D0838464

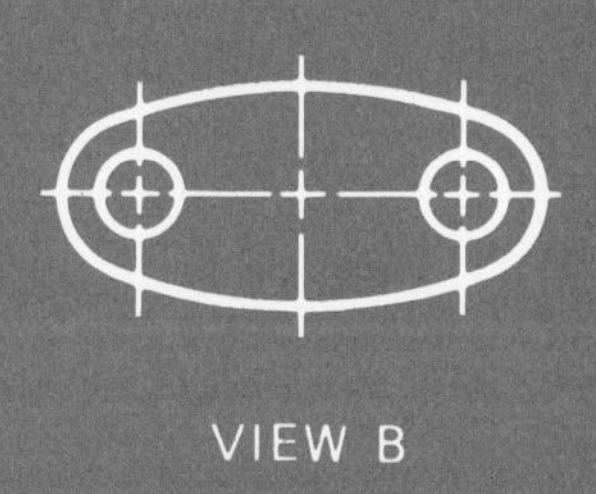
VIEW B

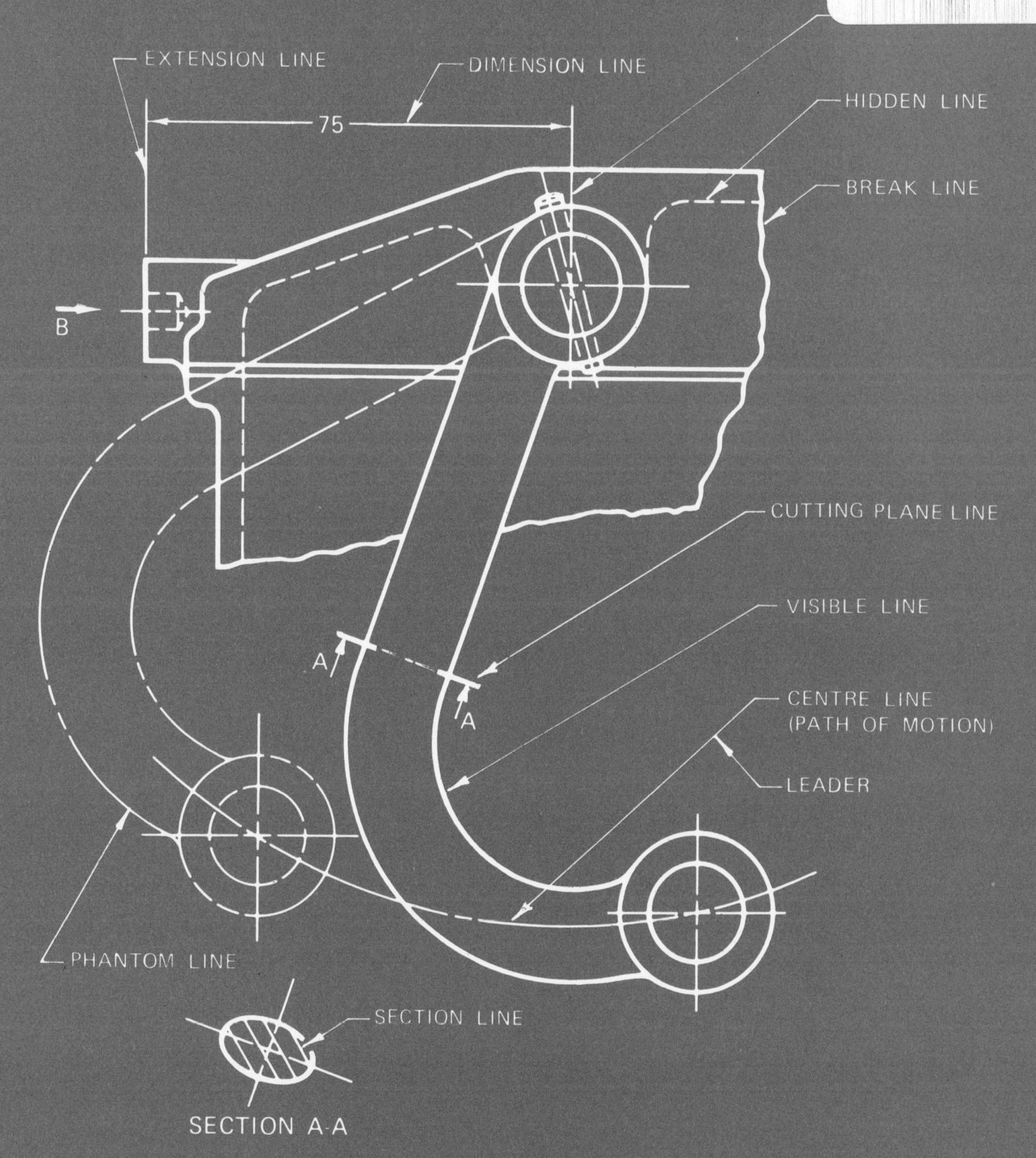
EXTENSION LINE
DIMENSION LINE
75
HIDDEN LINE
BREAK LINE
B
CUTTING PLANE LINE
VISIBLE LINE
A
A
CENTRE LINE
(PATH OF MOTION)
LEADER
PHANTOM LINE
SECTION LINE
SECTION A-A

DRAFTING FUNDAMENTALS

FIFTH EDITION
SI METRIC

CECIL JENSEN
Former Technical Director
R.S. McLaughlin Collegiate and Vocational Institute
Oshawa, Ontario

FRED MASON
Technical Director
Anderson Collegiate and Vocational Institute
Whitby, Ontario

McGraw-Hill Ryerson Limited

Toronto Montréal New York St. Louis San Francisco
Auckland Bogotá Guatemala Hamburg Johannesburg
Lisbon London Madrid Mexico New Delhi Panama
Paris San Juan Sao Paulo Singapore Sydney Tokyo

Drafting Fundamentals, Fifth edition

ISBN 0-07-548068-9

1 2 3 4 5 6 7 8 9 0 THB 1 9 8 7 6 5 4 3 2

Printed and bound in Canada

Canadian Cataloguing in Publication Data

Jensen, C.H. (Cecil Howard), Date -

Drafting fundamentals

ISBN 0-07-548068-9

1. Mechanical drawing. I. Mason, F.H.S. (Frederick Harry Sextus), Date - II. Title.

T353.J46
1982 604.2'4 C82-0949
94-9

Preface

Technical drawing, like all technical areas, is constantly changing. In this new edition of Drafting Fundamentals, the authors have made every effort to translate the most current technical information available into the most usable form from the standpoint of both teacher and student. The latest developments and current practices in all areas of graphic communication, functional drafting, materials representation, shop processes, numerical control, and modern engineering tolerancing have been added to the text in a manner that simplifies basic drawing standards and procedures into understandable instructional units.

Each chapter has been expanded and divided into single-concept units each with its own objective, instruction, examples, review and assignments. This organization provides the student with a logical sequence of experiences which can be adjusted to individual needs. It also provides for maximum efficiency in learning essential concepts. Development of each unit is from the simple to the complex and from the familiar to the unfamiliar.

The authors thank the many teachers who, over the years, have given valuable suggestions and assistance. Comments on this new edition will be most welcome.

Table of Contents

Chapter 1 The Language of Industry 1
Unit 1-1 The Language of Industry 1

Chapter 2 Developing Drafting Skills 5
Unit 2-1 Drafting instruments and equipment 5
2-2 Basic Linework and Lettering 12
2-3 Hidden Lines 18
2-4 Centre Lines 20
2-5 Drawing Circles and Arcs 21
2-6 Drawing Irregular Curves 24
2-7 Sketching 25

Chapter 3 Theory of Shape Description 28
Unit 3-1 Shape Description by Views 28
3-2 Arrangement of Views 33
3-3 All Surfaces Parallel to the Viewing Planes and All Edges and Lines Visible 36
3-4 All Surfaces Parallel to the Viewing Planes with Some Edges and Surfaces Hidden 38
3-5 Inclined Surfaces 41
3-6 Circular Features 44
3-7 Oblique Surfaces 46
3-8 Review of Chapter 3 48

Chapter 4 Basic Dimensioning 56
Unit 4-1 Dimensioning 46
4-2 Dimensioning Circular Features 67
4-3 Dimensioning Common Features 73
4-4 Dimensioning Methods 79
4-5 Surface Texture 83

Chapter 5 Working Drawings 88
Unit 5-1 Conventional Represent ation of Common Features 88
5-2 Special Views 95
5-3 Detail and Assembly Drawings 99
5-4 Drawing Reproduction 104

Chapter 6 Fastening Devices 107
Unit 6-1 Thread Forms and Their Pictorial Representation 107
6-2 Conventional Thread Representation 113
6-3 Common Threaded Fasteners 119
6-4 Special Fasteners 127
6-5 Keys and Pins 127

Chapter 7 Sections and Conventions 139
Unit 7-1 Sectional Views 139
7-2 Two or More Sectional Views on One Drawing 144
7-3 Half Sections 146
7-4 Threads in Section 148
7-5 Assemblies in Section 153
7-6 Offset Sections 153
7-7 Ribs, Holes, and Lugs in Section 156
7-8 Revolved and Removed Sections 159
7-9 Spokes and Arms in Section 162

7-10 Partial or Broken-out Sections 163
7-11 Phantom or Hidden Sections 165

Chapter 8 Auxiliary Views 166
Unit 8-1 Primary Auxiliary Views 166
8-2 Circular Features in Auxiliary Projection 169
8-3 Multi-auxiliary View Drawings 170

Chapter 9 Pictorial Drawings 173
Unit 9-1 Pictorial Drawings 173
9-2 Curved Surfaces in Isometric 179
9-3 Common Features in Isometric 183
9-4 Oblique Projection 187
9-5 Common Features in Oblique 190

Chapter 10 Development and Intersections 194
Unit 10-1 Straight-line Development 194
10-2 The Packaging Industry 200
10-3 Radial Line Development of Flat Surfaces 203
10-4 Parallel Line Development 207
10-5 Radial Line Development of Conical Surfaces . 209
10-6 Intersection of Flat Surfaces 213
10-7 Intersection of Cylindrical Surfaces 215

Chapter 11 Materials and Manufacturing Processes 216
Unit 11-1 Basic Metallurgy 216
11-2 Plastics 220
11-3 Manufacturing Processes 224

Chapter 12 Modern Engineering Tolerancing 231
Unit 12-1 Limits and Tolerances 231
12-2 Positional Tolerancing 237

Chapter 13 Drawing for Numerical Control 242
Unit 13-1 Two-axis Control Systems 242
13-2 Three-axis Control Systems 248

Chapter 14 Functional Drafting
Unit 14-1 Drawing Aids and Drawing Practices 252
14-2 Freehand Sketching 255
14-3 Simplified Drafting 256
14-4 Cut-and-Paste 261

Chapter 15 Architectural Drafting 264
Unit 15-1 Presentation Drawings 264
15-2 Construction Drawings 270
15-3 Developing a House Plan 285

Chapter 16 Applied Geometry 293
Unit 16-1 Straight Lines 293
16-2 Arcs and Circles 296
16-3 Polygons 299
16-4 The Ellipse 302
16-5 The Helix 304

Appendix 305

Index 323

About the Authors

Cecil Jensen is the author of many technical books, including Engineering Drawing and Design, Interpreting Engineering Drawings, Architectural Drawing and Design for Residential Construction, Residential Construction Drawings (4 sets), Home Planning and Design and Interior Design. Some of these books are printed in three languages and are used in many countries.

He has twenty-seven years teaching experience in mechanical and architectural drafting and was a technical director in the educational system of the Province of Ontario for twenty-three years. Mr. Jensen is Past President of the Ontario Technical Directors' Association and the Ontario Drafting Teachers' Association.

Before entering the teaching profession, Mr. Jensen gained several years of design experience in the industry. He has also been responsible for the supervision of the teaching of technical courses for General Motors apprentices in Oshawa, Canada.

He is a member of the Canadian (CSA) Metric Committee on Technical Drawings (which includes both mechanical and architectural drawing) and is chairman of the Committee on Dimensioning and Tolerancing. Mr. Jensen is Canada's representative on the American (ANSI) Standards for Dimensioning and Tolerancing and has recently represented this country at two world (ISO) conferences in Oslo, Norway and Paris, France on the standardization of technical drawings.

He took an early retirement from the teaching profession in 1979 in order to devote his full attention to writing and working on the committee for Canadian drawing standards.

Fred H.S. Mason entered the teaching profession after twelve years of industrial experience as a tool and machine designer and manufacturing methods planner. He has served the Durham Board of Education in Ontario, Canada, as a teacher of mechanical drafting and mathematics, guidance counsellor, technical director, and as an education officer. In this last capacity he was responsible for the co-ordination of technical subjects and facilities, co-operative education and Linkage programs, and the operation of adult training continuing education, summer school, driver education, and summer youth employment programs.

Mr. Mason has been an associate lecturer for the College of Education, University of Toronto in summer programs for technical teacher training. He has served on the drafting curriculum committee for the Ontario Ministry of Education and the executive committee for the Ontario Ministry of Education and the executive committee of the Durham Organization For Industrial Training. Mr. Mason is a Past President of the Ontario Technical Directors' Association.

Chapter 1

Language of Industry

UNIT 1-1
Drafting as a Language

Since earliest times people have used drawings to communicate ideas to each other, and to record those ideas so that they would not be forgotten. The earliest forms of writing, such as the Egyptian hieroglyphics, were picture forms.

The word graphic means dealing with the expression of ideas by lines or marks im-

Celotex Corporation

Fig. 1-1-1 Early Greek use of post and lintel construction

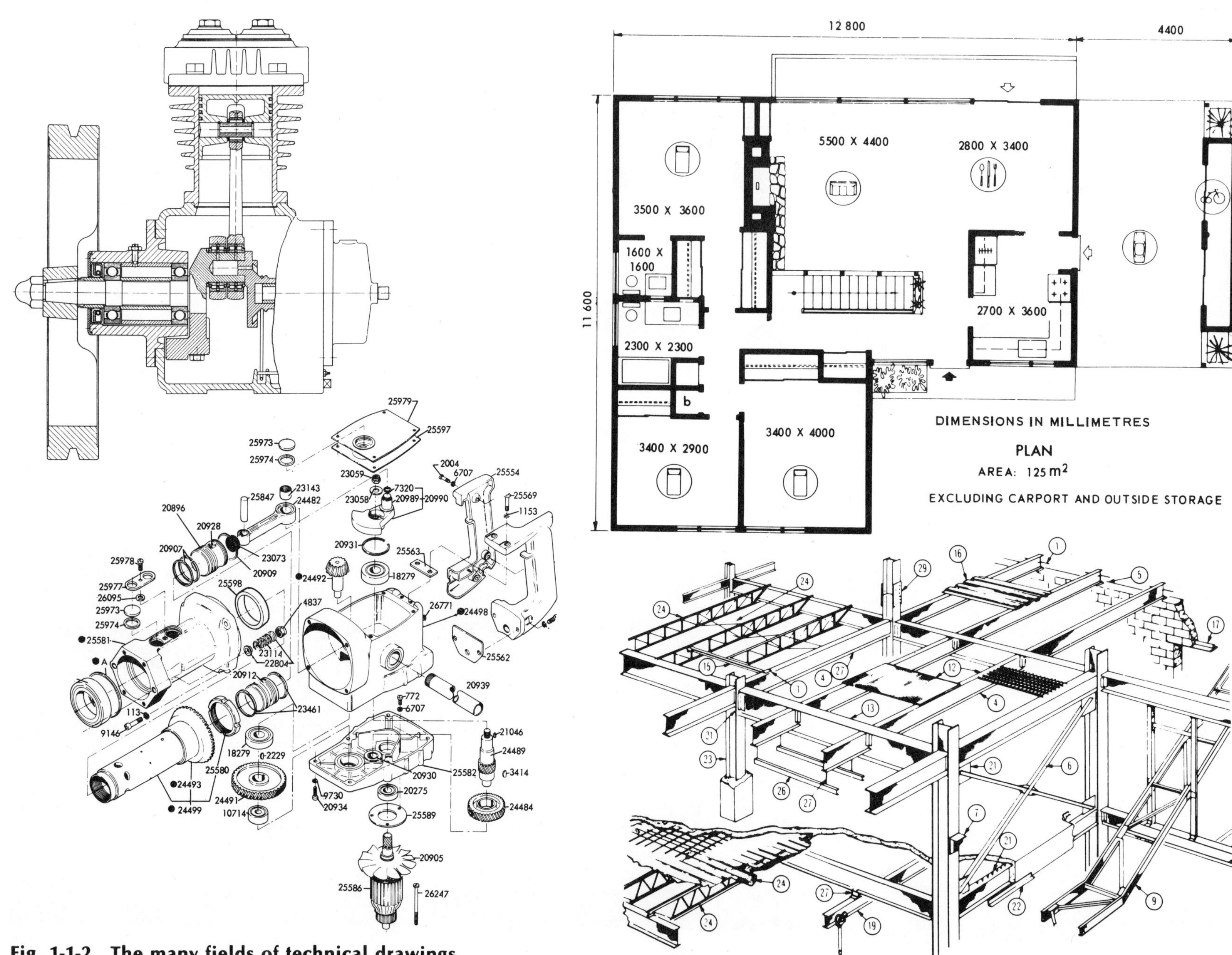

Fig. 1-1-2 The many fields of technical drawings

pressed on a surface. A drawing is a graphic representation of a real thing. Drafting, therefore, is a graphic language, because it uses pictures to communicate thoughts and ideas.

Drawing has developed along two distinct lines, with each form having a different purpose. Artistic drawing is concerned mainly with the expression of real or imagined ideas of a cultural nature. Technical drawing is concerned with the expression of technical ideas of a practical nature, and is the method used in all branches of technical industry.

Even highly developed word languages are inadequate for describing the size, shape, and relationship of physical objects. For every manufactured object there are drawings that describe its physical shape completely and accurately, communicating the drafter's ideas to the worker. For this reason, drafting is referred to as the language of industry.

A Universal Language

Throughout the long history of drafting, many drawing conventions, terms, abbreviations, symbols, and practices have come into common use. It is essential that all drafters use the same practices if drafting is to serve as a reliable means of communicating technical theories and ideas.

Discussions among representatives of the industrialized nations of the world have led to the creation and adoption of the SI Metric System and the standardization of drawing conventions. For such nations, drafting has truly become a universal language.

The Drafting Student

While students are learning basic drafting skills, they will also be increasing their general technical knowledge. Not all students will make drafting their career or occupation. However, an understanding of this graphic language is necessary for anyone who intends to work in industry. It is essential for anyone who plans to enter the skilled trades or become a technician, technologist or engineer.

Because a drawing is a set of instructions that the worker will follow, it must be accurate, clear, correct and complete. When technical drawings are made with the use of instruments, they are called instrumental drawings. When drawn without instruments, they are referred to as sketches. The ability to sketch ideas and designs and to make accurate instrumental drawings is a basic part of drafting skills.

In everyday life, a knowledge of drafting is helpful in understanding house plans, assembly, maintenance, and operating instructions for many manufactured products, as well as plans and specifications for hobby and spare-time activities.

Fig. 1-1-4 Typical drafting classroom

Bettman Archive, Inc.

(A) PAST

Charles Bruning Co.

(B) PRESENT

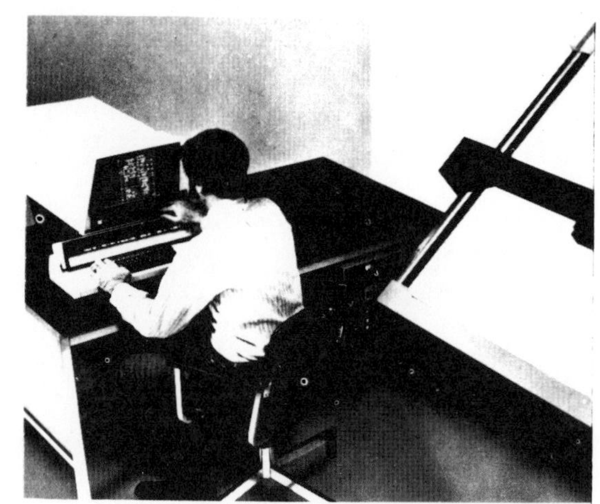
Gerber Scientific Instruments

(C) FUTURE

Fig. 1-1-5 The drafting office

Drafting Careers

The ability to draw does not in itself make a person a drafter. It is also necessary to have a good background knowledge in technology, mathematics, and physical sciences, plus a degree of creative ability and specialized knowledge and training in the particular field of technology of the employer.

Employment opportunities are greatest with manufacturing companies, engineering and architectural consulting firms, construction companies and public utilities. Drafters are also employed by federal, provincial and municipal branches of government in various departments such as housing, public works, and highways.

Depending on the employer's training program, you may be called an apprentice, a trainee, a beginner, or a junior drafter. Senior positions are held by those with the most ability, experience, and knowledge.

The Drafting Office

The drafting office is the starting point for all engineering work. The drawings produced are the main method of communication between all persons concerned with the design, manufacture, and assembly of parts.

Many improvements have taken place over the years in facilities and equipment. Most drafting classrooms have manual drafting equipment similar to that which is still used extensively in drafting offices.

Review Questions

1. Why is drafting a "graphic language"?
2. What is the main difference between artistic and technical drawing?
3. Why is drafting referred to as the "language of industry"?
4. Why are uniform drafting practices important?
5. In what other technical careers or occupations is knowledge of drafting essential?
6. Name four requirements for a good technical drawing.
7. What are the two forms in which technical drawings may be made?
8. In what ways is a knowledge of drafting helpful in personal life?
9. What are four other requirements for a drafter besides the ability to draw?
10. Name four areas of manufacturing in which drafters may be employed.

Chapter 2

Drafting Instruments and Equipment

UNIT 2-1

Drafting Instruments and Equipment

Straightedges

The straightedge is used in drawing horizontal lines and for supporting set squares when vertical and sloping lines are being drawn. See Figure 2-1-2. It is fastened on each end to cords which pass over pulleys. This arrangement permits movement of the

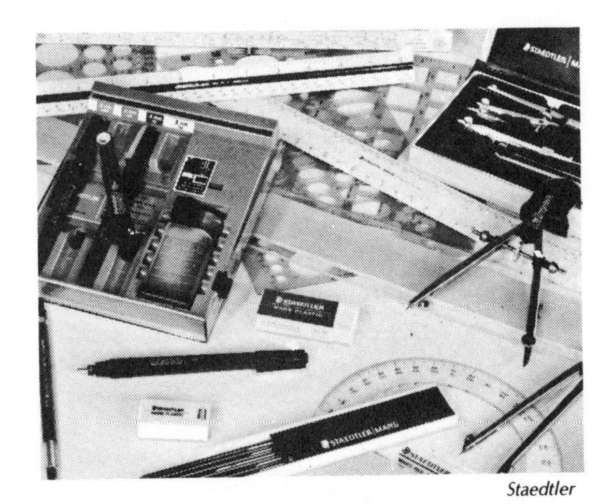

Staedtler

Fig. 2-1-1 Drafting equipment

Addressograph - Multigraph Corp.

Fig. 2-1-2 Drafting table with straightedge

straightedge up and down the board while maintaining the straightedge in a horizontal position.

T Squares

The T square (Fig. 2-1-3) performs the same function as the parallel straightedge. While you are using the T square, keep the head of the instrument firmly against the side of the board to ensure that the lines you draw will be parallel. The head will be on the left edge of the board if you are right-handed and on the right if you are left-handed.

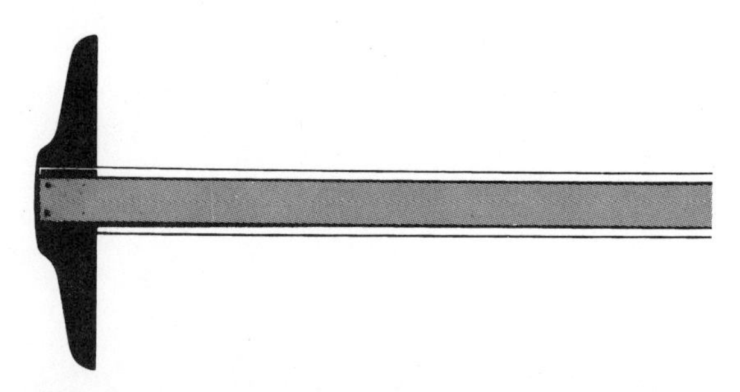

Fig. 2-1-3 T square

Set Squares

Set squares are used together with the parallel straightedge or T square when you are drawing vertical and sloping lines (Fig. 2-1-4). The set squares most commonly used are the 60-30° and the 45° set squares. Singly or in combination, these set squares can be used to form angles in all the multiples of 15°. For other angles, the protractor is used. All angles can be drawn with the adjustable

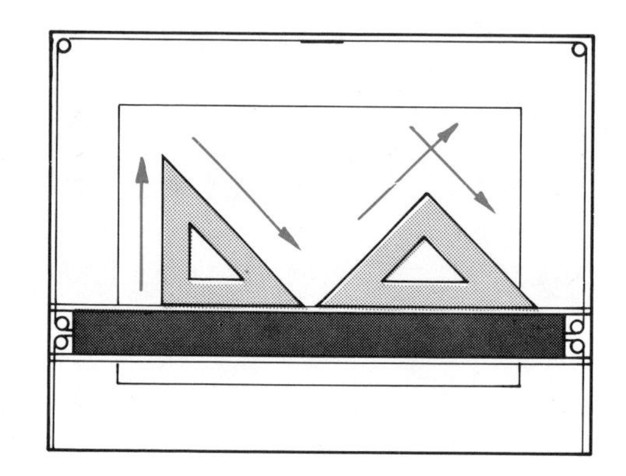

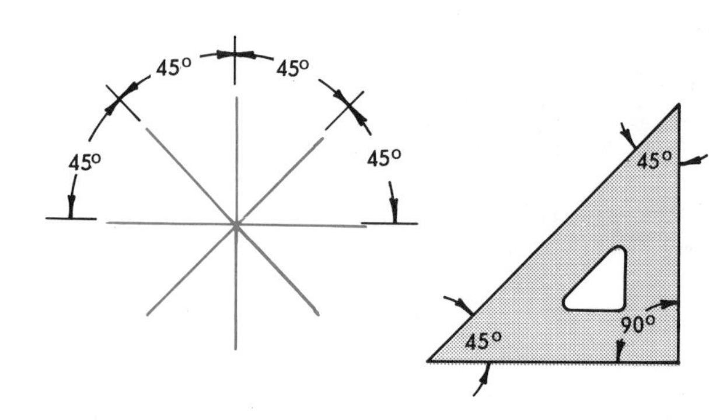

(A) THE 45° SET SQUARE

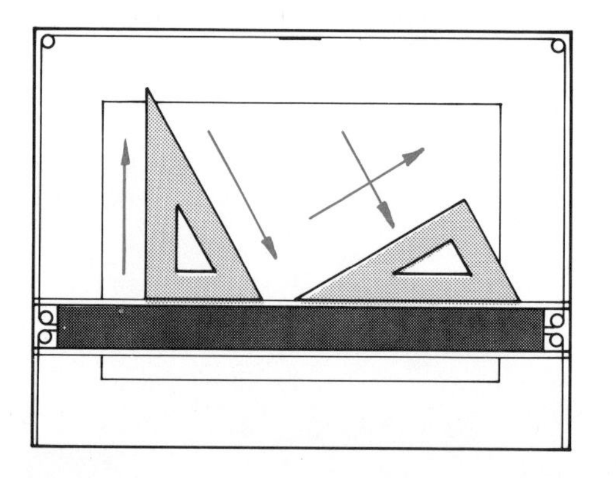

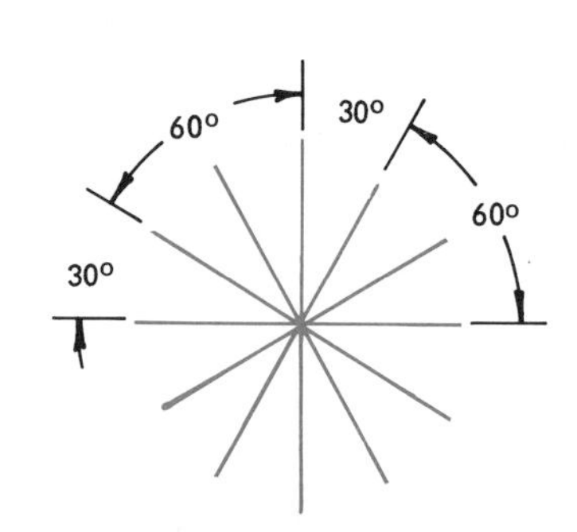

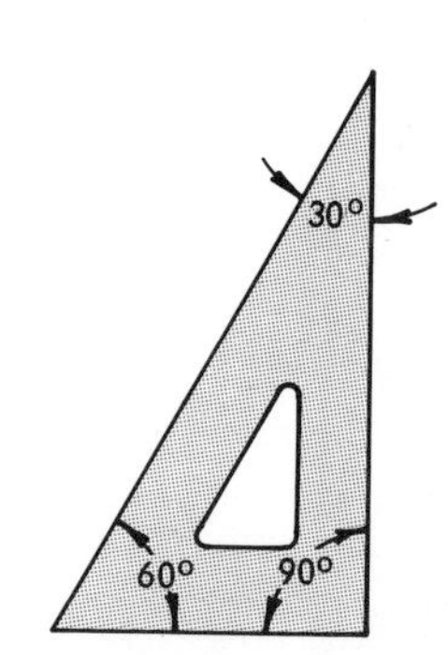

(B) THE 60° SET SQUARE

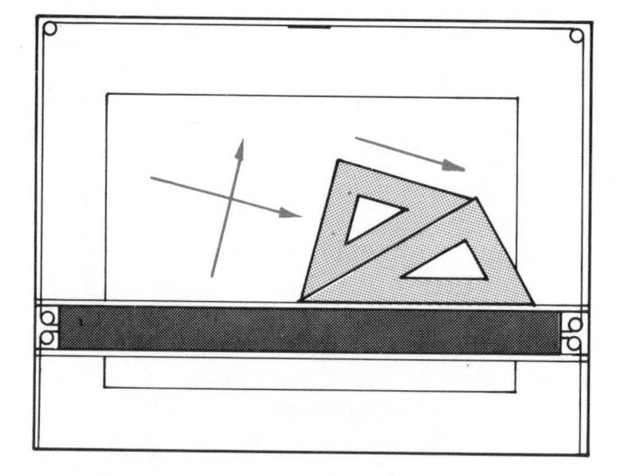

15°
75°
75°
15°

(C) THE SET SQUARES IN COMBINATION

Fig. 2-1-4 The set square

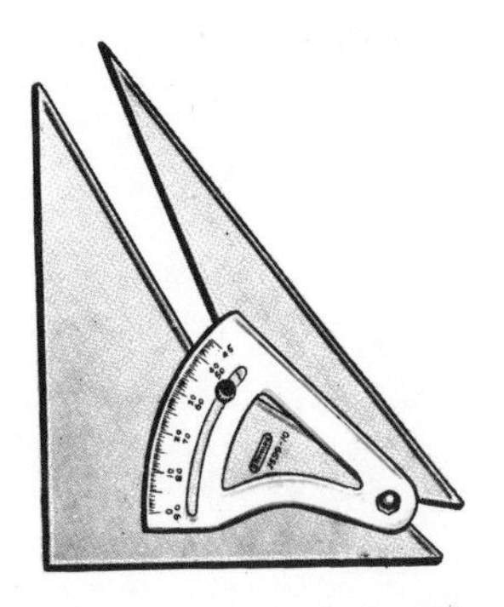

Fig. 2-1-5 Adjustable set square

set square (Fig. 2-1-5). This instrument replaces the two common set squares and the protractor.

Scales

Shown in Figure 2-1-6 are the common shapes of scales used by drafters to make measurements on their drawings. Scales are used only for measuring and are not to be used for drawing lines. It is important that drafters draw accurately to scale. The scale to which the drawing is made must be indicated in the title block or strip.

When objects are drawn at their actual

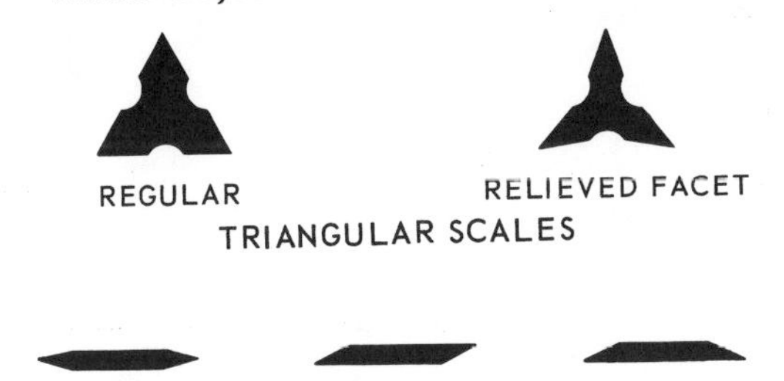

Fig. 2-1-6 End view shapes of scales

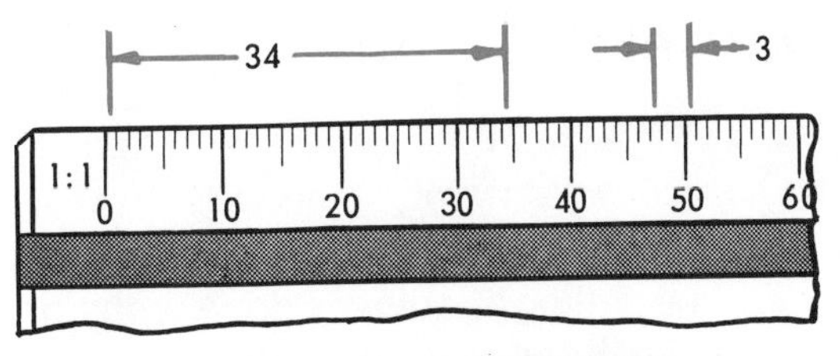

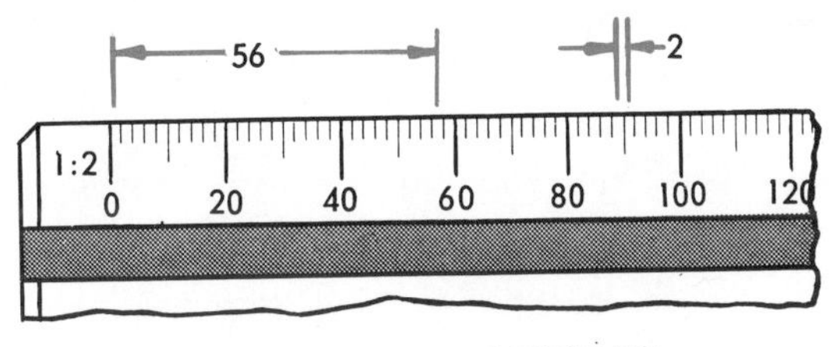

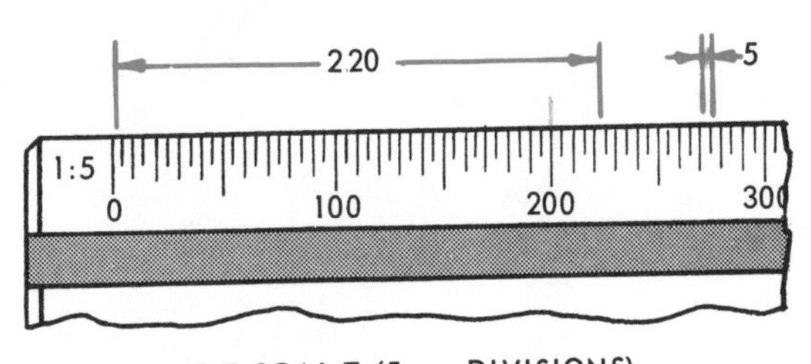

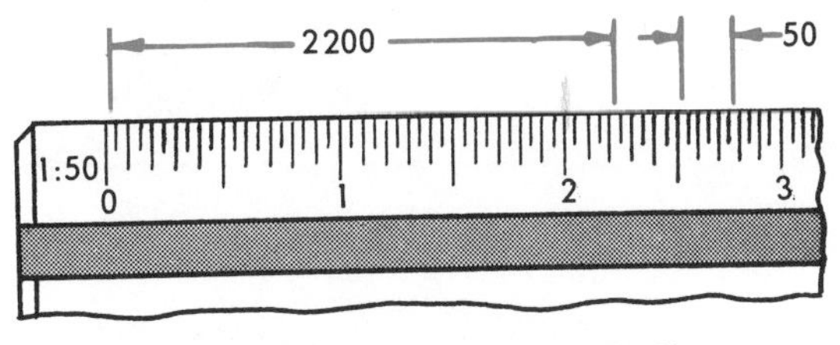

Fig. 2-1-7 Metric scales

size, the drawing is called **full scale** or **scale 1:1.** Many objects, however, such as buildings, ships, or airplanes, are too large to be drawn full scale, so they must be drawn to a reduced scale. An example would be the drawing of a house to a scale of 1:50.

Frequently, objects such as small watch parts are drawn larger than their actual size so that their shape can be seen clearly. Such a drawing has been drawn to an enlarged scale. The minute hand of a wristwatch, for example, could be drawn to scale 5:1.

Many mechanical parts are drawn to half scale, 1:2, and fifth scale, 1:5. Notice that the scale of the drawing is expressed as an equation. The left side of the equation represents a unit of the size drawn. The right side equals a unit of the actual object. Thus one unit of measurement on the drawing = 5 units of measurement on the actual object.

Scales are made with a variety of combined scales marked on their surfaces. This combination of scales spares the drafter the necessity of calculating the sizes to be drawn when working to a scale other than full size.

Metric Scales. The linear unit of measurement for mechanical drawings is the millimetre. Scale multipliers and divisors of 2 and 5 are recommended, which gives the scales shown in the accompanying table.

The numbers shown indicate the difference in size between the drawing and the actual part. For example, the ratio 10:1 shown on the drawing means that the drawing is 10 times the actual size of the part. A

METRIC SCALES		
ENLARGED	SIZE AS	REDUCED
1000:1	1:1	1:2
500:1		1:5
200:1		1:10
100:1		1:20
50:1		1:50
20:1		1:100
10:1		1:200
5:1		1:500
2:1		1:1000

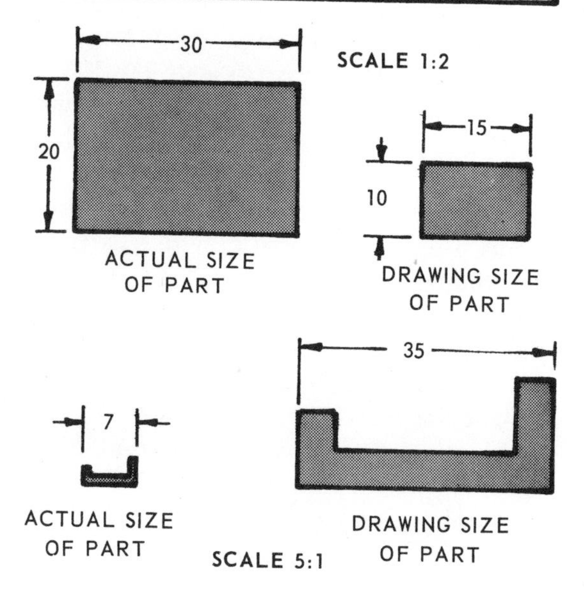

Fig. 2-1-8 Metric scales

ratio of 1:5 on the drawing means the object is 5 times as large as it is shown on the drawing.

The units of measurement for architectural drawings are the metre and millimetre. The same scale multipliers and divisors as used for mechanical drawings are used for architectural drawings.

Drafting Machines

This device, which combines the functions of T square, set squares, scale, and protractor, is estimated to save up to 50% of the user's time. All positioning is done with one hand, while the other is free to draw.

Drafting machines may be attached to any drafting board or table. Two types are currently available. In the track type, a vertical beam carrying the drafting instruments rides along a horizontal beam fastened to the top of the table. In the arm, or elbow type, two arms pivot from the top of the machine and relative to each other.

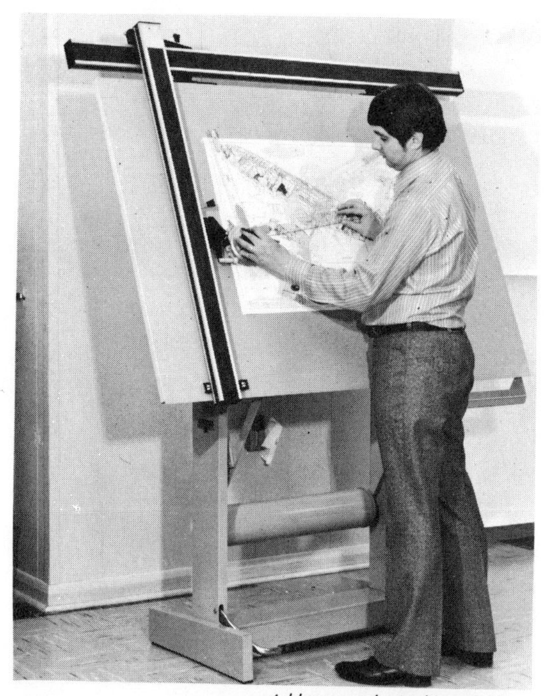

Addressograph - Multigraph Corp.

Fig. 2-1-9 Track type drafting machine

Keuffel and Esser Co.

Fig. 2-1-10 Arm type drafting machine

Compasses

The compass is used for drawing circles and arcs. Several basic types and sizes are available (Fig. 2-1-11).

- **Friction head compass,** standard in most drafting sets.
- **Bow compass,** which operates on jackscrew or ratchet principle by turning a large knurled nut.
- **Beam compass,** a bar with an adjustable needle and pencil-and-pen attachment for drawing large arcs or circles.

Dividers

Used for laying out or transferring measurements, dividers have a steel pin insert in each leg and come in a variety of sizes and designs, similar to the compass. See Figure 2-1-12. The compass can be used as a divider by replacing the lead point with a steel pin.

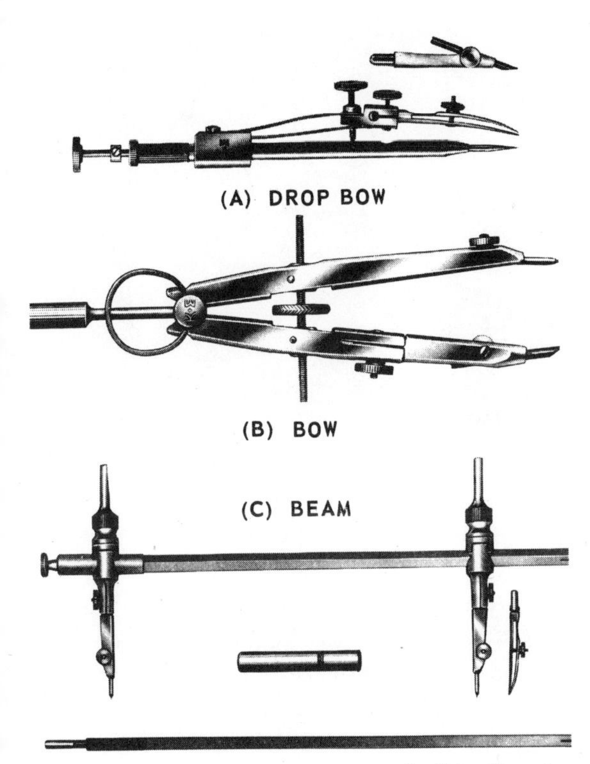

Fig. 2-1-11 Compasses *Keuffel and Esser Co.*

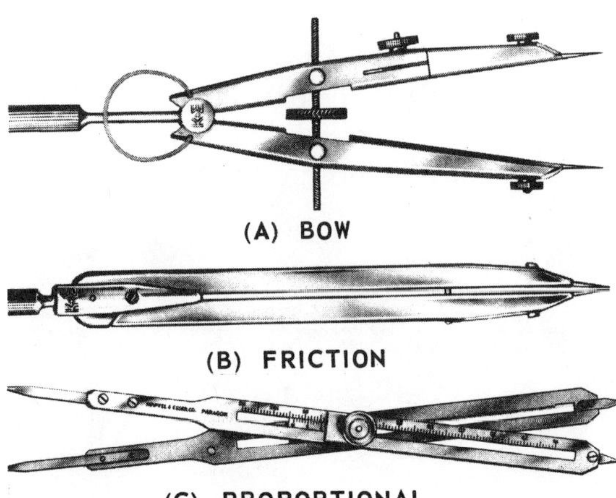

Fig. 2-1-12 Dividers *Keuffel and Esser Co.*

Drafting Leads and Pencils

Leads. Because of the drawing media used and the type of reproduction required, pencil manufacturers have marketed three types of lead for the preparation of engineering drawings.

Graphite Lead. This is the conventional type of lead which has been used for years. It is made from graphite, clay and resin. It is available in a variety of grades or hardness 9H, 8H, 7H, and 6H (hard); 5H and 4H (medium hard); 3H and 2H (medium); H and F (medium soft); and HB, B, 2B to 6B (very soft). Very soft lead is not recommended for use on paper, vellum, or initial layout on cloth. The selection of the proper grade of lead is important. A hard lead might penetrate the drawing while a soft lead will smear.

Plastic Lead. This type of lead is designed for use on film only. It has good microfilm reproduction characteristics.

Plastic-Graphite Lead. This is designed for use on film only, erases well, does not readily smear, and produces a good opaque line which is suitable for microfilm reproduction. The main drawback with this type of lead is that it does not hold a point well.

Fig. 2-1-13 Pencil point shapes

Drafting Pencils. The leads are held either in the conventional wood-bonded cases known as wooden pencils or in metal or plastic cases known as mechanical pencils. See Figure 2-1-14. With the latter, the lead is ejected to the desired length of projection from the clamping chuck and then pointed in the same manner as the wood-bonded pencil. Recently, disposable mechanical pencils became available. These operate just as any mechanical pencil, but they are discarded after the lead has been used.

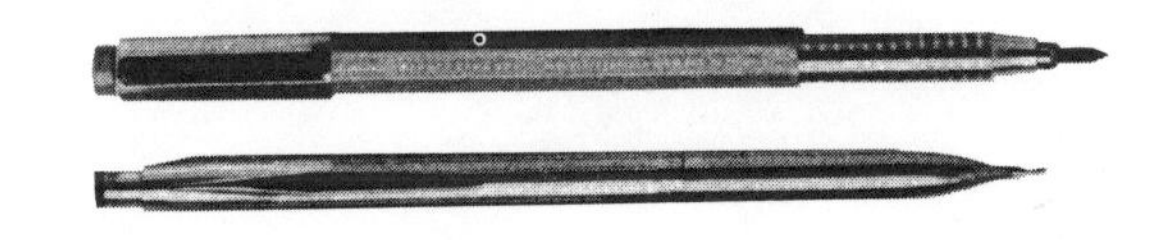

Fig. 2-1-14 Mechanical pencil-ejector type

Lead Pointers. A fast, convenient means of putting a clean drafter's point on refillable pencils and on the exposed leads of wood-cased pencils is with a lead pointer. See Figure 2-1-15. On some models short,

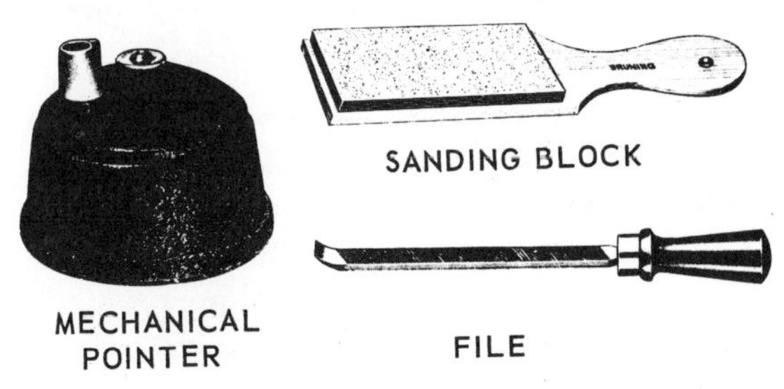

Fig. 2-1-15 Lead pointers

medium, or long tapered lead points can be made by merely adjusting the length of lead. A sandpaper block or file is used to keep sharp points on wooden pencils.

Brushes

A light brush (Fig. 2-1-16) is used to keep the drawing area clean. By using a brush to remove eraser particles and any accumulated dirt, the drafter avoids smudging the drawing.

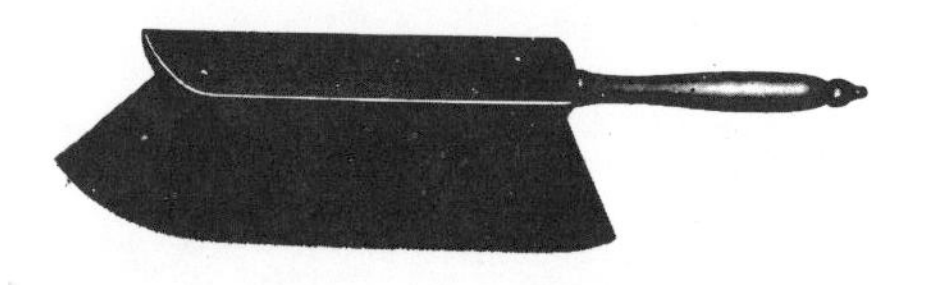

Charles Bruning Co. Canada Ltd.

Fig. 2-1-16 Brush

Lettering Aids

Lettering sets or guides (Fig. 2-1-17) are used when it is desirable to have more uniform and accurate letters and numerals than can be obtained by the freehand method. Lettering sets contain a number of guide templates that give a variety of letter shapes and sizes.

Instant Lettering is a dry transfer lettering which offers a wide variety of good quality lettering and speed. It adheres firmly to paper, wood, glass, and metal and is available in different colours. In case of errors, letters can be removed with cellophane tape or a pencil eraser.

Addressograph Multigraph Corp.

(A) MECHANICAL LETTERING

(B) APPLIQUES

Fig. 2-1-17 Lettering aids

Templates

To save time, many drafters now use templates for drawing small circles and arcs. Templates are also available for drawing standard square, hexagonal, triangular, and elliptical shapes and standard electrical and architectural symbols. See Figure 2-1-18.

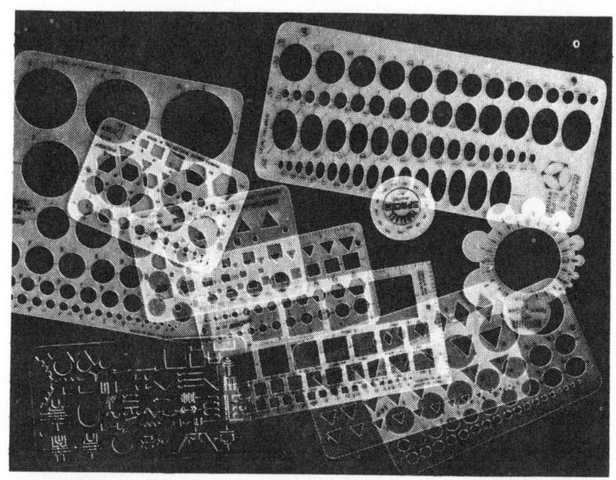

Rapid Design

Fig. 2-1-18 Templates

Irregular Curves

For drawing curved lines where (unlike circular arcs) the radius of curvature is not constant, an instrument known as an irregular or French curve (Fig. 2-1-19) is used.

The patterns for these curves are based on various combinations of ellipses, spirals, and other mathematical curves, and are available in a variety of shapes and sizes.

Generally, you plot a series of points of intersection along the desired path and then use the French curve to join these points so that a smooth-flowing curve results.

Teledyne Post

Fig. 2-1-19 Irregular curves

Calculators

Calculators, such as those shown in Figure 2-1-20, are used to make fast mathematical calculations, using division, multiplication, and extractions of square roots, and to solve problems involving areas, volumes, masses, strengths of materials, pressures, etc.

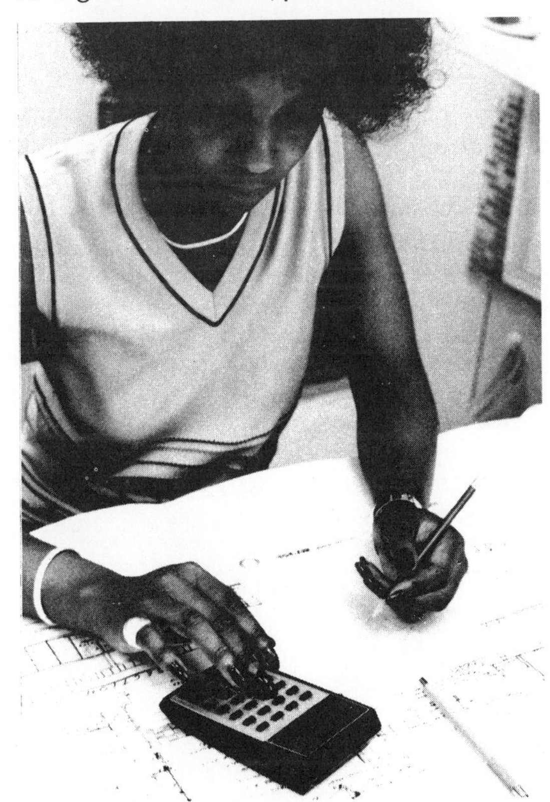

Fig. 2-1-20 Calculators

Erasing Shields

These thin pieces of metal have a variety of openings to permit the erasure of fine detail

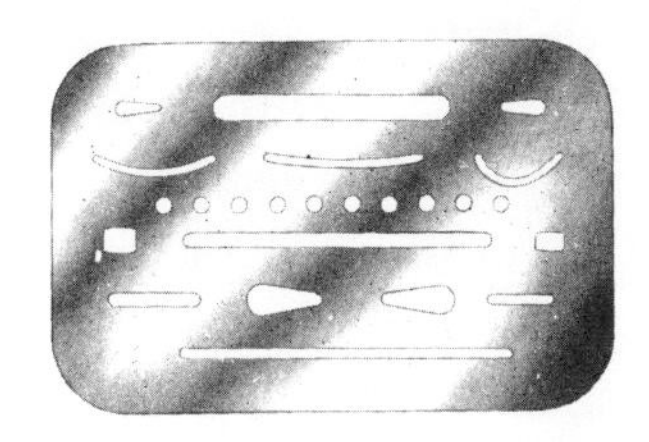

Keuffel and Esser Co.

Fig. 2-1-21 Erasing shield

line or note work without disturbing nearby work that is to be left on the drawing. Through the use of this device, erasures can be performed quickly and accurately.

Assignments

1. Scale measurement assignment Fig. 2-1-A.
2. Scale reading assignment Fig. 2-1-B.

Fig. 2-1-A Scale measurement assignment

Continued on page 12

USING THE SCALE

1:1 MEASURE DISTANCES A TO E
1:2 MEASURE DISTANCES F TO K (NO I)
1:5 MEASURE DISTANCES L TO P
1:10 MEASURE DISTANCES Q TO U
1:50 MEASURE DISTANCES V TO Z

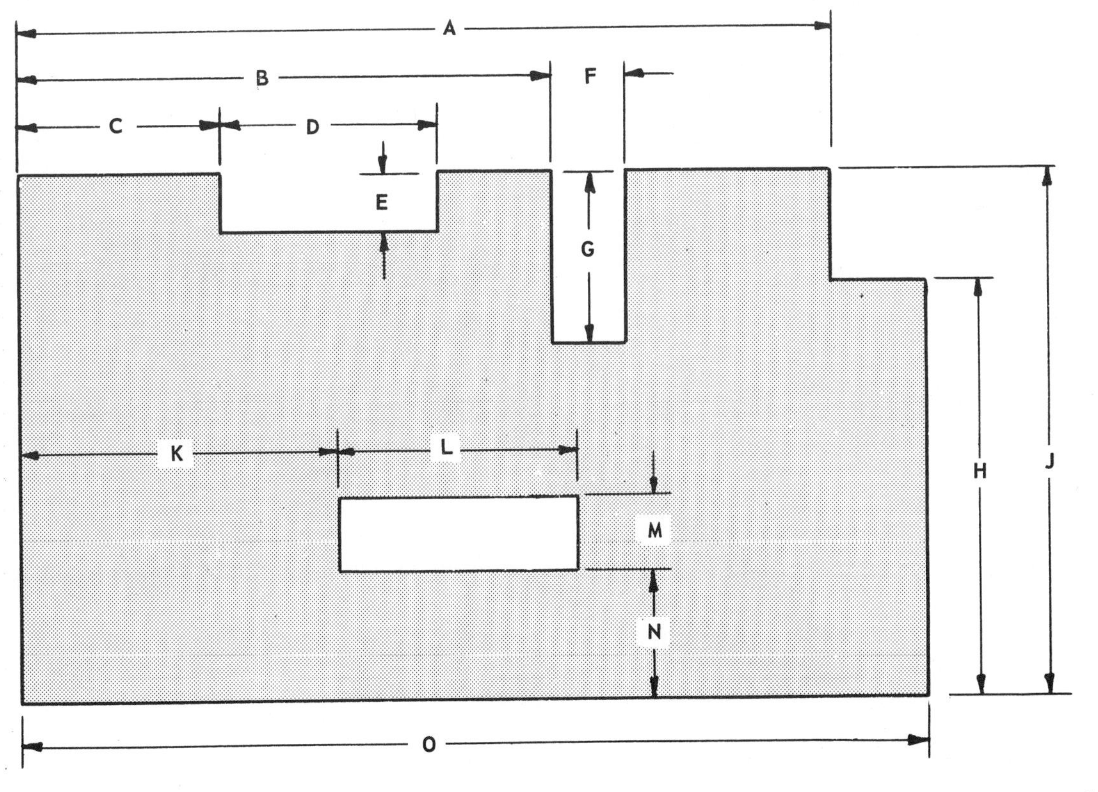

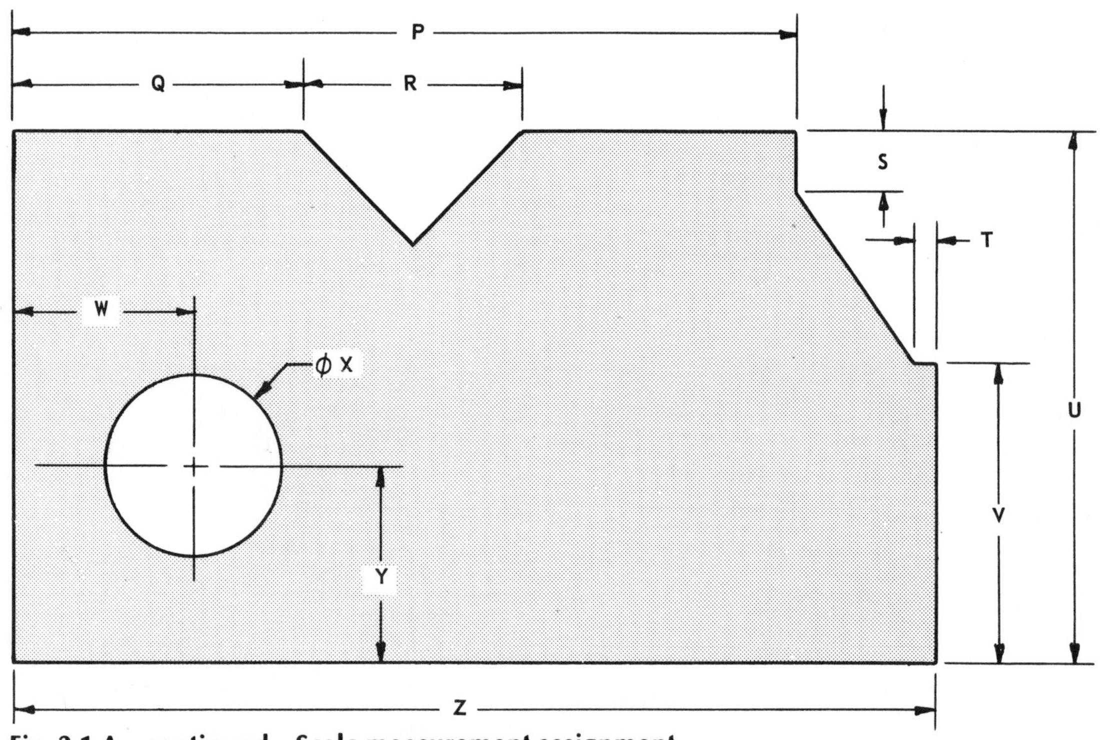

Fig. 2-1-A continued Scale measurement assignment

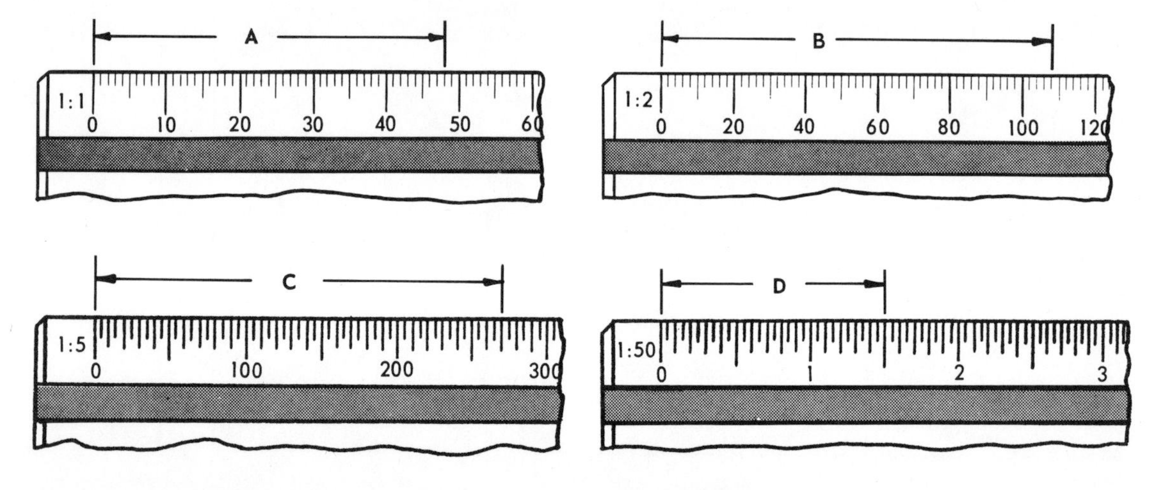

Fig. 2-1-B Scale reading assignment Determine distances A to D

UNIT 2-2

Basic Line Work and Lettering

Line Work

The various lines used in drawing form the alphabet of the drafting language: like letters of the alphabet, they are different in appearance. See inside front cover. The differences in the thickness and construction of all lines that are left on a drawing are most important. Lines must be clearly visible and stand out in sharp contrast to one another. This line contrast is necessary if the drawing is to be clear and easily understood.

First draw very light construction lines, setting out the main shape of the object in various views. Since these first lines are very light, they can be erased easily should changes or corrections be necessary. When you are satisfied that the layout is accurate, the construction lines are then changed to their proper type. Guidelines, used to ensure uniform lettering, are also drawn very lightly.

Line Widths

Two widths of lines, thick and thin, as shown in Figure 2-2-2, are recommended for use on drawings. Thick lines are 0.5 to 0.8 mm wide, thin lines between 0.3 and 0.5 mm wide. The actual width of each line is governed by the size and style of the drawing and the smallest size to which it is to be reduced. All lines of the same type

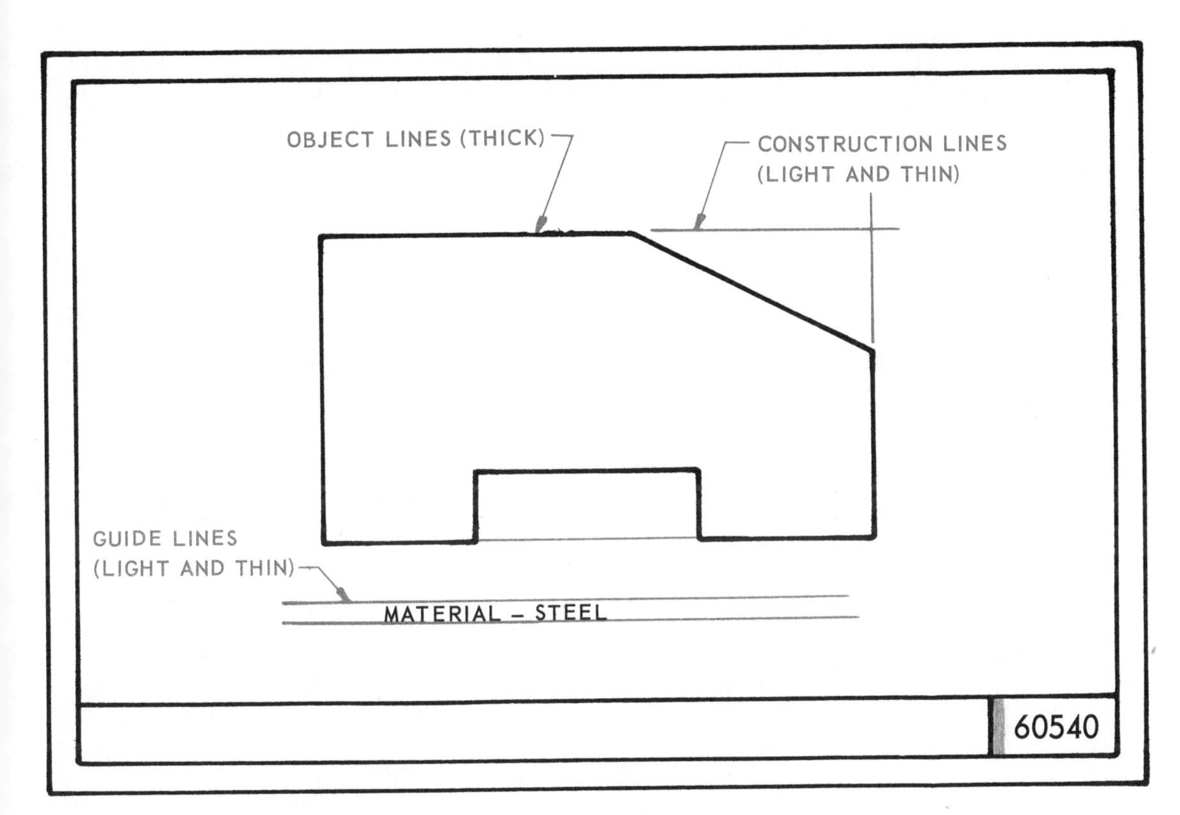

Fig. 2-2-1 Types of lines

THICK WIDTH 0.7 mm

THIN WIDTH 0.35 mm

Fig. 2-2-2 Thickness of lines

should be uniform throughout the drawing. Spacing between parallel lines should be such that there is no fill-in when the copy is reproduced by available photographic methods. Spacing of no less than 3 mm normally meets reproduction requirements.

Drawing Straight Lines

When using a conical-shaped lead, rotate the pencil slowly between your thumb and your forefinger when drawing lines. This keeps the lines uniform in width and the pencil sharp. Do not rotate a pencil having a bevel or wedge-shaped lead.

A general rule to follow when drawing lines is this: always draw in the direction in which the pencil is leaning.

Horizontal Lines. A right-handed person would lean the pencil to the right and draw horizontal lines from left to right. The left-

(A) DRAWING HORIZONTAL LINES

(B) DRAWING VERTICAL LINES

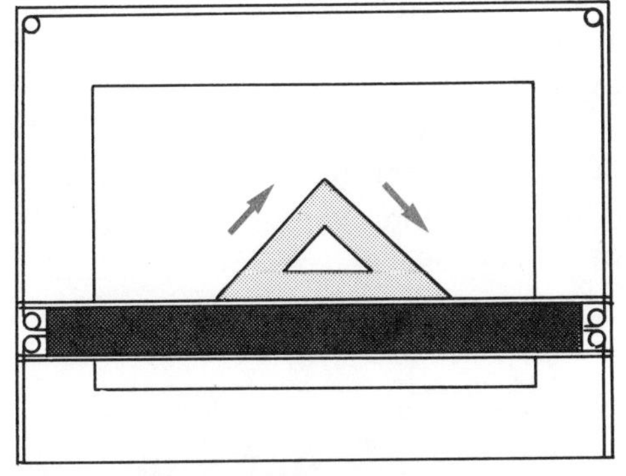

(C) DRAWING SLOPED LINES

Fig. 2-2-3 Drawing pencil lines

Fig. 2-2-4 **Drawing straight lines**

handed person would reverse this procedure and draw horizontal lines from right to left.

Vertical Lines. When drawing vertical lines, lean the pencil away from yourself, towards the top of the drafting board, and draw lines from bottom to top.

Sloped Lines. Lines sloping from the bottom to the top right are drawn from bottom to top; lines sloping from the bottom to the top left are drawn from top to bottom. This procedure for sloping lines would be reversed for a left-handed person.

Lettering

Styles of Letters and Numerals

The styles of lettering recommended for use on all technical drawings are single-stroke

ABCDEFGHIJKLMNOPQRSTUV
WXYZ& 1234567890

(A) UPPERCASE GOTHIC LETTERING

ABCDEFGHIJKLMNOPQR
STUVWXYZ 1234567890

(B) MICROFONT

Fig. 2-2-5 Recommended lettering for technical drawings

USE	MINIMUM LETTER HEIGHT		DRAWING SIZE
	FREEHAND	MECHANICAL	
DRAWING NUMBERS IN TITLE BLOCK AND DRAWING TITLE	7	7	ALL
DIMENSION, TOLERANCES, LIMITS, NOTES, SUBTITLES FOR SPECIAL VIEWS, TABLES, REVISIONS, AND ZONE LETTERS FOR THE BODY OF THE DRAWING	3.5	3.5	UP TO AND INCLUDING A2 (420 X 594)
	5	5	LARGER THAN 420 X 594

Fig. 2-2-6 Recommended lettering heights

upper-case Gothic and Microfont. These best meet the requirement that lettering be legible and easily executed. Note that in both styles the construction is vertical.

Lettering Heights

As microfilming and reduced print size of larger drawings becomes more widely used, the proper height of lettering has become essential to guarantee clarity on reproductions. The recommended lettering height for dimensions and notes is 3.5 mm for drawing sizes up to and including A2, and 5 mm in height for drawing sheet sizes A1 and larger. The height chosen must remain consistent on the drawing, except that titles and drawing numbers are made proportionately higher than the height of dimensions and notes.

Spacing

Letters should be close to each other but not touching, and the space between words the equivalent of one full width letter. The space between lines of lettering should be equal to the height of the lettering.

Guide Lines

Light guide lines should always be used to maintain a consistent height, and a consistent spacing between lines of lettering.

Drawing Sheets and Layout

Drawing Paper

Various types of drafting and tracing materials — paper, cloth, and film — are used in industry, depending on the reproduction process that will be used.

In most drafting classrooms, drawings are made on white or buff-coloured paper. When reproductions such as whiteprints are required, a type of translucent paper usually referred to as tracing paper, is used.

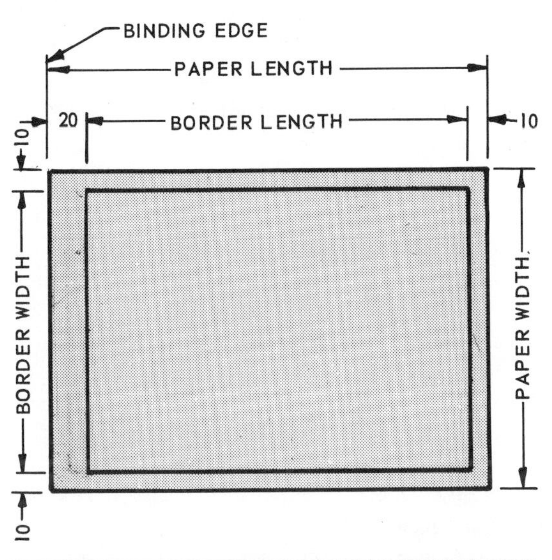

DRAWING SIZE	BORDER SIZE	OVERALL PAPER SIZE
A4	190 X 267	210 X 297
A3	277 X 390	297 X 420
A2	400 X 564	420 X 594
A1	574 X 811	594 X 841
A0	821 X 1159	841 X 1189

NOTE: PREFERRED BORDER WIDTHS DIFFER SLIGHTLY FOR MECHANICAL AND ARCHITECTURAL DRAWINGS

Fig. 2-2-7 Standard drawing sheet sizes

Standard Drawing Sheet Sizes

Metric drawing sizes are based on the A0 size — 821 × 1159. Each smaller size has an area half of the preceding size, and the length to width ratio remains constant.

Drawing sizes in the inch system are based on dimensions of commercial letterhead paper, 8.5 × 11. Each larger size has an area double the preceding size.

Title Strips and Blocks

Although the needs and wishes of individual industrial firms or schools vary, information printed in the title strip or title block usually includes: Title, Drawing Number, Name of Company or School, Scale Used, Name of Drafter, Date Drawn, Name of Checker or Supervisor.

In order to save drafting time, many firms purchase their drawing paper cut to the standard sheet sizes, with the border lines and the title strip or title block of their own design already printed on the sheets.

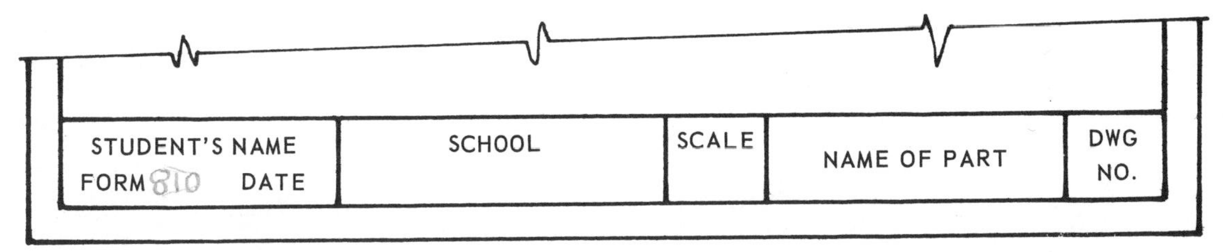

Fig. 2-2-8 Title strip

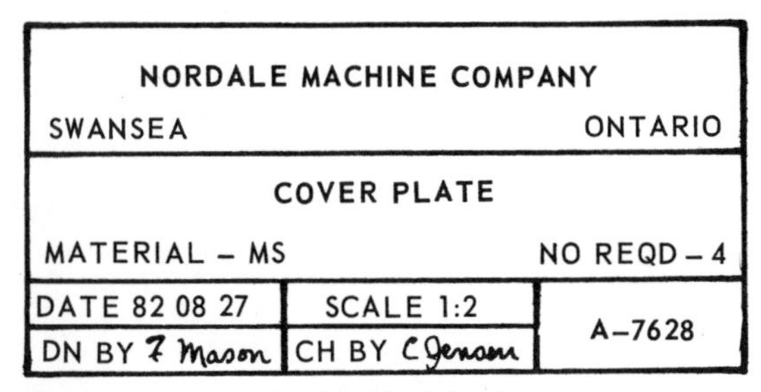

Fig. 2-2-9 Typical Title block

Fastening Paper to the Board

Drawing paper is usually fastened to the drawing surface by small pieces of tape across the four corners.

The top or bottom edge of the drawing sheet should be lined up with the top horizontal edge of the straightedge, T square or horizontal scale of the drafting machine.

When refastening a partially completed drawing, use horizontal lines already drawn for alignment.

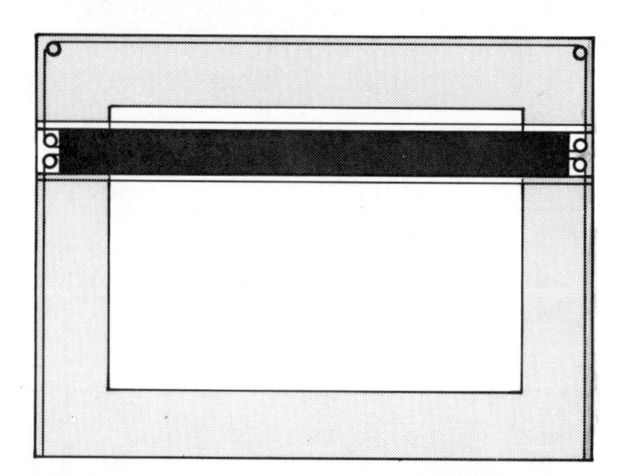

Fig. 2-2-10 Positioning the paper on the board

Assignments

1. Lettering Assignment. On an A4 sheet duplicate all of the lettering shown in Fig. 2-2-9 to the following heights: 3.5, 5 and 7 mm. Use very light guidelines and allow a 5 mm space between lines.

Sheet Size - A4, No Dimensions Required

2. Chess board, Fig. 2-2-A, scale 1:2.
3. Template #1, Fig. 2-2-B, scale 1:1.
4. Template #2, Fig. 2-2-C, scale 1:1.
5. Inlay designs, Fig. 2-2-D, scale 1:1. Do any two of A, B, C.
6. Stop sign, Fig. 2-2-E, scale 1:1.
7. Shuffleboard court, Fig. 2-2-F, scale 1:50. Letter on numbers and line names.

Review for Assignments

Unit 2-1 Drafting instruments and equipment

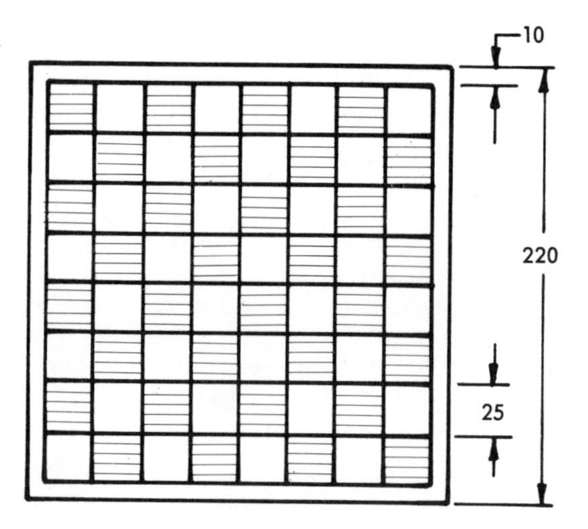

Fig. 2-2-A Chessboard

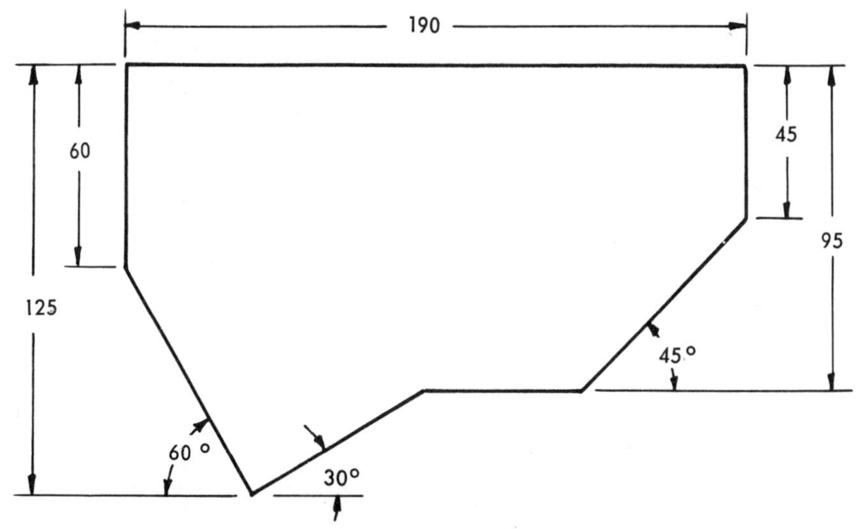

Fig. 2-2-B Template 1

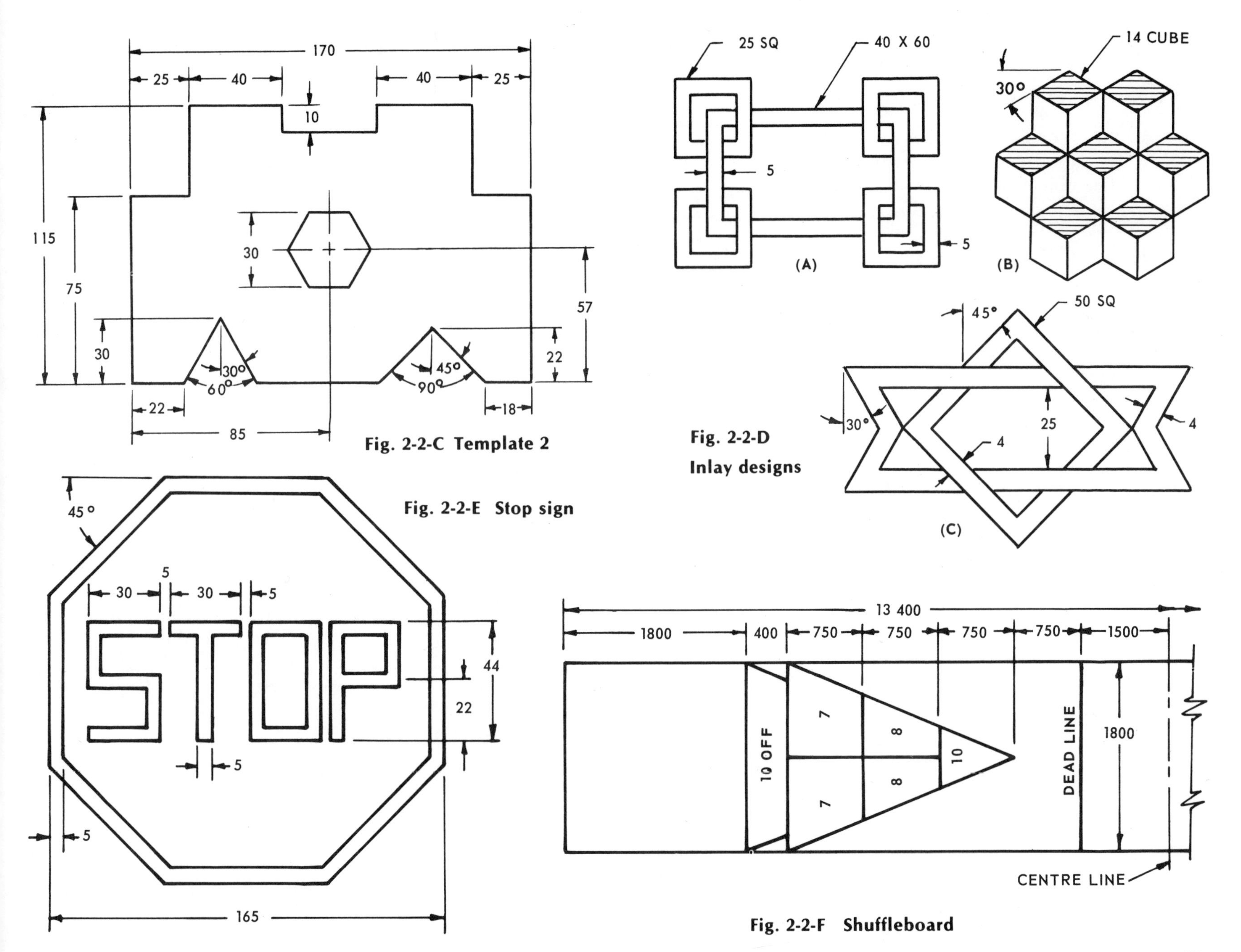

Fig. 2-2-C Template 2

Fig. 2-2-D Inlay designs

Fig. 2-2-E Stop sign

Fig. 2-2-F Shuffleboard

UNIT 2-3
Hidden Lines

Hidden lines consist of short, evenly-spaced dashes, and are used to show the hidden features of an object. They should be omitted when not required to preserve the clarity of the drawing. The length of dashes may vary slightly in relation to the size of the drawing. (Fig. 2-3-1).

Lines depicting hidden features and phantom details should always begin and end with a dash in contact with the line at which they start and end, except when such a dash would form a continuation of a visible detail line. Dashes should join at corners. Arcs should start with dashes at the tangent points. (Fig. 2-3-2)

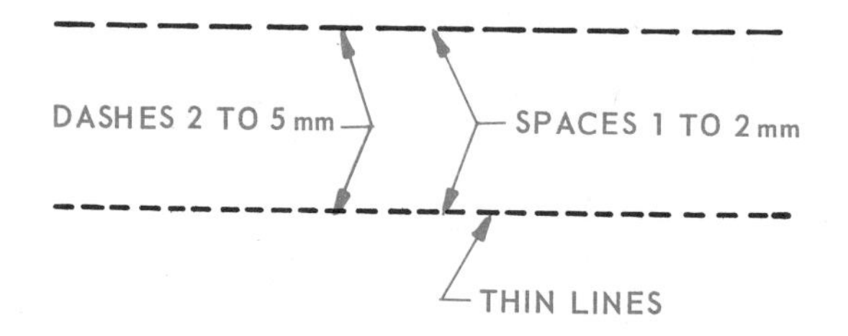

Fig. 2-3-1 Hidden line

Assignments

1. Garden gate Fig. 2-3-A, sheet size - A3, scale 1:10.
2. Roof truss Fig. 2-3-B, sheet size-A3, scale 1:20.

Review for Assignments
Unit 2-1 Scales

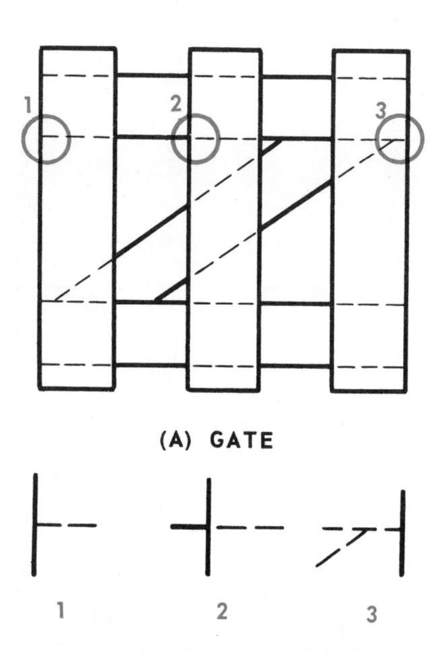

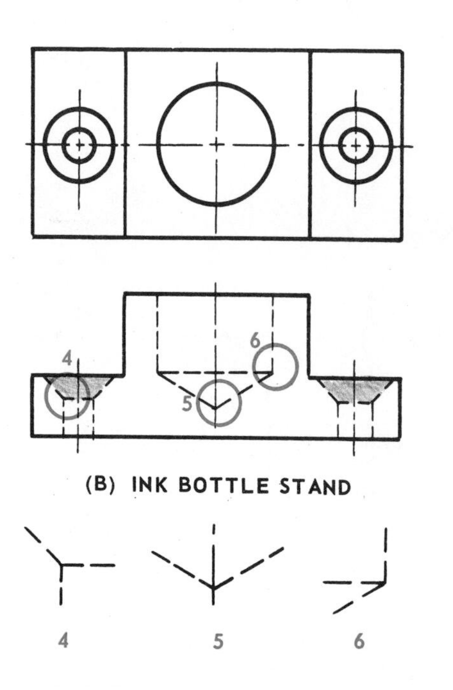

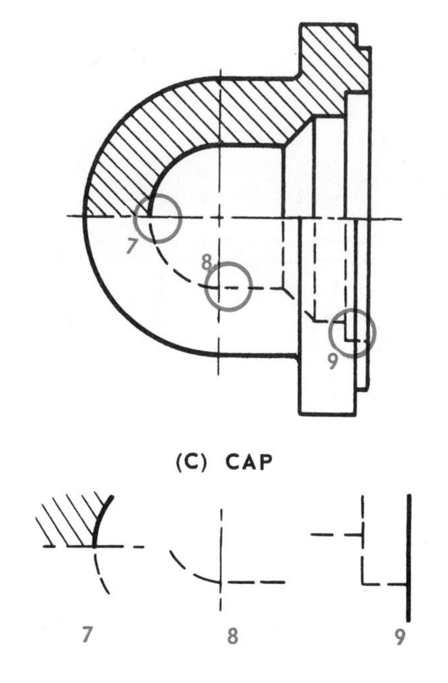

Fig. 2-3-2 Application of hidden lines

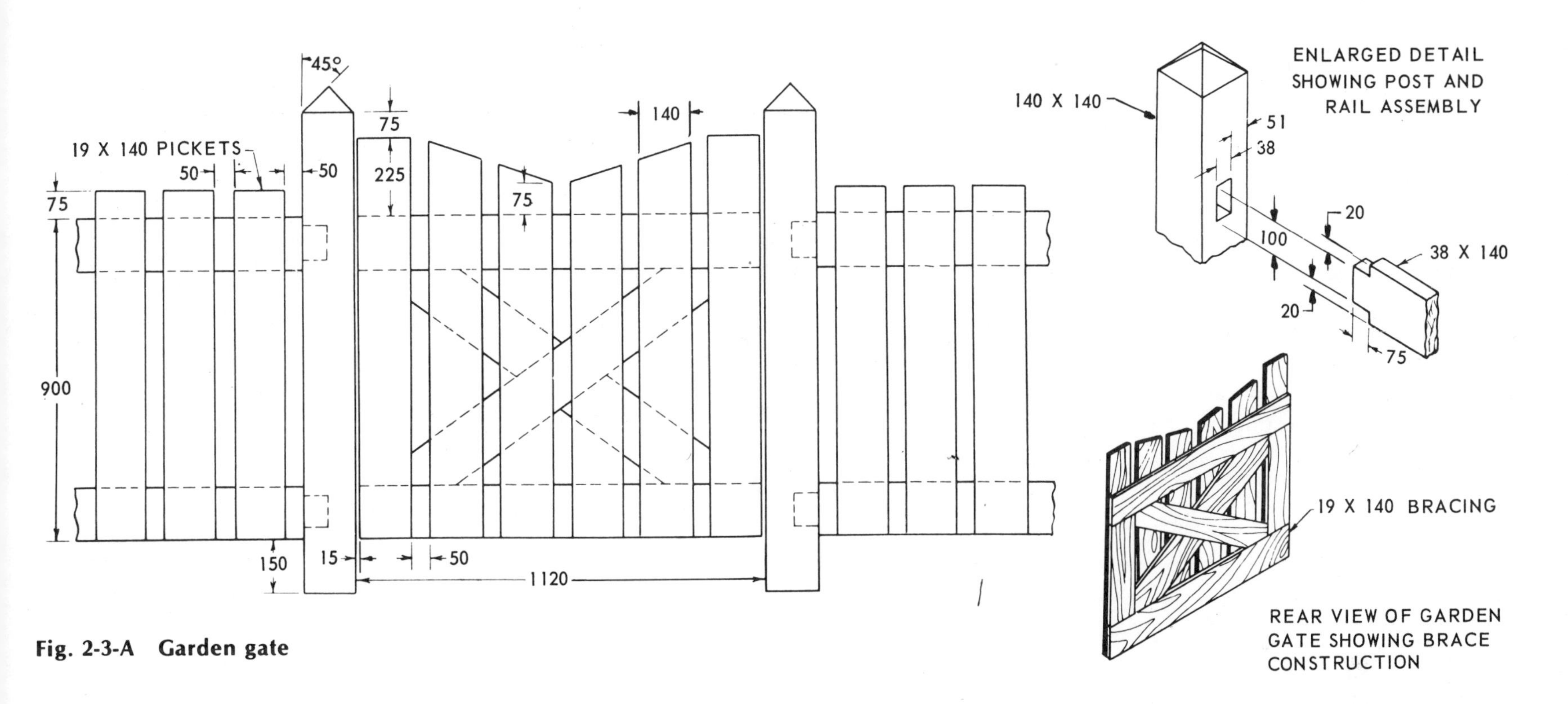

Fig. 2-3-A Garden gate

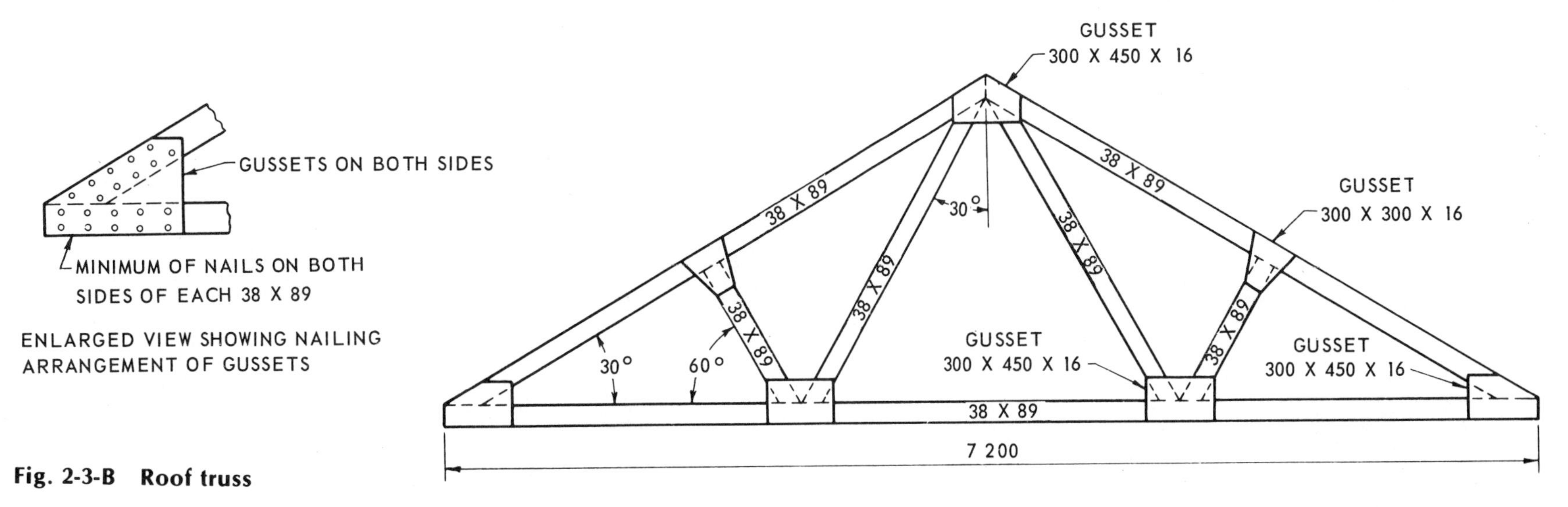

Fig. 2-3-B Roof truss

UNIT 2-4
Centre Lines

Centre lines are used to indicate centre points, axes of cylindrical parts, and axes of symmetry (Fig. 2-4-1). Centre lines should project for a short distance beyond the outline of the part or feature to which they refer.

A centre line is drawn as a thin, broken line of long and short dashes, alternately spaced and should project for a short distance beyond the outline of the part or feature to which they refer.

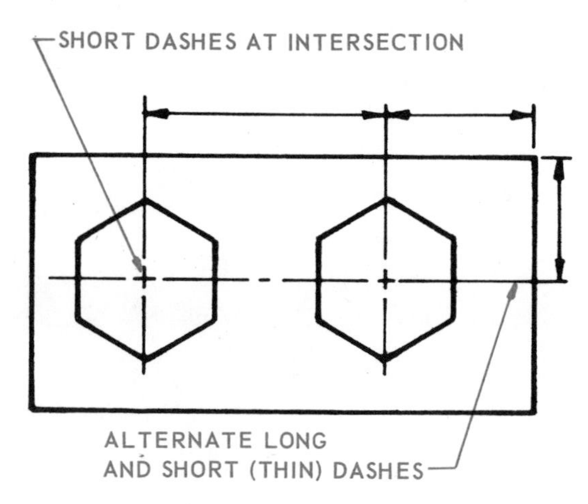

Fig. 2-4-1 Centre line construction

Assignments

Sheet Size - A4, Scale 1:1

1. Template #1, Fig. 2-4-A.
2. Template #2, Fig. 2-4-B.
3. Geometric Patterns, Fig. 2-4-C. Any two of A, B, C.

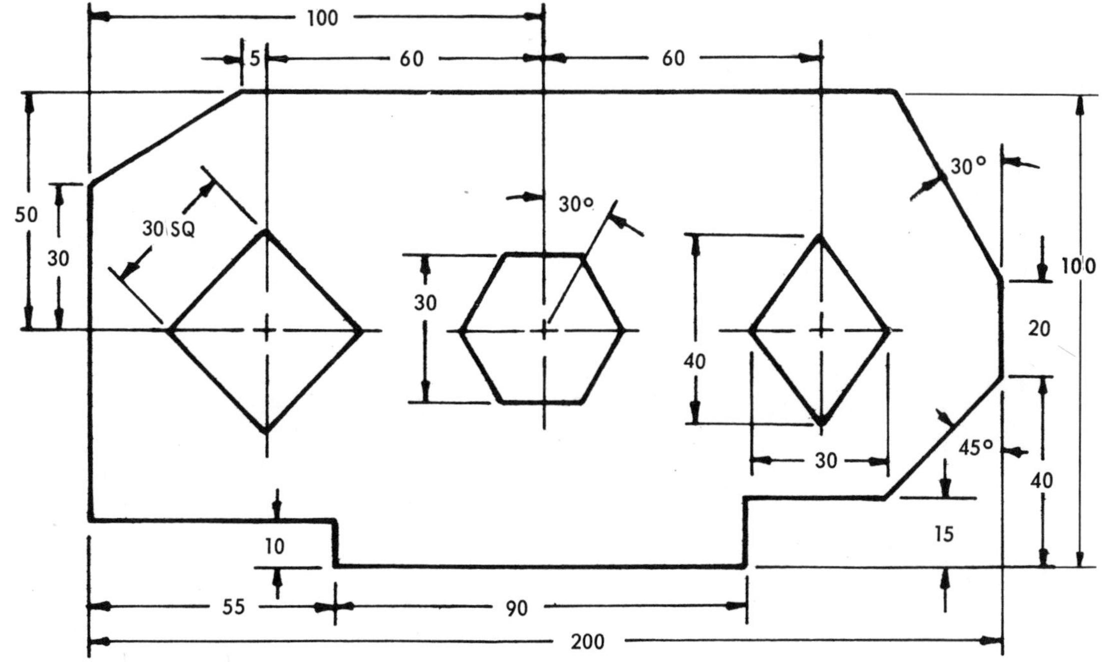

Fig. 2-4-A Template 1

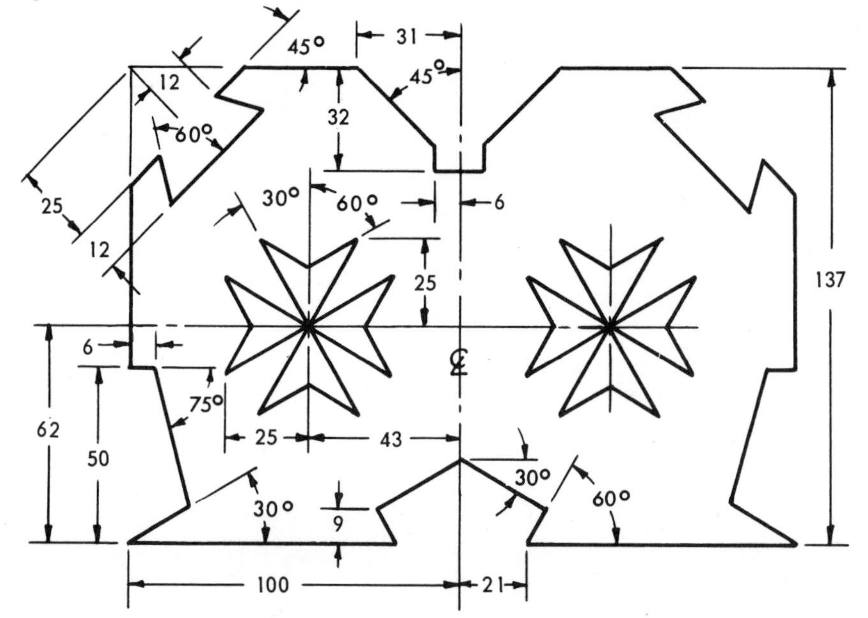

Fig. 2-4-B Template 2

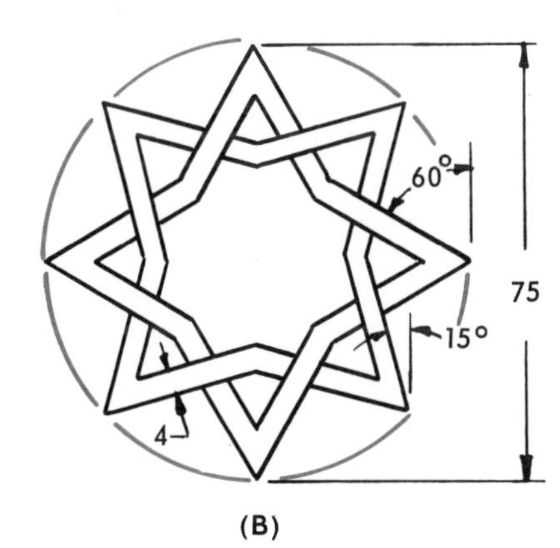

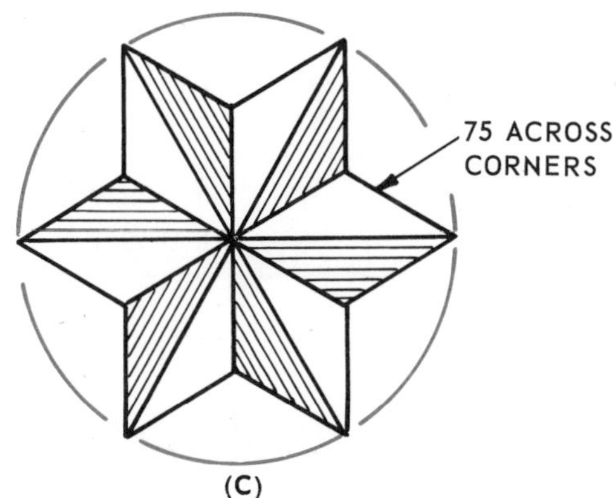

Fig. 2-4-C Geometric patterns

UNIT 2-5
Drawing Circles and Arcs

Circles and arcs are drawn with the aid of a compass or a template. When using a compass it is best to draw circles and arcs heavy the first time, since it is difficult to match up exactly a second time. Note the proper method of sharpening and setting the compass lead shown in Figure 2-5-2.

Centre lines are used to locate the centre of circles and arcs. They are first drawn as light construction lines, then finished as alternate long and short dashes, with the short dashes intersecting at the centre of the circle. (Fig. 2-5-3)

To draw a circle with a compass, locate and draw the centre lines, make a light mark

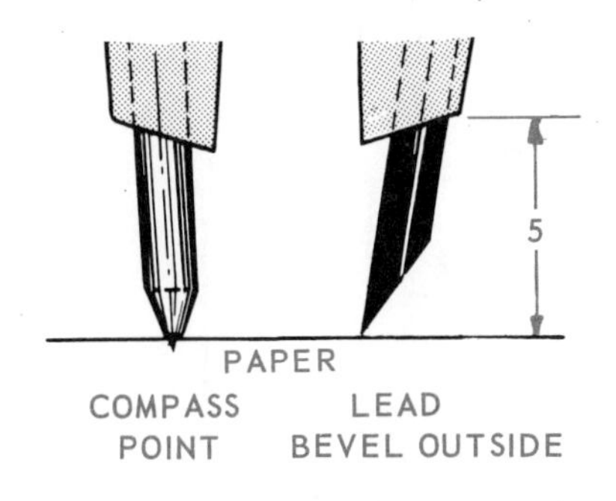

Fig. 2-5-2 Sharpening and setting the compass lead

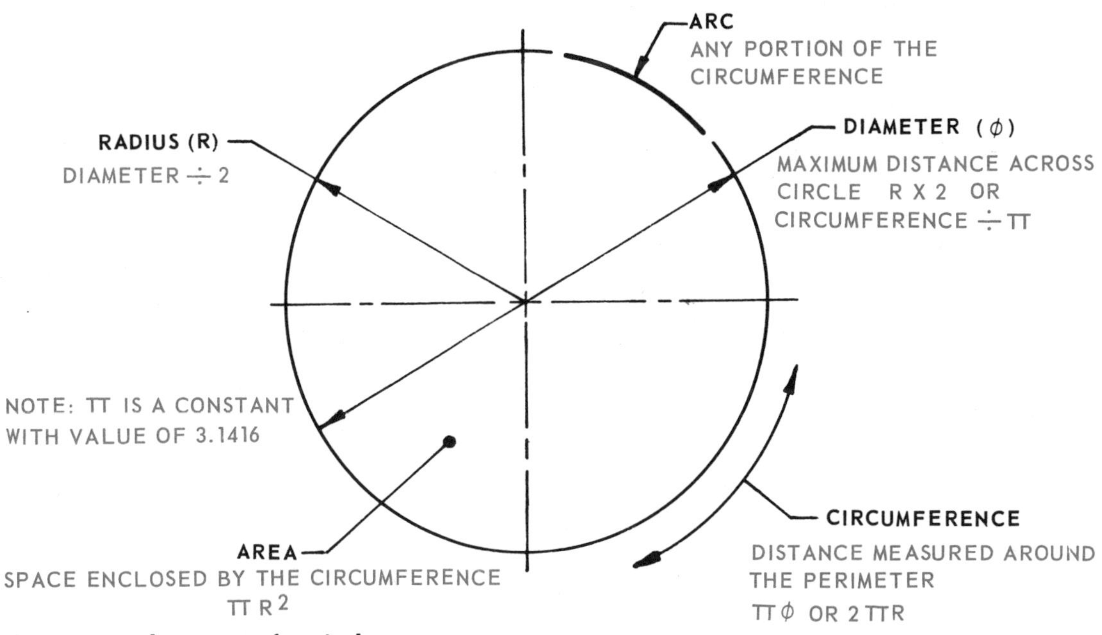

Fig. 2-5-1 Elements of a circle

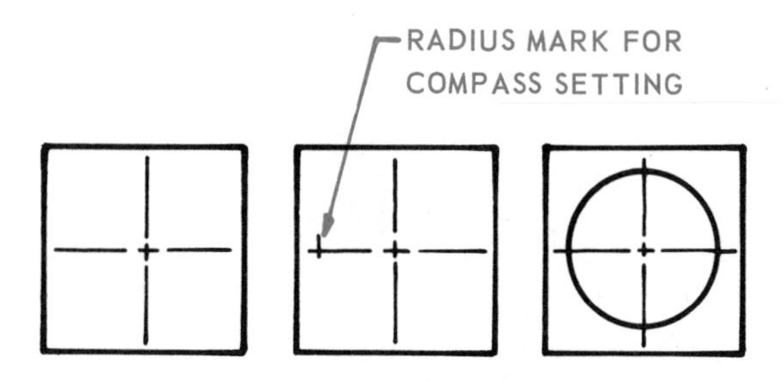

(A) DRAWING A CIRCLE

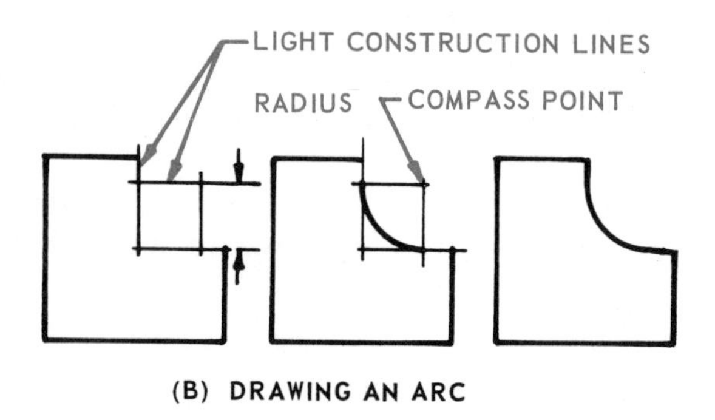

(B) DRAWING AN ARC

Fig. 2-5-3 Drawing circles and arcs

on one of the centre lines for the radius setting (one-half the diameter), locate the steel point carefully where the centre lines cross, adjust the lead to the mark and rotate the compass to produce a dark thick line.

When using a circle template, choose the correct diameter, line up the marks on the template with the centre lines and trace a dark thick line.

The drawing of arcs should be done before the tangent lines are made heavy. Draw light construction lines to establish the compass point and check to make certain that the compass lead meets properly with both tangent lines before drawing the arc.

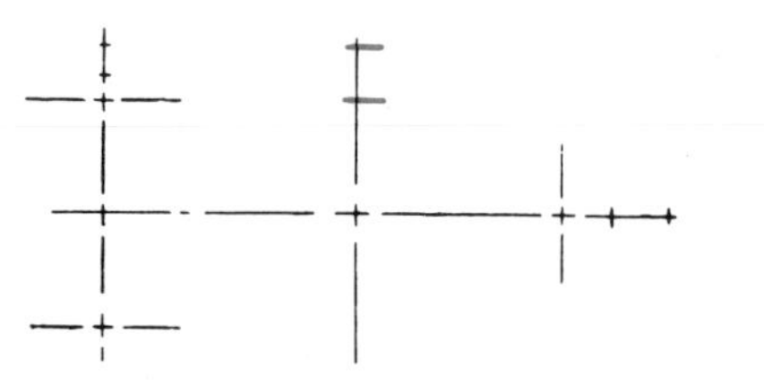

(A) ESTABLISH CENTRE LINES AND RADII MARKS

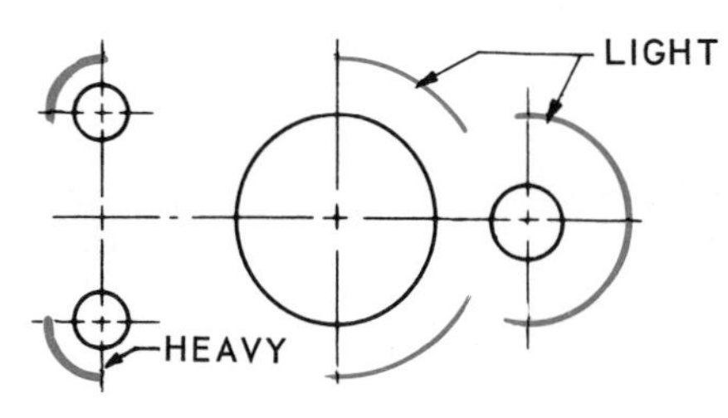

(B) DRAW CIRCLES AND ARCS

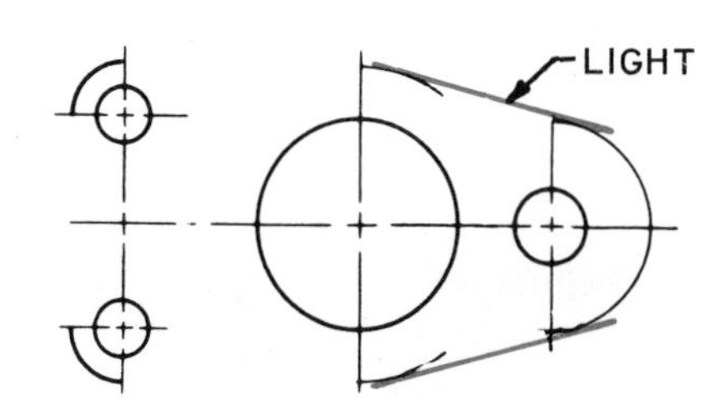

(C) DRAW TANGENT LINES

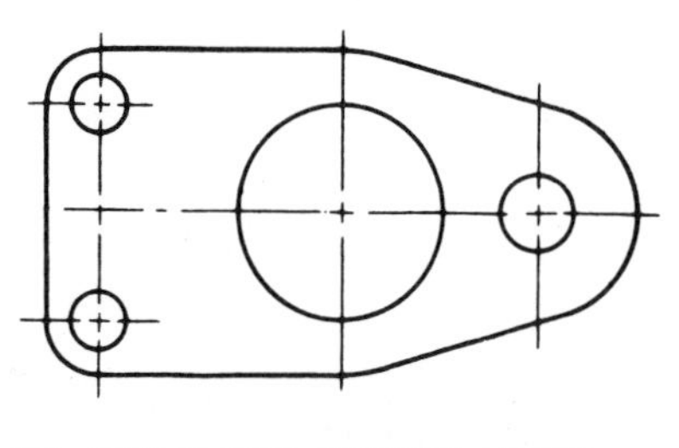

(D) COMPLETE OBJECT LINES

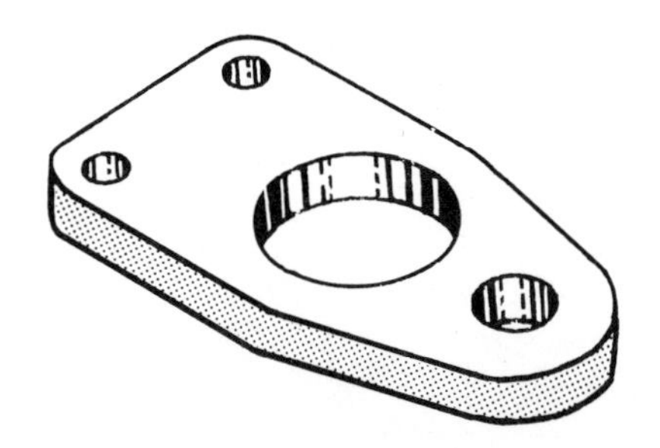

Fig. 2-5-4 Sequence of steps for drawing a view having circles and arcs

Assignments

1. Template, Fig. 2-5-A. Sheet size-A4, scale 1:1.
2. Shaft Support, Fig. 2-5-B. Sheet size-A4, scale 1:1.
3. Dial Indicator, Fig. 2-5-C. Sheet size-A4, scale 1:1.
4. Dart Board, Fig. 2-5-D. Sheet size-A4, scale 1:5.

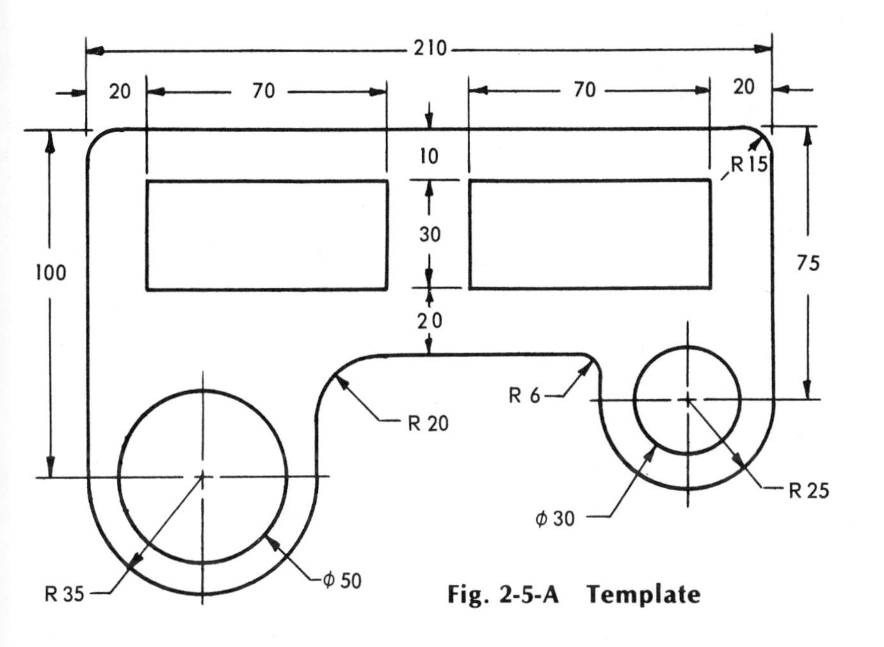

Fig. 2-5-A Template

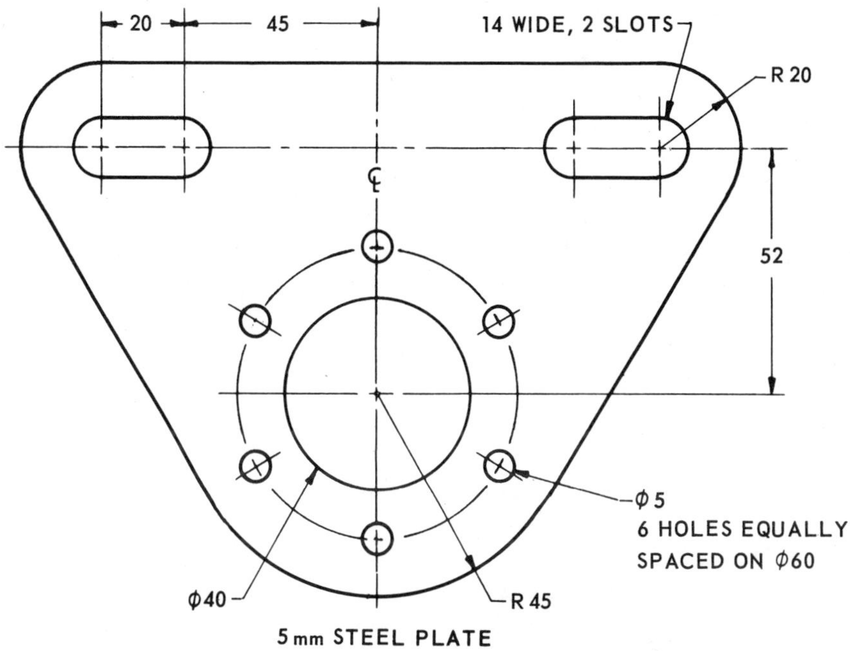

Fig. 2-5-B Shaft support

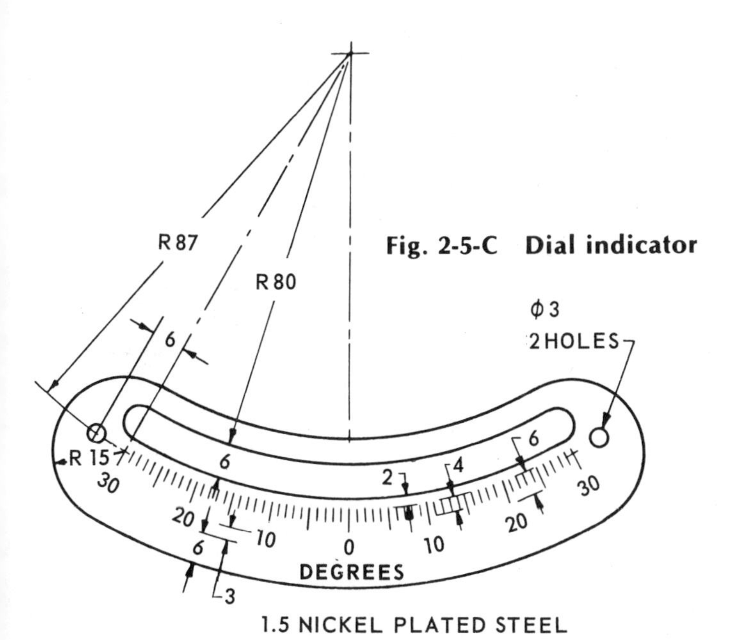

Fig. 2-5-C Dial indicator

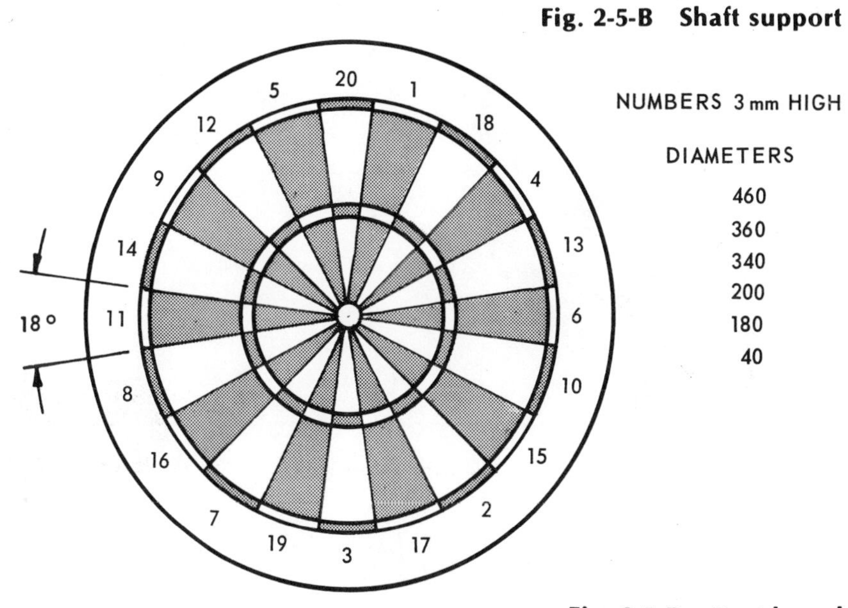

Fig. 2-5-D Dart board

UNIT 2-6
Drawing Irregular Curves

For drawing curved lines in which, unlike circular arcs, the radius of curvature is not constant, an instrument known as an irregular curve or French curve is used.

The patterns for these curves are based on various combinations of ellipses, spirals, and other mathematical curves. The curves are available in a variety of shapes and sizes.

Generally, you plot a series of points of intersection along the desired path, then use the French curve to join these points so that a smooth-flowing curve results.

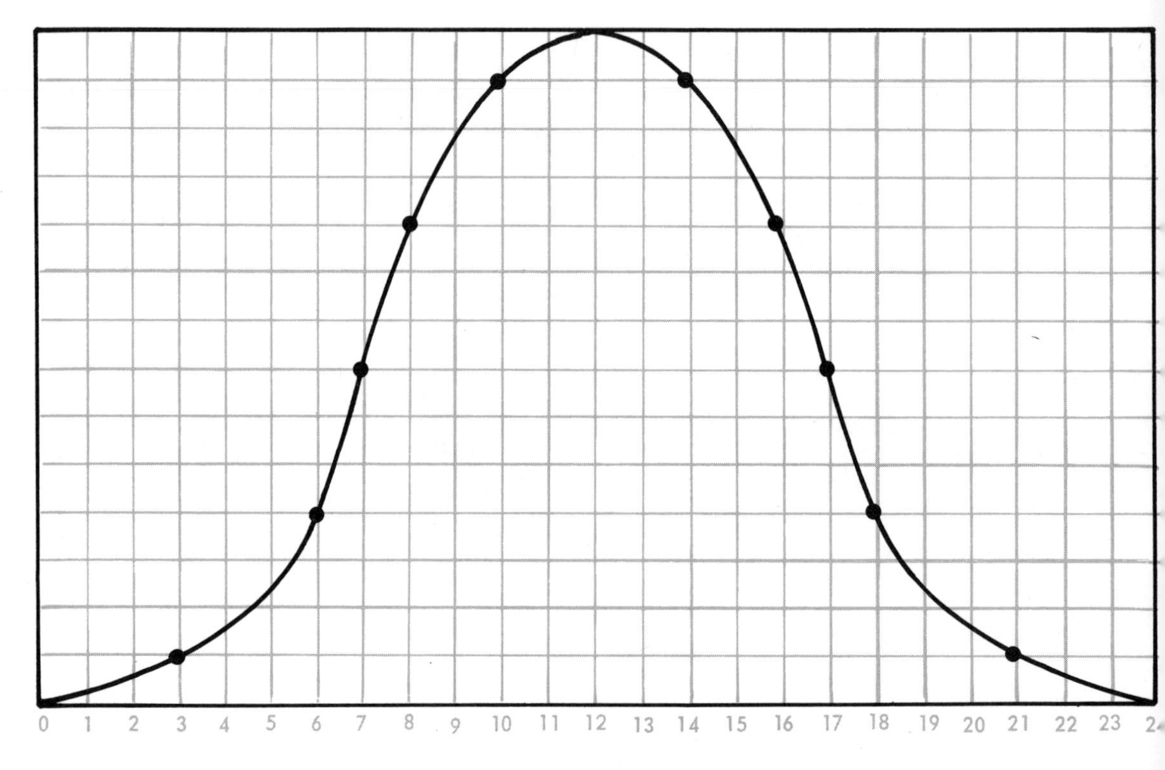

Fig. 2-6-A Line graph

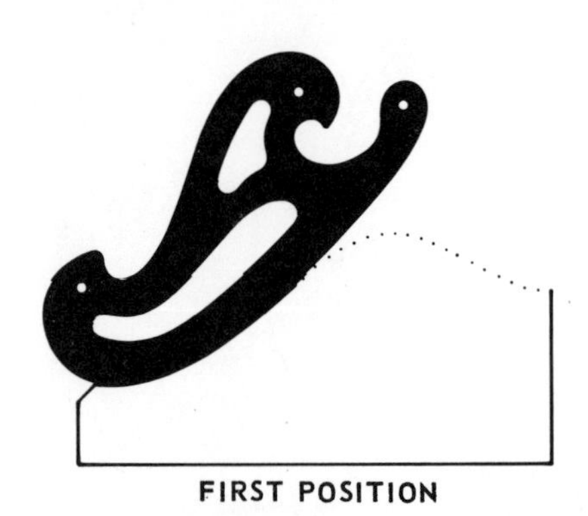

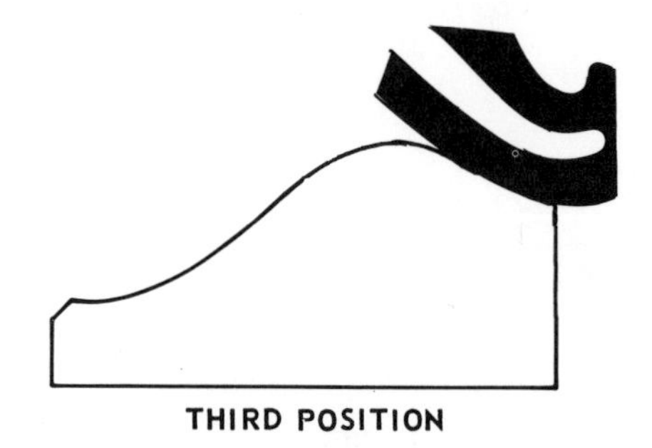

Fig. 2-6-1 Steps in drawing an irregular curve

Assignments

Use grid paper, sheet size-A4, or draw required grid lines.

1. Line graph, Fig. 2-6-A, scale 1:1. Plot and draw the graph line using an irregular curve.
2. Table Leg, Fig. 2-6-B, scale 1:2. Plot and draw shape using an irregular curve.

2 HOLES,
DRILL AND CSK FOR
10 WOOD SCREW
25 mm SQUARES

Fig. 2-6-B Table leg

UNIT 2-7
Sketching

Freehand sketching is a necessary part of drafting because the drafter in industry frequently sketches ideas and designs prior to making instrumental drawings.

The drafter may also use sketches to explain thoughts and ideas to other people in discussions of mechanical parts and mechanisms. Sketching, therefore, is an important method of communication.

Practice in sketching helps the student to develop a good sense of proportion and accuracy of observation. It can be used to advantage when you are learning the fundamentals of drafting practice and procedures.

A fairly soft (HB, F, or H) pencil should be used for preliminary practice. Many types of graph or ordinate paper are available and can be used to advantage when close accuracy to scale or proportion is desirable. The directions in which horizontal, vertical, and oblique lines are sketched are illustrated in Figure 2-7-2.

Since the shapes of objects are made up of flat and curved surfaces, the lines forming views of objects will be both straight and curved. Do not attempt to draw long lines with one continuous stroke. First plot points along the desired line path; then connect these points with a series of light strokes.

When you are sketching a view (or views), first lightly sketch the overall size as a rectangular or square shape, estimating its

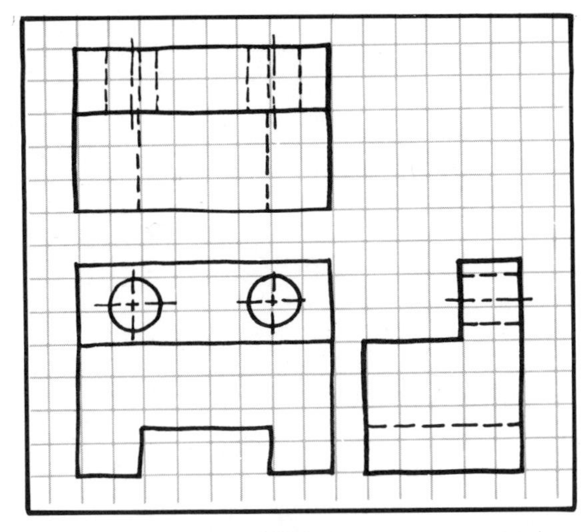

(A) CO–ORDINATE SKETCHING PAPER

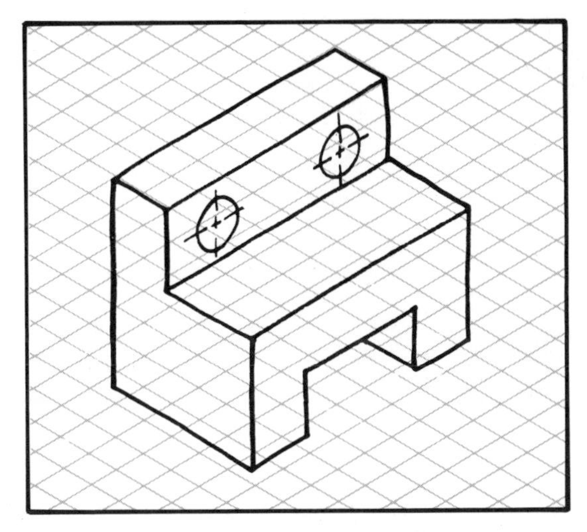

(B) ISOMETRIC SKETCHING PAPER

Fig. 2-7-1 Sketching paper

proportions carefully. Then add lines for the details of the shape, and thicken all lines forming a part of the view. See Figure 2-7-3.

Figure 2-7-4 shows two methods of sketching circles. Figure 2-7-5 illustrates, both pictorially and orthographically, the use of graph paper for the sketching of a machine part.

Assignment

1. On grid paper sketch the figures shown in Fig. 2-7-2.
2. Structural steel shapes, Fig. 2-7-A, grid paper, sheet size-A4. Sketch any two.

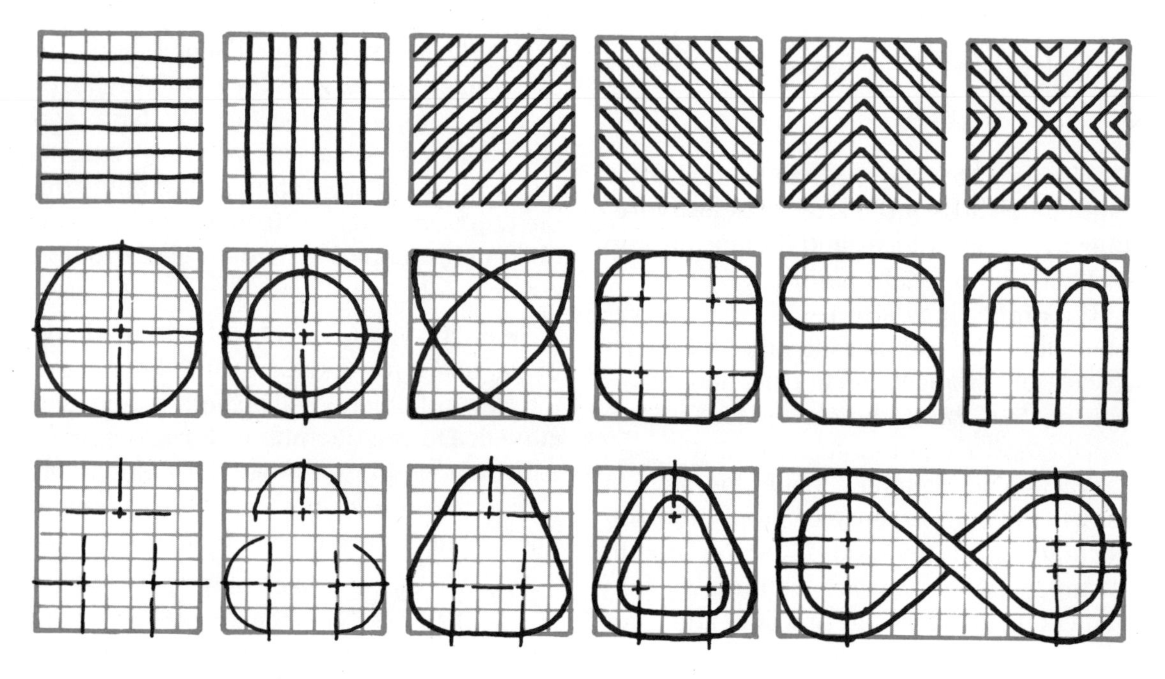

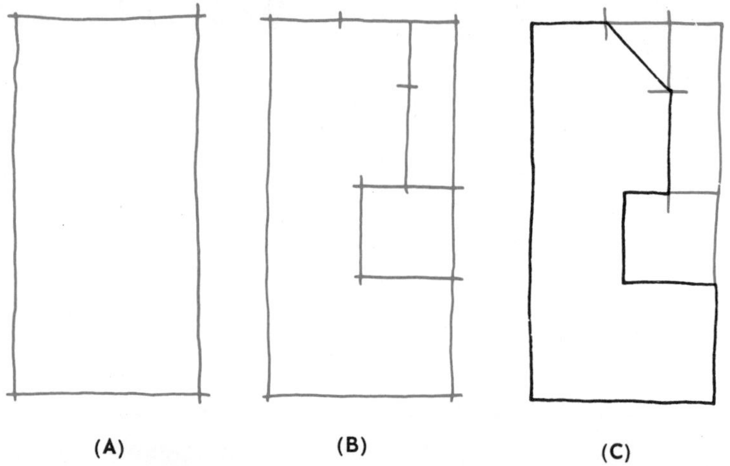

Fig. 2-7-2 Sketching lines and circles

(A) (B) (C)

Fig. 2-7-3 Sketching a view having straight lines

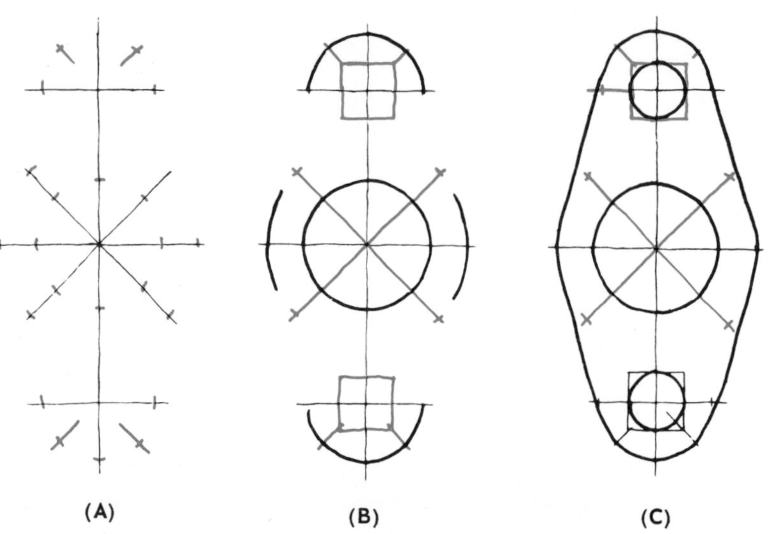

Fig. 2-7-4 Sketching a figure having circles and arcs

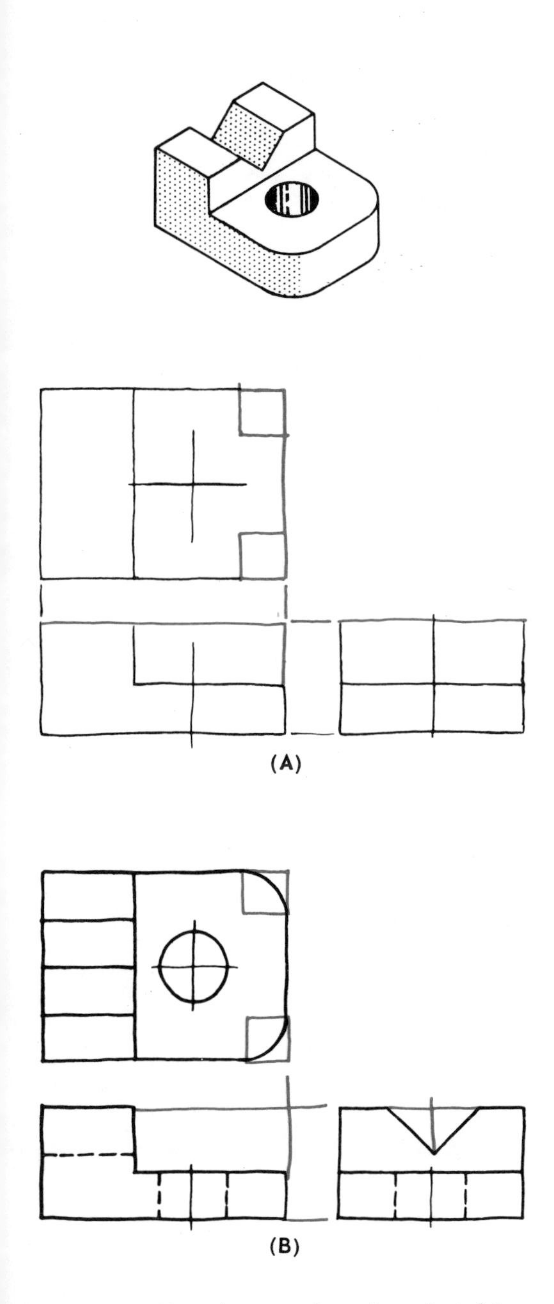

Fig. 2-7-5 Usual procedure for sketching three views

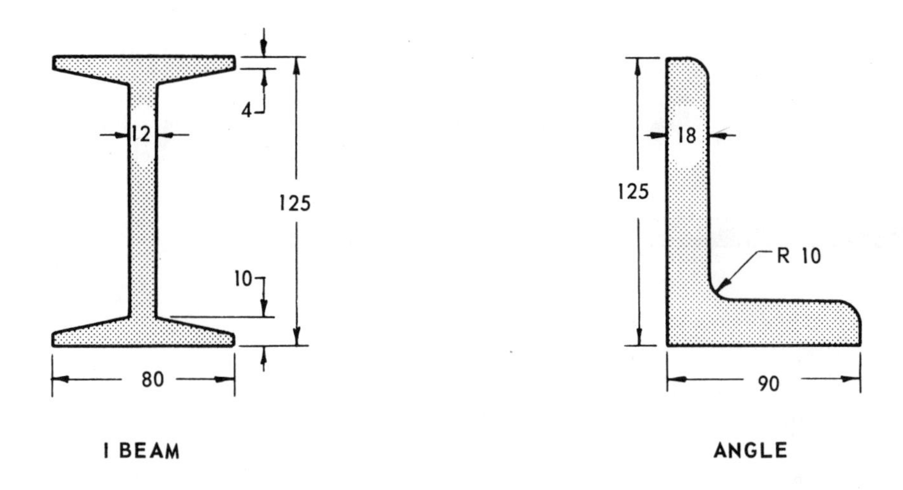

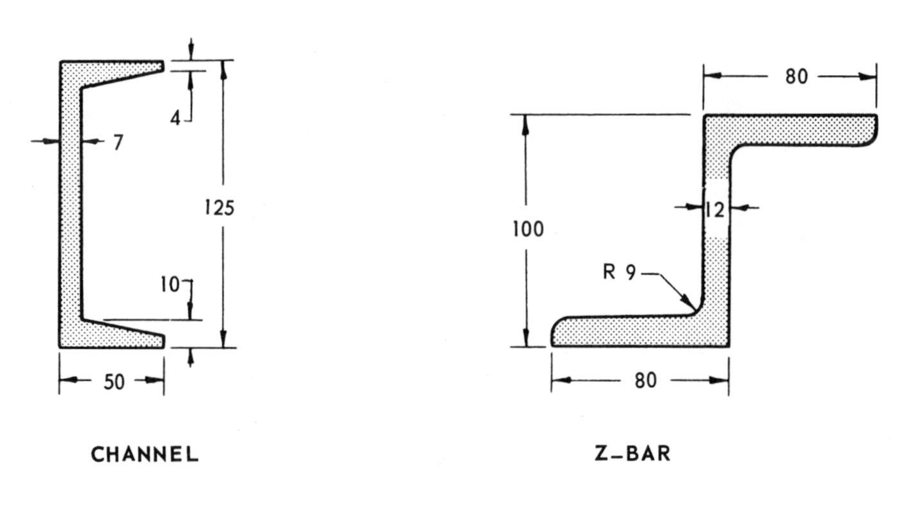

Fig. 2-7-A Structural steel shapes

Chapter 3

Theory of Shape Description

UNIT 3-1
Shape Description By Views

When looking at objects, we normally see them as three-dimensional; as having width, depth and height; or length, width and height. The choice of terms used depends on the shape and proportions of the object.

Spherical shapes, such as a basketball, would be described as having a certain **diameter.** (**one** term)

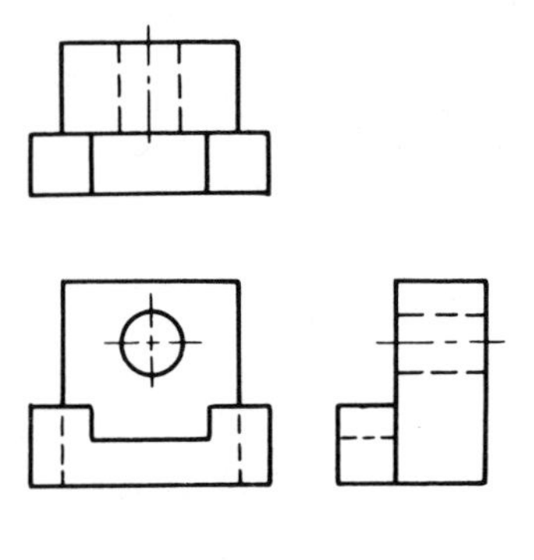

(A) ORTHOGRAPHIC PROJECTION

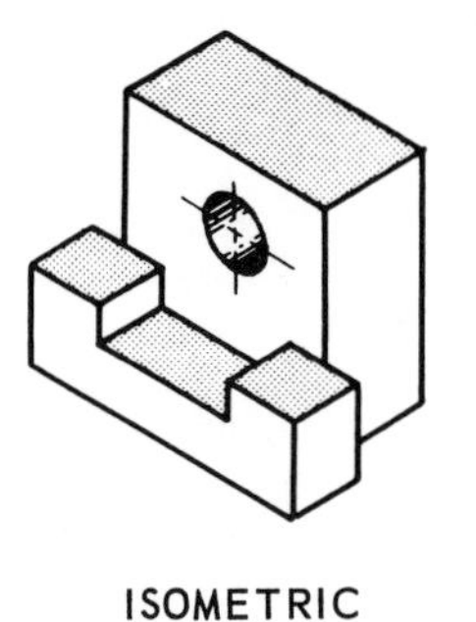

ISOMETRIC

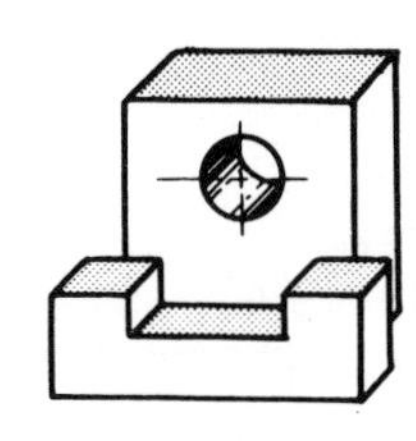

OBLIQUE

PERSPECTIVE

(B) PICTORIAL DRAWINGS

Fig. 3-1-1 Types of drawings

Cylindrical shapes, such as a baseball bat, would have **diameter** and **length.** However, a hockey puck would have **diameter** and **thickness.** (**two** terms)

Objects which are not spherical or cylindrical require **three** terms to describe their overall shape. The terms used for a car would probably be **length, width** and **height;** for a filing cabinet — **width, height** and **depth;** for a sheet of drawing paper — **length, width** and **thickness.** The terms used are interchangeable according to the **proportions** of the object being described, and the **position** it is in when being viewed. For example, a hydro pole lying on the ground would be described as having **diameter** and **length,** but when placed in a vertical position, its dimensions would be **diameter** and **height.**

In general, distances from left to right are referred to as width or length, distances from front to back as depth or width, and vertical distances (except when very small in proportion to the others) as height.

On drawings, the multi-dimensional shape is represented by a view or views on the flat surface of the drawing paper.

Pictorial Views

Pictorial drawings represent the shape with just one view, and are frequently used for illustrative purposes, for installation and maintenance drawings, and do-it-yourself projects for the general public. However, the majority of parts manufactured in industry are too complicated in shape and detail

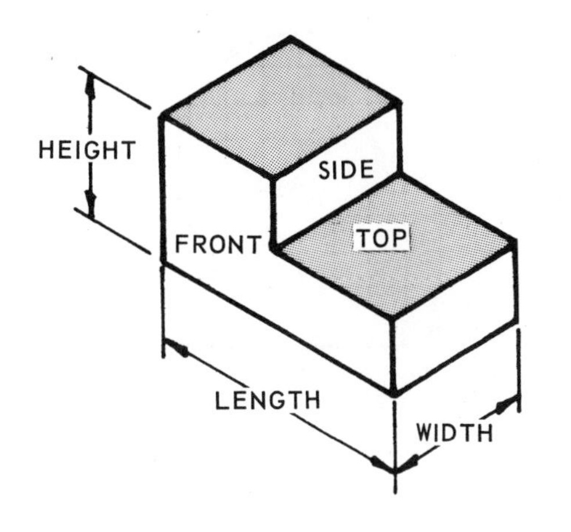

(A) PICTORIAL DRAWING (ISOMETRIC)

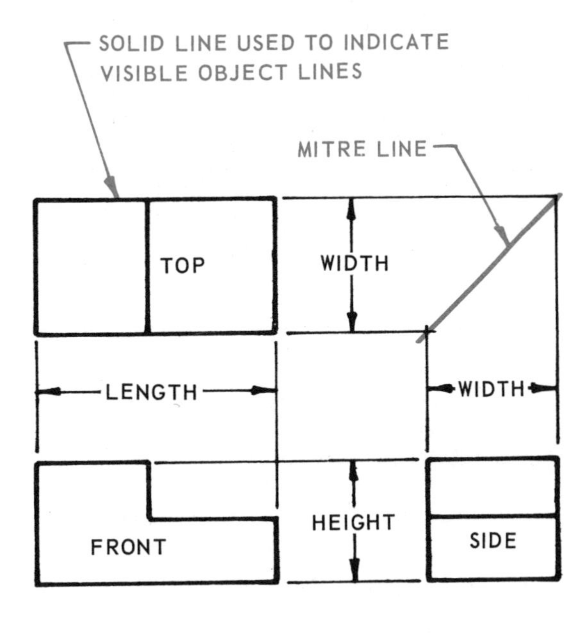

(B) ORTHOGRAPHIC PROJECTION DRAWING

Fig. 3-1-2 Description by views

to be described successfully by a pictorial view.

Orthographic Projection

In technical drawing, **orthographic** views are used to describe the shapes of objects completely and exactly. The word "orthographic" is derived from two Greek words: *orthos,* meaning straight, correct, at right angles to, and *graphikos,* to write or describe by drawing lines.

An orthographic view is what you would see looking directly at one side or "face" of the object. When looking directly at the front face, you would see width and height — two dimensions — but not the third dimension, depth. Each orthographic view gives two of the three major dimensions.

One View Drawings

Some objects, such as flat templates and parts which are mainly cylindrical in shape, only require one orthographic view. The third dimension, such as thickness, may be expressed by a note, or by descriptive words or symbols.

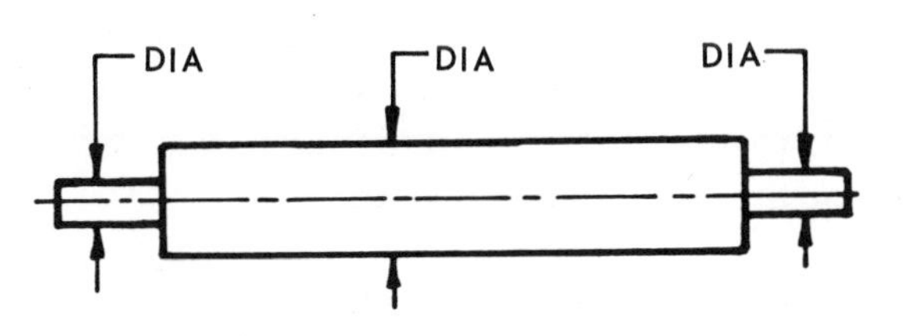

Fig. 3-1-3 One-view drawings

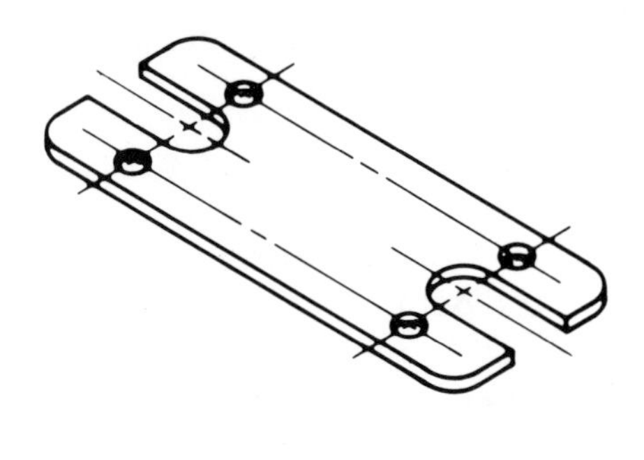

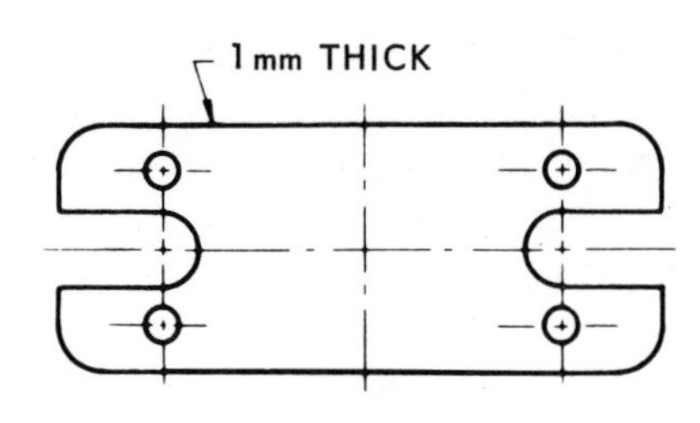

Fig. 3-1-4 One-view drawing

Two-View Drawings

Frequently, only two views are necessary to explain fully the shape of an object. For this reason, some drawings consist of top and front views only, or front and right side views only.

Two views are usually sufficient to explain fully the shape of cylindrical objects; if three views were used, two of them would be identical, or almost identical, depending on the detail structure of the part.

Multi-View Drawings

Except for complex objects of irregular shape, it is seldom necessary to draw more than three views. Each view represents a different side or face of the object, and the views are *projected* one from the other and arranged in a systematic manner, thus the term "orthographic projection".

The principles of orthographic projection can be applied in four different "angles" or systems: first -, second -, third -, and fourth-angle projection.

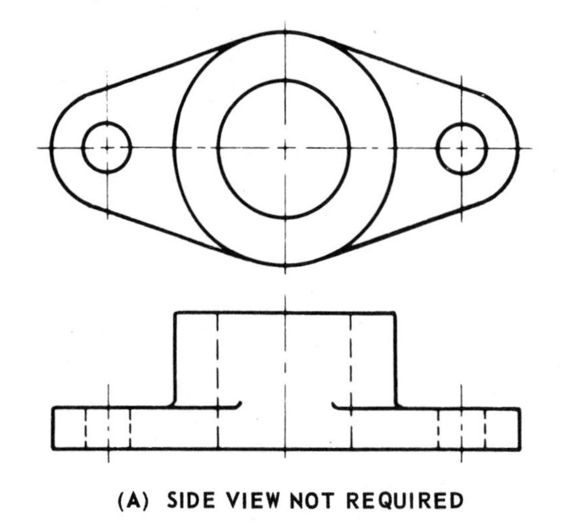

(A) SIDE VIEW NOT REQUIRED

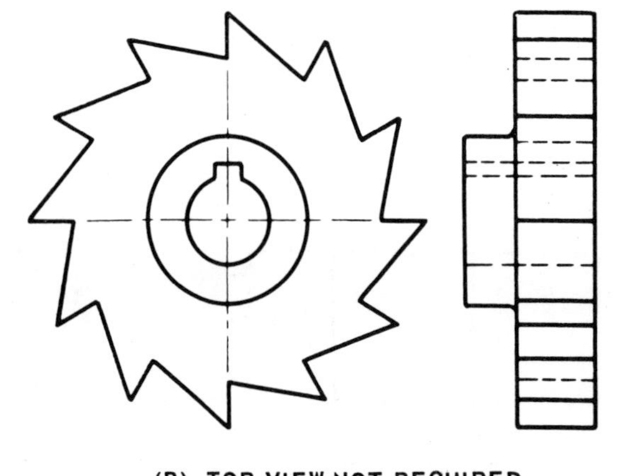

(B) TOP VIEW NOT REQUIRED

Fig. 3-1-5 Two-view drawings

Only two systems — first-angle and third-angle projection — are used. Third-angle projection is used in Canada, the United States, and many other countries throughout the world. First-angle projection is used mainly in European and Asiatic countries.

The basic rule for third-angle projection is this: *each view is a picture of the surface nearest to it in an adjacent view.* Applying this rule, the top view is placed above the front view, the right side view is to the right of the front view, etc.

Selection Of Views

Many mechanical parts do not have a definite "front" or "side" or "top," as do objects such as refrigerators, desks or houses, and their shapes vary from the simple to the complex. Decisions have to be made on how many views, and which views, will be drawn. Following are some basic guidelines.

1. Draw those views which are necessary to fully explain the shape.

2. The front view is usually the "key" view, and shows the width or length of the object and gives the most information about its shape. When the longest dimension is drawn in a horizontal position, the object will seem balanced.

3. Choose those views which will show most of the detailed features of the object as "visible," thus avoiding the extensive use of "hidden" feature lines.

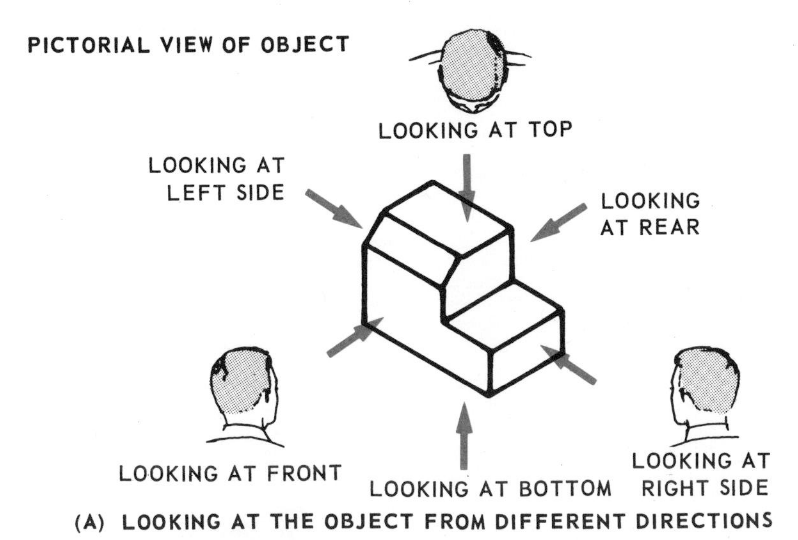

(A) LOOKING AT THE OBJECT FROM DIFFERENT DIRECTIONS

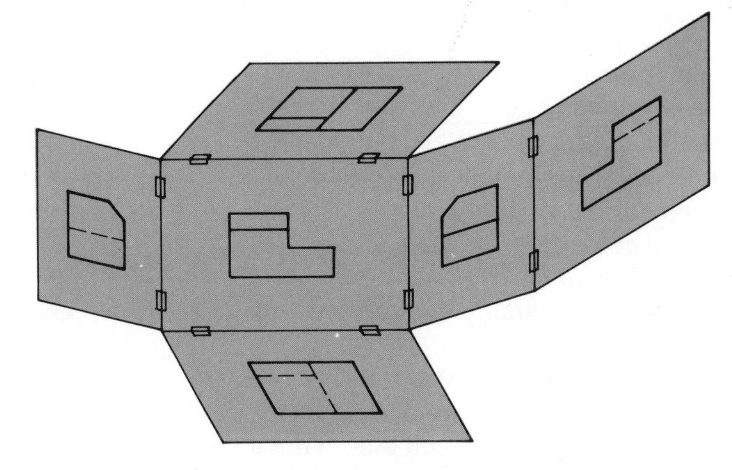

(C) UNFOLDING GLASS BOX TO GIVE THIRD ANGLE LAYOUT OF VIEWS

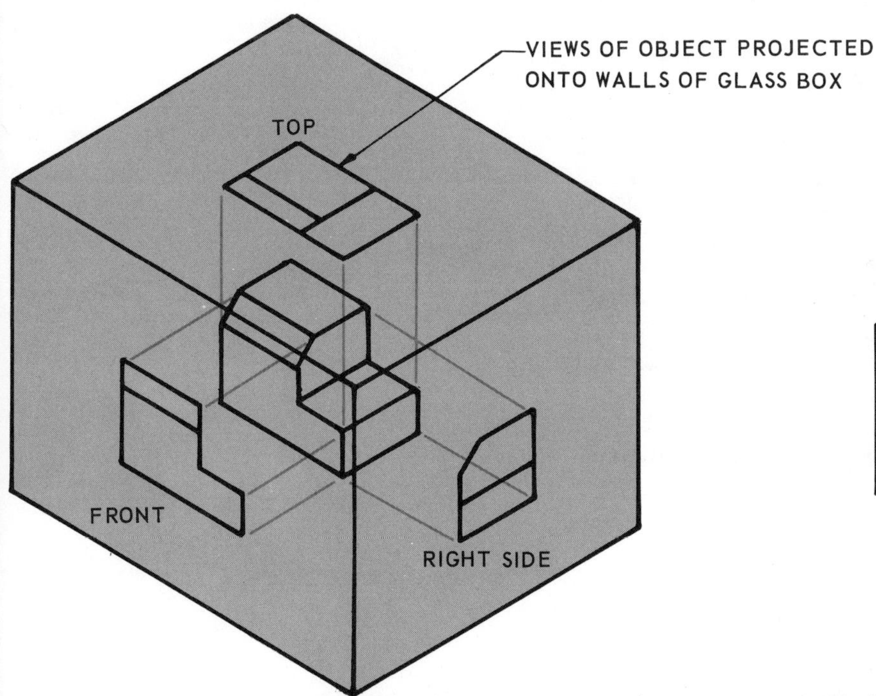

(B) OBJECT ENCLOSED IN GLASS BOX

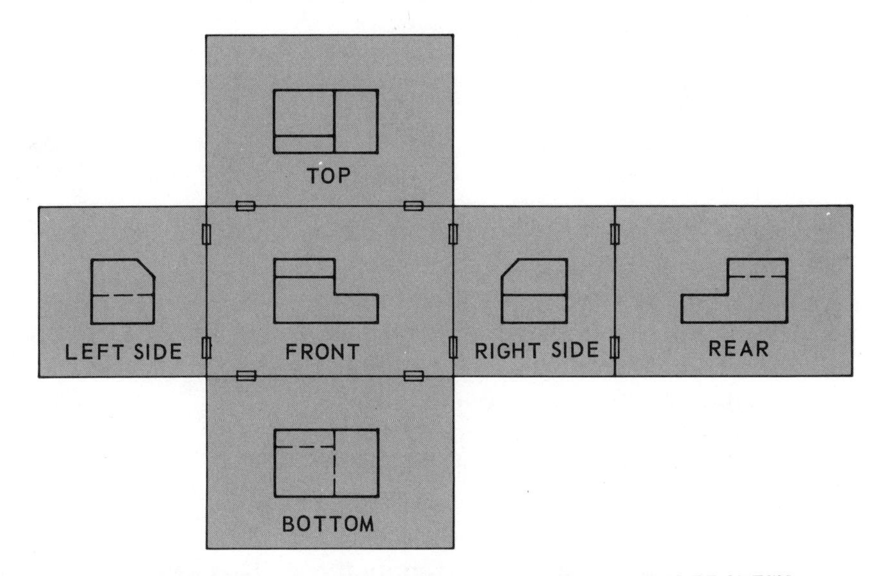

(D) GLASS BOX UNFOLDED SHOWING THE PROJECTION OF THREE VIEWS

Fig. 3-1-6 Systematic arrangement of views

Surface Terms

When describing the shape of an object, reference is often made to the types of *surfaces* found on the object, relative to the three principal viewing planes — horizontal plane, vertical plane, profile plane. These surfaces can be identified as:

Parallel— flat surfaces which are parallel to the three principal viewing planes;

Hidden — surfaces which are hidden in one or more reference planes;

Inclined— flat surfaces which are inclined in one plane and parallel to the other two planes;

Oblique— flat surfaces which are inclined in all three reference planes;

Circular— surfaces which have diameter or radius.

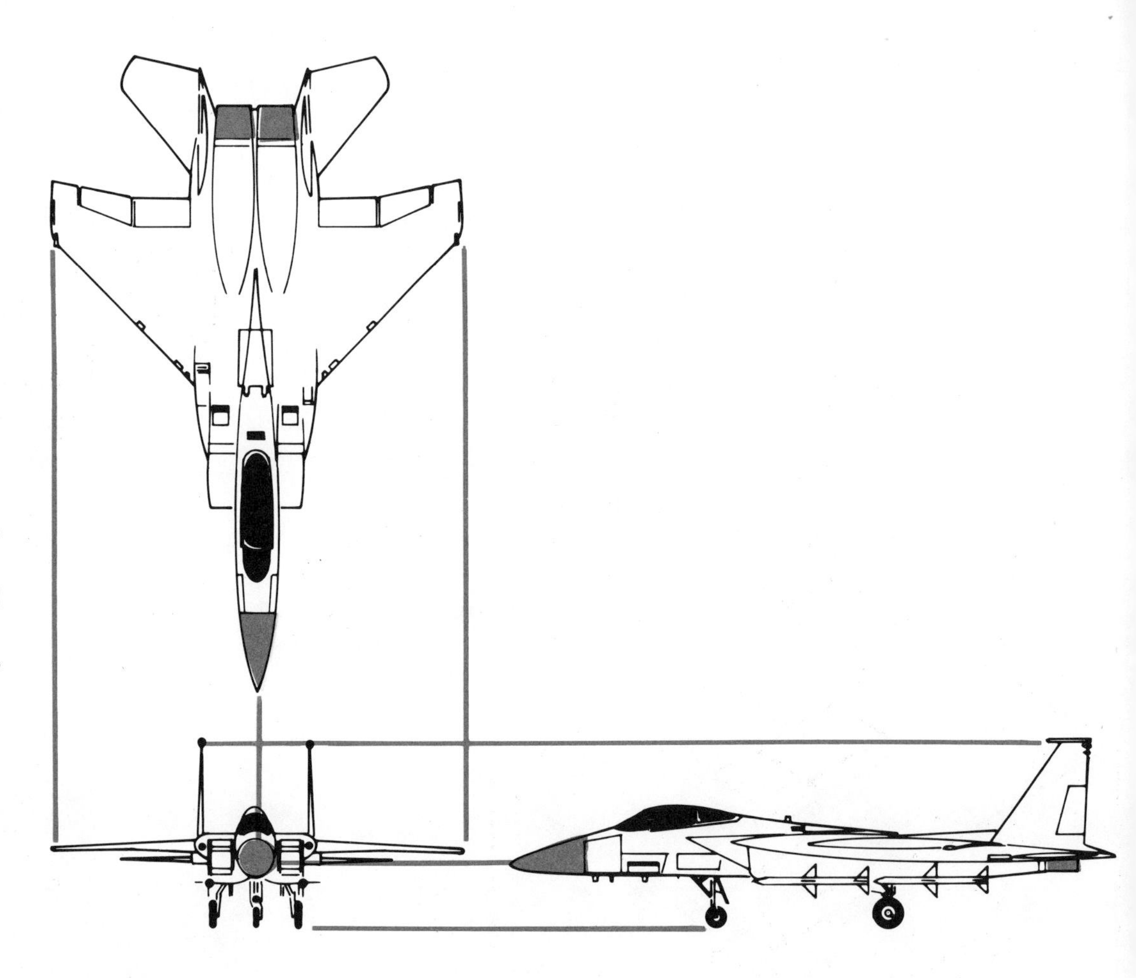

Fig. 3-1-7 Systematic arrangement of views

Assignments

1. On an A4 sheet of grid paper, make a two-view sketch, at a suitable scale, of one of the following: a coffee mug, a dinner plate, a drinking glass or mug. Identify the terms that would be used to best describe the object's overall size.
2. On an A4 sheet of grid paper, make a three-view sketch, at a suitable scale, of one of the following: a kitchen table, a filing cabinet, a writing desk, a car, a house, a chest of drawers. Identify the terms that would be used to best describe the object's overall size.

Review For Assignments

Unit 2-1 Scales
Unit 2-7 Sketching

UNIT 3 - 2

Arrangement of Views

Spacing The Views

It is important for clearness and good appearance that the views be well balanced on the drawing sheet, whether the drawing shows one view, two views, three views or more. You must plan the space required for the number of views to be drawn, then block them out on the drawing sheet to leave a margin that is reasonably equal all around the drawing. You have already done this for one-view drawings. Figure 3-2-1 shows how to balance the views for a three-view drawing. For a drawing with two or more views, follow these guidelines:

1. Decide on the views to be drawn and the scale to be used, eg. 1:1 or 1:2.

2. Make a sketch of the space required for each of the views to be drawn, showing these views in their correct location. (A simple rectangle for each view will be adequate, Figure 3-2-1(B).)

3. Put on the overall drawing sizes for each view. (These sizes are shown as W, D, and H, Figure 3-2-1(B))

4. Decide upon the space to be left between views, dimensions X and Y on Figure 3-2-1(B) (These spaces should be sufficient for the parallel dimension lines to be placed between views. For most drawing projects, 40 mm is sufficient.)

5. Total these dimensions to get the overall horizontal distance (A) and overall vertical distance (B).

6. Select the drawing sheet to best accommodate the overall size of the drawing with suitable open space around the views.

7. Measure the "drawing space" remaining after all border lines, title strip or title block, etc. are in place. (Fig. 3-2-1(C))

8. Take one-half of the difference between distance A and the horizontal "drawing space" to establish Plane 1.

9. Take one-half of the difference between distance B and the vertical "drawing space" to establish Plane 2.

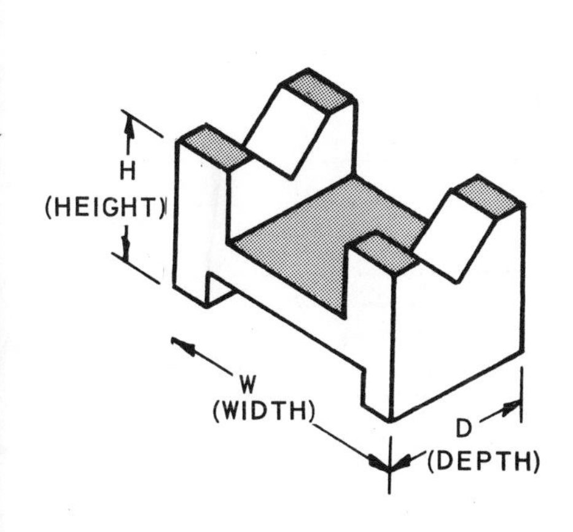

(A) DECIDING THE VIEWS TO BE DRAWN AND THE SCALE TO BE USED

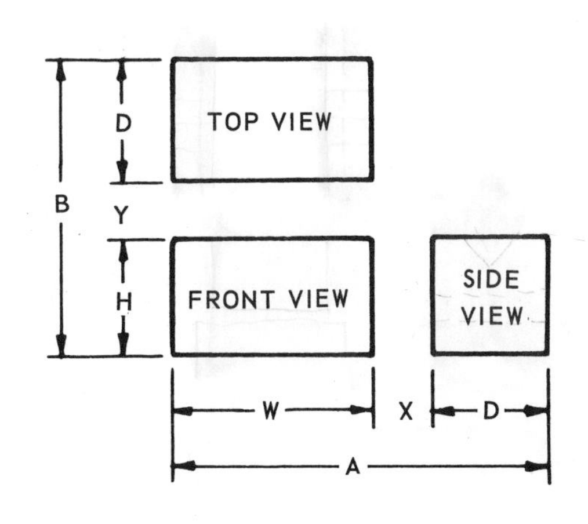

(B) CALCULATING DISTANCES A AND B

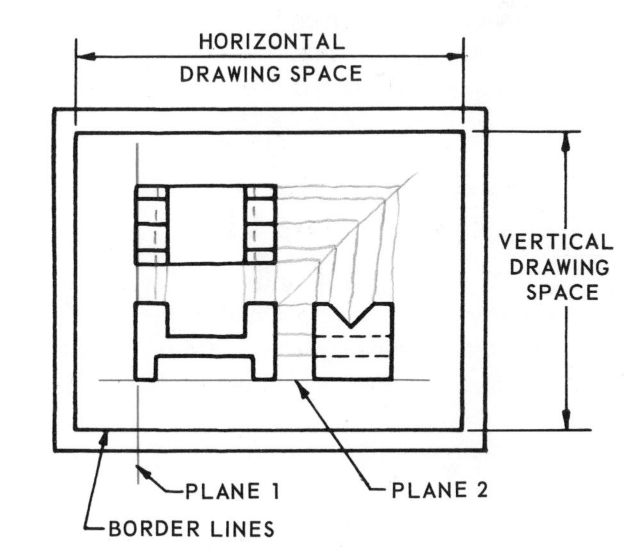

(C) ESTABLISHING LOCATION OF PLANES 1 AND 2

Fig. 3-2-1 Balancing the drawing on paper

Use of a Mitre Line

The use of a mitre line provides a convenient method of constructing the third view once two views are established. (Fig. 3-2-2)

Using A Mitre Line To Construct the Right Side View

1. Given the top and front views, project lines to the right of the top view.
2. Establish how far from the front view the side view is to be drawn. (Distance "D")
3. Construct the mitre line at 45° to the horizon.
4. Where the horizontal projection lines of the top view intersect the mitre line, drop vertical projection lines.
5. Project horizontal lines to the right of the front view and complete the side view.

Using A Mitre Line To Construct the Top View

1. Given the front and side views, project vertical lines up from the side view.
2. Establish how far away from the front view the top view is to be drawn. (Distance "D")
3. Construct the mitre line at 45° to the horizon.
4. Where the vertical projection lines of the side view intersect the mitre lines, project horizontal projection lines to the left.
5. Project vertical lines up from the front view and complete the top view.

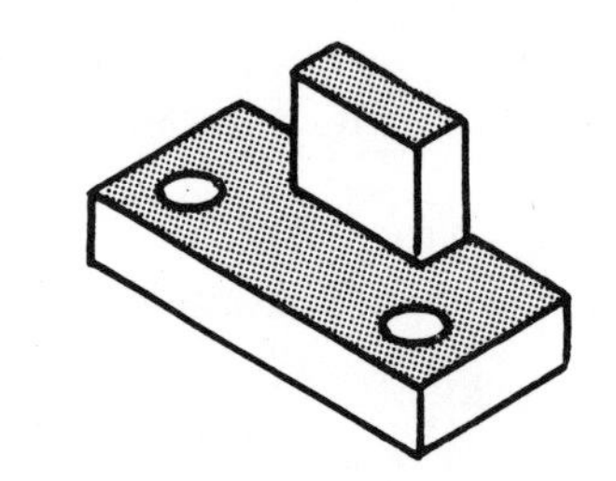

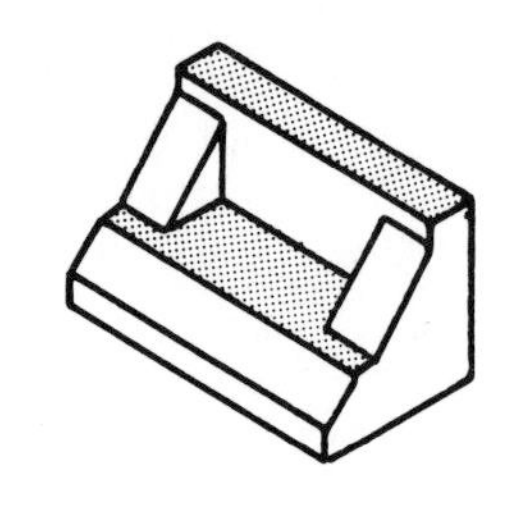

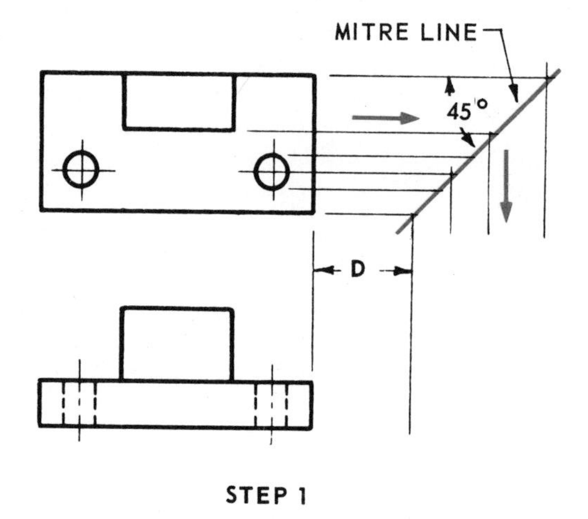

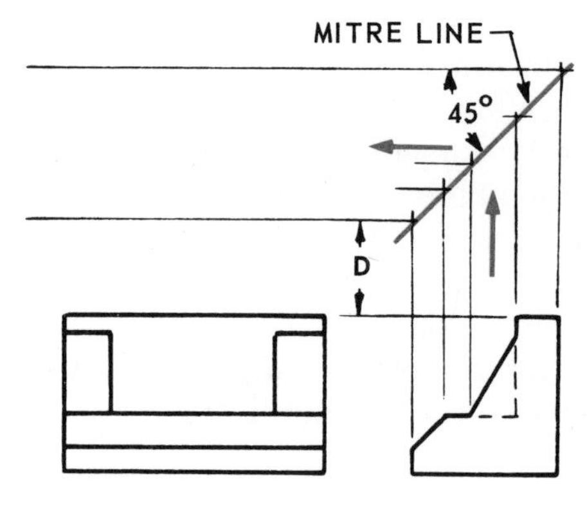

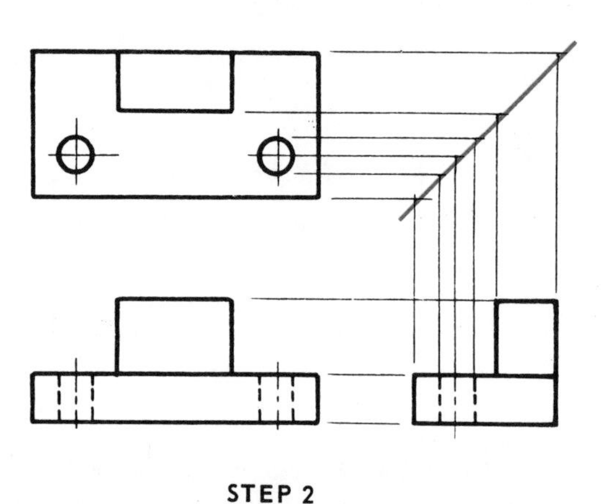

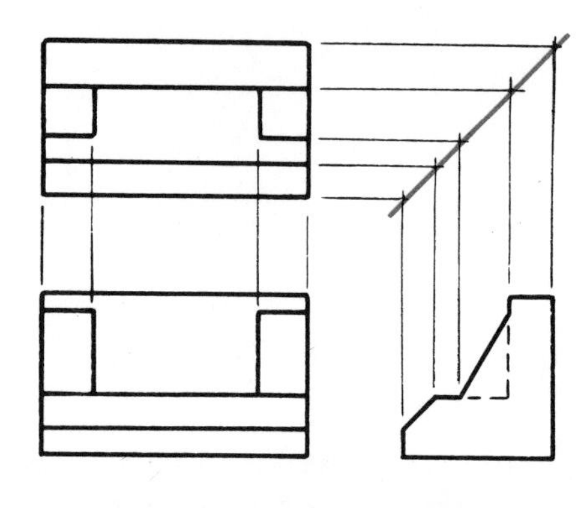

Fig. 3-2-2 Use of mitre line

Assignments

1. Make a sketch similar to Figure 3-2-1 B and C, and establish the distance between Plane 1 and the left border line, and between Plane 2 and the bottom border line, given the following:
   Top, Front and Right Side Views;
   Scale 1:1;
   Drawing Space 180 × 280;
   Width 104, Depth 56, Height 34;
   Space between views (X and Y) 40.
2. Make a sketch similar to Figure 3-2-1 B and C, and establish the distance between Plane 1 and the left border line, and between Plane 2 and the bottom border line, given the following:
   Top, Front and Right Side Views;
   Scale 1:2;
   Drawing Space 180 × 280;
   Width 240, Depth 60, Height 80;
   Space between views (X and Y) 40.
3. Stop block **#1**, Fig. 3-2-A, Sheet size-A4, scale 1:1. Make a three-view drawing using a mitre line to complete the right side view. Space between views to be 40 mm.
4. Stop block **#2**, Fig. 3-2-B, Sheet size-A4, scale 1:1. Make a three-view drawing using a mitre line to complete the top view. Space between views to be 40 mm.

Review for Assignments

Unit 2-2 Basic linework
Unit 2-7 Sketching

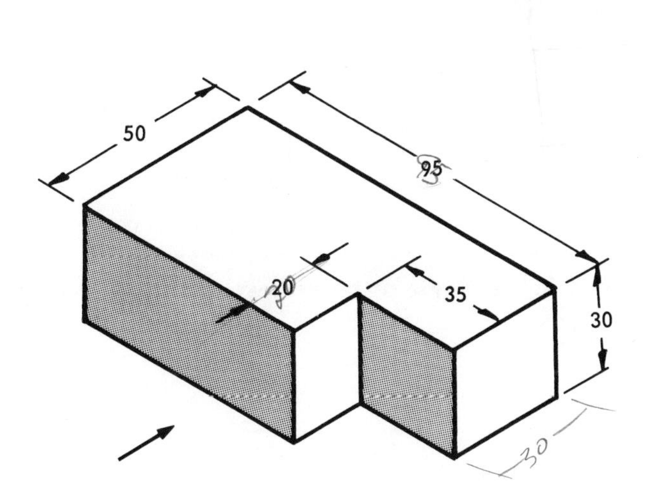

Fig. 3-2-A Step block

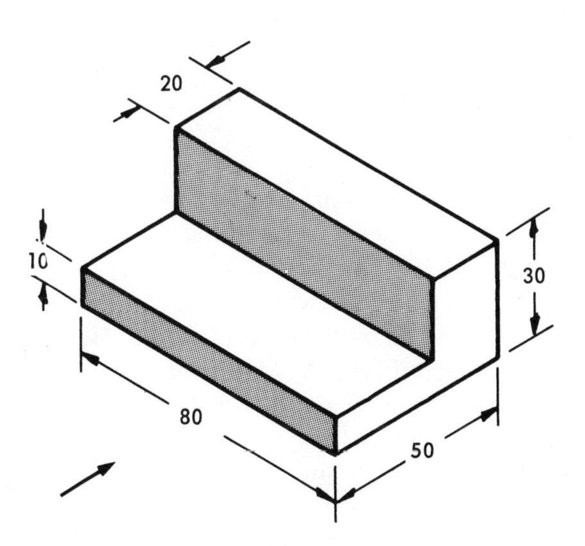

Fig. 3-2-B Stop block

UNIT 3-3

All Surfaces Parallel to the Viewing Planes and All Edges and Lines Visible.

To fully appreciate the shape and detail of views drawn in orthographic projection, the next five drawing units have been chosen according to the types of surfaces generally found on drawings.

When a surface is parallel to the viewing planes, that surface will show as a surface on one view and a line on the other views. The lengths of these lines are the same as the lines shown on the surface view. Figure 3-3-1 shows examples.

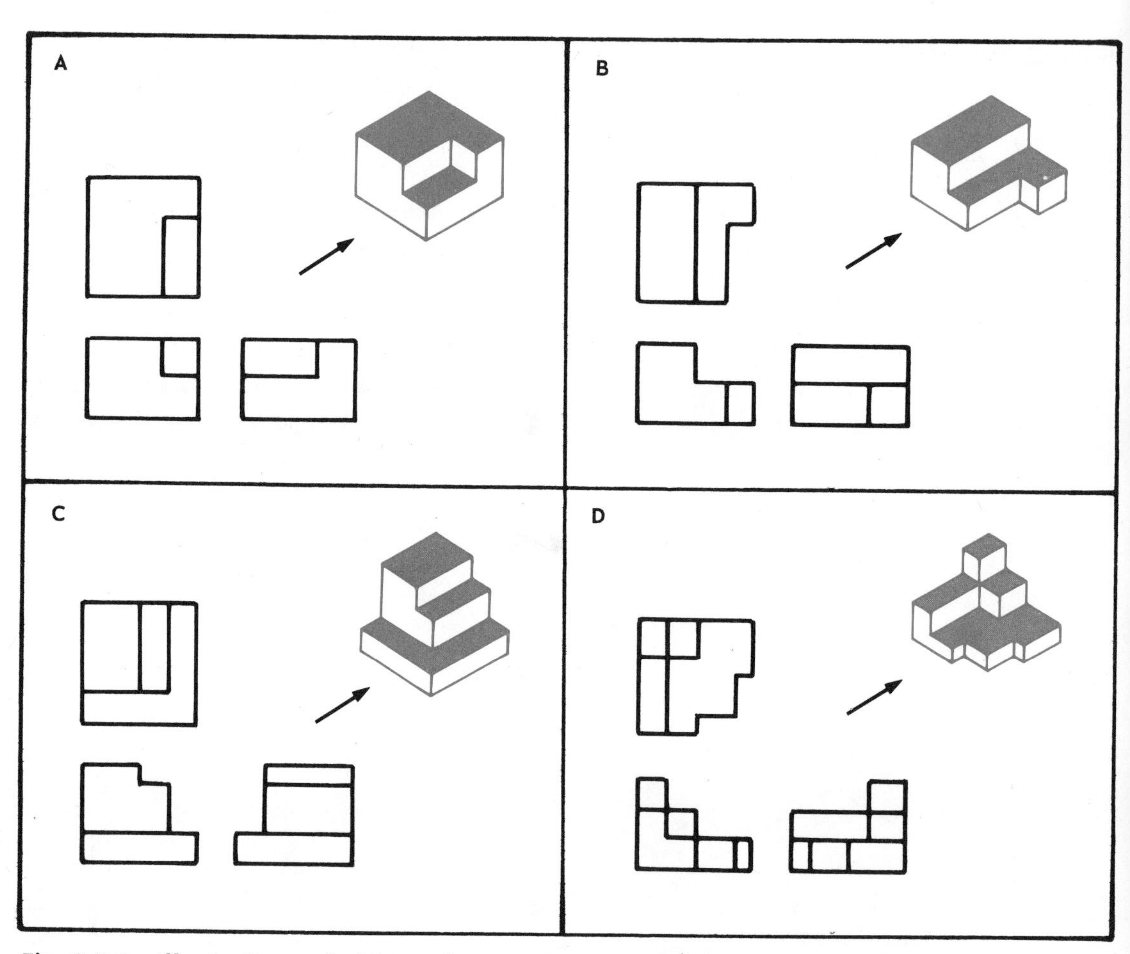

Fig. 3-3-1 Illustrations of objects drawn in 3rd angle ortho projection.

Assignments

1. On two A4-size sheets of preprinted grid paper, sketch three views of each of the objects shown in Figs. 3-3-A and 3- 3-B. Draw four objects on each sheet. Each square shown on the objects represents one square on the grid paper. Allow one grid space between views and two grid spaces between objects.
2. Step block, Fig. 3-3-C, Sheet size-A4, scale 1:1, three-view instrument drawing, space between views to be 30 mm.
3. Corner block, Fig. 3-3-D, Sheet size-A4, scale 1:1, three-view instrument drawing, space between views to be 30 mm.

Review for Assignments

Unit 2-7 Sketching
Unit 3-2 Arrangement of views

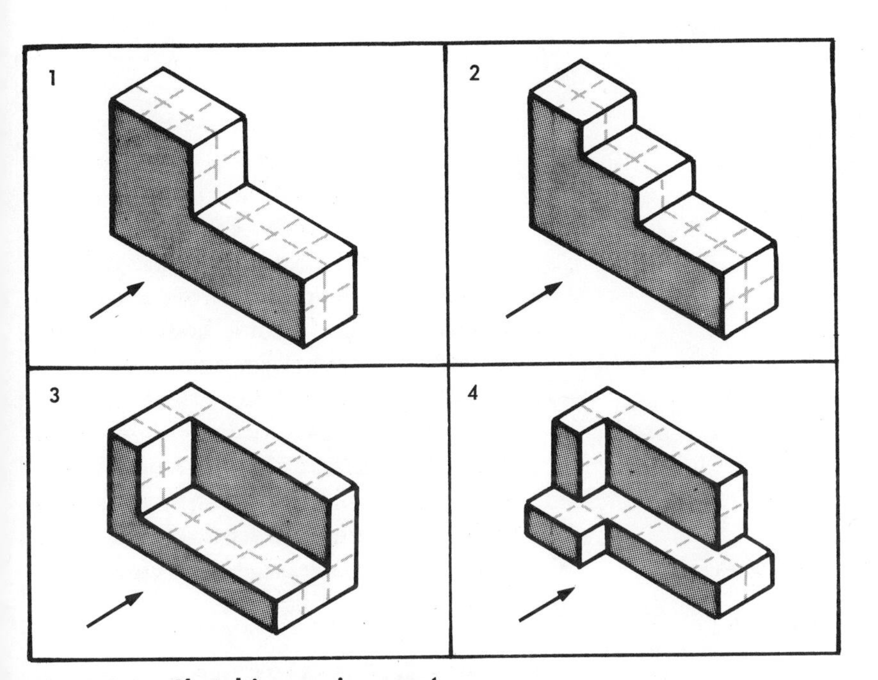

Fig. 3-3-A Sketching assignment

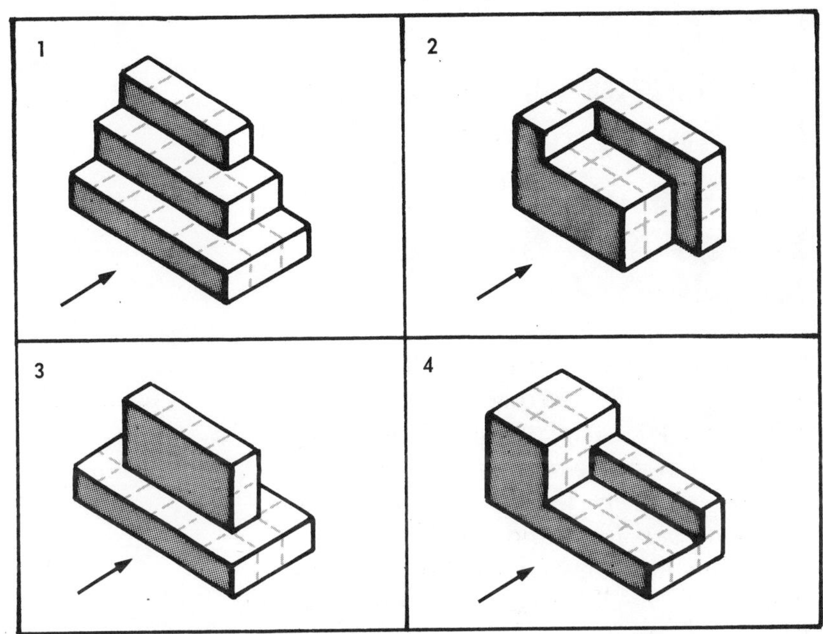

Fig. 3-3-B Sketching assignment

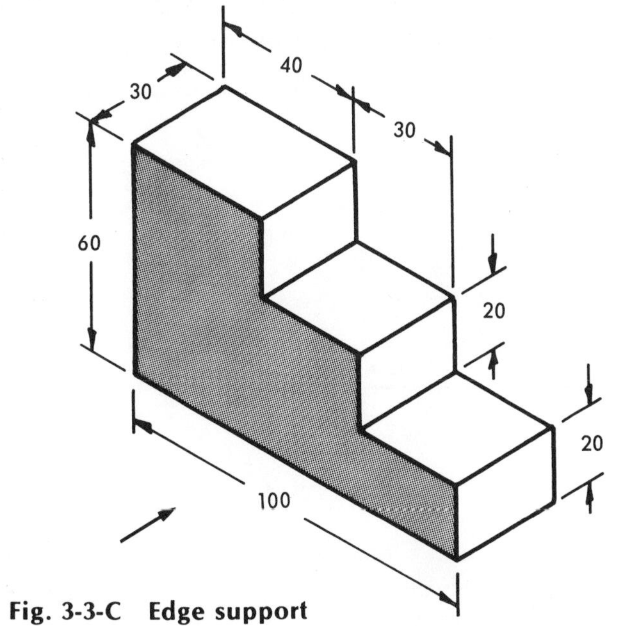

Fig. 3-3-C Edge support

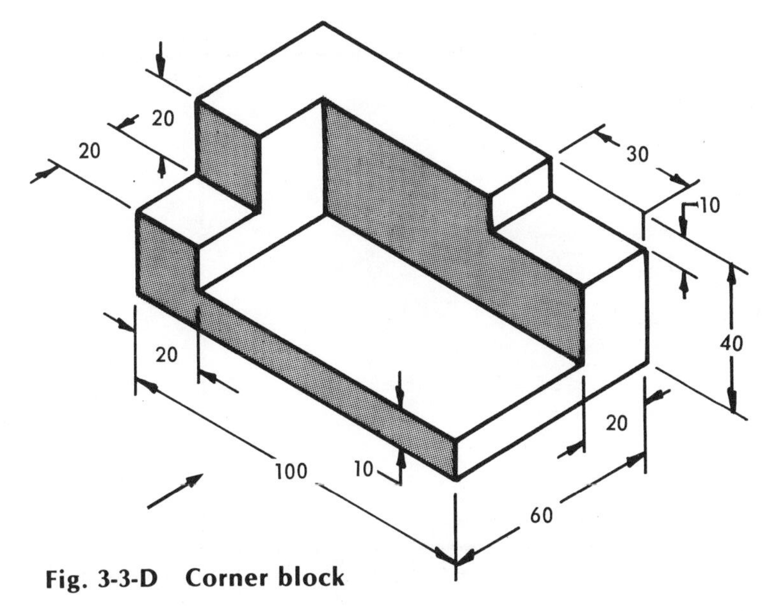

Fig. 3-3-D Corner block

UNIT 3-4
All Surfaces Parallel to the Viewing Planes with Some Edges and Surfaces Hidden

Most objects drawn in engineering offices are more complicated than the ones shown in Figure 3-4-1. Many features (lines, holes, etc.) cannot be seen when viewed from outside the piece. These hidden edges are called *hidden lines* and are normally required on the drawing to show the true shape of the object. Figure 3-4-2 shows additional examples of objects requiring hidden lines.

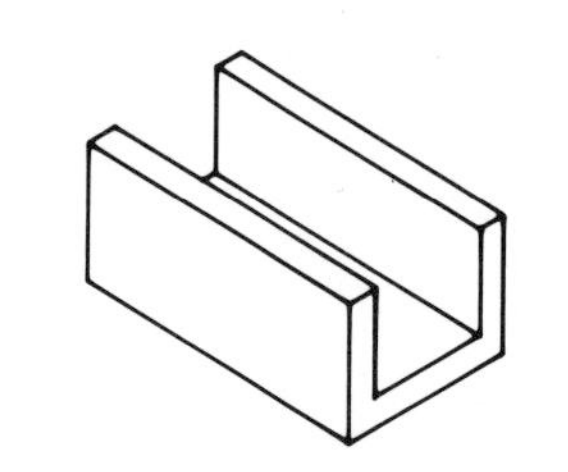

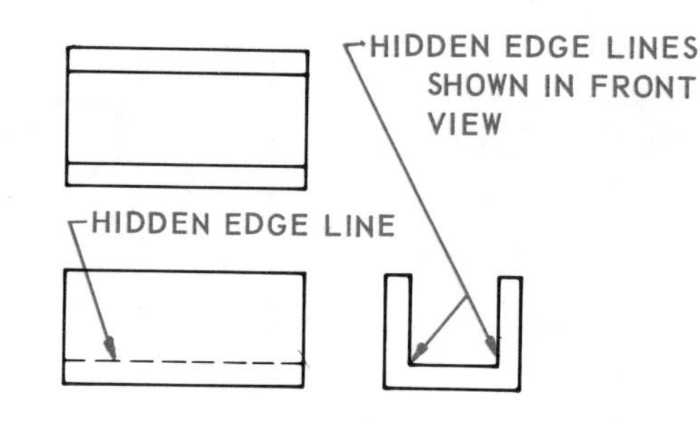

Fig. 3-4-1 Hidden lines

Assignments

1. On two A4-size sheets of preprinted grid paper sketch three views of each of the objects shown in Figs. 3-4-A and 3-4-B. Draw four objects on each sheet. Each square shown on the objects represents one square on the grid paper. Allow one grid space between views and two spaces between objects.
2. Adaptor, Fig. 3-4-C, Sheet size-A4, scale 1:1, three-view instrument drawing, 20 mm spacing between views.
3. Guide block, Fig. 3-4-D, Sheet size-A4, scale 1:1, three-view instrument drawing, 20 mm spacing between views.
4. Matching test, Fig. 3-4-E.

Review for Assignments

Unit 2-3 Hidden lines Unit 2-7 Sketching
Unit 3-2 Arrangement of views

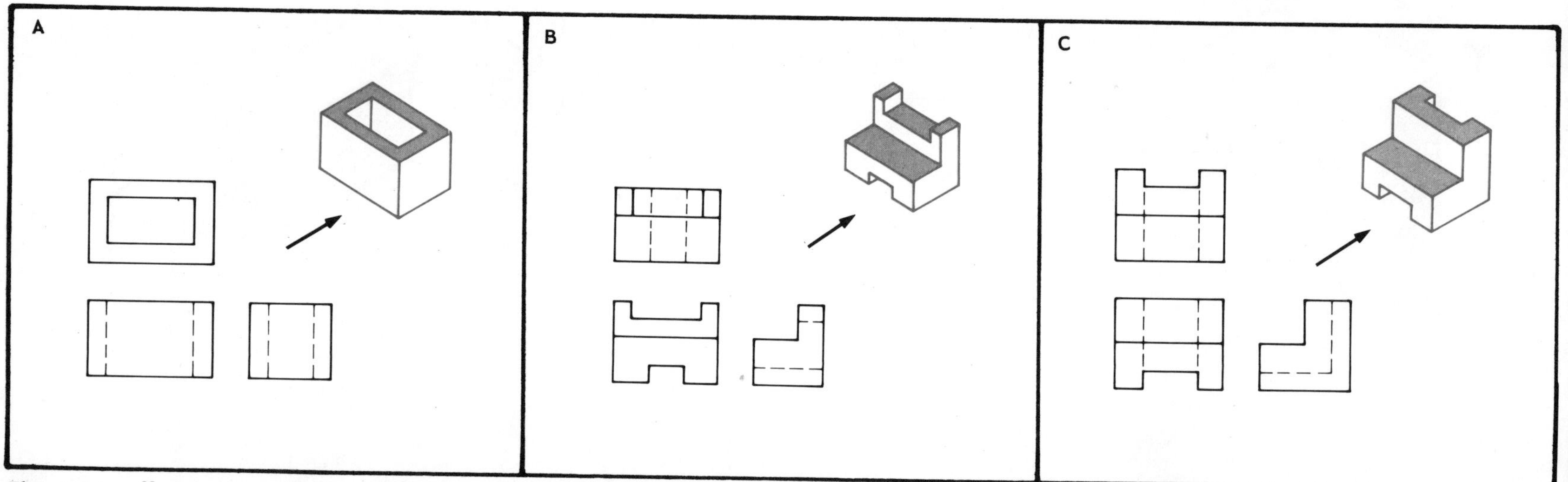

Fig. 3-4-2 Illustrations of objects having hidden features

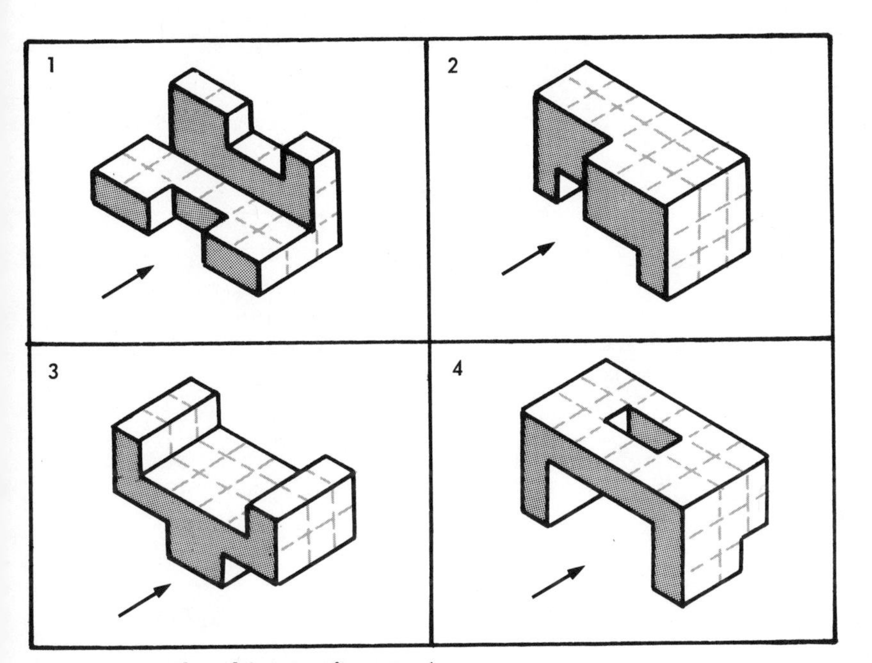

Fig. 3-4-A Sketching assignment

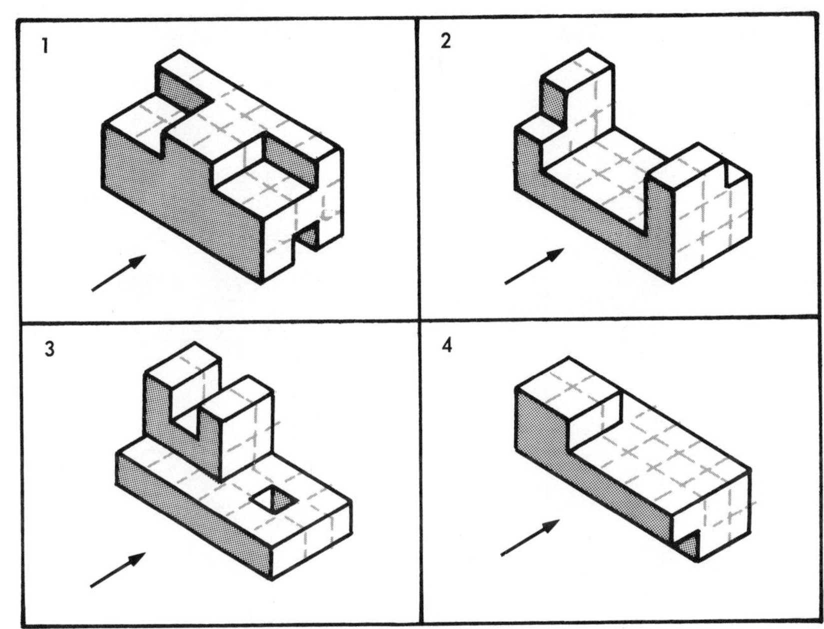

Fig. 3-4-B Sketching assignment

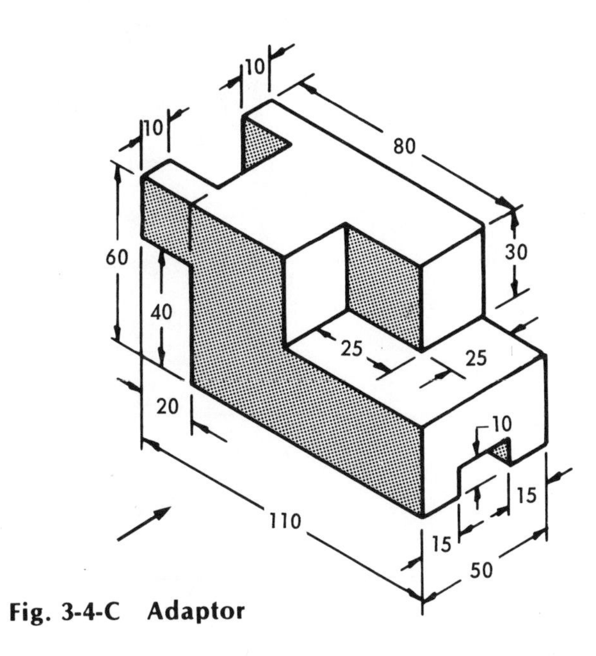

Fig. 3-4-C Adaptor

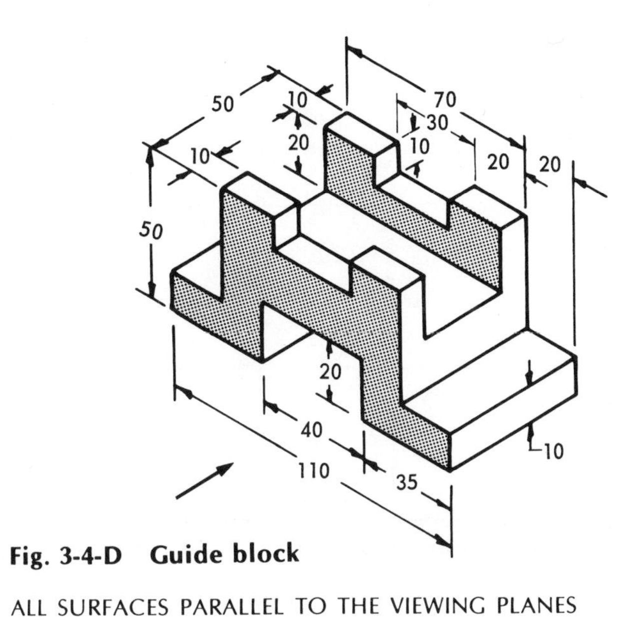

Fig. 3-4-D Guide block

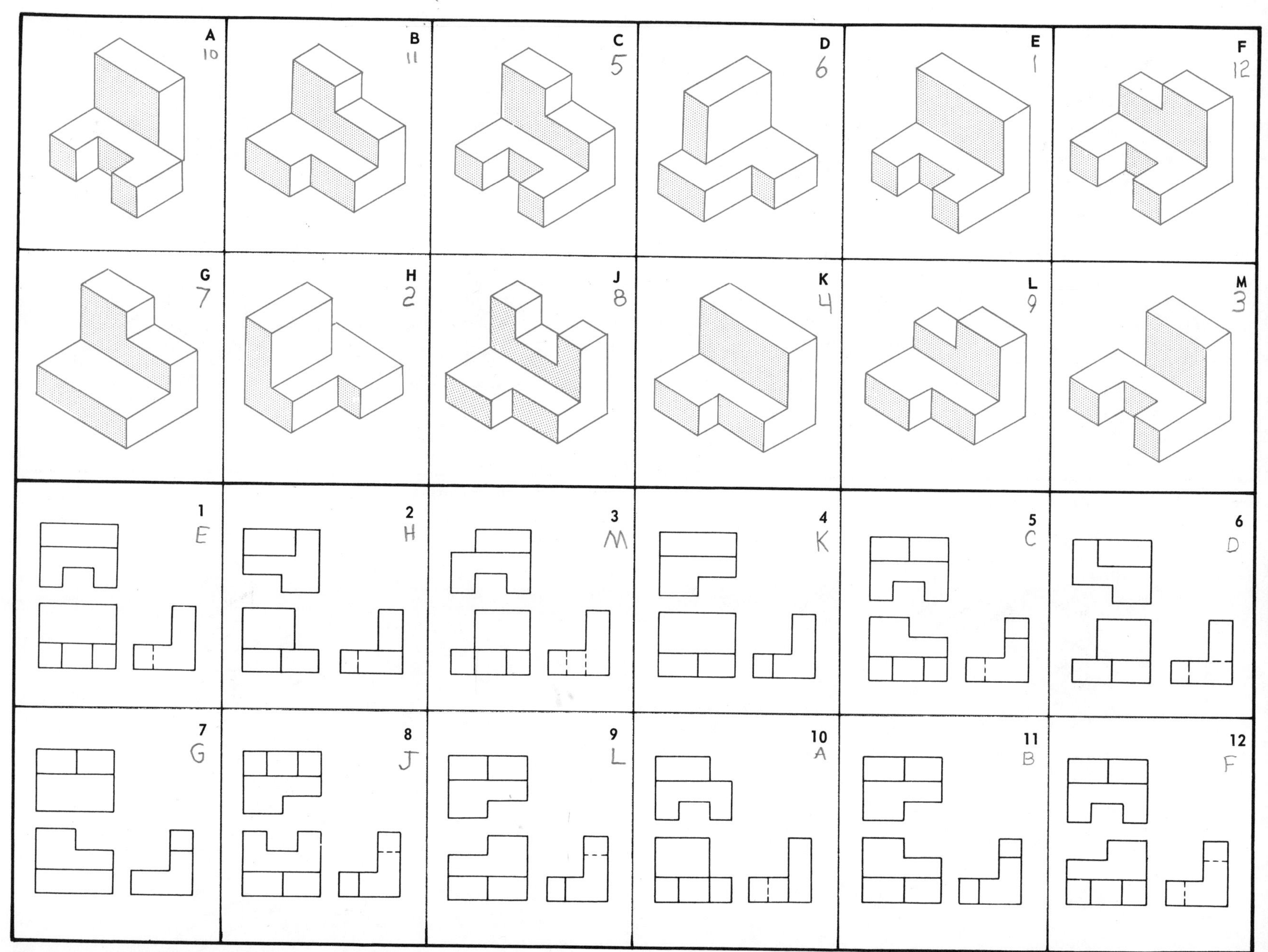

Fig. 3-4-E Match pictorial dwgs. A to M with ortho. dwg. 1 to 12

UNIT 3-5
Inclined Surfaces

If the surfaces of an object lie in either a horizontal or a vertical position, the surfaces appear in their true shapes in one of the three views, and these surfaces appear as a line in the other two views.

When a surface is inclined or sloped in only one direction, then that surface is not seen in its true shape in the top, front, or side view. It is, however, seen in two views as a distorted surface. On the third view it appears as a line.

The true length of surfaces A and B in Figure 3-5-1 is seen in the front view only. In the top and side views, only the width of surfaces A and B appears in its true size. The length of these surfaces is foreshortened. Figure 3-5-2 shows additional examples.

Where an inclined surface has important features that must be shown clearly and without distortion, an *auxiliary* or helper view must be used. This type of view will be discussed in detail in Chapter 8.

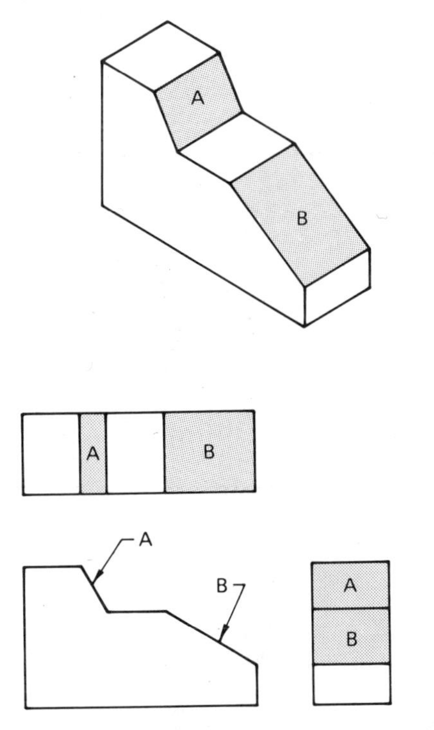

Fig. 3-5-1 Sloping surfaces

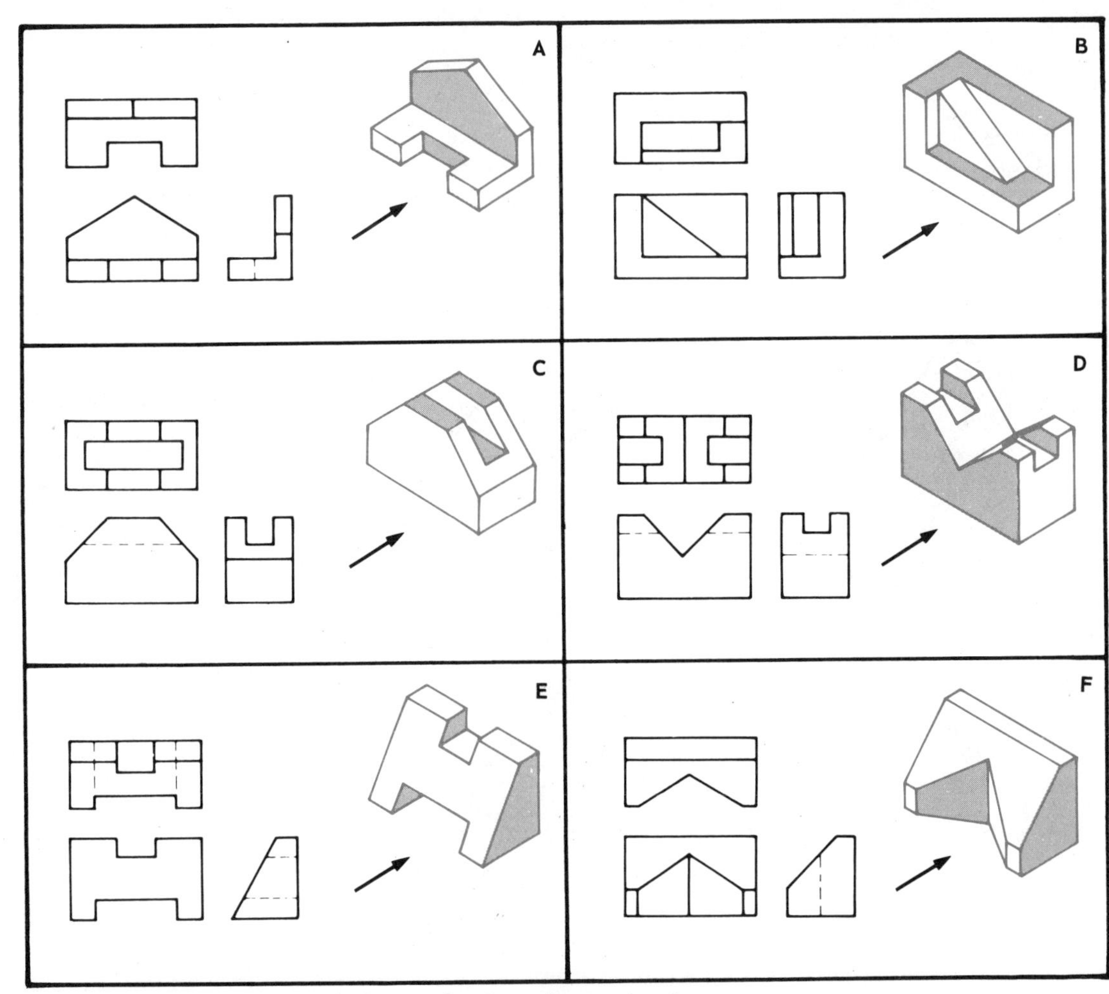

Fig. 3-5-2 Illustrations of objects having sloping surfaces

Assignments

1. On two A4-size sheets of preprinted grid paper sketch three views of each of the objects shown in Figs. 3-5-A and 3-5-B. Draw four objects on each sheet. Each square shown on the objects represents one square on the grid paper. Allow one grid space between views and two grid spaces between objects. The sloped (inclined) surfaces on each of the objects are identified by a letter. Identify the sloped surfaces on each of the three views with a corresponding letter. For the following instrument drawings — sheet size-A4, scale 1:1, three views, 20 mm spacing between views.
2. Adjusting guide, Fig. 3-5-C, sheet size A4, scale 1:1, three-view instrument drawing, 20 mm spacing between views.
3. Matching test, Fig. 3-5-D.

Review for Assignments

Unit 2-3 Hidden lines
Unit 2-7 Sketching
Unit 3-2 Arrangement of views

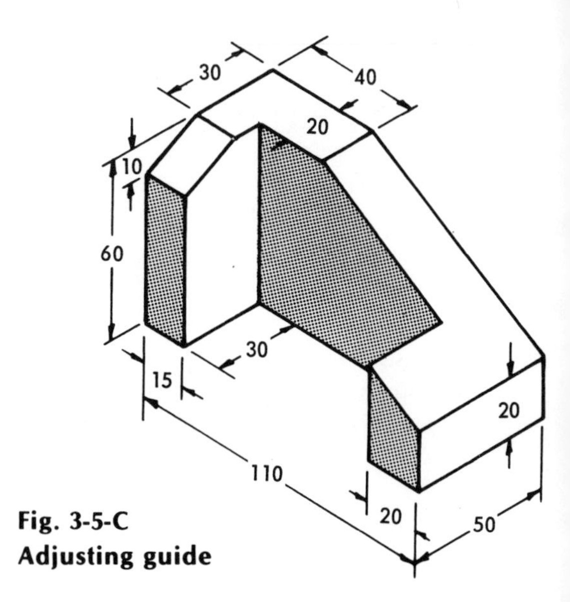

Fig. 3-5-C Adjusting guide

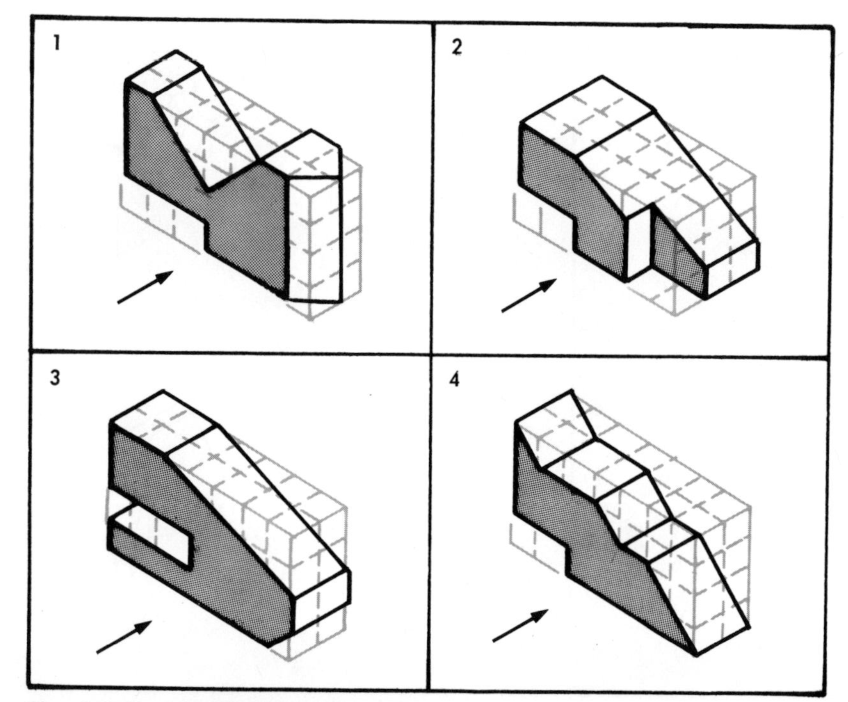

Fig. 3-5-A Sketching assignment

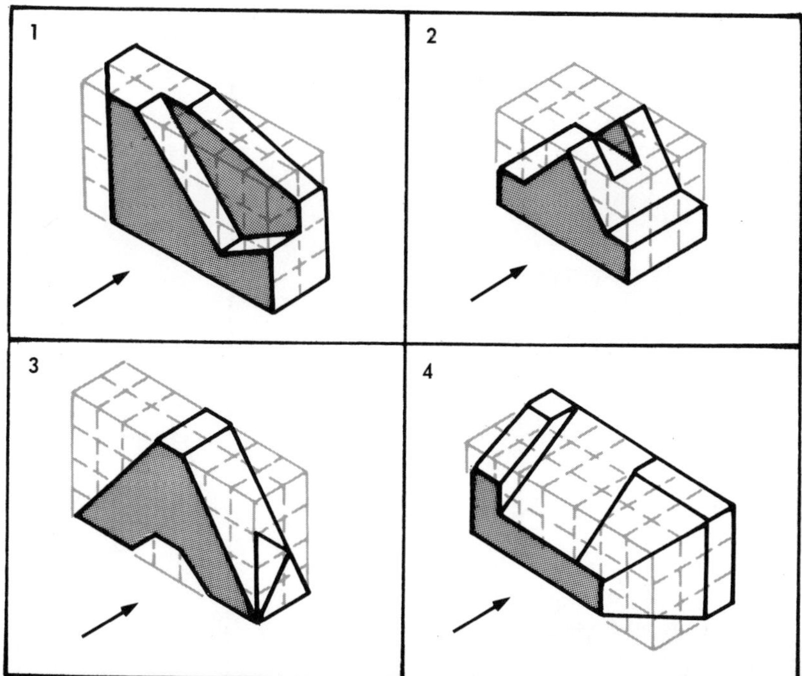

Fig. 3-5-B Sketching assignment

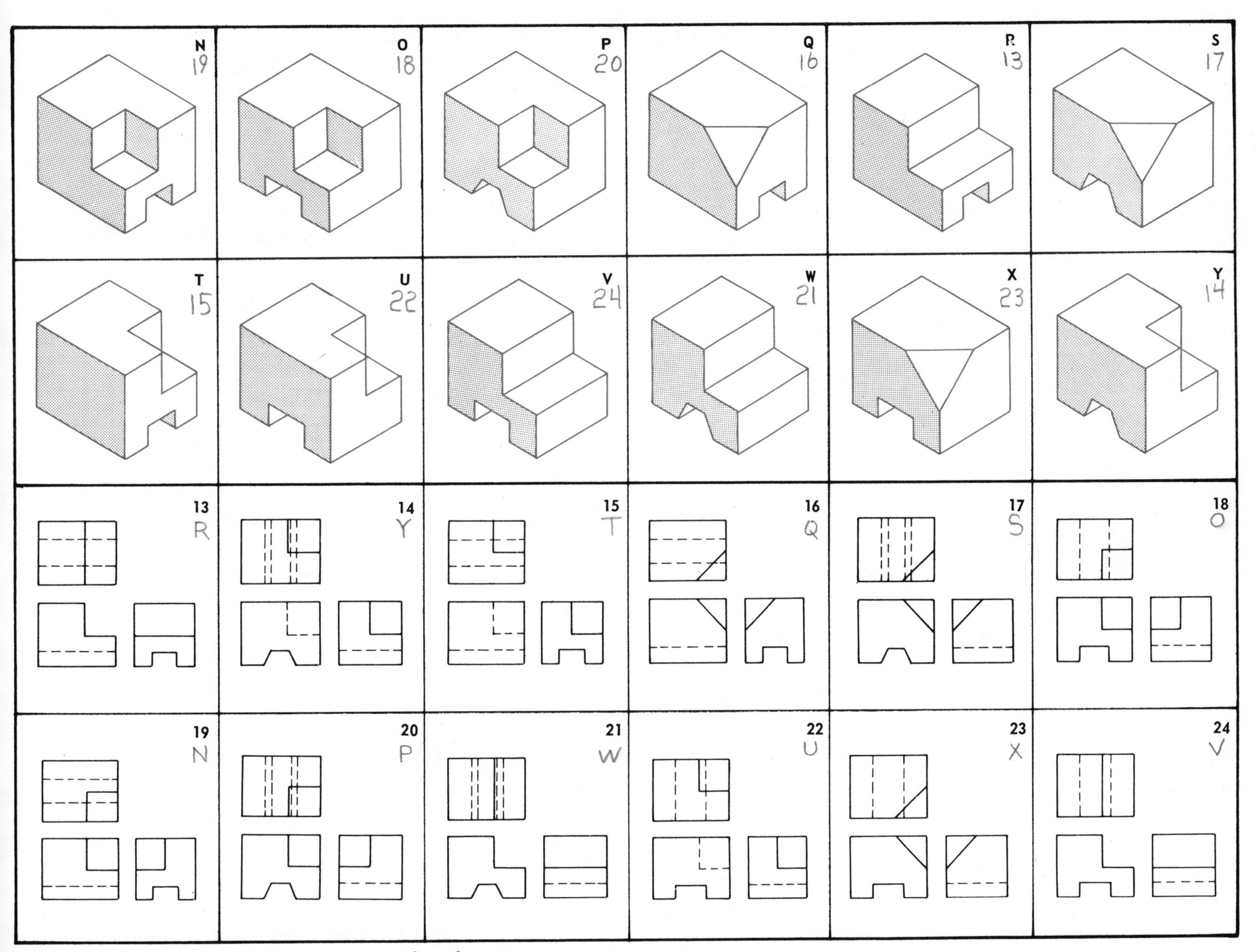

Fig. 3-5-D Match pictorial dwgs. N to Y with ortho. dwg. 13 to 24

UNIT 3-6
Circular Features

Typical parts with circular features are illustrated in Figure 3-6-1. Note that the circular feature appears circular in one view only and that no line is used to indicate where a curved surface joins a flat surface. Hidden circles, like hidden flat surfaces, are represented on drawings by a hidden line.

Centre Lines

A centre line is drawn as a thin, broken line of long and short dashes, spaced alternately. They may be used to indicate centre points, axes of cylindrical parts, and axes of symmetry, as shown in Fig. 3-6-2. Solid centre lines are often used when the circular features are small. Centre lines should project for a short distance beyond the outline of the part or feature to which they refer. They may be extended for use as extension lines for dimensioning purposes, but in this case the extended portion is not broken.

On views showing the circular features, the point of intersection of the two centre lines is shown by the two intersecting short dashes.

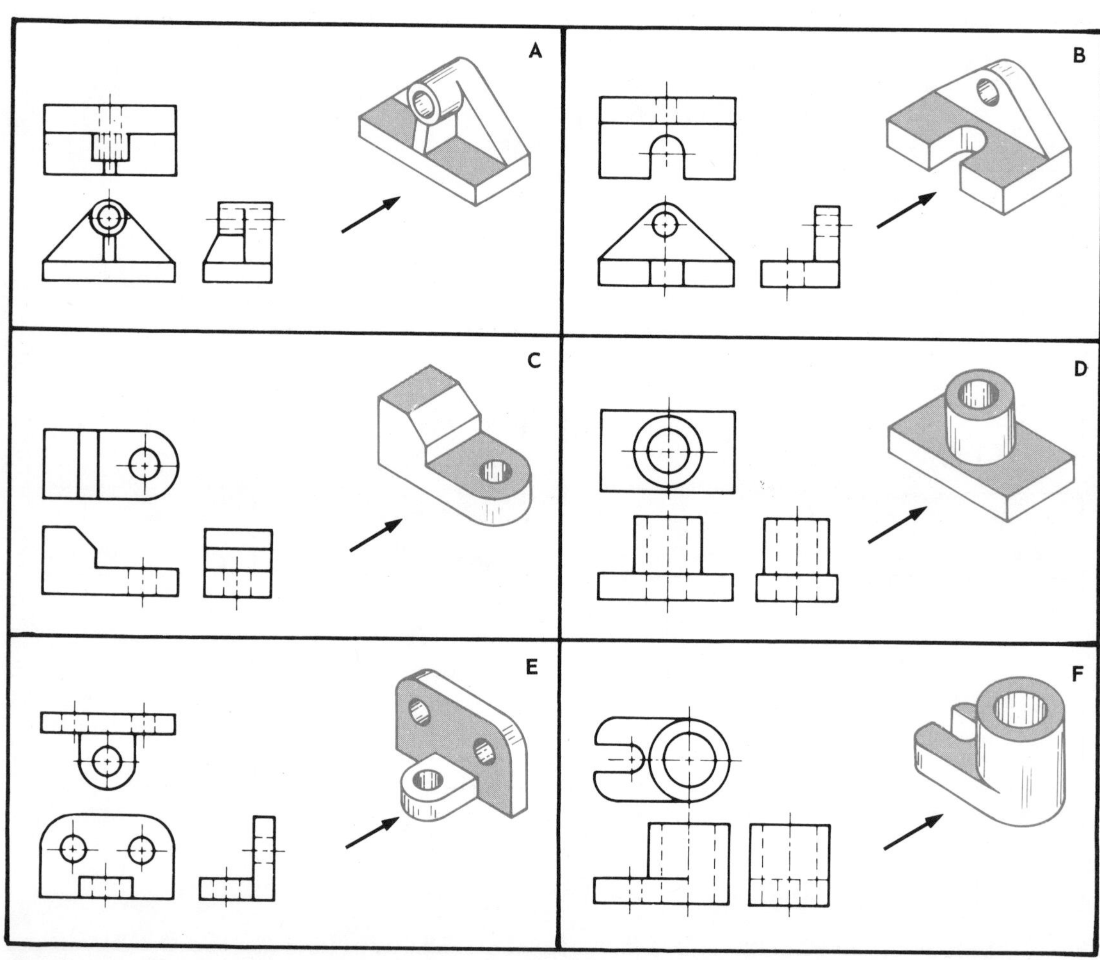

Fig. 3-6-1 Illustrations of objects having circular features

Assignments

1. On two A4-size sheets of preprinted grid paper sketch three views of each of

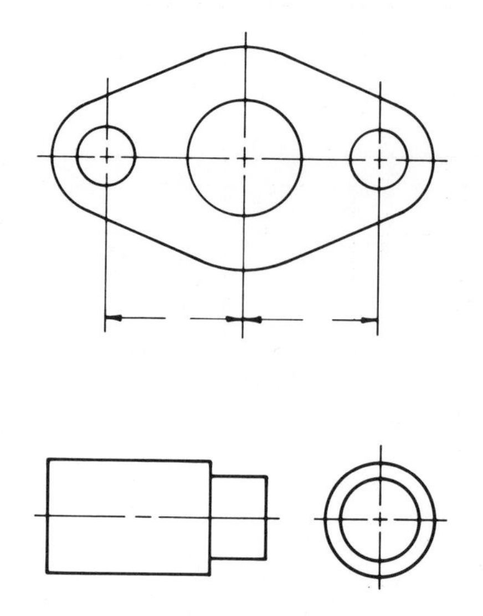

Fig. 3-6-2 Centre line application for circular features

the objects shown in Figs. 3-6-A and 3-6-B. Draw four objects on each sheet. Each square shown on the objects represents one square on the grid paper. Allow one grid space between views and two grid spaces between objects.

2. Cradle support, Fig. 3-6-C, sheet size-A4, scale 1:1, three-view instrument drawing, 20 mm spacing between views.
3. Pillow block, Fig. 3-6-D, sheet size-A4, scale 1:1, three-view instrument drawing, 20 mm spacing between views.

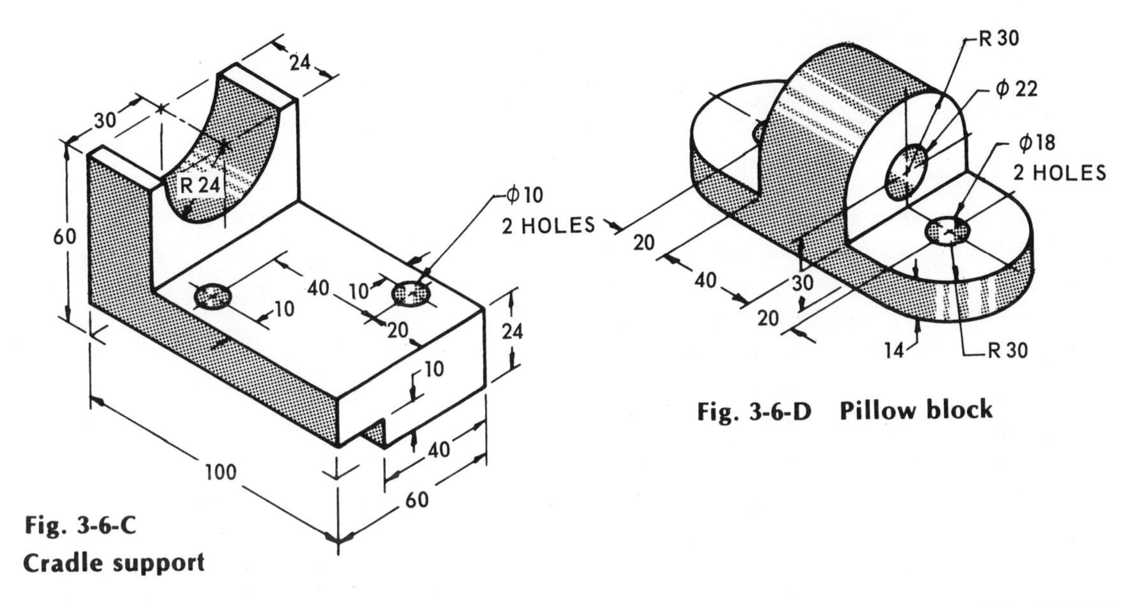

Fig. 3-6-C Cradle support

Fig. 3-6-D Pillow block

Review for Assignments

Unit 2-5 Drawing circles and arcs

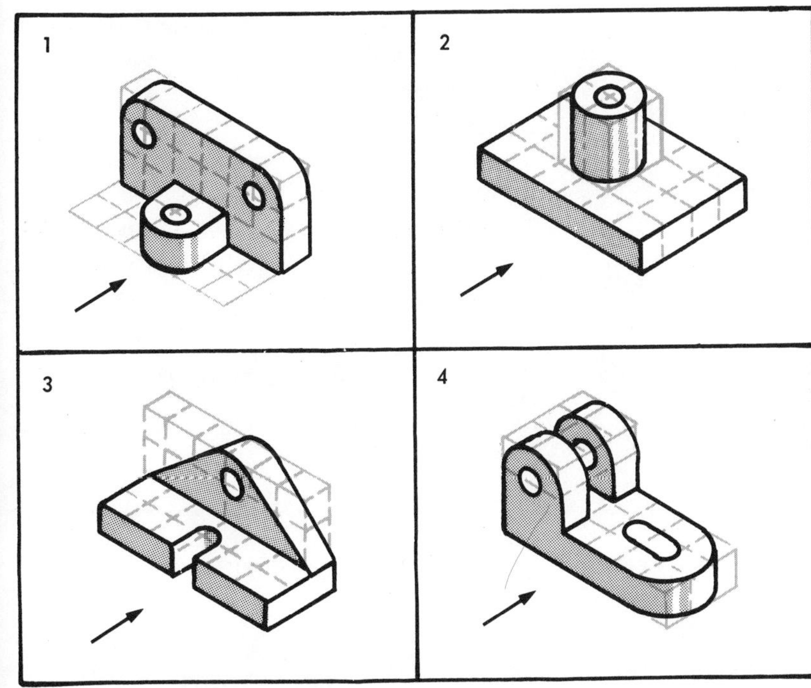

Fig. 3-6-A Sketching assignment

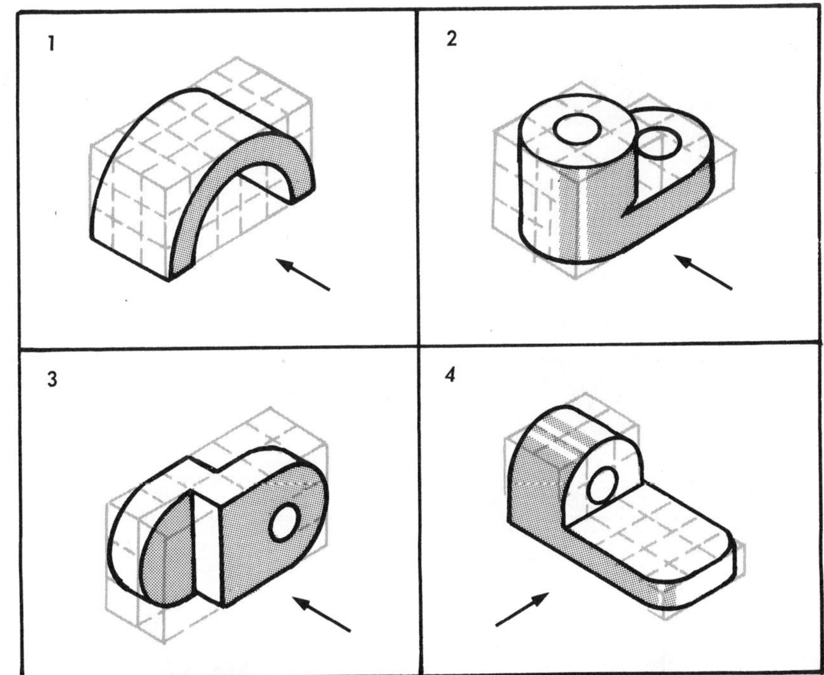

Fig. 3-6-B Sketching assignment

UNIT 3-7
Oblique Surfaces

When a surface is sloped so that it is not perpendicular to any of the three viewing planes, it will appear as a surface in all three views but never in its true shape. This is referred to as an **oblique surface** (Fig 3-7-1). Since the oblique surface is not perpendicular to the viewing planes, it cannot be parallel to them and consequently appears foreshortened. If a true view is required for this surface, two auxiliary views — a primary and a secondary view — need to be drawn. This is discussed in detail under Auxiliary Views in Chapter 8. Figure 3-7-2 shows additional examples of objects having oblique surfaces.

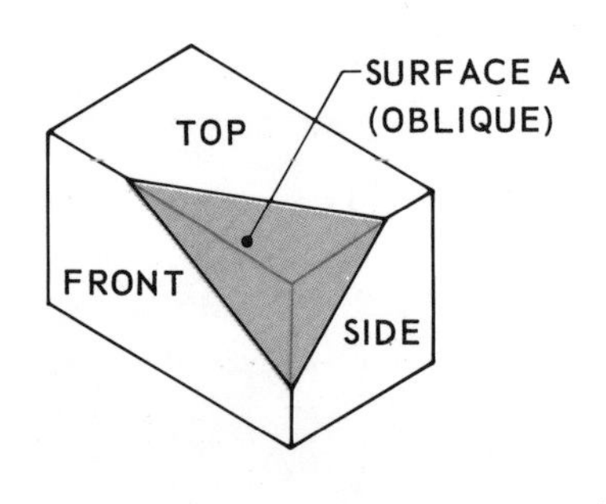

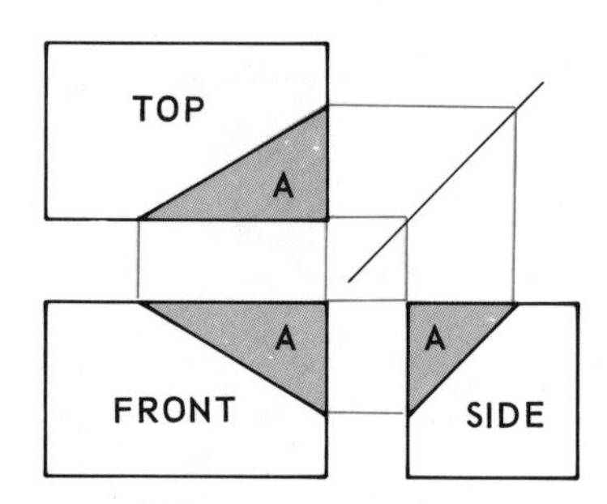

Fig. 3-7-1 Oblique surface A not true shape in any of the three views

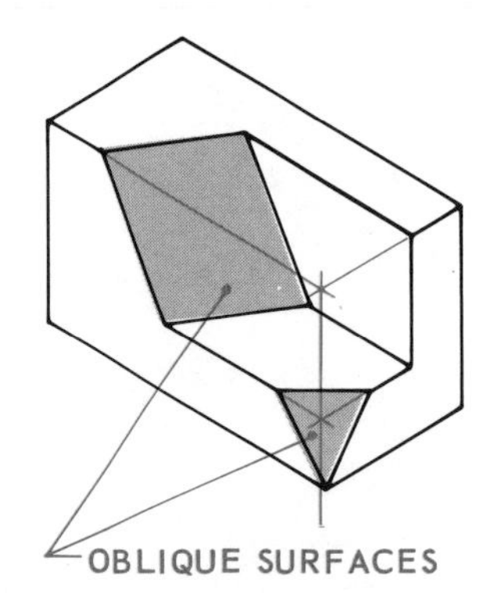

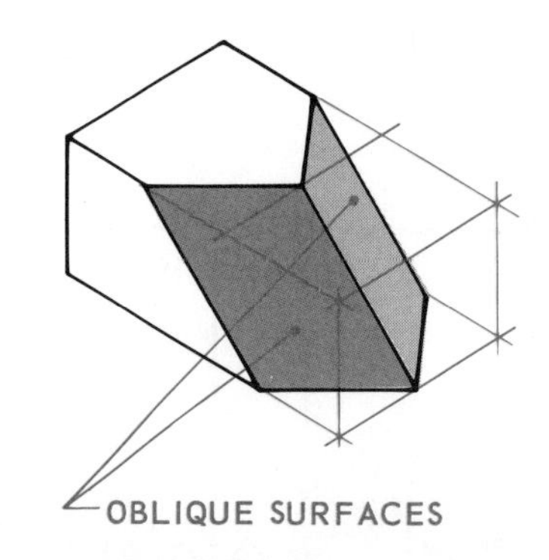

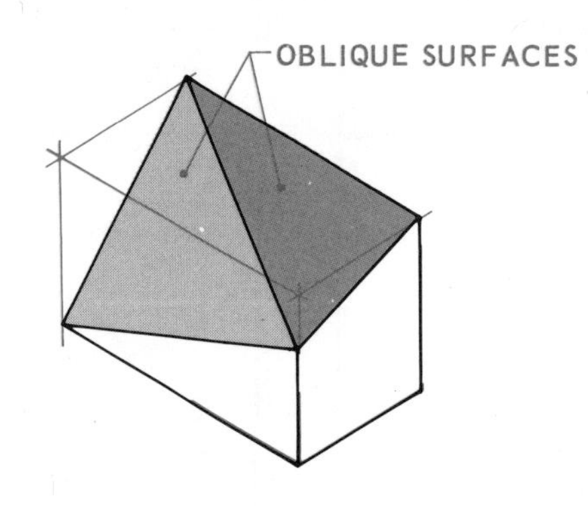

Fig. 3-7-2 Illustration of objects having oblique surfaces

Assignments

1. On two A4-size sheets of preprinted grid paper sketch three views of each of the objects shown in Figs. 3-7-A and 3-7-B. Draw four objects on each sheet. Each square on the objects represents one square on the grid paper. Allow one grid space between views and two grid spaces between objects. The oblique surfaces on the objects are identified by a letter. Identify the oblique surfaces on each of the three views with a corresponding letter.
2. Base plate, Fig. 3-7-C, sheet size-A4, scale 1:1, three-view instrument drawing, 20 mm spacing between views.
3. Locking base, Fig. 3-7-D, sheet size-A4, scale 1:1, three-view instrument drawing, 20 mm spacing between views.

Review for Assignment

Unit 2-3 Hidden lines
Unit 2-7 Sketching

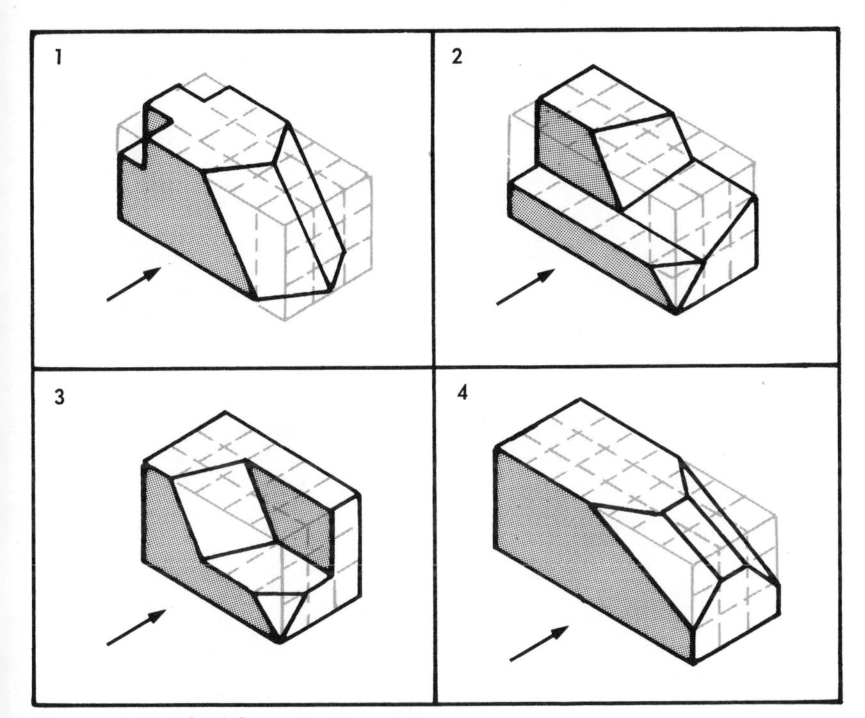

Fig. 3-7-A Sketching assignment

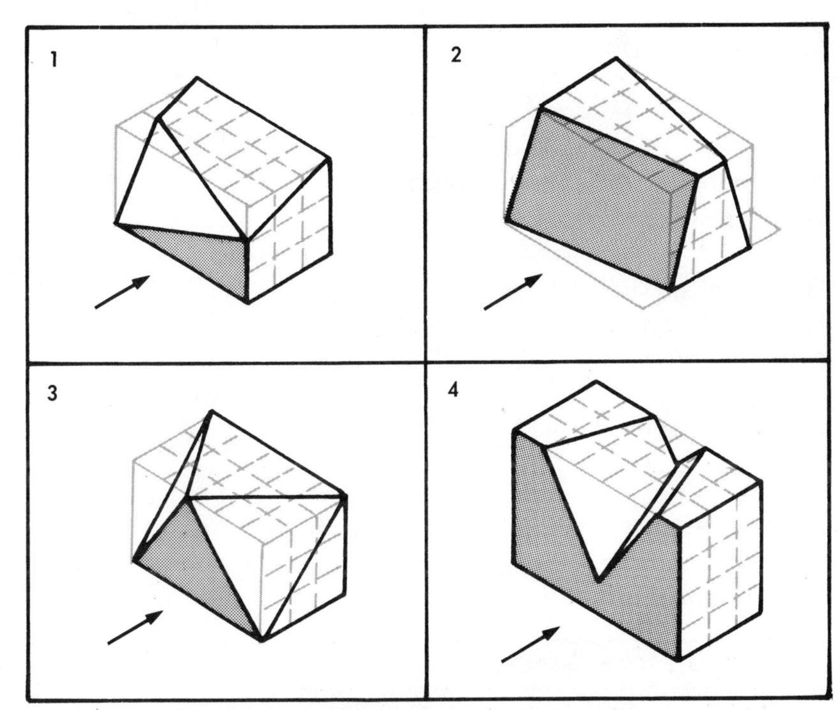

Fig. 3-7-B Sketching assignment

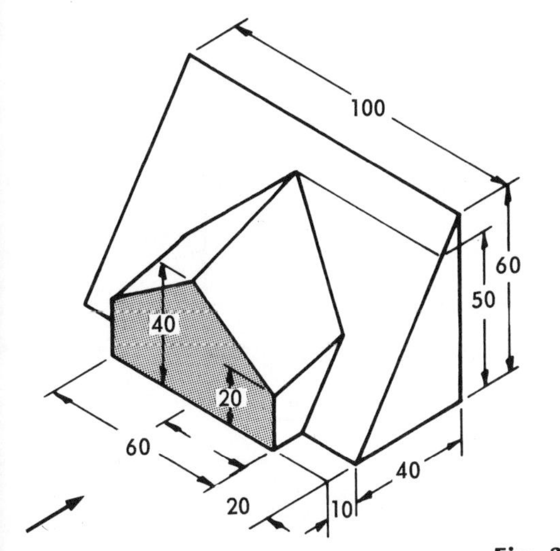

Fig. 3-7-C Base plate

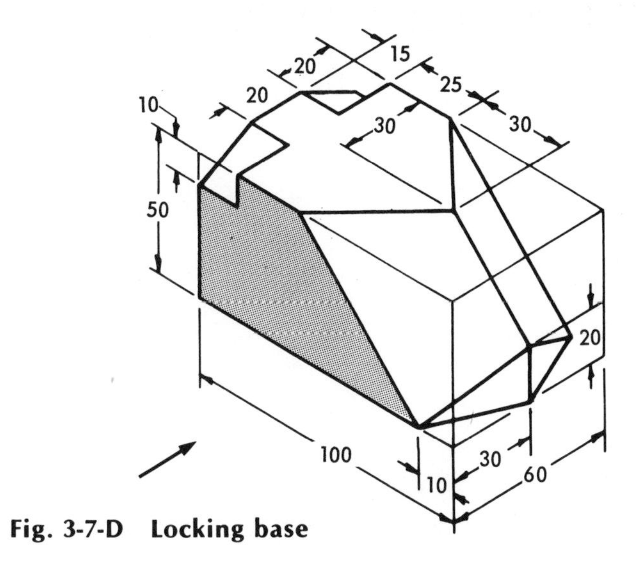

Fig. 3-7-D Locking base

UNIT 3-8
Review of Chapter 3

Surface Identification Problems 3-8-A to 3-8-E

Draw a chart similar to the one shown in Figure 3-8-1 and place in the chart the appropriate numbers corresponding to the letters shown in the pictorial drawing. The letters represent surfaces in the pictorial drawings.

The numbers enclosed by a circle refer to a hidden surface or a hidden line. In some instances an enclosed number may refer to more than one surface.

SURFACE IDENTIFICATION CHART			
PICTORIAL DRAWING LETTERS	ORTHOGRAPHIC DRAWING		
	TOP VIEW	FRONT VIEW	SIDE VIEW
A			
B			
C			
D			
E			
L			
M			
N			

Fig. 3-8-1 Surface identification chart

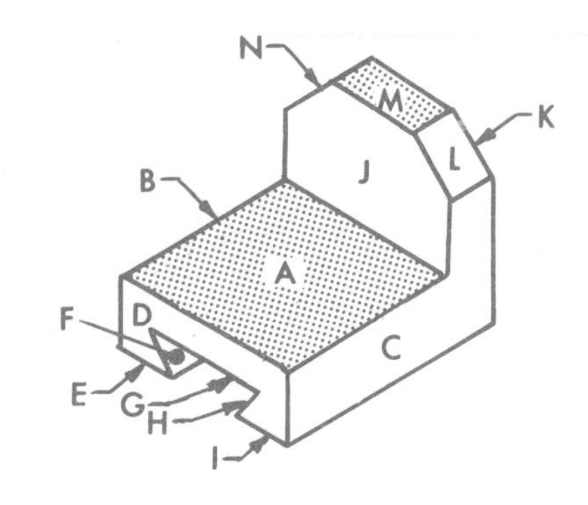

Dovetail Guide

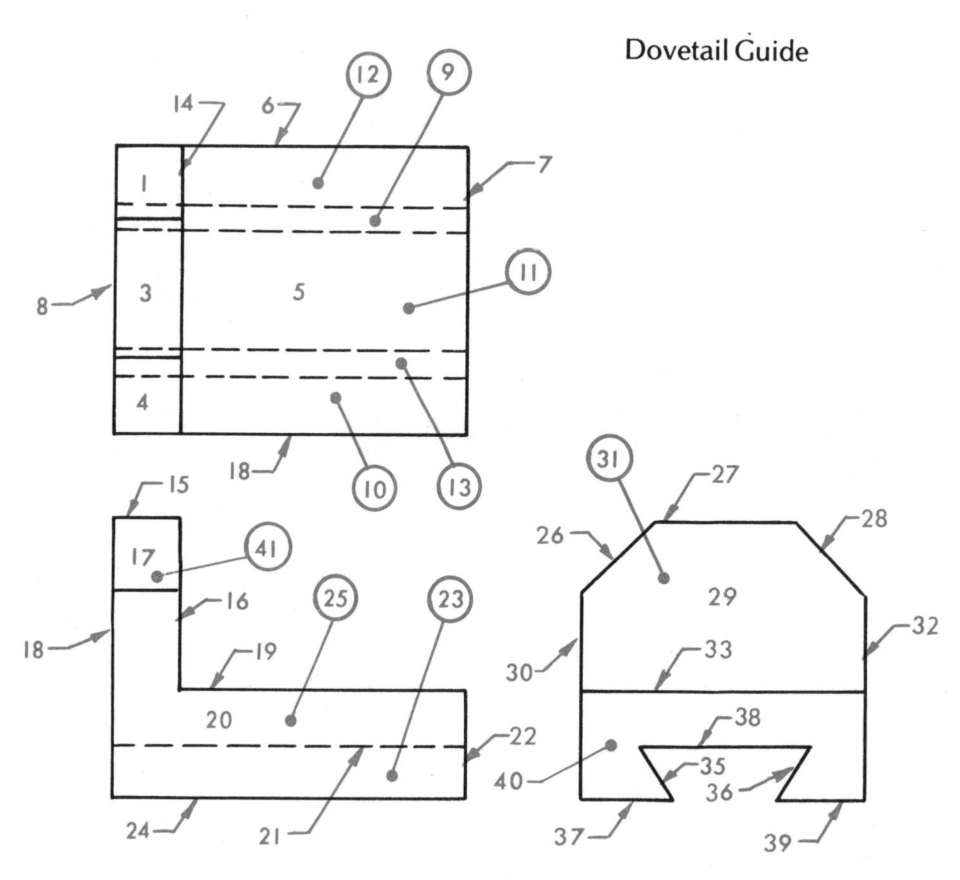

Fig. 3-8-A Surface identification problems

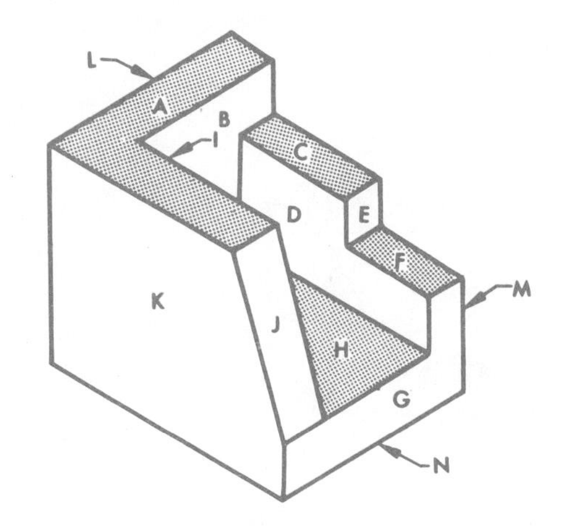

Bracket

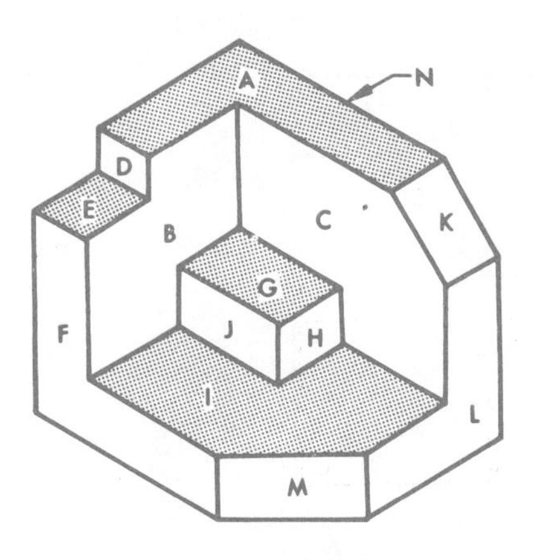

Corner Bracket

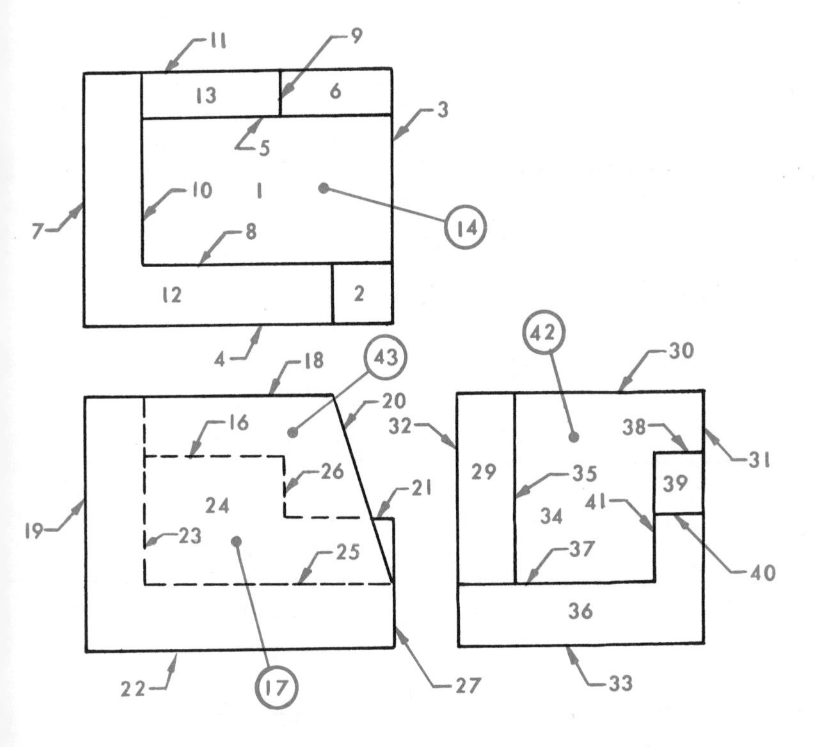

Fig. 3-8-B Surface identification problems

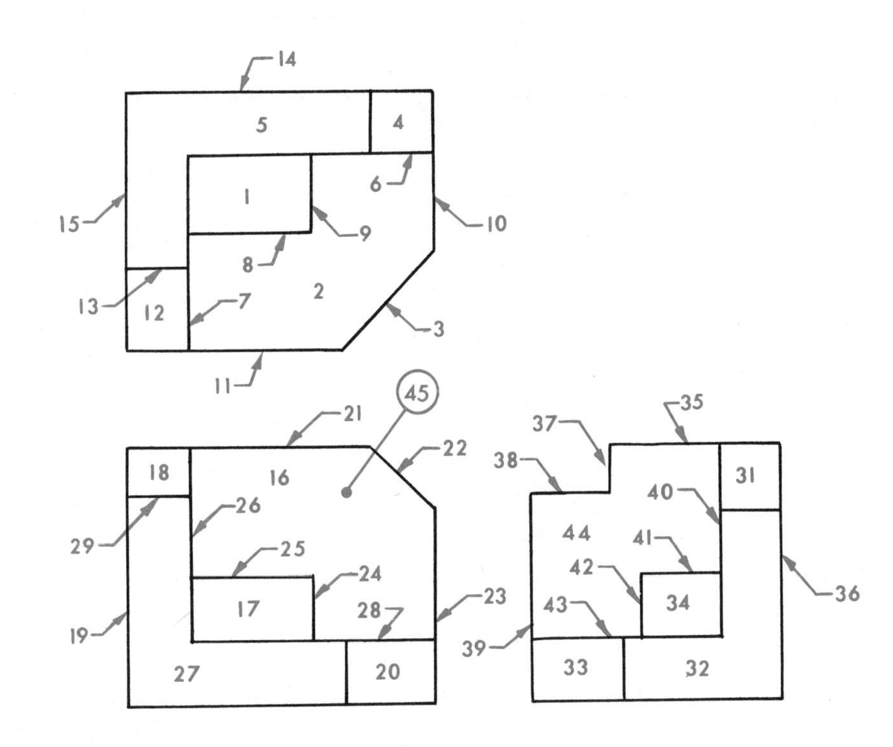

Fig. 3-8-C Surface identification problems

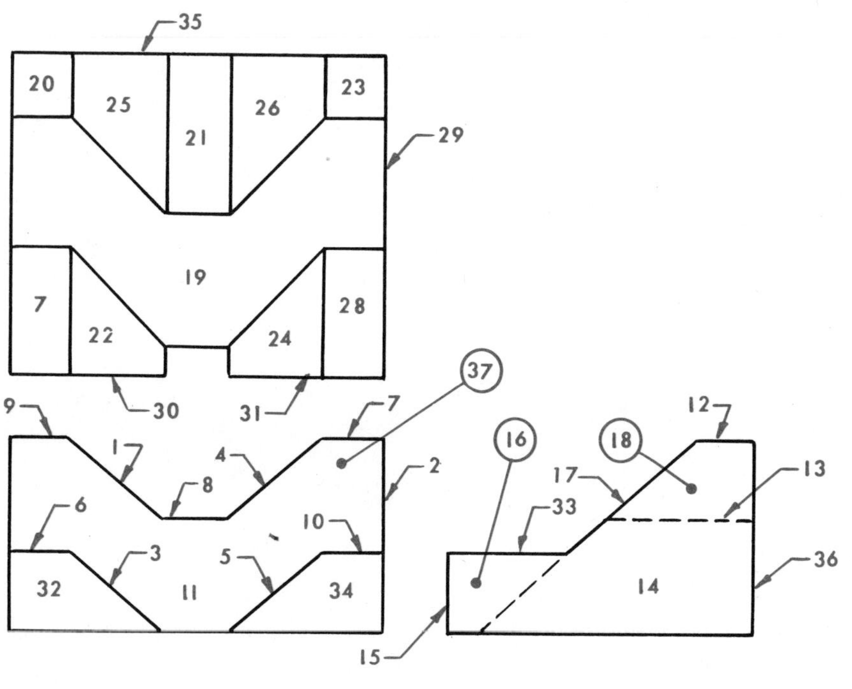

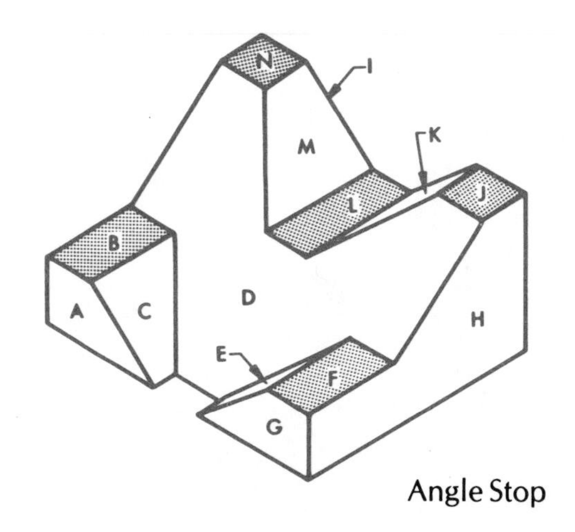

Fig. 3-8-D Surface identification problem

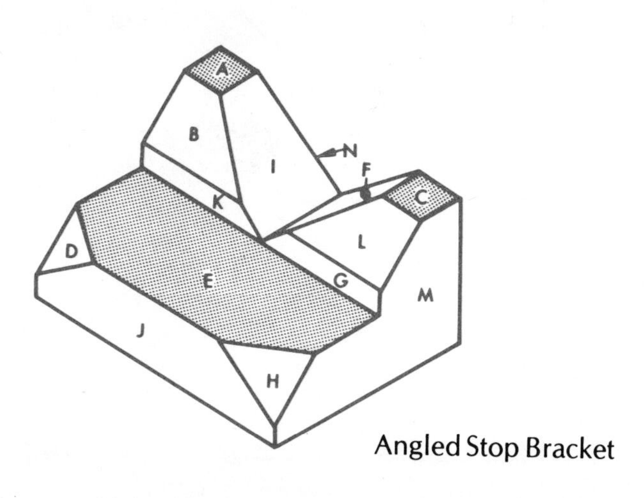

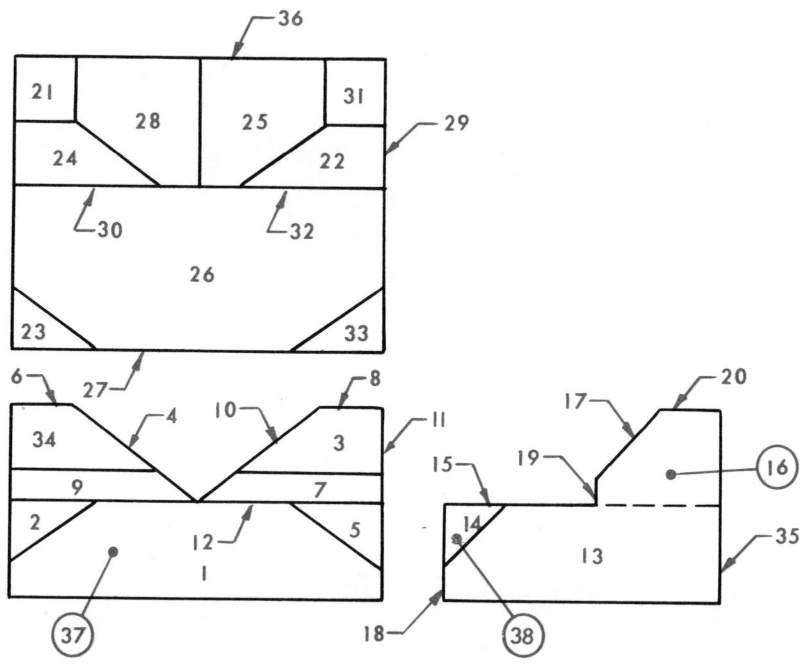

Fig. 3-8-E Surface identification problem

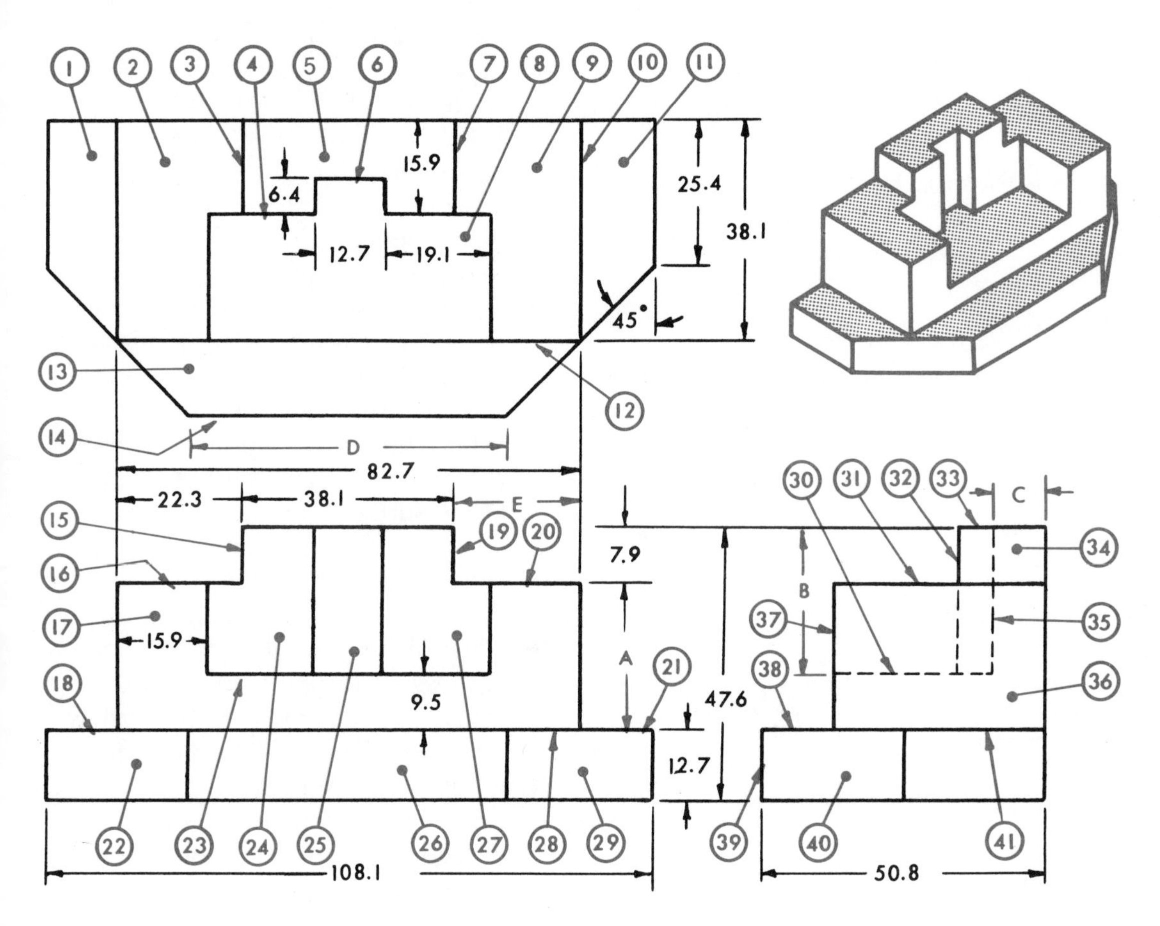

Fig. 3-8-F Step block

1. Calculate dimensions A to E.
2. What surfaces in the top view are represented by line 41 in the side view?
3. What line in the side view is represented by surface 5 in the top view?
4. What line in the side view is represented by line 4 in the top view?
5. What lines in the front view are represented by surface 34 in the side view?
6. What line in the side view is represented by surface 25 in the front view?
7. What surfaces in the front view are represented by surface 40 in the side view?
8. What surface in the top view is represented by line 30 in the side view?
9. What line in the side view is represented by surface 13 in the top view?
10. What lines in the front view are represented by line 41 in the side view?
11. What lines in the front view are represented by line 31 in the side view?
12. What line in the side view is represented by line 12 in the top view
13. What surface in the side view is represented by line 10 in the top view?
14. What line in the side view is represented by line 14 in the top view?
15. What surface in the front view is represented by line 12 in the top view?
16. What lines in the top view are represented by surface 34 in the side view?
17. What line in the front view is represented by surface 8 in the top view?
18. What line in the side view is represented by line 6 in the top view?
19. What line in the front view is represented by surface 13 in the top view?
20. What surfaces in the top view are represented by line 31 in the side view?
21. What line in the top view is represented by surface 25 in the front view?

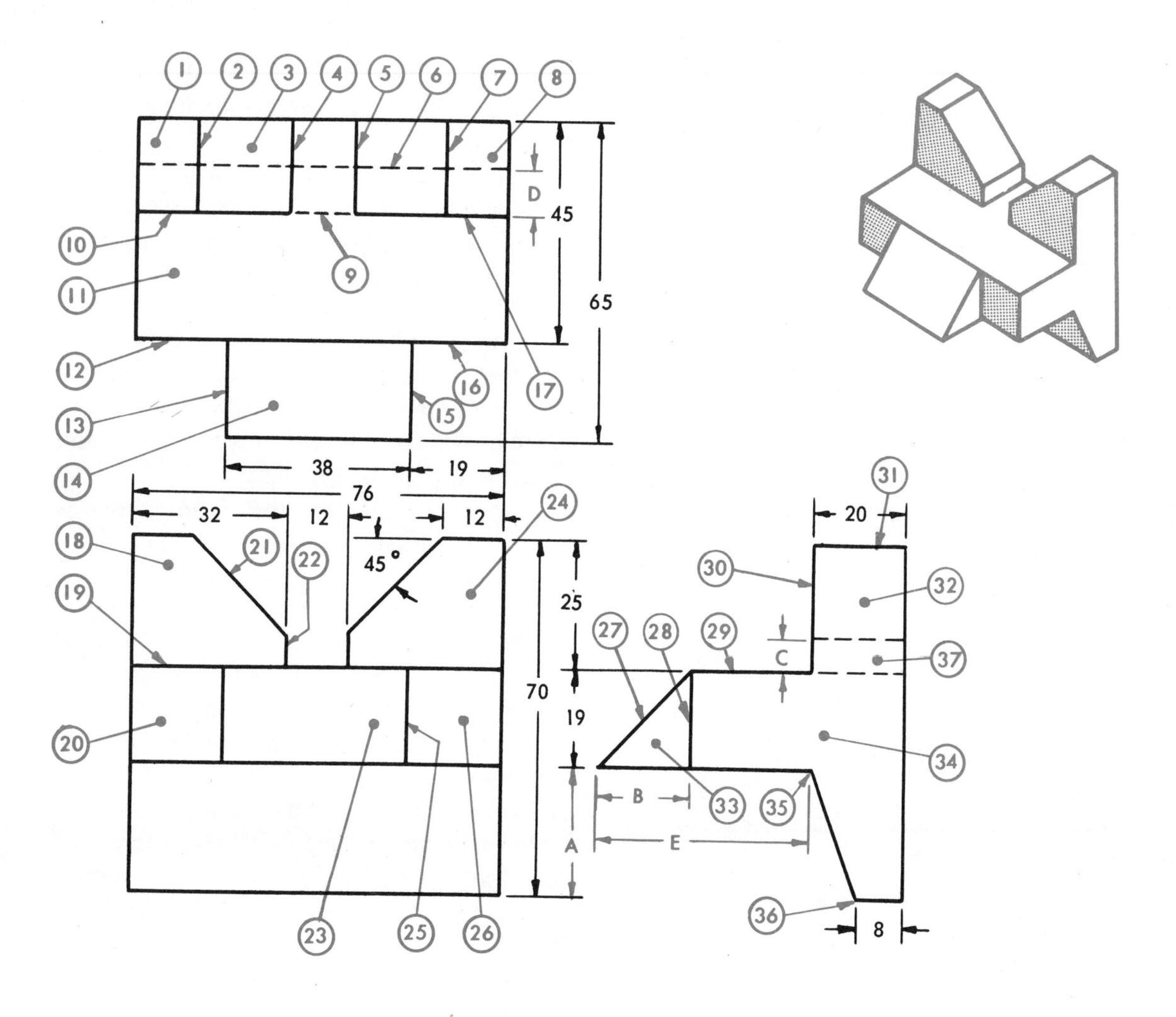

Fig. 3-8-G Slide bracket

1. Calculate dimensions A to E.
2. What line in the top view is represented by surface 18 in the front view?
3. What surfaces in the top view are represented by line 31 in the side view?
4. What line in the front view is represented by surface 3 in the top view?
5. What surface in the side view is represented by surface 3 in the top view?
6. What line in the side view is represented by surface 14 in the top view?
7. What line in the side view is represented by line 12 in the top view?
8. What line in the top view is represented by point 35 in the side view?
9. What line in the front view is represented by surface 11 in the top view?
10. What surface in the front view is represented by line 27 in the side view?
11. What surface in the top view is represented by line 29 in the side view?
12. What line in the side view is represented by surface 8 in the top view?
13. What lines in the top view are represented by line 28 in the side view?
14. What surfaces in the front view are represented by line 30 in the side view?
15. What line in the front view is represented by line 29 in the side view?
16. What lines in the top view are represented by line 30 in the side view?
17. What surface in the side view is represented by line 4 in the top view?
18. What surfaces in the front view are represented by line 28 in the side view?
19. What lines in the top view are represented by surface 33 in the side view?
20. What line in the top view is represented by point 36 in the side view?
21. What line in the front view is represented by line 4 in the top view?

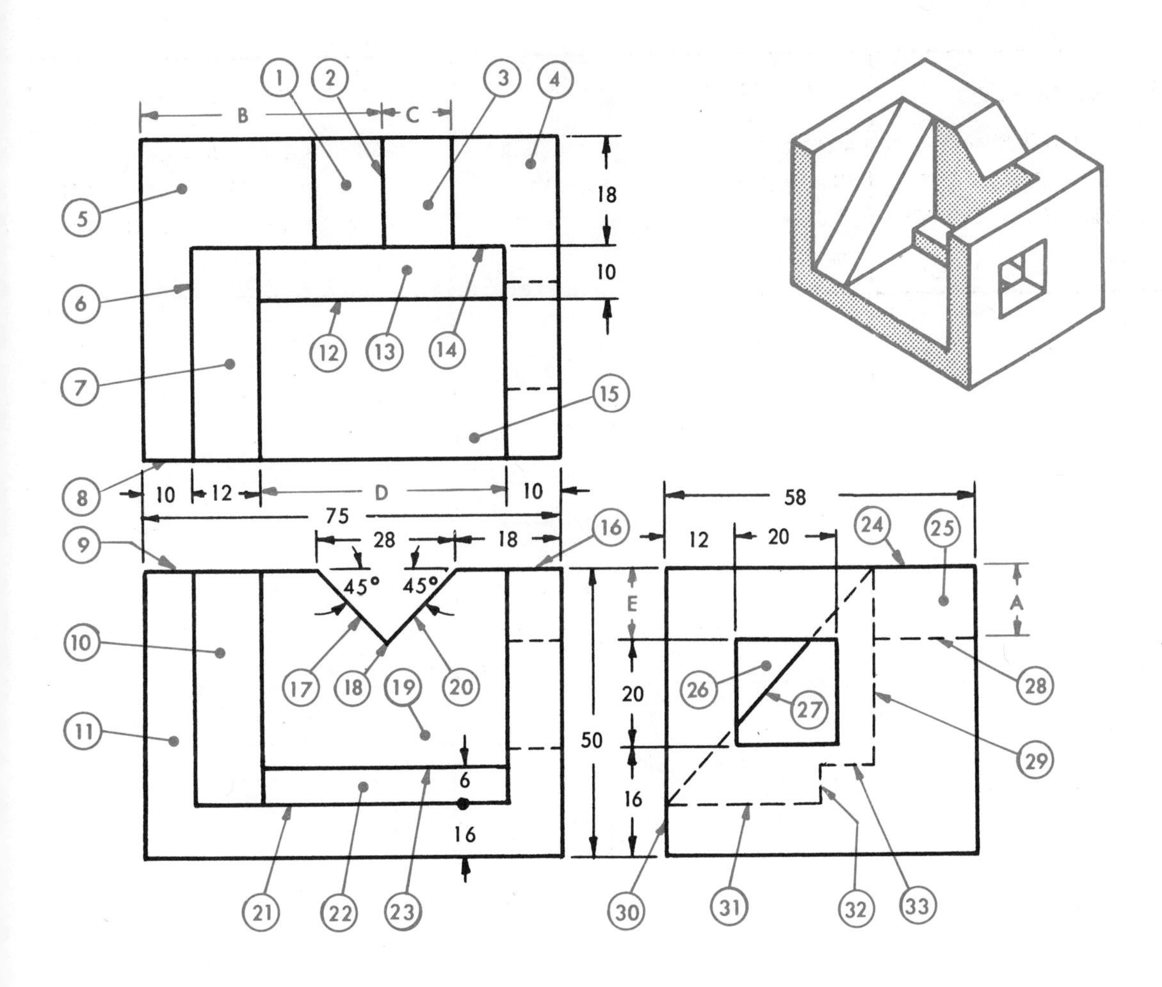

Fig. 3-8-H V bracket

1. Calculate dimensions A to E.
2. What line in the front view is represented by surface 1 in the top view?
3. What line in the side view is represented by surface 11 in the front view?
4. What surface in the top view is represented by line 31 in the side view?
5. What line in the side view is represented by line 12 in the top view?
6. What surface in the front view is represented by line 12 in the top view?
7. What lines in the front view are represented by line 24 in the side view?
8. What line in the side view is represented by point 18 in the front view?
9. What surface in the front view is represented by line 27 in the side view?
10. What line in the front view is represented by surface 3 in the top view?
11. What line in the side view is represented by line 8 in the top view?
12. What line in the top view is represented by line 32 in the side view?
13. What line in the side view is represented by line 21 in the front view?
14. What line in the top view is represented by line 29 in the side view?
15. What line in the side view is represented by surface 7 in the top view?
16. What surfaces in the top view are represented by line 24 in the side view?
17. What line in the front view is represented by surface 15 in the top view?
18. What surface in the front view is represented by line 29 in the side view?
19. What line in the top view is represented by surface 19 in the front view?
20. What surface in the side view is represented by line 6 in the top view?
21. What surface in the top view is represented by line 16 in the front view?

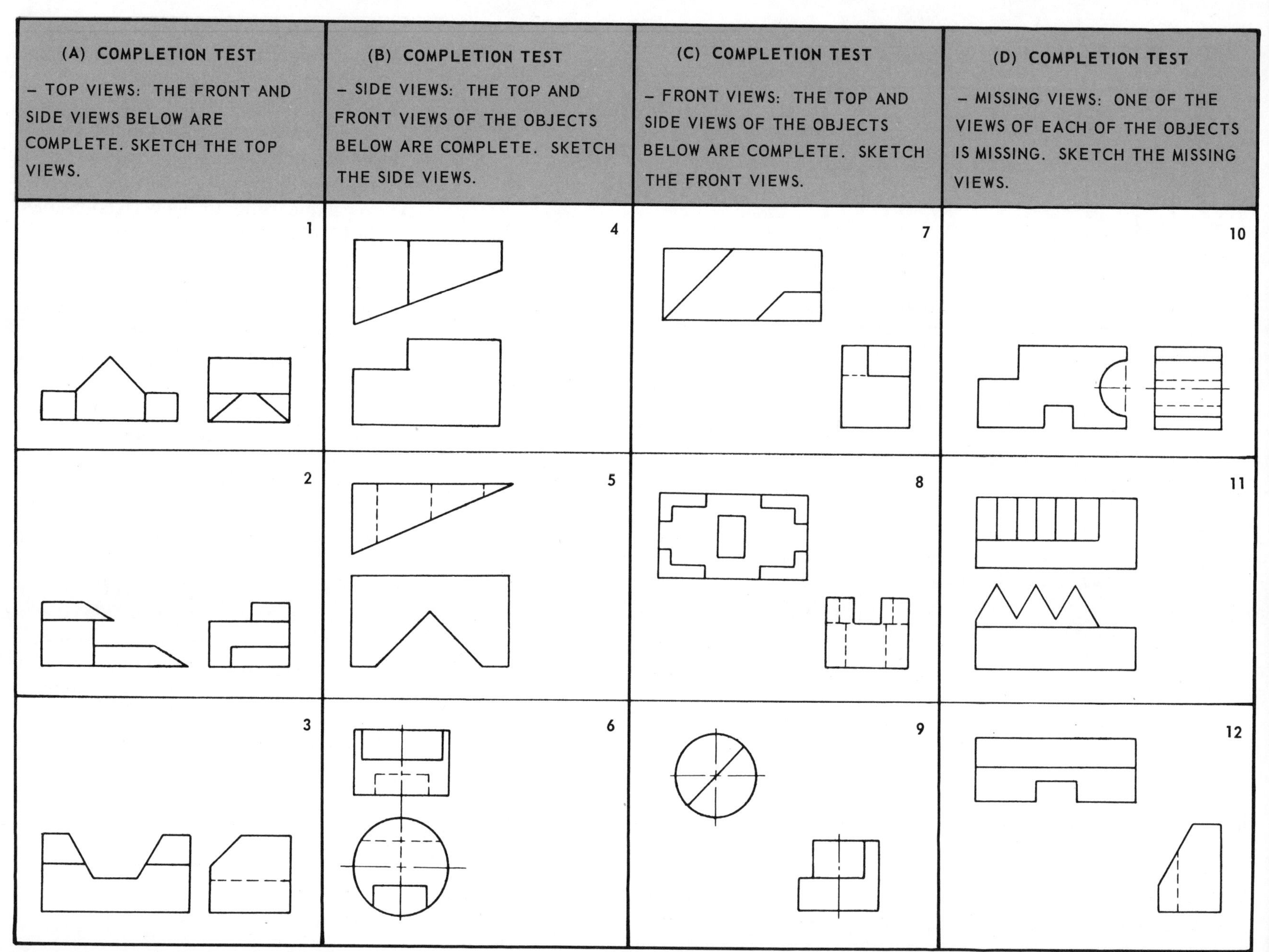

Fig. 3-8-J Completion tests 1 to 12

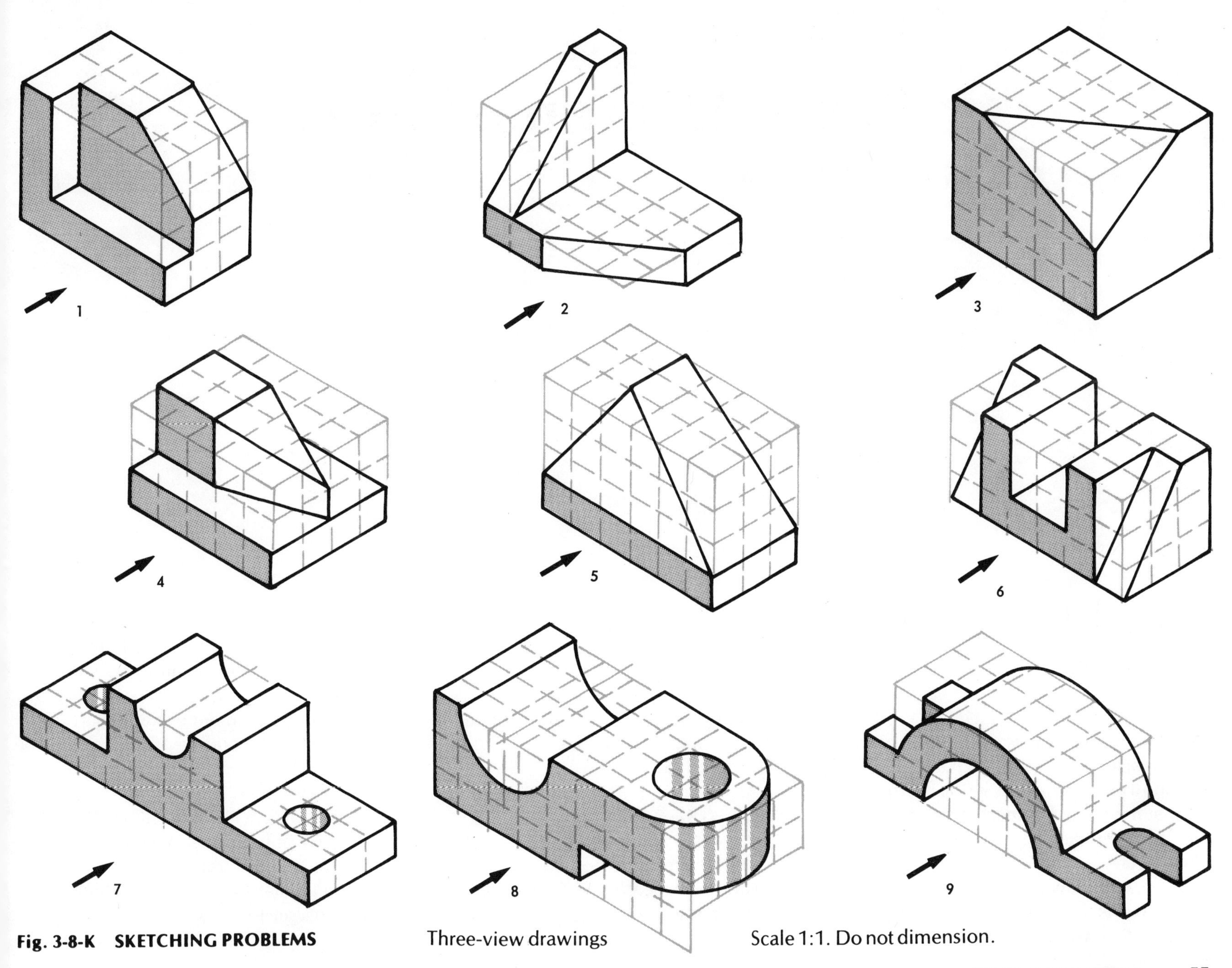

Fig. 3-8-K SKETCHING PROBLEMS Three-view drawings Scale 1:1. Do not dimension.

Chapter 4

Basic Dimensioning

UNIT 4-1

Working Drawings

A working drawing is one from which a tradesperson can produce a part. The drawing must be a complete set of instructions, so that it will not be necessary to give additional information to the person(s) making the object.

A working drawing consists of:

- The Views necessary to explain the shape;
- The Dimensions needed by the tradesperson to construct or assemble the part or parts;
- The required Specifications, such as material and quantity required. (This information may be in the form of notes on the drawing, or may be given in the title block or bill of material.)

Dimensioning

Dimensions are indicated on drawings by means of extension lines, dimension lines, leaders, arrowheads, figures, notes and symbols.

They define characteristics such as widths, heights, thicknesses, lengths, diameters, angles and locations of holes or slots.

The lines used in dimensioning are THIN in contrast to the outline of the object. The dimension must be clear and permit only one interpretation.

In general, each surface, line or point is located by only one set of dimensions. An exception to this basic rule is for arrowless and tabular dimensioning which is discussed in Unit 4-4.

Study the illustrations which demonstrate

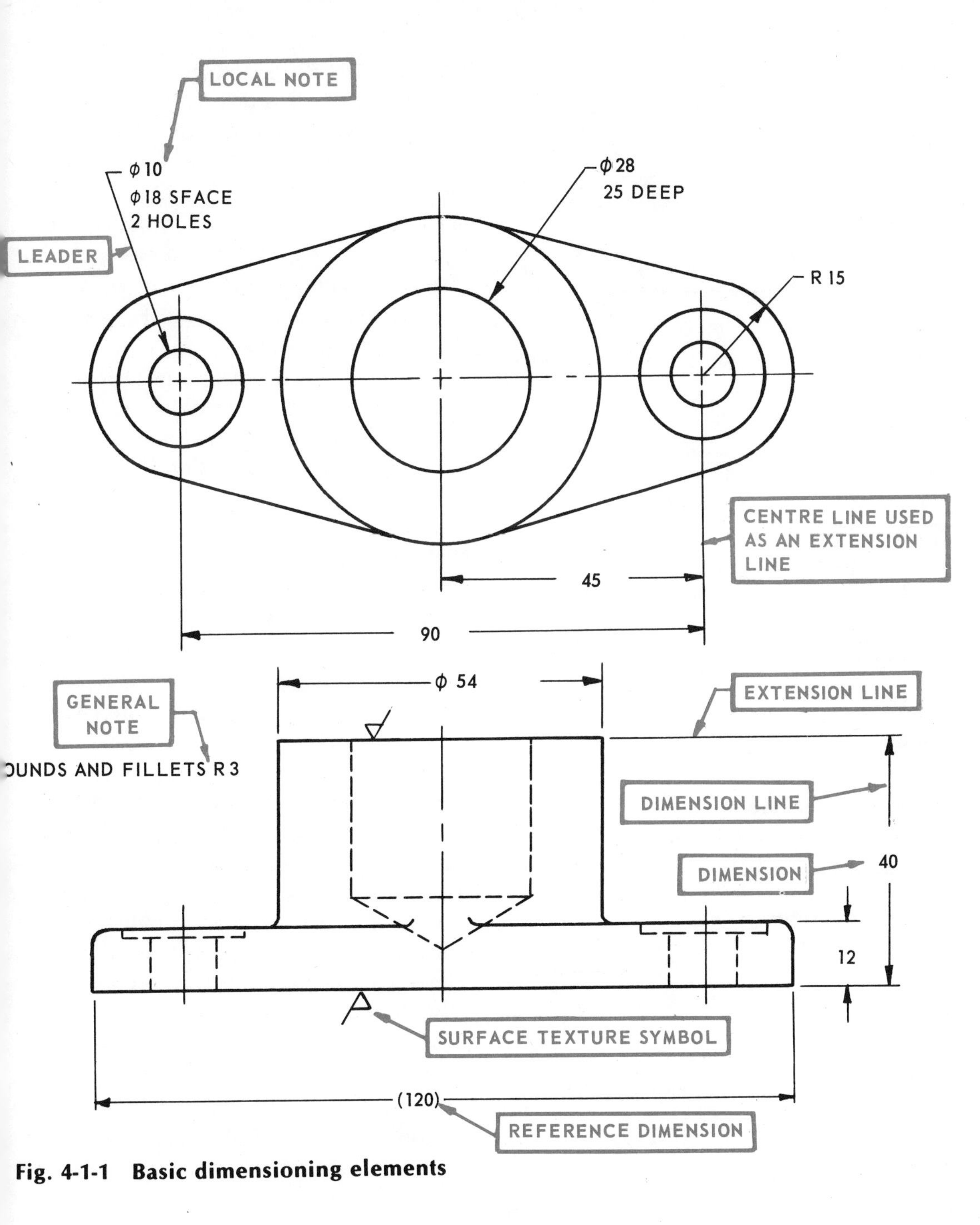

Fig. 4-1-1 **Basic dimensioning elements**

the basic dimensioning elements, rules, and techniques, and refer to them as you work on the assigned drawing projects.

Dimension and Extension Lines

Dimension lines are used to indicate the extent and direction of dimensions, and they are terminated by neatly made, uniform arrowheads, as shown in Figure 4-1-2. Arrowheads are usually drawn freehand, and the recommended length and the width should be in a ratio of 3:1 (Fig. 4-1-3*b*). The length of the arrowhead should be equal to the height of the dimension numerals. Where space is limited, a small circular dot may be used in lieu of an arrowhead (Fig. 4-1-3*d*).

Normally, a break is made near the centre of the dimension line for the insertion of the dimension which indicates the distance between the extension lines. When several dimension lines are directly above or next to

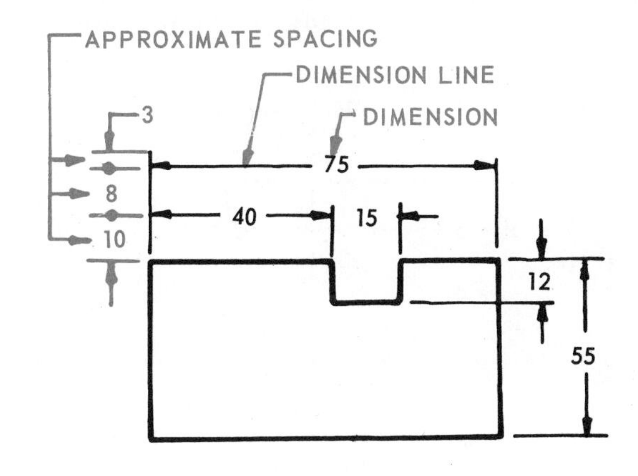

Fig. 4-1-2 **Dimension and extension lines**

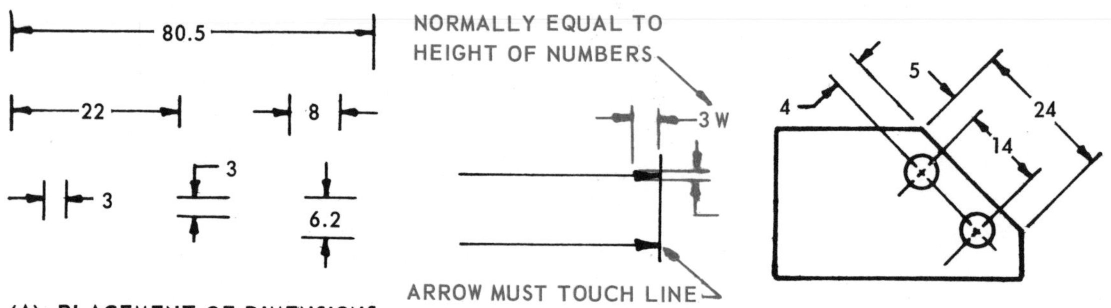

(A) PLACEMENT OF DIMENSIONS

(B) ARROWHEAD SIZE

(C) OBLIQUE DIMENSION LINES

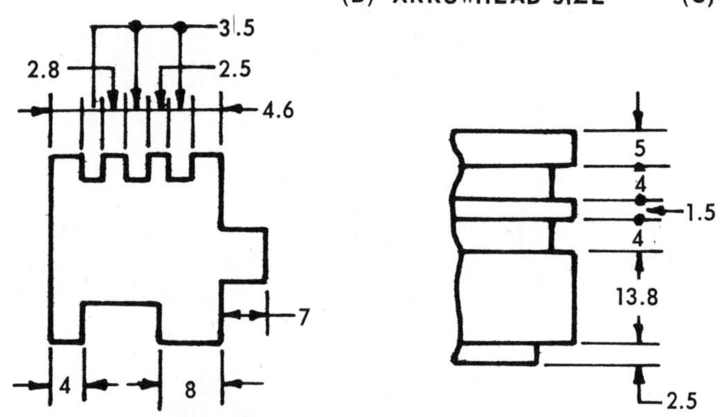

(D) DIMENSIONING IN RESTRICTED AREAS

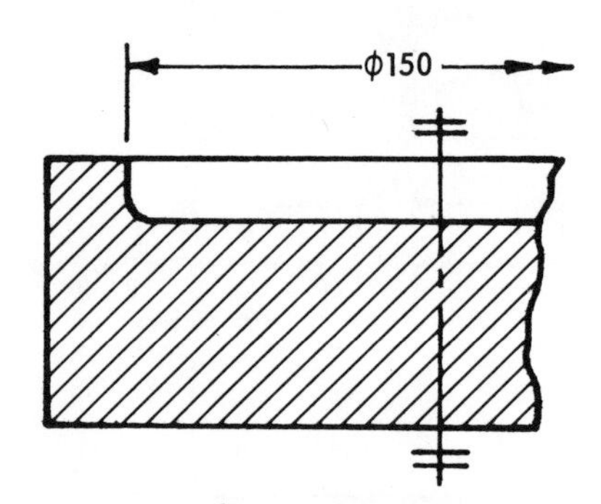

(E) DIMENSIONING PARTIAL (SYMMETRICAL) VIEWS

100

80

60

(F) STAGGERING DIMENSIONS FOR CLARITY

Fig. 4-1-3 Dimension lines

one another, it is good practice to stagger the dimensions in order to improve the clarity of the drawing. See Figure 4-1-3.

The spacing suitable for most drawings between parallel dimension lines is 8 mm, and the spacing between the outline of the object and the nearest dimension line should be approximately 10 mm.

When the space between the extension lines is too small to permit the placing of the dimension line complete with arrowheads and dimension, then the alternate methods of placing the dimension line, dimension, or both outside the extension lines are used (Fig. 4-1-3*d*).

Centre lines should never be used for dimension lines. Every effort should be made to avoid crossing dimension lines. This is accomplished by placing the shortest dimension closest to the outline (Fig. 4-3-1*c*).

Dimension lines should be placed outside the view where possible and should extend to extension lines rather than object lines. However, when readability is improved by avoiding either extra-long extension lines (Fig. 4-1-4) or the crowding of dimensions, placing of dimensions on views is permissible.

Avoid dimensioning to hidden lines. In order to do so, it may be necessary to use a sectional view or a broken-out section.

When the end of a dimension is not included, as when used on partial views, a double arrow is used. For symmetrical features the dimension line should extend beyond the centre line before the double arrowheads are added.

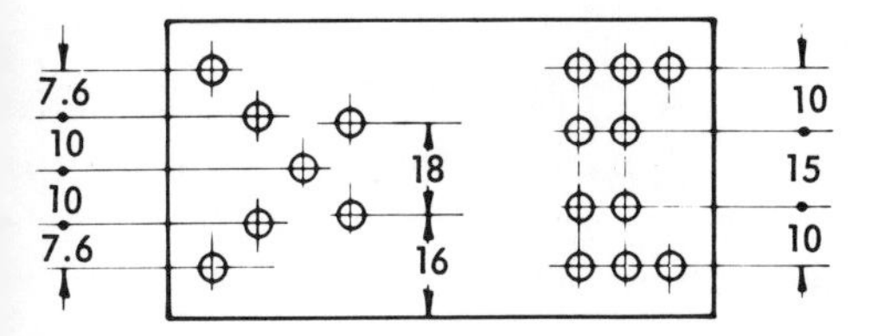

(A) IMPROVING READABILITY OF DRAWING

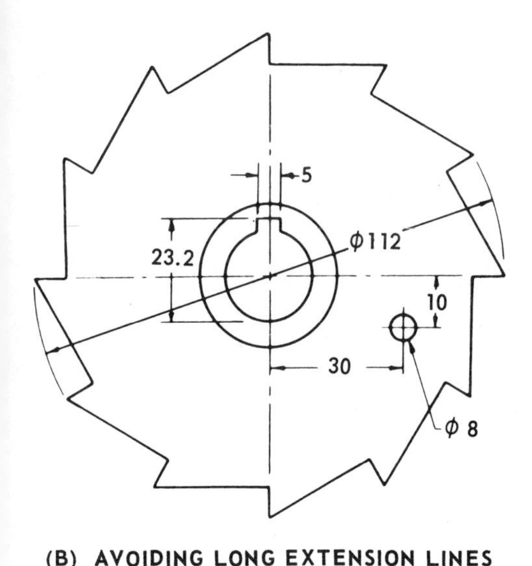

(B) AVOIDING LONG EXTENSION LINES

Fig. 4-1-4 Placing dimensions on view

Extension (witness) lines are used to indicate the point or line on the drawing to which the dimension applies (Fig. 4-1-5). A small gap of about 1 mm is left between the extension line and the outline to which it refers, and the extension line extends about 3 mm beyond the outermost dimension line. However, when extension lines refer to points, as in Figure 4-1-5*f*, they should extend through the points.

Extension lines are usually drawn perpendicular to dimension lines. However, to improve clarity or when there is overcrowding, extension lines may be drawn at an oblique angle but where they apply must be clearly illustrated.

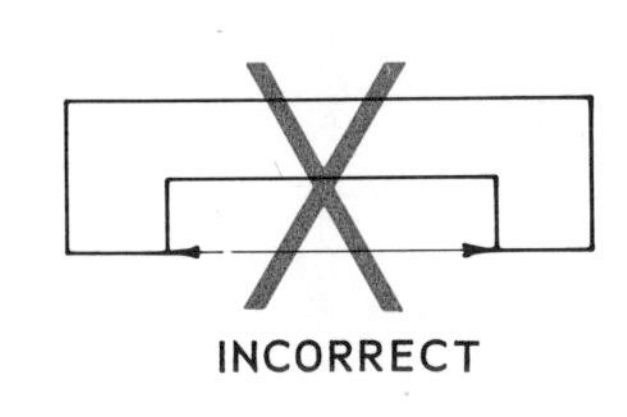

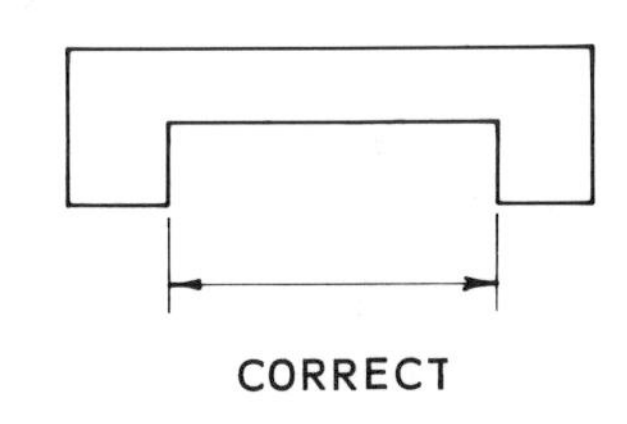

(A) USE OF EXTENSION LINES

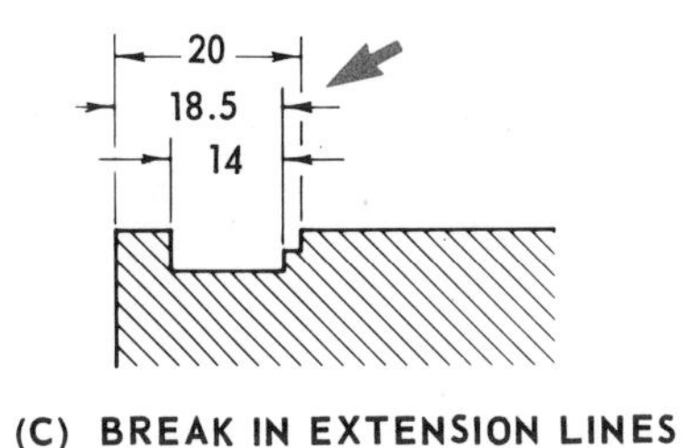

(C) BREAK IN EXTENSION LINES

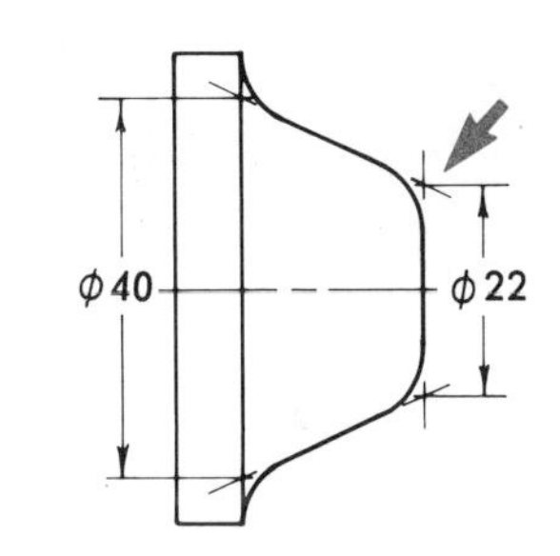

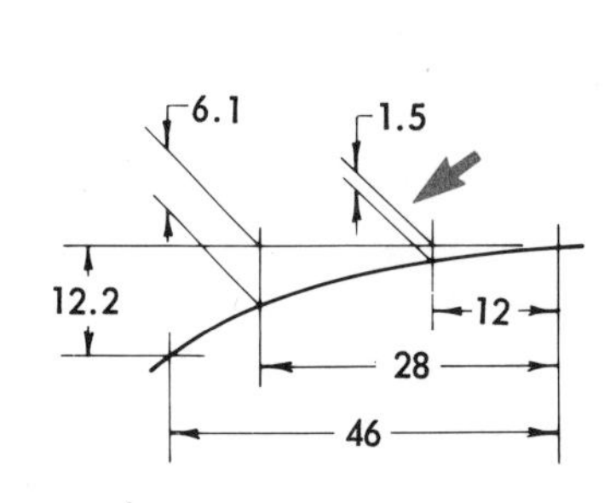

(D) OBLIQUE EXTENSION LINES

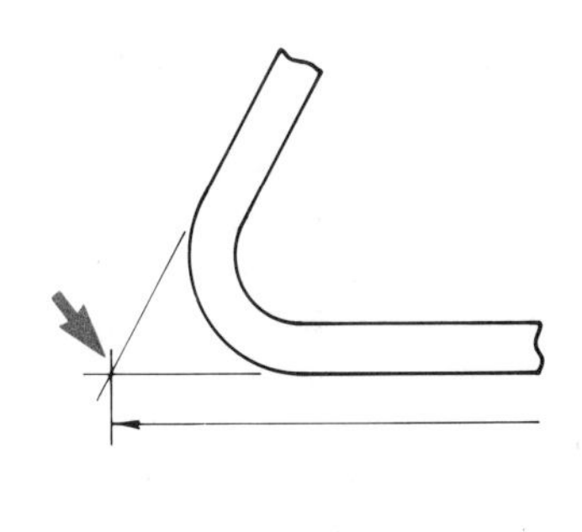

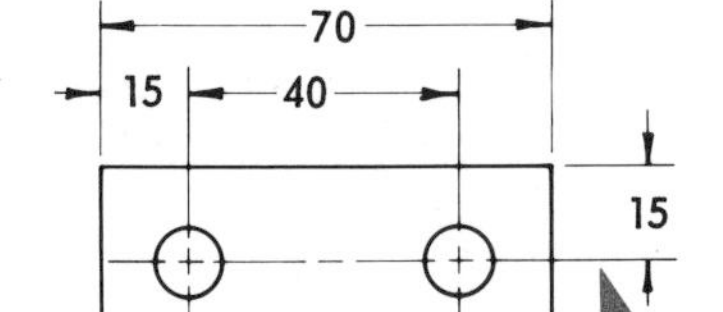

(B) CENTRE LINE USED AS AN EXTENSION LINE

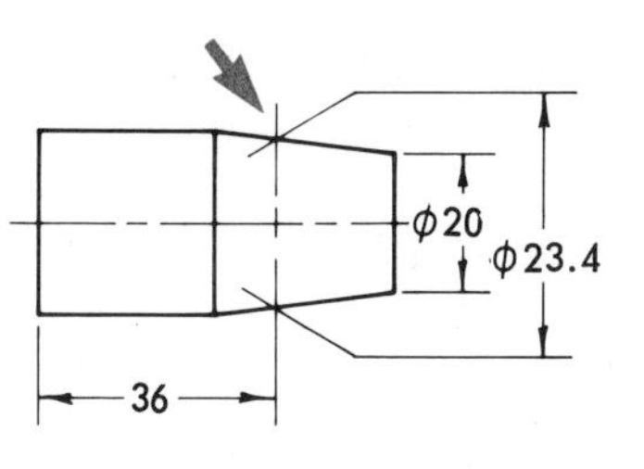

(E) EXTENSION LINES ON TAPERS

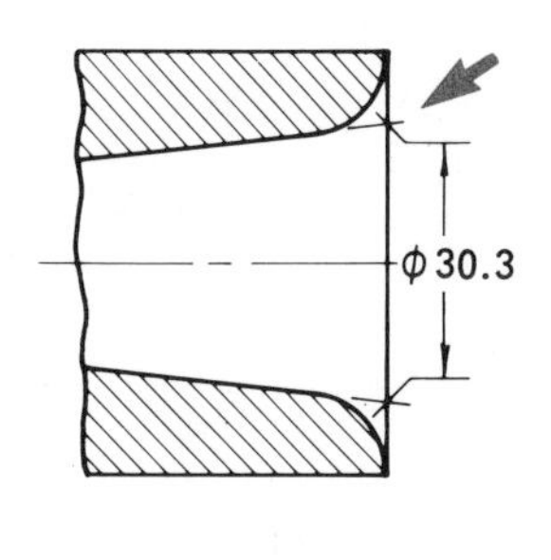

(F) EXTENSION LINES FROM POINTS

Fig. 4-1-5 Extension lines

Centre lines may be used as extension lines in dimensioning. The centre line extending past the circle is not broken.

Where extension lines cross other extension lines, dimension lines, or object lines, they are not broken. However, if extension lines cross arrowheads or dimension lines close to arrowheads, a break in the extension line is recommended.

Leaders

Leaders are used to direct notes, dimensions, symbols, item numbers or part numbers to features on the drawing (Fig. 4-1-6).

A leader should generally be a single straight inclined line (not vertical or horizontal) except for a short horizontal portion extending to the centre of the height of the first or last letter or digit of the note.

The leader is terminated by an arrowhead or a dot of at least 1.5 mm in diameter. Arrowheads should always terminate on a line; dots should be used within the outline of the object and rest on a surface.

Leaders should not be bent in any way unless it is unavoidable; they should not cross one another; and two or more leaders adjacent to one another should be drawn parallel if practicable. It is better to repeat dimensions or references than to use long leaders.

Where a leader is directed to a circle or circular arc, its direction should point to the centre of the arc or circle.

All notes and dimensions used with leaders are placed in a horizontal position.

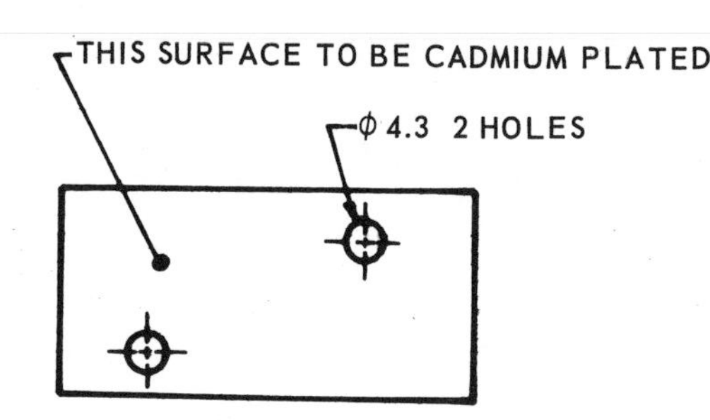

(A) DOT RESTS ON SURFACE

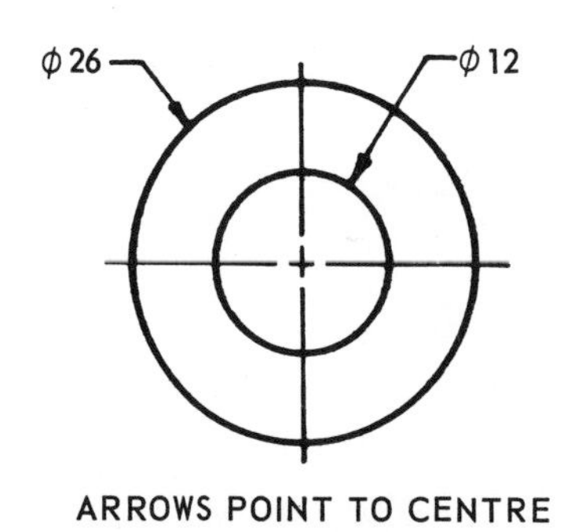

ARROWS POINT TO CENTRE

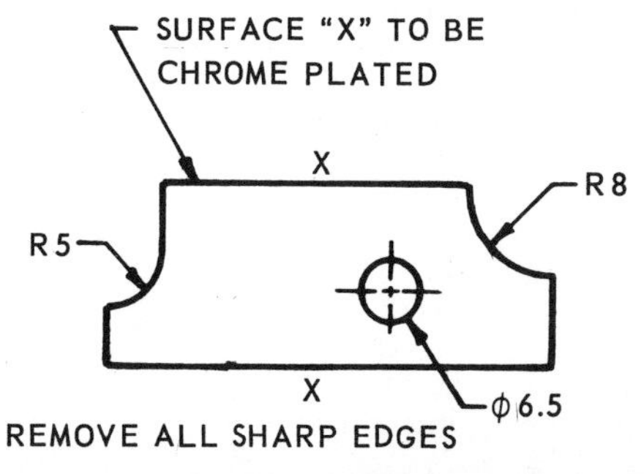

(B) ARROWHEADS POINT TO FEATURE

Fig. 4-1-6 Leaders

Notes

Notes are used to simplify or complement dimensioning by indicating information on a drawing in a condensed and systematic manner. They may be general or local notes, and should be in the present or future tense.

General Notes. These refer to the part or drawing as a whole. They should be shown in a central position below the view to which they apply or placed in a general note column. Typical examples of this type of note are

- FINISH ALL OVER
- ROUNDS AND FILLETS R 2
- REMOVE ALL SHARP EDGES

Local Notes. These apply to local requirements only and are connected by a leader to the point to which the note applies. Typical examples are

- ϕ6, 4 HOLES
- 2 × 45°
- ϕ3
  ϕ11.5 × 86° CSK
- M12 × 1.25

Linear Units of Measurements

Although the SI metric system of dimensioning and measurement has become the official standard, some drawings still in current use are dimensioned in inches or feet and inches. For this reason, you should be familiar with all the dimensioning systems which may be encountered.

SI Metric Units of Measurement

The standard metric units on engineering drawings are the millimetre (mm) for linear measure and micrometre (μm) for surface roughness (Fig. 4-1-7). For architectural drawings, metre and millimetre units are used. Unless otherwise specified, the dimensions in this book are in millimetres.

Whole numbers from 1 to 9 will be shown without a zero to the left of the number or a zero to the right of the decimal point.

2 *not* 02 or 2.0

A millimetre value of less than 1 is shown with a zero to the left of the decimal point.

0.2 *not* .2 or .20

0.26 *not* .26

Decimal points should be uniform and large enough to be clearly visible on reduced-size prints. They should be placed in line with the bottom of the associated numbers and be given adequate space.

Commas should not be used to separate groups of three numbers in either metric or inch values. A space should be used in place of the comma.

32 541 *not* 32,541

2.562 827 6 *not* 2.562827

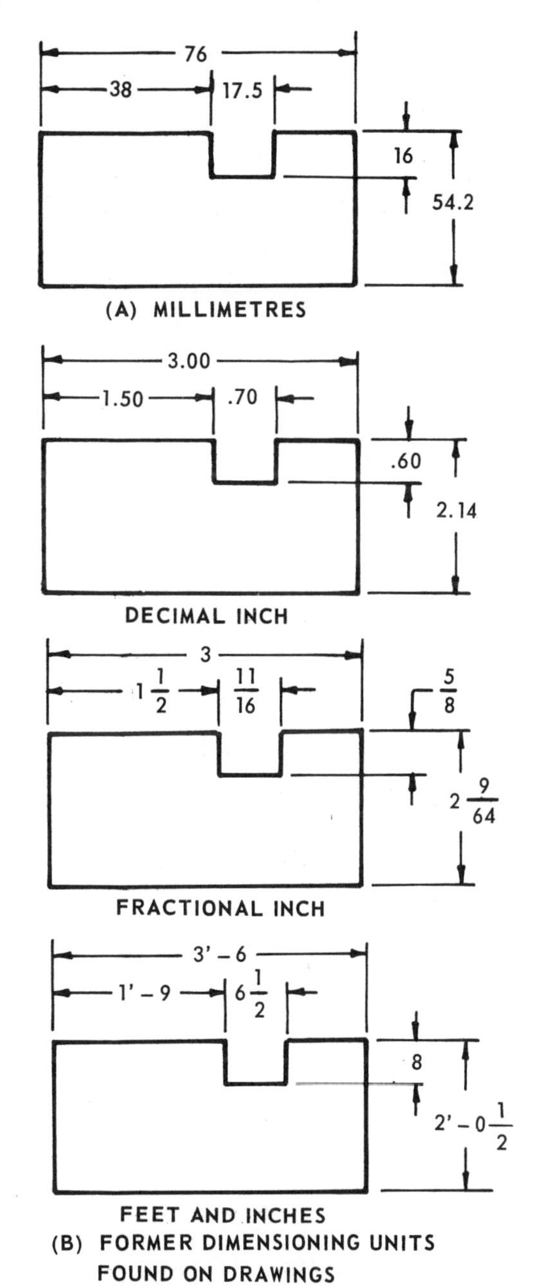

Fig. 4-1-7 Dimensioning units

Identification. A metric drawing should include a general note, such as UNLESS OTHERWISE SPECIFIED DIMENSIONS ARE IN MILLIMETRES and be identified by the word METRIC prominently displayed near the title block.

Inch Units of Measurement

Decimal Inch System. Parts were designed in basic decimal increments, preferably 0.02 in., and were expressed with a minimum of two figures to the right of the decimal point. Using the 0.02 module, the second decimal place (hundredths) is an even number or zero. By using the design modules having an even number for the last digit, dimensions could be halved for centre distances without increasing the number of decimal places. Decimal dimensions which were not multiples of 0.02, such as 0.01, 0.03, and 0.15, would be used only when it was essential to meet design requirements such as to provide clearance, strength, smooth curves, etc. When greater accuracy was required, sizes were expressed as three-or-four place decimal numbers, for example, 1.875.

Whole dimensions showed a minimum of two zeros to the right of the decimal point.

24.00 *not* 24

An inch value of less than 1 was shown *without* a zero to the left of the decimal point.

.44 *not* 0.44

In cases where parts have to be aligned with existing parts or commercial products, which are dimensioned in fractions, it may be necessary to use decimal equivalents of fractional dimensions.

Foot and Inch System. Feet and inches were often used for installation drawings, drawings of large objects, and floor plans associated with architectural work. In this case, all dimensions 12 in. or greater were specified in feet and inches. Parts of an inch were usually expressed as common fractions, rather than as decimals.

The inch marks (″) were not shown. The drawing should carry a note such as

DIMENSIONS ARE IN INCHES
UNLESS OTHERWISE SPECIFIED

A dash and space were to be left between the foot and inch values. For example, 1′ - 3, not 1′3.

Whole numbers of inches were expressed without decimals or fractions. For example, 24 in. was expressed as 2′ - 0 and 27 in. was expressed as 2′ - 3.

* N.B. The old inch (″) and foot (′) marks are now to be used only to indicate minutes and seconds of angular measure. (See Angular Units)

Angular Units of Measurements

Angles are measured in degrees. The decimal degree is now preferred to the use of degree, minutes, and seconds: For example, the use of 50.5° is preferred to the use of 50°30′. Where only minutes or seconds are specified, the number of minutes or seconds is preceded by 0°, or 0°0′, as applicable. Some examples follow.

Decimal Degree	Degrees, Minutes, and Seconds
10° ± 0.5°	10° ± 0°30′
0.75°	0°45′
0.004°	0°0′15″
90° ± 1.0°	90° ±1°
25.6° ± 0.2°	25°36′ ±0°12′
25.51°	25°30′40″

The dimension line of an angle is an arc drawn with the apex of the angle as the centre point for the arc, wherever practicable. The position of the dimension varies according to the size of the angle and appears in a horizontal position. Recommended arrangements are shown in Figure 4-1-8.

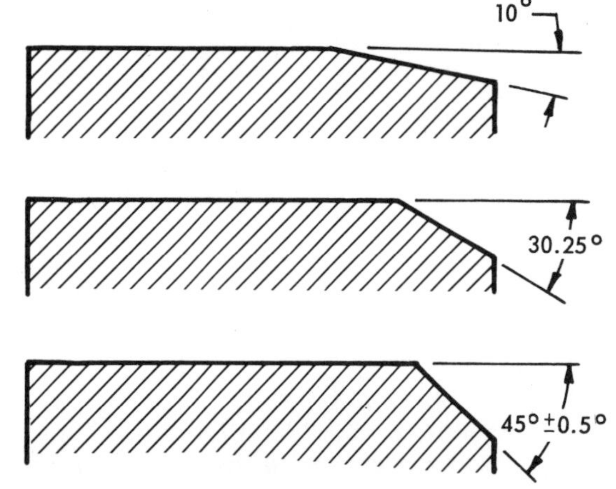

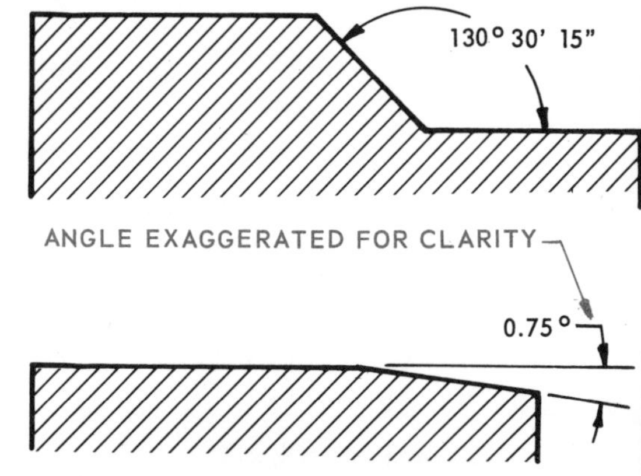

Fig. 4-1-8 Angular units

Reading Direction

Dimensions on drawings are placed to be read from the bottom of the drawing.

Angular dimensions and dimensions and notes shown with leaders should be aligned with the bottom of the drawing.

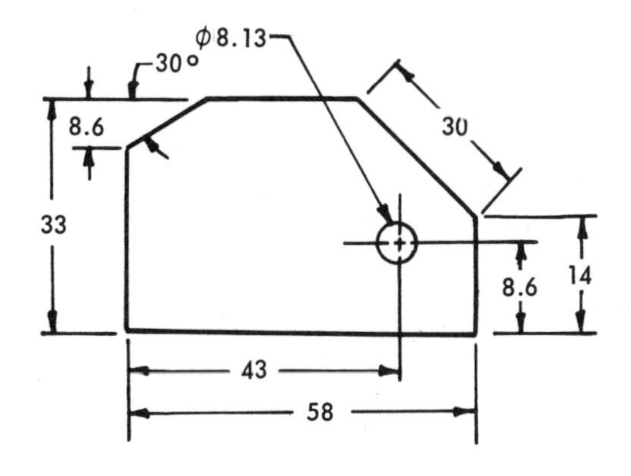

Fig. 4-1-9 Reading direction of dimensions

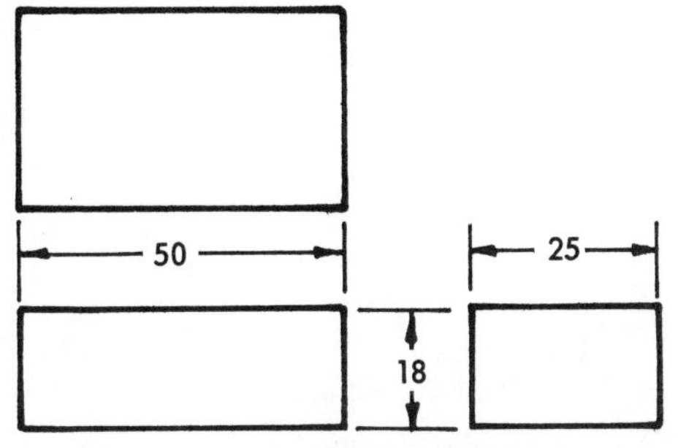

(A) PLACE DIMENSIONS BETWEEN VIEWS

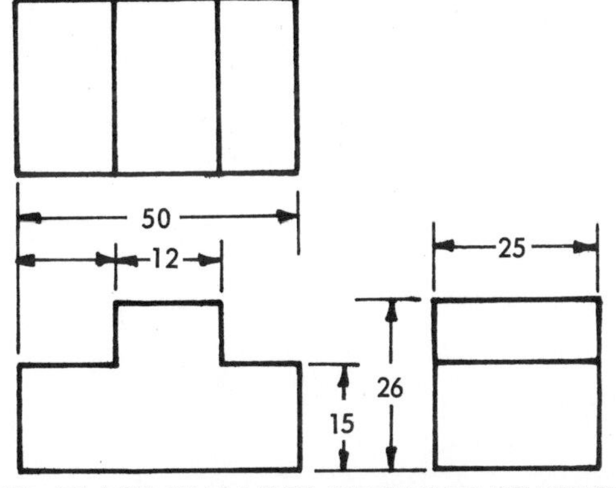

(B) PLACE SMALLEST DIMENSION NEAREST THE VIEW BEING DIMENSIONED

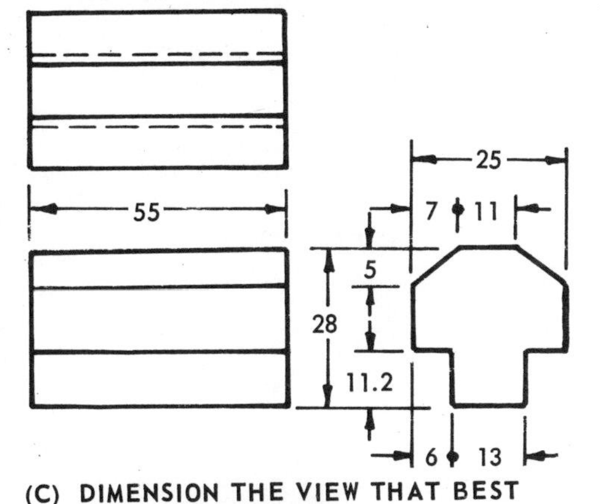

(C) DIMENSION THE VIEW THAT BEST SHOWS THE SHAPE

Fig. 4-1-10 Basic dimensioning rules

Basic Rules for Dimensioning

- Place dimensions between the views when possible.
- Place the dimension line for the shortest width, depth or height, nearest the outline of the object. Parallel dimension lines are placed in order of their size, making the longest dimension line the outermost.
- Place dimensions near the view that best shows the characteristic contour or shape of the object. In following this rule, dimensions will not always be between views.
- On large drawings, dimensions can be placed on the view to improve clarity.

Reference and Not-to-Scale Dimensions

When a reference dimension is shown on a drawing for information only, and is not necessary for the manufacture of that part, it is shown in parentheses, as shown in Figure 4-1-11.

When a dimension on a drawing is altered, making it not to scale, a freehand line is drawn below the dimension to indicate that the dimension is not drawn to scale. See Figure 4-1-12.

Abbreviations

Abbreviations and symbols are used on drawings to conserve space and time, but

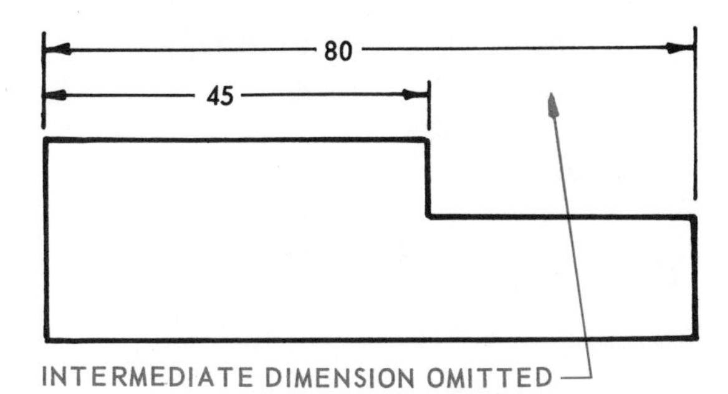

(A) NORMAL DIMENSIONING PRACTICE

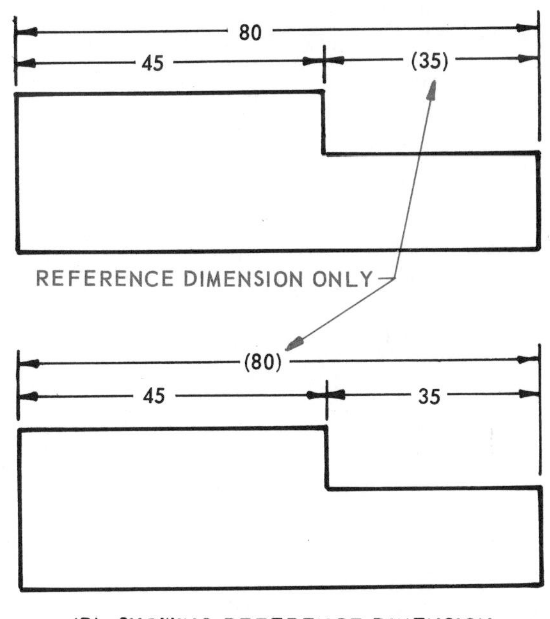

(B) SHOWING REFERENCE DIMENSION

Fig. 4-1-11 Reference dimension

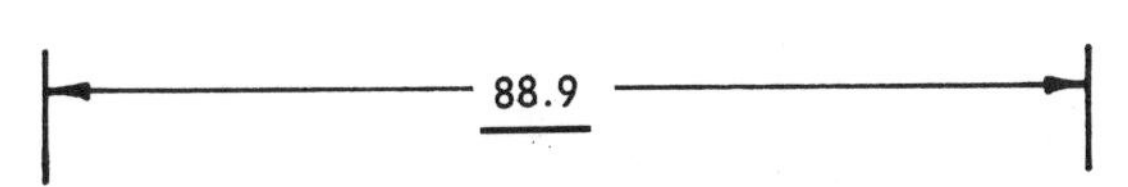

Fig. 4-1-12 Not-to-scale dimension

only where their meanings are quite clear. Therefore, only commonly accepted abbreviations such as those shown in the Appendix should be used on drawings.

Operational Names

The use of operational names of dimensions, such as turn, bore, grind, ream, tap, and thread, should be avoided. While you should be aware of the methods by which a part can be produced, the method of manufacture is better left to the producer.

If the completed part is adequately dimensioned and has surface texture symbols showing finish quality desired, it is then a shop problem to meet the drawing specifications.

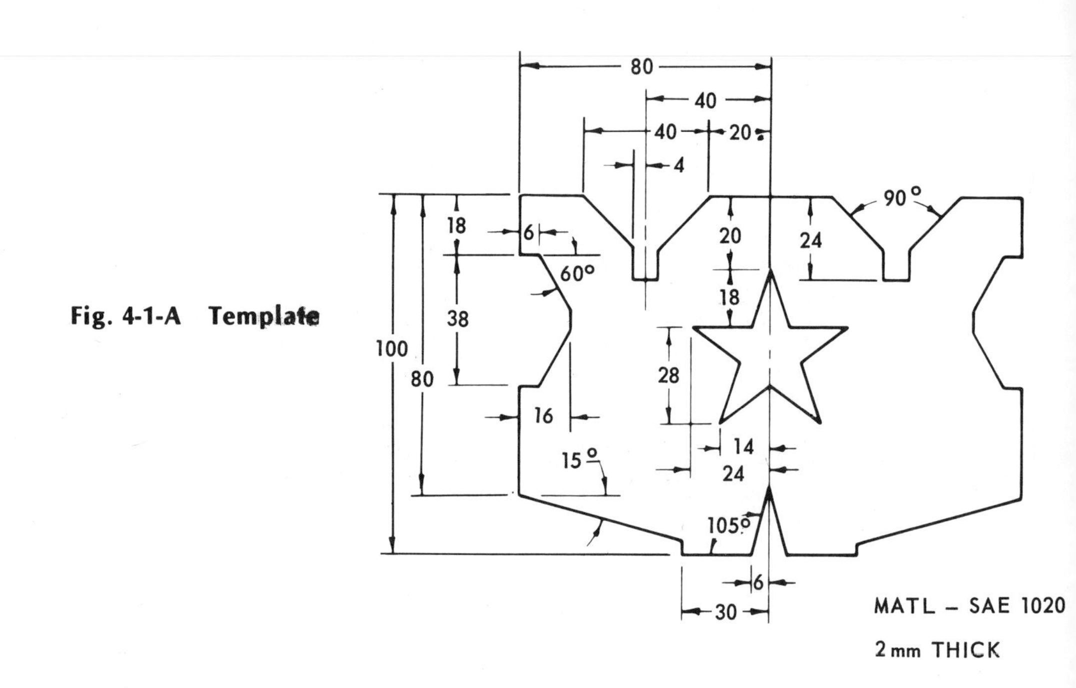

Fig. 4-1-A **Template**

Assignments

All drawings to be complete with dimensions, sheet size-A4, scale 1:1.

1. Template, Fig. 4-1-A.
2. Bracket, Fig. 4-1-B.
3. Step block, Fig. 4-1-C.
4. Guide block, Fig. 4-1-D.
5. Angle bracket, Fig. 4-1-E.
6. Angle stop, Fig. 4-1-F.
7. Control block, Fig. 4-1-G.
8. Sliding block, 4-1-H.

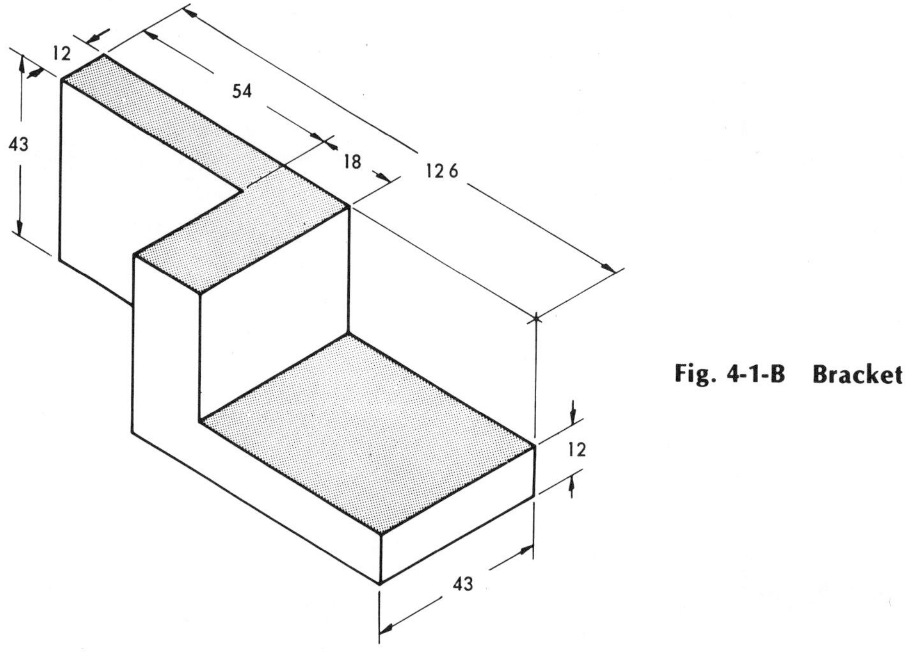

Fig. 4-1-B **Bracket**

Review for Assignment

Unit 3-7 Spacing the views

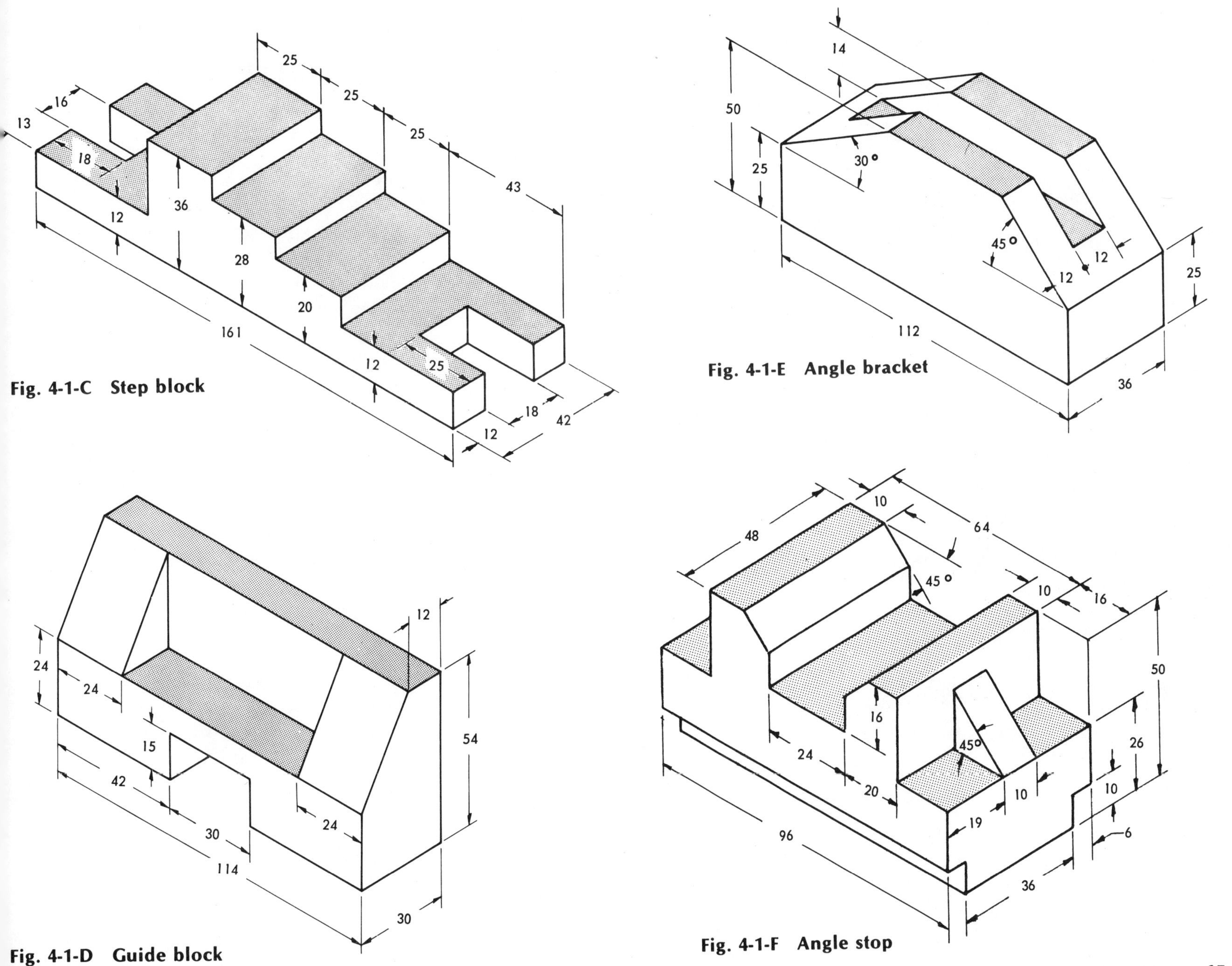

Fig. 4-1-C Step block

Fig. 4-1-E Angle bracket

Fig. 4-1-D Guide block

Fig. 4-1-F Angle stop

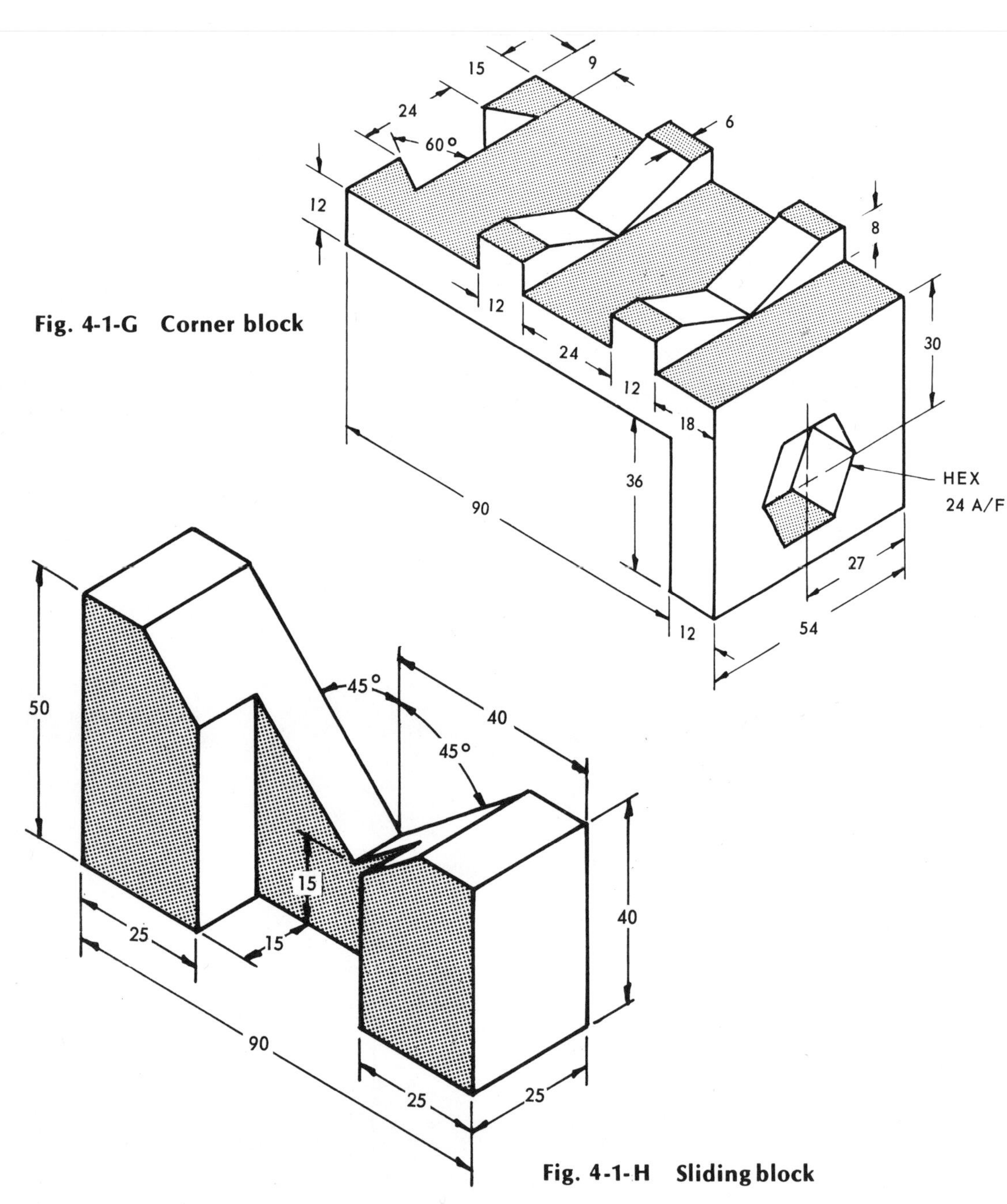

Fig. 4-1-G Corner block

Fig. 4-1-H Sliding block

UNIT 4-2
Dimensioning Circular Features

Diameters

Where the diameter of a single feature or the diameters of a number of concentric cylindrical features are to be specified, it is recommended that they be shown on the longitudinal view. (Fig. 4-2-1).

Where it is obvious from the views that the dimension refers to the diameter and a dimension line is used, the symbol ϕ or the abbreviation DIA may be omitted. It is recommended that the diameter symbol or abbreviation be used in conjunction with a leader.

Where the diameter is dimensioned on a single-view drawing which does not show that the feature is circular, the dimension for the diameter is preceded by the symbol ϕ or followed by the abbreviation DIA. The symbol is the preferred method since it is faster to draw, takes less space, and is universal in use.

Where space is restricted or when only a partial view is used, diameters may be dimensioned as illustrated in Figure 4-2-2.

Radii

The general method of dimensioning a circular arc is by giving its radius. A radius dimension line passes through, or is in line

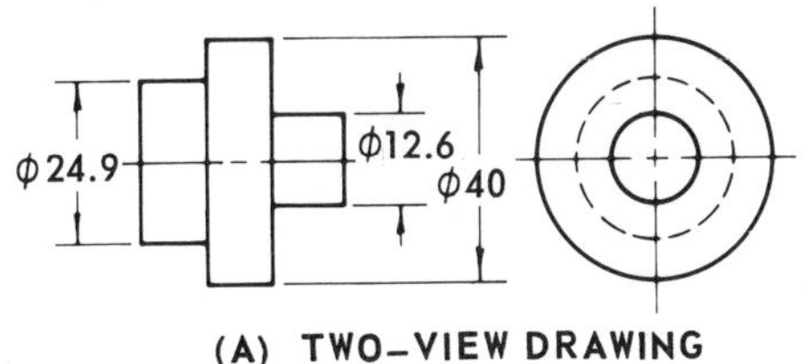

(A) TWO–VIEW DRAWING

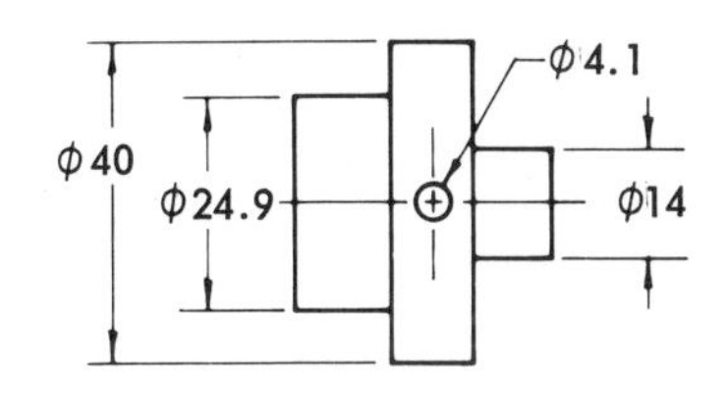

(B) ONE–VIEW DRAWING

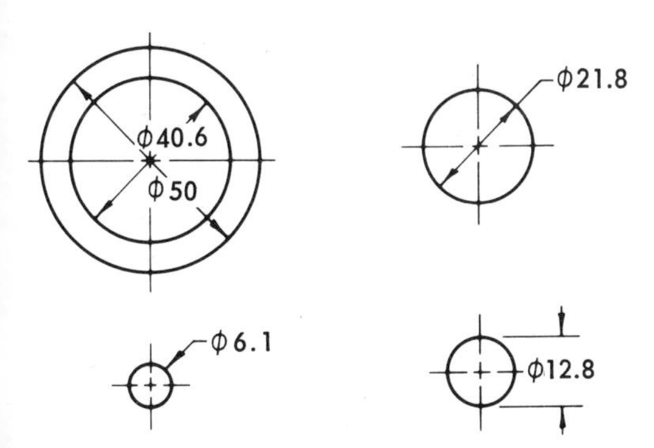

(C) DIMENSIONING DIAMETERS ON END VIEW

Fig. 4-2-1 Diameters

with, the radius centre and terminates with an arrowhead touching the arc (Fig. 4-2-3). An arrowhead is never used at the radius centre.

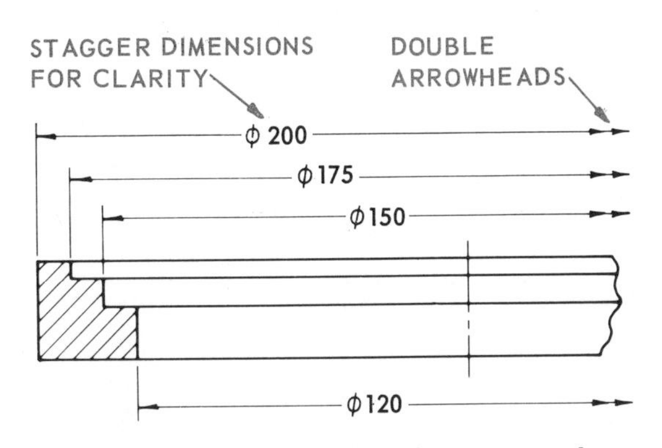

Fig. 4-2-2 Dimensioning diameters where space is restricted

The size of the dimension is preceded by the abbreviation R for metric dimensioning and followed by the abbreviation R for feet and inch dimensions.

Where space is limited, as for a small radius, the radial dimension line may extend through the radius centre. Where it is inconvenient to place the arrowhead between the radius centre and the arc, it may be placed outside the arc, or a leader may be used (Fig. 4-2-3*a*).

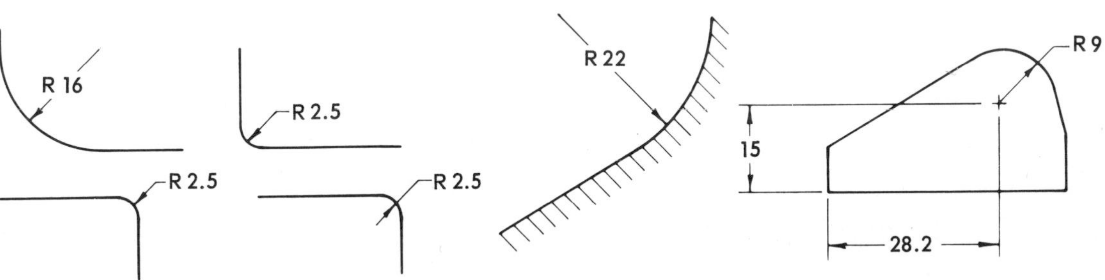

(A) RADII WHICH NEED NOT HAVE THEIR CENTRES LOCATED

(B) LOCATING RADIUS CENTRE

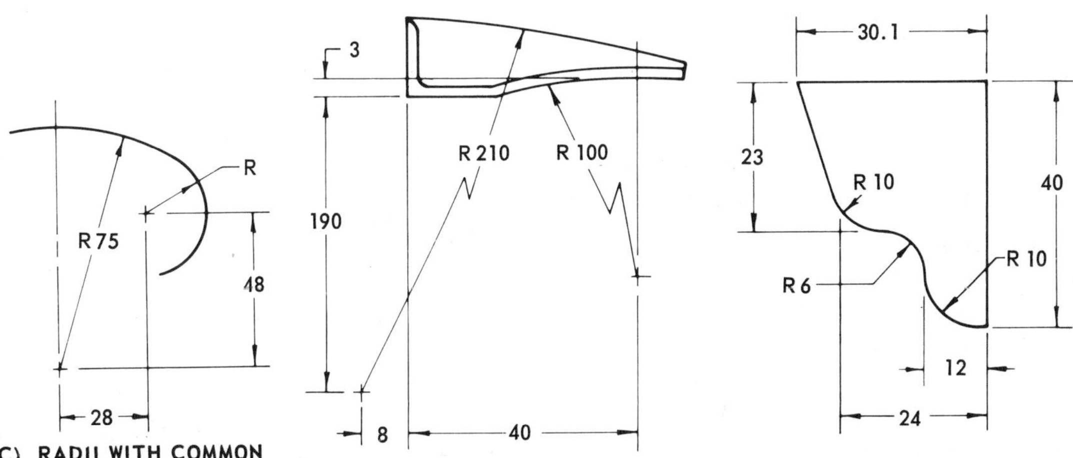

(C) RADII WITH COMMON TANGENT POINTS

(D) FORESHORTENED RADII

(E) RADII LOCATED BY TANGENTS

Fig. 4-2-3 Radii

Where a dimension is given to the centre of the radius, a small cross should be drawn at the centre (Fig. 4-2-3*b*). Extension lines and dimension lines are used to locate the centre. Where the location of the centre is unimportant, a radial arc may be located by tangent lines (Fig. 4-2-3*e*).

Where the centre of a radius is outside the drawing or interferes with another view, the radius dimension line may be foreshortened (Fig. 4-2-3*d*). The portion of the dimension line next to the arrowhead should be on the same line as the radius. Where the radius dimension line is foreshortened and the centre is located by coordinate dimensions, the dimensions locating the centre should be shown as foreshortened or as a dimension not to scale.

Simple fillet and corner radii may also be dimensioned by use of a general note, such as ALL ROUNDS AND FILLETS R 5 UNLESS OTHERWISE SPECIFIED or ALL RADII R 5.

Rounded Ends

Overall dimension should be used for parts or features having rounded ends. For fully rounded ends, the radius is indicated but not dimensioned (Fig. 4-2-4*a*).

For parts with partially rounded ends, the radius is dimensioned (Fig. 4-2-4*b*).

Where a hole and radius have the same centre and the hole location is more critical than the location of a radius, then either the radius or the overall length should be shown as a reference dimension (Fig. 4-2-4*c*).

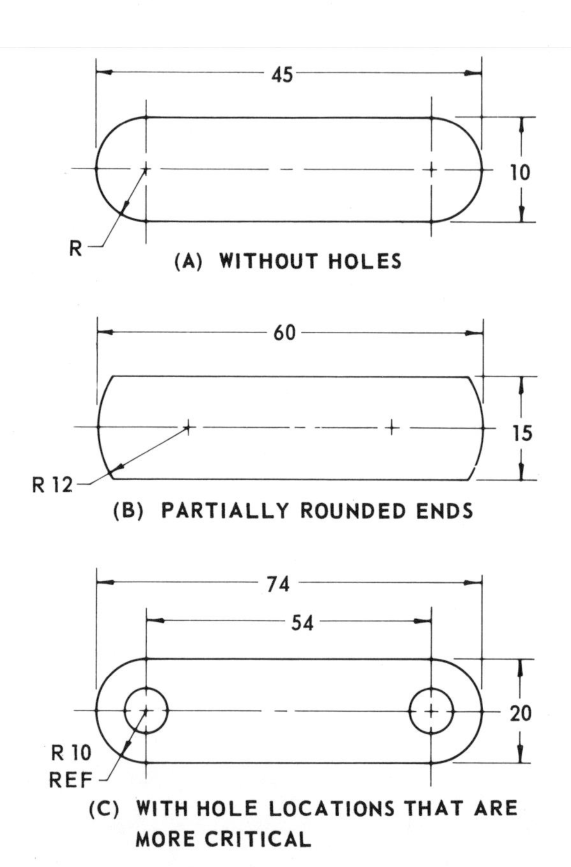

Fig. 4-2-4 External surfaces with rounded ends

Spherical Features

Spherical surfaces may be dimensioned as diameters or radii, but the dimension should be used with the abbreviations R SPHER, ϕ SPHER, or DIA SPHER.

The symbol ϕ or the abbreviation DIA may be omitted when a diameter is clearly indicated by the view and with the use of extension lines (Fig. 4-2-5).

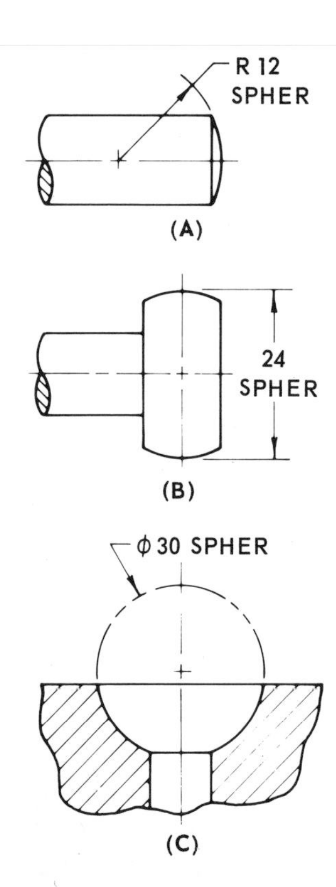

Fig. 4-2-5 Spherical surfaces

Cylindrical Holes

Plain, round holes are dimensioned in various ways, depending upon design and manufacturing requirements (Fig. 4-2-6). However, the leader is the method most commonly used.

When a leader is used to specify diameter sizes, as with small holes, the dimension is identified as a diameter by preceding the numerical value with the diameter symbol ϕ or by adding the abbreviation DIA after the numerical size.

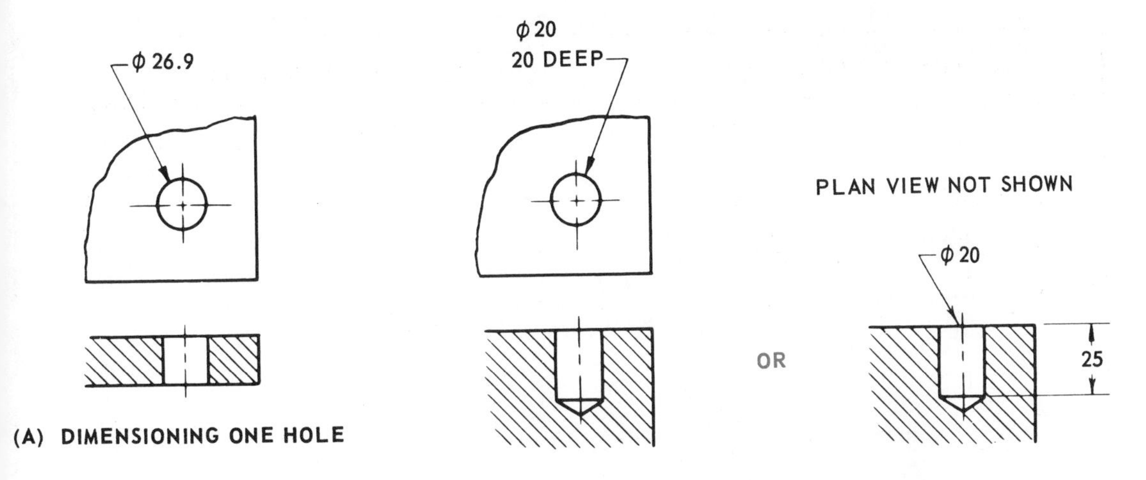

ϕ 16 THRU

ϕ 4.3
4 HOLES

ϕ 12
4 HOLES EQ SP
ϕ 60

(C) DIMENSIONING A THROUGH HOLE WHICH IS NOT SHOWN IN A LONGITUDINAL VIEW

(D) DIMENSIONING A GROUP OF HOLES

Fig. 4-2-6 Cylindrical holes

The size, quantity, and depth may be shown on a single line, or on several lines if preferable. For through holes, the abbreviation THRU should follow the dimension if the drawing does not make this clear. The depth dimension of a blind hole is the depth of the full diameter and is normally included as part of the dimensioning note. When the symbol for depth is used, it precedes the dimension which applies to the depth.

When more than one hole of a size is required, the number of holes should be specified. However, care must be taken to avoid placing two values together without adequate spacing. For example, it may be better to show the note on two or more lines than to use a line note which might be misread (Fig. 4-2-6*d*).

Slotted Holes

Elongated holes and slots allow for adjustment of parts. (Fig. 4-2-7). The method used to locate the slot would depend on how the slot was made.

The method shown in Figure 4-2-7*b* is used when the slot is punched out and the

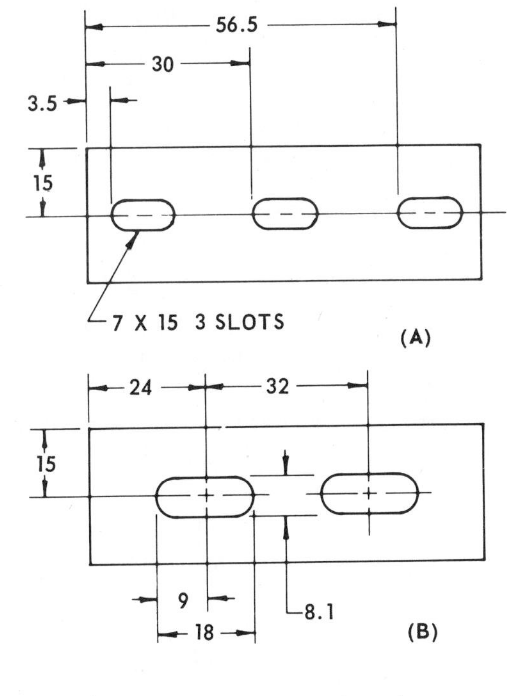

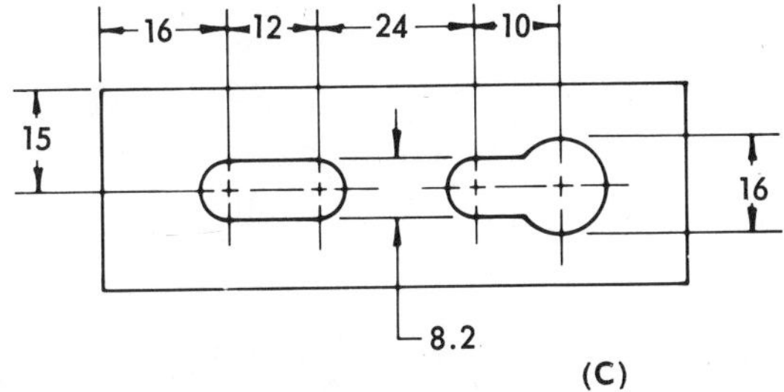

Fig. 4-2-7 Slotted holes

location of the punch is given. Figure 4-2-7*a* is the method of dimensioning used when the slot is machined out.

Countersink, Counterbore, and Spotface

The abbreviations CSK, CBORE, and SF represent countersink, counterbore, and spotface, respectively. These abbreviations, or their symbols, indicate the form of the surface only and not the methods used to produce that form. The dimensions for them are usually given as a note, preceded by the size of the through hole (Fig. 4-2-8).

A countersink is an angular-sided recess to accommodate the head of flathead screws, rivets, and similar items. The diameter at the surface and the included angle are given.

When the depth of the tapered section of the countersink is important, this depth is specified in the note or by dimension. For counterdrilled holes, the diameter, depth, and included angle of the counterdrill are given.

A counterbore is a flat-bottomed, cylindrical recess which permits the head of a fastening device, such as a bolt, to lie recessed into the part. The diameter, depth, and corner radius are specified in the note. In some cases, the thickness of the remaining stock may be dimensioned rather than the depth of the counterbore.

A spotface is an area where the surface is machined just enough to provide smooth, level seating for a bolthead, nut, or washer.

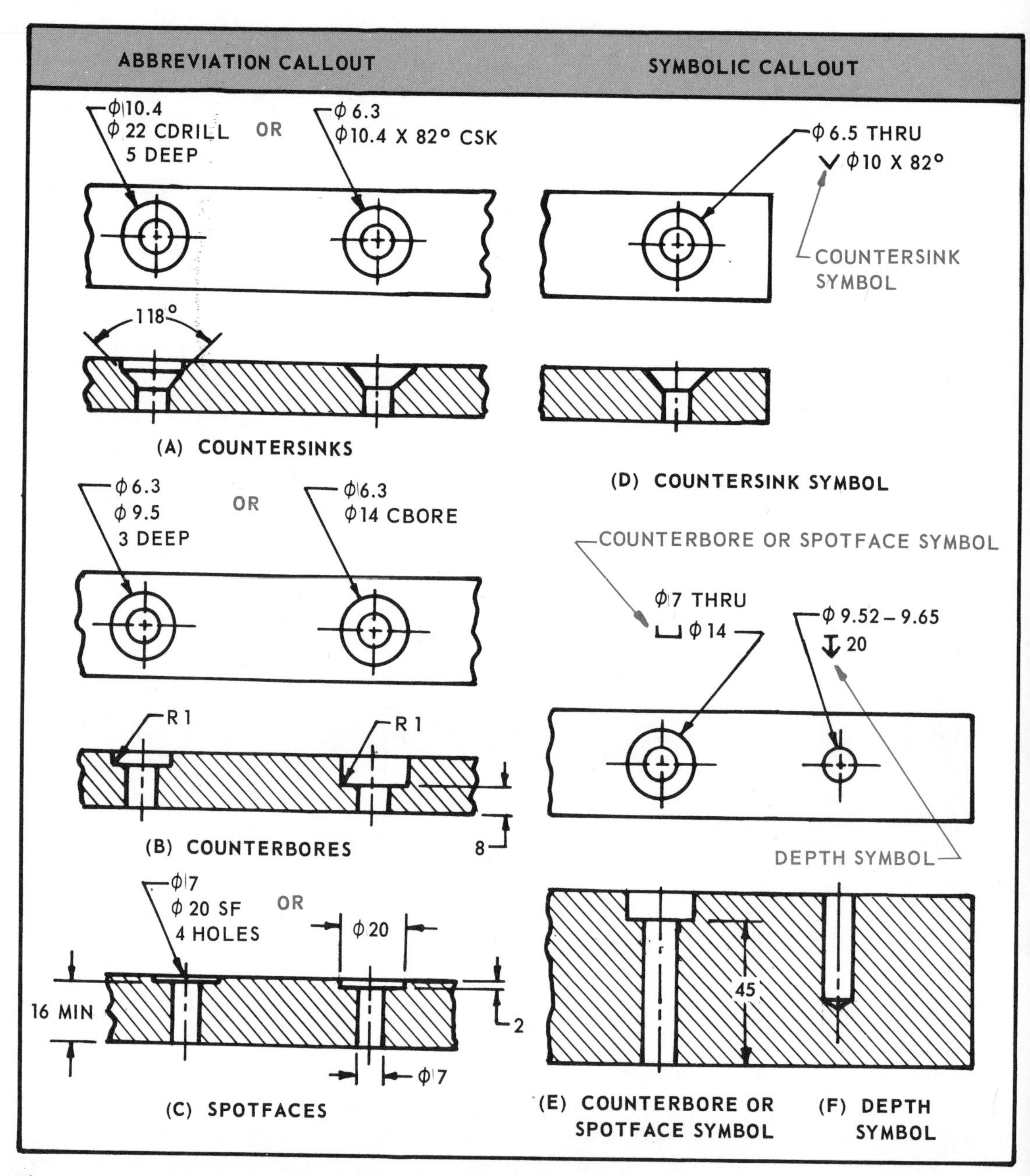

Fig. 4-2-8 Countersink, counterbores and spotfaces

The diameter of the faced area and either the depth or the remaining thickness are given. A spotface may be specified by a note only, and not measured on the drawing.

If no depth or remaining thickness is specified, it is implied that the spot-facing is the minimum depth necessary to clean up the surface to the specified diameter.

Assignments

All drawings to be complete with dimensions.

1. Dial indicator, Fig. 4-2-A, sheet size-A4, scale 1:1, one view.
2. Element plate, Fig. 4-2-B, sheet size-A3, scale 1:1, one view.
3. Ink bottle stand, Fig. 4-2-C, sheet size-A4, scale 1:1, three views.
4. Yoke, Fig. 4-2-D, sheet size-A4, scale 1:1, three views.
5. Offset plate, Fig. 4-2-E, sheet size-A4, scale 1:1, two views
6. Guide bracket, Fig. 4-2-F, sheet size-A4, scale 1:1, three views.
7. Ratchet wheel, fig. 4-2-G, sheet size-A4, scale 1:1, two views.

Review for Assignments

Unit 3-5 Circular features
Unit 4-1 Dimensioning

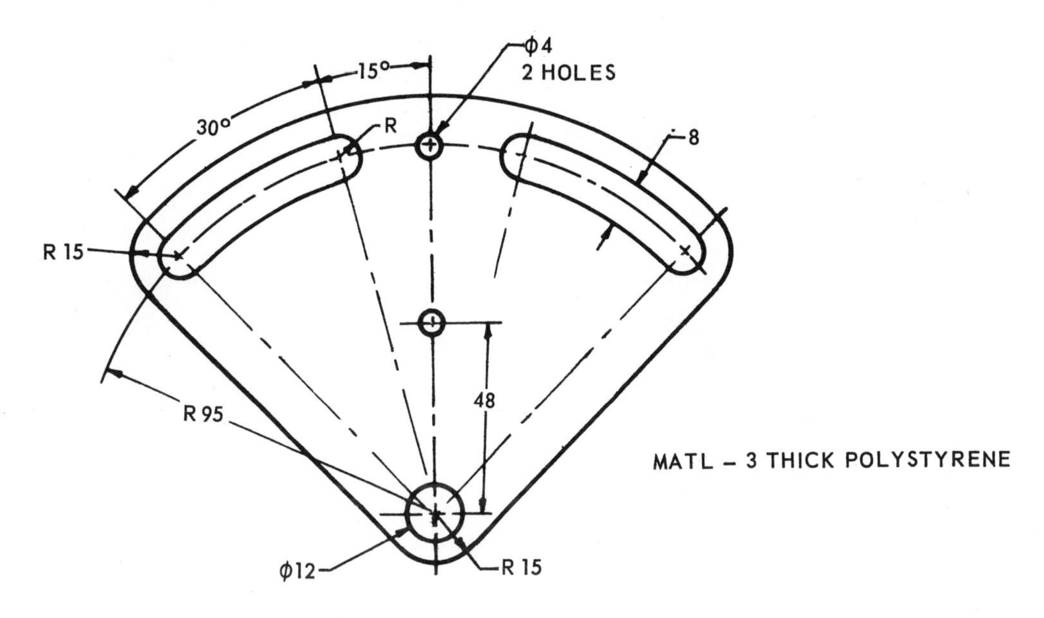

Fig. 4-2-A Dial indicator

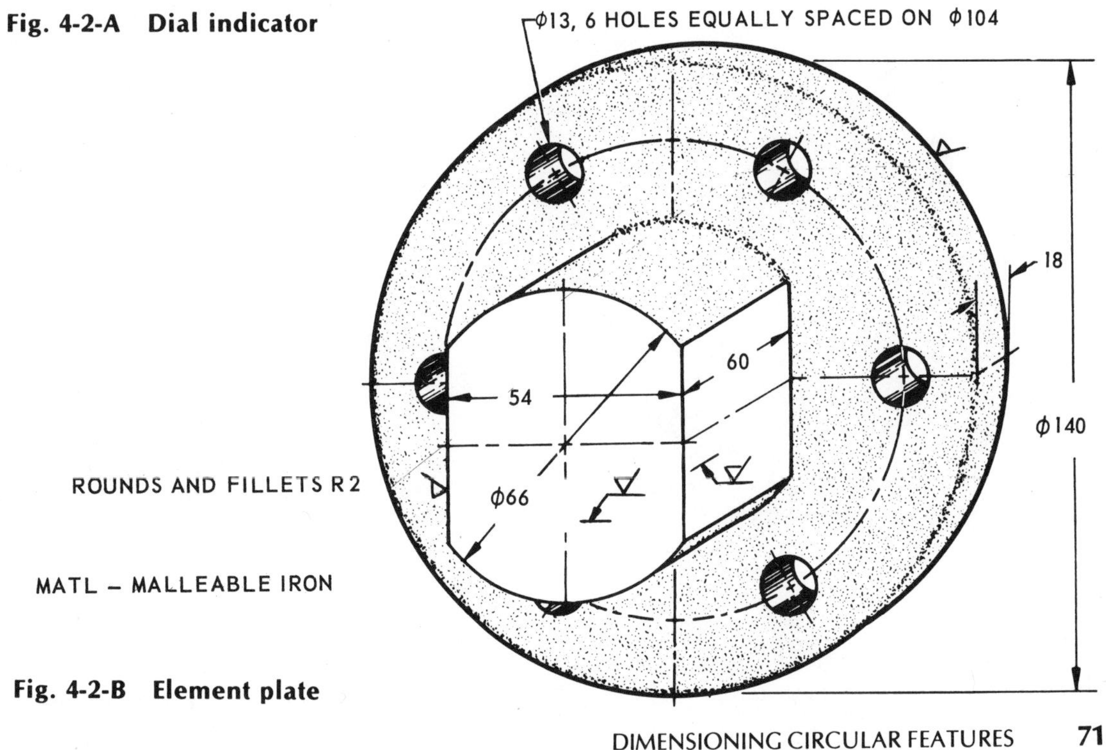

Fig. 4-2-B Element plate

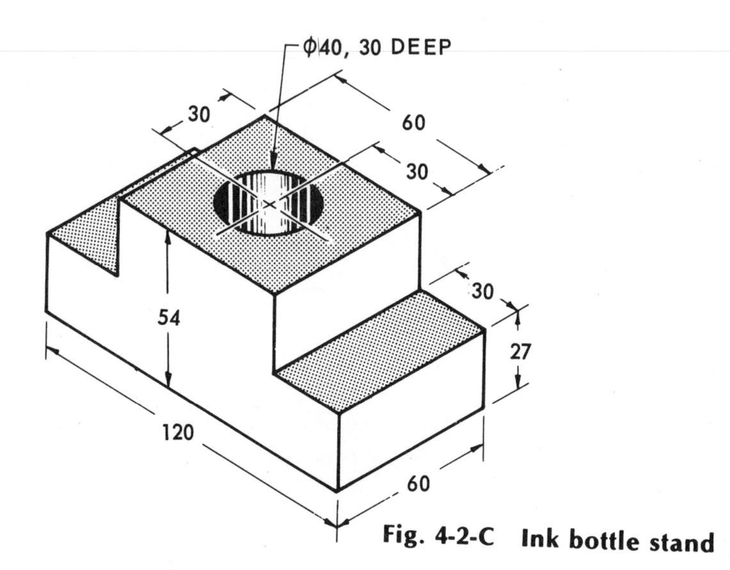

Fig. 4-2-C Ink bottle stand

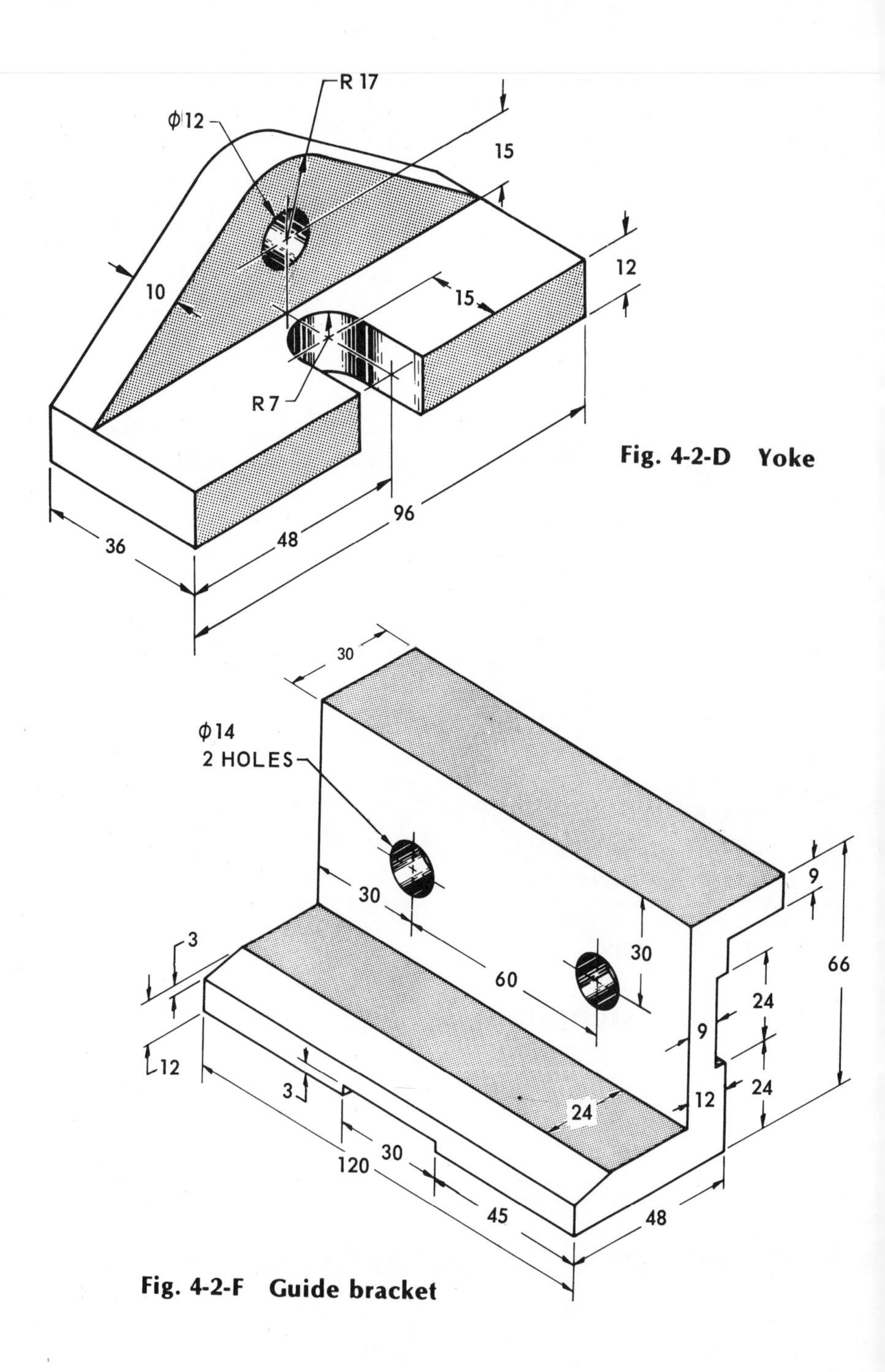

Fig. 4-2-D Yoke

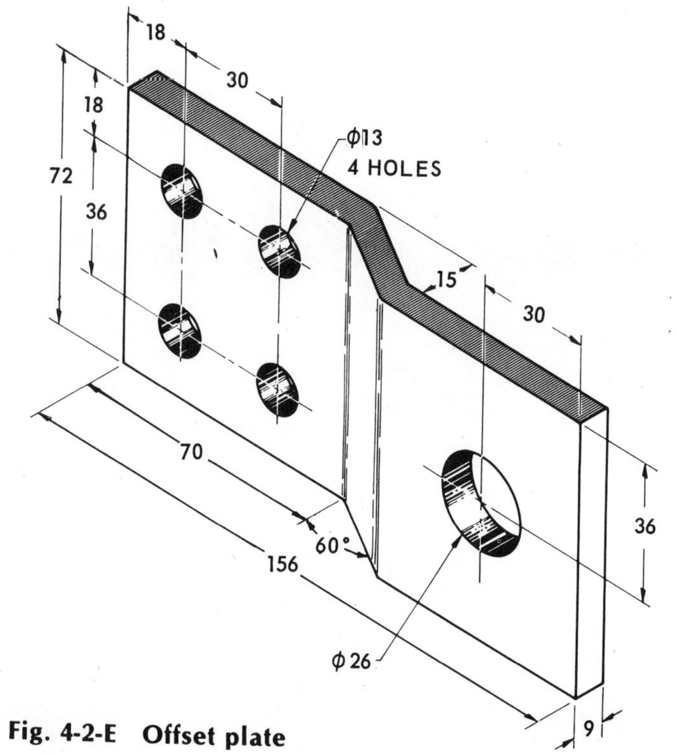

Fig. 4-2-E Offset plate

Fig. 4-2-F Guide bracket

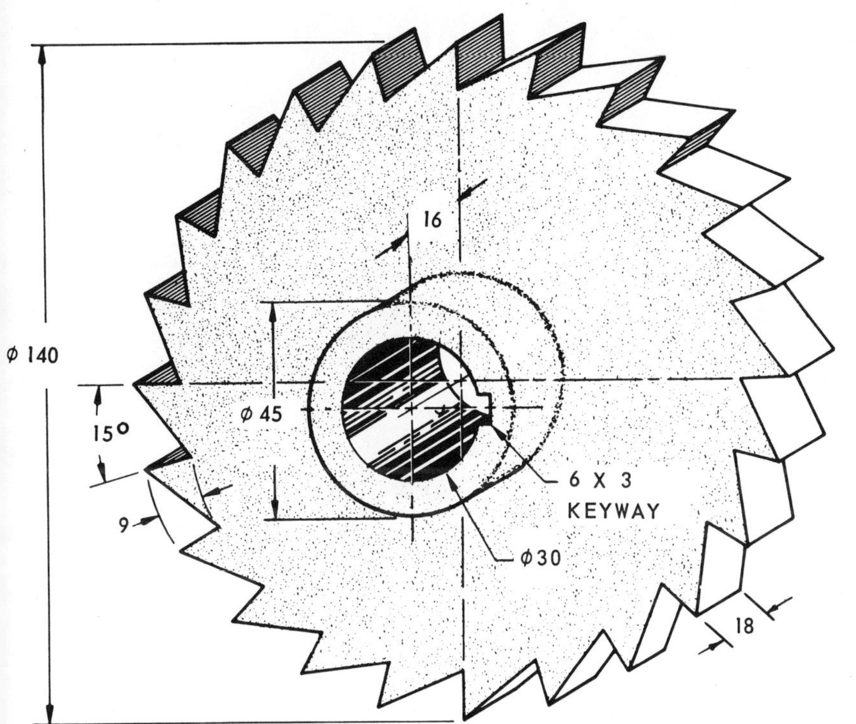

Fig. 4-2-G Ratchet wheel

UNIT 4-3 Dimensioning Common Features

Tapers

Circular Tapers. Taper shanks are used on many small tools, such as drills, reamers, counterbores, and spotfaces, to hold them accurately in the machine spindle (Fig. 4-3-2). **Taper** means the difference in diameter or width in a given length. There are many standard tapers; the Morse taper and the Brown and Sharpe taper are the most common.

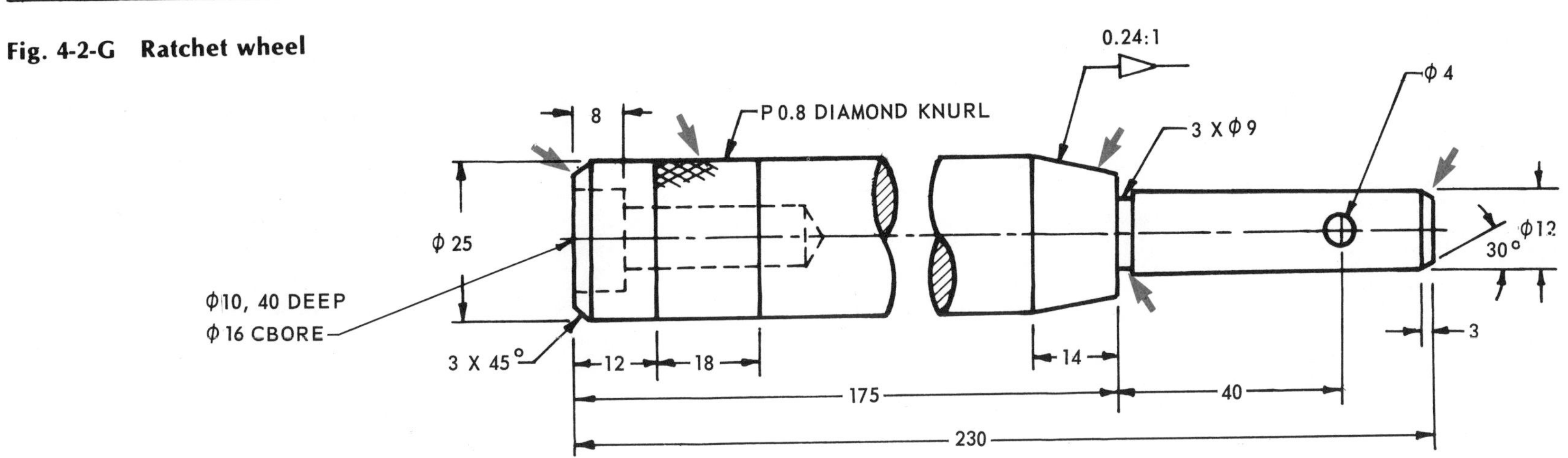

Fig. 4-3-1 Typical detail drawing

The following dimensions may be used, in suitable combinations, to define the size and form of tapered features:

- The diameter (or width) at one end of the tapered feature
- The length of the tapered feature
- The rate of taper
- The included angle
- Taper ratio

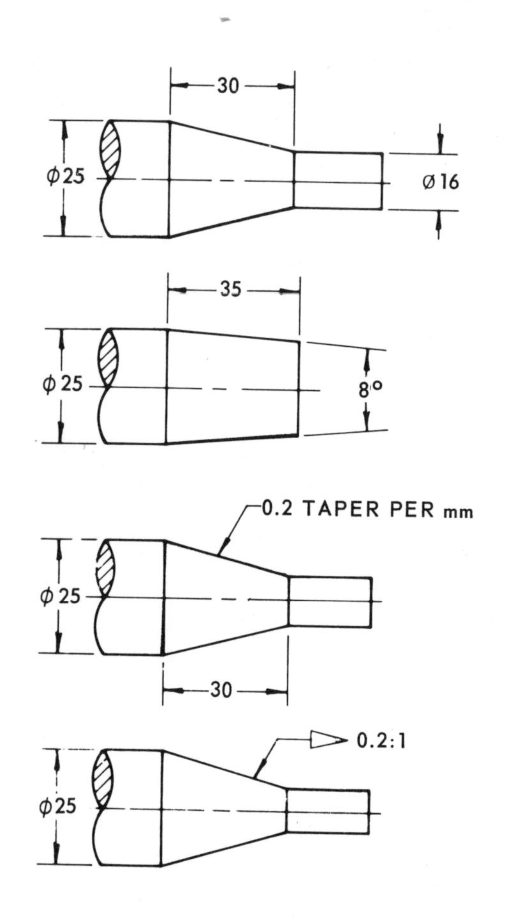

Fig. 4-3-2 Dimensioning circular tapers

In dimensioning a taper by means of the taper ratio, the taper symbol ⊳ should precede the ratio figures.

Flat Tapers. Flat tapers are used as locking devices such as taper keys and adjusting shims. The methods recommended for dimensioning flat tapers are shown in Figure 4-3-3.

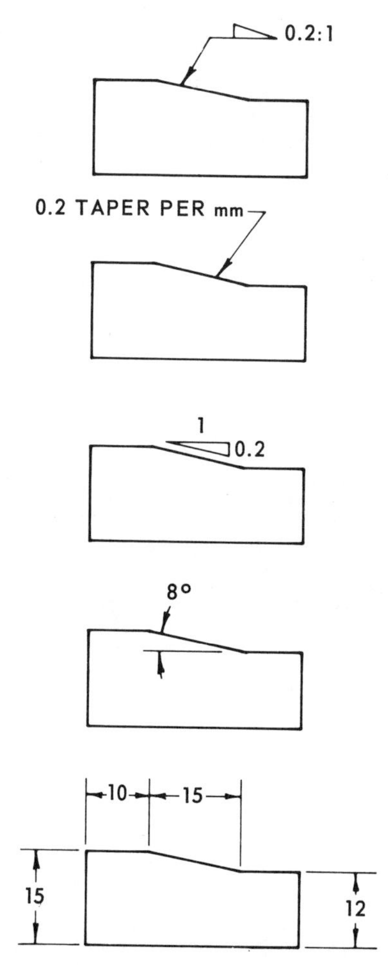

Fig. 4-3-3 Dimensioning flat tapers

Chamfers

The process of chamfering, that is, cutting away the inside or outside piece, is done to facilitate assembly. Chamfers are normally dimensioned by giving their angle and length (Fig. 4-3-4). When the chamfer is 45°, it may be specified as a note.

When a very small chamfer is permissible, primarily to break a sharp corner, it may be dimensioned but not drawn, as in Figure 4-3-4*c*. If not otherwise specified, an angle of 45° is understood.

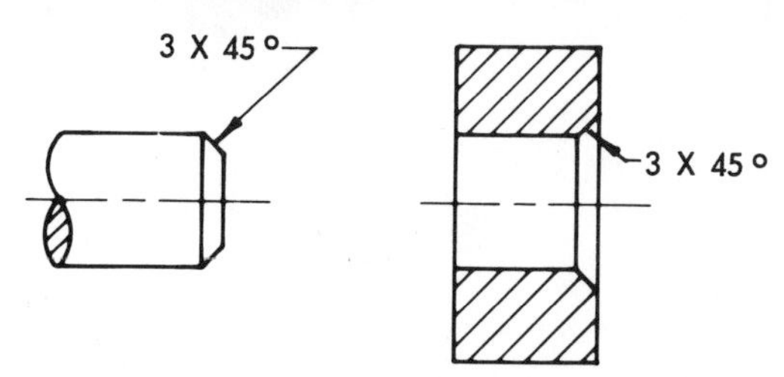

(A) FOR 45° CHAMFERS ONLY

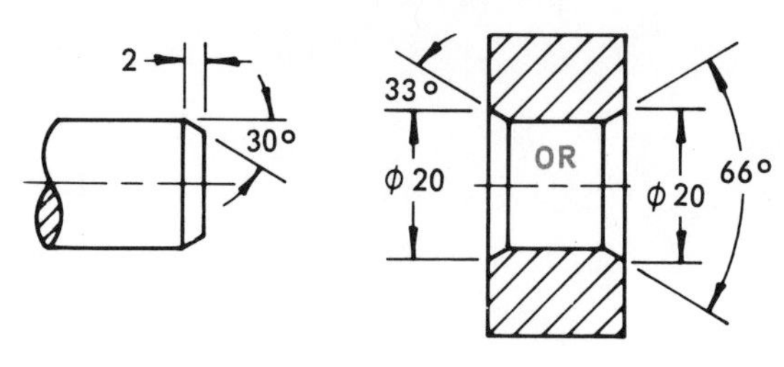

(B) FOR ALL CHAMFERS

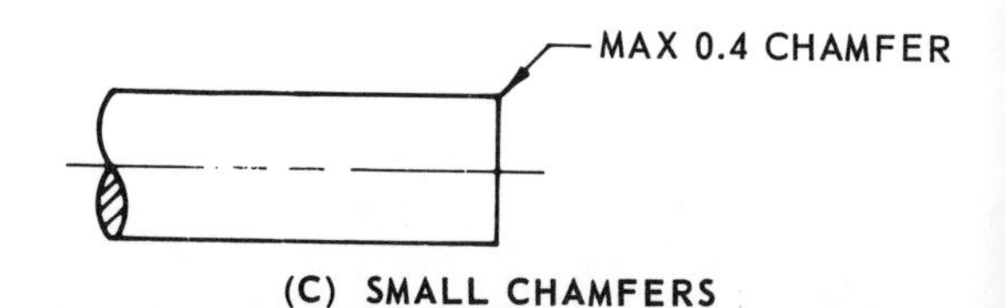

(C) SMALL CHAMFERS

Fig. 4-3-4 Dimensioning chamfers

Internal chamfers may be dimensioned in the same manner, but it is often desirable to give the diameter over the chamfer. The angle may also be given as the included angle if this is a design requirement. This type of dimensioning is generally necessary for larger diameters, especially those over 50 mm, whereas chamfers on small holes are usually expressed as countersinks.

Chamfers are never measured along the angular surface.

Knurls

Knurling is specified in terms of type, pitch, and diameter before and after knurling (Fig. 4-3-5). The letter P precedes the pitch number. Where control is not required, the diameter after knurling is omitted. Where only portions of a feature require knurling, axial dimension must be provided. Where required to provide a press fit between parts, knurling is specified by a note on the drawing which includes the type of knurl required, the pitch, the toleranced diameter of the feature prior to knurling, and the minimum acceptable diameter after knurling. Commonly used types are straight, diagonal, spiral, convex, raised diamond, depressed diamond, and radial. The pitch is usually expressed in terms of teeth per millimetre and may be the straight pitch, circular pitch, or diametral pitch. For cylindrical surfaces, the latter is preferred.

The knurling symbol is optional and is used only to improve clarity on working drawings.

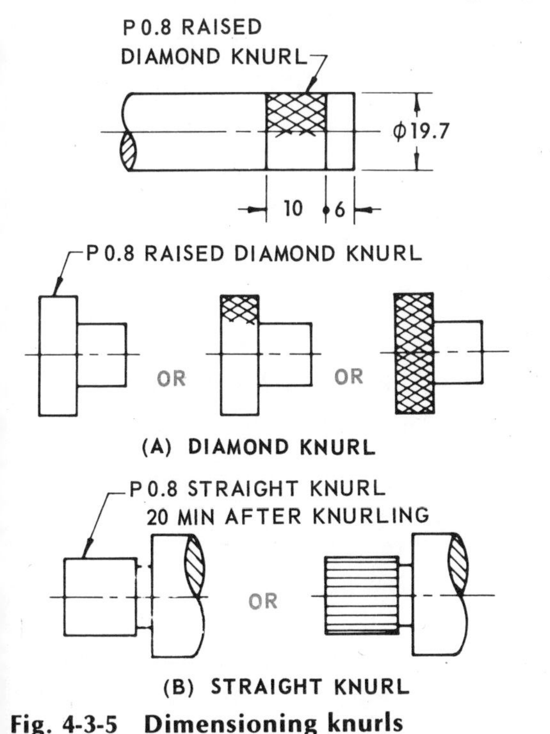

Fig. 4-3-5 Dimensioning knurls

Undercuts

The operation of undercutting or necking, that is, cutting a recess in a diameter, is done to permit two parts to come together, as illustrated in Figure 4-3-6*a*. It is indicated on the drawing by a note listing the width first and then the diameter. If the radius is shown at the bottom of the undercut, it will be assumed that the radius is equal to half the width unless otherwise specified, and the diameter will apply to the centre of the undercut. When the size of the undercut is unimportant, the dimension may be left off the drawing.

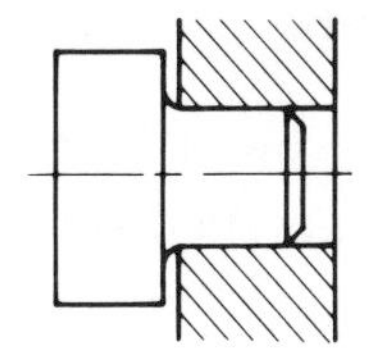

PART CANNOT FIT FLUSH IN HOLE BECAUSE OF SHOULDER

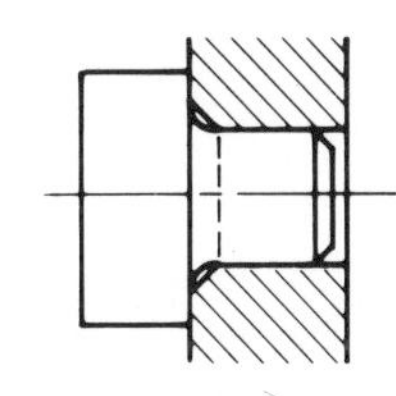

CHAMFER ADDED TO HOLE TO ACCEPT SHOULDER OF PART

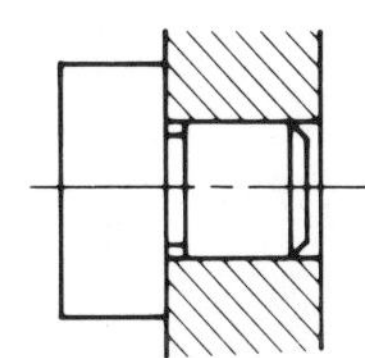

SAME PART WITH UNDERCUT ADDED PERMITS PART TO FIT FLUSH

(A) CHAMFER AND UNDERCUT APPLICATION

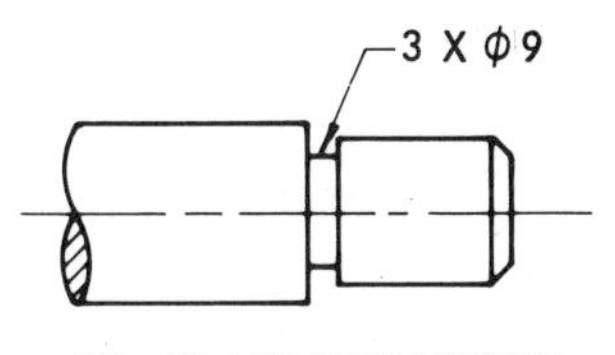

(B) PLAIN UNDERCUT

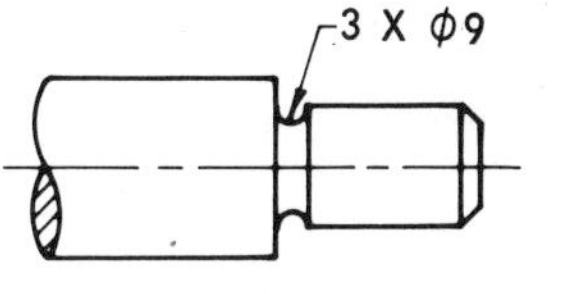

(C) RADIUSED UNDERCUT

Fig. 4-3-6 Dimensioning undercuts

Formed Parts

In dimensioning formed parts, the inside radius is usually specified, rather than the outside radius, but all forming dimensions should be shown on the same side if possible. Dimensions apply to the side on which the dimensions are shown unless otherwise specified (Fig. 4-3-7).

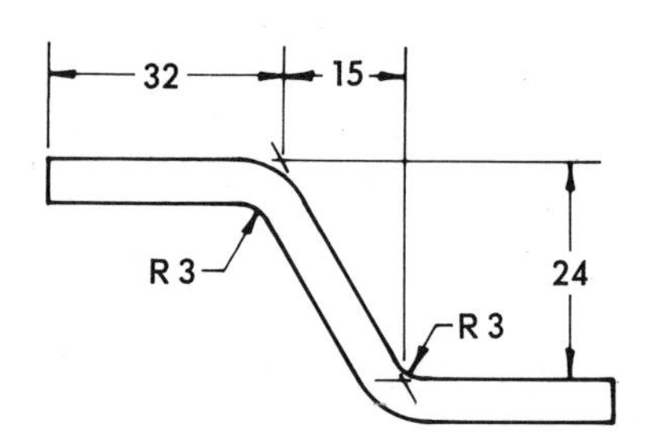

Fig. 4-3-7 Dimensioning theoretical points of intersection

Fillets and Rounds

A *round,* or radius, is put on the outside of a piece to improve its appearance and to avoid forming a sharp edge that might injure anyone handling it.

A *fillet* is additional metal allowed in the inner intersection of two surfaces for casting practices and to improve the strength of that part.

The dimensioning of fillets and rounds normally takes the form of a note. Where fillets and rounds vary in size, the individual dimensions must be shown (Fig. 4-3-8).

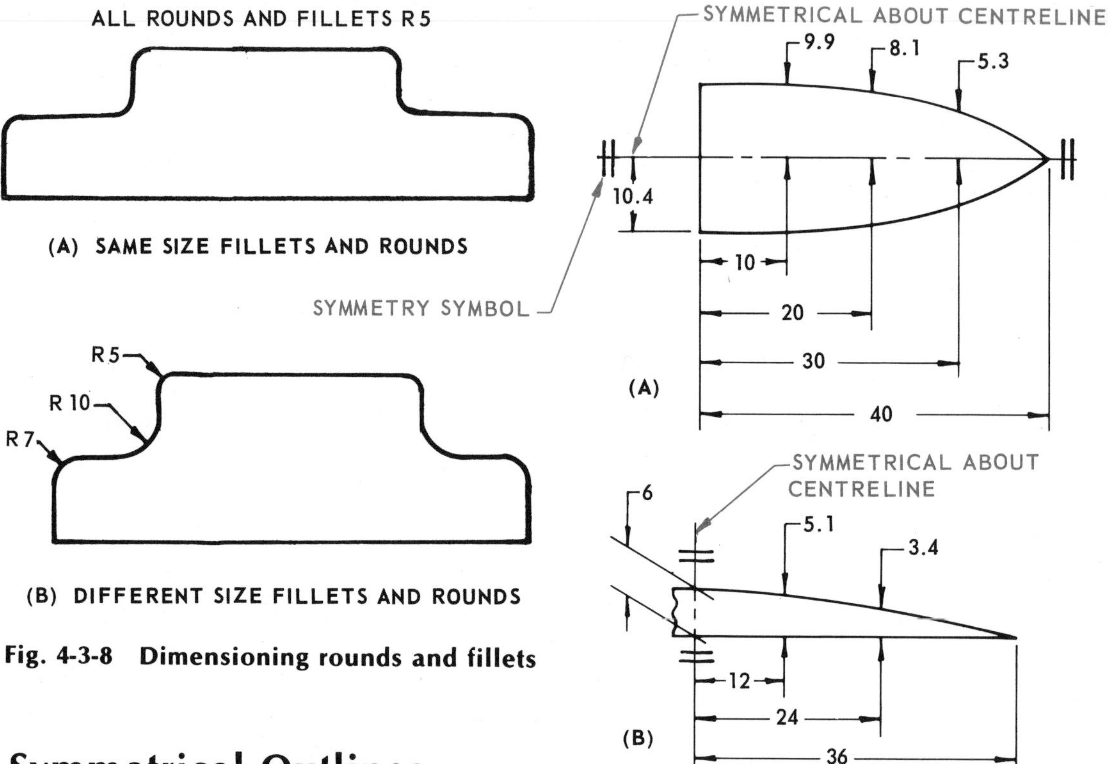

Fig. 4-3-8 Dimensioning rounds and fillets

Fig. 4-3-9 Dimensioning symmetrical features

Symmetrical Outlines

Symmetrical outlines may be dimensioned on one side of the axis of symmetry only (Fig. 4-3-9). Where only part of the outline is shown — because of functional drafting procedures, the size of the part, or space limitations — symmetrical shapes may be shown by only half of the outline and symmetry is indicated by the symmetry symbol. In such cases, the outline of the part should extend slightly beyond the centre line and terminate with a break line. Note the dimensioning method of extending the dimension lines to act as extension lines for the perpendicular dimension.

Repetitive Features and Dimensions

Where a series of holes or other features are spaced equally, the methods shown in Figure 4-3-10 may be used. Alternately, repetitive features and dimensions may be specified by the use of an "x" in conjunction with the numeral to indicate the "number of times" or "places" they are required.

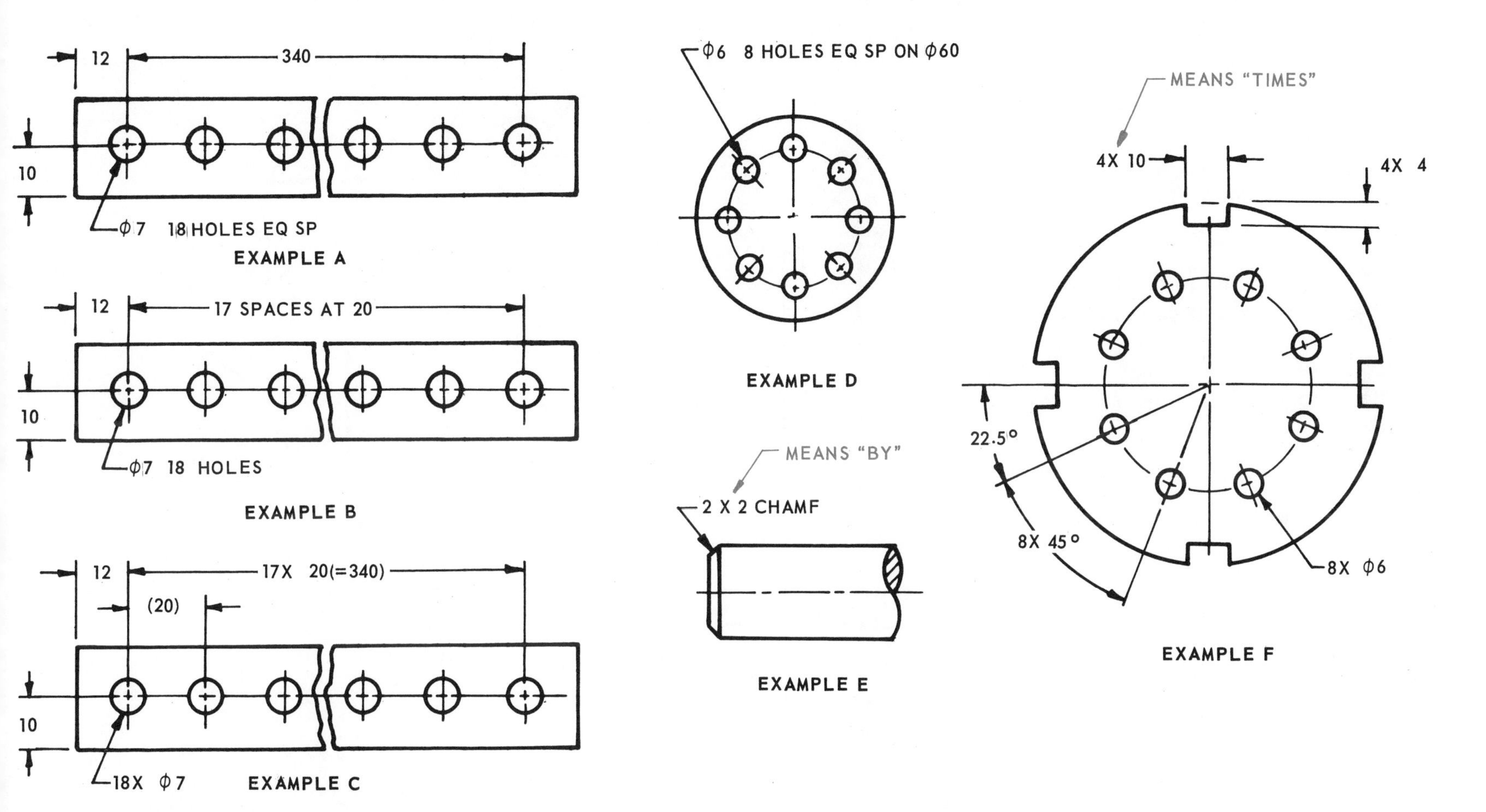

Fig. 4-3-10 Dimensioning repetitive detail

The number of spaces or features is given first, followed by an "x", then a space, then the dimension.

When it is difficult to distinguish between the dimension and the number of spaces, as shown in Figure 4-3-10, one space is dimensioned and identified as reference.

The "x" is sometimes used to indicate "BY" for dimensions shown as a note. Where both uses are found on a drawing, make certain the meaning is clear.

Wire, Sheet Metal, and Drill Rod

Wire, sheet metal, and drill rod, which are manufactured to gauge or code sizes, should be shown by their decimal dimensions; but gauge numbers, drill letters, etc., may be shown in parentheses following those dimensions.

Examples:

Sheet — 3.57 (No. 10 USS GA)

— 2.05 (No. 12 B & S GA)

Assignments

To be complete working drawings, sheet size-A4, Scale 1:1.

1. Screwdriver, Fig. 4-3-A
2. Indicator rod, Fig. 4-3-B.
3. Gasket, Fig. 4-3-C, can be drawn two ways.
   (1) Use technique for symmetrical outlines and draw and dimension just one quarter of the gasket.
   (2) Draw all of the gasket and use the dimensioning technique for repetitive features.

Review for Assignment

Unit 4-2 Dimensioning circular features

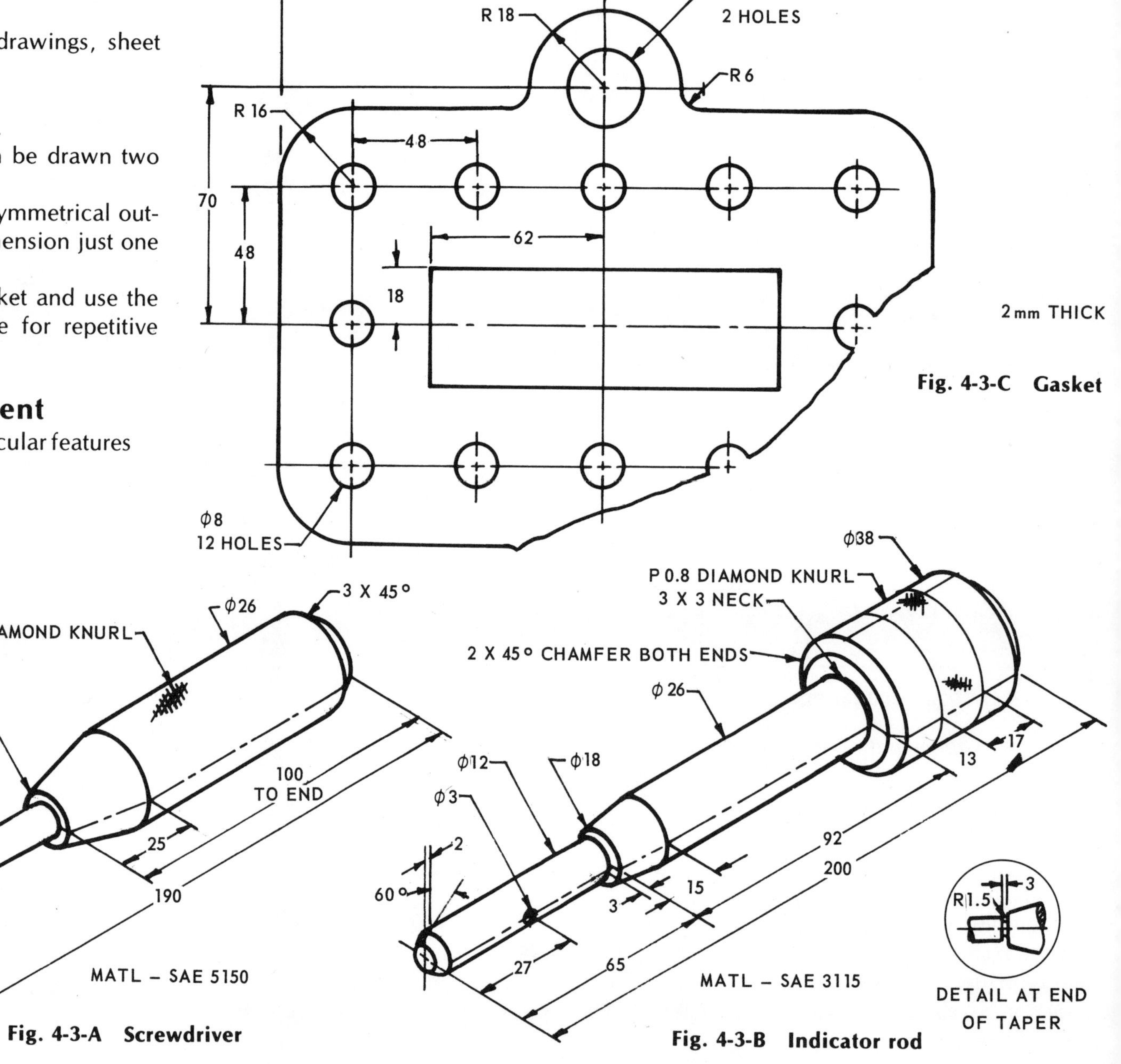

Fig. 4-3-C Gasket

Fig. 4-3-A Screwdriver

Fig. 4-3-B Indicator rod

UNIT 4-4

Dimensioning Methods

The choice of the most suitable dimensions and dimensioning methods will depend, to some extent, on how the part will be produced and whether the drawings are intended for unit or mass production.

Unit production refers to cases where each part is to be made separately, using general-purpose tools and machines.

Mass production refers to parts produced in quantity, where special tools and gauges are usually provided.

Either linear or angular dimensions may locate features with respect to one another (point-to-point) or from a datum. Point-to-point dimensions may be adequate for describing simple parts. Dimensions from a datum may be necessary if a part with more than one critical dimension must mate with another part.

The following systems of dimensioning are used more commonly for engineering drawings.

Rectangular Coordinate Dimensioning

This is a method for indicating distance, location, and size by means of linear dimensions. These are measured parallel or perpendicular to reference axes or datum planes that are perpendicular to one another.

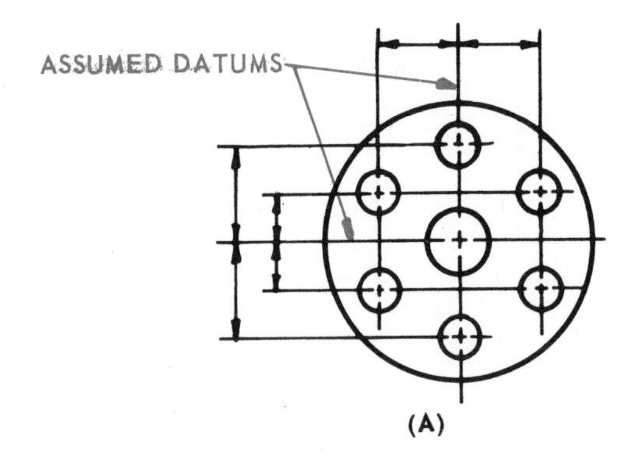

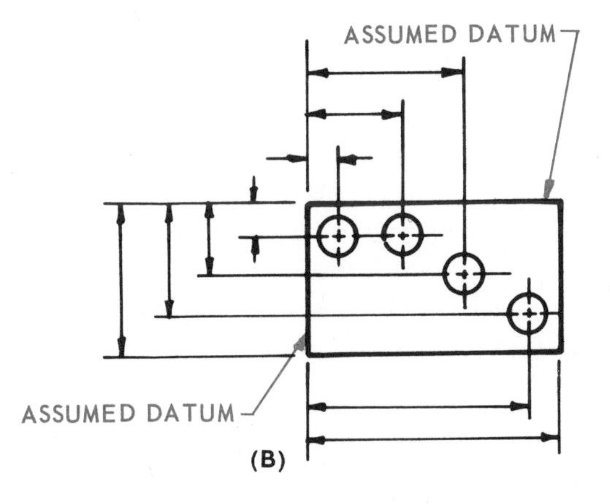

Fig. 4-4-1 Rectangular coordinate dimensioning

Coordinate dimensioning with dimension lines must clearly identify the datum features from which the dimensions originate (Fig. 4-4-1).

True Position Dimensioning

True position dimensioning (Fig. 4-4-2) has many advantages over the coordinate dimensioning system. See Unit 12-1.

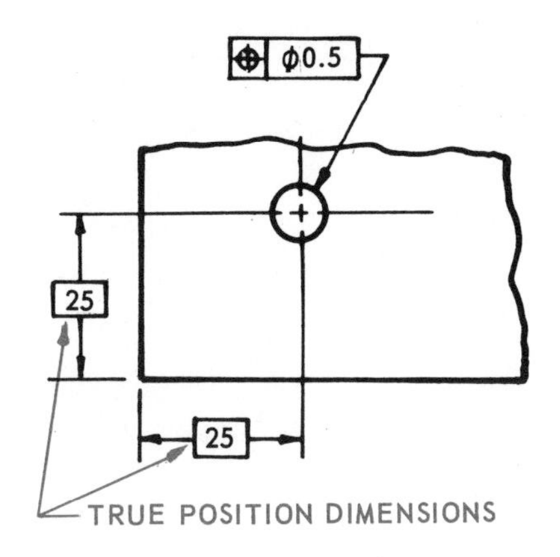

Fig. 4-4-2 True position dimensioning

Polar Coordinate Dimensioning

Polar coordinate dimensioning is commonly used in circular planes or circular configurations of features. It is a method of indicating the position of a point, line, or surface by means of a linear dimension and an angle, other than 90°, that is implied by the vertical and horizontal centre lines (Fig. 4-4-3).

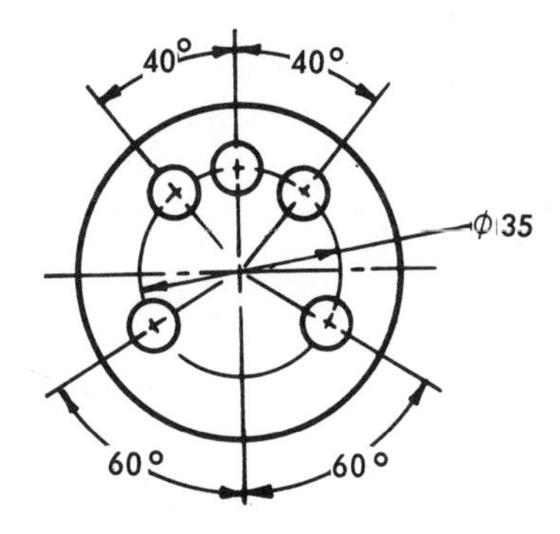

Fig. 4-4-3 Polar coordinate dimensioning

Chordal Dimensioning

The chordal dimensioning system may also be used for the spacing of points on the circumference of a circle relative to a datum, where manufacturing methods indicate that this will be more convenient (Fig. 4-4-4).

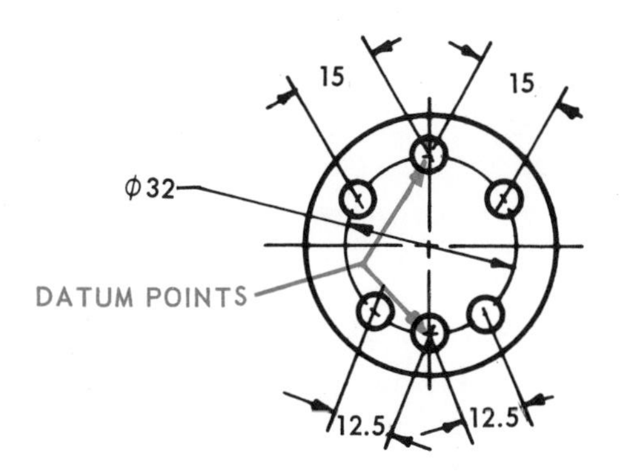

Fig. 4-4-4 Chordal dimensioning

Arrowless Dimensioning

Arrowless dimensioning is coordinate dimensioning without dimension lines (Fig. 4-4-5). The dimensions originate from datum planes which are indicated by zero coordinates. Dimensions from them are shown on extension lines. There should never be more than one zero line in each direction.

Tabular Dimensioning

Tabular dimensioning is a type of rectangular coordinate dimensioning in which dimensions from mutually perpendicular datum planes are listed in a table on the drawing rather than on the pictorial delineation (Fig. 4-4-6). Tables may be prepared in any suitable manner which will adequately locate the features.

HOLE SYMBOL	HOLE DIA
A	6
B	5
C	4
D	3

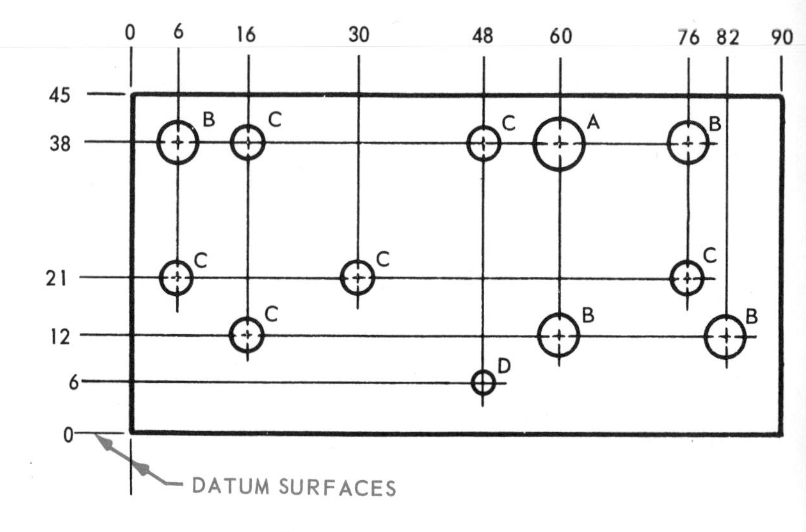

Fig. 4-4-5
Arrowless dimensioning

Chain Dimensioning

When a series of dimensions is applied on a point-to-point basis, it is called **chain dimensioning** (Fig. 4-4-7). A possible disadvantage of this system is that it may result in an undesirable accumulation of tolerances between individual features (see Unit 4-5).

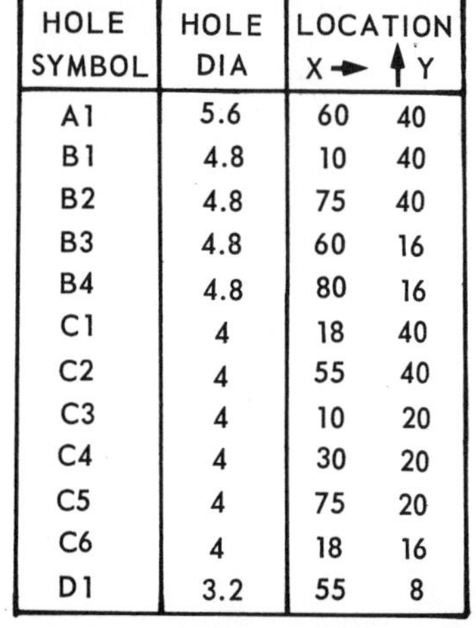

HOLE SYMBOL	HOLE DIA	LOCATION X →	↑ Y
A1	5.6	60	40
B1	4.8	10	40
B2	4.8	75	40
B3	4.8	60	16
B4	4.8	80	16
C1	4	18	40
C2	4	55	40
C3	4	10	20
C4	4	30	20
C5	4	75	20
C6	4	18	16
D1	3.2	55	8

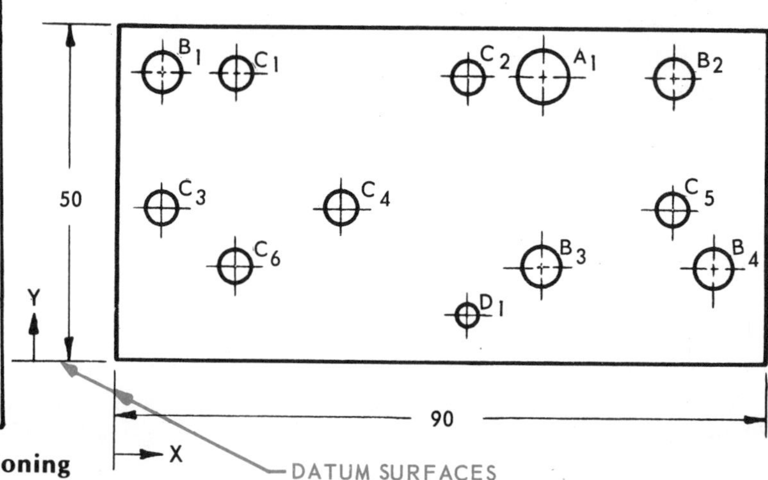

Fig. 4-4-6 Tabular dimensioning

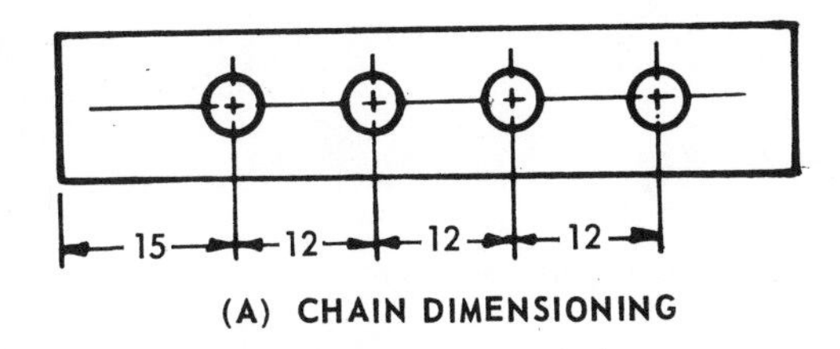

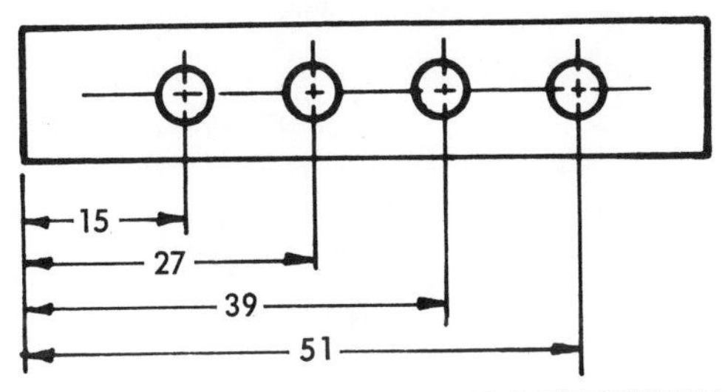

Fig. 4-4-7 A comparison between chain and datum dimensioning

Datum or Common-Point Dimensioning

When several dimensions emanate from a common reference point or line, the method is called **common-point** or **datum dimensioning.**

Assignments

1. Select one of the problems shown in Figs. 4-4-A or 4-4-B, and on an A3-size sheet make a working drawing of the part. The arrowless dimensioning shown is to be replaced with rectangular coordinate dimensioning and has the following dimensioning changes for Fig. 4-4-A.

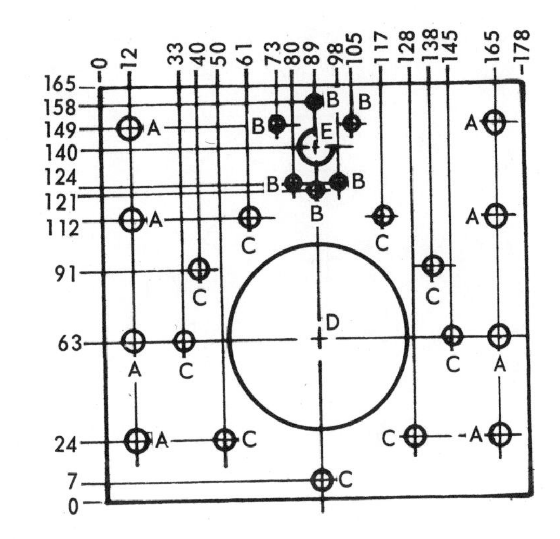

MATL – SAE 1006
3 mm THICK

HOLE	SIZE
A	8
B	4
C	5
D	76
E	12

Fig. 4-4-A Cover plate

- Holes A, E, and D are located from the zero coordinates.
- Holes B are located from centre of hole E.
- Hole C is located from centre of hole D.

For Fig. 4-4-B,

- Holes E and D are located from left and bottom edges.
- Holes A and C are located from centre of hole D.

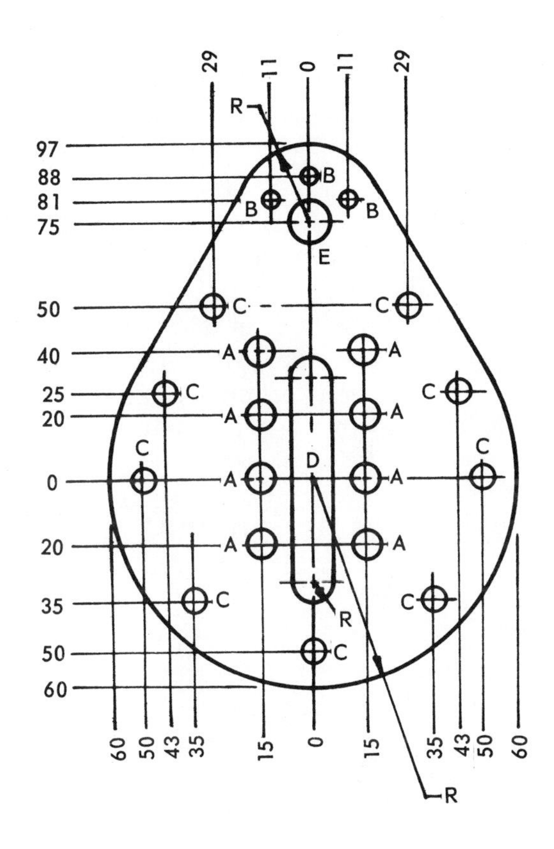

MATL – SAE 1008
3 mm THICK

HOLE	SIZE
A	8
B	4
C	6
D	10 X 70
E	12

Fig. 4-4-B Transmission cover

- Holes B are located from centre of hole E.

For the sake of clarity, some dimensions may best be shown on the part. Scale for drawings is 1:1.

2. On an A3-size sheet make a working drawing of the gasket (Fig. 4-4-C.) Dimensioning is to be either arrowless or tabular.

3. Divide an A3 - size sheet into 4 quadrants by bisecting the vertical and horizontal sides. In each quadrant draw the adaptor plate shown in Fig. 4-4-D. Different methods of dimensioning are to be used for each drawing. The methods are rectangular coordinate, chordal, arrowless and tabular. Scale 1:1.

Review for Assignment

Unit 2-6 Drafting skills

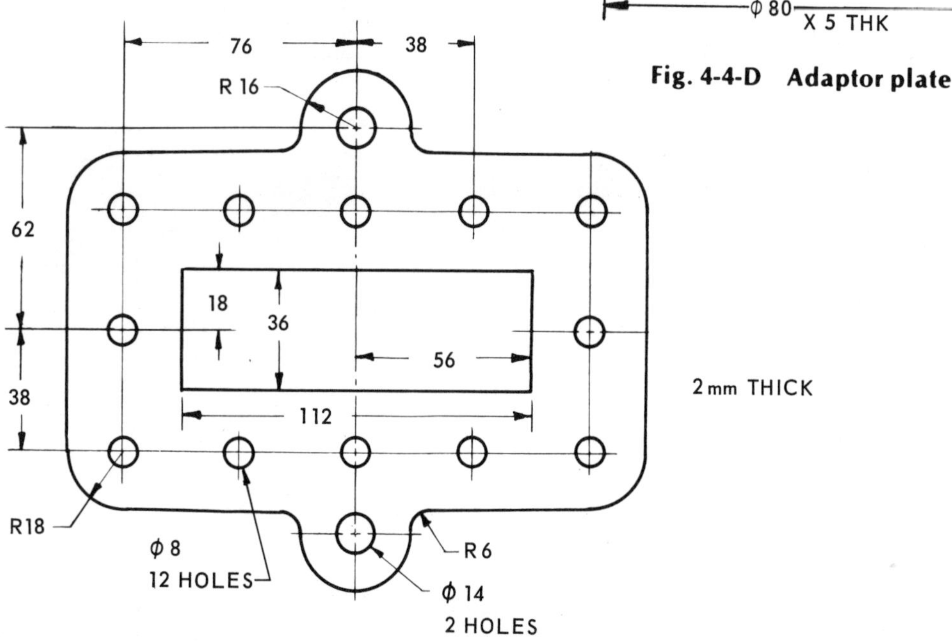

Fig. 4-4-C Gasket

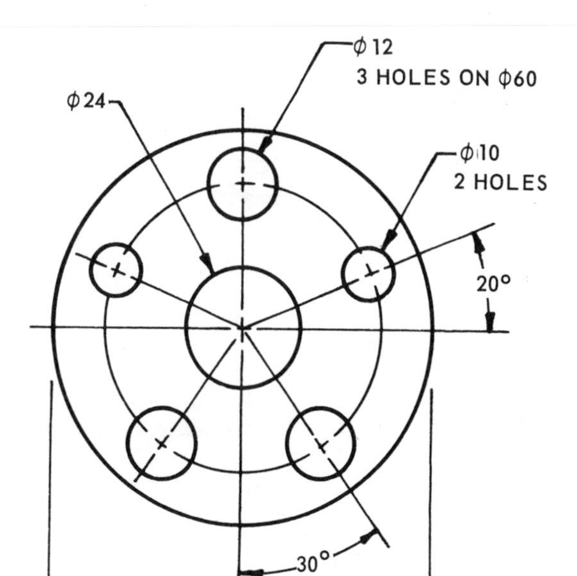

Fig. 4-4-D Adaptor plate

UNIT 4-5

Surface Texture

There are many reasons to control the roughness of a surface. Surfaces coming in contact with one another to prevent leakage of gases or liquids must be sufficiently smooth to act as a seal. For a part to withstand severe operating conditions, such as friction and wear, a particular surface finish is essential. Therefore, the degree of surface finish control must be specified on the drawing.

In selecting the required surface texture for any particular part, the drafter must consider factors such as size and function of the part, loading, speeds, operating conditions, and materials.

To meet the requirements for proper surface texture designation and control, a system for accurately describing the desired surface is necessary. Only the height and width of surface irregularities will be covered in this text.

Surface Texture Symbol

The basic symbol consists of two legs of unequal length inclined at approximately

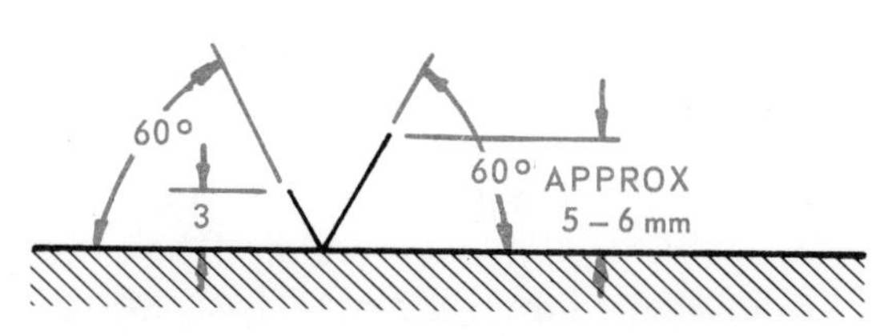

Fig. 4-5-1 Basic surface texture symbol

60° to the line representing the surface being controlled, as shown in Figure 4-5-1. The basic symbol alone has no meaning. The symbol should point from outside the material of the piece.

Removal of Material by Machining

If the removal of material by machining is required, a bar is added to the basic symbol, as shown in Figure 4-5-2. This symbol, when placed on a line indicating a surface, means that this surface is to be machined. It does not indicate the desired smoothness required or the process to be used.

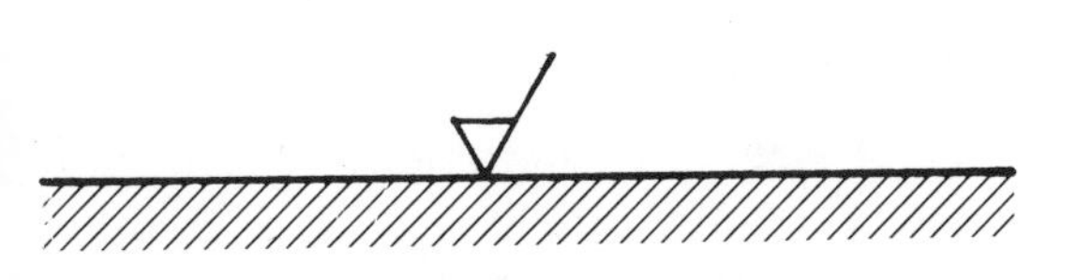

Fig. 4-5-2 Symbol used to indicate removal of material by machining

Parts to be cast, molded, or forged are examples of where some surfaces may require a greater degree of smoothness than can be obtained by the production process used to shape the part.

Where all the surfaces are to be machined, a general note such as "∇ ALL OVER" or "FAO" (Finish All Over) may be used and the symbols on the drawing omitted. See Figure 4-5-3.

Surface symbols, like dimensions, are not normally duplicated. They should be used on the same view as the dimensions that give the size or location of the surfaces concerned. The symbol is placed on the line representing the surface or, where desirable, on the extension line locating the surface. Figures 4-5-4 and 4-5-5 show examples of the use of symbols to indicate surfaces requiring machining.

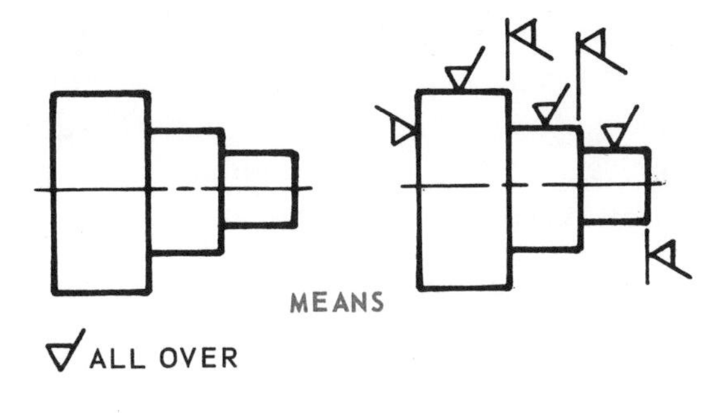

Fig. 4-5-3 Specifying surface finish by a note

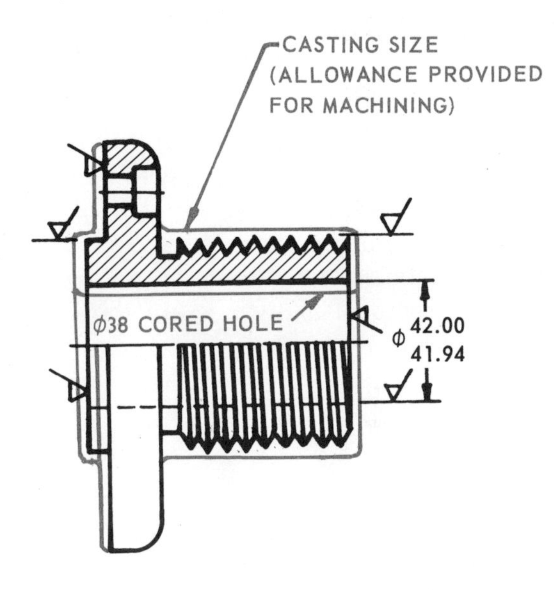

Fig. 4-5-5 Extra metal allowance required for machined surfaces

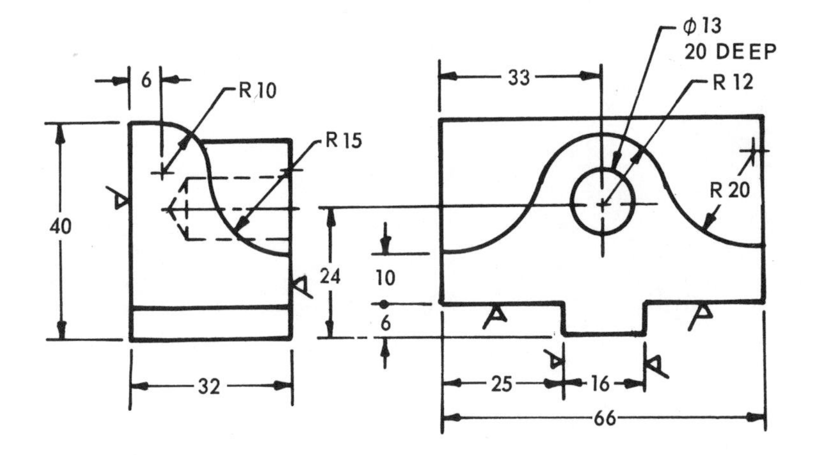

Fig. 4-5-4 Application of machining symbols

Material Removal Allowance

When it is desirable to indicate the amount of material to be removed, the amount of material in millimetres is shown to the left of the symbol. In conforming with the general rule of dimensioning, the symbol and its values should be placed to read from the bottom of the drawing. Illustrations showing material removal allowance are shown in Figures 4-5-6 and 4-5-7.

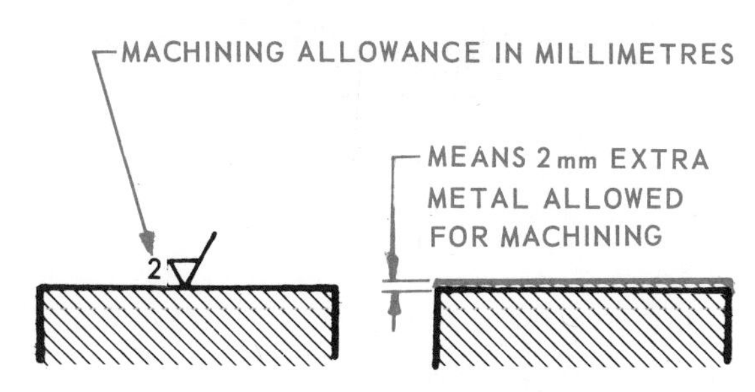

Fig. 4-5-6 Indication of machining allowance

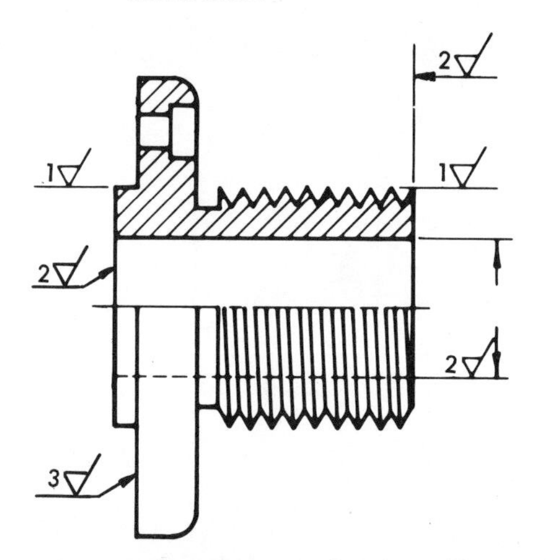

Fig. 4-5-7 Indicating machining allowance on drawings

Material Removal Prohibited

When it is necessary to indicate that a surface must be produced without material removal, the machining prohibited symbol shown in Figure 4-5-8 must be used.

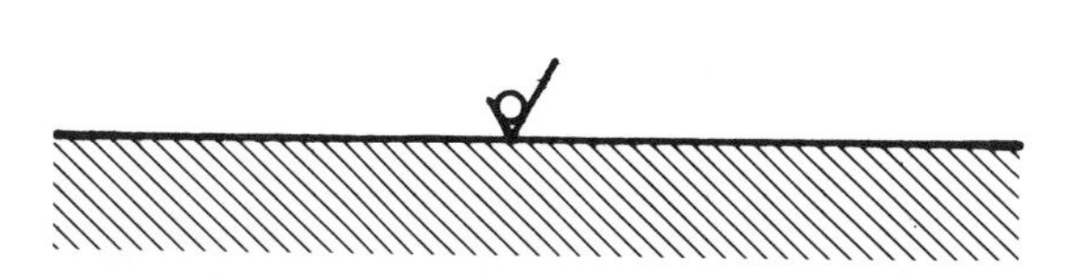

Fig. 4-5-8 Symbol used to indicate that the surface texture must be obtained without the removal of material

Surface Texture Control

All surface finish control starts in the drafting room. The designer has the responsibility of specifying the right surface to give maximum performance and service life at the lowest cost. In selecting the required surface finish for any particular part, the designer bases his or her decision on past experience with similar parts, on field service data, or on engineering tests. Such factors as size and function of the parts, type of loading, speed and direction of movement, operating conditions, physical characteristics of both materials in contact, type and amount of lubricant, contaminants, temperature, etc., influence the choice.

The two principal reasons for surface finish control are to reduce friction and to control wear.

Surface Texture Characteristics

Refer to Figure 4-5-9.

Micrometre. A micrometre is one millionth of a metre (0.000 001 m). The SI unit symbol for the micrometre, μm, is used

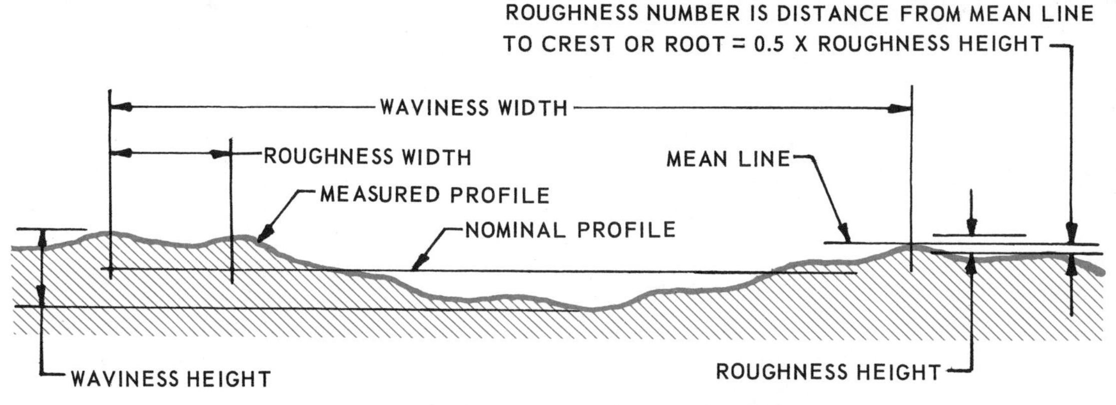

Fig. 4-5-9 Surface finish characteristics

with the appropriate number(s) for written specifications or reference to surface roughness requirements.

Roughness. Roughness consists of the finer irregularities in the surface texture usually including those which result from the production process.

Roughness-Height Value. Roughness-height value is rated as the arithmetic average (AA) deviation expressed in micrometres measured normal to the centre line.

Roughness Spacing. Roughness spacing is the distance parallel to the nominal surface between successive peaks or ridges which constitute the predominant pattern of the roughness. Roughness spacing is rated in millimetres.

Waviness. Waviness is the usually widely spaced component of surface texture. Waviness may result from such factors as machine or work deflections, vibration, chatter, heat treatment, or warping strains. Roughness may be considered as superimposed on a "wavy" surface.

Roughness Control

Surface characteristics of roughness may be controlled by applying the desired values to the surface texture symbol, shown in Figure 4-5-10, in a general note, or both. The point of the symbol should be located on the line indicating the surface, on an extension line to the surface or extension line (Fig. 4-5-11). The symbol should be in an upright position in order to be readable from the bottom.

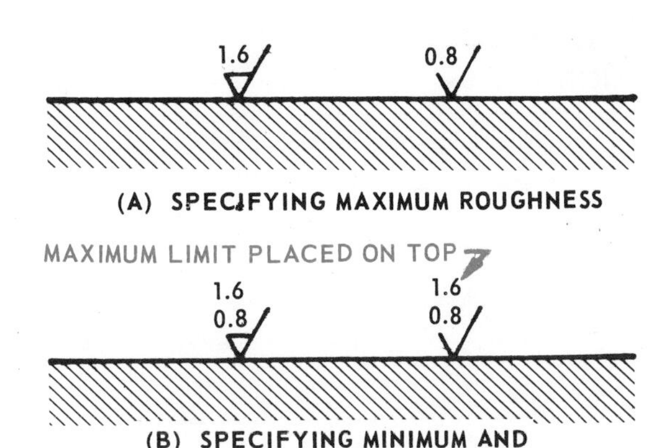

Fig. 4-5-10 Location of roughness numbers on symbol

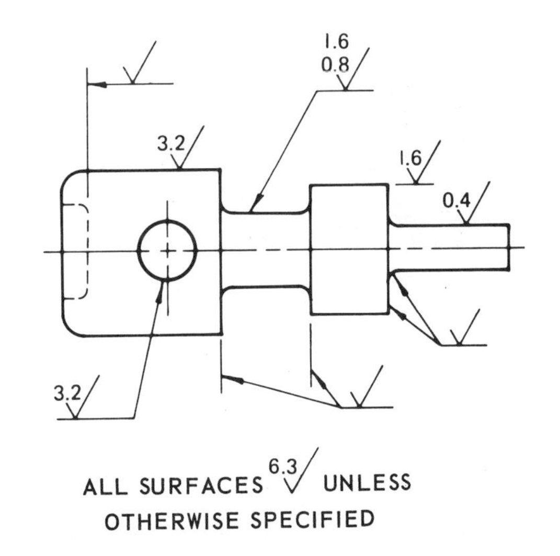

Fig. 4-5-11 Application of surface texture symbols and notes

This means that the long leg is always on the right. The symbol applies to the entire surface, unless otherwise specified.

Showing only one roughness rating indicates the maximum value, and any lesser value is acceptable. Showing two ratings indicates the minimum and maximum values, and anything lying within that range is acceptable.

Figure 4-5-12 shows the surface textures which are obtainable by the common production methods used in the industry.

Assignments

1. On an A3 sheet make a working drawing of the cut-off stop shown in Fig. 4-5-A. The four sliding surfaces are to have a maximum roughness of 1.6 μm and a machining allowance of 2 mm. The stop-surface is to have a maximum roughness of 3.2 μm and a machining allowance of 3 mm. Scale 1:1.
2. On an A3 sheet make a working drawing of the sparker bracket shown in Fig. 4-5-B. The end faces for the ϕ12 holes are to have a maximum roughness of 3.2 μm and a machining allowance of 2 mm. The back face and two mounting surfaces are to have a maximum roughness of 1.6 μm and a machining allowance of 3 mm. Scale 1:1.
3. On an A3 sheet make a working drawing of the cross slide shown in Fig. 4-5-C. The dovetail slot is to have a maximum roughness of 3.2 μm and a machining allowance of 2 mm. The T-slot is to have a maximum roughness of 1.6 μm and a machining allowance of 3 mm. The top surface is to have a maximum roughness of 6.3 μm and a machining allowance of 3 mm. Scale 1:1.

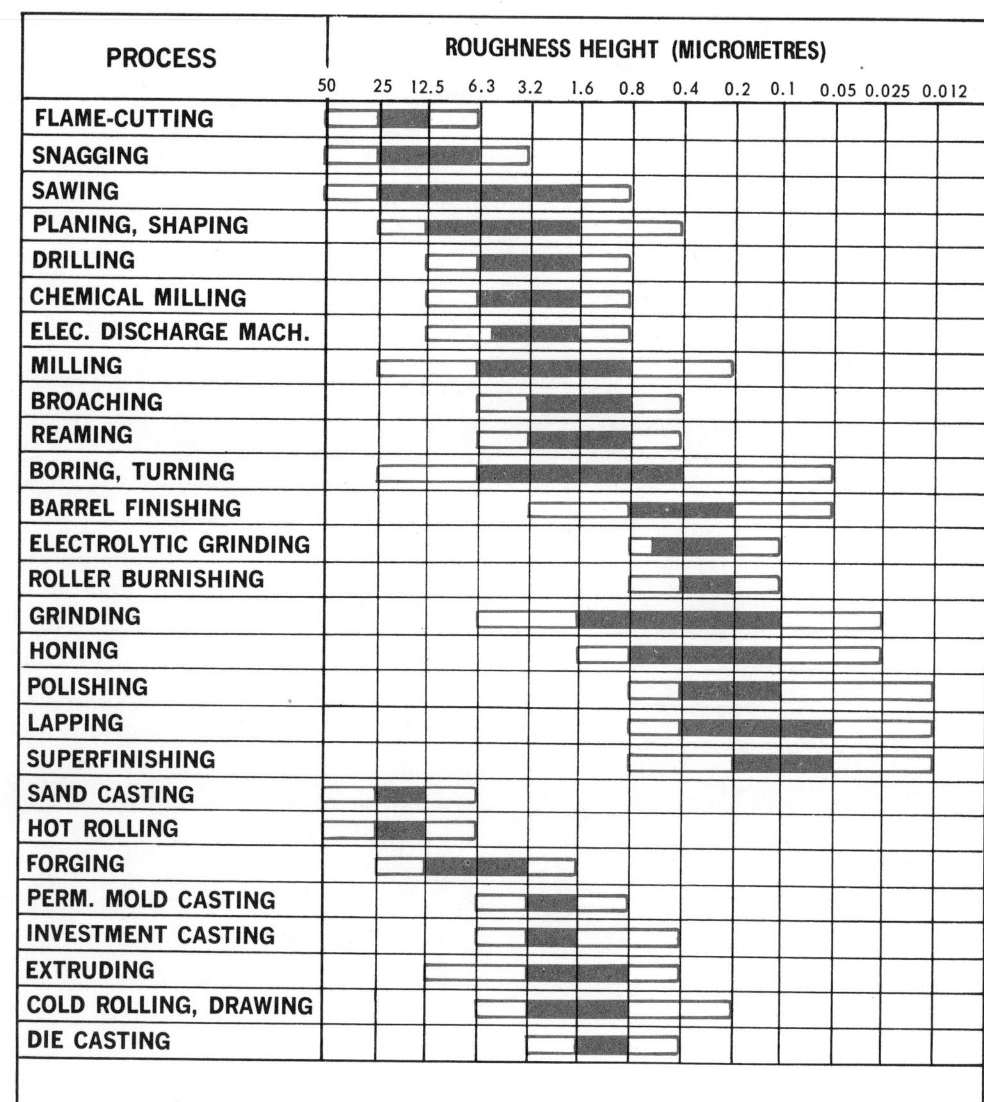

Fig. 4-5-12 Surface roughness range for common production

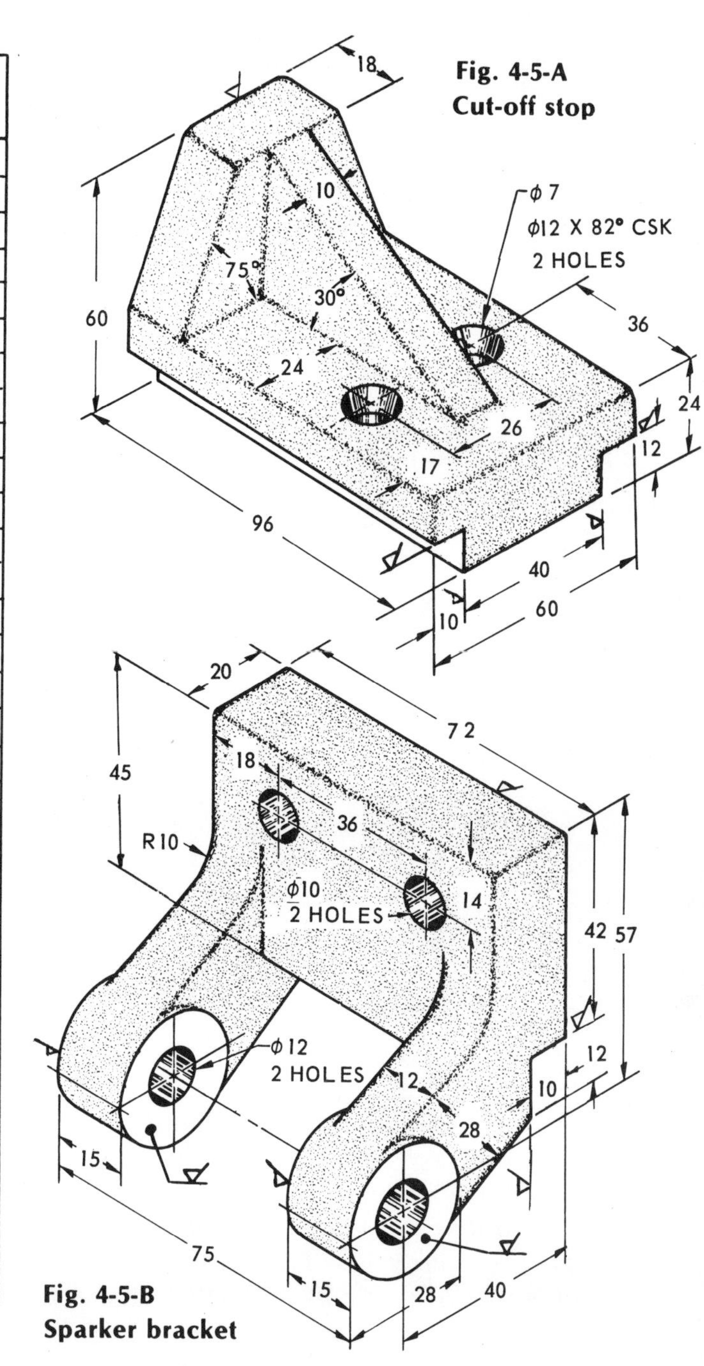

Fig. 4-5-A
Cut-off stop

Fig. 4-5-B
Sparker bracket

4. On an A3 sheet make a working drawing of the column bracket shown in Fig. 4-5-D. The bottom of the base is to have a maximum roughness value of 3.2 μm and a machining allowance of 3 mm.

The tops of the bosses are to have a maximum roughness value of 6.3 μm and a machining allowance of 2 mm.

The end surfaces of the hubs supporting the shafts are to have maximum and minimum roughness values of 1.6 and 0.8 μm and a machining allowance of 2 mm.

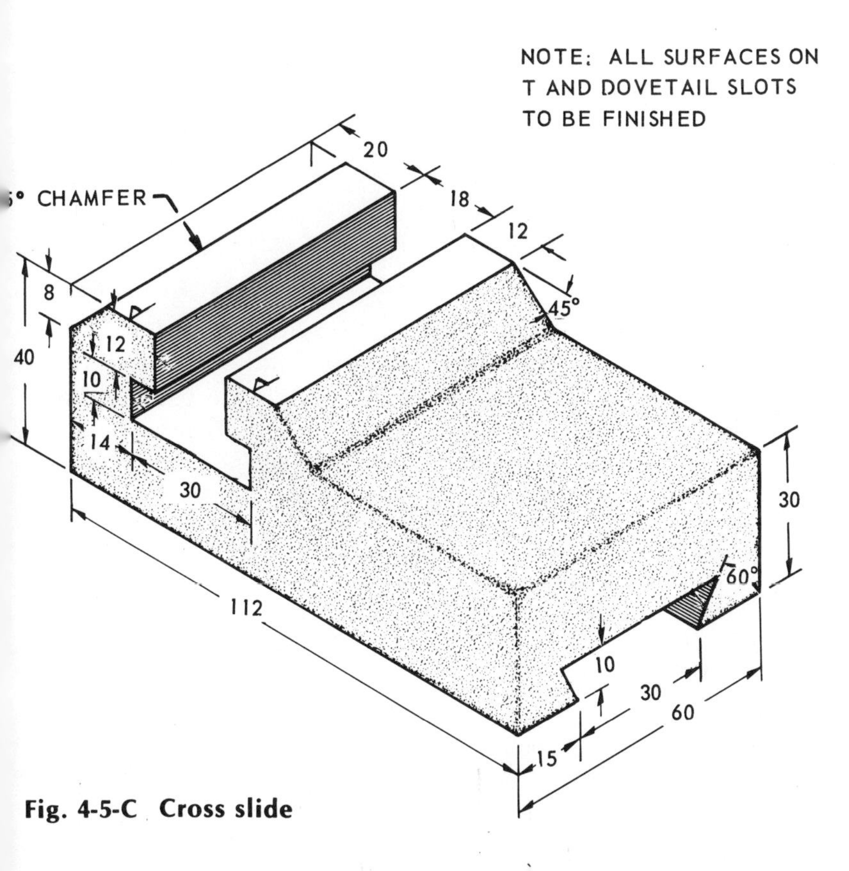

Fig. 4-5-C Cross slide

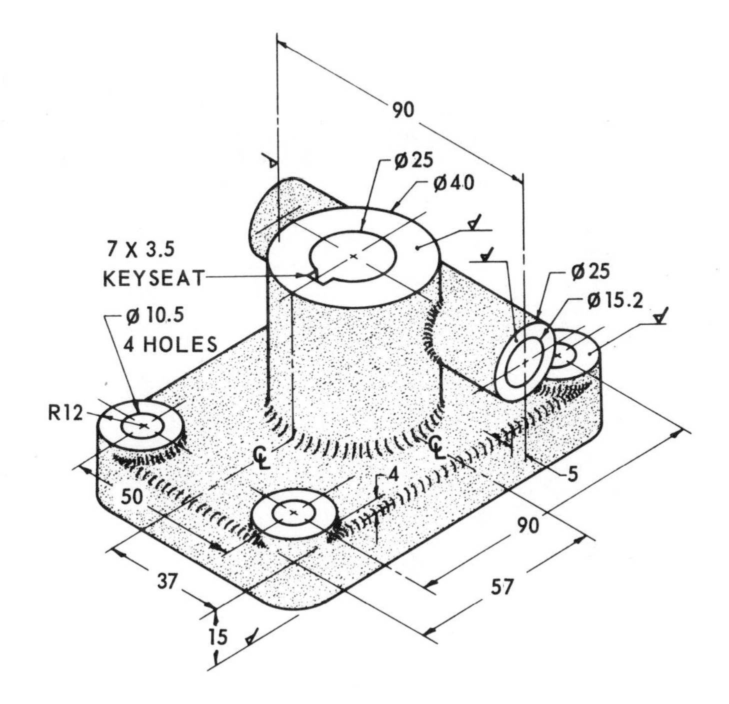

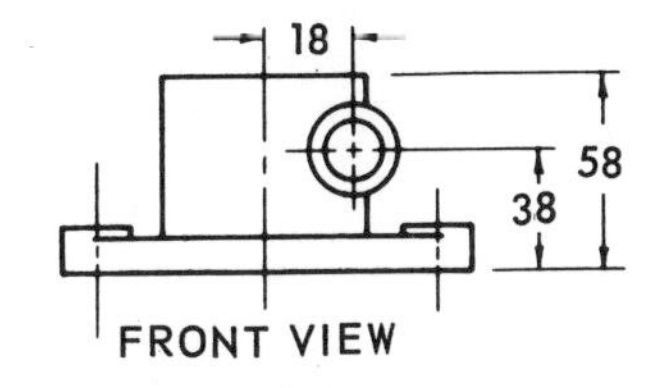

Fig. 4-5-D Column bracket

Chapter 5

Working Drawings

UNIT 5-1

Conventional Representation of Common Features

To simplify the representation of common features a number of conventional drawing practices are used. Many conventions are deviations from "true" projection for the purpose of clarity; others are used for the purpose of saving time (Fig. 5-1-1). These conventions must be executed carefully, for clarity is even more important than speed.

Intersection of Unfinished Surfaces

The intersection of unfinished surfaces that are rounded or filleted at the point of theoretical intersection may be indicated conventionally by a line coinciding with the theoretical point of intersection.

The need for this convention is shown by the examples illustrated in Figure 5-1-2, where the upper top views are shown in true projection. Note that in each example the true projection would be misleading. In the case of the large radius such as shown in Figure 5-1-2(D) no line is drawn.

Members such as ribs and arms that blend into other features end in curves called **runouts.** Small runouts are usually drawn freehand. Large runouts are drawn with an irregular curve, template, or compass. (See Fig. 5-1-3).

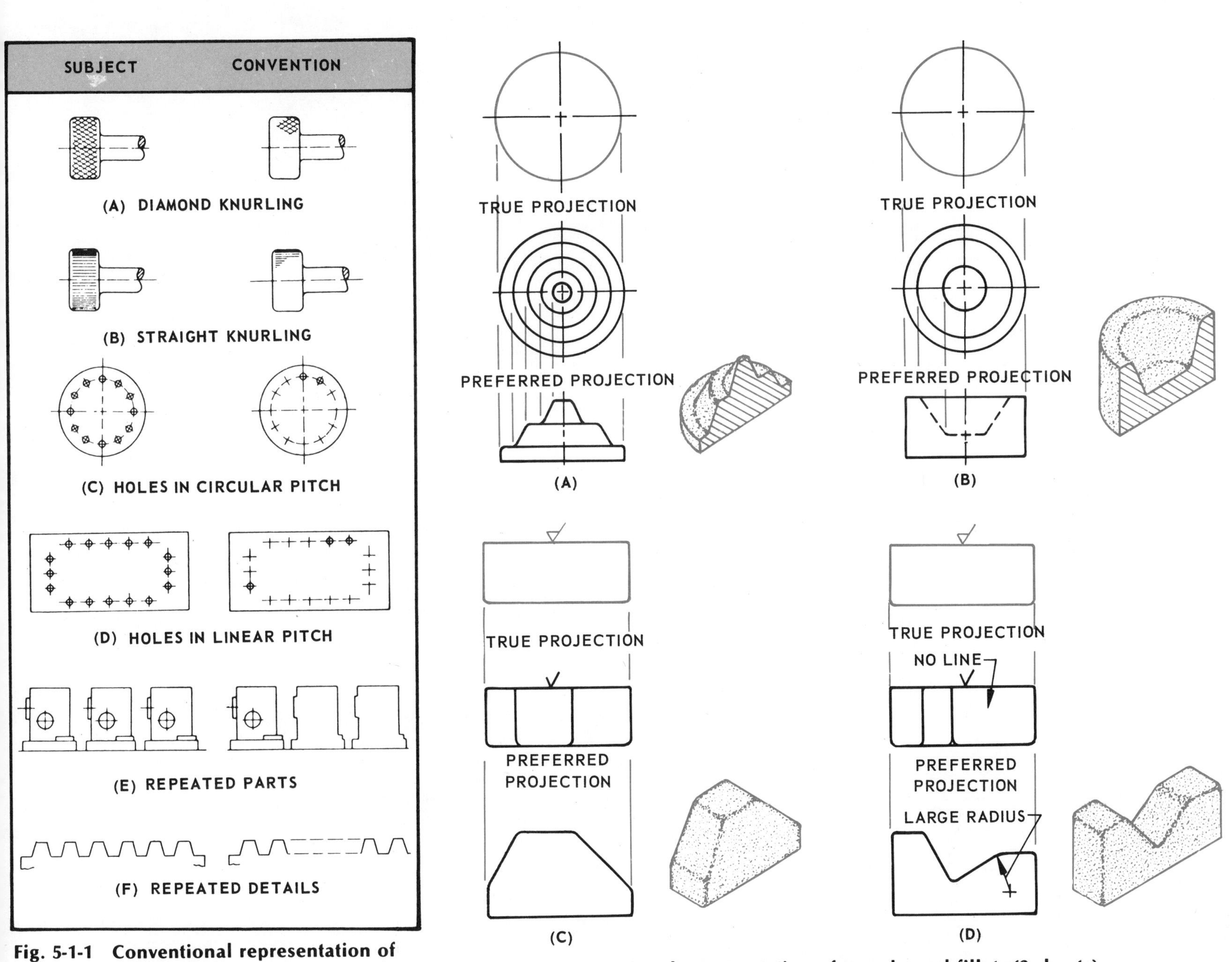

Fig. 5-1-1 Conventional representation of common features

Fig. 5-1-2 Conventional representation of rounds and fillets (2 sheets)

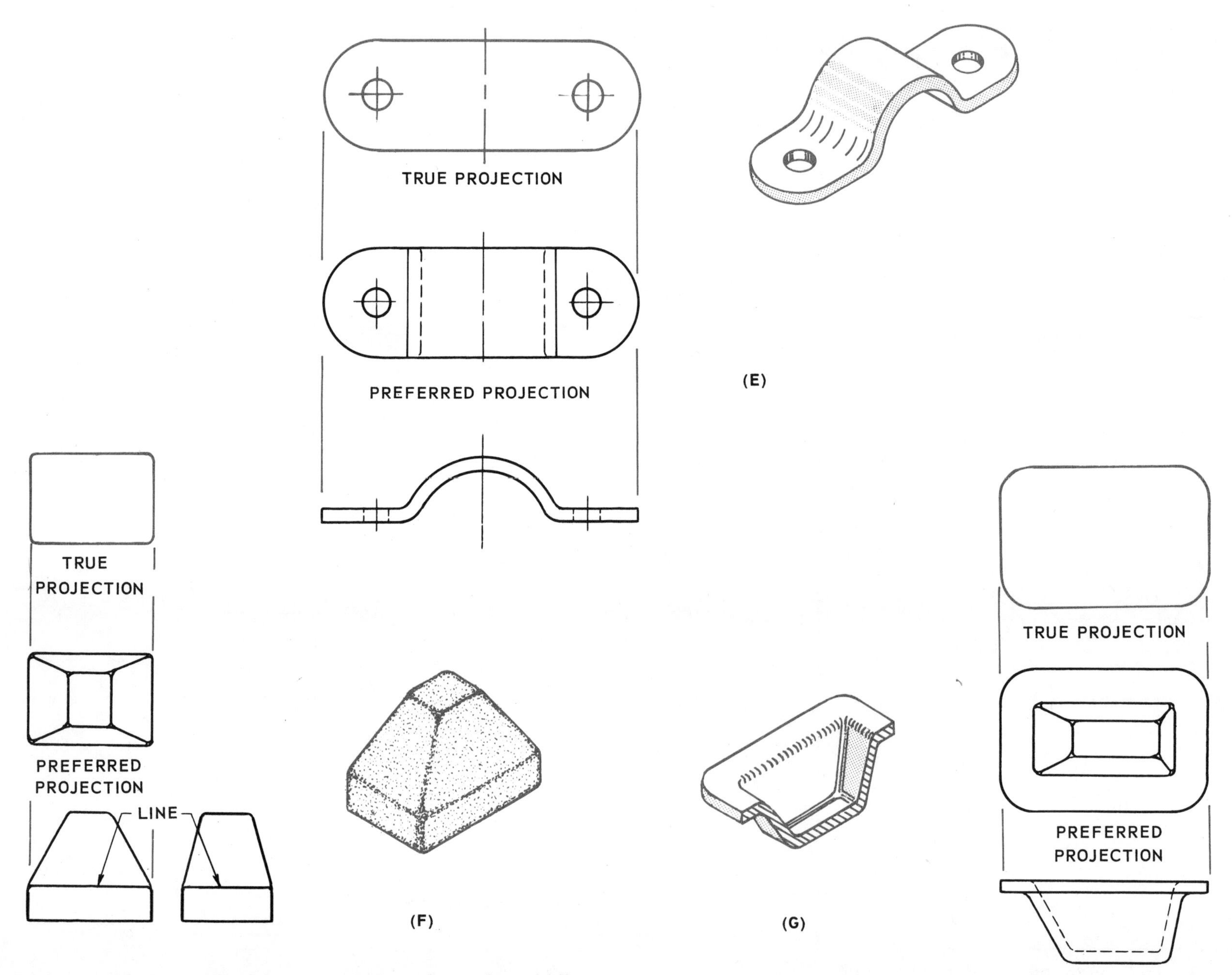

Fig. 5-1-2 cont'd. Conventional representation of rounds and fillets

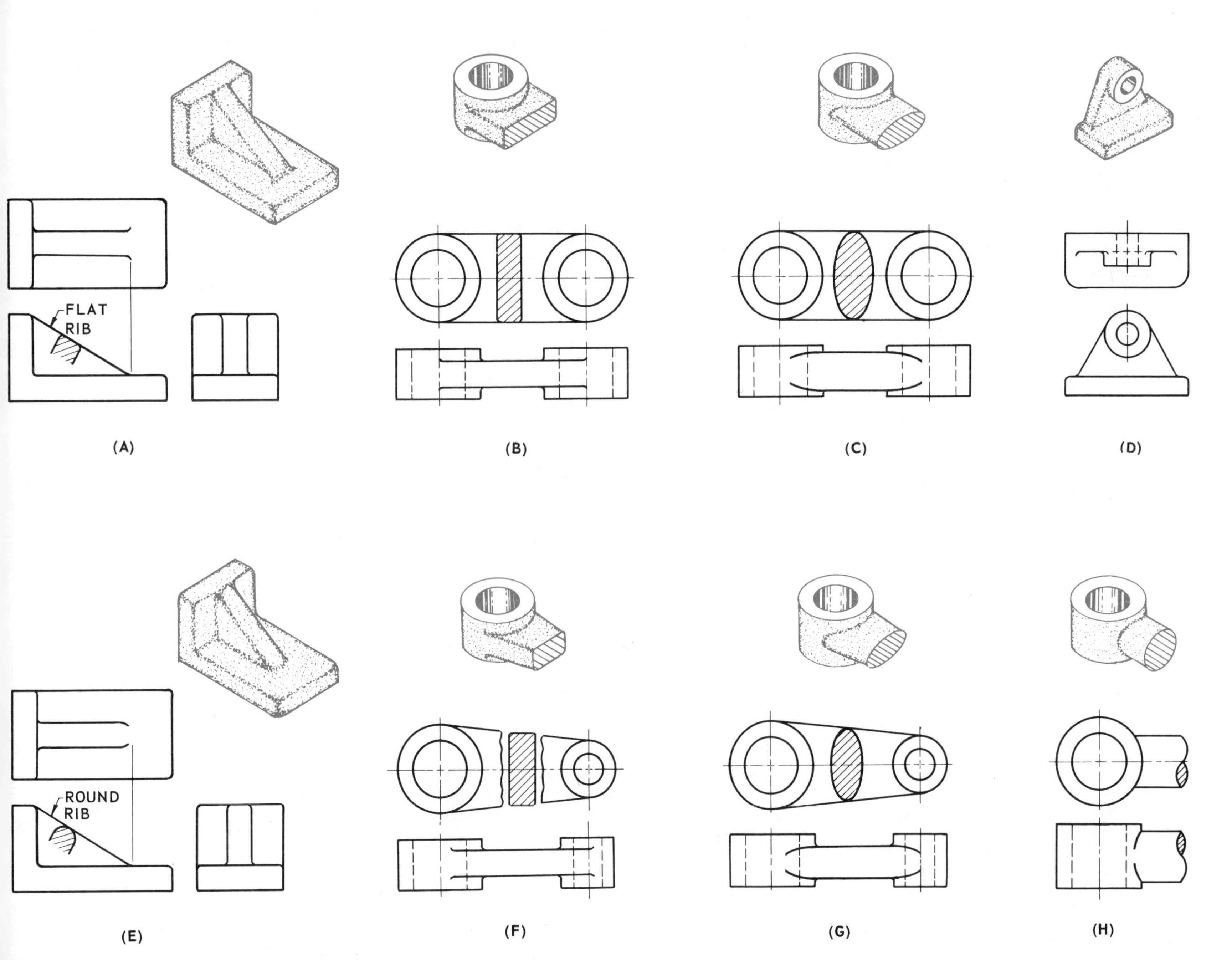

Fig. 5-1-3 Conventional representation of runouts

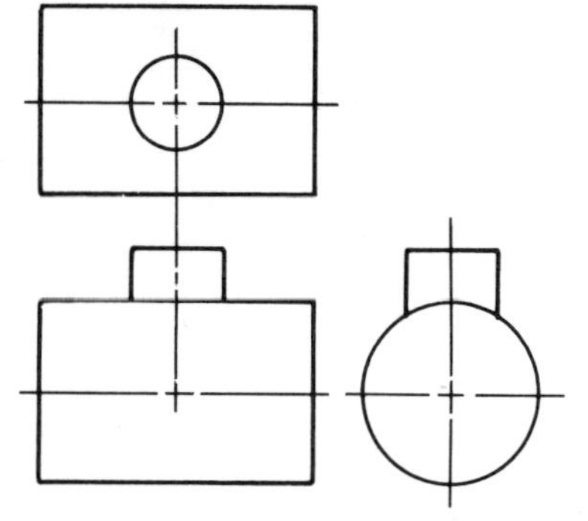

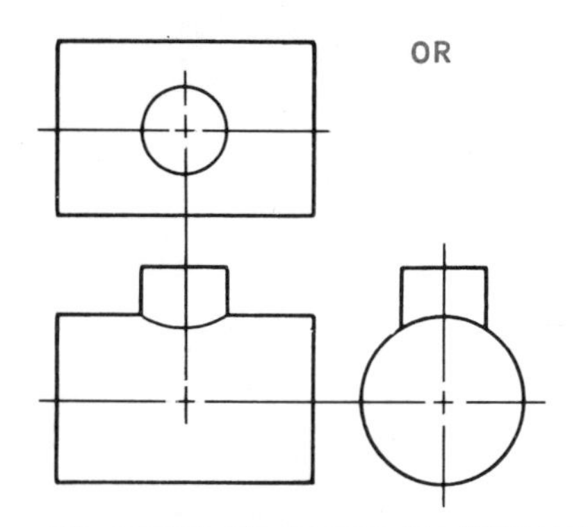

Fig. 5-1-4 Conventional representation of external intersections

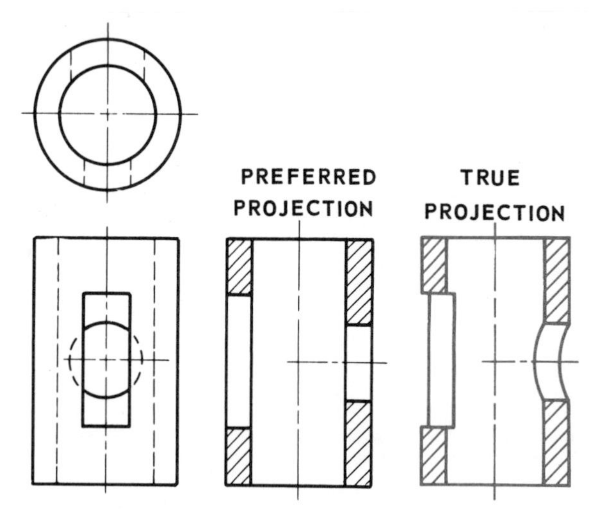

Fig. 5-1-5 Conventional representation of cylindrical intersections

Cylindrical Intersections

The intersection of rectangular and circular contours, unless they are very large, is shown conventionally as in Figures 5-1-4 and 5-1-5. The same convention may be used to show the intersection of two cylindrical contours; or the curve of intersection may be shown as a circular arc.

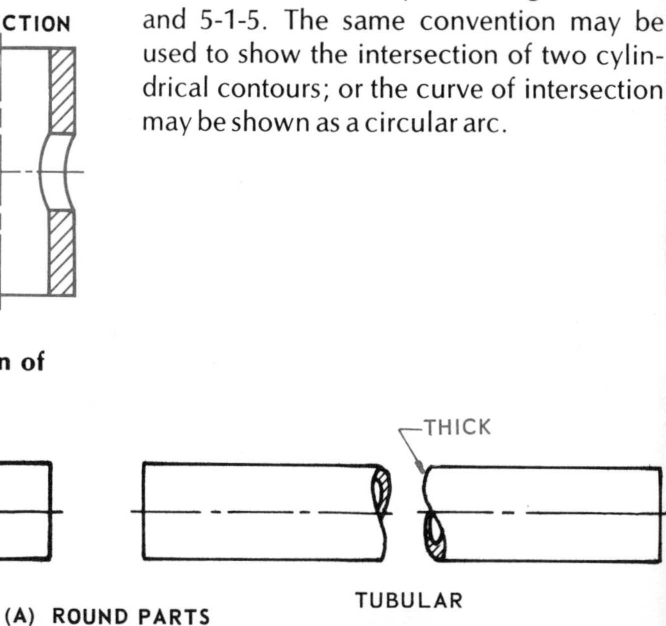

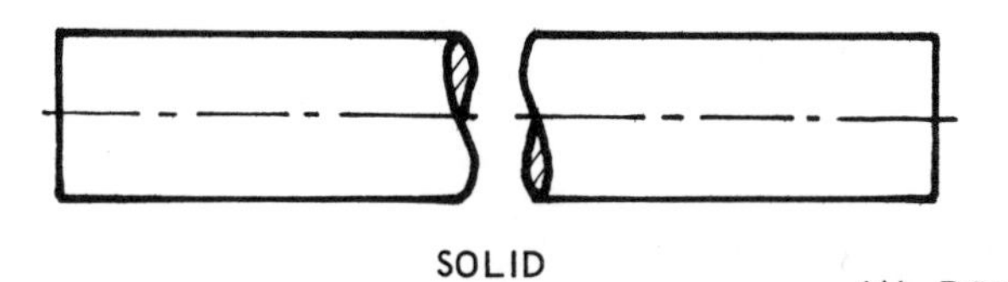

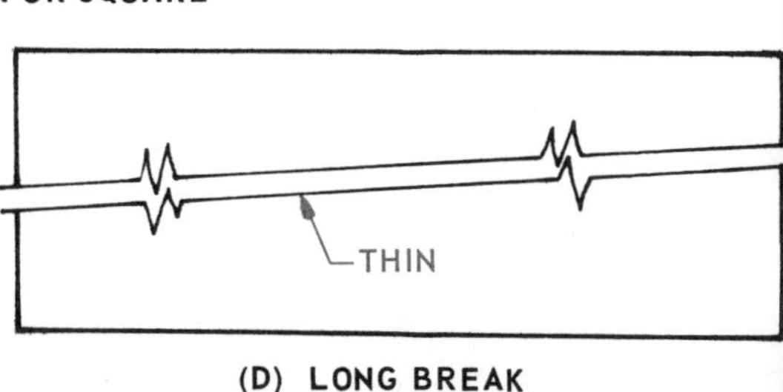

Fig. 5-1-6 Conventional breaks

Conventional Breaks

Long, simple parts such as shafts, bars, tubes, and arms need not be drawn to their entire length. Conventional breaks located at a convenient position may be used and the true length indicated by a dimension (Fig. 5-1-6).

Often a part can be drawn to a larger scale to produce a clearer drawing if a conventional break is used. The breaks used on circular objects, known as "S breaks," may be drawn freehand, with an irregular curve, template, or compass.

Foreshortened Projection

When the true projection of a piece would result in confusing foreshortening, parts such as ribs or arms should be rotated until they are parallel to the line of the section or projection (Fig. 5-1-8(A)(B)(C)).

Holes Revolved to Show True Distance from Centre

Drilled flanges in elevation or section should show the holes at their true distance from the centre, rather than the true projection (Fig. 5-1-8(D)).

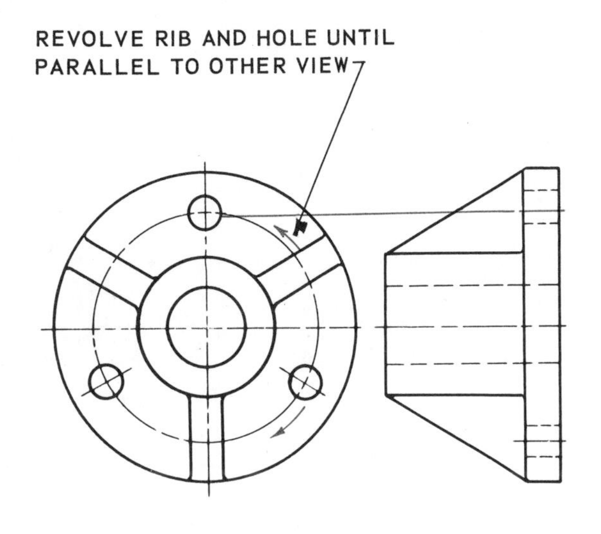

(A) ALIGNMENT OF RIB AND HOLES

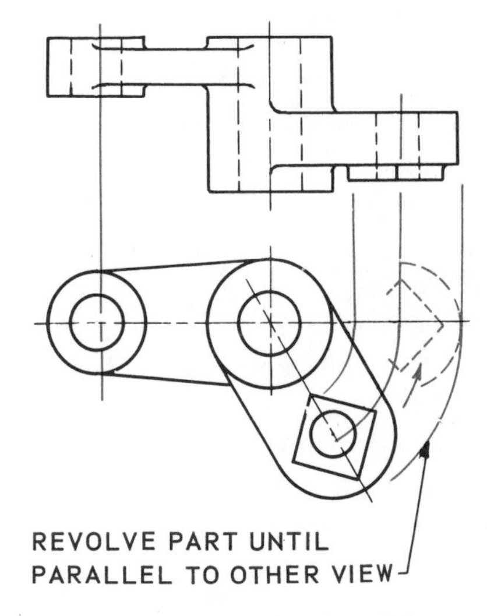

(B) ALIGNMENT OF PART

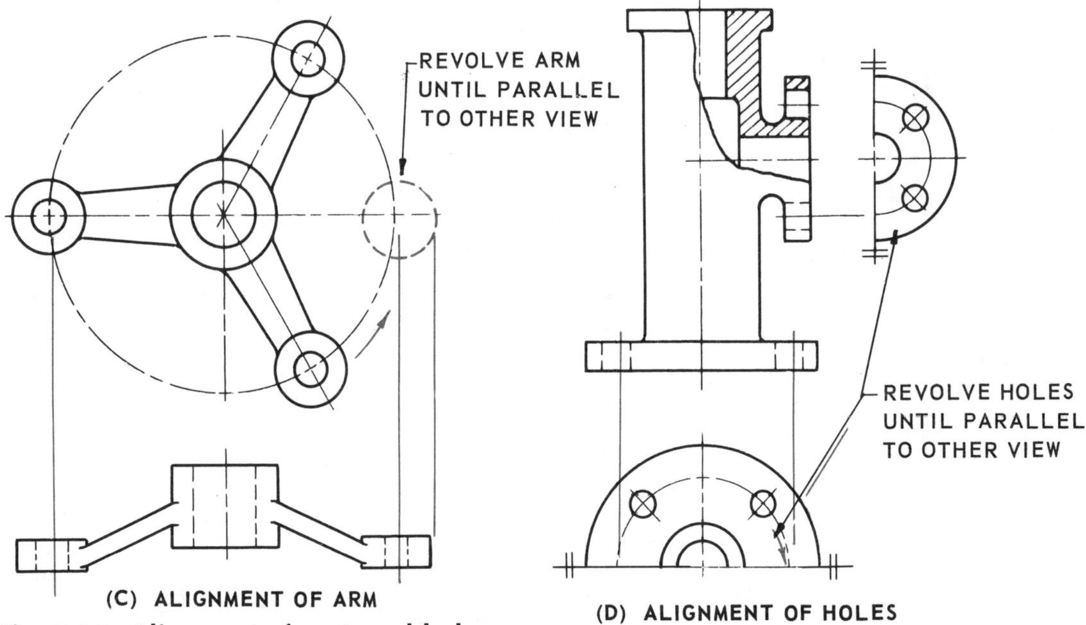

(C) ALIGNMENT OF ARM

(D) ALIGNMENT OF HOLES

Fig. 5-1-7 Alignment of parts and holes

Assignments

1. Cross slide, Fig. 5-1-A, sheet size-A4, scale 1:1. Make a three-view working drawing.
2. Shift lever, Fig. 5-1-B, sheet size-A4, scale 1:1. Make a two-view drawing using a conventional break for the ϕ20 connecting piece.
3. Mounting bracket, Fig. 5-1-C, sheet size-A3, scale 1:1. Make a two-view working drawing following the techniques for foreshortened projection and revolving of holes to show true distance from centre.

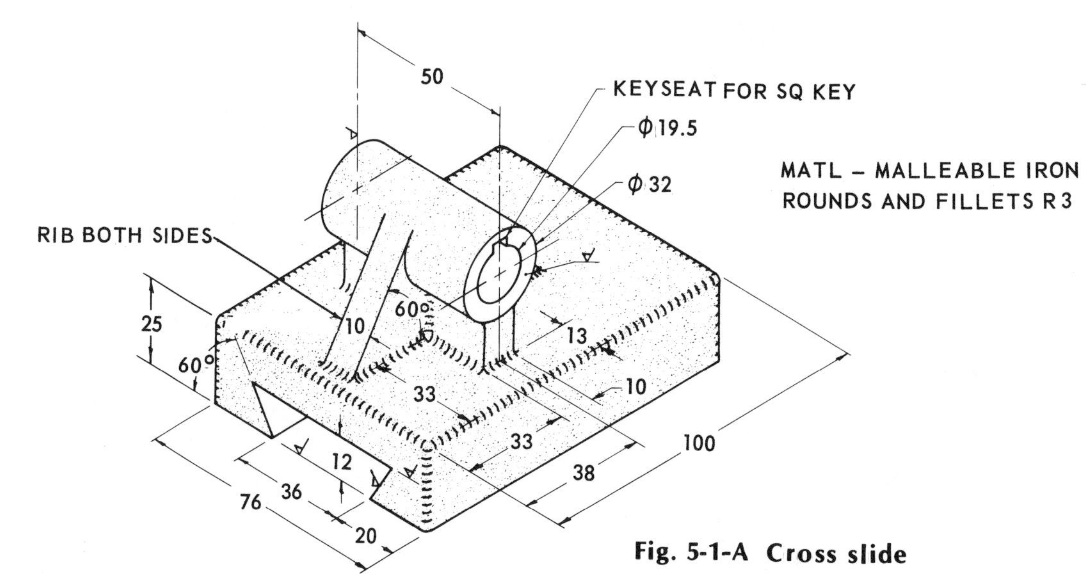

Fig. 5-1-A Cross slide

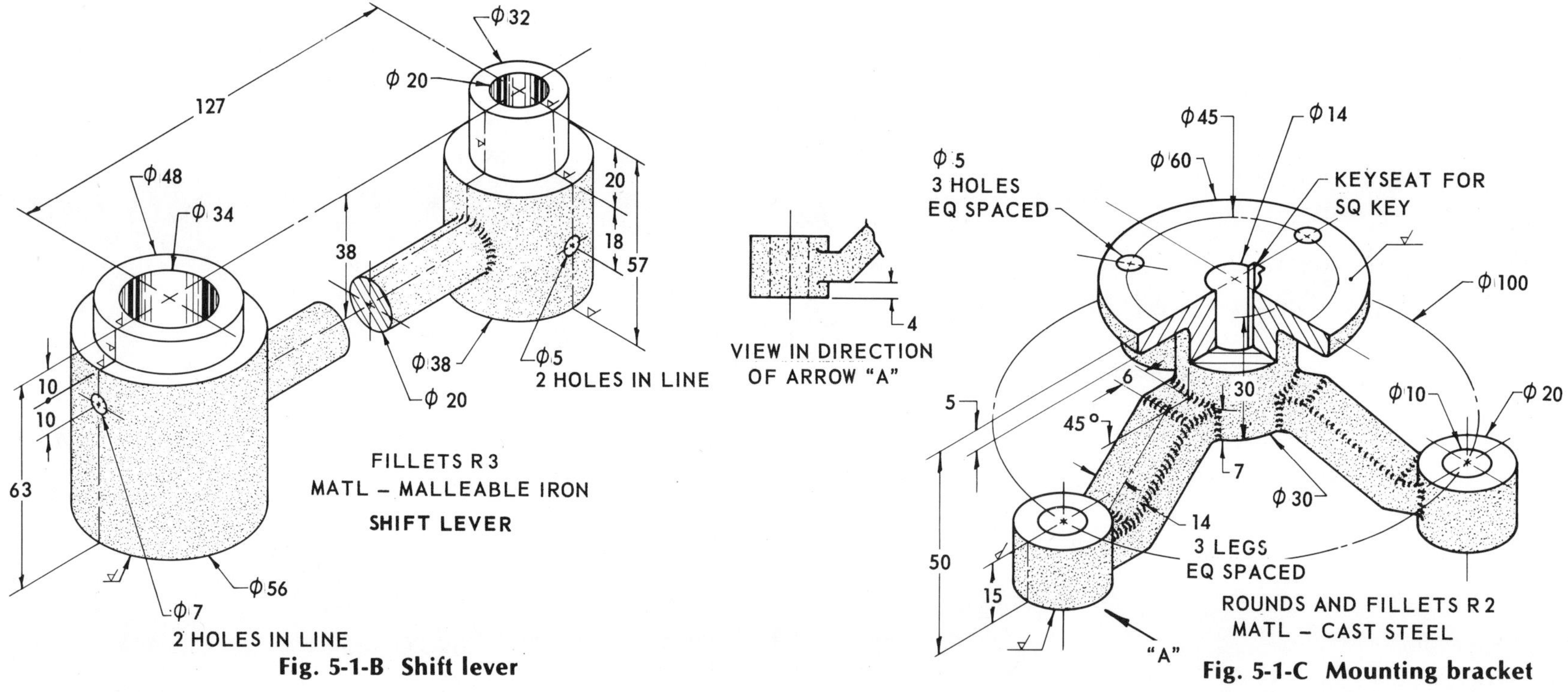

Fig. 5-1-B Shift lever

Fig. 5-1-C Mounting bracket

UNIT 5-2
Special Views

Placement of Views

When views are placed in their regular projected positions shown in Figure 5-2-1, it is rarely necessary to identify them. When they are placed in other than the regular projected position, the removed view must be clearly identified.

Whenever appropriate, the orientation of the main view on a detail drawing should be the same as on the assembly drawing. To avoid the crowding of dimensions and notes, ample space must be provided between views.

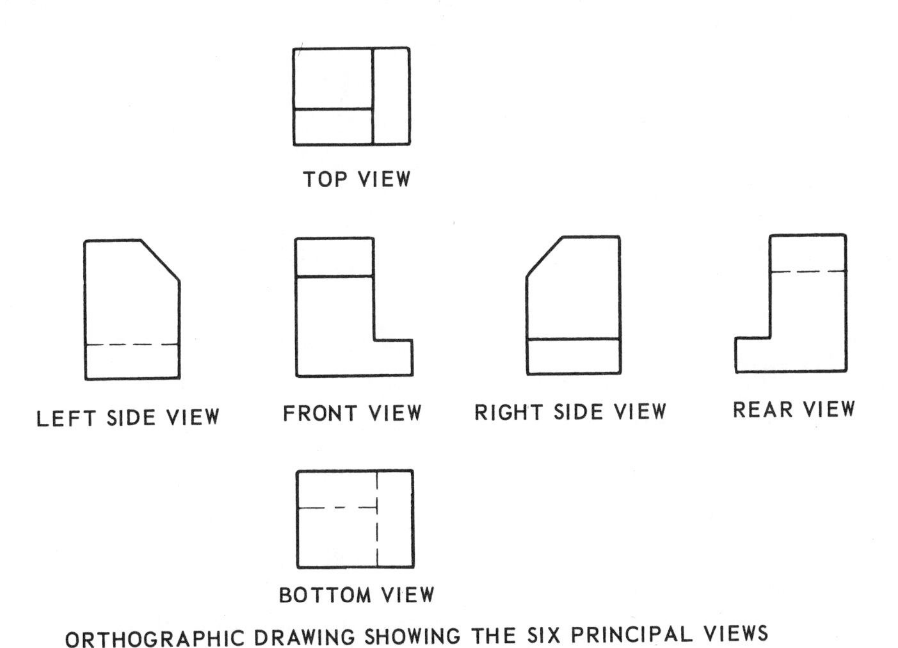

Fig. 5-2-1 Regular projected position of views

Enlarged Views

Enlarged views are used when it is desirable to show a feature in greater detail or to eliminate the crowding of details or dimensions (Fig. 5-2-2). The enlarged view should be oriented in the same manner as the main view. However, if an enlarged view is rotated, state the direction and the amount of rotation of the detail. The scale of enlargement must be shown, and both views should be identified by one of the three methods shown.

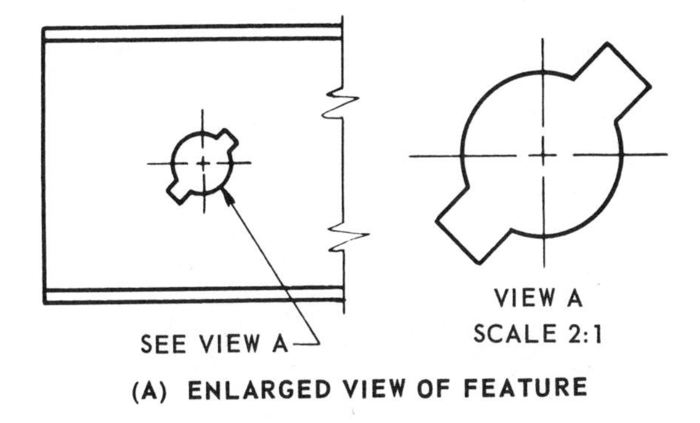

(A) ENLARGED VIEW OF FEATURE

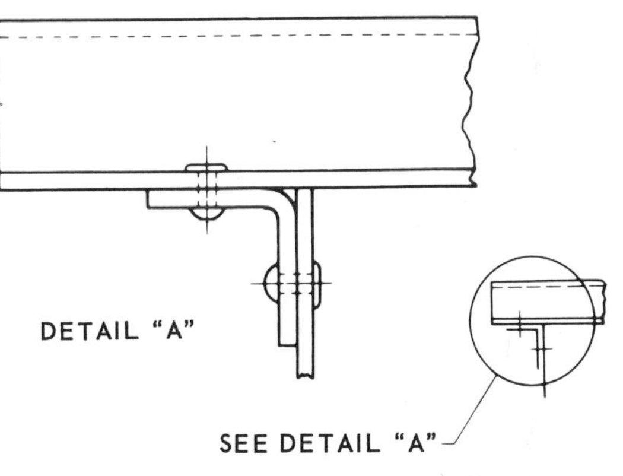

(B) ENLARGED VIEW OF ASSEMBLY

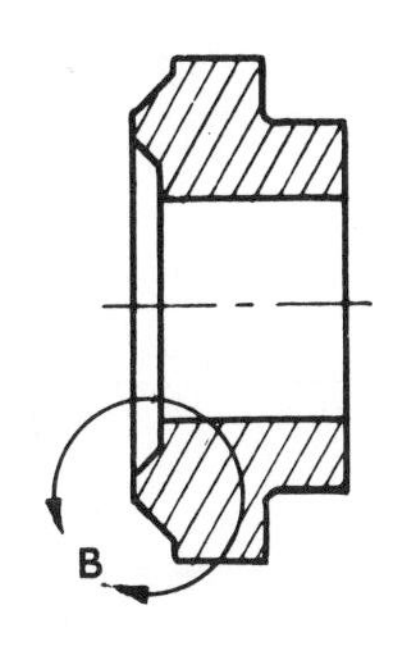

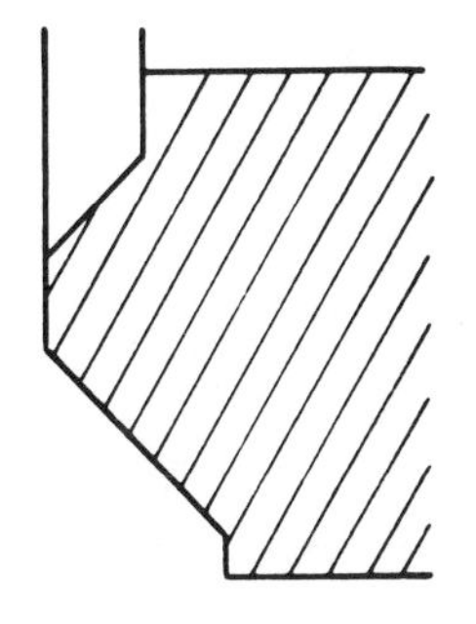

(C) ENLARGED REMOVED VIEW

Fig. 5-2-2 Enlarged views

Rear Views

Rear views are normally projected to the right or left. When this projection is not practical because of the length of the part, particularly for panels and mounting plates, the rear view must not be projected up or down. Doing so would result in the part's being shown upside down. Instead, the view should be drawn as if it were projected sideways but located in some other position, and it should be clearly labelled REAR VIEW REMOVED (5-2-3).

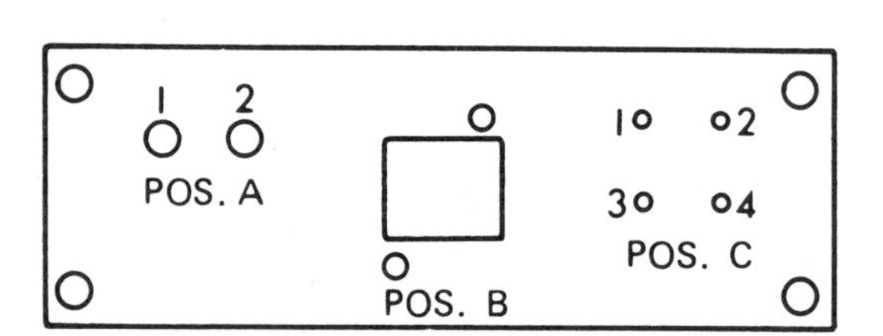

Fig. 5-2-3 Removed rear view

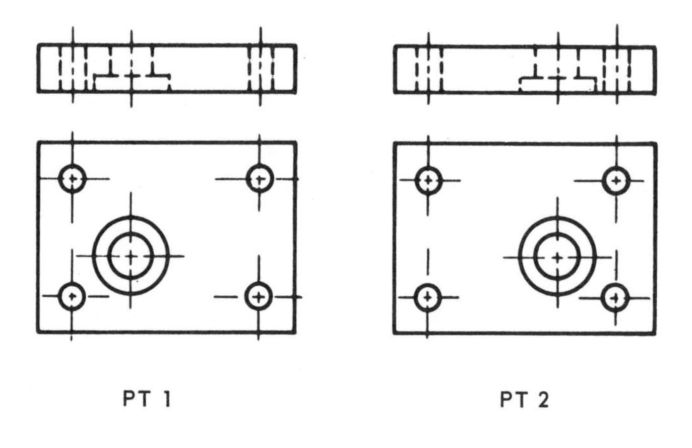

Fig. 5-2-4 Opposite hand views

Opposite-Hand Views

Where parts are symmetrically opposite, such as for right- and left-hand usage, one part is drawn in detail and the other is indicated by a note such as PART B SAME EXCEPT OPPOSITE HAND. It is preferable to show both part numbers on the same drawing (Fig. 5-2-4).

Partial Views

Symmetrical objects may often be adequately portrayed by half or quarter views (Fig. 5-2-5A and B). Partial views, which show only a limited portion of the object with remote details omitted, should be used, when necessary, to clarify the meaning of the drawing (Fig. 5-2-5(C)). Such views are used to avoid the necessity of drawing many hidden features.

On drawings of objects where two side views can be used to better advantage than one, each need not be complete if together they depict the shape. Show only the hidden lines of features immediately behind the view (Fig. 5-2-5(D)).

Assignments

1. Radio cover plate, Fig. 5-2-A, sheet size - A4, scale 1:1. Make a working drawing of the part showing both front and rear views.
2. Crescent truss, Fig. 5-2-B, sheet size - A3, scale 1:10. Draw the enlarged views of the gusset assemblies.
3. Flanged coupling, Fig. 5-2-C, sheet size - A3, scale 1:1. Make a working drawing using only those views (full and partial) which are necessary to fully describe the part.

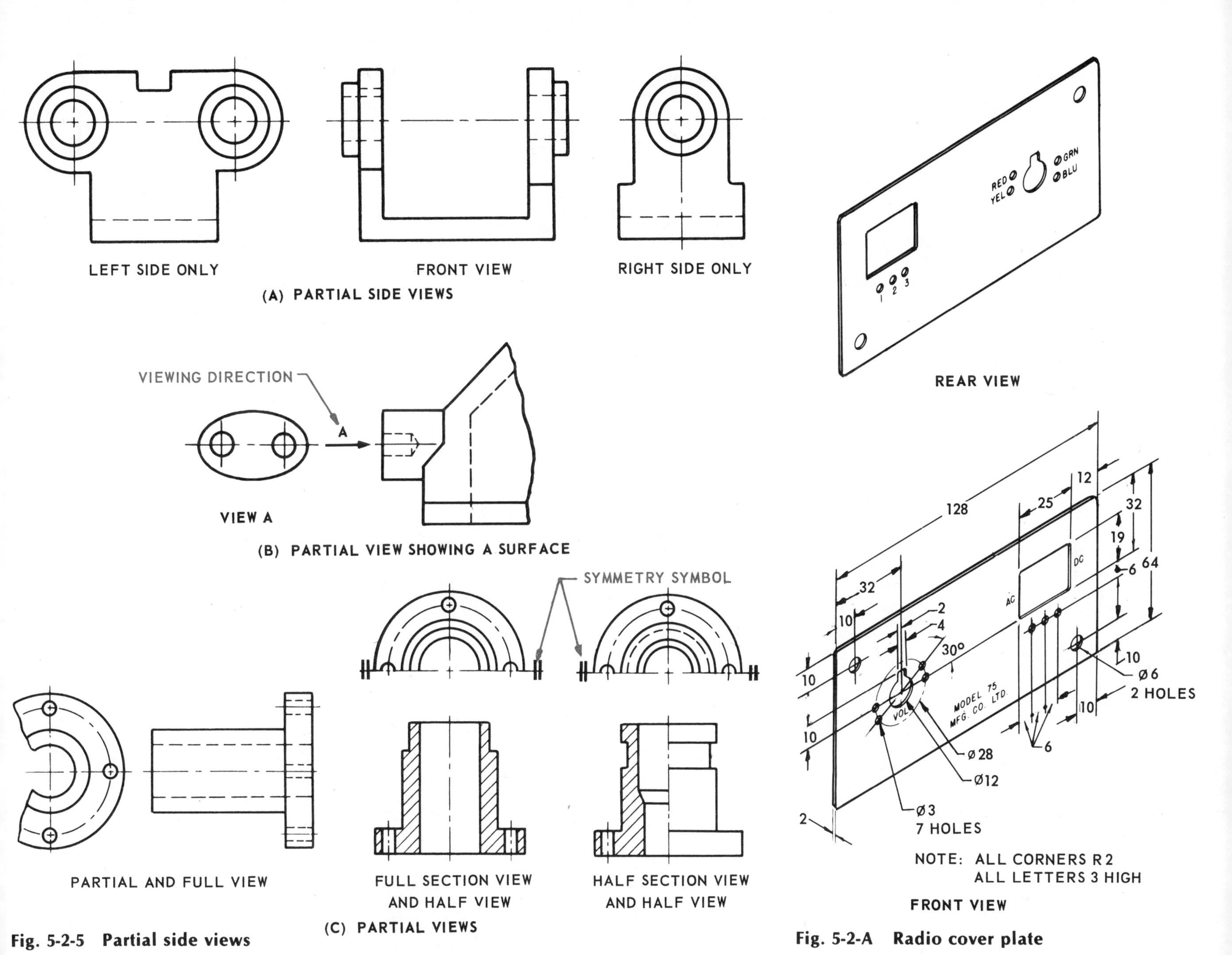

Fig. 5-2-5 Partial side views

Fig. 5-2-A Radio cover plate

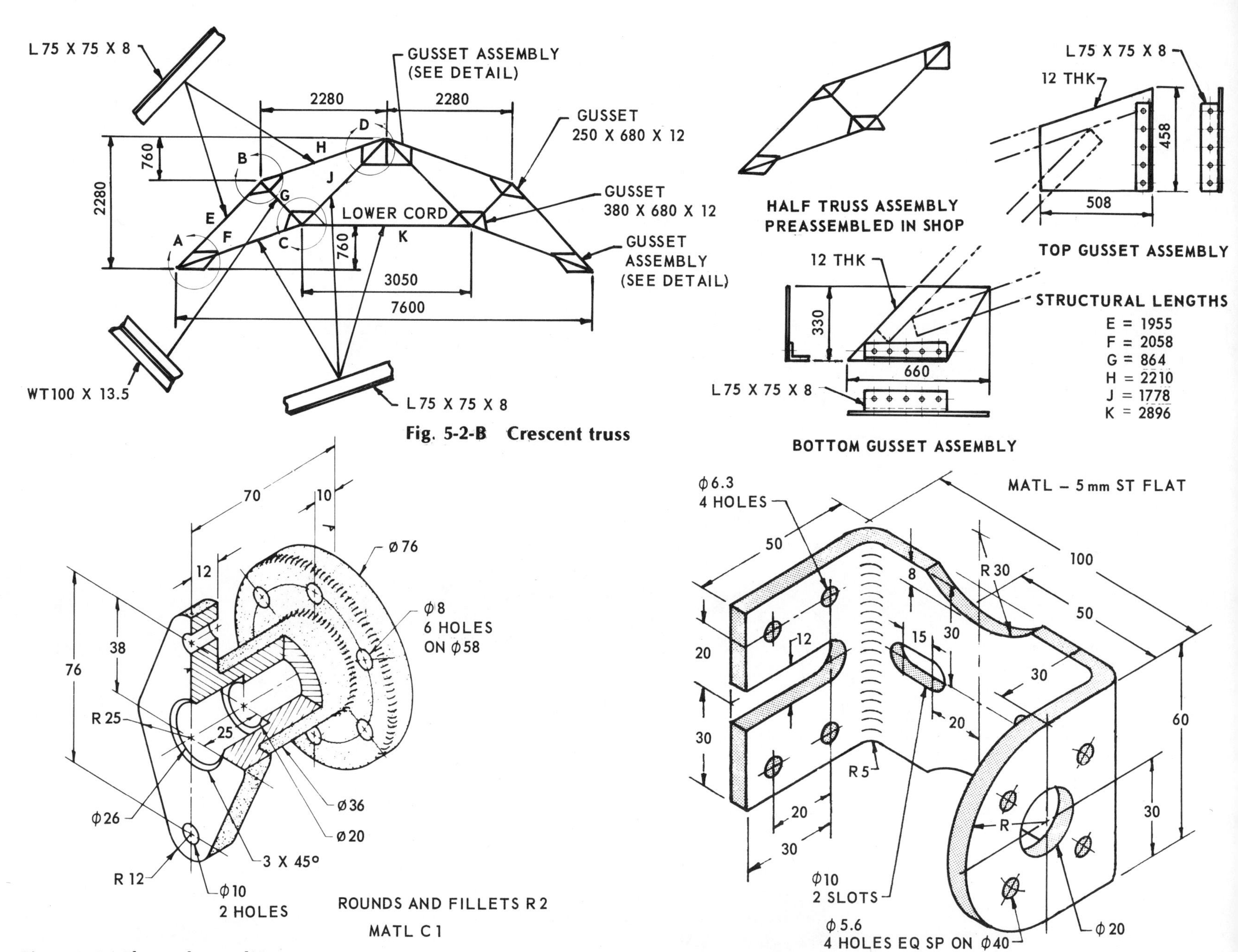

Fig. 5-2-B Crescent truss

Fig. 5-2-C Flanged coupling

Fig. 5-2-D Connector

UNIT 5-3
Detail and Assembly Drawings

A *working drawing* is a drawing that supplies information and instructions for the manufacture or construction of machines or structures. Generally, working drawings may be classified into two groups: detail drawings, which provide the necessary information for the manufacture of the parts, and assembly drawings, which supply the necessary information for their assembly.

Detail Drawings

The working drawings of each of the parts are called *detail drawings*, since each part is a "detail" of the complete machine or mechanism. Very often the working drawing for each detail is made on a separate drawing sheet (Fig. 5-3-1). When a number of the details are to be made of the same material and will be manufactured similarly, they can be grouped on a large common sheet (Fig. 5-3-2).

Several drawings may be made of the same part, each one giving only the information necessary for a particular step in the manufacture of the part. A part made from cast iron, for example, may have one detail drawing for the patternmaker and one detail drawing for the machinist, each drawing having only the dimensions and specifications necessary for that particular step in its manufacture.

The nature of the parts, the manufacturing techniques to be employed, and the drafting practices of the individual engineering office will determine the procedures for making detail drawings.

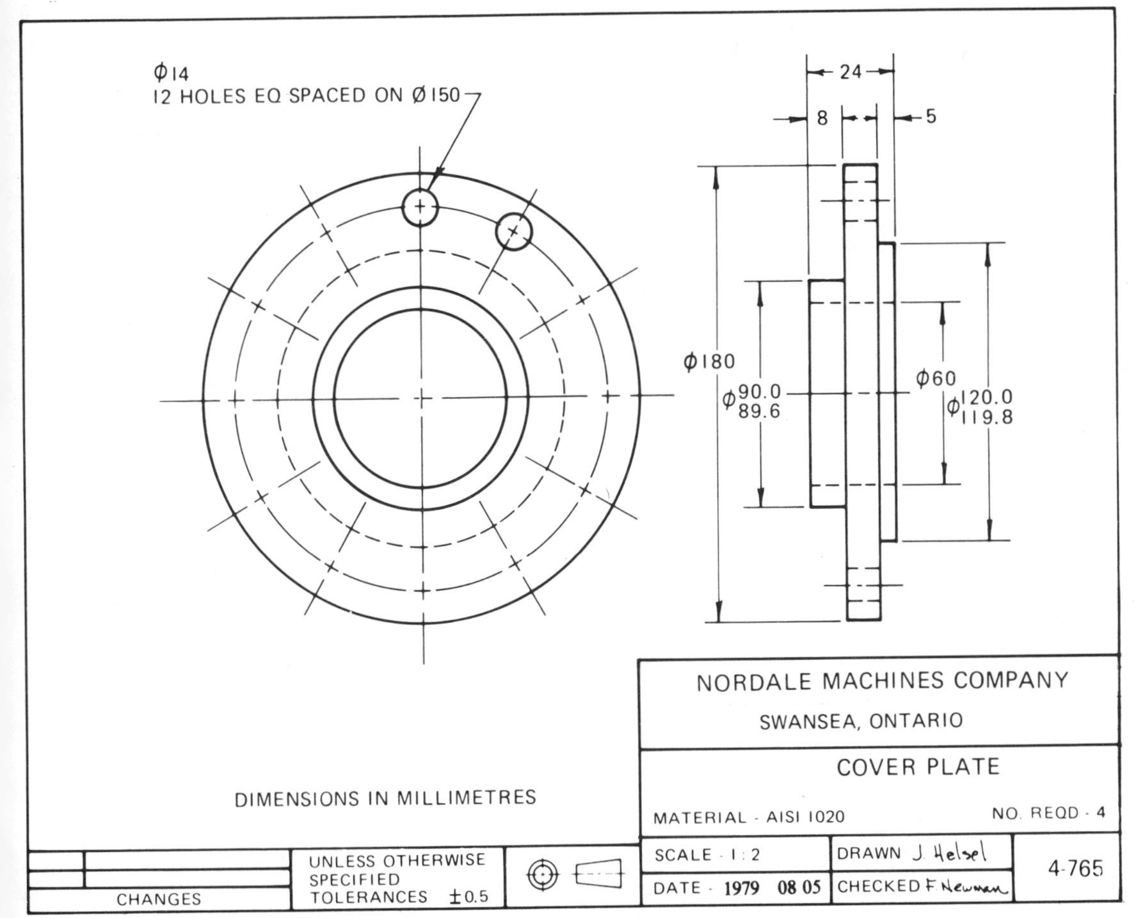

Fig. 5-3-1 A simple detail drawing

Assembly Drawings

All machines and mechanisms are made up of numerous parts. A drawing showing the product in its completed state is called an *assembly drawing* (Fig. 5-3-3).

Assembly drawings vary greatly in the amount and type of information they give, depending on the nature of the machine or mechanism they depict. The primary functions of the assembly drawings are to

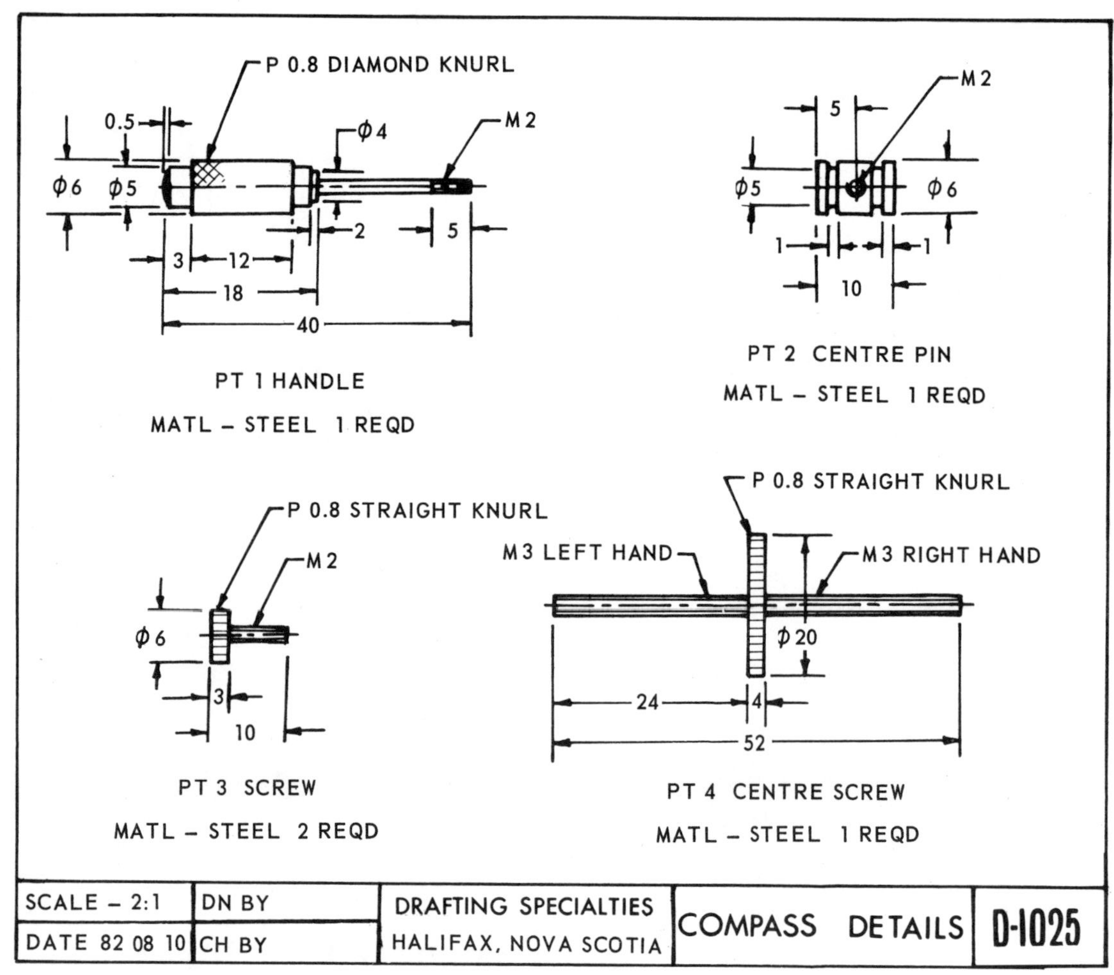

Fig. 5-3-2 Detail drawing showing many details on one drawing

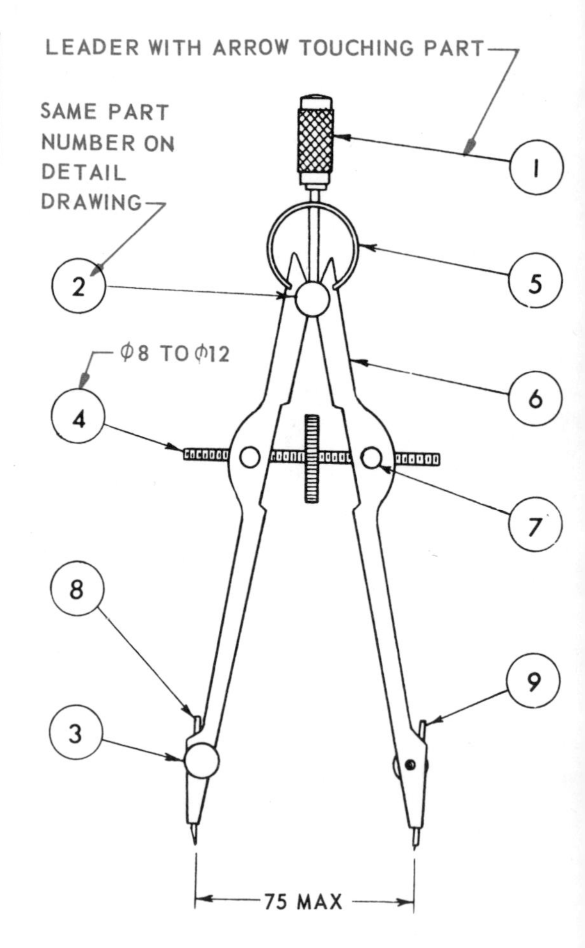

Fig. 5-3-3 An assembly drawing

show the product in its completed shape to indicate the relationship of its various components, and to designate those components by a part or detail number. Other information that might be given includes: over-all dimensions, capacity dimensions, relationship dimensions between parts (necessary information for assembly), operating instructions, and data on design characteristics. Some mechanisms are assembled units in themselves, but also form part of a total machine. Such mechanisms are often referred to as *sub-assemblies*. The transmission of an automobile is an example of a sub-assembly.

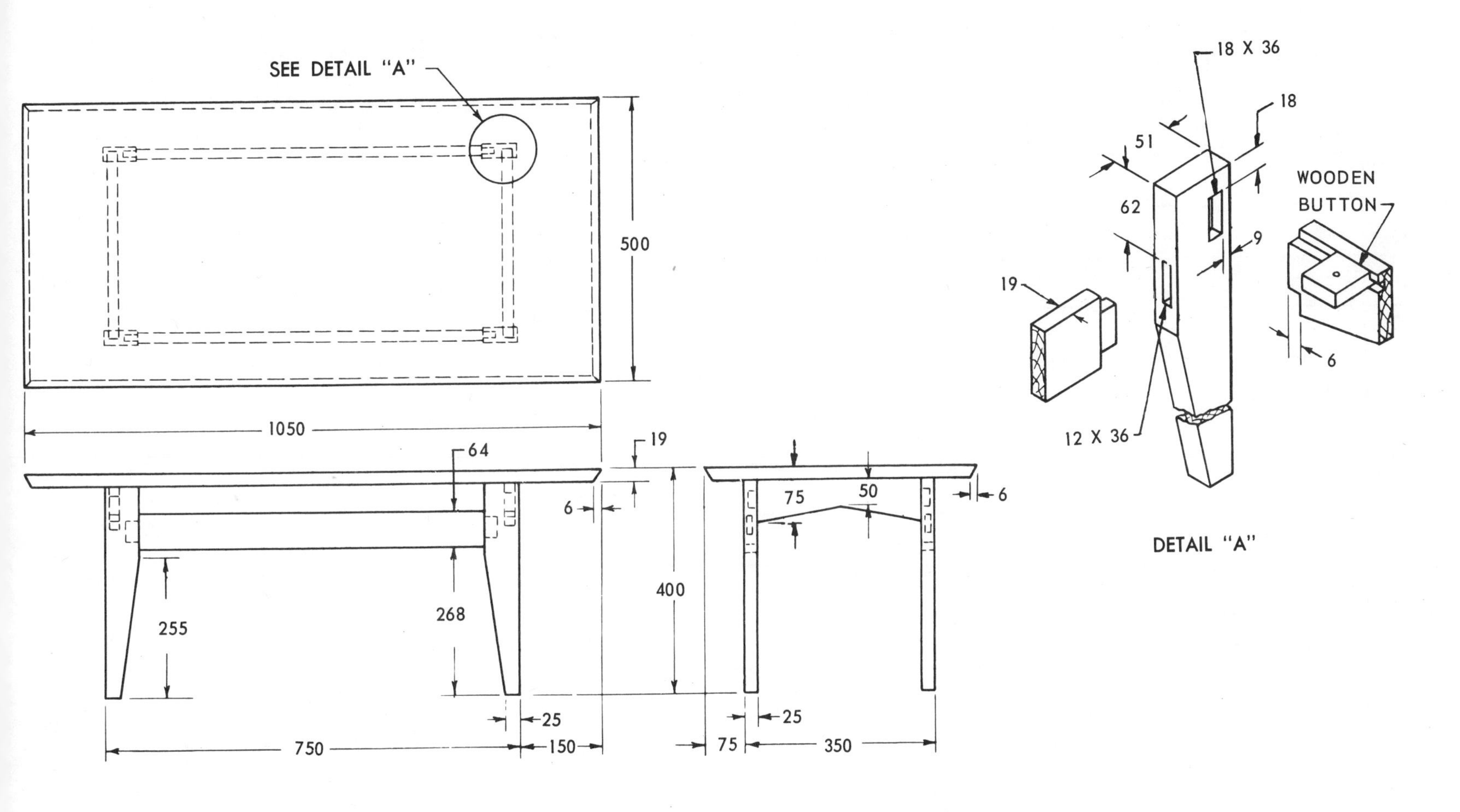

Fig. 5-3-4 Combined detail and assembly drawing

Detail Assembly Drawings

Detail assembly drawings are often made for fairly simple objects, such as pieces of furniture, when the parts are few in number and not intricate in shape. All the dimensions and information necessary for the construction of each part and for the assembly of the parts are given directly on the assembly drawing. Separate views of specific parts, or enlargements showing the fitting together of parts, may also be drawn in addition to the regular drawing. Note that in Figure 5-3-4 the enlarged views are drawn in picture form and not as regular orthographic views. This method is peculiar to the cabinet-making trade and is not normally used in mechanical drawing.

QTY	ITEM	MATL	DESCRIPTION	PT NO
1	BASE	CI	PATTERN # A3154	1
1	CAP	CI	PATTERN # B87156	2
1	SUPPORT	MS	10 X 50 X 110	3
1	BRACE	MS	6 X 25 X 50	4
1	COVER	ST	3.57 (10 USS GA) X 150	5
1	SHAFT	CRS	ϕ 25 X 160	6
2	BEARINGS	SKF	RADIAL BALL 62002	7
2	RETAINING RING	TRUARC	N5000 - 725	8
1	KEY	ST	WOODRUFF 608	9
1	SET SCREW CUP		HEX SOCKET M6 X 10 LG	10
4	BOLT - HEX HD - REG	SEMI - FIN	M10 X 40 LG	11
4	NUT - REG - HEX	ST	M10	12

(A) TYPICAL BILL OF MATERIAL
PTS 7 TO 12 ARE PURCHASED ITEMS

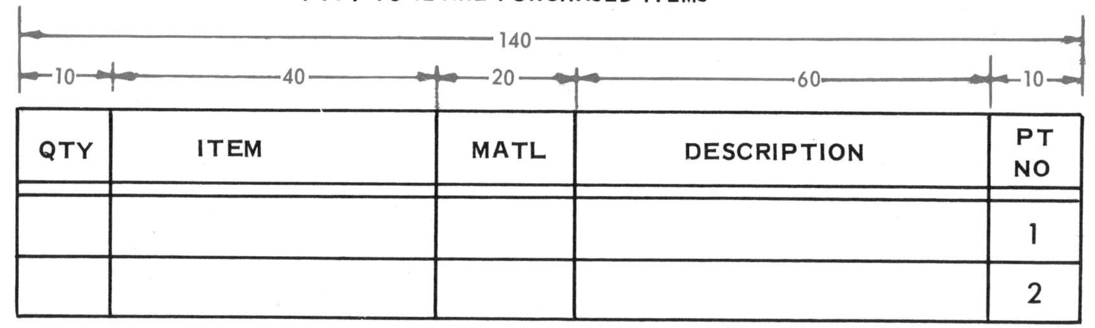

QTY	ITEM	MATL	DESCRIPTION	PT NO
				1
				2

(B) SAMPLE SIZES

Fig. 5-3-5 Bill of material

Bills of Materials

A bill of material is an itemized list of all the components shown on an assembly drawing or on a detail drawing. (See Fig. 5-3-5). Often a bill of material is placed on a separate sheet for ease of handling and duplicating. As the bill of material is used by the purchasing department to order the necessary material for the design, it is necessary to show in the bill of material the raw material size rather than the finished size of the part.

For castings, a pattern number should appear in the size column in lieu of the physical size of the part.

Standard components, such as bolts, nuts and bearings, which are purchased ready-made, should have a part number and appear in the bill of material. Sufficient information should be shown in the descriptive column to enable the purchasing agent to order correct parts.

Parts lists for bills of materials placed on the bottom of the drawing should read from bottom to top, while bills of materials placed on the top of the drawings should read from top to bottom. This practice allows additions to be made at a later date.

Assignments

1. Book rack, Fig. 5-3-A, sheet size-A3, scale 1:5. Make a detailed assembly drawing, complete with a bill of materials.

2. Saw horse, Fig. 5-3-B, sheet size-A3, scale 1:5. Make a detailed assembly drawing, complete with a bill of materials.
3. Caster, Fig. 5-3-C, sheet size-A3, scale 1:1. Make detailed drawings of parts 1, 2, 3, 4 and 5 on a common sheet.

Review for Assignment

Unit 2-2 Linework and lettering
Unit 3-1 Shape description by views
Unit 4-1 Dimensioning

Fig. 5-3-A Book rack

Fig. 5-3-B Saw horse

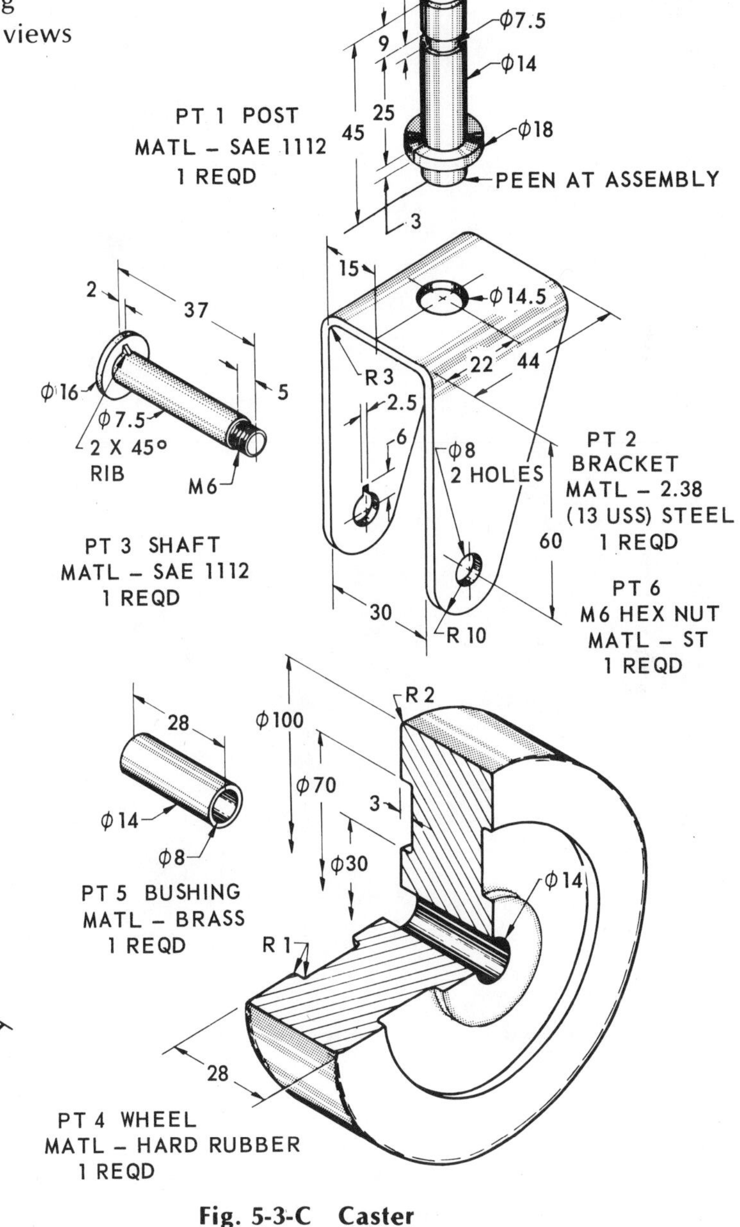

Fig. 5-3-C Caster

UNIT 5-4
Drawing Reproduction

A revolution in reproduction technologies and methods began in the 1940s and 1950s. It brought with it new equipment and supplies which have made quick copying commonplace. The new technologies make it possible to apply improved systems approaches and new information handling techniques to all types of files ranging from small documents to large engineering drawings. (See Fig. 5-4-1).

For many years, the most popular reproduction process was blueprinting. Although durable and waterproof, blueprints had to be washed and dried, and duplicates could only be made as negatives of the originals.

Diazo (Whiteprint)

The demand for a faster and more versatile type of reproduction was responsible for the introduction of the diazo process (white prints) of duplicating drawings. The diazo process was faster than blueprinting and produced copies which were direct duplicates of the originals. It was reliable, inexpensive and, as a result, enjoyed rapid acceptance (Fig. 5-4-2 and 5-4-3).

With this type of duplicating process the drawing is made on a translucent material . . . paper, cloth or plastic. The drawing, which is referred to as the tracing, is placed on sensitized print paper and, while being held still, is exposed to a strong light for a predetermined length of time. The length of time depends on the strength of the light and the printing speed of the print paper.

The light penetrates through the tracing, except where the ink or lead shields the chemicals on the print paper from the light,

Fig. 5-4-1 Drawing reproduction

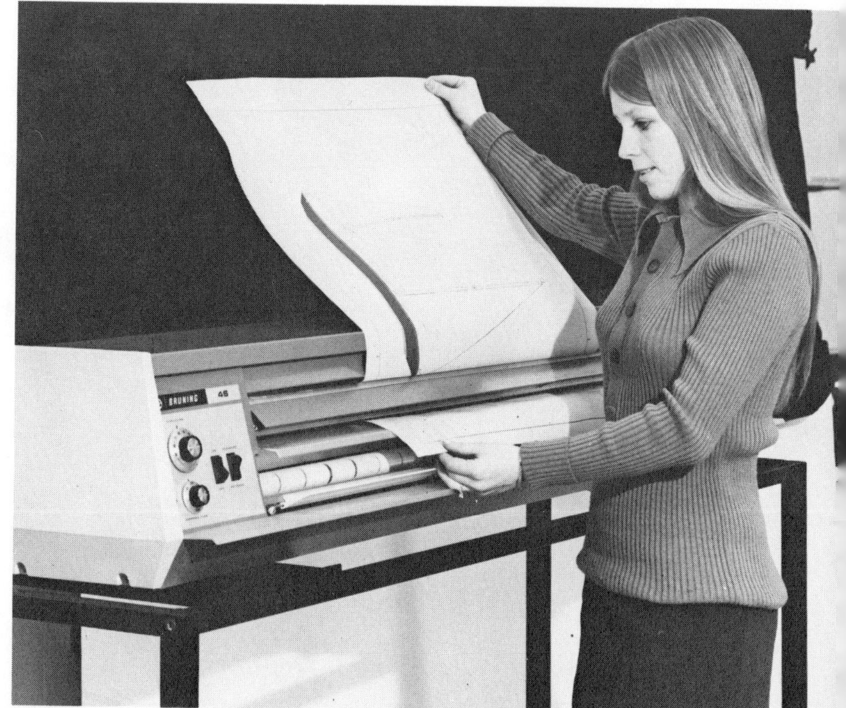

Fig. 5-4-2 A whiteprint machine in use

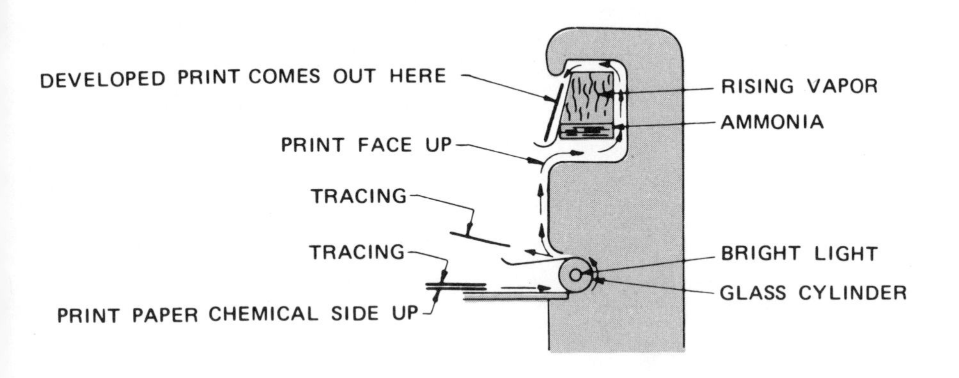

Fig. 5-4-3 The Diazo printing process

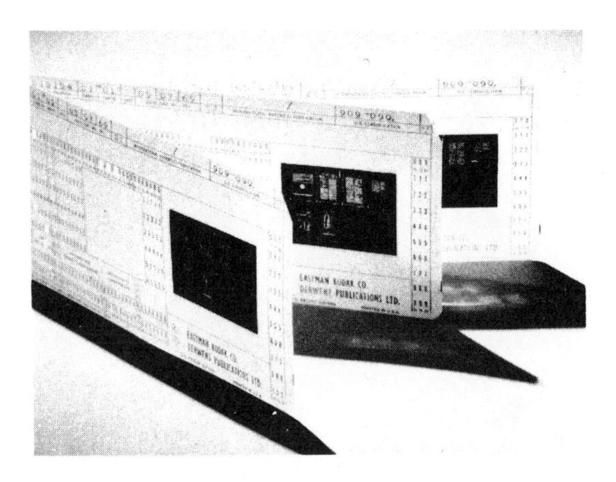

Fig. 5-4-4 Aperature cards

and dissolves the chemical on the print paper, creating a latent image on the copy paper.

There are several methods for developing the latent image . . . (1) by conveying the copy paper across ammonia vapour (dry process), (2) by passing the copy through developer rollers (moist process) where a chemical solution is applied to the emulsion, or (3) by pressure diazo.

Microfilm

Microfilming of engineering drawings is now an established practice in many drafting offices. (See Fig. 5-4-4). This has come about because of the primary savings in lower transportation, labour, and storage costs of microfilm.

Fig. 5-4-5 Reader-printer for microfilm

Another advantage is that the drafter can draw on practically any type of paper since the drawing is photographed and prints or tracings can be made from the original work.

Such photo-copying equipment is excellent for scissors and paste-up drafting.

A drawing can be stored in a fraction of the space needed for filing the full-size paper originals. It can be referenced on the screen of a reader, or, if a print is required, it can be blown back on an enlarger-printer.

Photo-reproduction Process

Photo-reproduction has proven to be another of the newest and most versatile methods for reproducing engineering drawings. One such printer offers, in addition to 1:1 printing, five reduction settings by which a drawing, or part of a drawing, can be physically reduced in size. Drawing reductions allow such advantages as less paper use, lower handling and mailing costs, and smaller filing space requirements.

Fig. 5-4-6 Photoreproduction process. This machine will reproduce, reduce, fold and sort prints

Chapter 6

Fastening Devices

UNIT 6-1

Thread Forms and Their Pictorial Representation

Fastening devices are important in the construction of manufactured products, in the machines and devices used in manufacturing processes, and in the construction of all types of buildings. Fastening devices are used in the smallest watch and the largest ocean liner. (See Fig. 6-1-1).

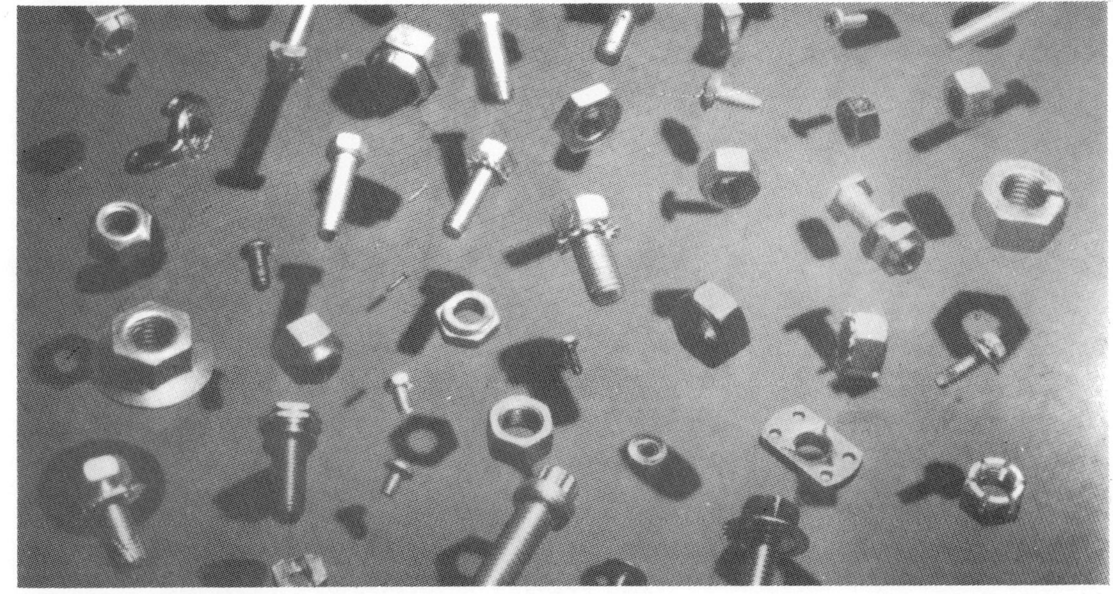

Fig. 6-1-1 Fasteners

There are two basic kinds of fastening. Rivets and welds are **permanent fastenings;** bolts, screws, studs, nuts, pins and keys are **removable fastenings.**

As industry progressed, fastening devices became standardized, and they have definite characteristics and names. A thorough knowledge of the design and graphic representation of common fasteners is an essential part of drafting.

Screw Threads

Threaded fastenings, such as screws, bolts, studs, and nuts, are manufactured in a great variety of forms and sizes. See the charts in the appendix for the dimensions and specifications of standard threaded fasteners, such as machine screws, cap screws, bolts, nuts, and studs.

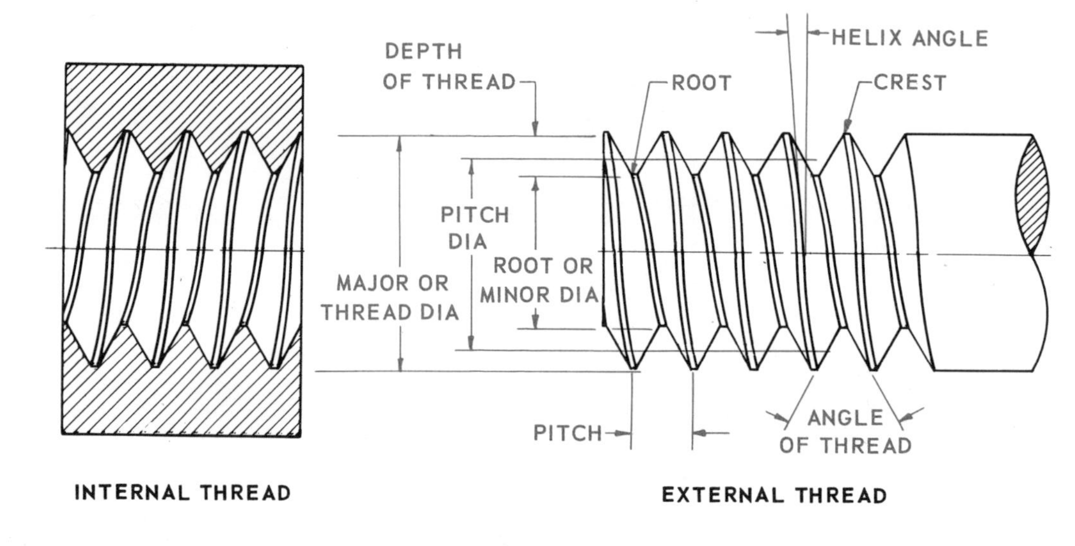

Fig. 6-1-3 Screw thread terms

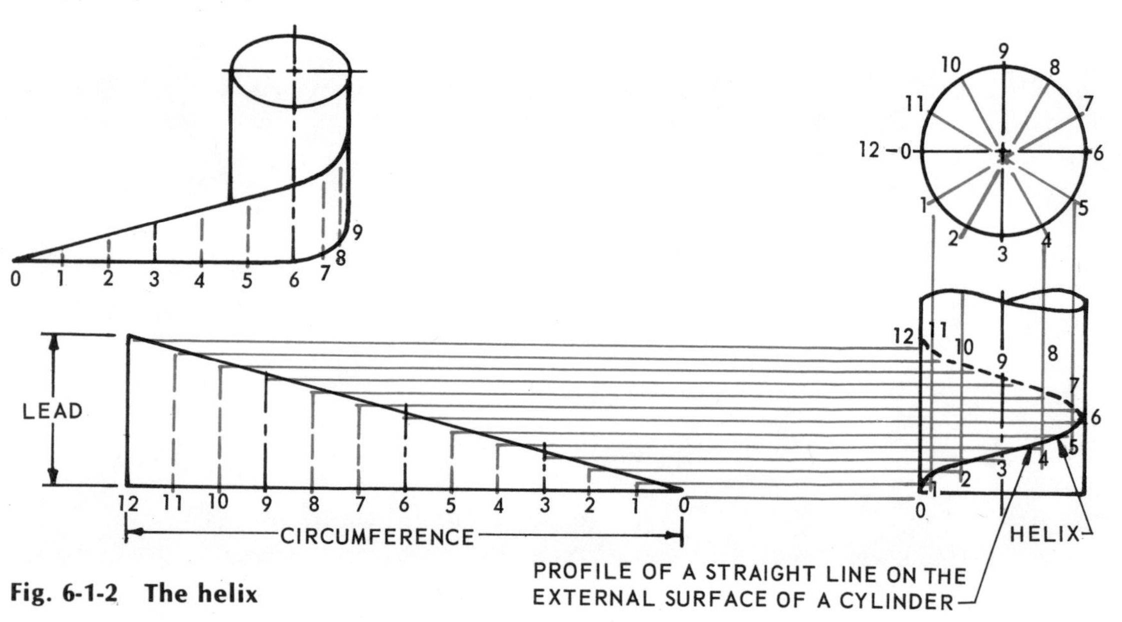

Fig. 6-1-2 The helix

A screw thread is a continuous ridge of uniform section in the form of a helix on the external or internal surface of a cylinder.

A **helix** (plural, **helices**) is a curve generated by a point moving uniformly about a cylinder and uniformly parallel to the axis of the cylinder. The principle of the helix curve can be demonstrated by winding a piece of string about a transparent glass cylinder, such as a test tube or a straight-sided drinking glass (See Fig. 6-1-2).

The **pitch** of a thread **P** is the distance from a point on the thread form to the corresponding point on the next form, measured parallel to the axis (Fig. 6-1-3). The **lead L** is the distance the threaded part would move parallel to the axis during one complete rotation in relation to a fixed mating part (the distance a screw would enter a threaded hole in one turn).

Thread Forms

Figure 6-1-4 shows some of the more common thread forms in use today. The ISO metric thread form shown in Figure 6-1-4 has now replaced all the former V-shaped metric and inch threads. As for the other thread forms shown, the proportions will be the same for both metric and inch-size threads.

The sharp V thread form is seldom used for general fastening purposes, but it is sometimes used in precision instruments.

Fig. 6-1-5 Application of a knuckle thread

The knuckle thread is usually rolled or cast. A familiar example of this form is seen on electric light bulbs and sockets.

The square and acme forms are designed to transmit motion or power, as on the lead screw of a lathe.

The buttress thread takes pressure in only one direction — against the surface perpendicular to the axis.

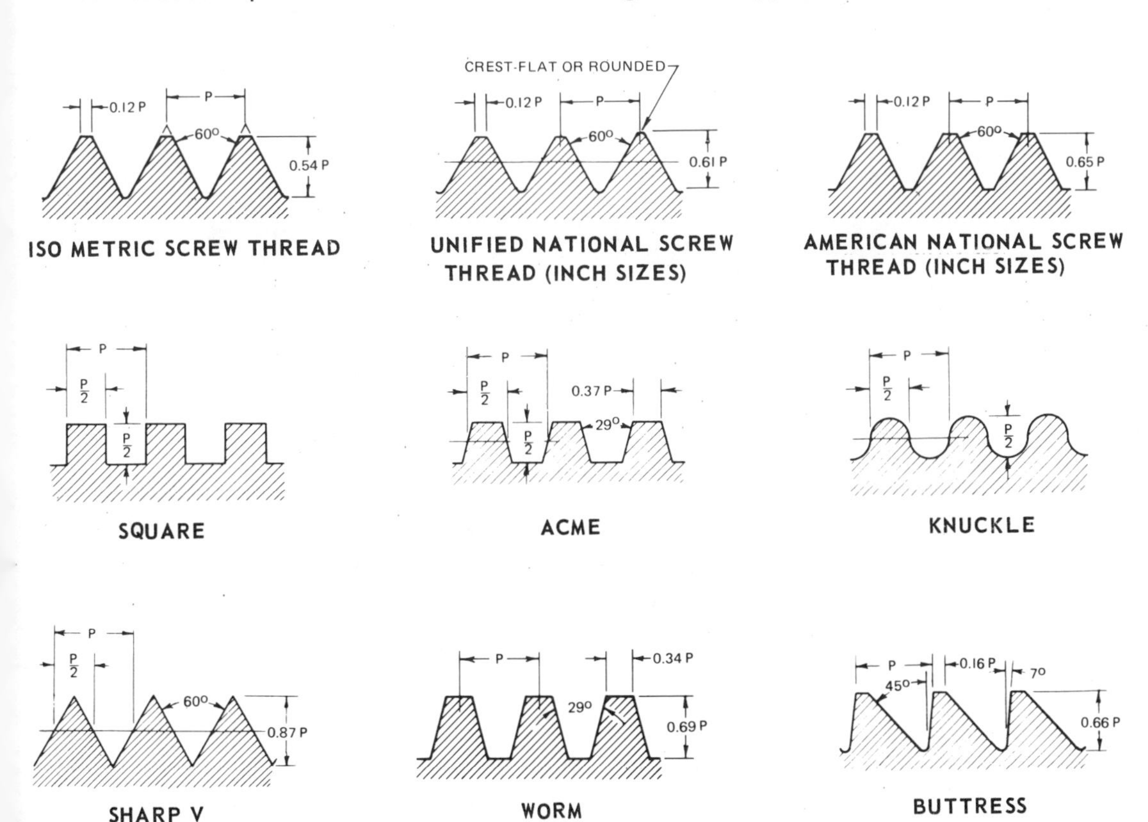

Fig. 6-1-4 Standard thread forms

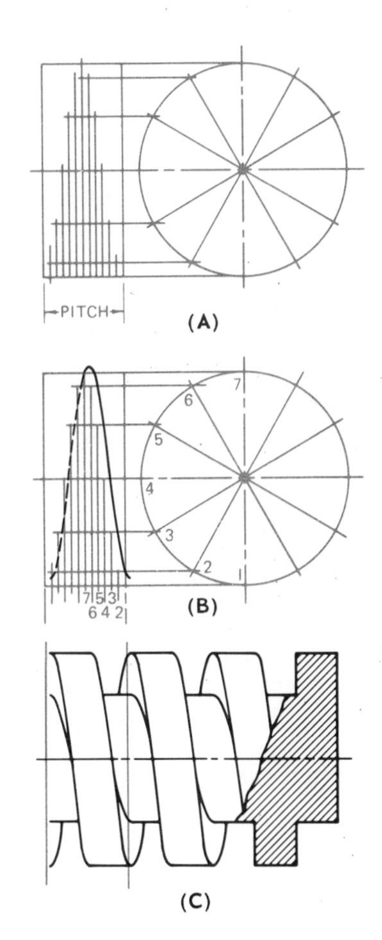

Fig. 6-1-6 The helix of a square thread

Pictorial Representation

A close approximation to the actual appearance of the screw thread is shown in Figure 6-1-7 and 6-1-8. It is simplified in that crests and roots for full threads are shown sharp, with a single straight line, instead of a double curved line which would be required for flat crests and roots.

Pictorial representation should be used only for enlarged detail and other special applications.

Pictorial Representation of Square Threads. The depth of the square thread is one-half the pitch. In Figure 6-1-7A, lay off spaces equal to *P*/2 along the diameter and add light lines which indicate the depth of thread. At B draw the crest lines. At C draw the root lines, as shown. At D the internal square thread is drawn in section. Note the reverse direction of the lines.

Pictorial Representation of Acme Threads. The depth of the acme thread is one-half the pitch (Fig. 6-1-7). The stages in drawing acme threads are shown in E. The pitch diameter is midway between the outside diameter and the root diameter and locates the pitch line. On the pitch line, lay off half-pitch spaces and draw the root lines to complete the view. The construction shown at F is enlarged.

A sectional view of an internal acme thread is shown at G. Other representations used for internal threads are hidden lines and sections. These are shown at H.

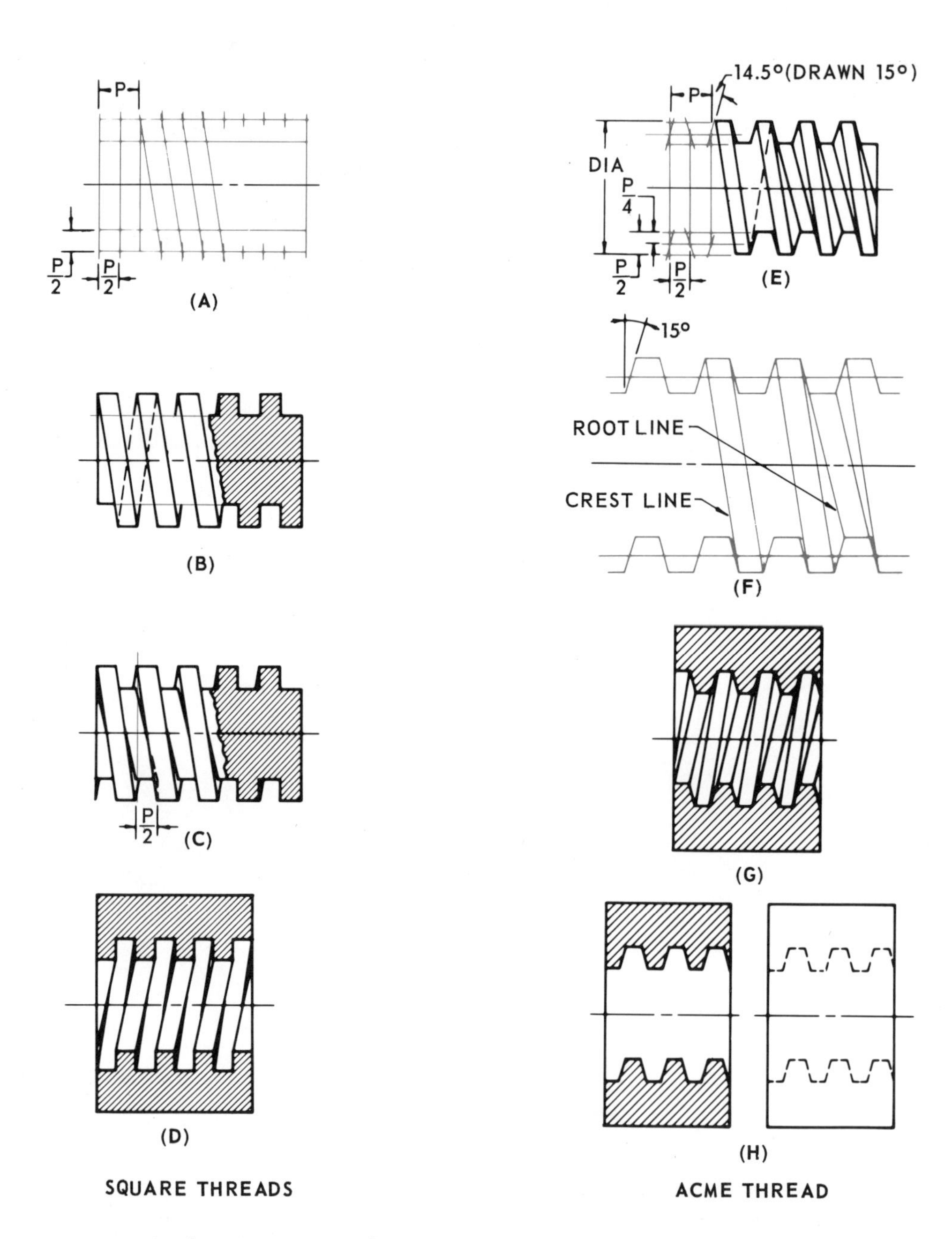

Fig. 6-1-7 Steps in drawing pictorial representation of square and Acme threads

Pictorial Representation of Screw Threads. The pictorial representation uses the sharp-V profile. Straight lines are used to represent the helices of the crest and root lines.

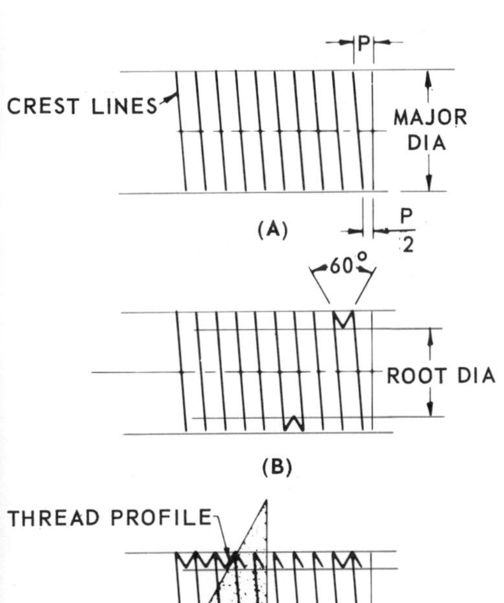

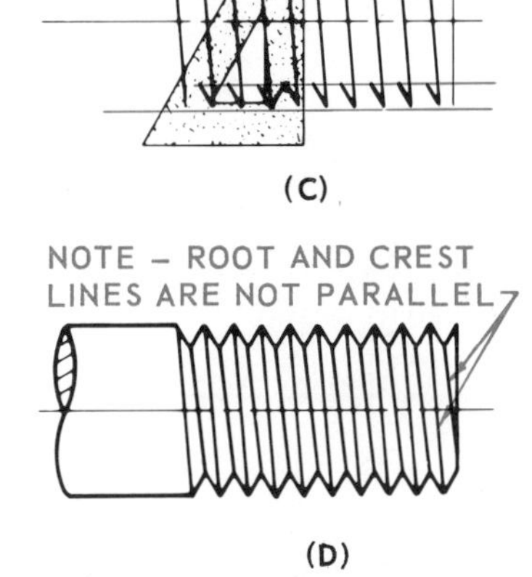

Fig. 6-1-8 Steps in drawing pictorial representation of V-shaped threads

The order of drawing the V-form thread is shown in Figure 6-1-8. The pitch is seldom drawn to scale. Lay off the pitch *P* and the half-pitch *P*/2, as shown. Adjust the set square to the slope, and draw the crest lines (if a drafting machine is used, set the ruling arm to the slope of the crest line). At B draw the V profile for one thread, top and bottom, locating the root diameter. Draw light construction lines for the root diameter. At C set the ruling face of the 30° set square and draw one side of the remaining V's (thread profile). Reverse the set square and draw the other sides of the V's, completing the thread profile. At D, draw the root lines, which complete the pictorial representation of the threads.

Single and Multiple Threads

Most screws have single threads; it is assumed that, unless the thread is designated otherwise, it is a single thread. The single thread has a single ridge in the form of a helix. (See Fig. 6-1-9). The **lead** of a thread is the distance travelled parallel to the axis in one rotation of a part in relation to a fixed mating part (the distance a nut would travel along the axis of a bolt with one rotation of the nut). In single threads the lead is equal to the pitch.

A double thread has two ridges, started 180° apart, in the form of helices, and the lead is twice the pitch.

A triple thread has three ridges, started 120° apart, in the form of helices, and the lead is three times the pitch.

Double and triple threads are used where fast movement is desired with a minimum number of rotations, such as on threaded mechanisms for opening and closing windows.

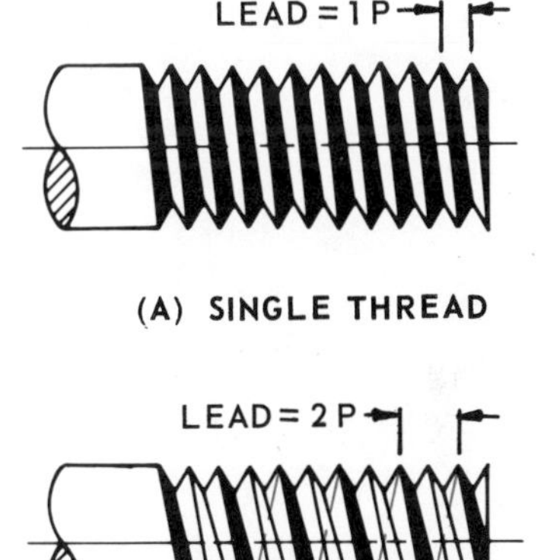

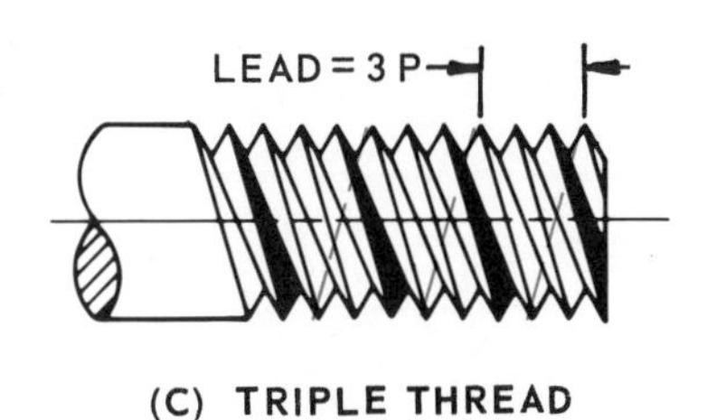

Fig. 6-1-9 Single and multiple threads

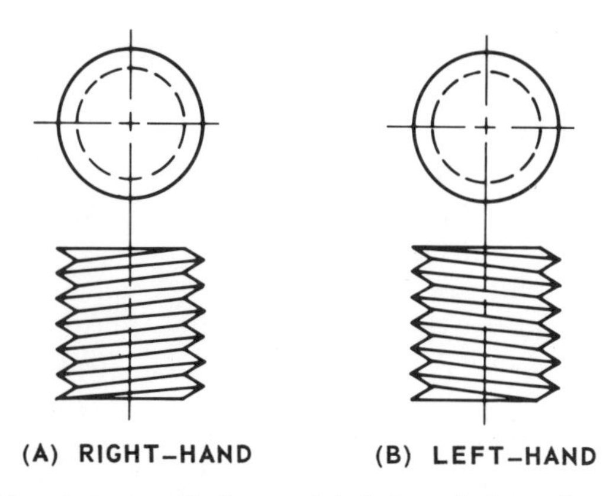

Fig. 6-1-10 Right- and left-hand threads

Right- and Left-Handed Threads

Unless designated otherwise, threads are assumed to be right-hand. A bolt being threaded into a tapped hole would be turned in a right-hand (clockwise) direction. (See Fig. 6-1-10).

For some special applications, such as turnbuckles, left-hand threads are required. When such a thread is necessary, the letters LH are added after the thread designation.

Assignments

Prepare working drawings showing pictorial representation of the threads. Sheet size-A4, scale 1:1.

1. Plug, Fig. 6-1-A. (See Fig. 6-1-4 and Fig. 6-1-7 and Unit 16 for drawing a hexagon).
2. Fuse assembly, Fig. 6-1-B. (See Fig. 6-1-4).
3. Jack screw, Fig. 6-1-C. (See Fig. 6-1-4). Use a conventional break to reduce the length on the drawing.
4. Guide rod, Fig. 6-1-D. (See Fig. 6-1-4 and Fig. 6-1-8). Use a conventional break to reduce the length on the drawing.

Review for Assignments

Chapter 3 Theory of shape description
Chapter 4 Basic dimensioning
Chapter 5 Working drawings

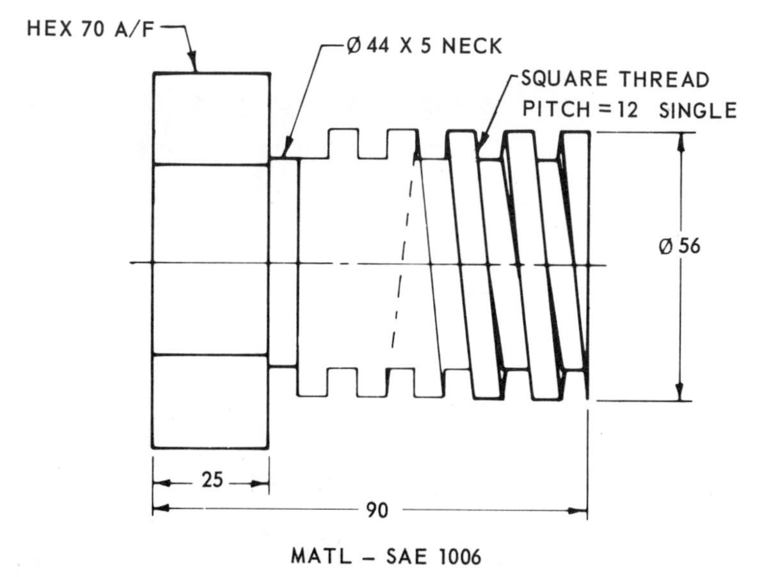

Fig. 6-1-A Plug

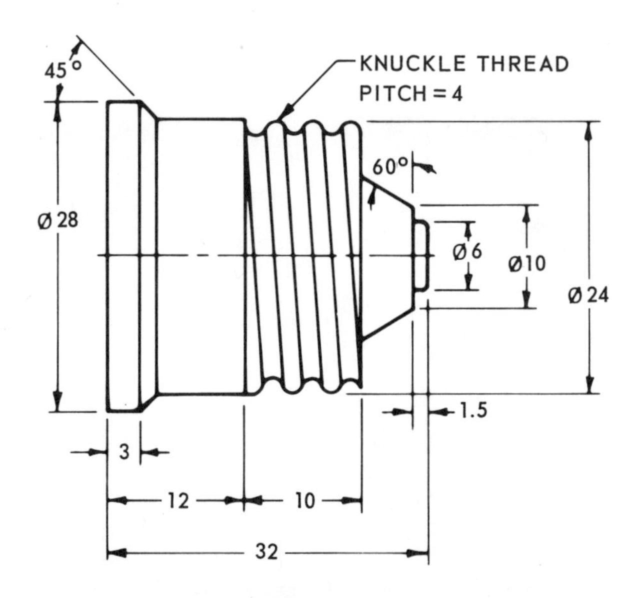

Fig. 6-1-B Fuse assembly

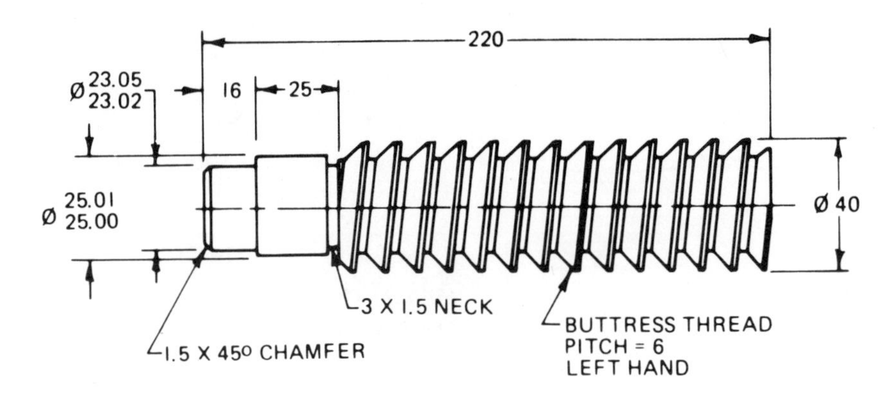

Fig. 6-1-C Jack screw

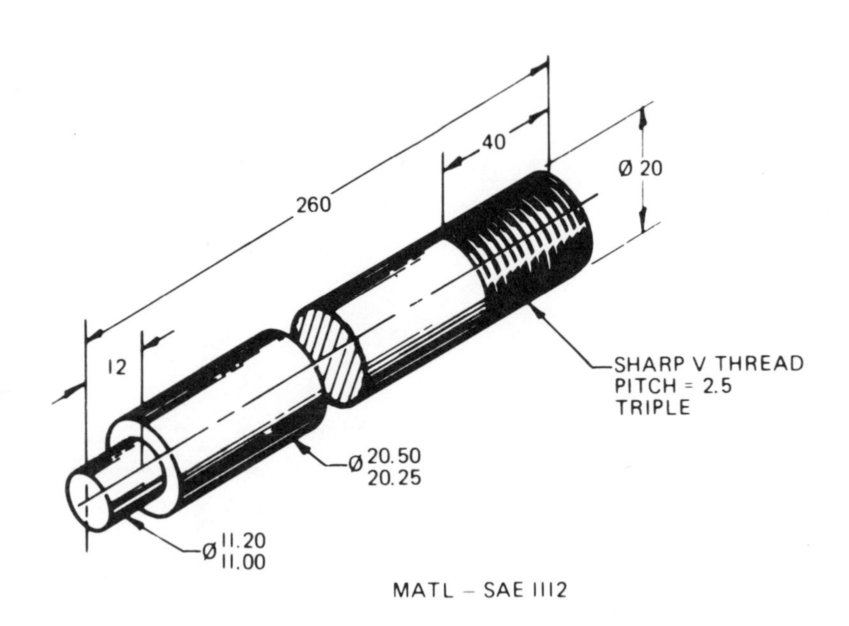

Fig. 6-1-D Guide rod

UNIT 6-2
Conventional Thread Representation

True representation of a screw thread is never shown on working drawings as it would require very laborious and accurate drawing, involving repetitious development of the helix curve of the thread. A conventional representation for showing threads on drawings should be used whenever it conveys the required information without confusion, as it requires the least amount of drafting effort.

Conventional Representation

The conventional representation of threaded parts on technical drawings is shown in Figure 6-2-1. This method may be used for any type of screw thread required. The type of screw thread and its dimensions are to be indicated by the designations shown later in this Unit.

In this system the thread crests, except in hidden views, are represented by a thick outline, and the thread roots by a thin continuous line. In the end views the root line extends approximately 270° or three quarters of a complete circle but should not start or end on a centre line.

The end of the full form thread is indicated by a thick line across the part, and imperfect or runout threads are shown beyond this line by running the root line at an

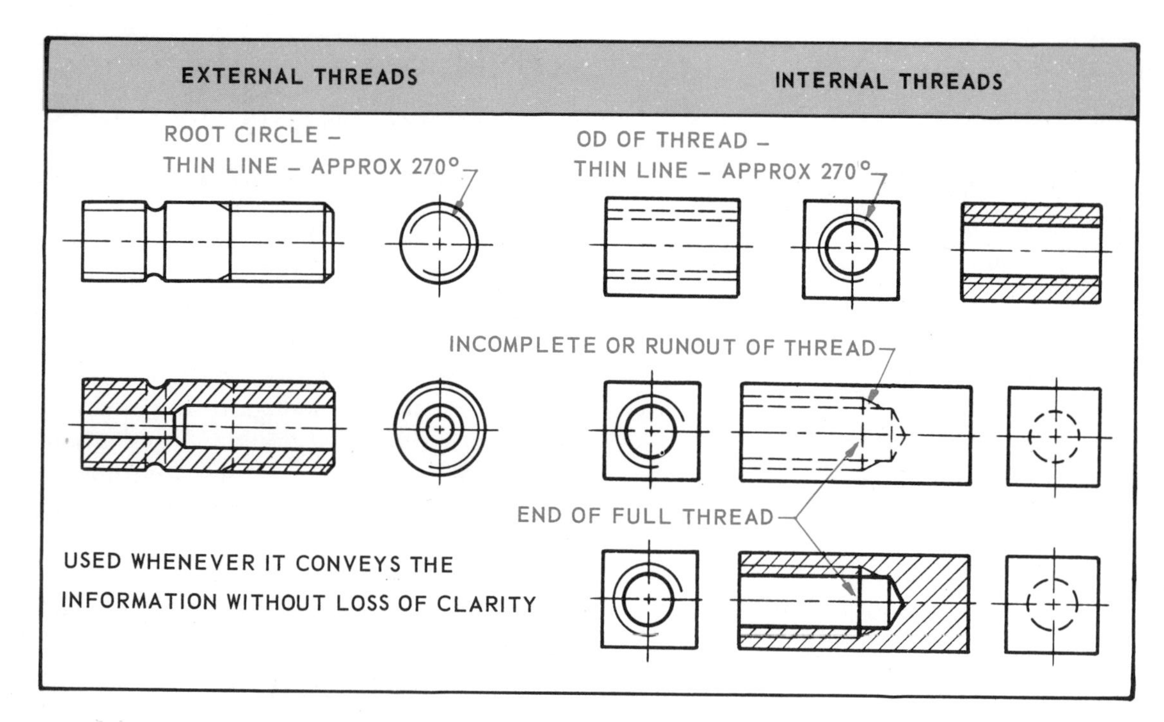

Fig. 6-2-1 Conventional thread representation

INTERIOR VIEW

EXTERIOR VIEW

USED ON ENLARGED DETAIL AND OTHER SPECIAL APPLICATIONS

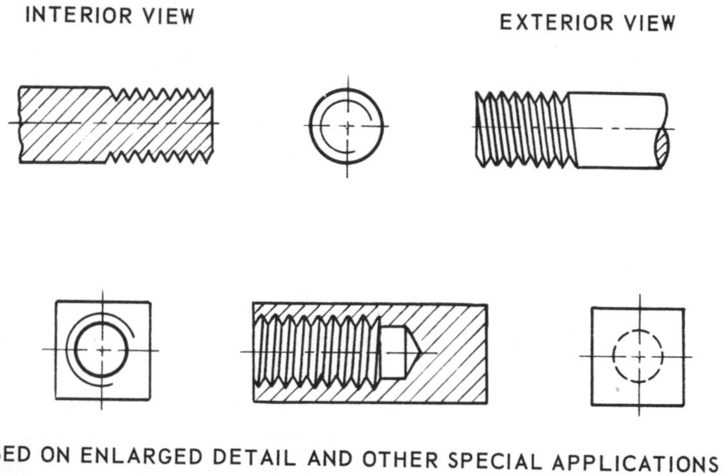

Fig. 6-2-2 Alternate representation (pictorial)

angle to meet the crest line. If the length of runout thread is unimportant, this portion of the thread representation may be omitted.

Hidden threads are indicated on drawings by showing both the crest and root lines as thin broken lines.

Alternative Representation (Pictorial)

If it is desirable for clarity, for example on certain assembly drawings and other special applications, the method shown in Figure 6-2-2 may be used.

This represents the actual appearance of the screw thread. It is simplified in that the crest and roots for full threads are shown sharp, with single straight lines, instead of double curved lines which would be required for flat crests and roots. It is not necessary to draw the pitch to the exact scale.

Threaded Assemblies

In sectional views, the externally threaded part is always shown covering the internally threaded part, as shown in Figure 6-2-3. If it is desirable for clarity (eg. on certain assembly drawings), the alternate representation may be used. Both methods of indicating threads may be used on the same drawing.

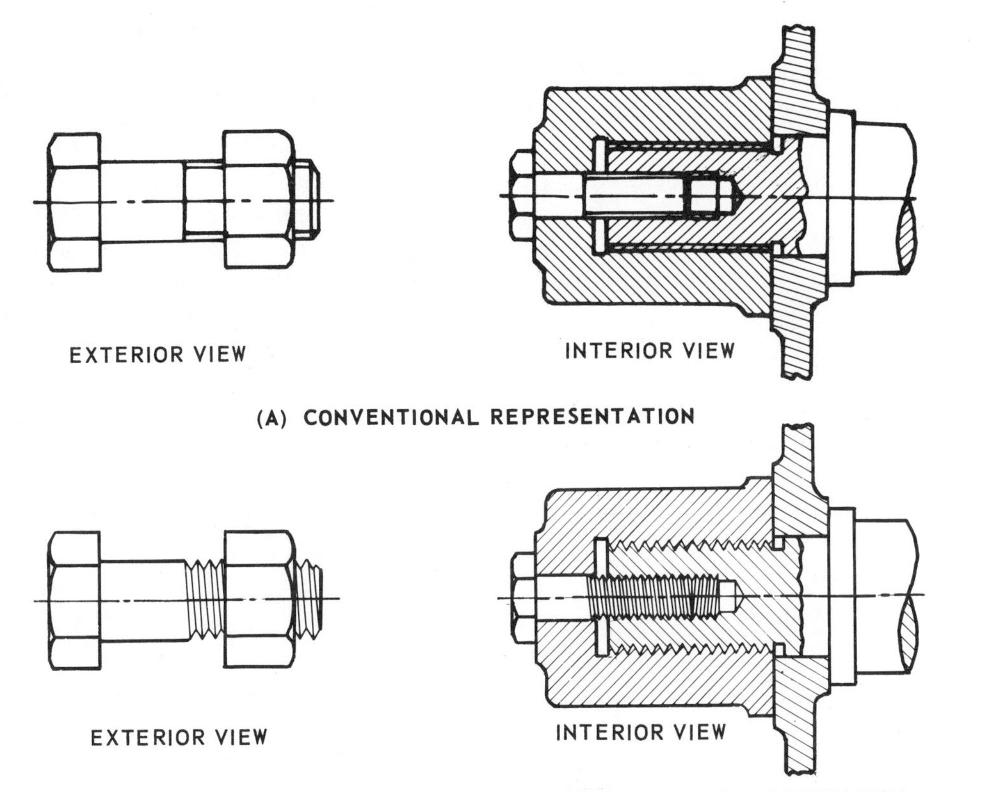

Fig. 6-2-3 Conventional and alternate representation of threads in assembly drawings

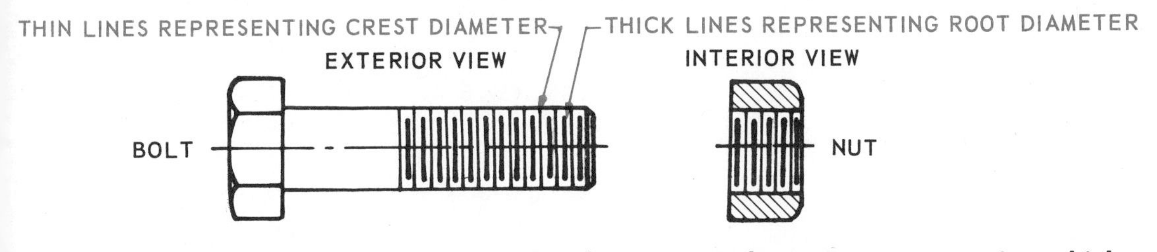

Figure 6-2-4 shows the thread representation known as *schematic representation* which was used in the past. As it took longer to draw than the conventional representation, the Canadian Standards Association recommended to discontinue its use.

Thread Standards

With the progress and growth of industry, a need has grown for uniform, interchangeable threaded fasteners. The factors that influence standards, aside from the threaded forms previously mentioned, are the **pitch** of the thread and the **major diameter.**

Metric Threads. Metric threads are grouped into diameter-pitch combinations distinguished from each other by the pitch applied to specific diameters. The "pitch" for metric threads is the distance between corresponding points on adjacent teeth. In addition to a coarse and fine pitch series, a series of constant pitches is available.

Inch Threads. Up to 1976, practically all threaded assemblies on this continent were designed using inch-sized threads. In this system the pitch is equal to

$$\frac{1}{\text{Number of threads per inch.}}$$

The number of threads per inch is set for different diameters in what is called a thread "series." For the Unified National system there is the coarse thread series and the fine thread series.

In addition, there is an extra fine thread series, UNEF, for use where a small pitch is desirable, such as on thin-walled tubing. For special work and for diameters larger than those specified in the coarse and fine series, the Unified thread system has three series that provide for the same number of threads per inch regardless of the diameter. These are the 8, 12, and 16-thread series.

Thread Grades and Classes

The **fit** of a screw thread is the amount of clearance between the screw and the nut when they are assembled together.

Metric Threads. For each of the two main thread elements, pitch diameter and crest diameter, a number of tolerance grades have been established. The number of the tolerance grades reflects the size of the tolerance. For example: Grade 4 tolerances are smaller than Grade 6 tolerances and Grade 8 tolerances are larger than Grade 6 tolerances.

In each case, Grade 6 tolerances should be used for medium quality length of engagement applications. The tolerance grades below Grade 6 are intended for applications involving fine quality and/or short lengths of engagement. Tolerance grades above Grade 6 are intended for coarse quality and/or long lengths of engagement.

In addition to the tolerance grade a positional tolerance is required. This defines the maximum-material limits of the pitch and crest diameters of the external and internal threads and indicates their relationship to the basic profile.

Two factors affecting tolerances are current coating (or plating) thickness requirements and the demand for ease of assembly. Therefore a series of tolerance positions which require varying amounts of allowance has been established as follows:

For external threads:

Tolerance position "e" (large allowance)
Tolerance position "g" (small allowance)
Tolerance position "h" (no allowance)

For internal threads:

Tolerance position "G" (small allowance)
Tolerance position "H" (no allowance)

Inch Threads. Three classes of external thread (Classes 1A, 2A, and 3A) and three classes of internal thread (Class 1B, 2B, and 3B) are provided. These classes differ from each other in the amount of the allowances and tolerances provided in each class.

The general characteristics and uses of the various classes are as follows:

Classes 1A and 1B. These classes produce the loosest fit, that is, the greatest amount of play in assembly. They are useful for work where ease of assembly and disassembly is essential, such as for some ordnance work and for stove bolts and other rough bolts and nuts.

Classes 2A and 2B. These classes are designed for the ordinary good grade of commercial products, such as machine screws and fasteners, and for most interchangeable parts.

Classes 3A and 3B. These classes are intended for exceptionally high-grade commercial products, where a particularly close or snug fit is essential and the high cost of precision tools and machines is warranted.

Thread Designation

Metric Threads

ISO metric screw threads are defined by the nominal size (basic major diameter) and pitch, both expressed in millimetres (Fig. 6-2-5). An "M" specifying an ISO metric screw thread precedes the nominal size and an "x" separates the nominal size from the pitch.

For the course thread series only, the pitch is not shown unless the dimension for the length of the thread is required. When specifying the length of thread an "x" is used to separate the length of thread from the rest of the designations.

For external threads, the length of thread may be given as a dimension on the drawing.

For example, a 10 mm diameter, 1.25 pitch, fine thread series is expressed as M10 × 1.25. A 10 mm diameter, 1.5 pitch, coarse thread series is expressed as M10; the pitch is not shown unless the length of thread is required. If the latter thread was 25 mm long and this information was required on the drawing then the thread callout would be M10 × 1.5 × 25.

A complete designation for an ISO metric screw thread comprises in addition to the basic designation, an identification for the tolerance class. The tolerance class designation is separated from the basic designation by a dash and includes the symbol for the pitch diameter tolerance followed immediately by the symbol for crest diameter tolerance.

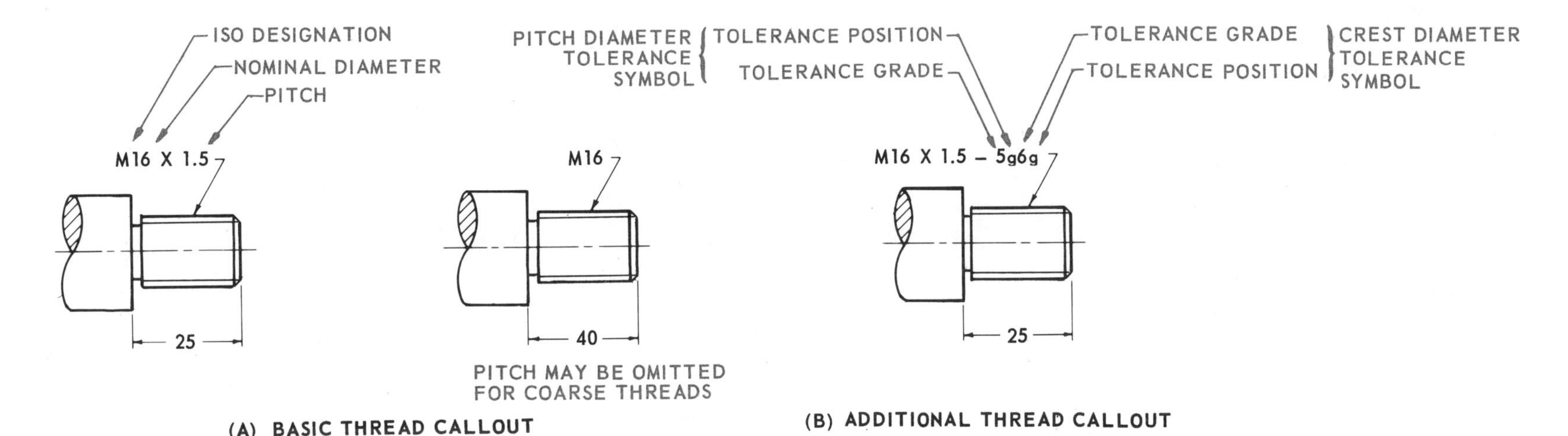

Fig. 6-2-5 Metric thread designation

Each of these symbols consists of first a numeral indicating the grade tolerance followed by a letter indicating the tolerance position (a capital letter for internal threads and a lower case letter for external threads).

Where the pitch and crest diameter symbols are identical, the symbol need only be given once and not repeated.

Inch Threads

Thread designation for inch threads, whether external or internal, is expressed in this order: diameter (nominal or major diameter), number of threads per inch, thread form and series, and class of fit (Fig. 6-2-6).

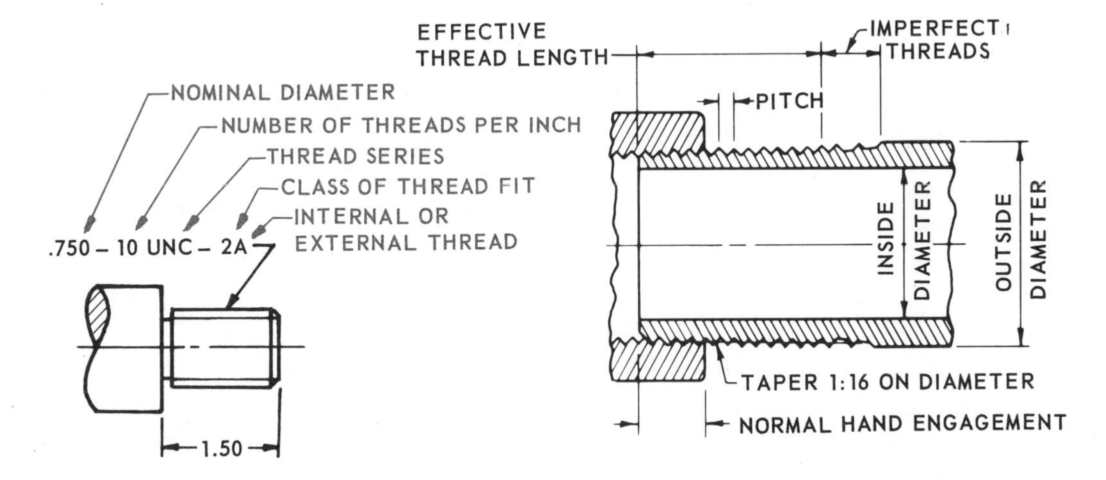

Fig. 6-2-6 Inch thread designation

Fig. 6-2-7 Pipe thread terminology

Pipe Threads

Pipe sizes and pipe threads have not been changed with metric conversion, although the outside diameter and wall thickness are now given in millimetres. In calling for the size of thread, the note used is similar to that for screw threads. See Figure 6-2-7.

Example 1. 4 × 8NPT

where 4 = nominal diameter of pipe, in inches
8 = number of threads per inch
N = American Standard
P = pipe
T = taper thread

Assignments

Use conventional representation for all threads.

1. Parallel clamp, Fig. 6-2-A, sheet size-A4, scale 1:1. Make a two-view assembly drawing showing the clamp at maximum jaw opening.
2. Parallel clamp, Fig. 6-2-A, sheet size - A3, scale 1:1. Make detail drawings of each of the parts shown.
3. Turnbuckle, Fig. 6-2-B, sheet size - A3, scale 1:1. Make detail drawings of each of the parts shown. Draw only one drawing of the eye bolts and use the opposite-hand convention of showing their differences.
4. Pipe plug, Fig. 6-2-C, sheet size-A4, scale 1:2. Make a two-view working drawing.

Review for Assignment

Chapter 3 Theory of shape description
Chapter 4 Basic dimensioning
Chapter 5 Working drawings

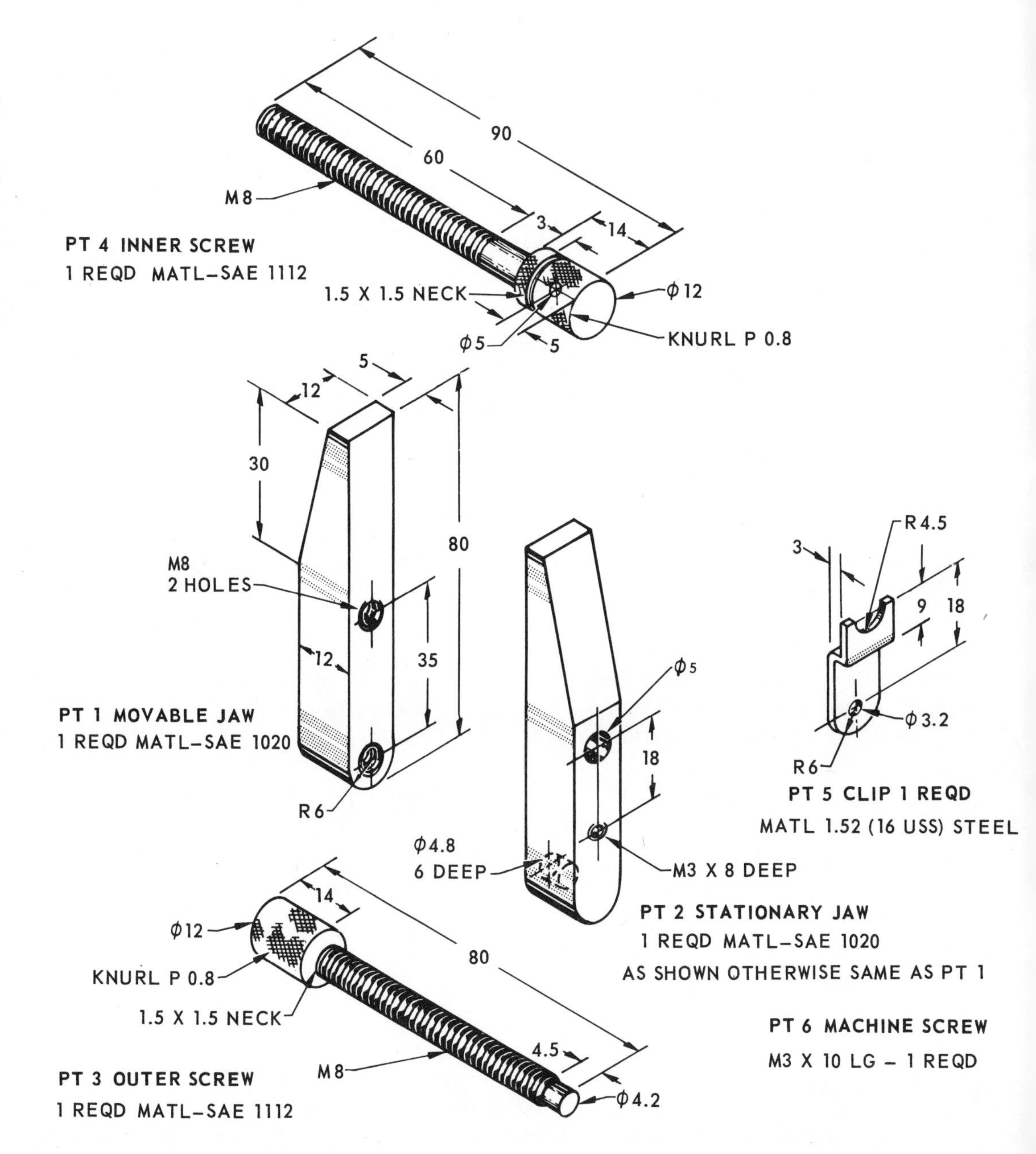

Fig. 6-2-A Parallel clamps

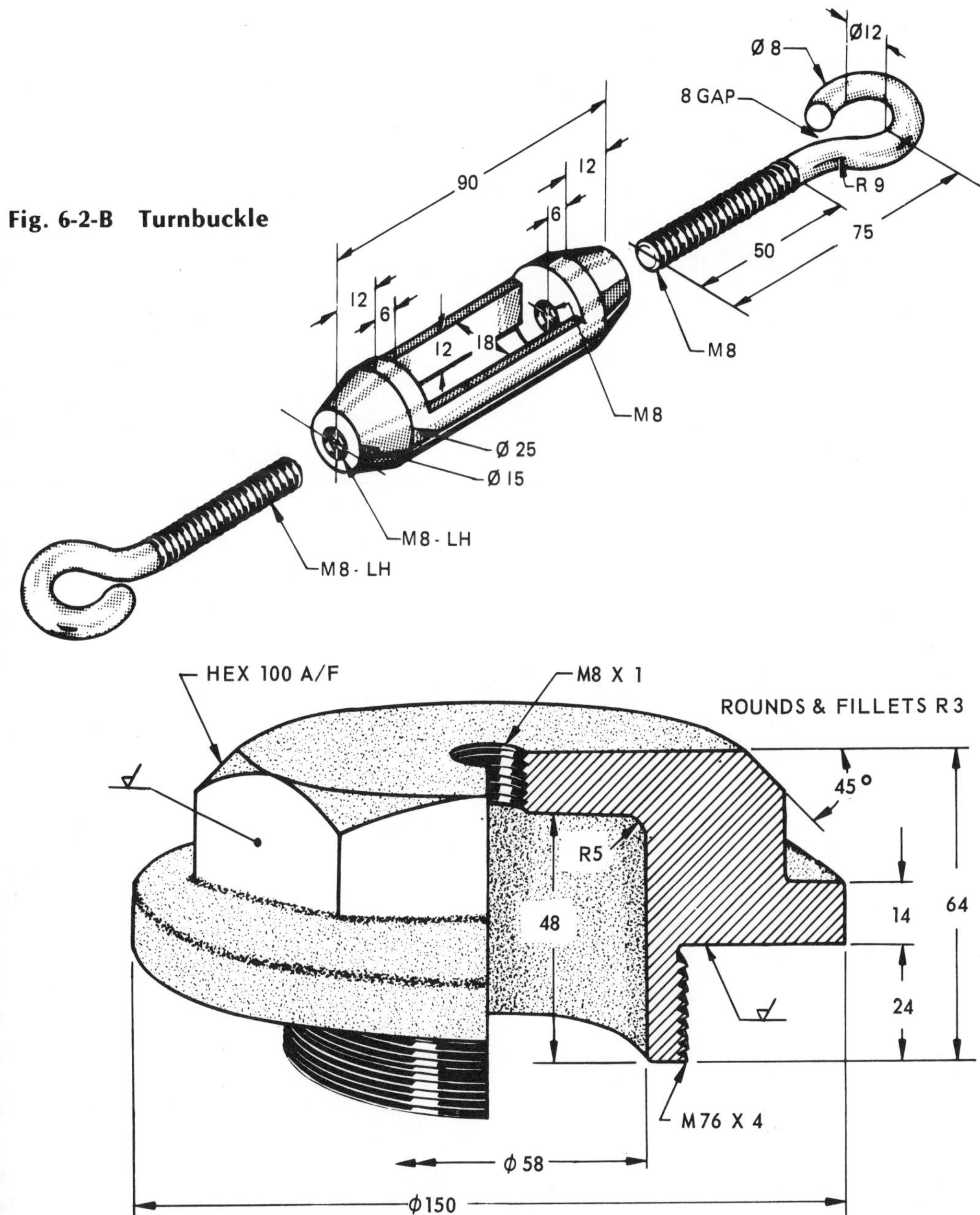

Fig. 6-2-B Turnbuckle

Fig. 6-2-C Pipe plug

UNIT 6-3 Common Threaded Fasteners

Fastener Definitions

Machine Screws. Machine screws are available in a range of sizes and thread classes, and are available in a variety of head shapes. They may be used in tapped holes or with nuts.

Cap Screws. A cap screw is a threaded fastener which joins two or more parts together by passing through a clearance hole in one part and screwing into a tapped hole in the other. They are tightened or released by torquing the head.

Cap screws range in size starting from 6.0 mm in diameter and are available in five basic types of head.

Bolts. A bolt is a threaded fastener which passes through clearance holes in assembled parts and threads into a nut. Bolts and nuts are available in a variety of shapes and sizes. The square and hexagon head are the two most popular designs and range in size from 6.0 to 72 mm in diameter.

Studs. Studs are shafts threaded at both ends and are used in assemblies. One end of the stud is threaded into one of the parts being assembled; other assembly parts, such as washers and covers, are guided over the studs through clearance holes and are held together by means of a nut which is threaded over the exposed end of the stud.

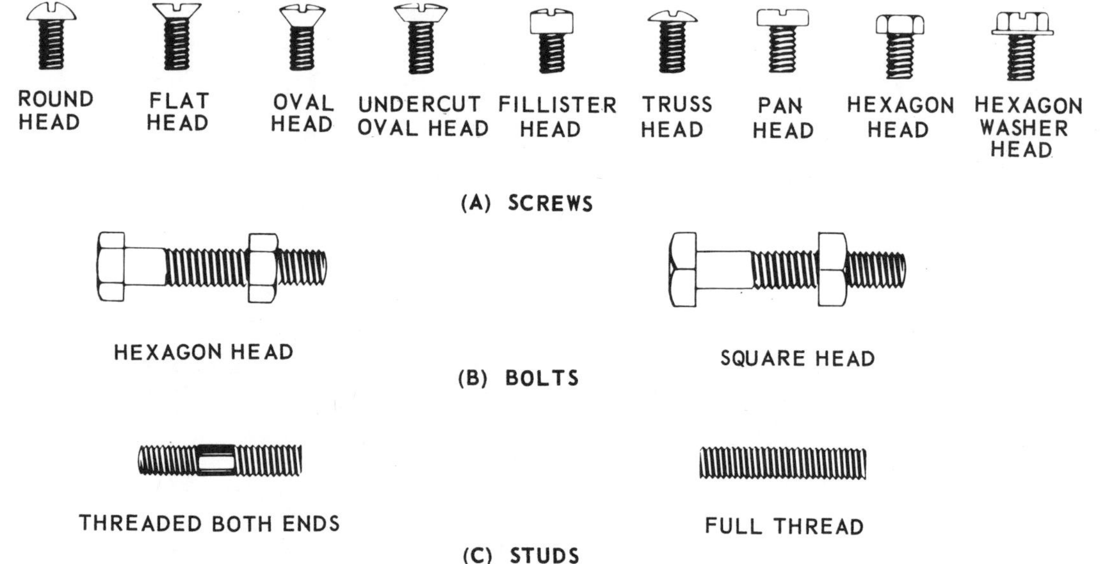

Fig. 6-3-1 Threaded fasteners

Note

A bolt is designed for assembly with a nut. A screw is designed to be used in a hole that has been tapped or preformed in another way. Because of basic design, it is possible to use certain types of screws together with a nut. However, any externally threaded fastener which, by its design can be used properly in a tapped or preformed hole, is called a **screw.** This is true no matter how it actually may be used.

Head Styles

Figure 6-3-3 shows some of the head styles which are common to bolts and screws alike. Of these basic styles, five (the pan, round, binding, truss, and hex cap) are designed to do the same job.

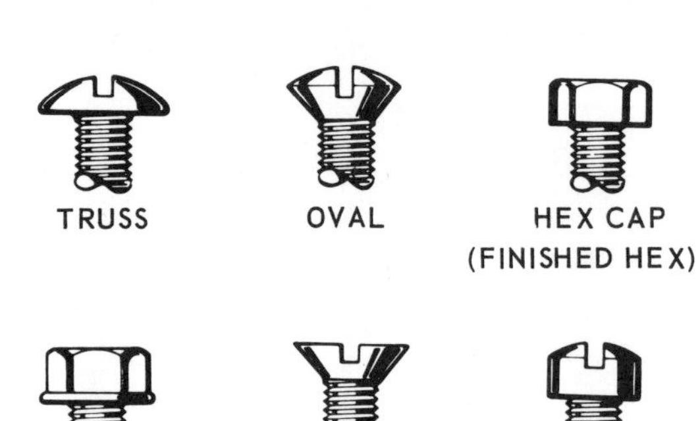

Fig. 6-3-2 Common head styles

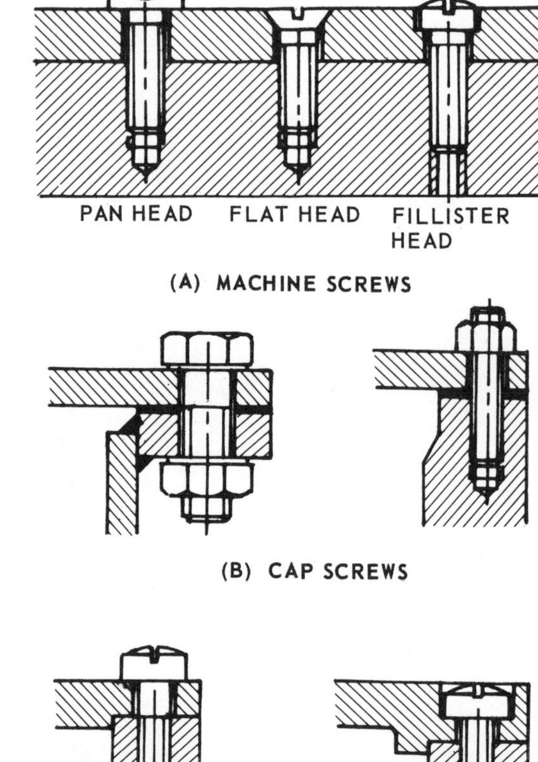

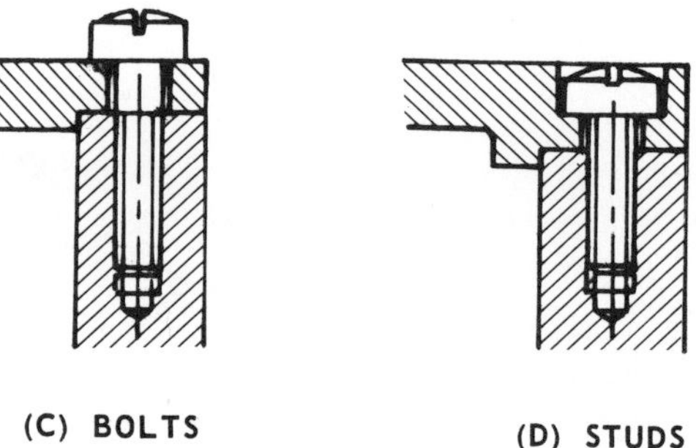

Fig. 6-3-3 Fastener application

The pan, however, is the most versatile. The Industrial Fasteners Institute of the United States has recommended that the pan replace round and binding head styles. They also suggest that pan heads can substitute for truss and fillister heads in most assemblies. Economically, it makes good sense. A switch to a uniform head style means that fewer types of screws can be manufactured.

Drive Configurations or Shapes. Figure 6-3-4 indicates five popular driving designs.

Fig. 6-3-4 Drive configurations

Shoulders and Necks. The *shoulder* of a fastener is the enlarged portion of the body of a threaded fastener or the shank of an unthreaded fastener. See Figure 6-3-5.

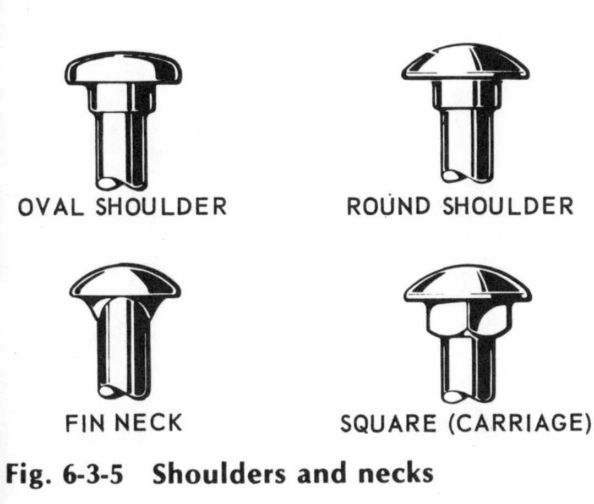

Fig. 6-3-5 Shoulders and necks

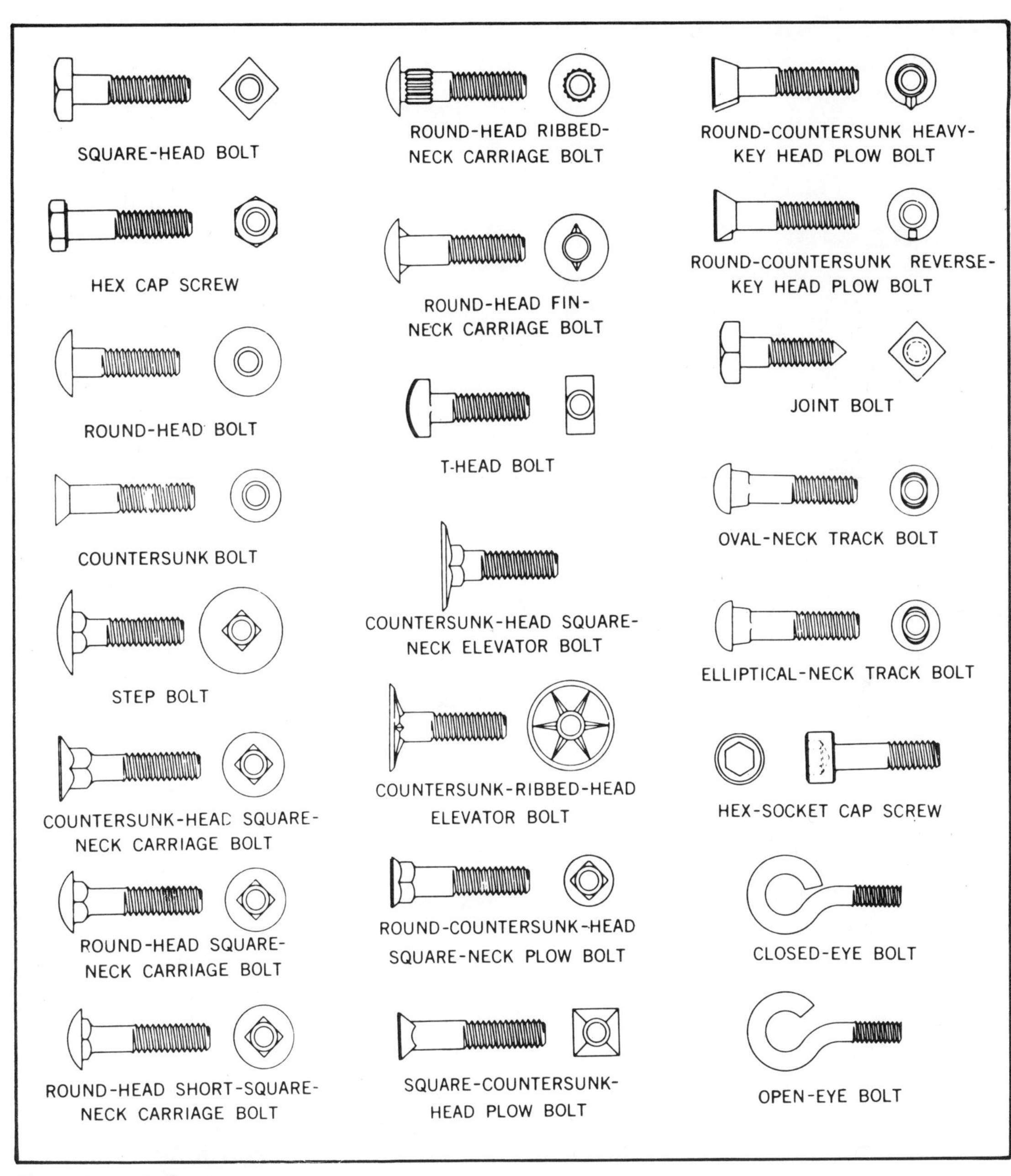

Fig. 6-3-6 Bolts and cap screws

Point Designs. The *point* of a fastener is the configuration of the end of the shank of a headed fastener, or of each end of a headless fastener. Points are usually either roll-formed or cold-headed. They may also be formed by pinching or milling. See Figure 6-3-6.

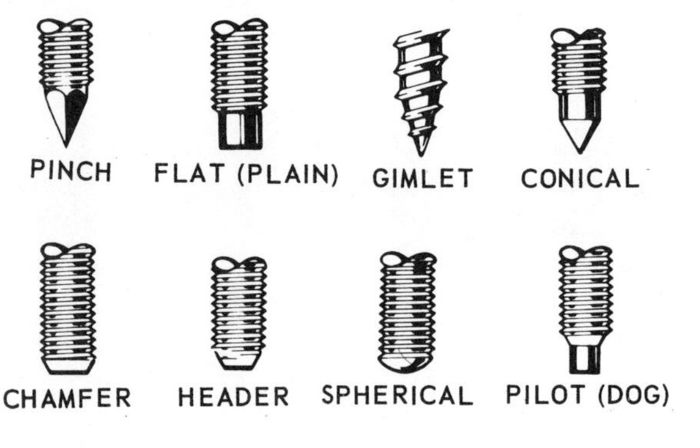

Fig. 6-3-7 Point styles

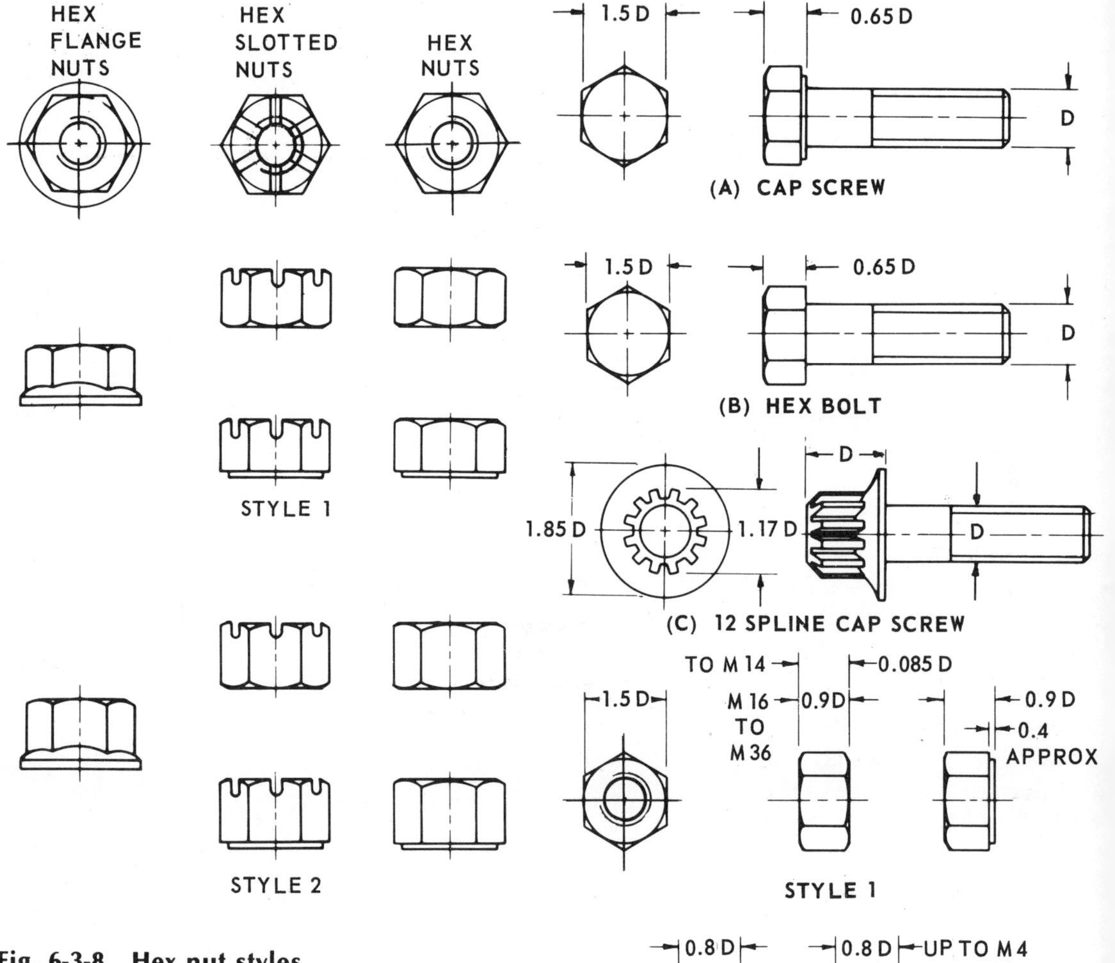

Fig. 6-3-8 Hex nut styles

Nuts

The terms **regular** and **thick** for describing nut thicknesses have been replaced by the terms **style 1** and **style 2.** The design of style 1 and 2 steel nuts shown in Figure 6-3-8 is based on providing sufficient nut strength to reduce the possibility of thread stripping.

Drawing a Bolt and Nut

Bolts and nuts are not normally drawn on detail drawings unless they are of a special size or have been modified slightly. On some assembly drawings it may be necessary to draw the nut and bolt. Approximate nut and bolt sizes are shown in Figure 6-3-9. Actual sizes are found in the Appendix. Nut and bolt templates are also available and are recommended as a cost-saving device. Conventional drawing practice is to draw the nuts and bolt heads in the across-corners position in all views.

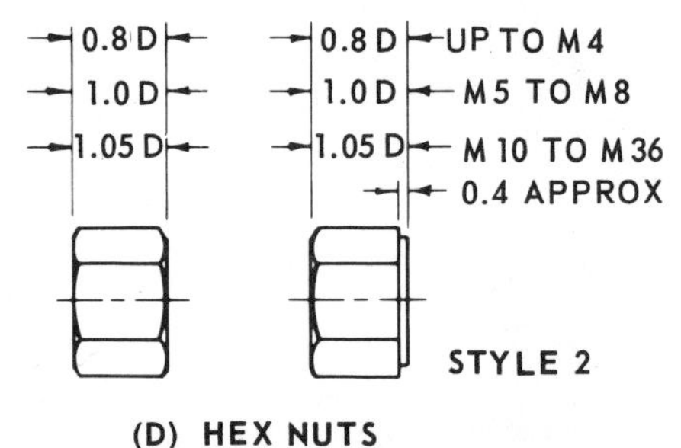

Fig. 6-3-9 Approximate head sizes for screws, bolts and nuts

Washers

Washers are one of the most common forms of hardware and have many uses in mechanically fastened assemblies. They may be required to span an oversize clearance hole, to give better bearing for nuts or screw faces, or to distribute loads over a greater area. Often, they serve to lock screws tight. They are also used to maintain a spring-resistance pressure, to guard surfaces against marring, to provide a seal, or an electrical connection.

Classification of Washers

Flat Washers. Plain, or flat washers are used primarily to provide a bearing surface for a nut or a screw head, to cover large clearance holes, and to distribute fastener loads over a larger area — particularly on soft materials such as aluminum or wood. See Figure 6-3-10.

Conical Washers. These washers are used with screws to effectively add spring take-up to the screw elongation. They are usually made of hardened spring steel.

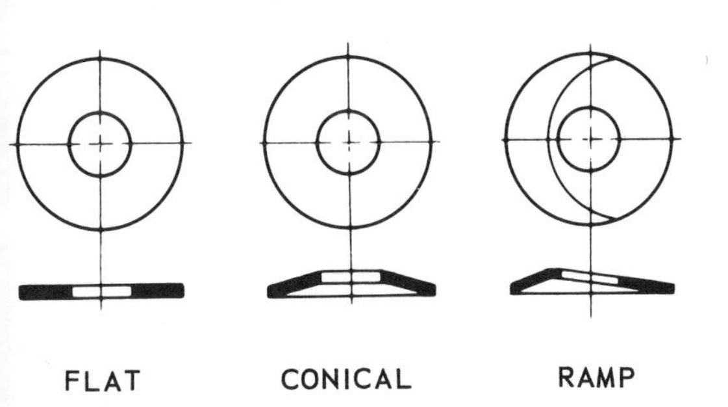

Fig. 6-3-10 Flat & conical washers

Belleville is a term commonly, but incorrectly, applied to coned (conical) washers. Cone washers do not have any auxiliary locking features other than friction, and in the flattened position, they are the equivalent of any flat washer as far as locking action is concerned.

Helical Spring Washers. These washers are formed into a helix of one coil so that the free height is approximately twice the thickness of the washer section. See Figure 6-3-11.

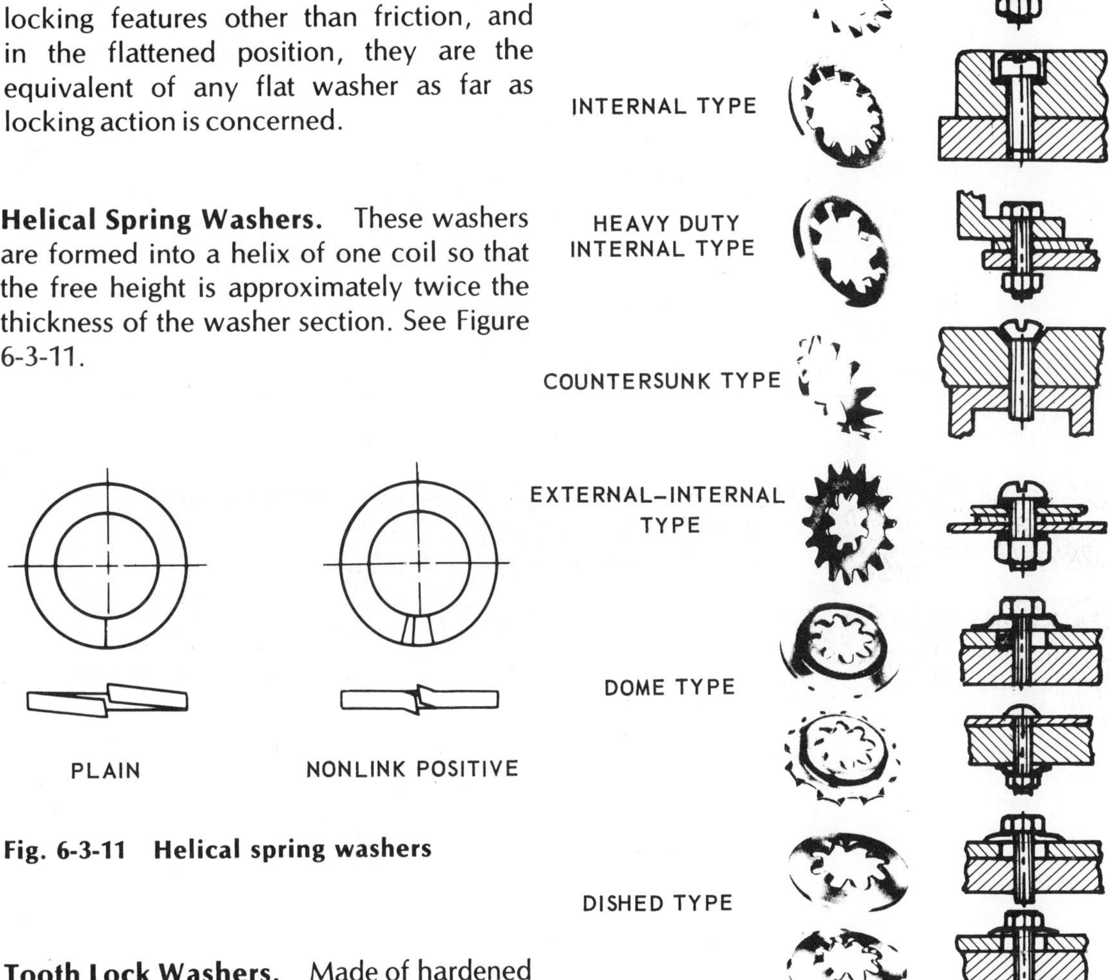

Fig. 6-3-11 Helical spring washers

Tooth Lock Washers. Made of hardened carbon steel, a tooth lock washer has teeth that are twisted or bent out of the plane of the washer face so that sharp cutting edges are presented to both the workpiece and the bearing face of the screwhead or nut. See Figure 6-3-12.

Fig. 6-3-12 Tooth lock washers

Spring Washers. There are no standard designs for spring washers. See Figure 6-3-13. They are made in a great variety of sizes and shapes and are usually selected from a manufacturer's catalogue for some specific purpose.

Special Purpose Washers. Molded or stamped nonmetallic washers are available in many materials and may be used as seals, electrical insulators, or for protection of the surface of assembled parts.

Many plain, cone, or tooth washers are available with special mastic sealing compounds firmly attached to the washer. These washers are used for sealing and vibration isolation in high-production industries.

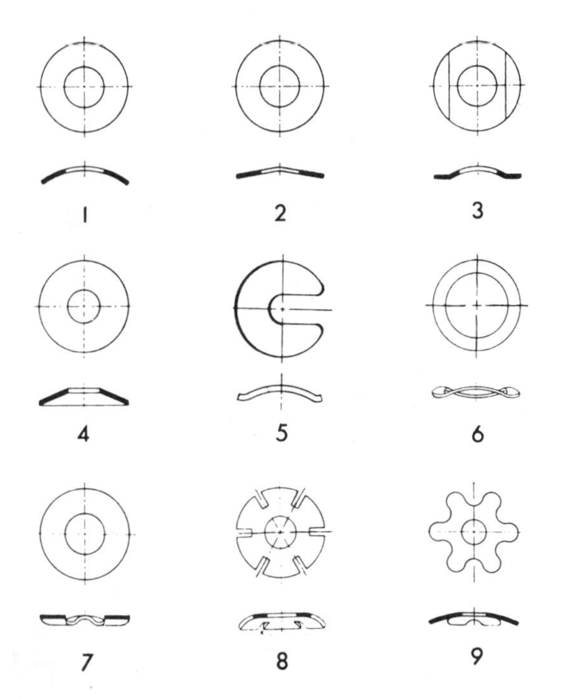

Fig. 6-3-13 Spring washer designs

Setscrews

Setscrews are used to hold a collar, sheave, or gear on a shaft against the forces of motion of the parts. Unlike most fastening devices, the setscrew is essentially a compression device. Forces developed by the screw point on tightening produce a strong clamping action that resists motion between assembled parts. The basic problem in setscrew selection is to find the best combination of setscrew form, size, and point style that provides the required holding power.

Setscrews can be categorized in two ways: by their forms and by the point style desired. See Figure 6-3-14. Each of the standardized setscrew forms is available in any one of five point styles.

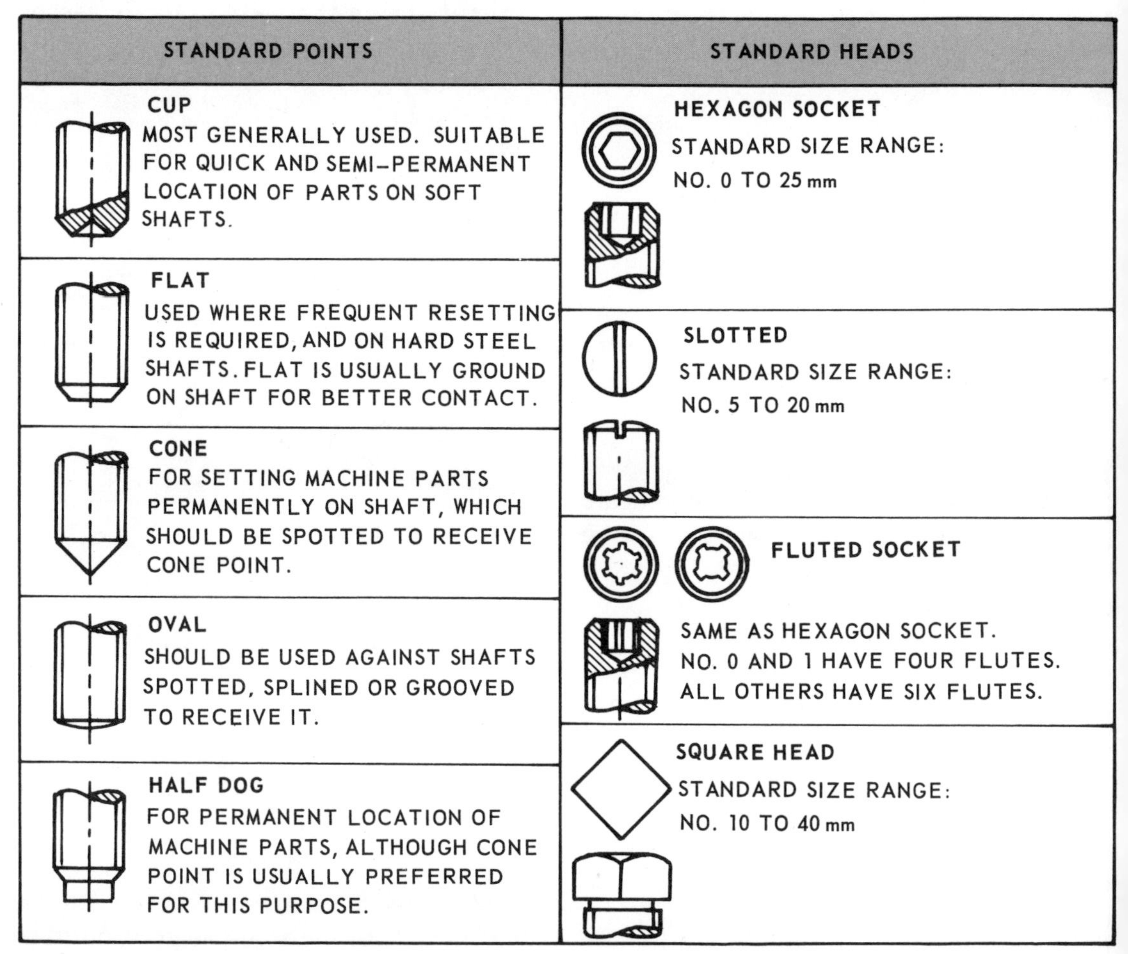

STANDARD POINTS
CUP MOST GENERALLY USED. SUITABLE FOR QUICK AND SEMI-PERMANENT LOCATION OF PARTS ON SOFT SHAFTS.
FLAT USED WHERE FREQUENT RESETTING IS REQUIRED, AND ON HARD STEEL SHAFTS. FLAT IS USUALLY GROUND ON SHAFT FOR BETTER CONTACT.
CONE FOR SETTING MACHINE PARTS PERMANENTLY ON SHAFT, WHICH SHOULD BE SPOTTED TO RECEIVE CONE POINT.
OVAL SHOULD BE USED AGAINST SHAFTS SPOTTED, SPLINED OR GROOVED TO RECEIVE IT.
HALF DOG FOR PERMANENT LOCATION OF MACHINE PARTS, ALTHOUGH CONE POINT IS USUALLY PREFERRED FOR THIS PURPOSE.

STANDARD HEADS
HEXAGON SOCKET STANDARD SIZE RANGE: NO. 0 TO 25 mm
SLOTTED STANDARD SIZE RANGE: NO. 5 TO 20 mm
FLUTED SOCKET SAME AS HEXAGON SOCKET. NO. 0 AND 1 HAVE FOUR FLUTES. ALL OTHERS HAVE SIX FLUTES.
SQUARE HEAD STANDARD SIZE RANGE: NO. 10 TO 40 mm

Fig. 6-3-14 Setscrews

Often setscrews are selected using a general rule-of-thumb: the setscrew diameter should be roughly equal to one-half the shaft diameter. This rule of thumb often gives satisfactory results, but its range of usefulness is limited.

Setscrews and Keyways. When a setscrew is used with a key, the screw diameter should be equal to the width of the key. In this combination the setscrew is locating the parts in an axial direction only. Torsional load on the parts is carried by the key.

The key should be tight-fitting, so that no motion is transmitted to the screw. Under high reversing or alternating loads, a poorly fitted key will cause the screw to back out and lose its clamping force.

Terms Related to Threaded Fasteners

The **tap drill size** for a threaded (tapped) hole is a diameter equal to the minor diameter of the thread. The **clearance drill size,** which permits the free passage of a bolt, is a diameter slightly greater than the major diameter of the bolt. See Figure 6-3-15. A **counterbored hole** is a circular, flat-bottomed recess that permits the head of a bolt or cap screw to rest below the surface of the part. A **countersunk hole** is an angular-sided recess that fits the shape of a flat-head cap screw or machine screw or an oval-head machine screw. **Spot-facing** is a machine operation that provides a smooth, flat surface where a bolt head or a nut will rest.

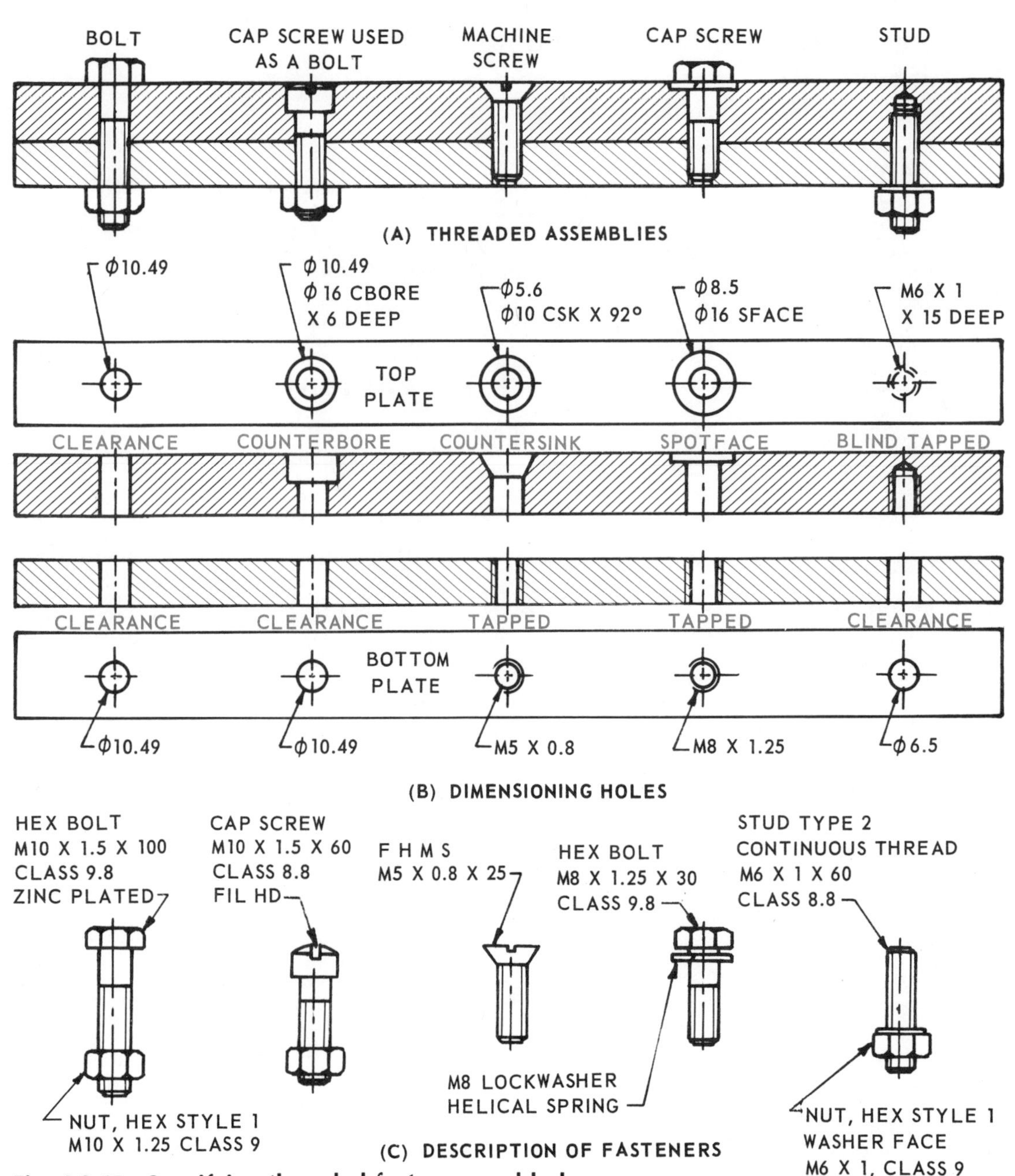

Fig. 6-3-15 Specifying threaded fasteners and holes

Counterbore, Countersink and Depth Symbols

The symbolic symbols for counterbore or spotface, countersink and depth are shown in Fig. 6-3-16. In each case the symbol precedes the dimension.

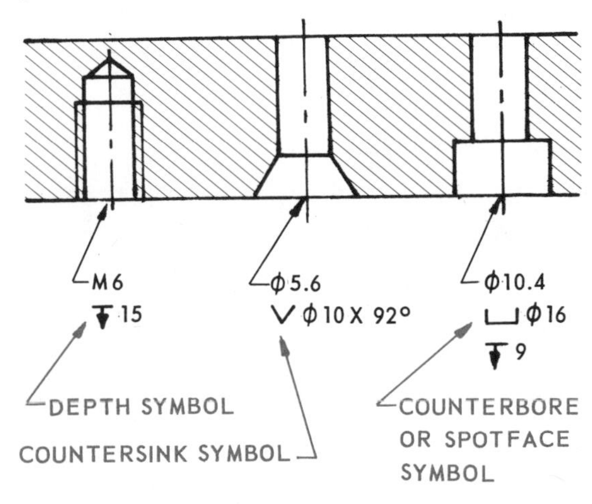

Fig. 6-3-16 Counterbore, countersink and depth symbols

Specifying Fasteners

In order for the purchasing department to properly order the fastening device which has been selected in the design, the following information is required. (Note: The information listed will not apply to all types of fasteners):

1. Type of fastener
2. Thread specifications
3. Fastener length
4. Type of driving recess
5. Point type (setscrews only)
6. Material
7. Head style
8. Finish

Examples:

M8 × 1.25 × 100, HEX BOLT, ZINC PLATED

M10 × 1.5 × 50, 12-SPLINE FLANGE SCREW

NUT, HEX, STYLE 1, M12 STEEL

MACH SCREW, PHILLIPS ROUND HD M4 × 0.7 × 25 LG, BRASS

WASHER, FLAT 8.4 ID × 17 OD × 2 THK

Assignments

Use conventional thread symbols. See Fig. 6-3-15 and consult the appendix for actual sizes.

1. Standard fasteners, Fig. 6-3-A, sheet size-A4, scale 1:1. Draw the four standard fasteners as designated. See Chapter 7 for section lining.
2. Brace assembly, Fig. 6-3-B, sheet size-A4, scale 1:1. Make a one-view drawing complete with the fasteners designated. No dimensions required.

Review for Assignments

Unit 6-2 Thread representation
Appendix Nut, bolt and washer sizes
Unit 16-3 Drawing a hexagon

12 12 10 40

CONNECTION A
M 10 X 30 LG
HEX HD CAP SCREW

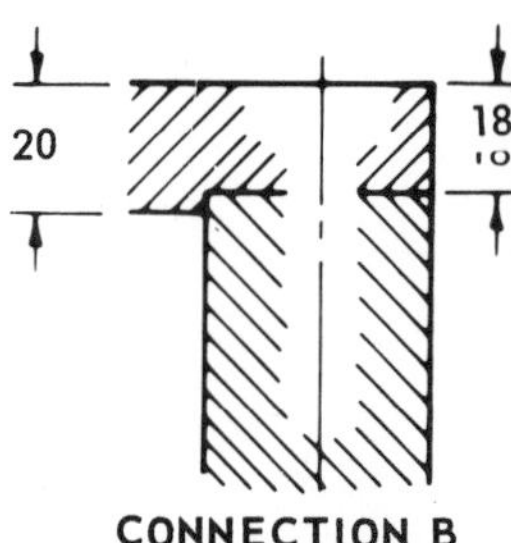

CONNECTION B
M 10 X 40 LG STUD
THREAD EACH END 20 LG
HEX NUT STYLE 1 AND
SPRING LOCKWASHER

CONNECTION C
M 10 X 30 LG
FL HD CAP SCREW

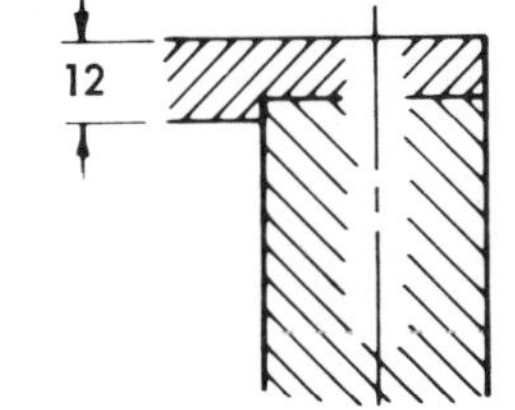

CONNECTION D
M 10 X 1.25 X 25 LG
SOCKET HEAD CAP
SCREW AND SPRING
LOCKWASHER

Fig. 6-3-A Standard fasteners

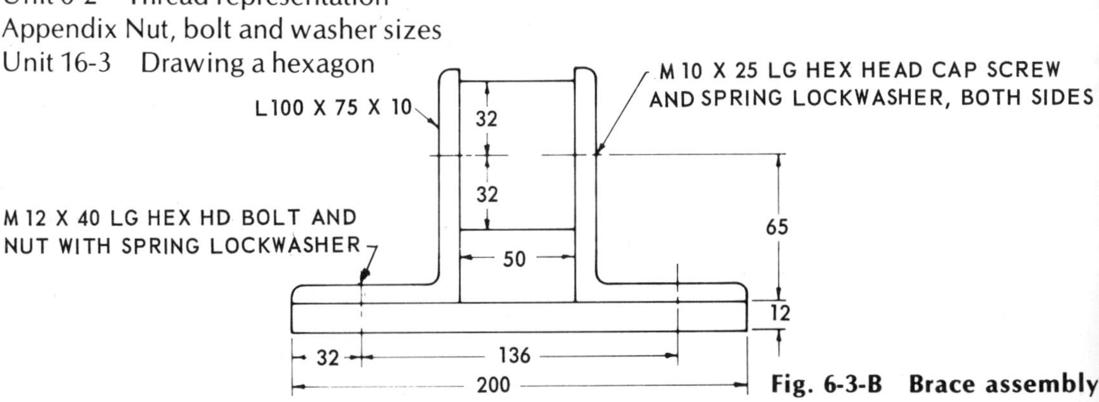

Fig. 6-3-B Brace assembly

UNIT 6-4
Special Fasteners

Keeping Fasteners Tight

Fasteners are inexpensive, but the cost of installing them can be high. Probably the simplest way to cut assembly costs is to make sure that, once installed, fasteners stay tight.

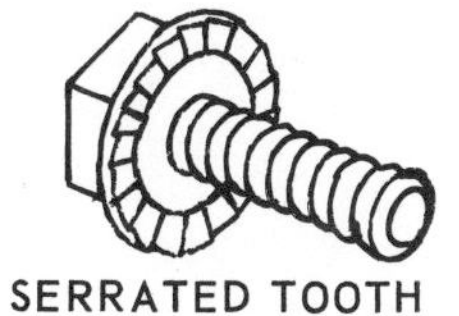

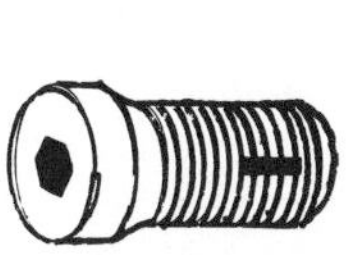

Fig. 6-4-1 Locking methods

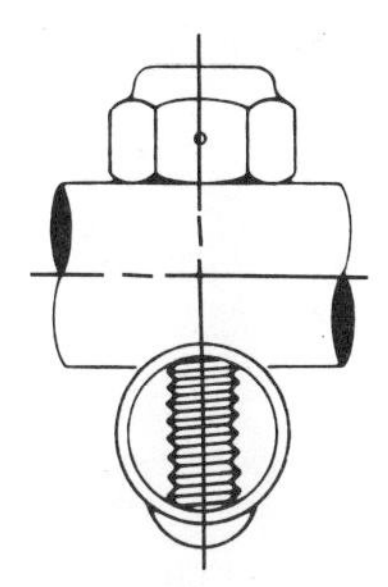

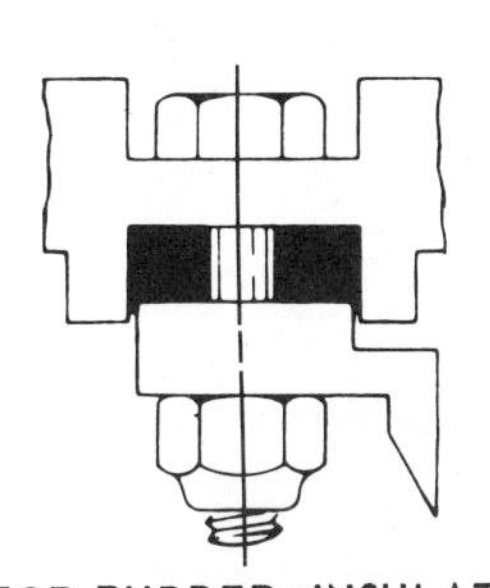

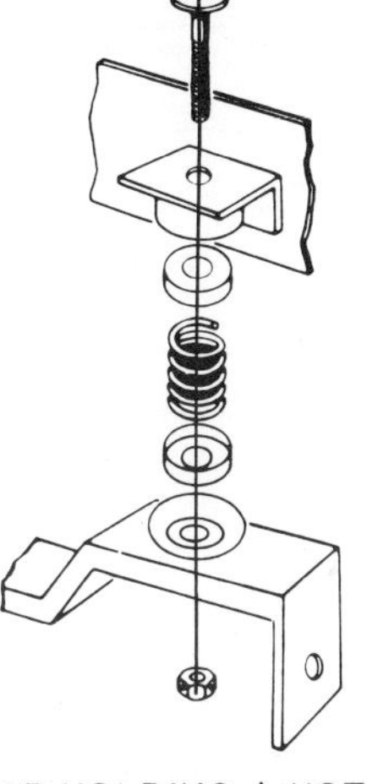

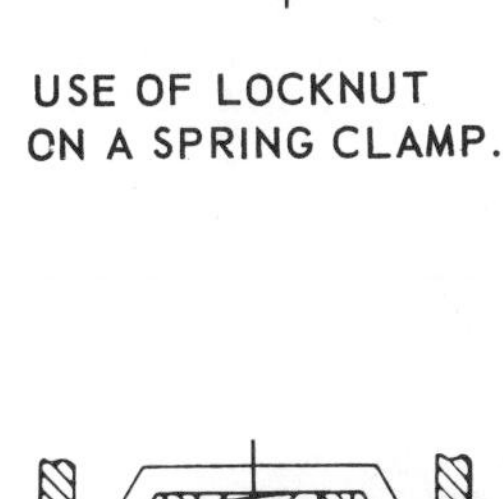

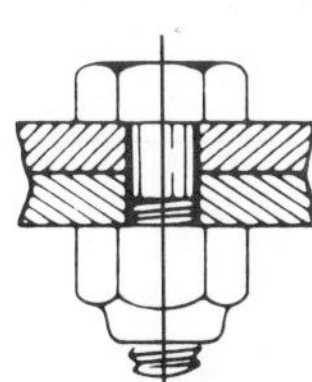

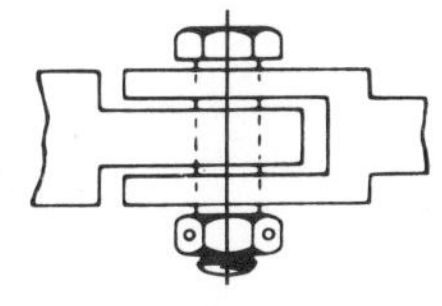

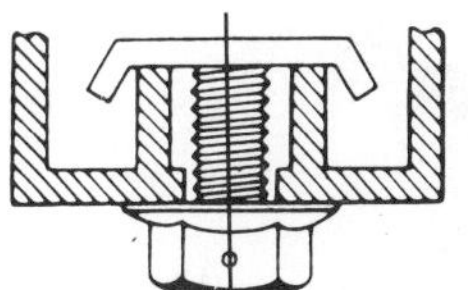

Fig. 6-4-2 Locknut applications

The American National Standards Institute has identified three basic locking methods; free-spinning, prevailing-torque, and chemical locking. Each has its own advantages and disadvantages. See Figure 6-4-1.

Free-spinning devices include toothed and spring lockwashers and screws and bolts with washerlike heads. With these arrangements, the fasteners spin free in the clamping direction, which makes them easy to assemble, and the break-loose torque is greater than the seating torque. However, once break-loose torque is exceeded, free-spinning washers have no prevailing torque to prevent further loosening.

Prevailing-torque methods make use of increased friction between nut and bolt. Metallic types usually have deformed threads or contoured thread profiles that jam the threads on assembly. Nonmetallic types make use of nylon or polyester insert elements that produce interference fits on assembly.

Chemical locking is achieved by coating the fastener with an adhesive. These adhesive chemicals can be formulated for

(A) PREVAILING TORQUE LOCKNUTS

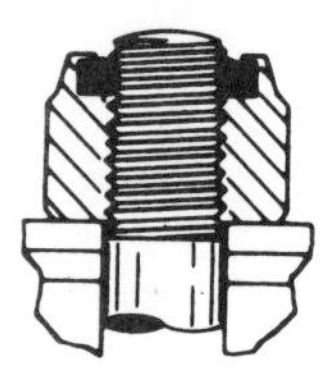

NONMETALLIC COLLAR CLAMPED IN THE TOP OF THIS NUT PRODUCES LOCKING ACTION.

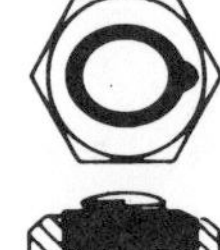

THREADED ELLIPTICAL SPRING-STEEL INSERT GRIPS THE BOLT AND PREVENTS TURNING.

SLOTTED SECTION OF THIS PREVAILING-TORQUE NUT FORMS BEAMS WHICH ARE DEFLECTED INWARD AND GRIP THE BOLT.

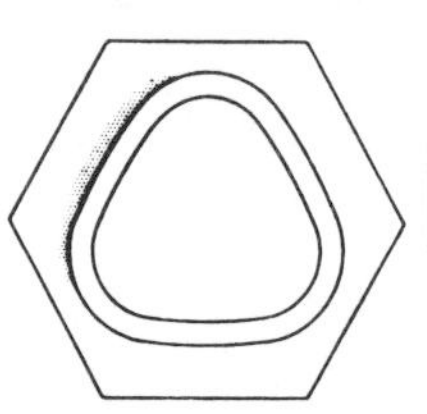

THREE SECTORS OF TAPERED CONE, PREFORMED INWARDLY, ARE ELASTICALLY RETURNED TO CIRCULAR FORM WHEN THE NUT IS APPLIED.

(B) FREE-SPINNING LOCKNUTS

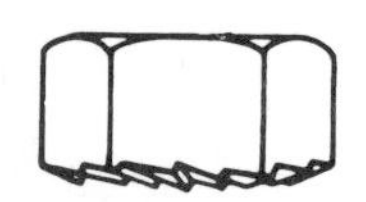

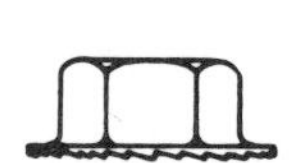

DEFORMED BEARING SURFACE. TEETH ON THE BEARING SURFACE "BITE" INTO WORK TO PROVIDE A RATCHET LOCKING ACTION.

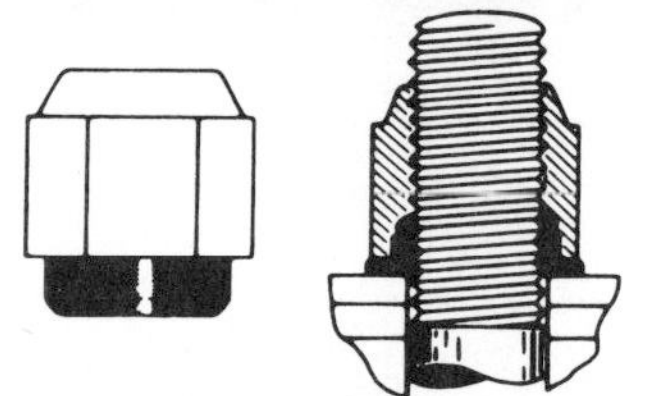

NYLON INSERT FLOWS AROUND THE BOLT RATHER THAN BEING CUT BY THE BOLT THREADS TO PROVIDE LOCKING ACTION AND AN EFFECTIVE SEAL.

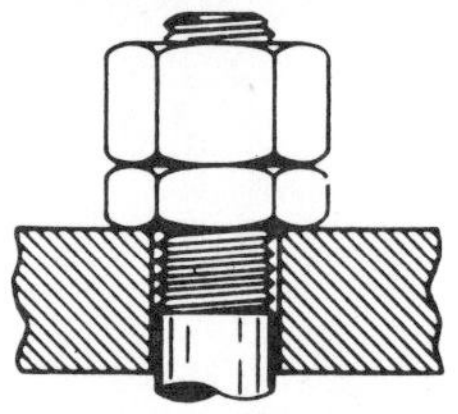

JAM NUT, APPLIED UNDER A LARGE REGULAR NUT, IS ELASTICALLY DEFORMED AGAINST BOLT THREADS WHEN THE LARGE NUT IS TIGHTENED.

NUT WITH A CAPTIVE-TOOTHED WASHER. WHEN TIGHTENED, THE CAPTIVE WASHER PROVIDES THE LOCKING MEANS WITH SPRING ACTION BETWEEN THE NUT AND WORKING SURFACE.

(C) OTHER TYPES

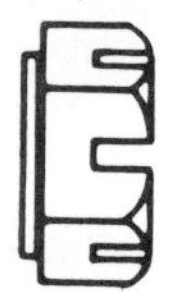

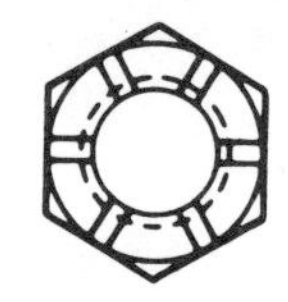

SLOTTED NUT USES A COTTER PIN THROUGH A HOLE IN THE BOLT FOR LOCKING ACTION.

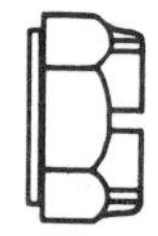

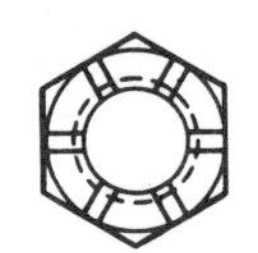

CASTLE NUT IS BASICALLY A SLOTTED NUT WITH A CROWN OF REDUCED DIAMETER.

SINGLE-THREAD LOCKNUT, WHICH IS SPEEDILY APPLIED, LOCKS BY GRIP OF ARCHED PRONGS WHEN BOLT OR SCREW IS TIGHTENED.

Fig. 6-4-3 Locknuts

different strengths, can be used on any size fastener, and can reduce the number of fasteners that must be carried in inventory.

Locknuts

A **locknut** is a nut with special internal means for gripping a threaded fastener or connected material to prevent rotation in use. Generally it has the dimensions, mechanical requirements, and other specifications of a standard nut, but with a locking feature added.

Locknuts are divided into two general classifications: prevailing-torque and free-spinning types. However, there are some types which do not fall exactly into these categories. These are shown in Figure 6-4-3.

Prevailing-Torque Locknuts

Prevailing-torque locknuts spin freely for a few turns, and then must be wrenched to final position. The maximum holding and locking power is reached as soon as the threads and the locking feature are engaged. Locking action is maintained until the nut is removed. Prevailing-torque locknuts are classified by basic design principles:

1. Thread deflection causes friction to develop when the threads are mated; thus the nut resists loosening.

2. Out-of-round top portion of the tapped nut grips the bolt threads and resists rotation.

3. Slotted section of locknut is pressed inward to provide a spring frictional grip on the bolt.

4. Inserts, either nonmetallic or of soft metal, are plastically deformed by the bolt threads to produce a frictional interfence fit.

5. Spring wire or pin engages the bolt threads to produce a wedging or ratchet-locking action.

Free-Spinning Locknuts

Free-spinning locknuts are free to spin on the bolt until seated. Additional tightening locks the nuts.

Free-spinning locknuts are often specified when long travel of nut on bolt is unavoidable. Since most free-spinning locknuts depend on clamping force for their locking action, they are usually not recommended for joints that might relax through plastic deformation or for fastening materials that might crack or crumble when subjected to preload.

Other Locknut Types

Jam nuts are thin nuts used under full-sized nuts to develop locking action. The large nut has sufficient strength to elastically deform the lead threads of the bolt and jam nut. Thus, a considerable resistance against loosening is built up. The use of jam nuts is decreasing; a one-piece, prevailing-torque locknut usually is used instead at a savings in assembled cost.

The jam nut is considered ideal for assemblies where long travel of nut on bolt under load is necessary to bring mating parts into position.

Slotted and castle nuts have slots which receive a cotter pin that passes through a drilled hole in the bolt and thus serves to lock the nut. These nuts are essentially free-spinning nuts with the locking feature added after the preload condition is developed. Castle nuts differ from slotted nuts in that they have a circular crown of a reduced diameter.

From an assembled-cost viewpoint, these nut forms are expensive because of the extra operations involved in their assembly. Usually a one-piece, prevailing-torque locknut is a better solution.

Single-thread engaging locknuts are spring steel fasteners which may be speedily applied. Locking action is provided by the grip of the thread-engaging prongs and the reaction of the arched base. Their use is limited to nonstructural assemblies and usually to screw sizes below 6 mm in diameter. Compared with multiple-thread locknuts, these types are less expensive and lighter in mass. The spring steel will absorb a certain amount of motion. Tightening torques for these nuts, however, must be lower than for multiple-thread locknuts.

Captive or Self-Retaining Nuts

Self-retained or captive nuts provide a permanent, strong, multiple-threaded fastener in many types of thin materials. See Figure 6-4-4. They are especially good where there are blind locations, and they can generally be attached without damaging finishes. Methods of attaching these types of nuts vary and tools required for assembly are

(A) PLATE NUT (B) CAGED NUT

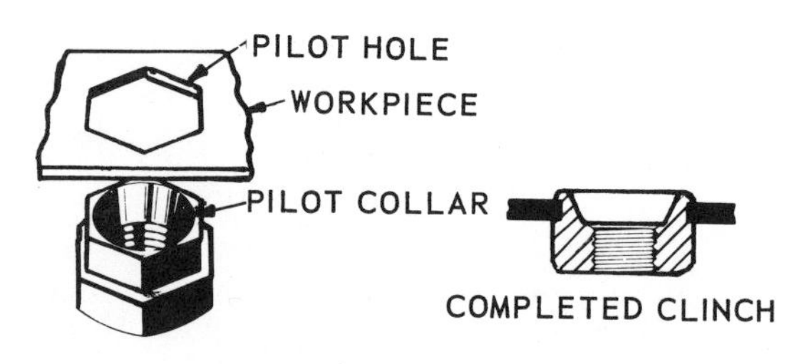

(C) CLINCH NUT

(1) UNIVERSAL PIERCE NUT

(2) HIGH–STRESS PIERCE NUT

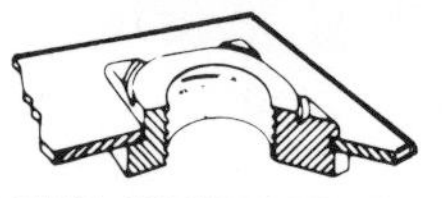

(3) PIERCE NUT WITH CORNERS CLINCHED

(D) PIERCE NUTS

Fig. 6-4-4 Captive or self retaining nuts

generally uncomplicated and inexpensive. In this section, the self-retained nuts are grouped according to four means of attachment:

1. Plate or anchor nuts: These nuts have mounting lugs which can be riveted, welded, or screwed to the part.

2. Caged nuts: A spring-steel cage retains a standard nut. The cage snaps into a hole or clips over an edge to hold the nut in position.

3. Clinch nuts: They are specially designed nuts with pilot collars which are clinched or staked into the parent part through a precut hole.

4. Self-piercing nuts: A form of clinch nut that cuts its own hole.

Single-Thread Engaging Nuts

Single-thread engaging nuts are formed by stamping a thread-engaging impression in a flat piece of metal. The stamped impression can take a number of shapes. For example, shear-formed helical prongs engage and lock on the screw-thread root diameter (Fig.6-4-5a), or a protruding truncated cone (Fig. 6-4-5b) stamped into the metal provides a ramp that the screw climbs as it turns. In another type of impression, the part for engaging the thread is formed in a spiral to match the pitch of the screw threads (Fig. 6-4-5c).

Inserts

Inserts are a special form of nut designed to serve as a tapped hole in blind or through-hole locations. See Figure 6-4-6. They are sometimes referred to as **solid bushings.** Another basic type of screw-thread insert consists of precision-formed wire, spirally coiled to provide threads of proper form for

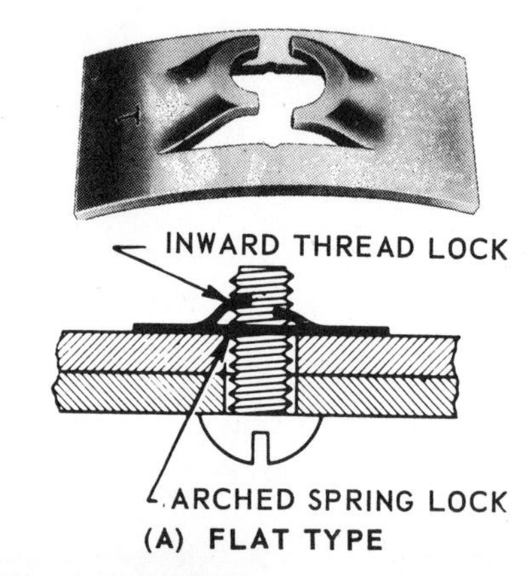

(A) FLAT TYPE

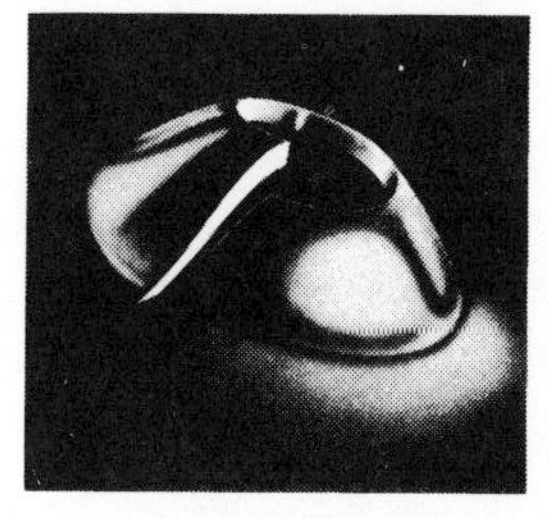

(B) FLAT–TYPE CONICAL THREAD

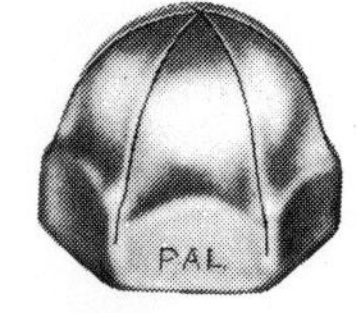

(C) SPIRAL–FORMED THREAD

Fig. 6-4-5 Single-thread engaging nuts

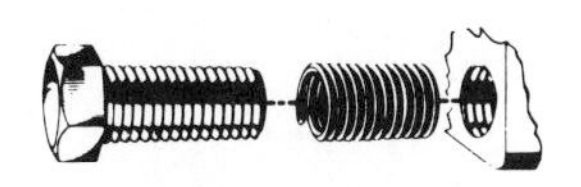

(A) EXTERNAL–INTERNAL THREADED INSERT

(B) SELF–TAPPING INSERT

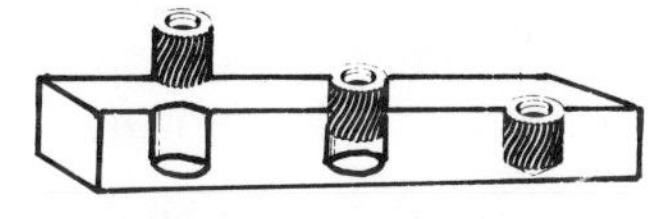

(C) PRESSED–IN INSERT

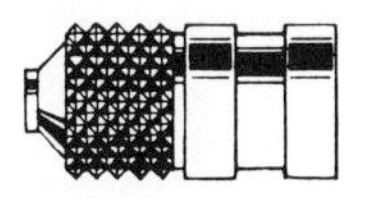

(D) MOLDED–IN INSERT

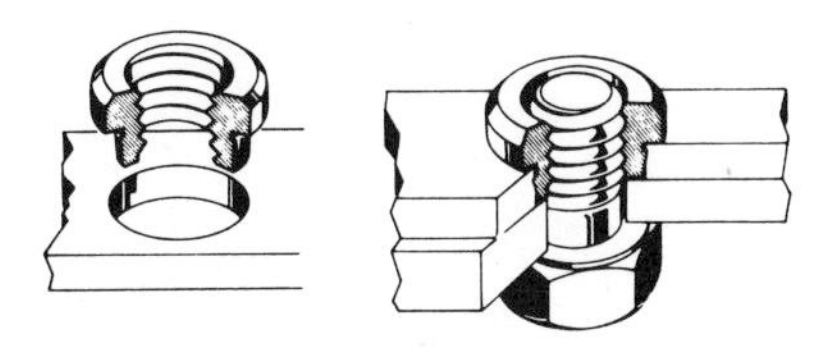

(E) THIN MATERIAL INSERT

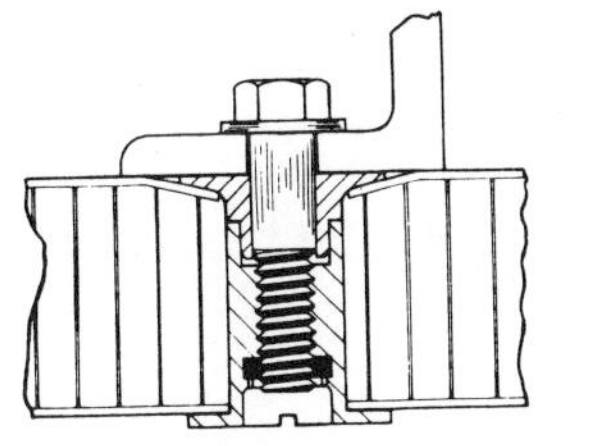

(F) SANDWICH PANEL INSERT

Fig. 6-4-6 Inserts

installation in a tapped hole. It is known as a **wire insert.**

The internal thread of the insert is usually of standard size and thread form as governed by the screw-thread industry. The external configuration of the insert is designed to suit its particular purpose and may use threads, flanges, grooves, knurls, or other shapes for holding strength.

Sealing Fasteners

Fasteners hold two or more parts together, but they can perform other functions as well. One important auxiliary function is that of sealing gases and liquids against leakage.

Two types of scaled-joint constructions are possible with fasteners. See Figure 6-4-7. In one approach, the fasteners enter the sealed medium and are separately sealed. A number of fastener designs with built-in sealing elements have been developed for this purpose.

The second approach uses a separate sealing element which is held in place by the clamping forces produced by conventional fasteners, such as rivets or bolts.

FASTENERS SEPARATELY SEALED

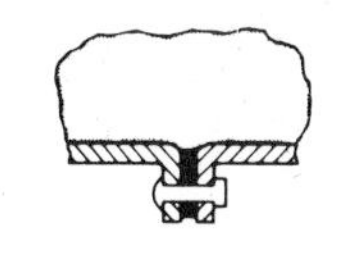

SEALING ELEMENT CLAMPED IN PLACE

Fig. 6-4-7 Types of sealed-joint construction

Sealing Fastener Types

There are many methods of obtaining a seal using sealing fasteners, as shown in Figure 6-4-8. The more common sealing fasteners are screws, rivets, nuts, and washers.

Screws. Most sealing screws provide a sealing action by the use of a resilient material added to the screw. It may be rubber, plastic, mastic, or metal. In addition, some screws make use of an interference fit in the threads.

Rivets. Sealing rivets are used extensively in aircraft and ships. They seal integral tanks, pressurized cabins, and pontoons and are used in other critical applications.

Nuts. Starting with the simple device of adding a soft metal washer under a cap nut to effect a seal, the range of sealing nuts available today has steadily increased. Mastic, rubber, plastic, and soft metals are all used as seals in a wide variety of types.

Washers. Sealing washers are used under either the screw head or the nut. Cut rubber rings, O-rings, or molded-in sections are most commonly used. More recently, flowed-in sealants and nylon rings or sleeves have been made available.

Assignment

List and describe six different examples of the use of special fasteners as used in automobiles, machines, home appliances, etc.

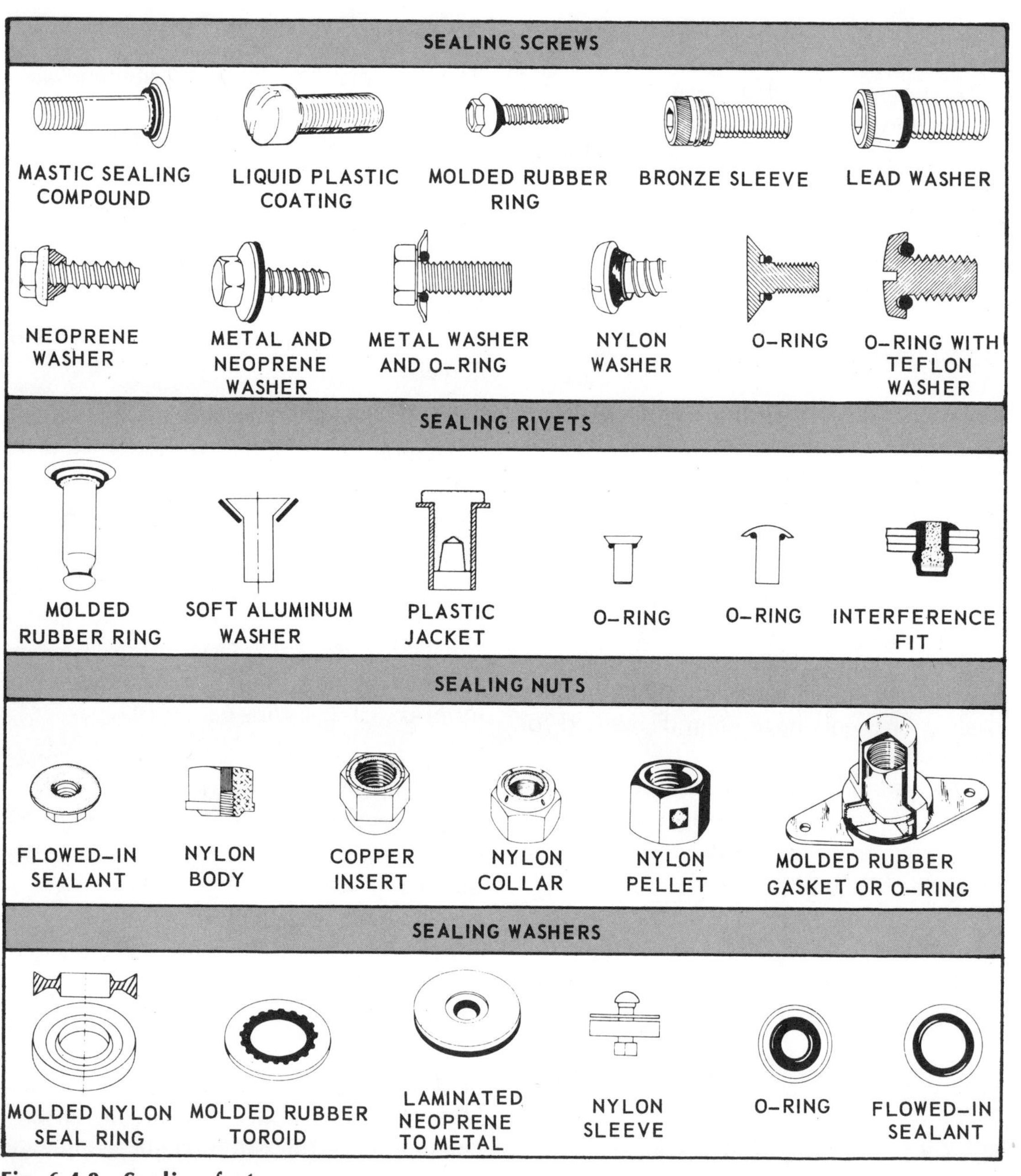

Fig. 6-4-8 Sealing fasteners

UNIT 6-5 Keys and Pins

Keys

A **key** is a piece of steel lying partly in a groove in the shaft, and extending into another groove in the hub. It is used to secure gears, pulleys, cranks, handles, and similar machine parts to shafts. In this way the motion of the part is transmitted to the shaft, or the motion of the shaft to the part, without slippage. The key may also act in a safety capacity; its size is generally calculated so that when overloading takes place, the key will shear or break before the part or shaft breaks.

There are many kinds of keys. The most common types are shown in Figure 6-5-1. Square and flat keys are widely used in industry. The width of the square and flat key should be approximately one-quarter the shaft diameter, but for proper key selection refer to the Appendix. These keys are also available with a 1:100 taper on their top surfaces and are know as **square taper** or **flat tapered** keys. The hole in the hub is tapered to accommodate the taper on the key.

The gibhead key is the same as the square or flat tapered key but has a head added for easy removal.

The Pratt and Whitney key is rectangular with rounded ends. Two-thirds of this key sits in the shaft, one-third sits in the hub.

The Woodruff key is semicircular and fits into a semicircular keyseat in the shaft and

TYPE OF KEY	ASSEMBLY SHOWING KEY, SHAFT AND HUB	SPECIFICATION
SQUARE		6 SQUARE KEY, 30 LG OR 6 SQUARE TAPERED KEY, 30 LG
FLAT		5 X 3 FLAT KEY, 25 LG OR 5 X 3 FLAT TAPERED KEY, 25 LG
GIB–HEAD		10 SQUARE GIB–HEAD KEY, 50 LG
PRATT AND WHITNEY		NO. 15 PRATT AND WHITNEY KEY
WOODRUFF		NO. 1210 WOODRUFF KEY

Fig. 6-5-1 Common keys

a rectangular keyseat in the hub. The width of the key should be approximately one-quarter the diameter of the shaft, and its diameter should approximate the diameter of the shaft. Half the width of the key extends above the shaft and into the hub. Refer to the Appendix for exact sizes.

Woodruff keys are identified by a number which indicates the nominal dimensions of the key. The numbering system which originated many years ago is identified with the fractional-inch system of measurement. The last two digits of the number give the normal diameter in eights of an inch, and the digits preceding the last two give the nominal width in thirty-seconds of an inch. For example, a no. 1210 Woodruff key indicates a key 12/32 × 10/8 in., or a 3/8 × 1¼ in. key.

In calling up keys on a bill of material, only the information shown in the column "Specifications" in Figure 6-5-1 need be given.

Dimensioning of Keyseats

Keyseats are dimensioned by width, depth, location and, if required, length. The depth is dimensioned from the opposite side of the shaft or hole (Fig 6-5-2).

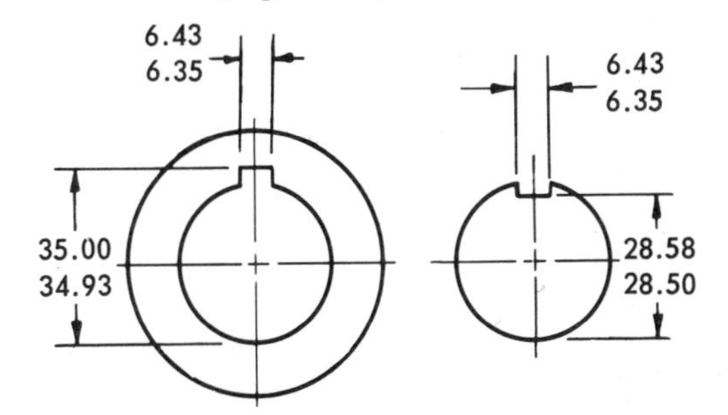

Fig. 6-5-2 Dimensioning keyseats

Since standard milling cutters for Woodruff keys have the same appropriate number, it is possible to call for a Woodruff keyseat by the number only.

Where it is desirable to detail Woodruff keyseats on a drawing, all dimensions are given in the form of a note in the following order: width, depth, and radius of cutter (Fig. 6-5-3).

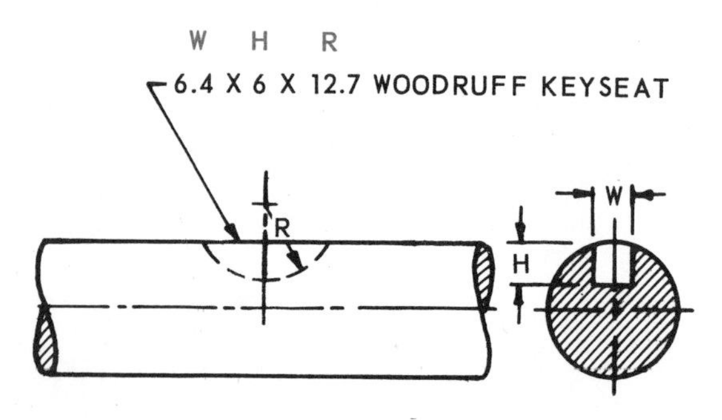

Fig. 6-5-3 Alternate method of detailing a Woodruff keyseat

Pin Fasteners

Pin fasteners are an inexpensive and effective approach to assembly where loading is primarily in shear. They can be separated into two groups: **semipermanent** and **quick-release.**

Semipermanent Pins

Semipermanent pin fasteners require application of pressure or the aid of tools for installation or removal. The two basic types are machine pins and radial-locking pins.

The following general design rules apply to all types of semipermanent pins:

- Avoid conditions where the direction of vibration parallels the axis of the pin.
- Keep the shear plane of the pin a minimum distance of 1 diameter from the end of the pin.
- In applications where engaged length is at a minimum and appearance is not critical, allow pins to protrude the length of the chamfer at each end for maximum locking effect.

Machine Pins. Four types are generally considered to be most important: **hardened and ground dowel pins and commercial straight pins, taper pins, clevis pins,** and **standard cotter pins.** Descriptive data and recommended assembly practices for these four traditional types of machine pins are presented in Figure 6-5-4. For proper size selection of cotter pins, refer to Figure 6-5-5.

Radial Locking Pins. Two basic pin forms are employed: **solid with grooved surfaces** and **hollow spring pins,** which may be either slotted or spiral-wrapped. In assembly, radial forces produced by elastic action at the pin surface develop a secure,

HARDENED AND GROUND DOWEL PIN	TAPER PIN	CLEVIS PIN	COTTER PIN
STANDARDIZED IN NOMINAL DIAMETERS RANGING FROM 3 TO 22 mm. 1. HOLDING LAMINATED SECTIONS TOGETHER WITH SURFACES EITHER DRAWN UP TIGHT OR SEPARATED. 2. FASTENING MACHINE PARTS WHERE ACCURACY OF ALIGNMENT IS A PRIMARY CONSIDERATION. 3. LOCKING COMPONENTS ON SHAFT	STANDARD PINS HAVE A TAPER OF 1:48 MEASURED ON THE DIAMETER. BASIC DIMENSION IS THE DIAMETER OF THE LARGE END. USED FOR LIGHT DUTY SERVICE IN THE ATTACHMENT OF WHEELS, LEVERS AND SIMILAR COMPONENTS TO SHAFTS.	STANDARD NOMINAL DIAMETERS FOR CLEVIS PINS RANGE FROM 5 TO 25 mm. BASIC FUNCTION OF THE CLEVIS PIN IS TO CONNECT MATING YOKE, OR FORK, AND EYE MEMBERS IN KNUCKLE–JOINT ASSEMBLIES. HELD IN PLACE BY A SMALL COTTER PIN OR OTHER FASTENER WHICH CAN BE READILY DISCONNECTED FOR ADJUSTMENT OR MAINTENANCE.	SIZES 1 TO 20 mm IN DIAMETER. LOCKING DEVICE FOR OTHER FASTENERS. USED WITH A CASTLE OR SLOTTED NUT ON BOLT, SCREWS, OR STUDS, IT PROVIDES A CONVENIENT, LOW–COST LOCKNUT ASSEMBLY CAN BE USED WITH OR WITHOUT A PLAIN WASHER.

Fig. 6-5-4 Machine pins

THREAD SIZE (mm)	COTTER PIN SIZE	COTTER PIN HOLE	END CLEAR-ANCE
6	1.5	1.9	3
8	2	2.4	3
10	2.5	2.8	4
12	3	3.4	5
14	3	3.4	5
16	4	4.5	6
20	4	4.5	7
24	5	5.6	8
27	5	5.6	8
30	6	6.3	10
36	6	6.3	11

Fig. 6-5-5 Recommended cotter pin sizes

frictional locking grip against the hole wall. These pins are reusable and can be removed and reassembled many times without appreciable loss of fastening effectiveness. Spring action at the pin surface also prevents loosening under shock and vibration and allows for variations in hole size.

Figure 6-5-6 shows six of the grooved-pin constructions and their applications.

Grooved Straight Pins. Locking action of the groove pin is provided by parallel, longitudinal grooves uniformly spaced around the pin surface. Rolled or pressed into solid pin stock, the grooves expand the effective pin diameter. When the pin is driven into a drilled hole corresponding in size to nominal pin diameter, elastic deformation of the raised groove edges produces a secure force-fit with the hole wall. For typical grooved pin size selection, refer to Fig. 6-5-7.

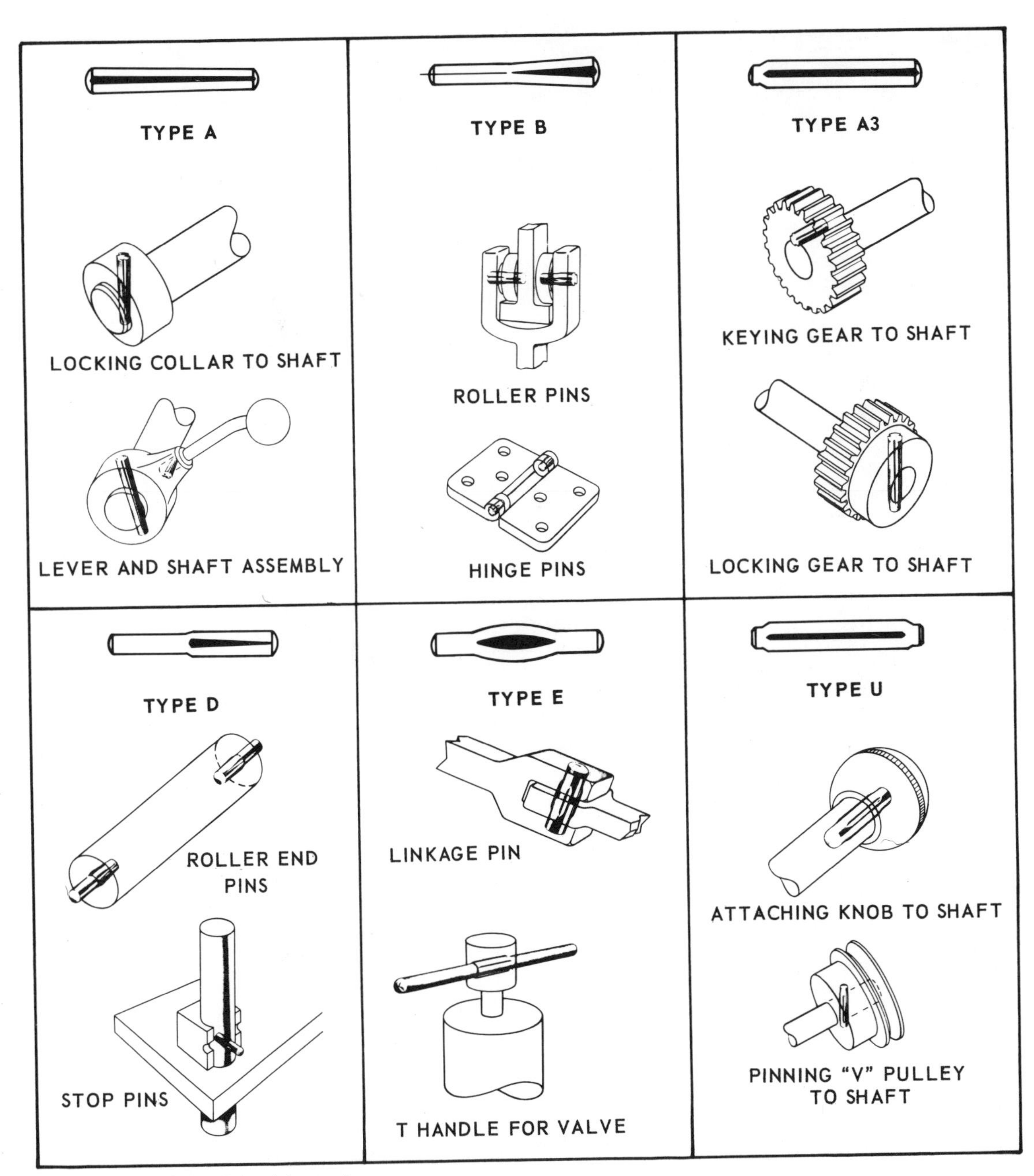

Fig. 6-5-6 Radial locking pins

SHAFT DIA (mm)	TRANSVERSE KEY		LONGITUDINAL KEY PIN DIA (mm)
	STRAIGHT PIN DIA	TAPER PIN NO.	
10	3	3/0	2.5
12	4	0	3
14	5	2	4
16	5	2	4
18	6	3	5
20	6	4	5
22	6	4	6
24	8	5	6
26	8	6	6
28	10	7	—
30	10	7	—
32	10	7	8
34	11	7	10

Fig. 6-5-7 Recommended grooved pin sizes

Hollow Spring Pins. Resilience of hollow cylinder walls under radial compression forces is the principle of spiral-wrapped and slotted tubular pins. Both pin forms are made to controlled diameters greater than the holes into which they are pressed. Compressed when driven into the hole, the pins exert spring pressure against the hole wall along their entire engaged length to develop locking action.

Locking force of a sprial-wrapped pin is a function of length of engagement, pin diameter, and wall thickness.

Standard slotted tubular pins are designed so that several sizes can be used inside one another. In such combinations, shear strengths of the individual pins are additive. For spring pin application refer to Figure 6-5-8.

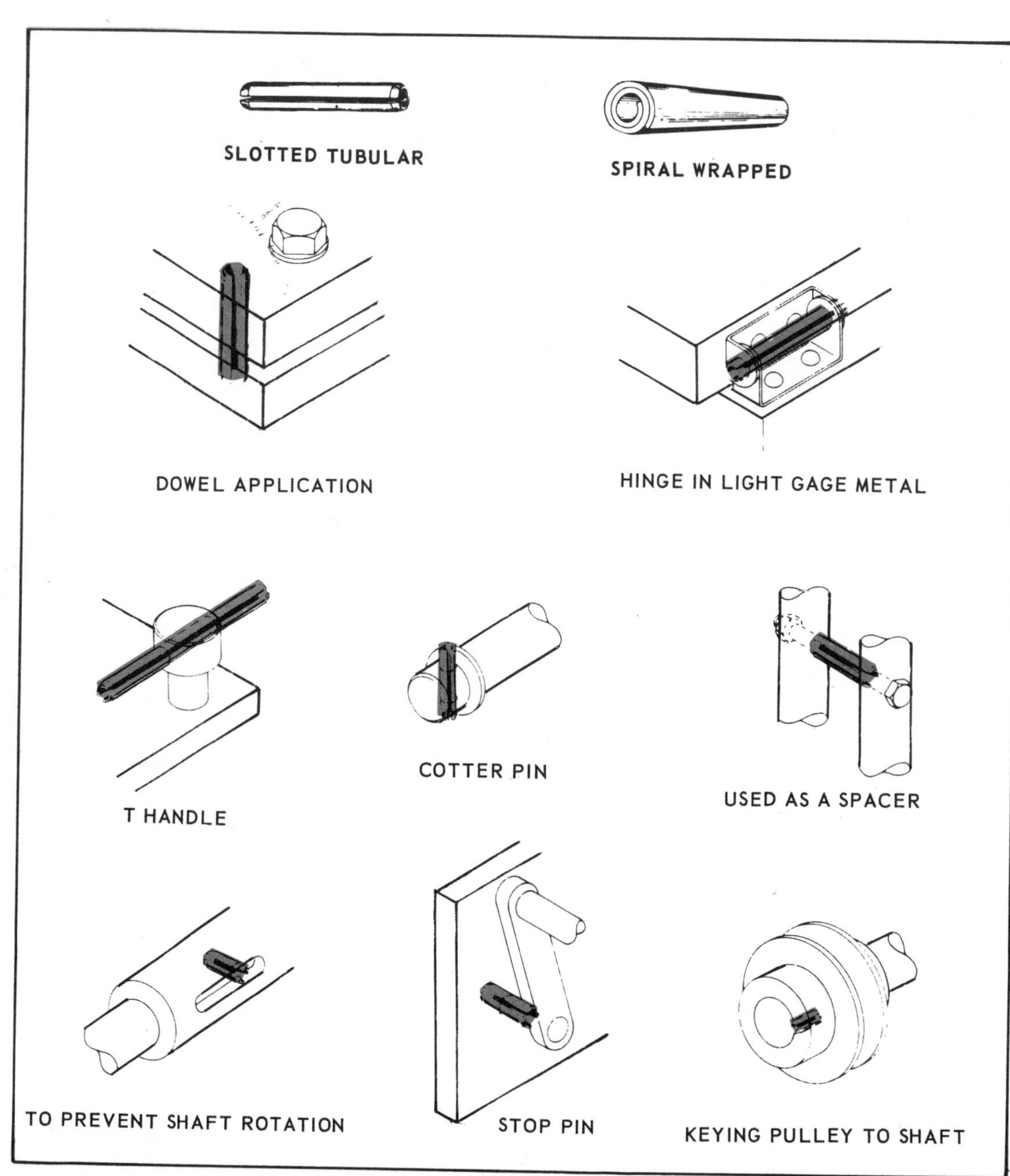

Fig. 6-5-8 Spring pin application

Assignments

1. On an A3 sheet, lay out the following four fastener assemblies, scale 1:1. Refer to the appendix for sizes and use your judgement for dimensions not shown.
   Pin and bolt fasteners, Fig. 6-5-A.
   Grooved pin fastener, Fig. 6-5-B.
   Square key fastener, Fig. 6-5-C.
   Woodruff key fastener, Fig. 6-5-D.

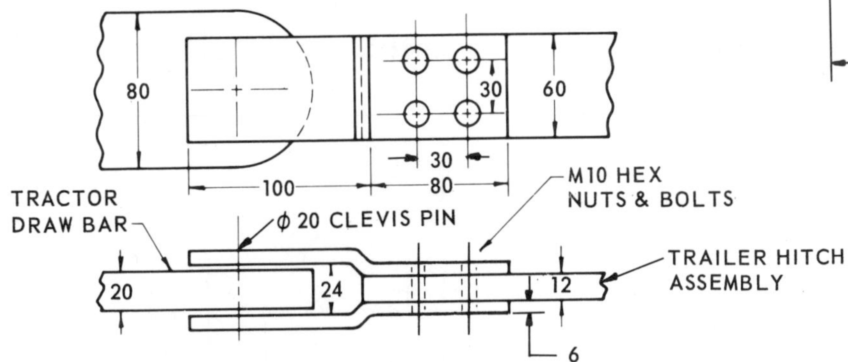

Fig. 6-5-A **Pin and bolt fasteners**

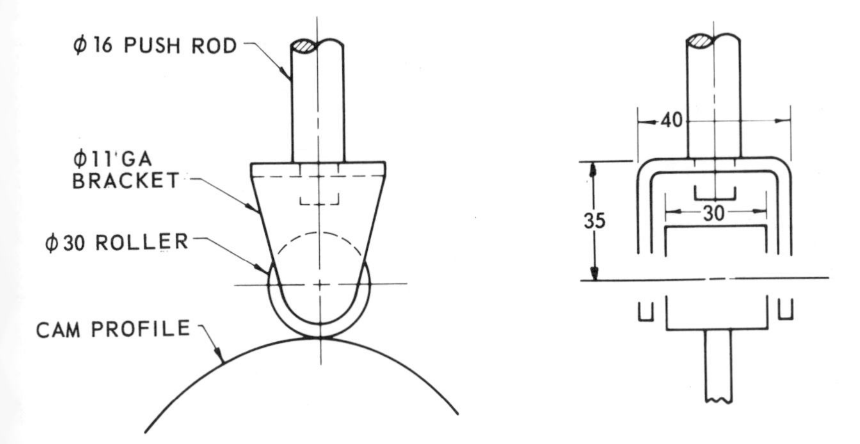

Fig. 6-5-B **Grooved pin fastener**

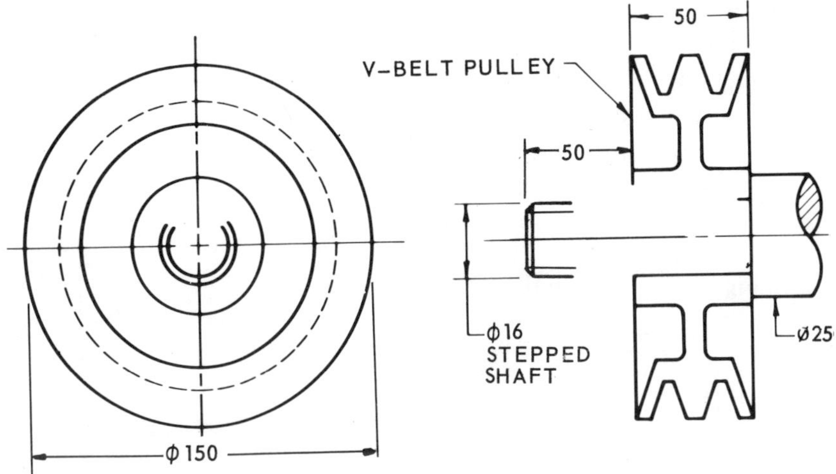

Fig. 6-5-C **Square key fastener**

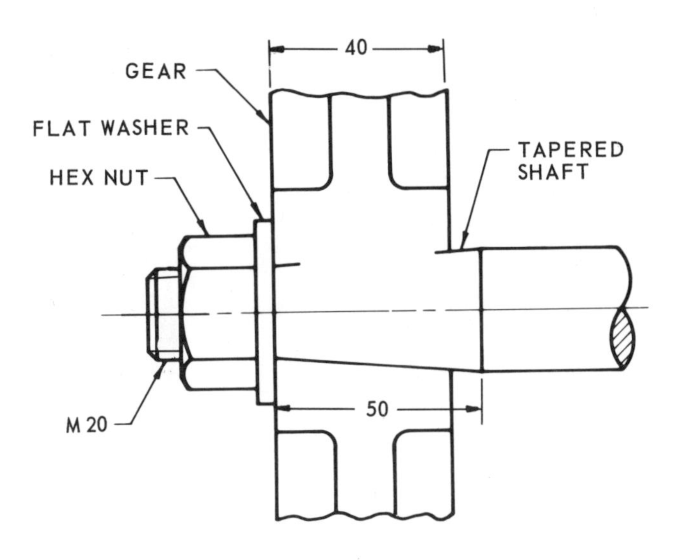

Fig. 6-5-D **Woodruff key fastener**

2. Crane hook assembly, Fig. 6-5-D, sheet size-A3, scale 1:1. Make a two view drawing. The hook is to be held to the U-frame with a slotted locknut. A spring pin is inserted through the locknut to prevent the nut from turning. A clevis pin with washer and cotter pin holds the pulley to the frame. Include on the drawing a bill of material.

Review for Assignments

Unit 2-5 Drawing circles and arcs

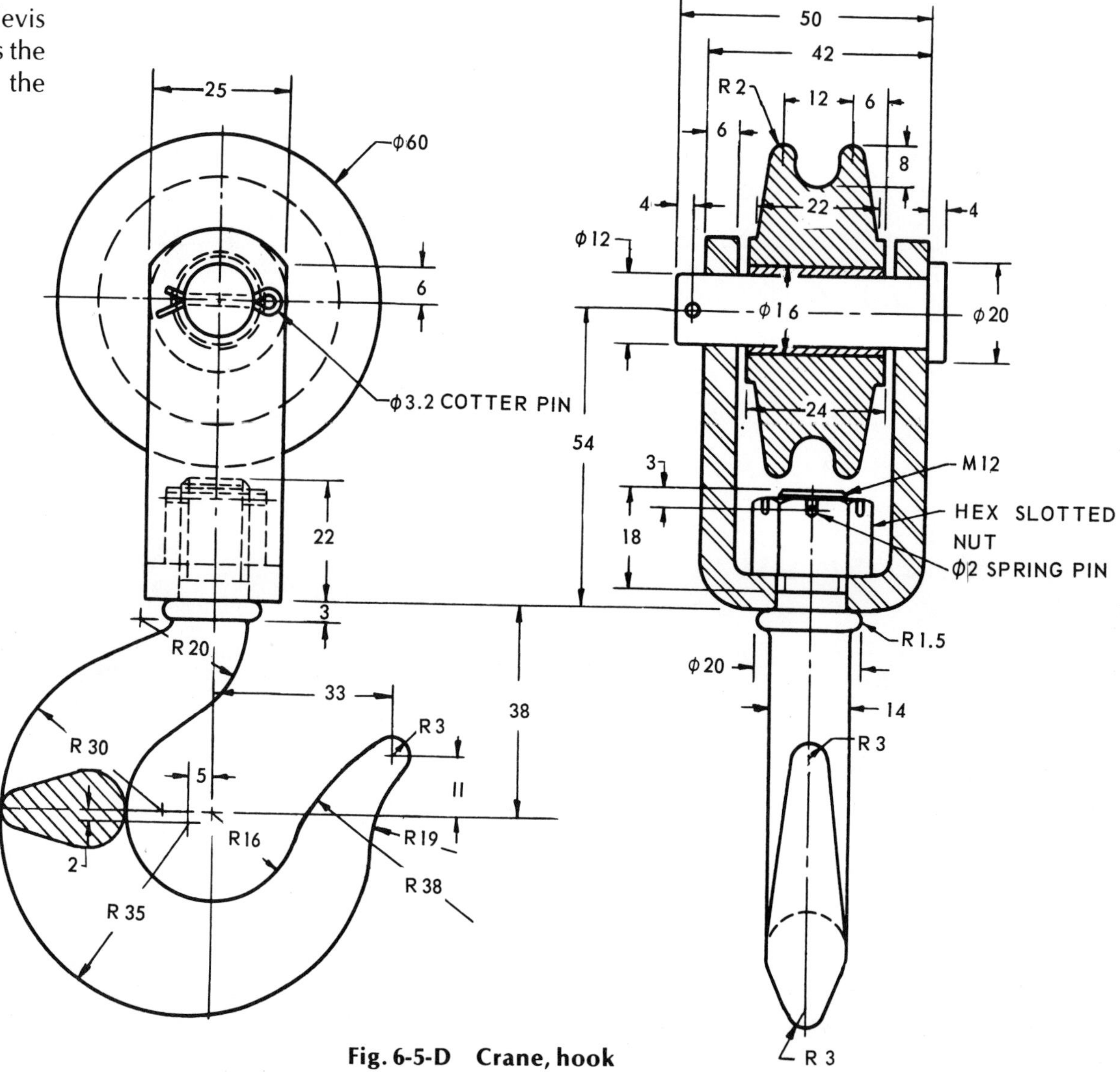

Fig. 6-5-D Crane, hook

Chapter 7

Sections and Conventions

UNIT 7-1
Sectional Views

Sectional views, commonly called **sections,** are used to show interior detail that is too complicated to be shown clearly by regular views that contain many hidden lines. For some assembly drawings, they indicate a difference in materials.

A sectional view is obtained by supposing the nearest part of the object to be cut or broken away on an imaginary cutting plane. The exposed or cut surfaces are identified by cross-hatching.

Hidden lines and details behind the cutting-plane line are usually omitted unless they are required for clarity or dimensioning. It should be understood that only in the sectional view is any part of the object shown as having been removed.

A sectional view frequently replaces one of the regular views. For example, a regular front view is replaced by a front view in section, as shown in Figure 7-1-1.

Whenever practical, except for revolved sections, sectional views should be projected perpendicular to the cutting plane and be placed in the normal position for third-angle projection.

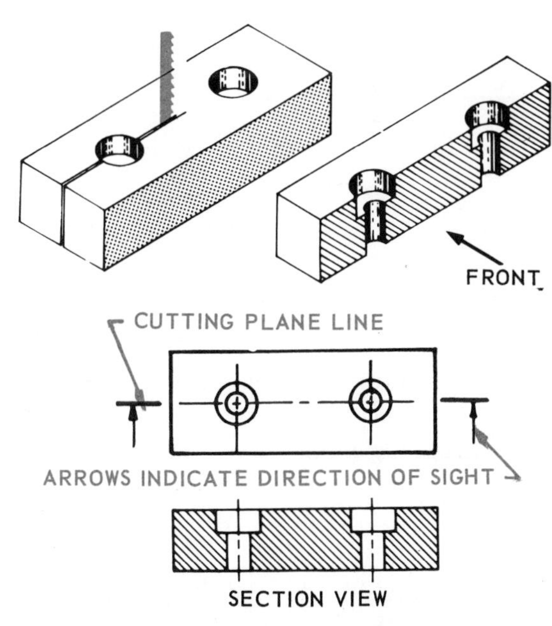

Fig. 7-1-1 A full section drawing

When the preferred placement is not practical, the sectional view may be removed to some other convenient position on the drawing, but it must be clearly identified, usually by two capital letters, and labelled.

Cutting-Plane Lines

Cutting-plane lines (Fig. 7-1-2) are used to indicate the location of cutting planes for sectional views and the viewing position for removed partial views.

(A) STRAIGHT

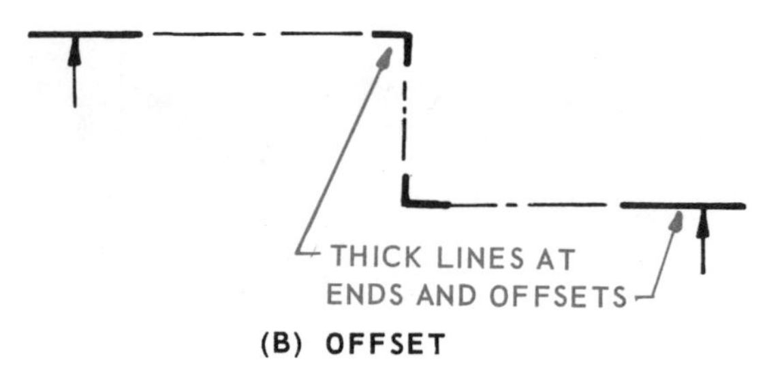

(B) OFFSET

Fig. 7-1-2 Cutting-plane lines

The cutting-plane line can be omitted when it corresponds to the centre line of the part and it is obvious where the cutting plane lies. On drawings with complicated line work, cutting-plane lines may be modified by omitting the dashes between the line ends for the purpose of simplicity.

Full Sections

When the cutting plane extends entirely through the object in a straight line and the front half of the object is theoretically removed, a full section is obtained (Figs. 7-1-3 and 7-1-4). This type of section is used for both detail and assembly drawings.

When the cutting-plane line falls on the centre line, it is not necessary to show its location (Fig. 7-1-5). However, it may be identified and indicated in the normal manner to increase clarity, if so desired.

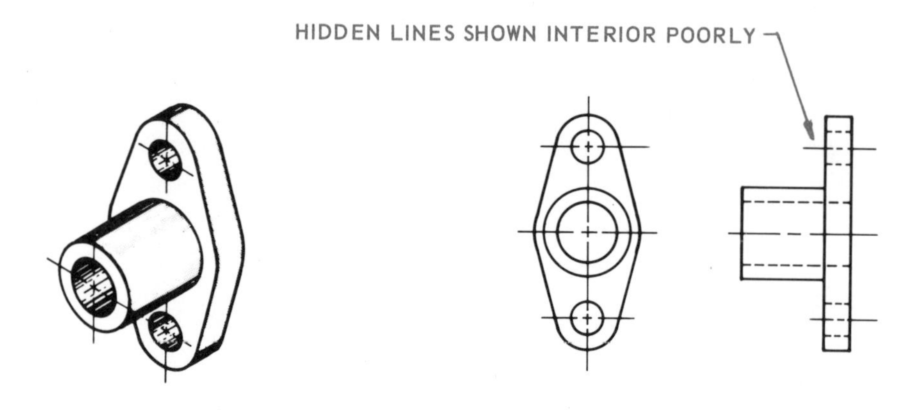

(A) SIDE VIEW NOT SECTIONED

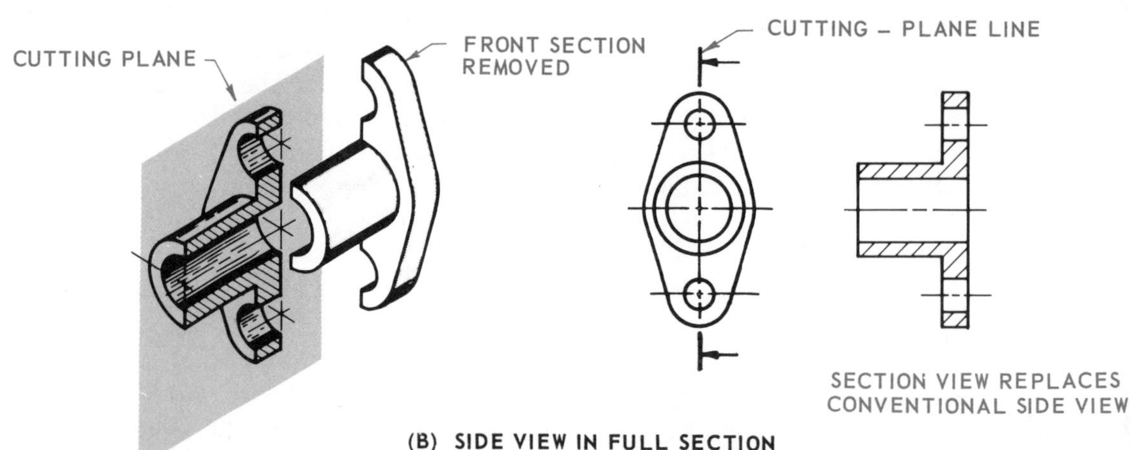

(B) SIDE VIEW IN FULL SECTION

Fig. 7-1-3 Full section view

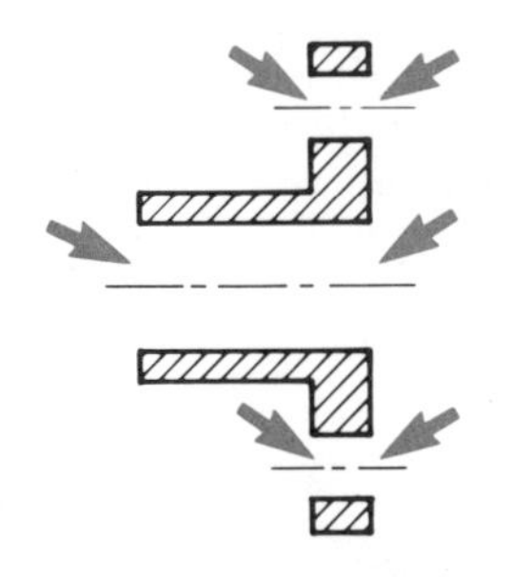

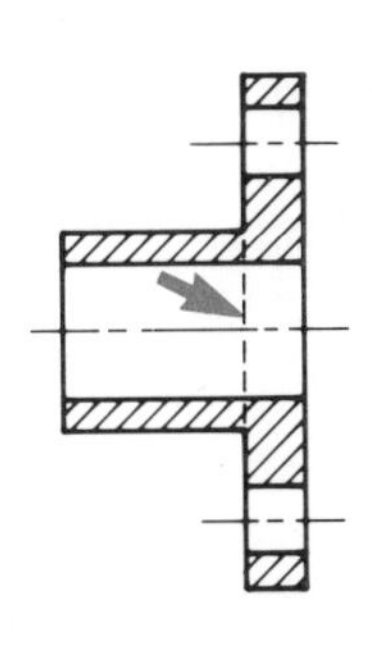

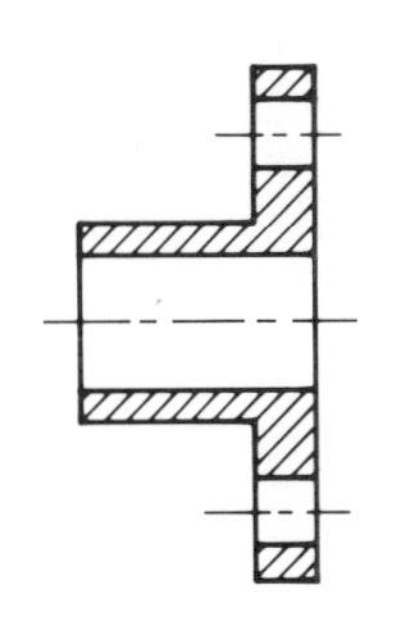

Fig. 7-1-4 Visible and hidden lines on section views

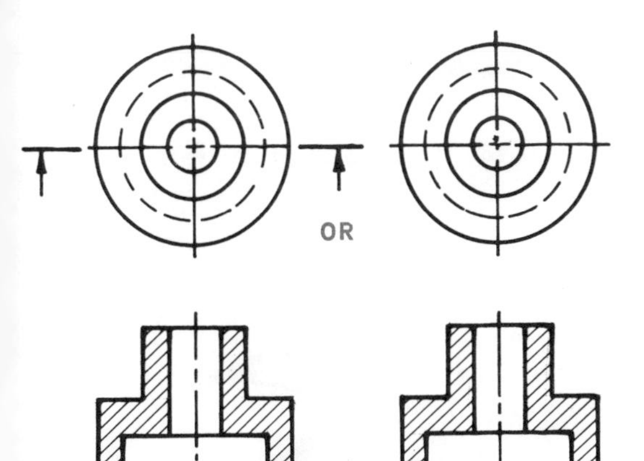

Fig. 7-1-5 Cutting-plane line may be omitted when it corresponds with the centre line

Hatching Lines

Hatching lines can serve a double purpose. They indicate the surface that has been theoretically cut and make it stand out clearly, thus helping the observer to understand the shape of the object. Hatching lines may also indicate the material from which the object is to be made, when the section lining symbols shown in Figure 7-1-6 are used.

Hatching Lines for Detail Drawings.

Since the exact material specifications for a part are usually given elsewhere on the drawing, the general-purpose hatching lines symbol is recommended for most detail drawings. An exception may be made for wood when it is desirable to show the direction of the grain.

IRON AND GENERAL USE FOR ALL MATERIAL

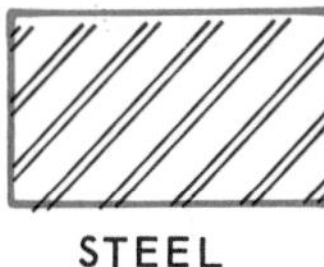

SOUND INSULATION

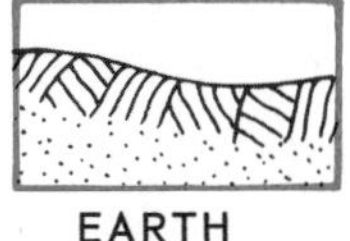

WHITE METAL, ZINC, LEAD, BABBITT, AND ALLOYS

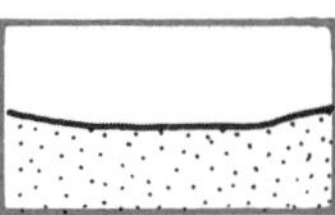

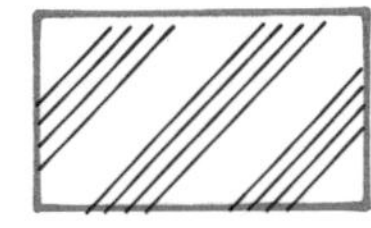

CONCRETE

Fig. 7-1-6 Symbolic hatching lines

Hatching lines are thin and are usually drawn at an angle of 45° to the major outline of the object. The same angle is used for the whole "cut" surface of the object. If the part shape would cause hatching lines to be parallel, or nearly so, to one of the sides of the part, then some angle other than 45° should be chosen (Fig. 7-1-7). The spacing of the hatching lines should be reasonably uniform to give a good appearance to the drawing. The pitch, or distance between lines, normally varies between 1 and 3 mm depending on the size of the area to be sectioned.

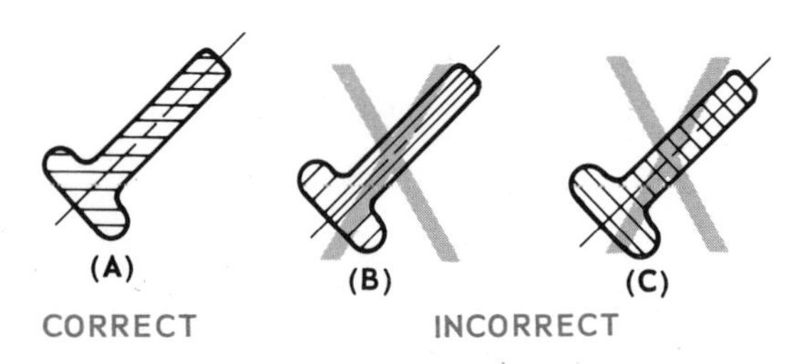

Fig. 7-1-7 Direction of hatching lines

Large areas shown in section need not be entirely lined (Fig. 7-1-8). Hatching lines around the outline will usually be sufficient, providing clarity is not lost.

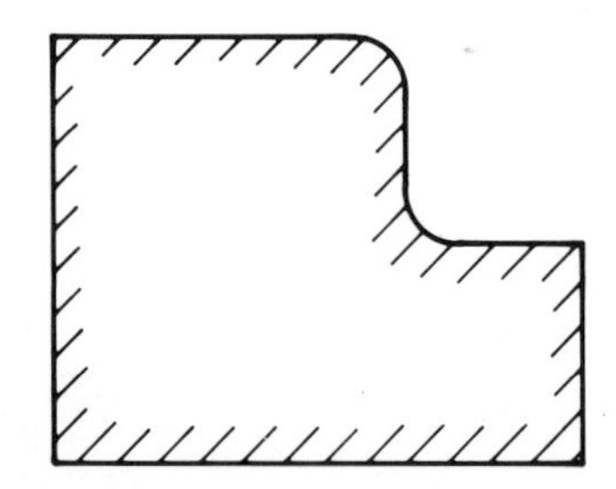

Fig. 7-1-8 Hatching lines outlining the part

Dimensions or other lettering should not be placed in sectioned areas. When this is unavoidable, the hatching lines should be omitted for the numerals or lettering (Fig. 7-1-9).

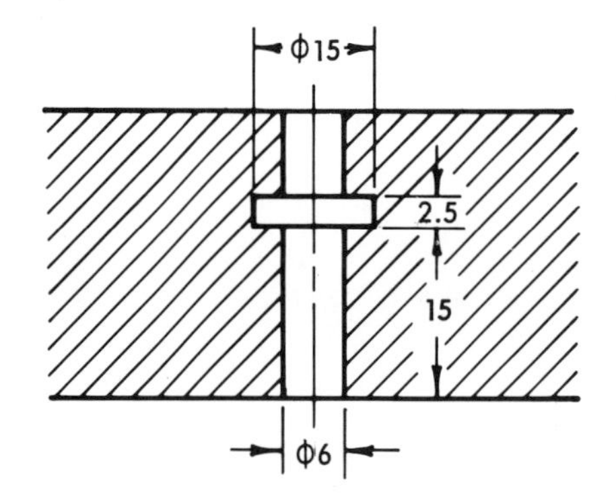

Fig. 7-1-9 Hatching lines omitted to accommodate dimensions

Sections which are too thin for hatching lines, such as sheet-metal items, packing, and gaskets, may be shown without these lines, or the area may be filled in completely (Fig. 7-1-10).

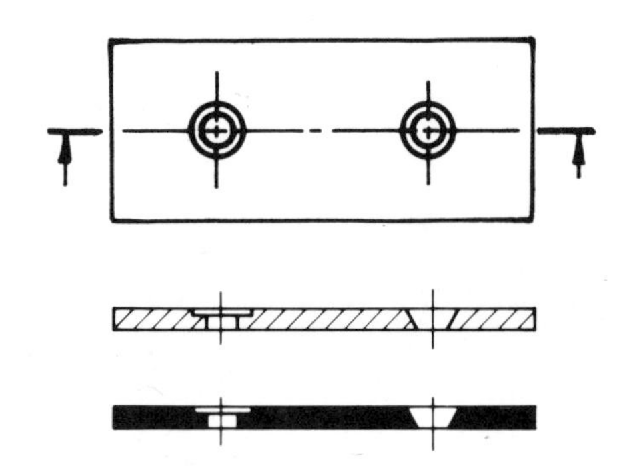

Fig. 7-1-10 Thin parts in section

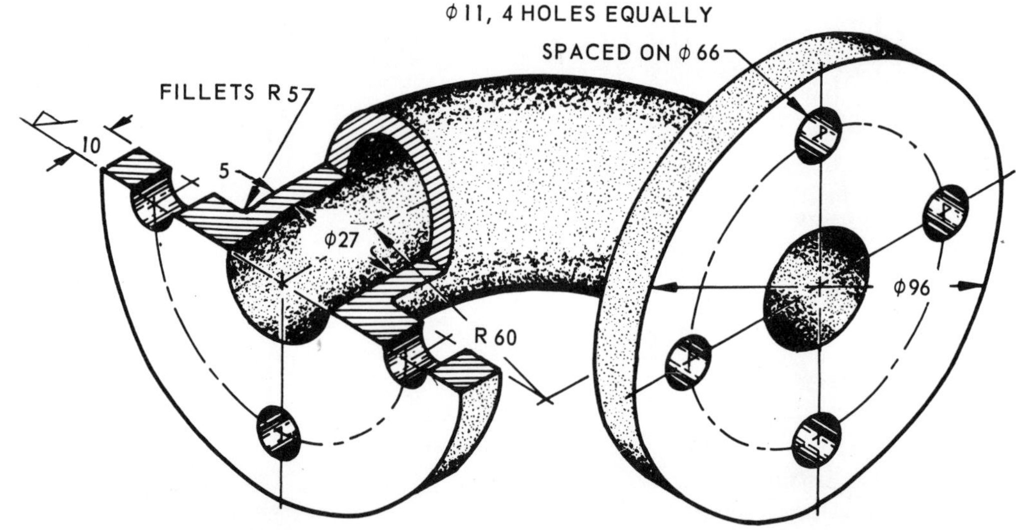

Fig. 7-1-A Flanged elbow

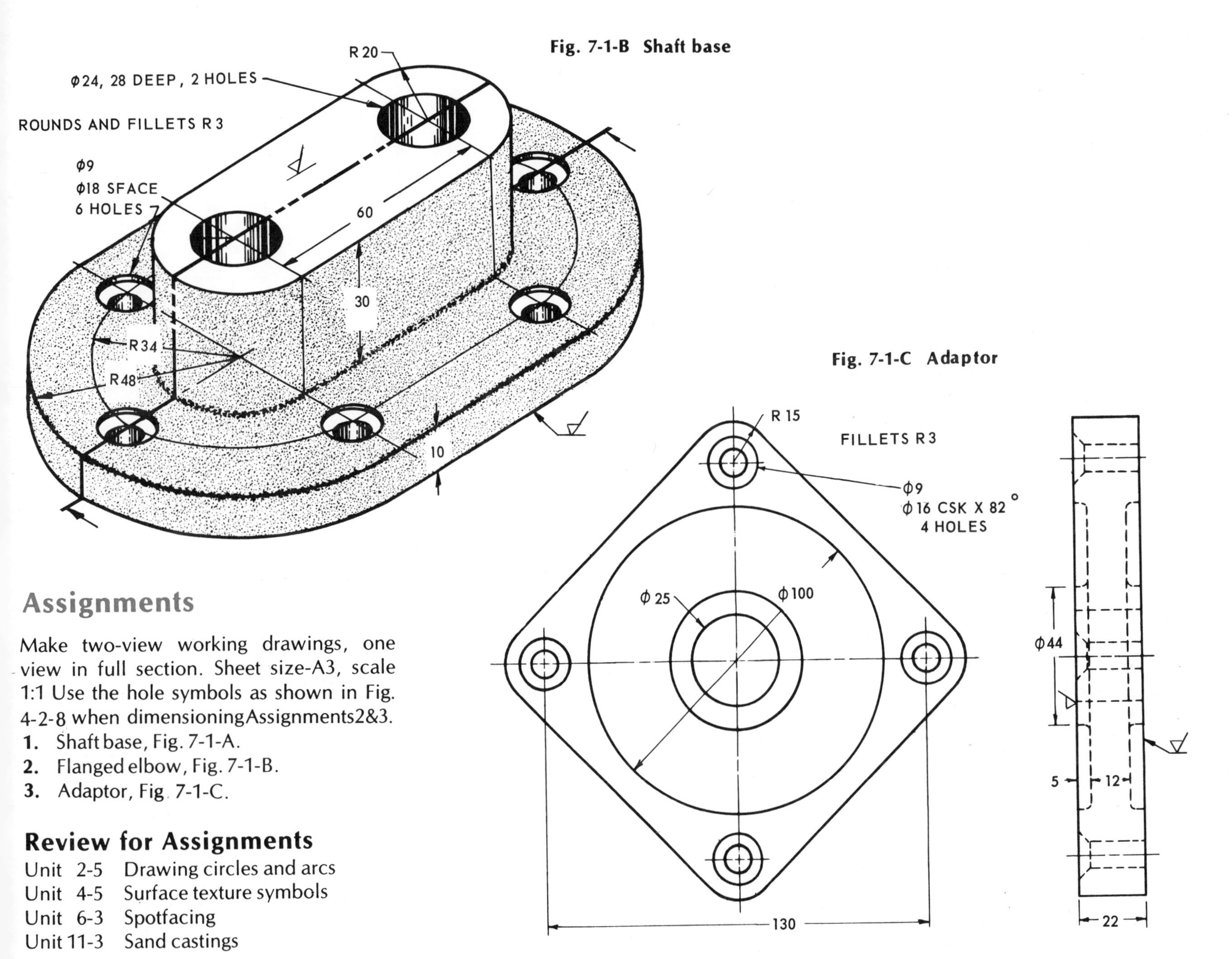

Fig. 7-1-B Shaft base

Fig. 7-1-C Adaptor

Assignments

Make two-view working drawings, one view in full section. Sheet size-A3, scale 1:1 Use the hole symbols as shown in Fig. 4-2-8 when dimensioning Assignments 2&3.

1. Shaft base, Fig. 7-1-A.
2. Flanged elbow, Fig. 7-1-B.
3. Adaptor, Fig. 7-1-C.

Review for Assignments

Unit 2-5 Drawing circles and arcs
Unit 4-5 Surface texture symbols
Unit 6-3 Spotfacing
Unit 11-3 Sand castings

UNIT 7-2
Two or More Sectional Views on One Drawing

If two or more sections appear on the same drawing, the cutting-plane lines are identified by two identical large, Gothic letters. One is placed at each end of the line, behind or beside the arrowhead so that the arrow points away from the letter. The identification letters should not include I, O, and Q (Fig. 7-2-1).

Sectional view subtitles are given when identification letters are used. They appear directly below the view, incorporating the letters at each end of the cutting-plane line thus: **SECTION A-A**, or abbreviated, **SECT. B-B**. When the scale is different from the main view, it is stated below the subtitle thus:

SECTION A-A
SCALE 1:5

NOTE: HIDDEN LINES SHOWN ON SECTION VIEWS, OTHERWISE FEATURES D AND E MAY BE MISTAKEN AS BEING SOLID.

LETTER PLACED BEHIND OR TO THE SIDE OF THE ARROW

SECTION A–A

SECTION B–B

Fig. 7-2-1 Detail drawing having two section views

Assignments

Make working drawings with section views as indicated. Select appropriate sheet size and scale.

1. Guide block, Fig. 7-2-A.
2. Housing, Fig. 7-2-B.
3. Casing, Fig. 7-2-C.

Review for Assignments

Unit 4-5 Surface texture symbols
Unit 7-1 Hatching lines

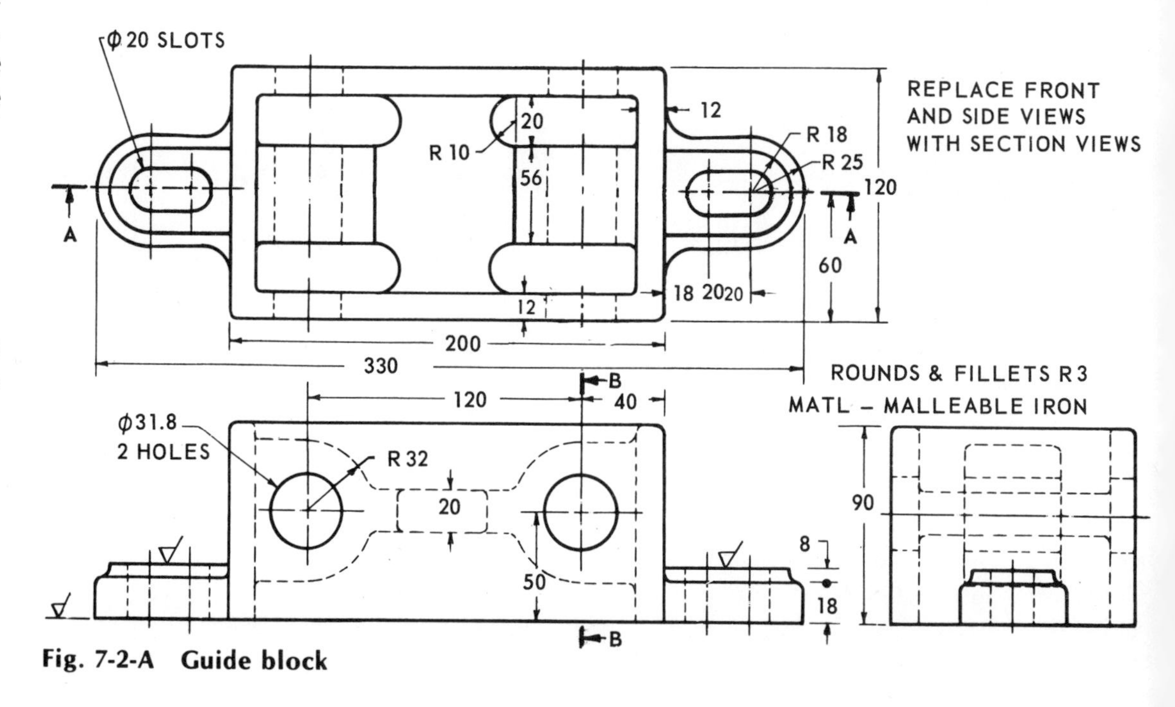

Fig. 7-2-A Guide block

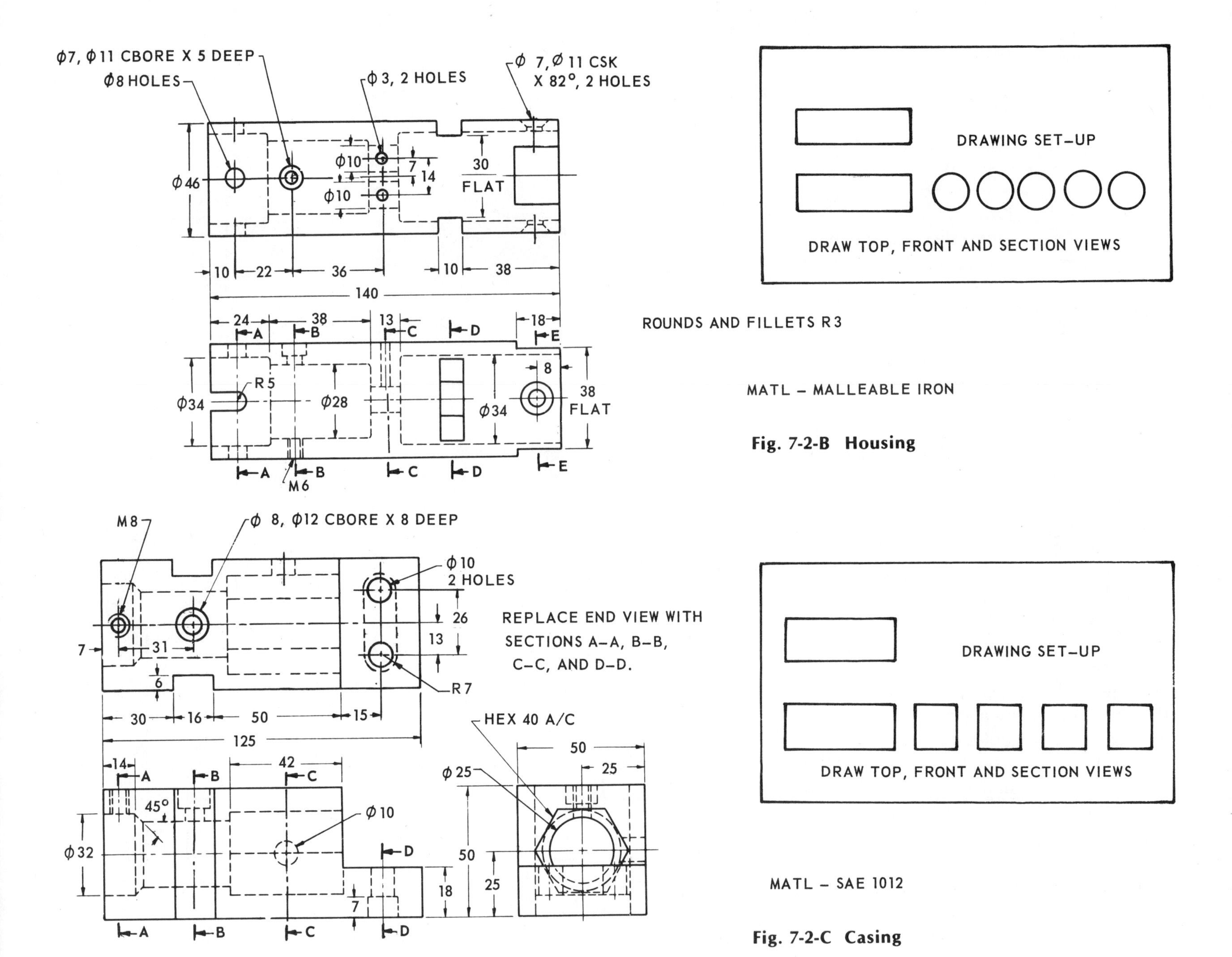

Fig. 7-2-B Housing

Fig. 7-2-C Casing

UNIT 7-3
Half Sections

A **half section** is a view of an assembly or object, usually symmetrical, showing one half of the view in section. See Figures 7-3-1 and 7-3-2. Two cutting-plane lines, perpendicular to each other, extend halfway through the view, and one quarter of the view is considered removed and the interior is exposed to view.

Similar to the practice followed for full-section drawings, the cutting-plane line need not be drawn for half sections when it is obvious where the cutting plane took place. Instead, centre lines may be used. When a cutting plane is used, the common practice is to show only one end of the cutting-plane line, terminating with an arrow to indicate the direction of sight for viewing the section.

On the sectional view a centre line or a visible object line may be used to divide the sectioned half from the unsectioned half of the drawing. This type of sectional drawing is best suited for assembly drawings where both internal and external construction is shown on one view and where only overall and centre-to-centre dimensions are required. The main disadvantage of using this type of sectional drawing for detail drawings is the difficulty in dimensioning internal features without adding hidden lines. However, hidden lines may be added for dimensioning, as shown in Figure 7-3-3.

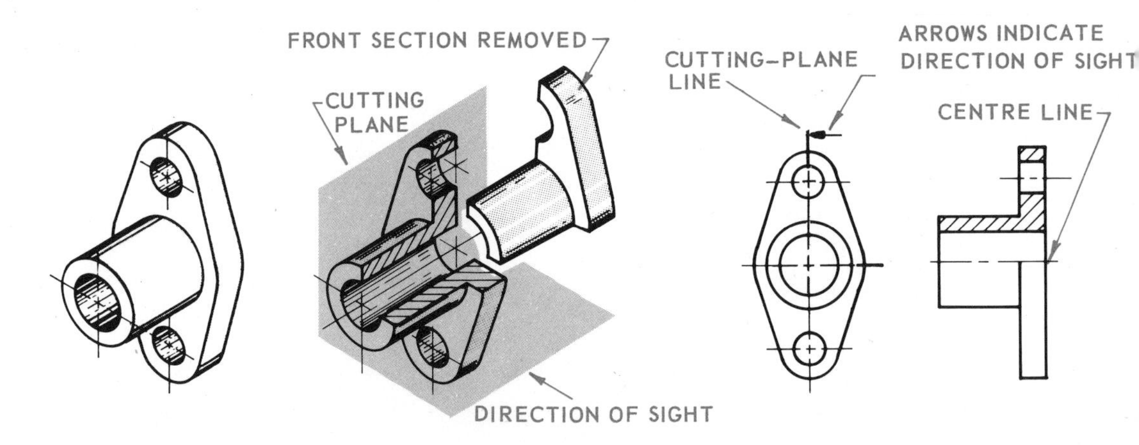

Fig. 7-3-1 A half-section drawing

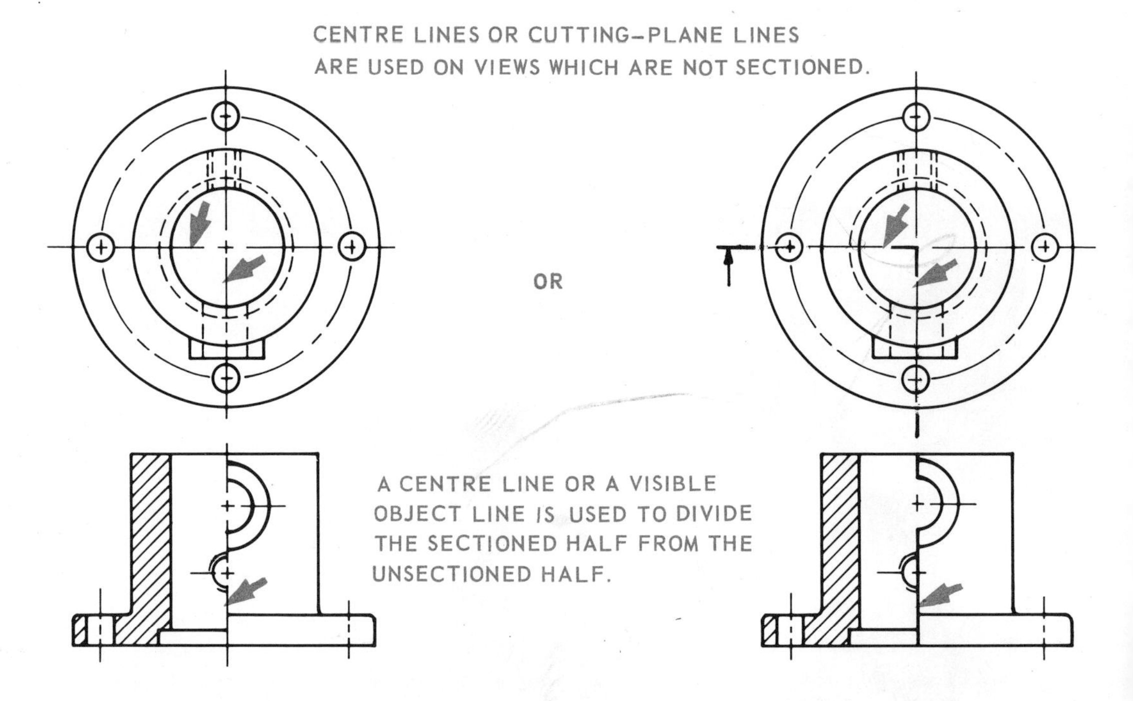

Fig. 7-3-2 Half-section views

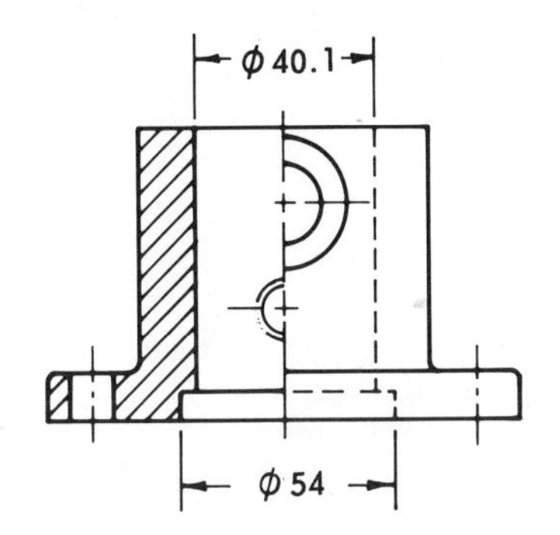

HIDDEN LINES ADDED FOR DIMENSIONING

Fig. 7-3-3 Dimensioning half section views

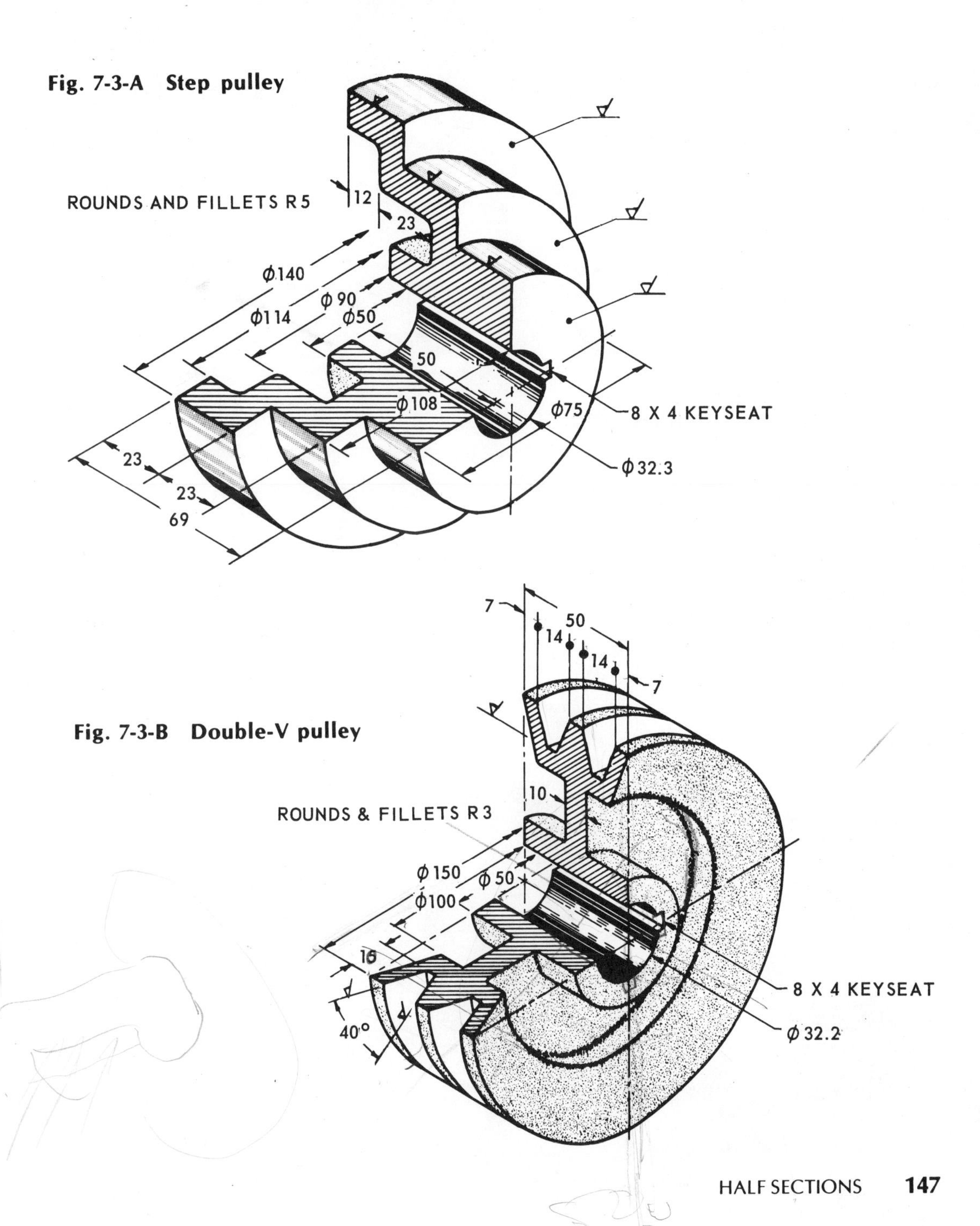

Fig. 7-3-A Step pulley

Fig. 7-3-B Double-V pulley

Assignments

Make two-view working drawings showing the side view in section. Sheet size-A3, scale 1:1.

1. Step pulley, Fig. 7-3-A.
2. Double-V pulley, Fig. 7-3-B.

Review for Assignments

Unit 4-5 Surface texture symbols
Unit 6-5 Keys

UNIT 7-4
Threads in Section

True representation of a screw thread is seldom provided on working drawings. It would require very laborious and accurate drawing and repetitious development of the helix curve of the thread. A symbolic representation of threads is now standard practice.

Two types of conventions are in general use for screw thread representation (Fig. 7-4-1). These are known as conventional, and alternative representation.

Conventional representation should be used whenever possible.

Alternative (pictorial) representation requires more drafting time, but is sometimes necessary to avoid confusion with other parallel lines or to more clearly portray particular aspects of the threads.

Threaded Assemblies

Any of the thread conventions shown here may be used for assemblies of threaded parts, and two or more methods may be used on the same drawing, as shown in Figure 7-4-2. In sectional views, the externally threaded part is always shown covering the internally threaded part, as illustrated in Figure 7-4-3.

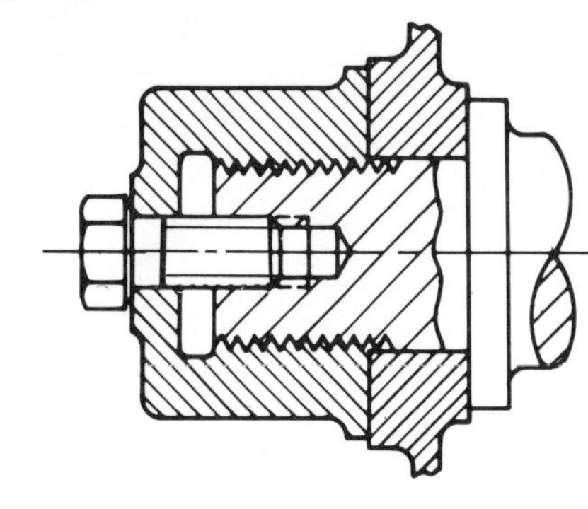

Fig. 7-4-2 Threaded assembly

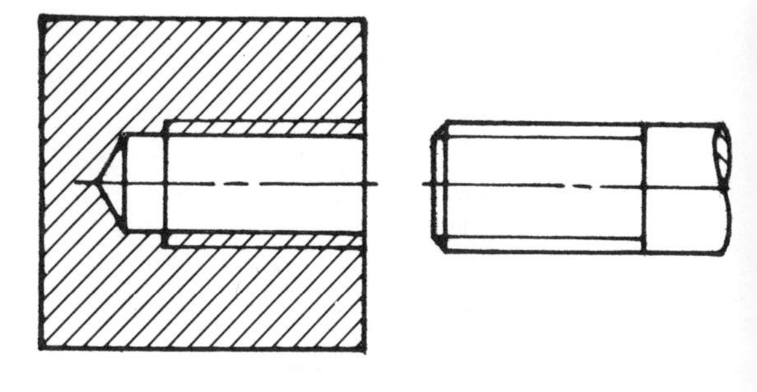

(A) BEFORE ASSEMBLY

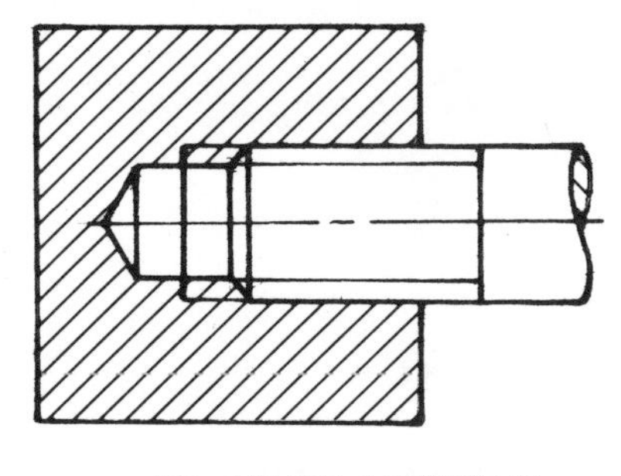

(B) AFTER ASSEMBLY

Fig. 7-4-3 Drawing threads in section

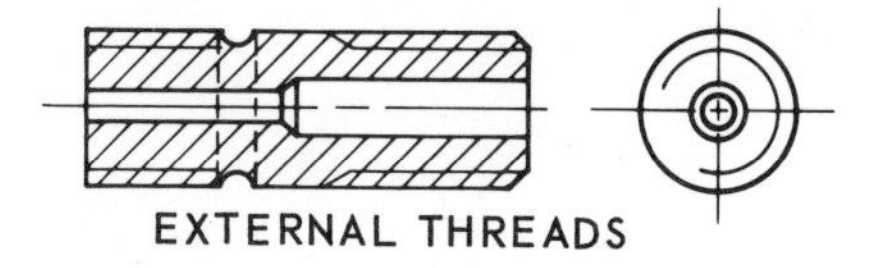

EXTERNAL THREADS

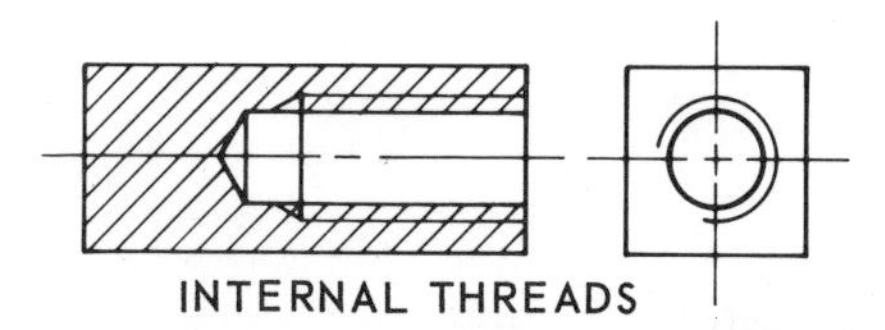

INTERNAL THREADS

(A) CONVENTIONAL THREAD CONVENTION

EXTERNAL THREADS

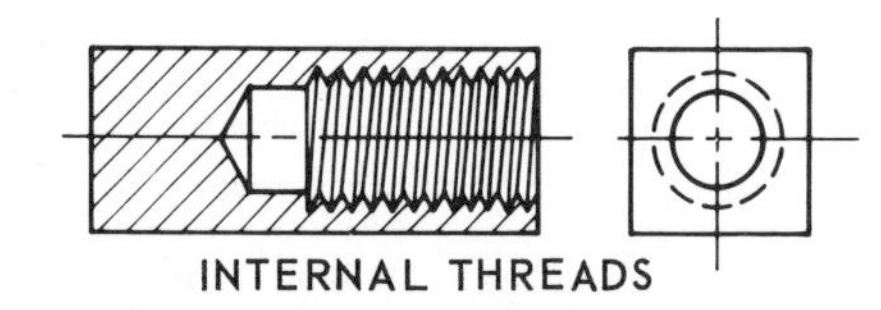

INTERNAL THREADS

(B) ALTERNATIVE (PICTORIAL) THREAD CONVENTION

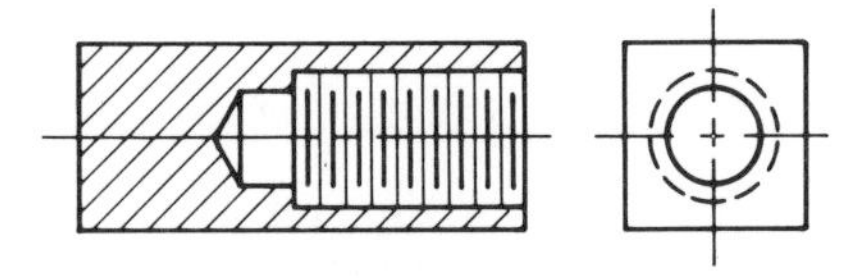

(C) FORMER THREAD (SCHEMATIC) CONVENTION

Fig. 7-4-1 Threads in section

Assignments

Make working drawings, selecting the number of views and the best type of section view which will clearly describe the part. Select suitable sheet size and scale.

1. Valve body, Fig. 7-4-A.
2. Pipe plug, Fig. 7-4-B.
3. End plate, Fig. 7-4-C.

Review for Assignments

Unit 6-1 Thread forms

Unit 6-2 Conventional thread representation

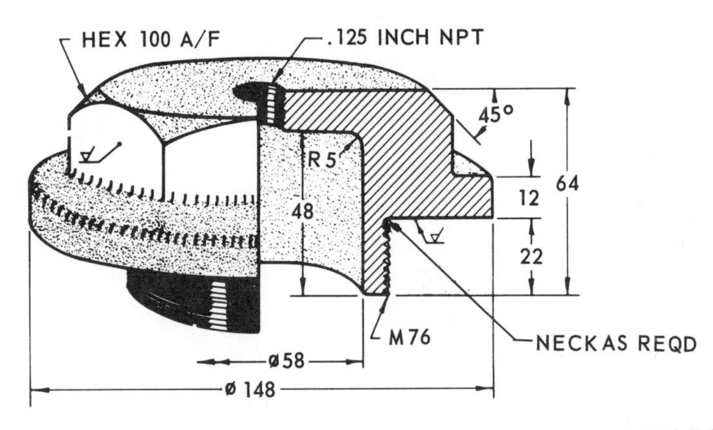

Fig. 7-4-B Pipe plug

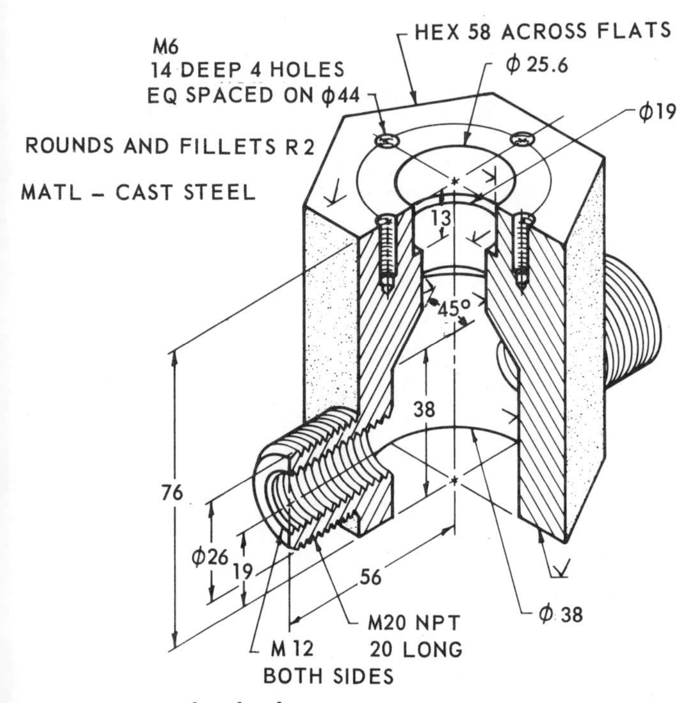

Fig. 7-4-A Valve body

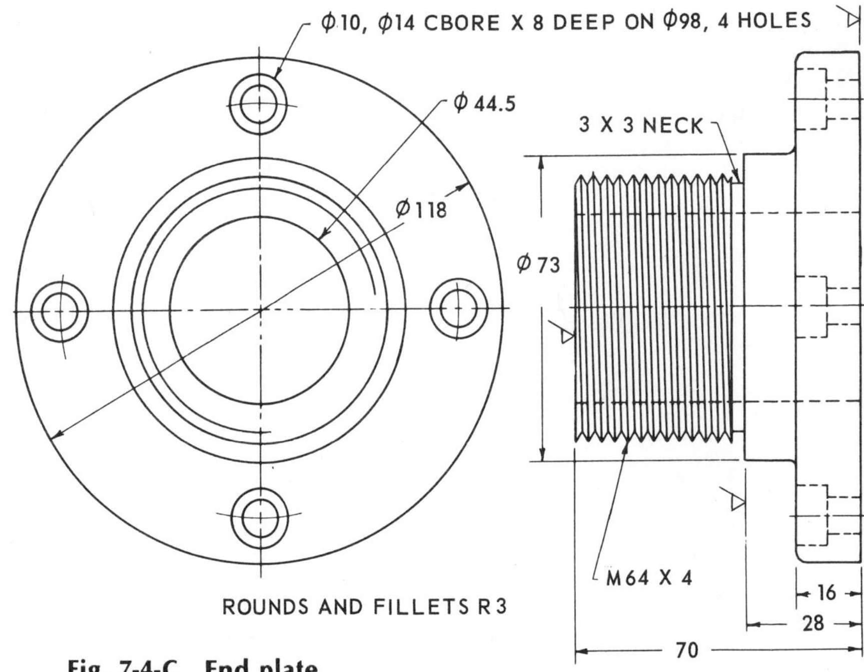

Fig. 7-4-C End plate

UNIT 7-5
Assemblies in Section

Hatching Lines on Assembly Drawings

General-purpose hatching lines are recommended for most assembly drawings, especially if the detail is small. Symbolic hatching lines are generally not recommended for drawings that will be microfilmed.

General-purpose hatching lines should be drawn at an angle of 45° with the main outlines of the view. On adjacent parts, the hatching lines should be drawn in the opposite direction, as shown in Figures 7-5-1 and 7-5-2.

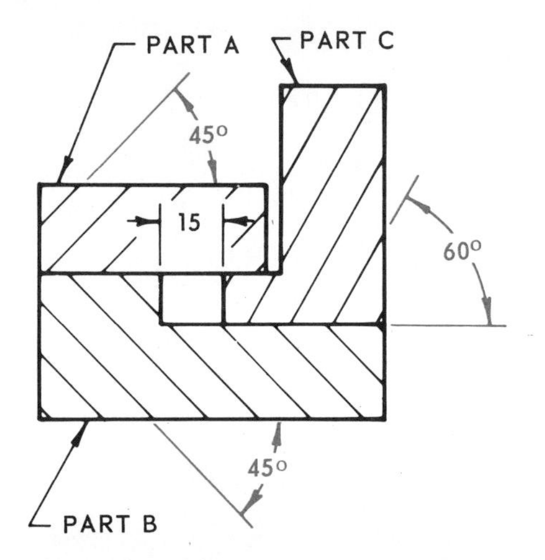

Fig. 7-5-1 Direction of hatching lines

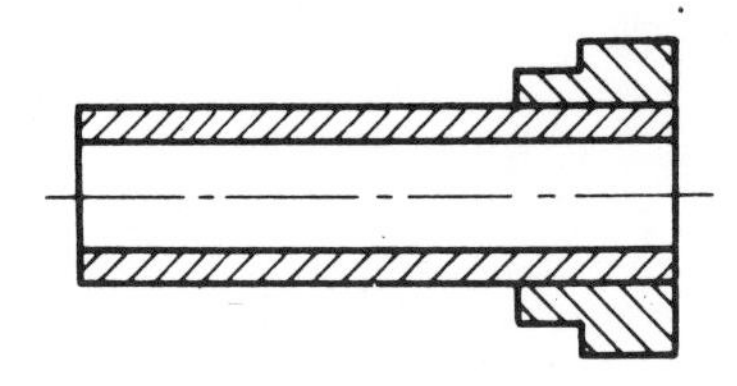

(A) ADJACENT PARTS

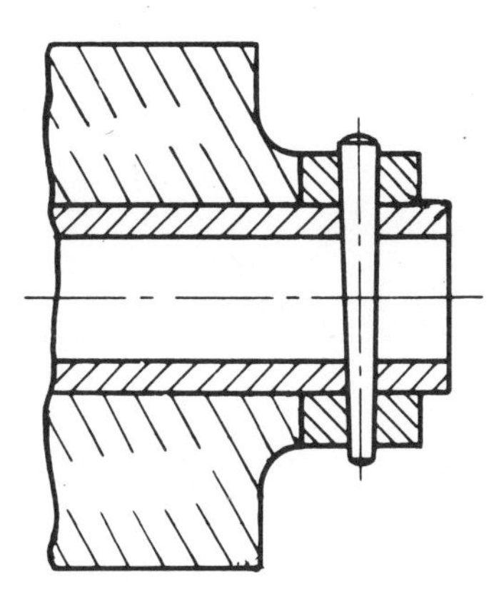

(B) SPACING OF HATCHING LINES ACCORDING TO THE AREA TO BE SECTIONED

Fig. 7-5-2 Arrangement of hatching lines

For additional adjacent parts, any suitable angle may be used to make each part stand out separately and clearly. Hatching lines should not be purposely drawn to meet at common boundaries.

When two or more thin adjacent parts are filled in, a space is left between them, as shown in Figure 7-5-3.

Symbolic hatching lines are used on special-purpose assembly drawings such as illustrations for parts catalogues, display assemblies, promotional materials, etc., when it is desirable to distinguish between different materials (Fig. 7-1-6).

All assemblies and subassemblies pertaining to one particular set of drawings should use the same symbolic conventions.

(A) STEEL PLATES

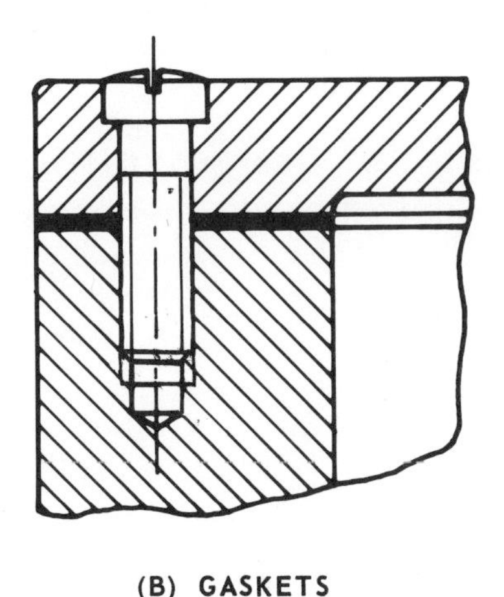

(B) GASKETS

Fig. 7-5-3 Assembly of thin parts in section

Shafts, Bolts, Pins, Keyways, etc., in Section. Shafts, bolts, nuts, rods, rivets, keys, pins, and similar solid parts, the axes of which lie in the cutting plane, should not be sectioned. Only a broken-out section of the shaft may be used to indicate clearly the key, keyseat, or pin (Fig. 7-5-4).

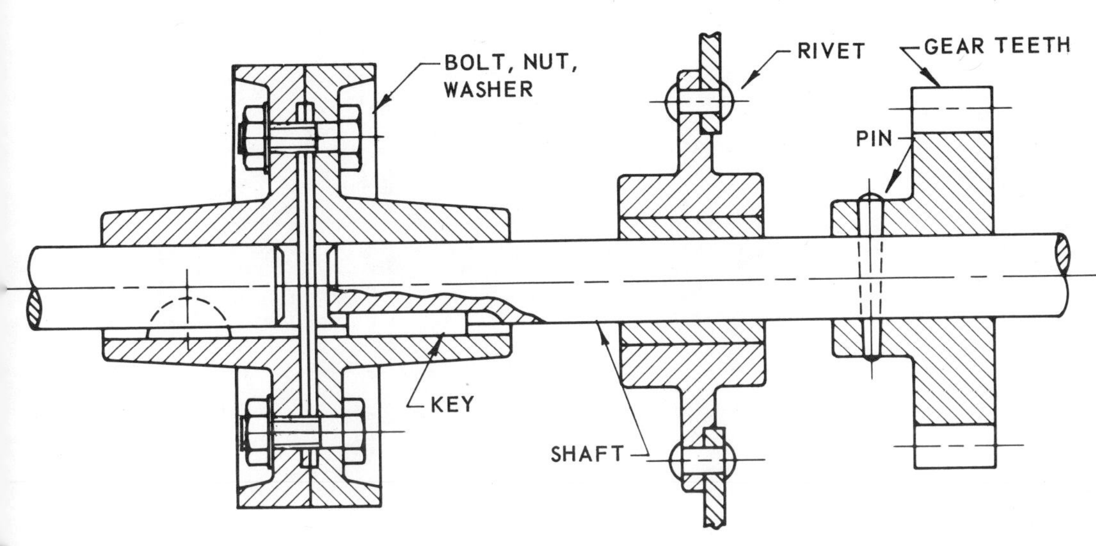

Fig. 7-5-4
Parts that are not hatched even though the cutting plane line passes through them

ϕ13, 4 HOLES EQUALLY SPACED
14
ϕ116
2
8 SQ KEY
ϕ110
ϕ32
ϕ90
ϕ60
ϕ140
32
72

FLANGES HELD TOGETHER BY M12 X 1.75 X 50 LG
HEX HD BOLTS WITH LOCKWASHERS
2 mm NEOPRENE GASKET BETWEEN FLANGES

Fig. 7-5-A Flanged coupling

Assignments

Make a one-view assembly drawing in section. Include a bill of material and identify the parts on the assembly. Sheet size and scale to suit.

1. Flanged coupling, Fig. 7-5-A.
2. Connecting link, Fig. 7-5-B.
3. Caster, Fig. 7-5-C.

Review for Assignments

Unit 5-3 Assembly drawings
Unit 5-3 Bills of materials
Unit 6-3 Threaded fasteners
Unit 6-3 Spotfacing

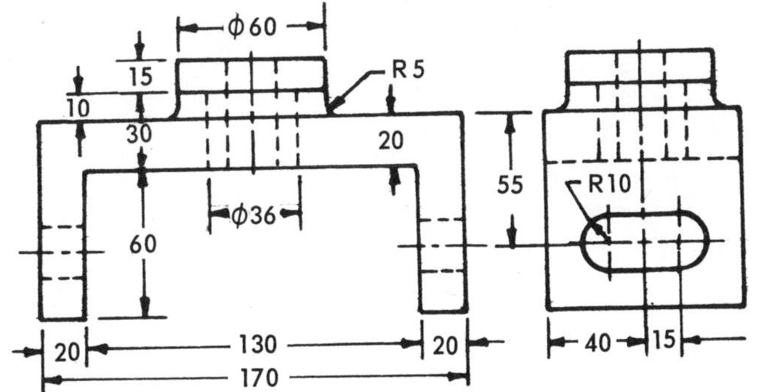

BUSHING
BASE

Fig. 7-5-B Connecting link

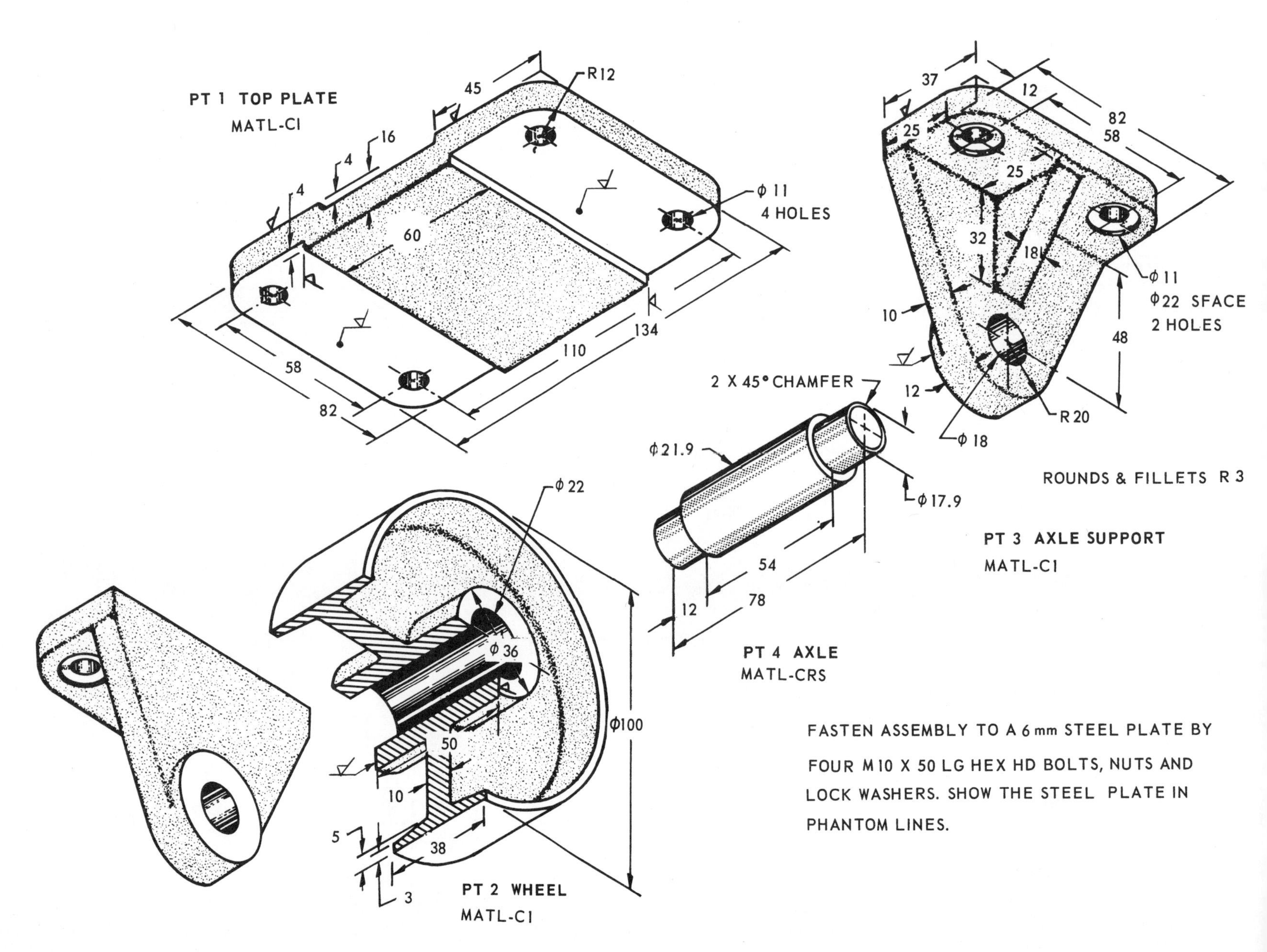

FASTEN ASSEMBLY TO A 6 mm STEEL PLATE BY FOUR M10 X 50 LG HEX HD BOLTS, NUTS AND LOCK WASHERS. SHOW THE STEEL PLATE IN PHANTOM LINES.

Fig. 7-5-C Caster

UNIT 7-6

Offset Sections

In order to include features that are not in a straight line, the cutting plane may be offset or bent, so as to include several planes or curved surfaces (Figs. 7-6-1 and 7-6-2).

An offset section is similar to a full section in that the cutting-plane line extends through the object from one side to the

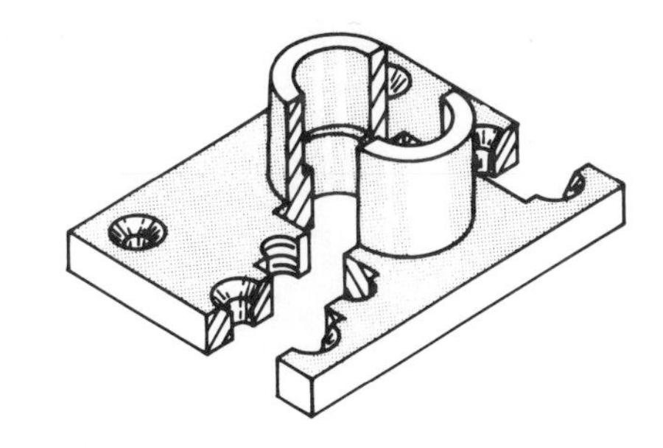

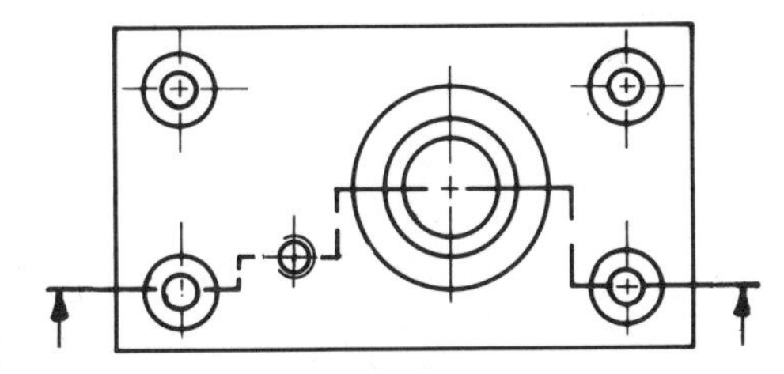

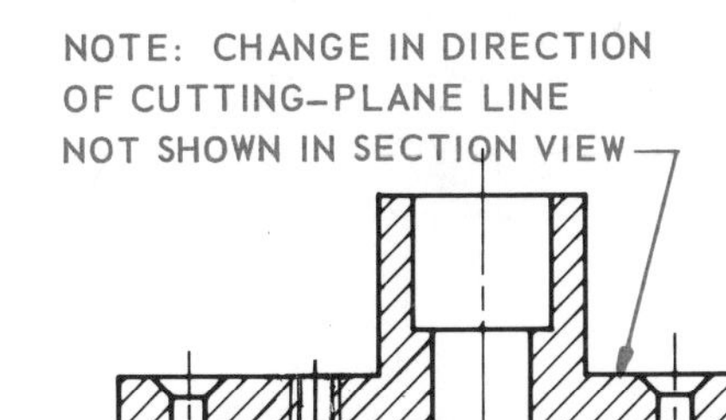

Fig. 7-6-1 An offset section

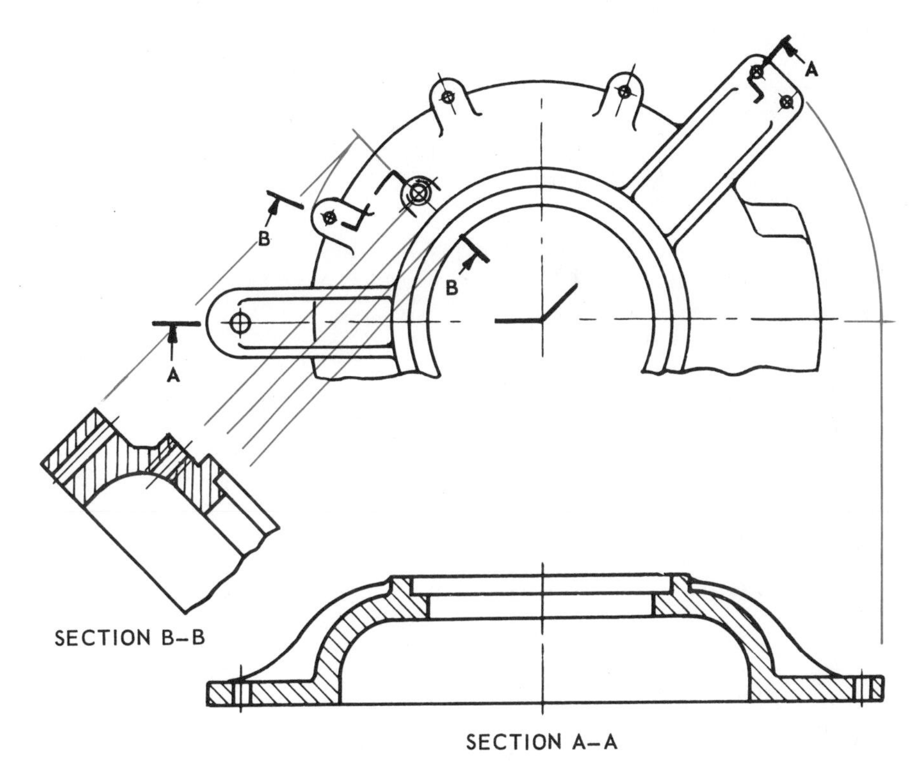

Fig. 7-6-2 Positioning offset sections

other. The change in direction of the cutting-plane line is not shown in the sectional view.

Assignments

Make working drawings with section views as indicated. Sheet size and scale to suit.

1. Clamp guard, Fig. 7-6-A.
2. Locating block, Fig. 7-6-B.
3. Index block, Fig. 7-6-C.
4. Jacket, Fig. 7-6-D.
5. Base plate, Fig. 7-6-E.
6. Mounting plate, Fig. 7-6-F.

Review for Assignments

Unit 4-4 Arrowless and tabular dimensioning

Unit 14-3 Simplified drafting

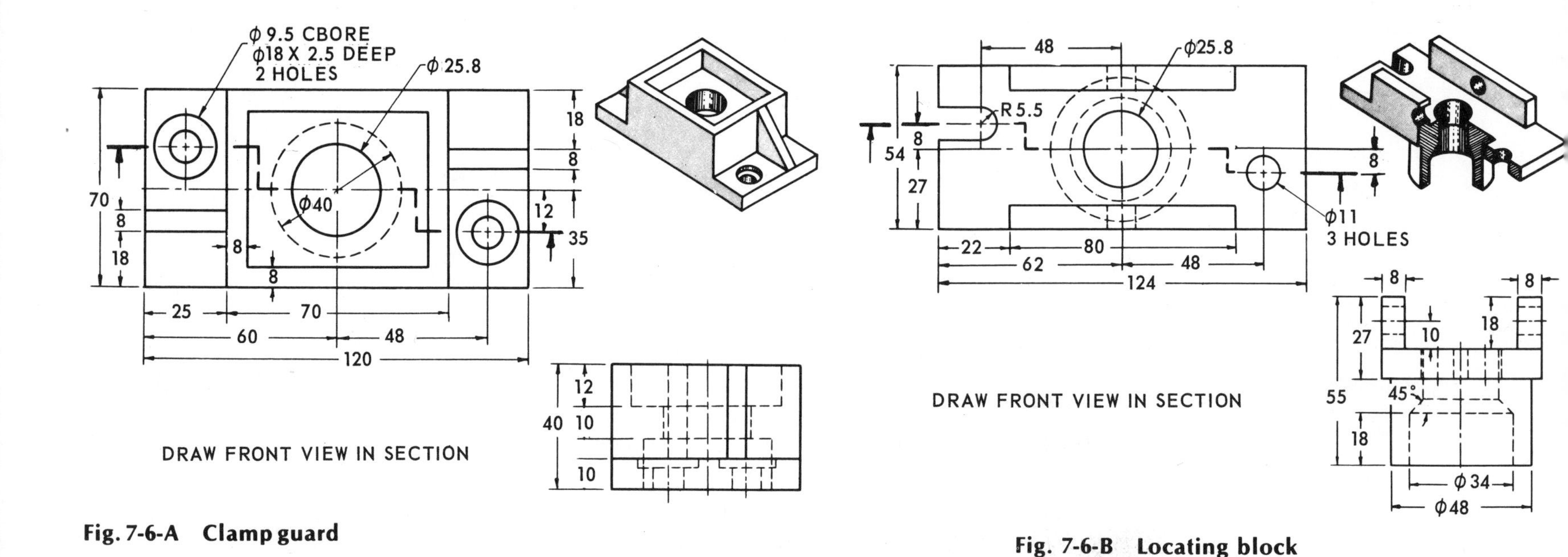

Fig. 7-6-A Clamp guard

Fig. 7-6-B Locating block

M10 X 1.25

φ11
4 HOLES

R 10

DRAW FRONT VIEW IN SECTION

Fig. 7-6-C Index block

φ12 SLOT

M 10 X 1.5

φ10 SLOT

φ 10
3 HOLES

DRAW THE FRONT VIEW IN SECTION

φ 22

φ 12.5

Fig. 7-6-D Jacket

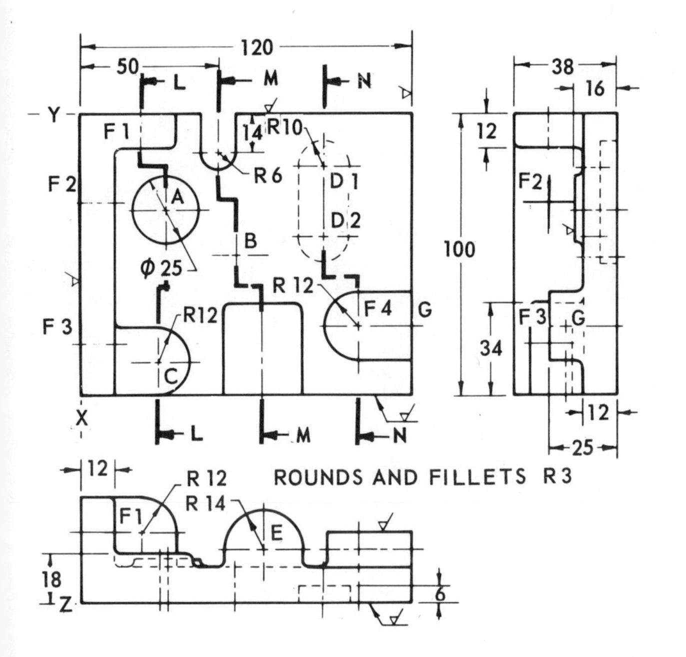

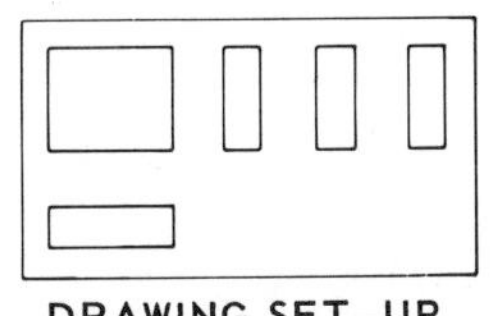

DRAW TOP, FRONT AND 3 SECTION VIEWS

MATL – MALLEABLE IRON

HOLE	HOLE SIZE	LOCATION		
		X	Y	Z
A	M 12	32	35	
B	Ø 8 CSK Ø 12 X 82°	58	50	
C	Ø 8 CBORE Ø 12 X 6 DEEP	28	88	
D_1	Ø 8	88	19	
D_2	Ø 8	88	44	
E	M 12 X 20 DEEP	66		19
F_1	Ø 12	22		25
F_2	Ø 12		32	25
F_3	Ø 12		82	25
F_4	Ø 12	100	76	
G	Ø 3 THROUGH		76	19

Fig. 7-6-E Base plate

HOLE	HOLE SIZE	LOCATION		
		X	Y	Z
A_1	Ø12	16	9	
A_2	Ø12	100	9	
A_3	Ø12	30	92	
A_4	Ø12	87	92	
B_1	Ø8	38	32	
B_2	Ø8	80	32	
C_1	M6 X 12 DEEP	12	50	
C_2	M6	104	52	
D	Ø6 CBORE Ø12 X 6 DEEP	58	70	
E	Ø10 X 12 DEEP	58		11
F_1	Ø6		32	20
F_2	Ø6		70	20

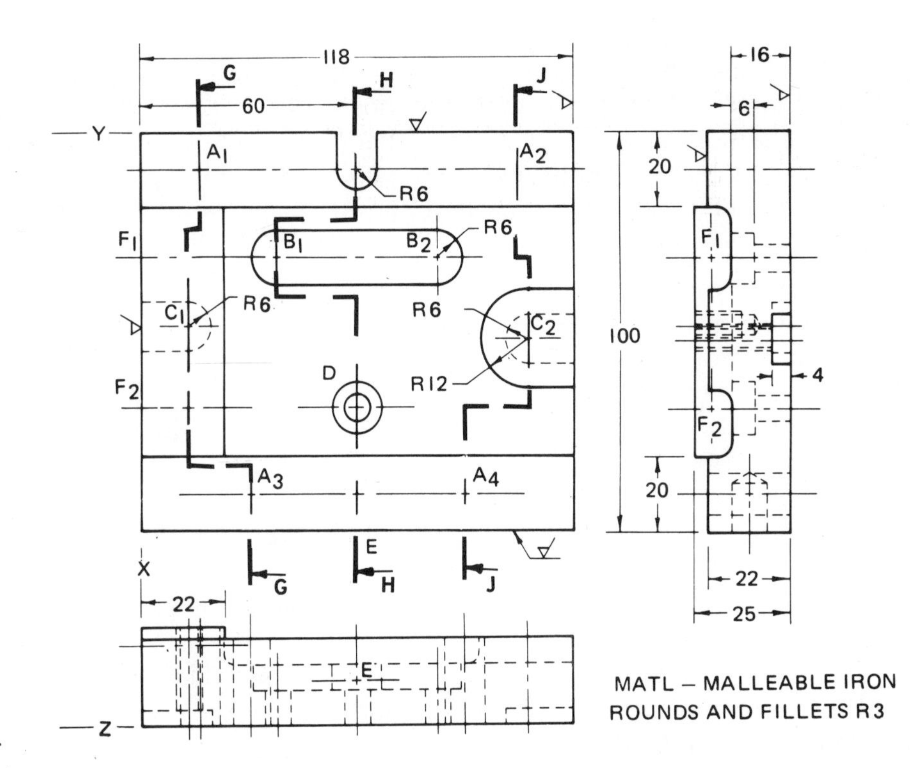

REPLACE END VIEW WITH SECTIONS G–G, H–H, AND J–J.

Fig. 7-6-F Mounting plate

UNIT 7-7
Ribs, Holes, and Lugs in Section

Ribs in Sections

A true-projection sectional view of a part, such as shown in Figure 7-7-1, would be misleading when the cutting plane passes longitudinally through the centre of the rib. To avoid this impression of solidity, a section not showing the ribs cross-hatched is preferred.

When there is an odd number of ribs, such as those shown in Figure 7-7-1*b*, the top rib is aligned with the bottom rib to show its true relationship with the hub and flange. If the rib is not aligned or revolved, it appears distorted on the sectional view and is therefore misleading.

At times it may be necessary to use an alternative method of identifying ribs in a sectional view. Figure 7-7-2 shows a base and a pulley in section. If rib A of the base were not sectioned as previously mentioned, it would appear exactly like rib B in the sectional view and would be misleading. Similarly, rib C shown on the pulley may be overlooked.

To distinguish between the ribs on the base and the ribs and spaces on the pulley, alternate hatching lines on the ribs are used. The line between the rib and solid portions is shown as a broken line.

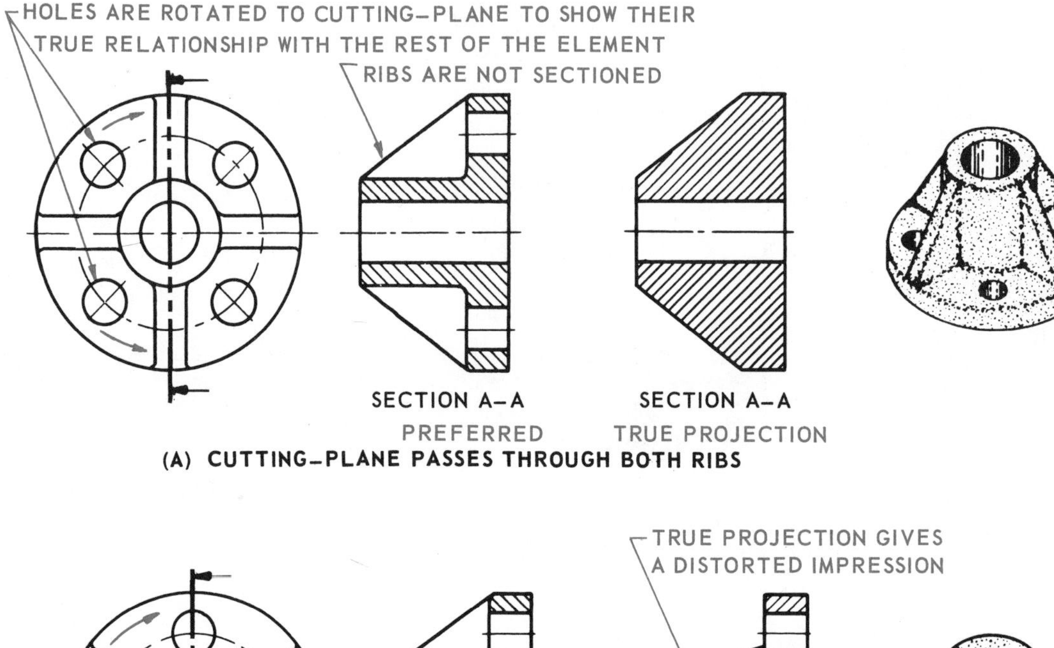

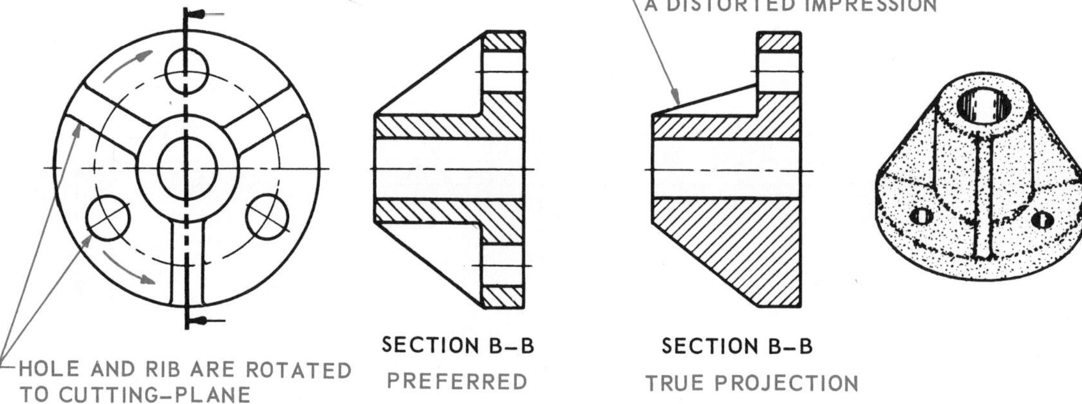

Fig. 7-7-1 Preferred and true projection through ribs and holes

Holes in Sections

Holes, like ribs, are aligned as shown in Figure 7-7-1 to show their true relationship to the rest of the part.

Lugs in Section

Lugs, like ribs and spokes, are also aligned to show their true relationship to the rest of the part, because true projection may be

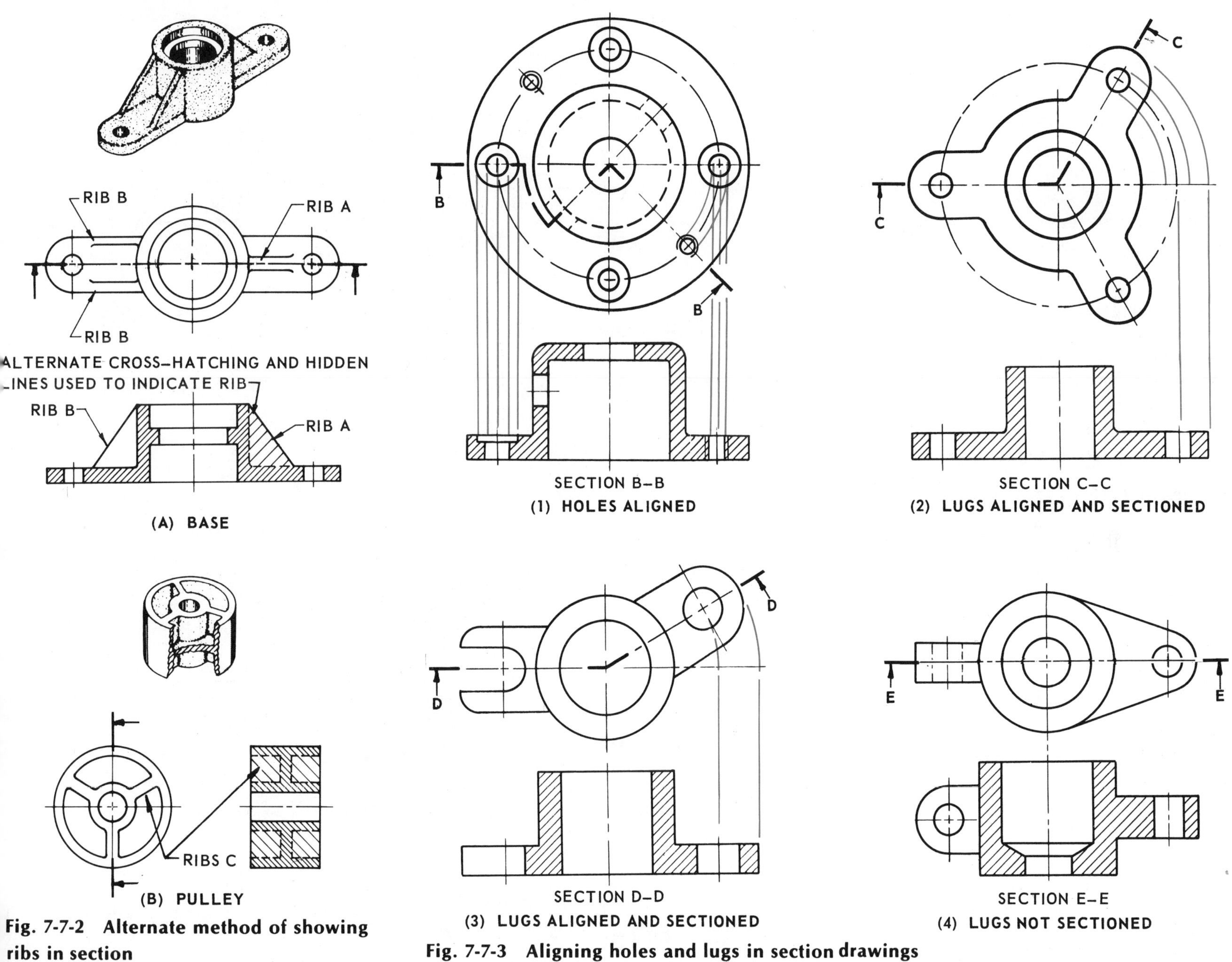

Fig. 7-7-2 Alternate method of showing ribs in section

Fig. 7-7-3 Aligning holes and lugs in section drawings

misleading. Figure 7-7-3 shows several examples of lugs in section.

Note how the cutting plane line is bent or offset so that the features may be clearly shown in the sectional view.

Some lugs are shown in section, and some are not. When the cutting plane passes through the lug crosswise, the lug is sectioned; otherwise, the lugs are treated in the same manner as ribs.

Assignments

Make a three-view working drawing, showing the front and side views in section. Sheet size-A3, scale 1:1.

1. Shaft support, Fig. 7-7-A.
2. Bracket bearing, Fig. 7-7-B.

Review for Assignment

Unit 4-5 Surface texture symbols
Unit 5-1 Conventional representation of common features

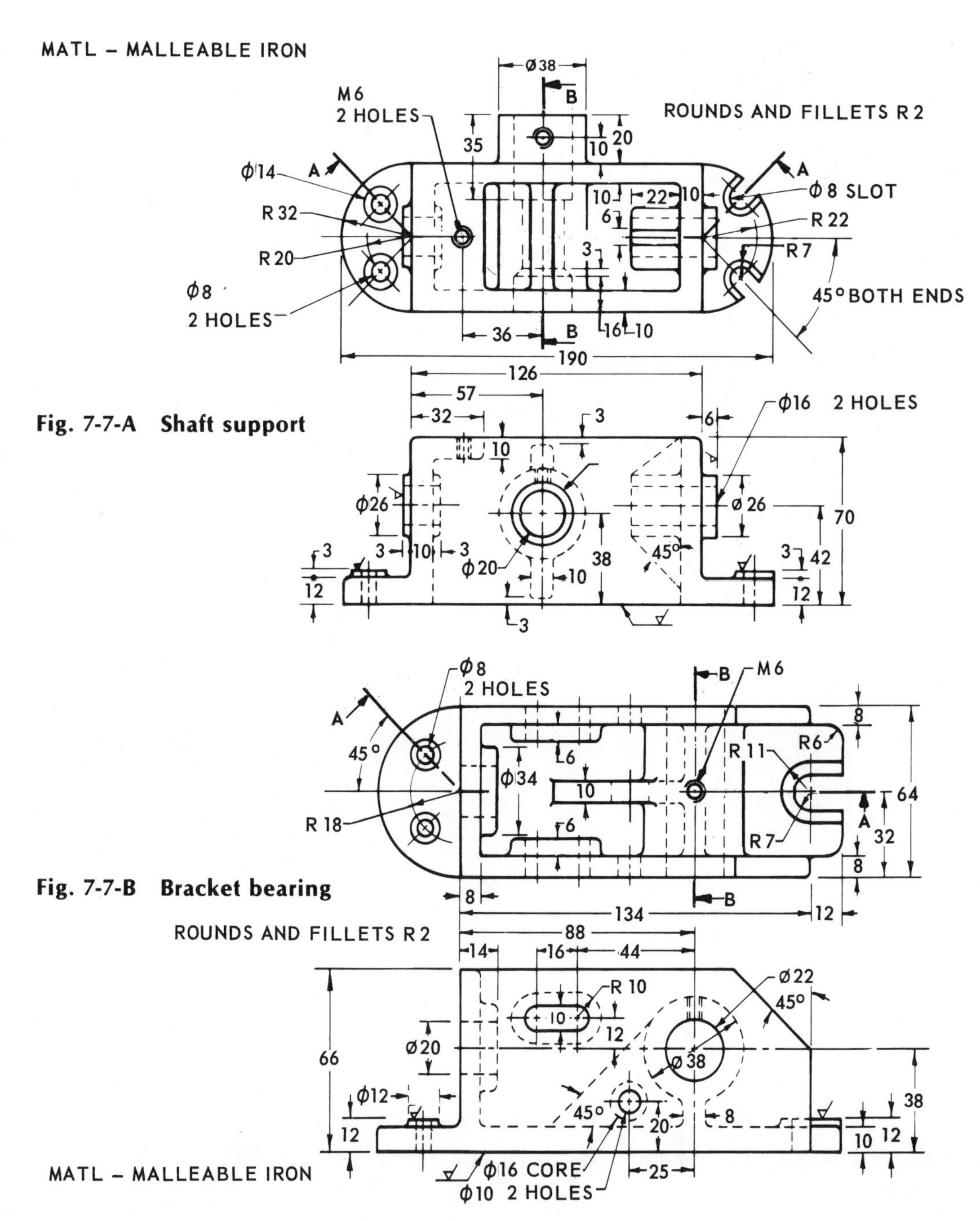

Fig. 7-7-A **Shaft support**

Fig. 7-7-B **Bracket bearing**

UNIT 7-8

Revolved and Removed Sections

Revolved and removed sections are used to show the cross-sectional shape of ribs, spokes, or arms when the shape is not obvious in the regular views (Figs. 7-8-1 to 7-8-3). Often, end views are not needed when a revolved section is used.

For a revolved section, draw a centre line through the shape on the plane to be described. Imagine the part to be rotated 90°. Superimpose on the view the shape that would be seen when rotated (Figs. 7-8-1 and 7-8-3).

If the revolved section does not interfere with the view on which it is revolved, then the view is not broken unless it would provide for clearer dimensioning. When the revolved section interferes or passes through lines on the view on which it is revolved, then the general practice is to break the view (Fig. 7-8-3). Often the break is used to shorten the length of the object.

In no circumstances should the lines on the view pass through the section. When superimposed on the view, the outline of the revolved section is a thin, continuous line.

The removed section differs in that the section, instead of being drawn right on the view, is removed to an open area on the drawing (Fig. 7-8-2).

Frequently the removed section is drawn to an enlarged scale for clarification and

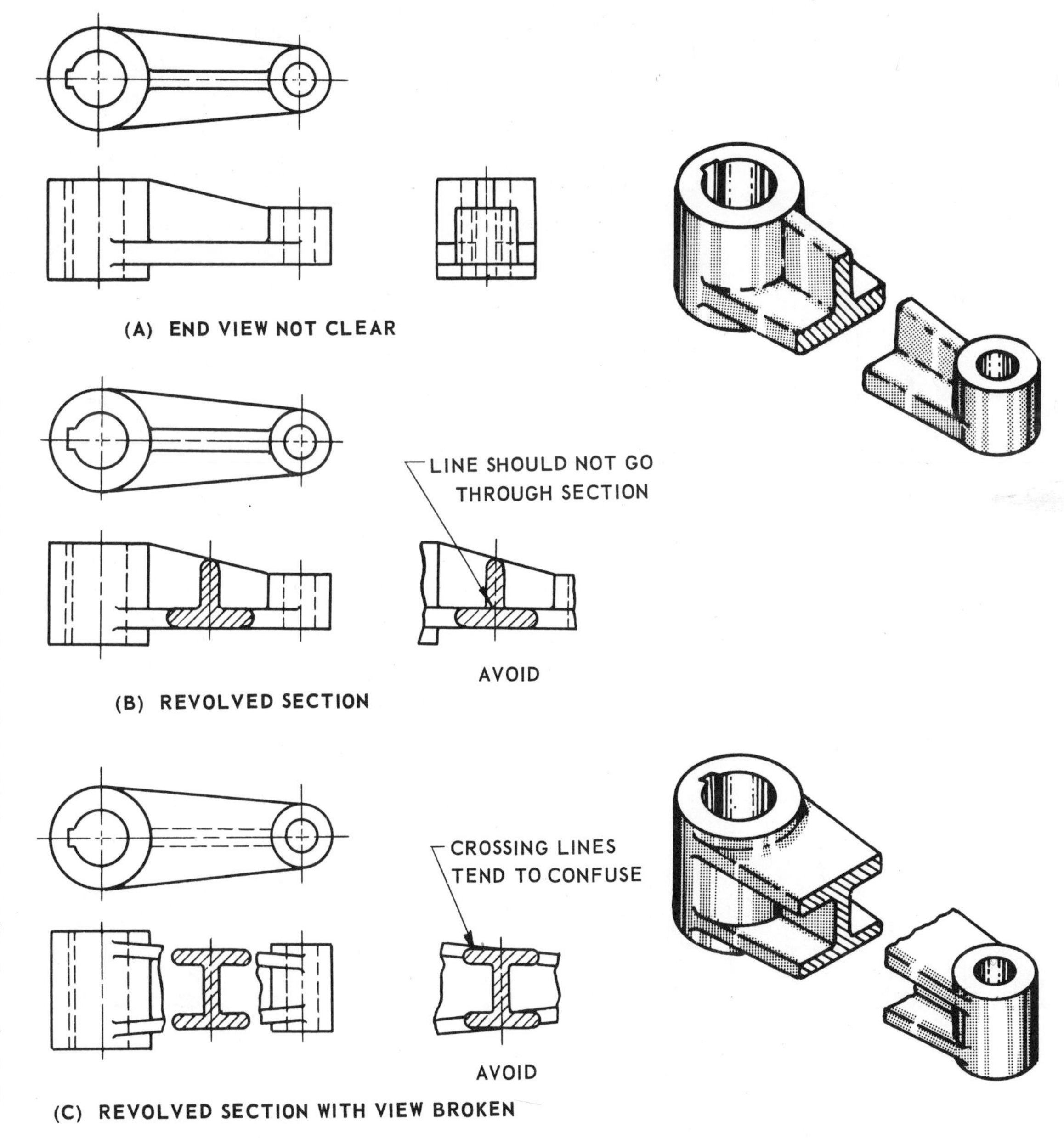

Fig. 7-8-1 Revolved sections

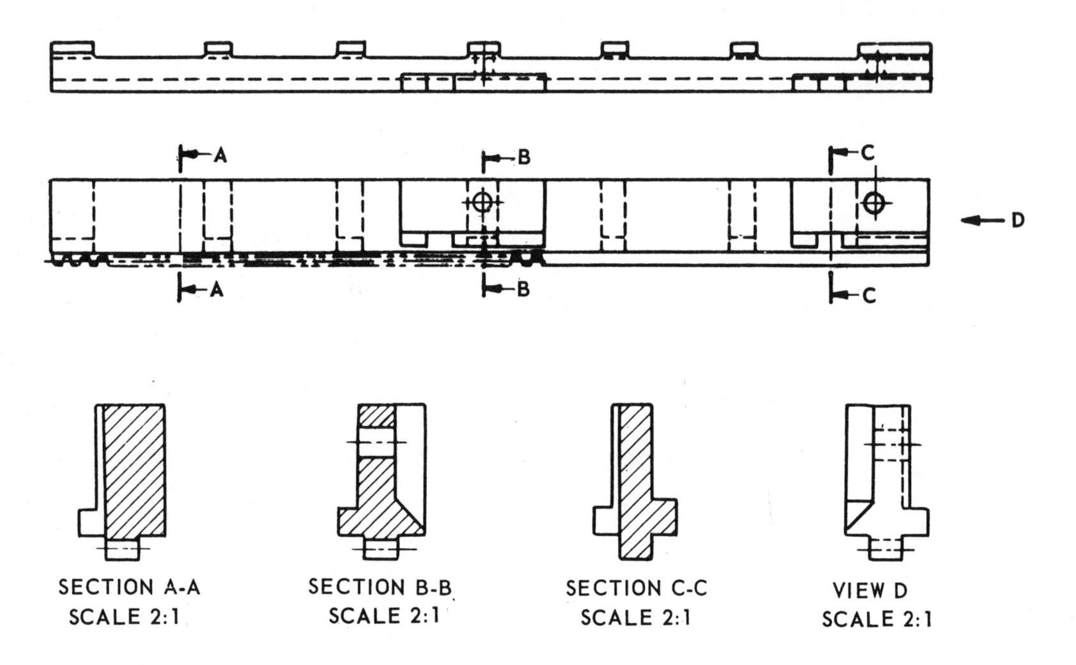

(A) REMOVED SECTIONS AND REMOVED VIEW

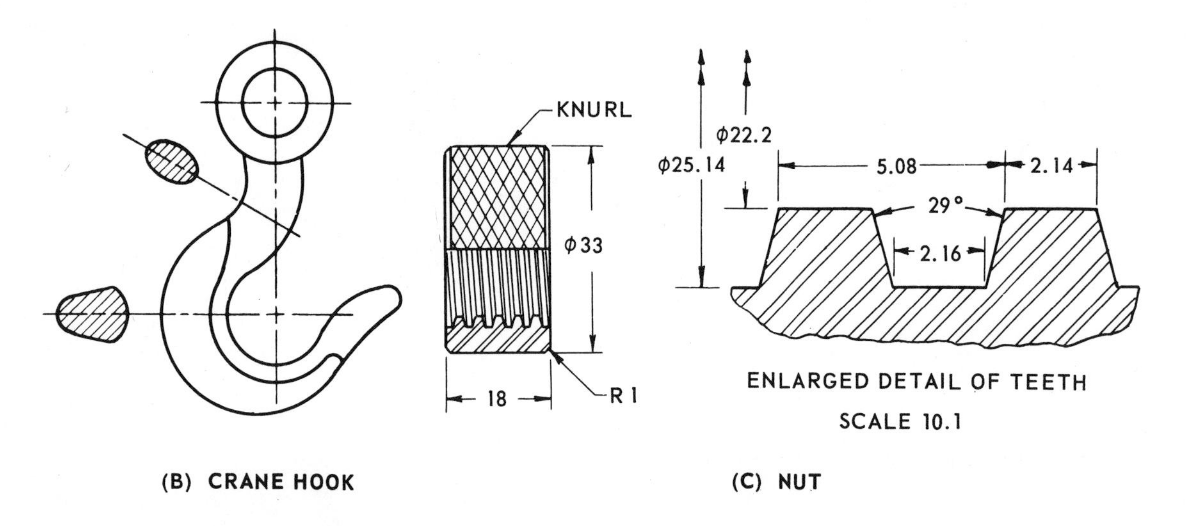

(B) CRANE HOOK

(C) NUT

Fig. 7-8-2 Removed sections

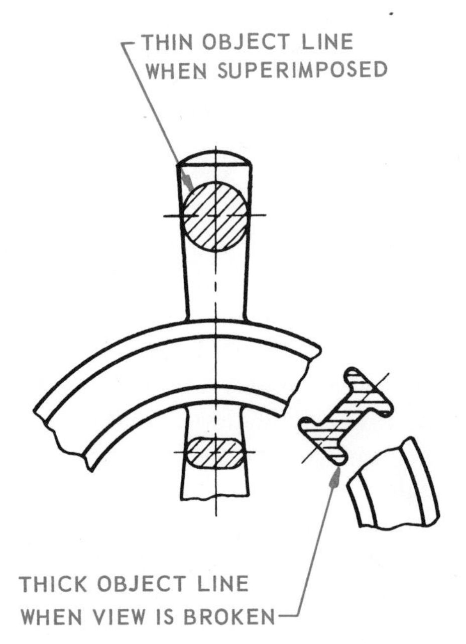

Fig. 7-8-3 Revolved (superimposed) sections

easier dimensioning. Removed sections of symmetrical parts should be placed, whenever possible, on the extension of the centre line (Fig. 7-8-2*b*).

Placement of Sectional Views

Whenever practical, except for revolved sections, section views should be projected perpendicular to the cutting plane and be placed in the normal position for third-angle projection (Fig. 7-8-4).

When the preferred placement is not practical, the sectional view may be removed to some other convenient position on the drawing, but it must be clearly identified, usually by two capital letters, excluding I, O, Q, and Z, and be labelled.

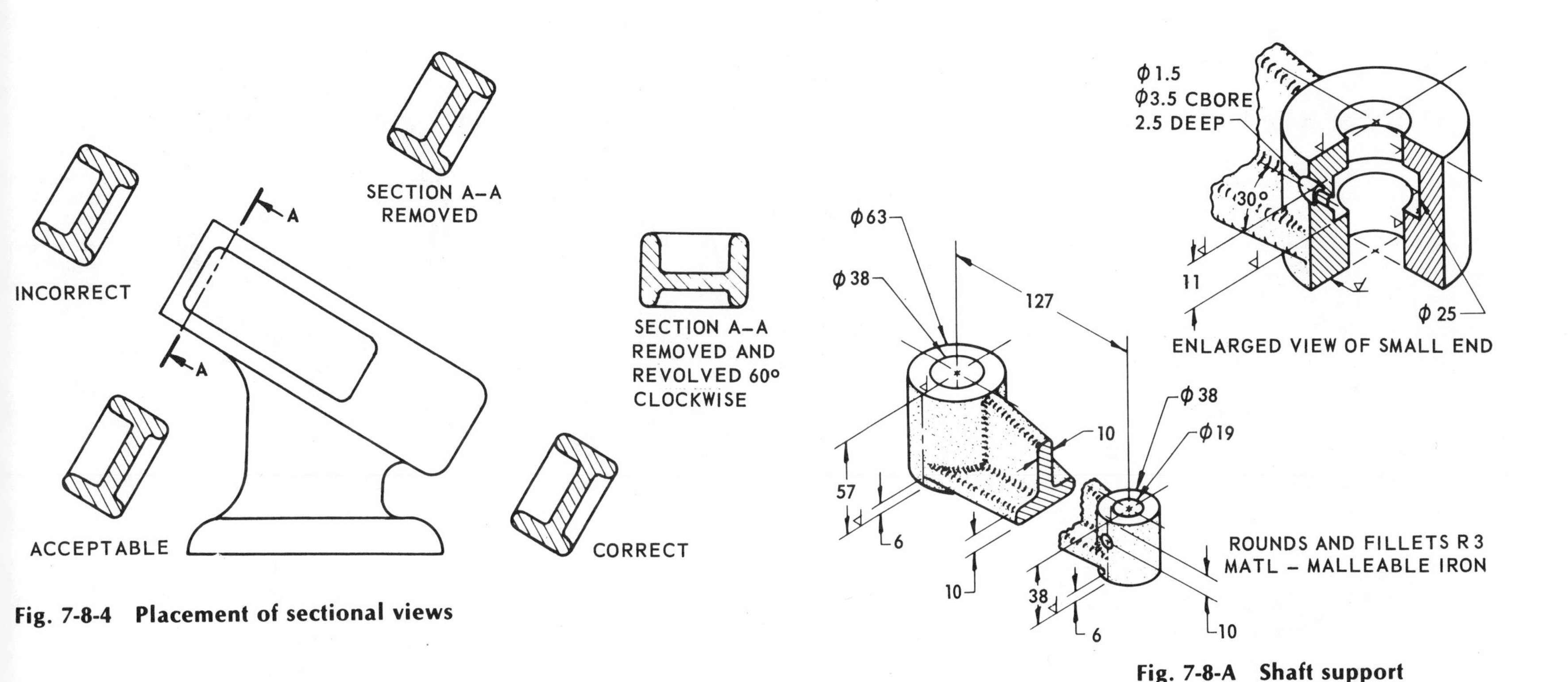

Fig. 7-8-4 Placement of sectional views

Fig. 7-8-A Shaft support

Assignments

Make a working drawing with a revolved section and an enlarged removed section as indicated. Sheet size and scale to suit.

1. Shaft support, Fig. 7-8-A.
2. Idler support, Fig. 7-8-B.

Review for Assignments

Unit 4-5 Surface texture symbols
Unit 5-1 Cylindrical intersections

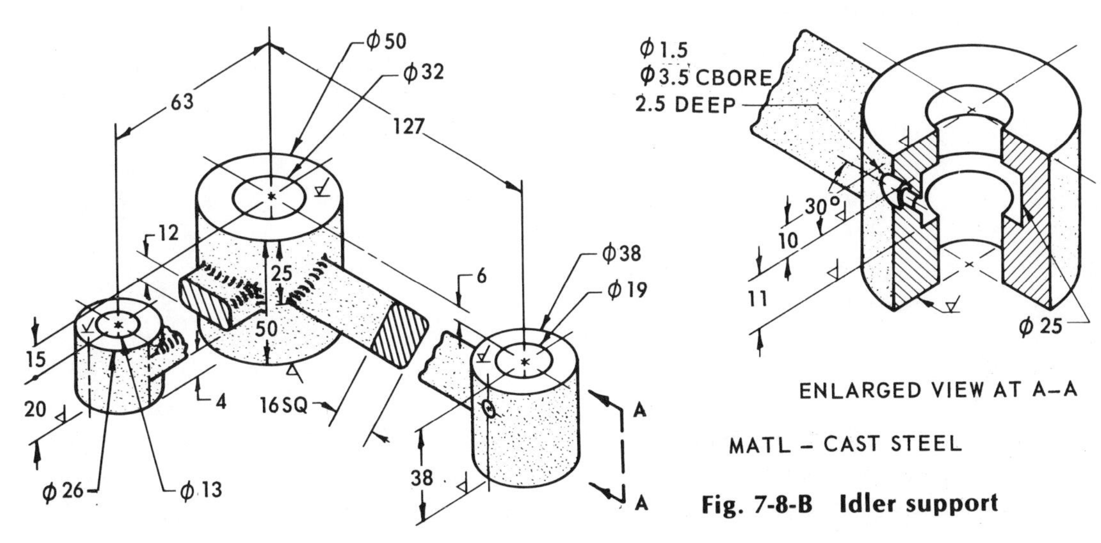

Fig. 7-8-B Idler support

UNIT 7-9

Spokes and Arms in Section

A comparison of the true projection of the wheel with spokes and the wheel with a web is made in Figures 7-9-1*a* and *b*. This comparison shows that a preferred section for the wheel and spokes is desirable so that it will not appear to be a wheel with a solid web.

In preferred sectioning, any part that is not solid or continuous around the hub is drawn without hatching lines, even though the cutting plane passes through the spoke.

When there is an odd number of spokes, as shown in Figure 7-9-1*c*, the bottom spoke is aligned with the top spoke to show its true relationship to the wheel and to the hub. If the spoke were not revolved or aligned, it would appear distorted in the sectional view.

Assignments

Make a two-view working drawing, with the side view in full section, and a revolved section of the spoke on the front view. Sheet size-A3, scale 1:1.

1. Offset handwheel, Fig. 7-9-A.
2. Handwheel, Fig. 7-9-B.

Review for Assignments

Unit 5-1 Cylindrical intersections
Unit 7-8 Revolved sections

R
R
SECTION R–R
WEB IS SECTIONED
TRUE PROJECTION

(A) FLAT PULLEY WITH WEB

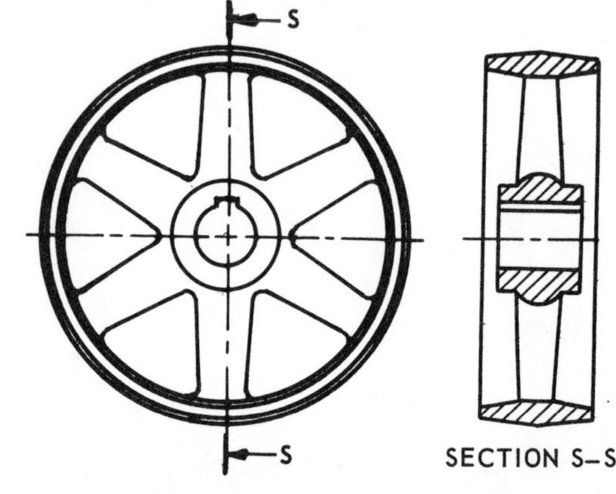

SPOKES ARE NOT SECTIONED
PREFERRED PROJECTION

(B) CROWNED PULLEY WITH EVEN NUMBER OF SPOKES

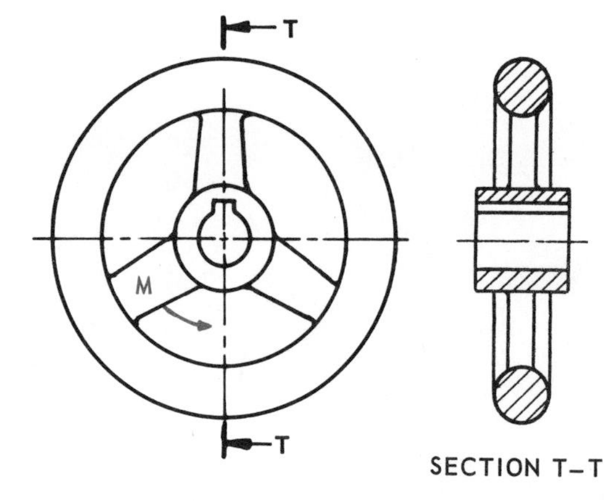

SPOKE M IS ROTATED TO CUTTING–PLANE LINE BUT NOT SECTIONED
PREFERRED PROJECTION

(C) HANDWHEEL WITH ODD NUMBER OF SPOKES

U
U
N
SECTION U–U
SPOKE N IS ROTATED TO CUTTING–PLANE LINE BUT NOT SECTIONED
PREFERRED

(D) HANDWHEEL WITH ODD NUMBER OF OFFSET SPOKES

Fig. 7-9-1 Preferred and true projection through spokes

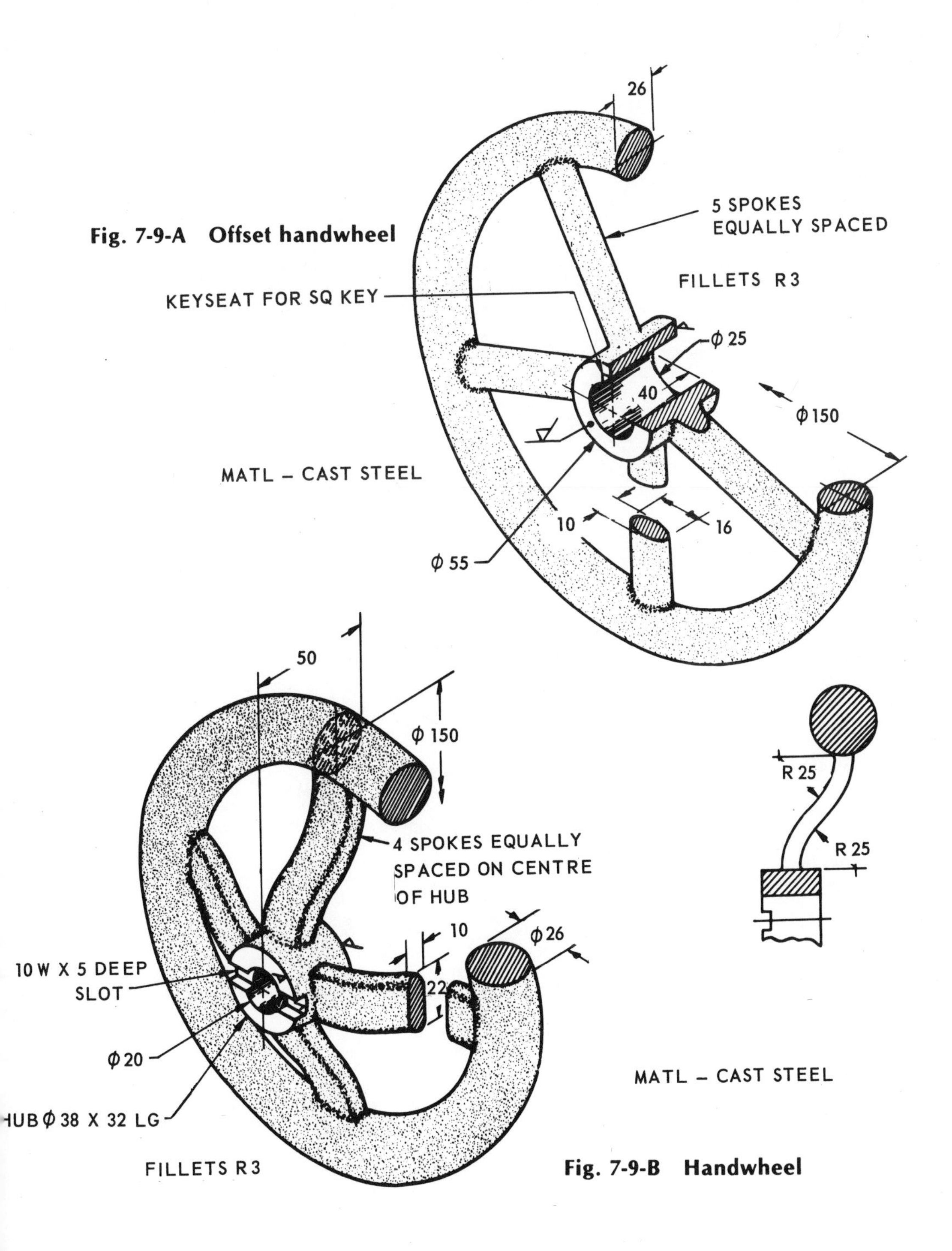

Fig. 7-9-A Offset handwheel

Fig. 7-9-B Handwheel

UNIT 7-10
Partial or Broken-Out Sections

Where a sectional view of only a portion of the object is needed, partial sections may be used (Fig. 7-10-1). An irregular break line is used to show the extent of the section. With this type of section, a cutting-plane is not required.

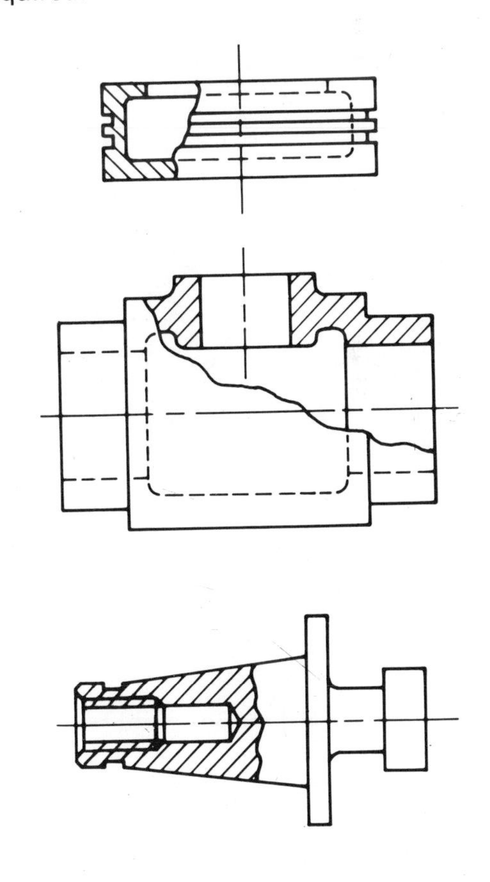

Fig. 7-10-1 Partial or broken-out sections

Assignments

Make a two-view working drawing, using partial sections where necessary to fully explain the shape. Sheet size and scale to suit.

1. Hold-down bracket, Fig. 7-10-A.
2. Tumble box, Fig. 7-10-B.

Review for Assignments

Unit 4-5 Surface texture symbols
Unit 5-1 Intersection of unfinished surfaces

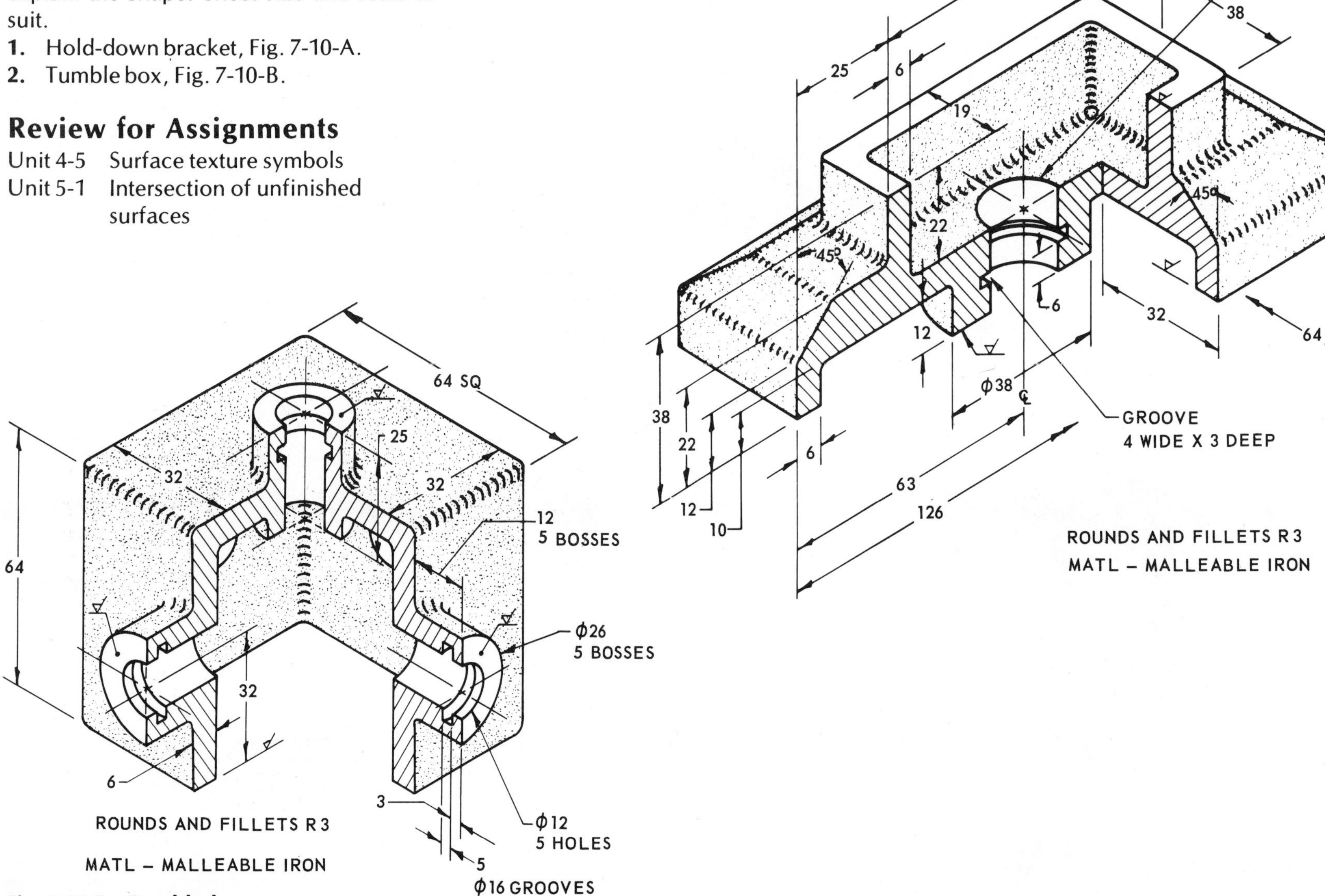

Fig. 7-10-A Offset handwheel

Fig. 7-10-B Tumble box

UNIT 7-11

Phantom or Hidden Sections

A phantom section is used to show the typical interior shapes of an object in one view when the part is not truly symmetrical in shape. It is also used to show mating parts in an assembly drawing (Fig. 7-11-1). It is a sectional view superimposed on the regular view without the removal of the front portion of the object. The hatching lines used for phantom sections consist of light, evenly spaced, broken lines.

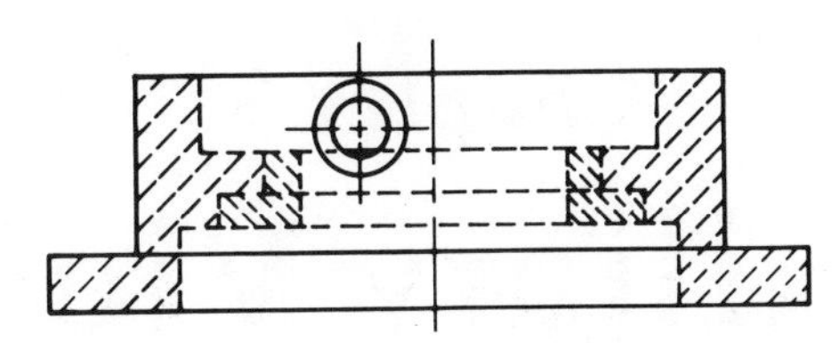

Fig. 7-11-1 Phantom or hidden sections

Assignments

Make a two-view assembly drawing showing the front view as a phantom section. Dimension only the bushing hole sizes and their location. Sheet size and scale to suit.

1. Drill-jig assembly, Fig. 7-11-A.
2. Lock assembly, Fig. 7-11-B.

Review for Assignments

Unit 5-3 Assembly drawings
Unit 7-1 Hatching lines

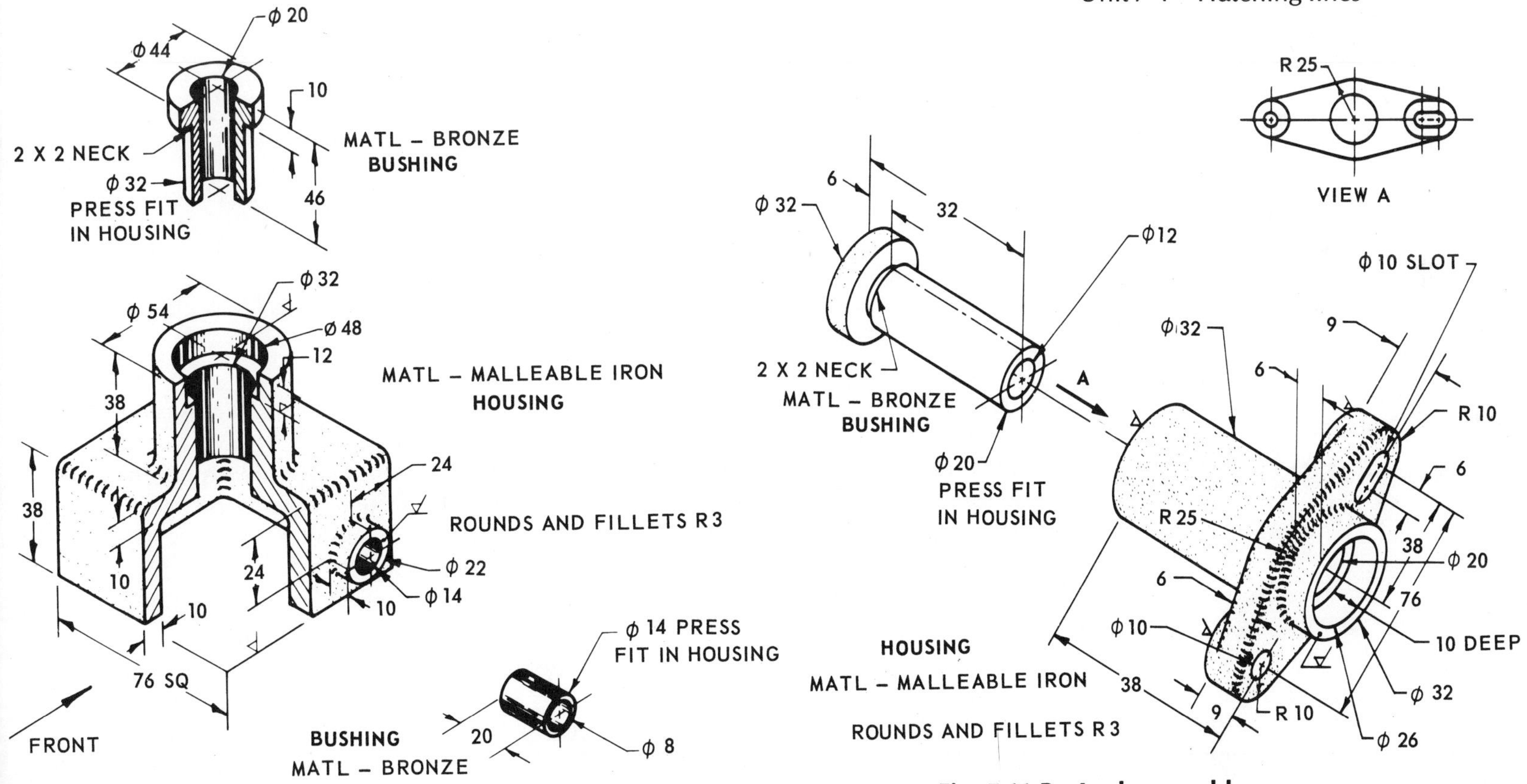

Fig. 7-11-A Drill jig assembly

Fig. 7-11-B Lock assembly

Chapter 8

Auxiliary Views

UNIT 8-1

Primary Auxiliary Views

Many machine parts have surfaces that are not perpendicular, or at right angles, to the plane of projection. These are referred to as **sloping** or **inclined** surfaces.

In the regular orthographic views, such surfaces appear to be foreshortened, and their true shape is not shown.

When an inclined surface has important characteristics that should be shown clearly and without distortion, an auxiliary view is used so that the drawing completely and clearly explains the shape of the object.

In many cases, the auxiliary view will replace one of the regular views on the drawing, as illustrated in Figure 8-1-1.

One of the regular orthographic views will have an edge line representing the inclined surface. The auxiliary view is projected from this edge line, at right angles, and is drawn parallel to the edge line.

Only the true shape features on the views need be drawn, as shown in Figure 8-1-2. Since the auxiliary view shows only the true shape and detail of the inclined surface or features, a partial auxiliary view is all that is necessary.

Likewise, the distorted features on the regular views may be omitted. Hidden lines are usually omitted unless required for clarity. This procedure is recommended for functional and production drafting where drafting costs are an important consideration. However, the drafter may be called

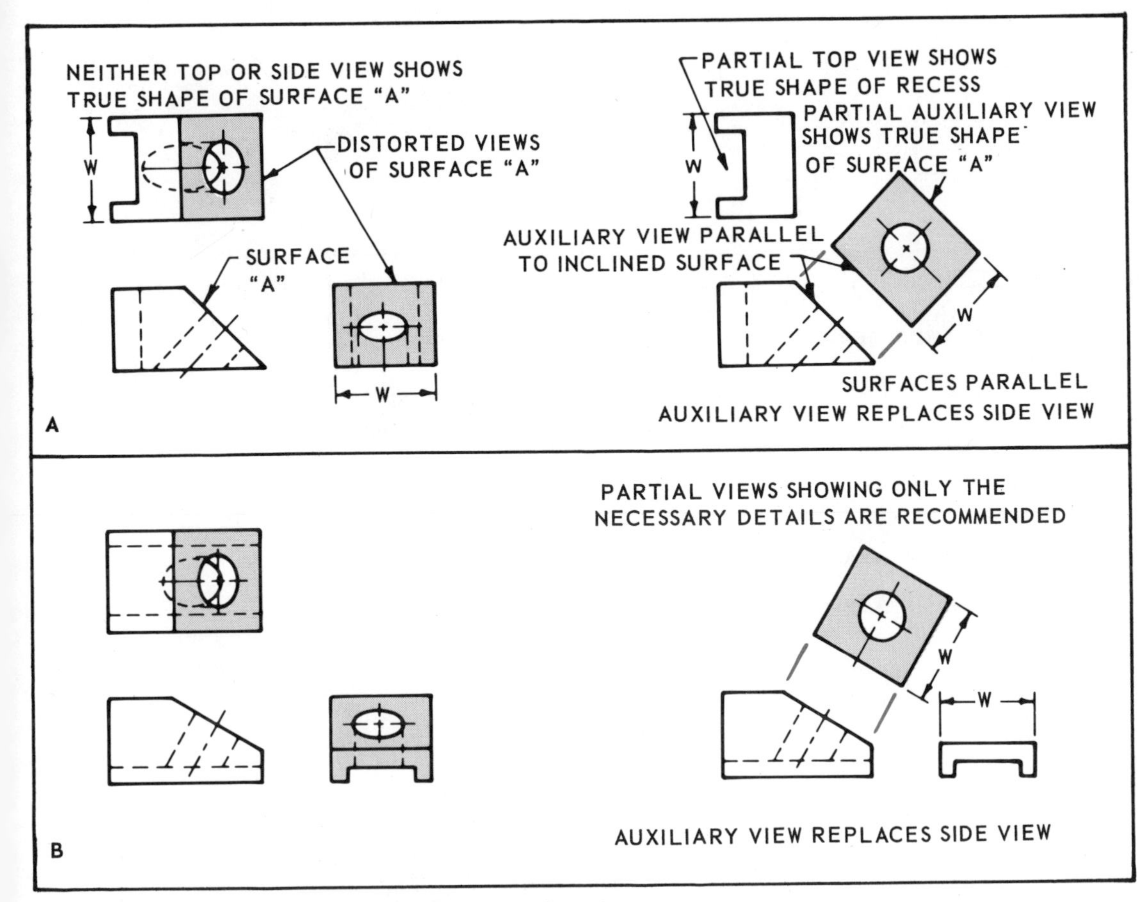

Fig. 8-1-1 Auxiliary views replacing regular views

upon to draw the complete views of the part. This type of drawing is often used for catalogue and standard parts drawings.

Dimensioning Auxiliary Views

One of the basic rules of dimensioning is to dimension the feature where it can be seen in its true shape and size. Thus the auxiliary view will show only the dimensions pertaining to those features for which the auxiliary view was drawn.

The recommended dimensioning method for engineering drawings is the unidirectional system. Not only is this method of dimensioning easier to prepare and read, but it is readily adaptable to mechanical lettering. See Figure 8-1-3.

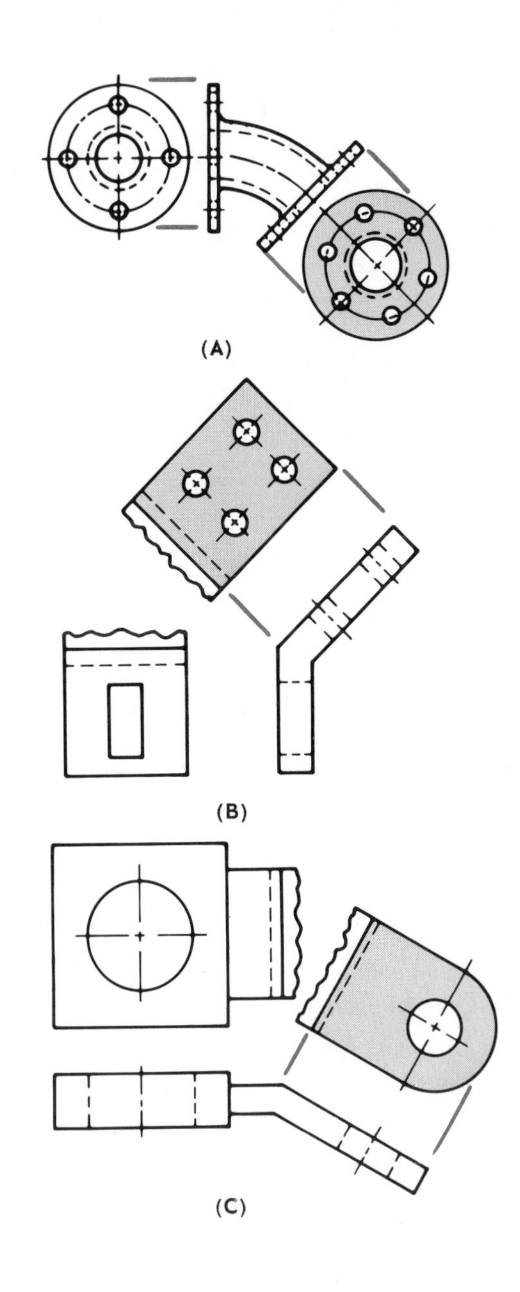

Fig. 8-1-2 Examples of auxiliary view drawings

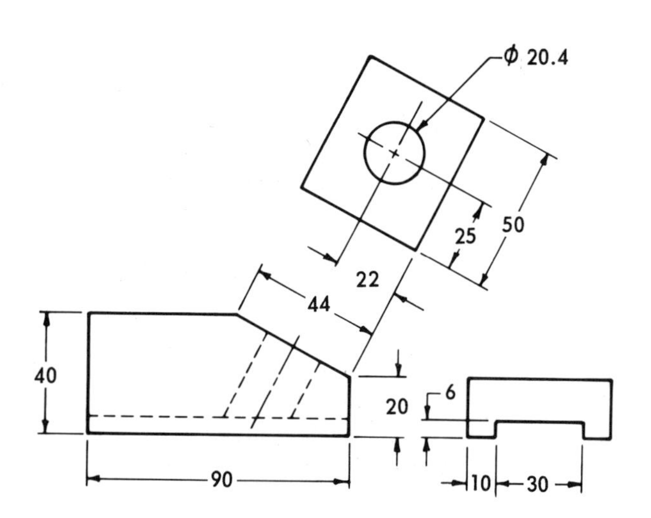

Fig. 8-1-3 Dimensioning auxiliary view drawings

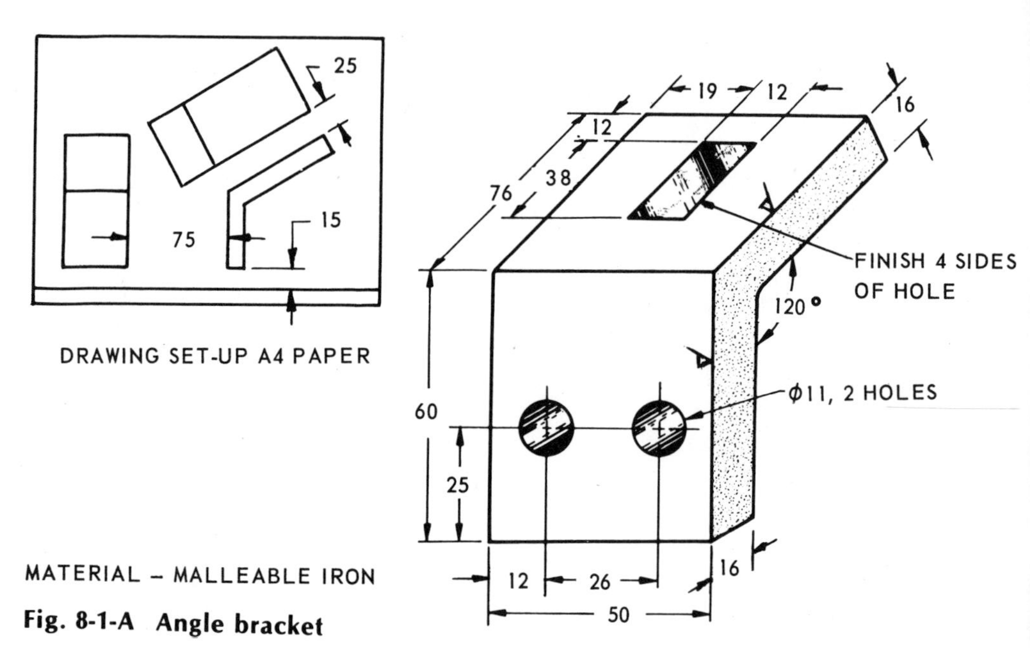

Fig. 8-1-A Angle bracket

Assignments

1. On an A3 sheet, make a working drawing of the angle bracket shown in Fig. 8-1-A. Replace the top view with an auxiliary view. Draw complete views with hidden lines. Scale is 1:1.
2. On an A3 sheet, make a working drawing of the cross slot bracket shown in Fig. 8-1-B. Replace the side view with an auxiliary view. Only partial views need be drawn, and hidden lines may be added to improve clarity. Scale is 1:1.

Review for Assignments

Unit 4-3 Fillets and rounds
Unit 4-5 Surface texture symbols

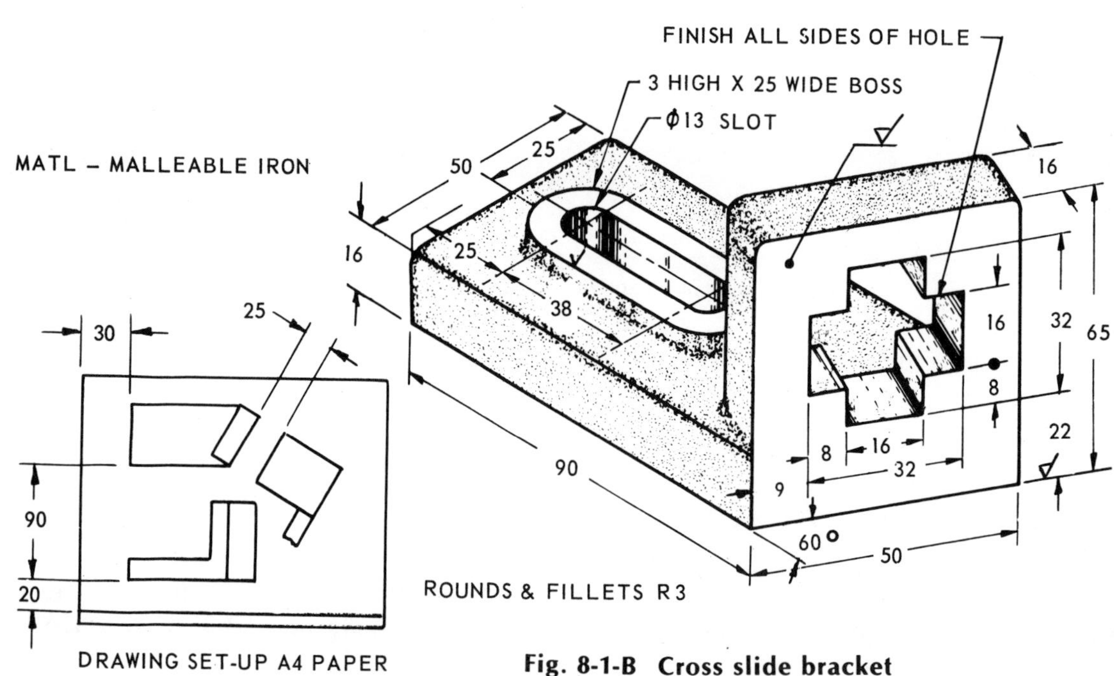

Fig. 8-1-B Cross slide bracket

UNIT 8-2

Circular Features in Auxiliary Projection

As mentioned in Unit 8-1, at times it is necessary to show the complete views of an object. If circular features are involved in auxiliary projection, then the surfaces appear elliptical, not circular, in one of the views.

The method most commonly used to draw the true-shape projection of the curved surface is the plotting of a series of points on the line, the number of points being governed by the accuracy of the curved line required.

Figure 8-2-1 illustrates an auxiliary view of a truncated cylinder. The shape seen in the auxiliary view is an **ellipse.** This shape is drawn by plotting **lines of intersection.** The perimeter of the circle in the top view is divided to give a number of equally spaced points — in this case, 12 points, *A* to *M*, spaced 30° apart (360°/12 = 30°).

These points are projected down to the edge line on the front view, then at right angles to the edge line to the area where the auxiliary view will be drawn. A centre line for the auxiliary view is drawn parallel to the edge line, and width settings taken from the top view are transferred to the auxiliary view. Note width setting *R* for point *L*.

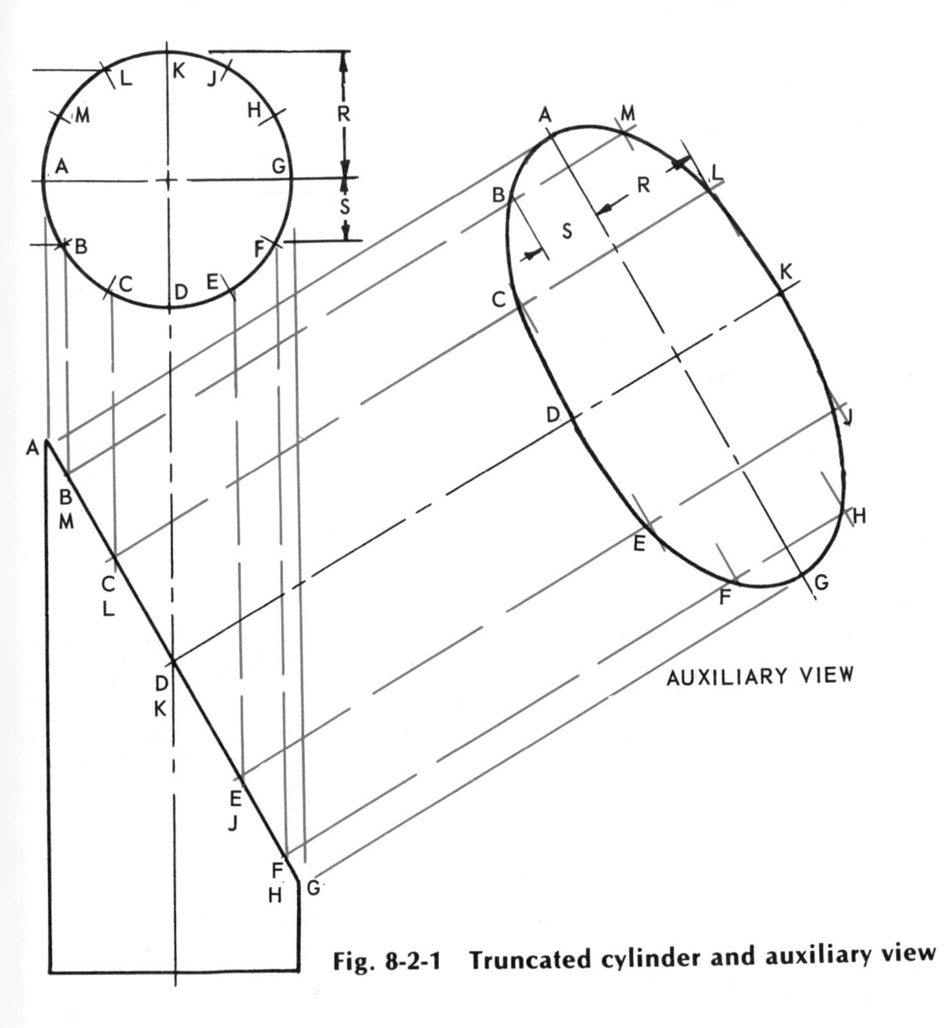

Fig. 8-2-1 **Truncated cylinder and auxiliary view**

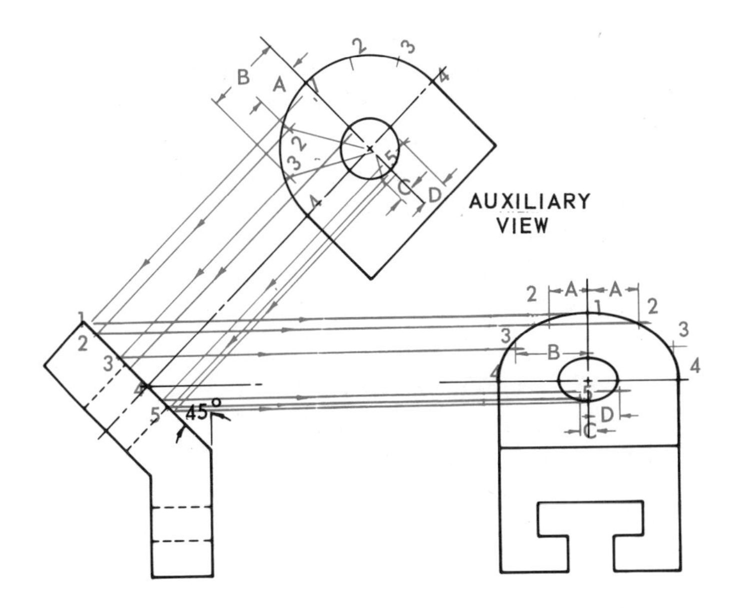

Fig. 8-2-2 **Constructing the true shape of a curved surface by the plotting method**

Because the illustration shows a true cylinder and the point divisions in the top view are all equal, the width setting *R* taken at *L* is also the correct width setting for *C, E,* and *J*. Width setting *S* for *B* is also the correct width setting for *F, H,* and *M*.

When all the width settings have been transferred to the auxiliary view, the resulting points of intersection are connected with the use of a French curve to give the desired elliptical shape.

It is often necessary to construct the auxiliary view first in order to complete the regular views. This is shown in Figure 8-2-2.

Assignments

Make a working drawing, sheet size-A3, scale 1:1. Add hidden lines when required for clarity. Location dimensions are to the centre of auxiliary views.

1. Link, Fig. 8-2-A. Draw full top and front views, and partial auxiliary view.
2. Control block, Fig. 8-2-B. Draw full top, front and side views, and partial auxiliary view.

Review for Assignments

Unit 4-5 Surface texture symbols
Unit 5-1 Intersections of unfinished surfaces
Unit 8-1 Dimensioning auxiliary views

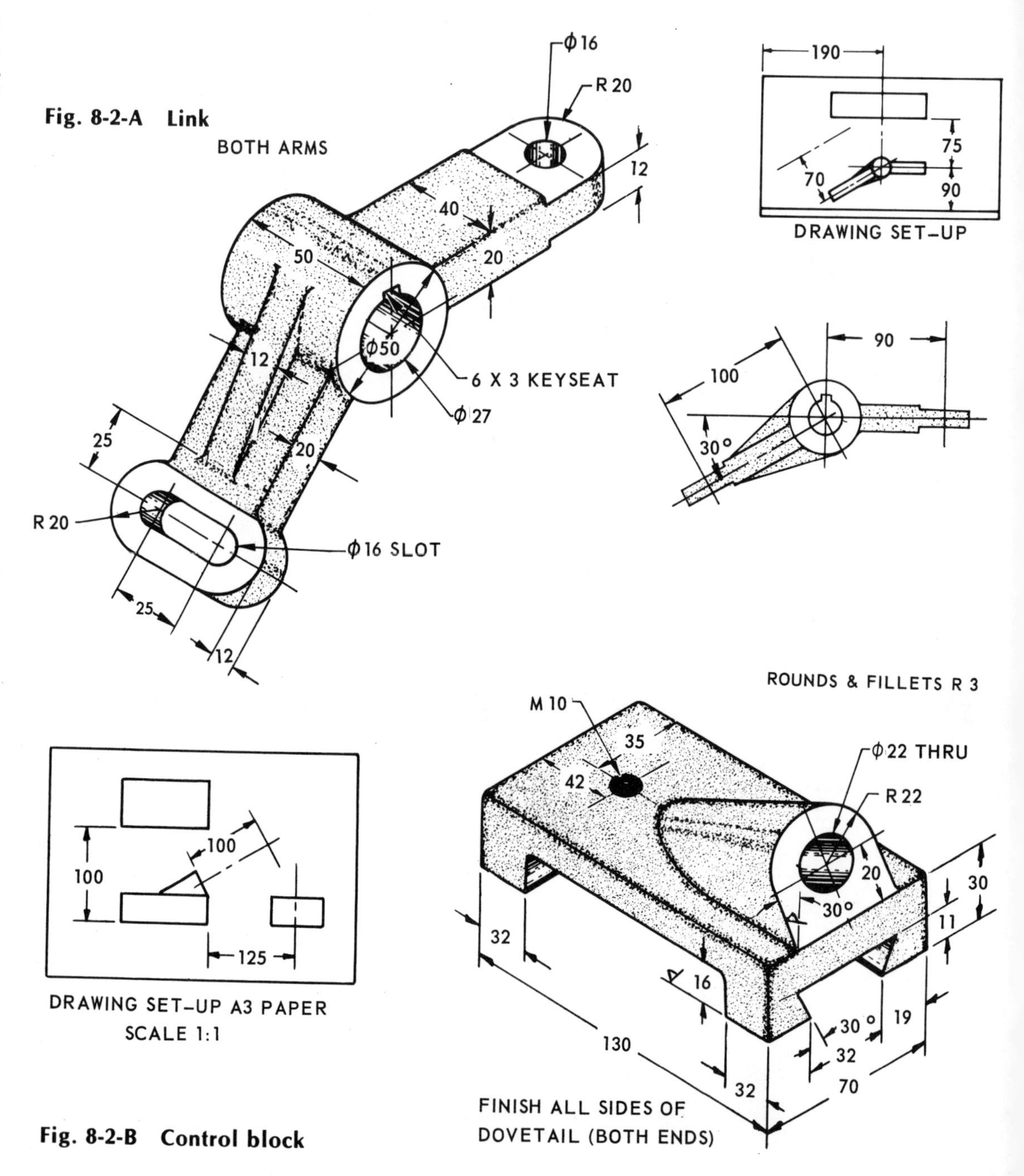

Fig. 8-2-A Link

Fig. 8-2-B Control block

UNIT 8-3
Multi-Auxiliary-View Drawings

Some objects have more than one surface not perpendicular to the plane of projection. In preparing working drawings of these objects, an auxiliary view may be required for each surface.

Naturally, this would depend upon the amount and type of detail lying on these surfaces. This type of drawing is often referred to as a **multi-auxiliary-view** drawing. See Figure 8-3-1.

One can readily see the advantage of using the unidirectional system of dimensioning for dimensioning an object such as shown in Figure 8-3-2.

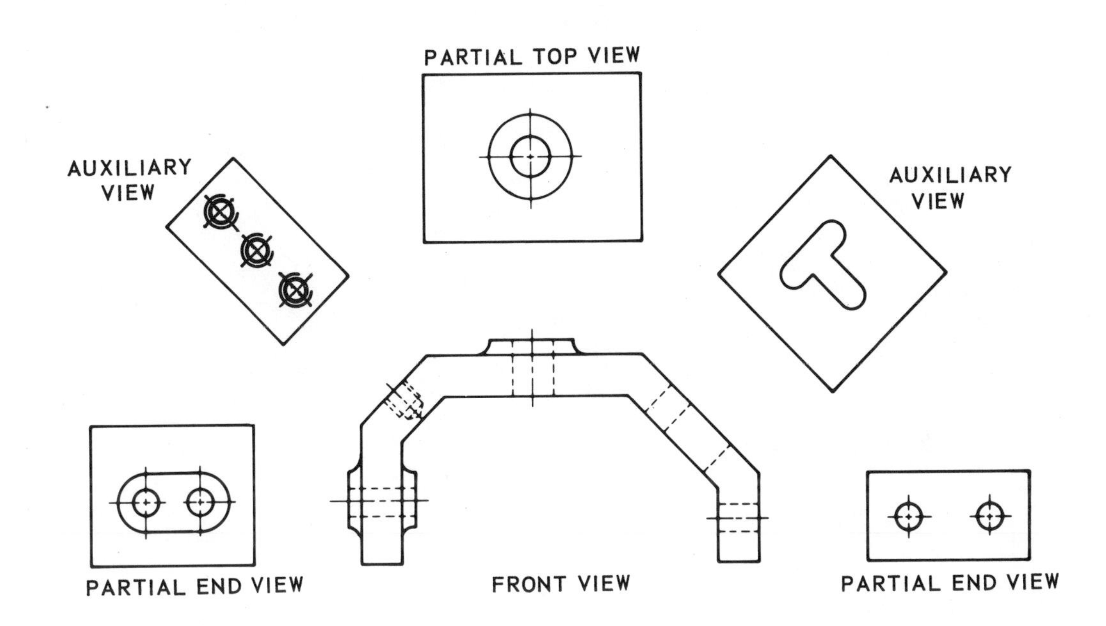

Fig. 8-3-1 Auxiliary views added to regular views to show true shape of features

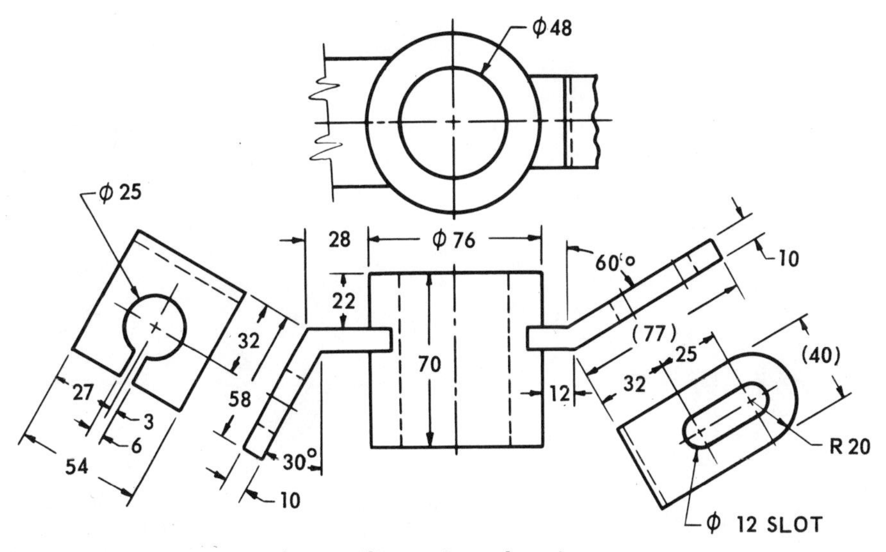

Fig. 8-3-2 Dimensioning a multi-auxiliary view drawing

Assignments

Make a working drawing following the instructions for selection and placement of views. Only partial views need be drawn except where noted. Hidden lines may be added to improve clarity. Sheet size-A3, scale 1:1.

1. Connecting bar, Fig. 8-3-A.
2. Angle slide, Fig. 8-3-B.
3. Bracket, Fig. 8-3-C.

Review for Assignments

Unit 4-5 Surface texture symbols
Unit 5-1 Intersection of unfinished surfaces
Unit 8-1 Dimensioning auxiliary views

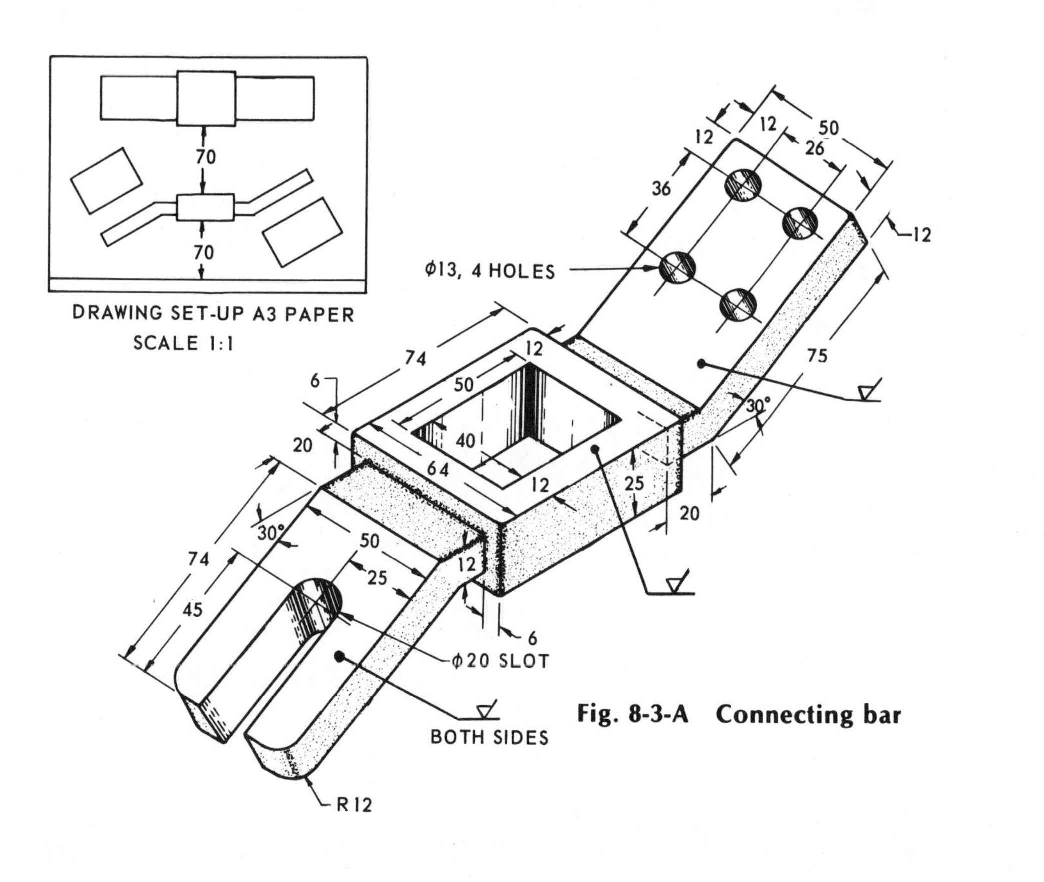

Fig. 8-3-A Connecting bar

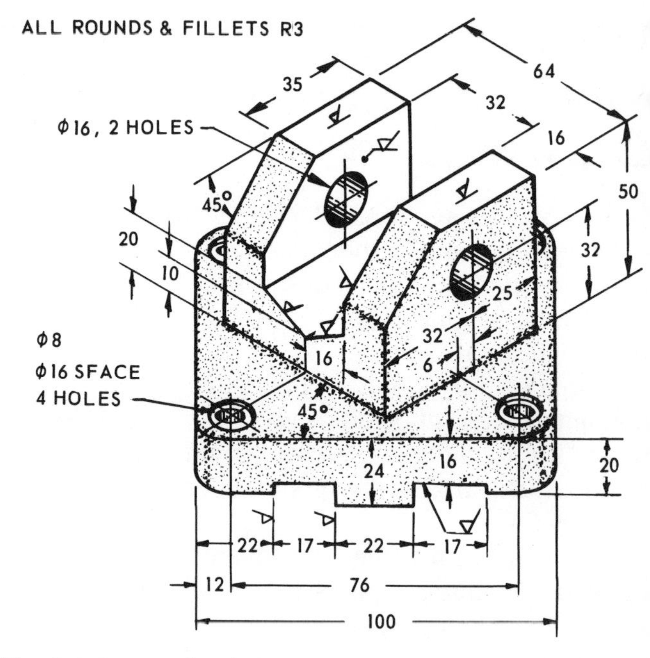

Fig. 8-3-B Angle slide

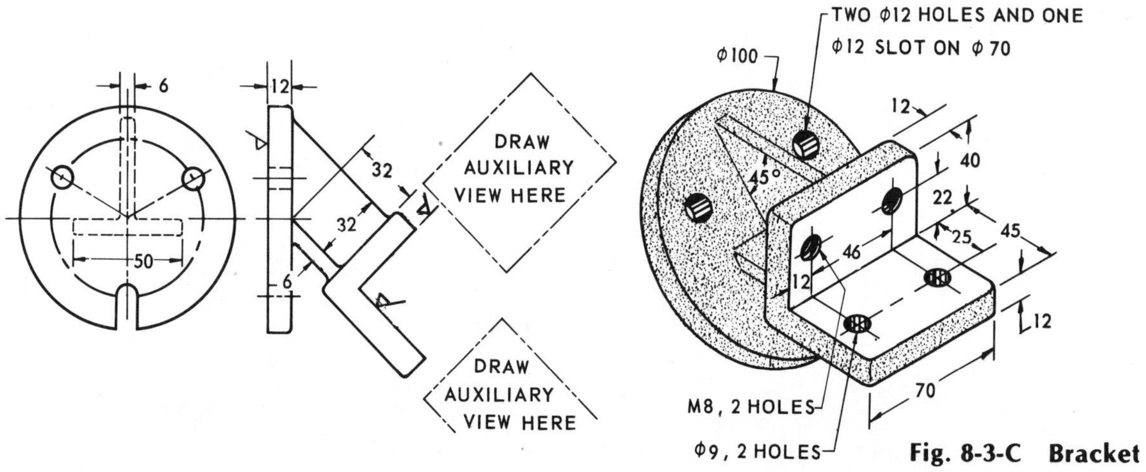

Fig. 8-3-C Bracket

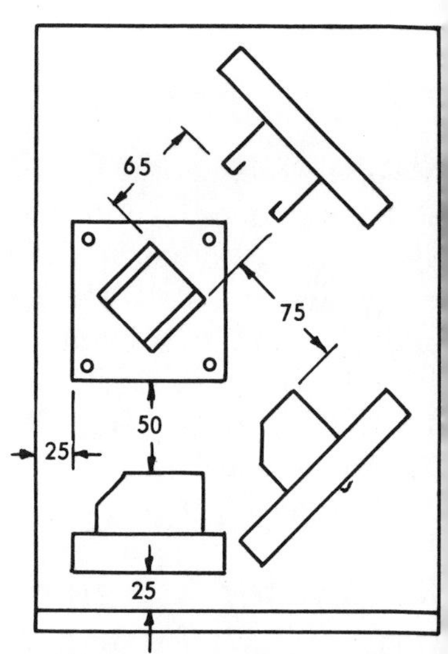

Chapter 9
Pictorial Drawings

UNIT 9-1
Pictorial Drawings

Pictorial drawing is the oldest written method of communication known, but the character of pictorial drawing has continually changed with the advance of civilization. In this text only those kinds of pictorial drawings commonly used by the engineer and drafter are considered.

Pictorial drawings are useful in design, construction or production, erection or assembly, service or repairs, and sales. The type of pictorial drawing used depends on the purpose for which it is drawn.

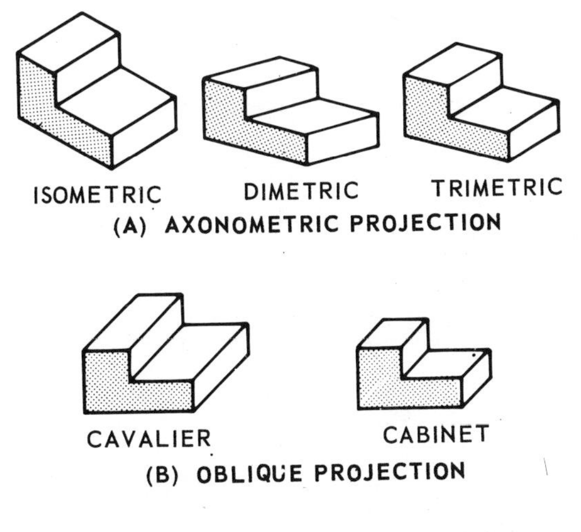

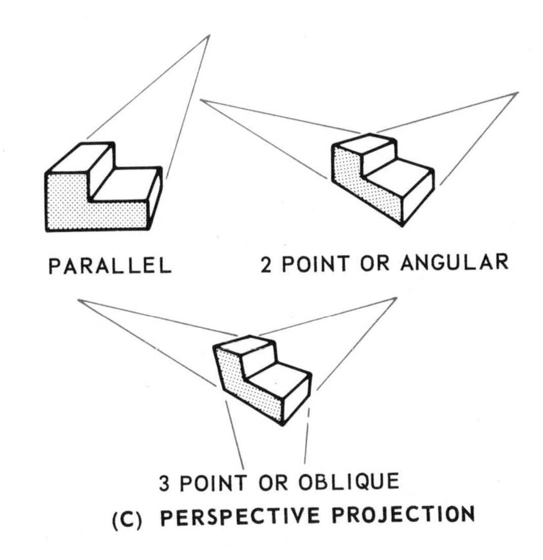

Fig. 9-1-1 Types of pictorial drawings

There are three general types into which pictorial drawings may be divided: axonometric, oblique, and perspective. These three differ from one another with regard to projection, as shown in Figure 9-1-1.

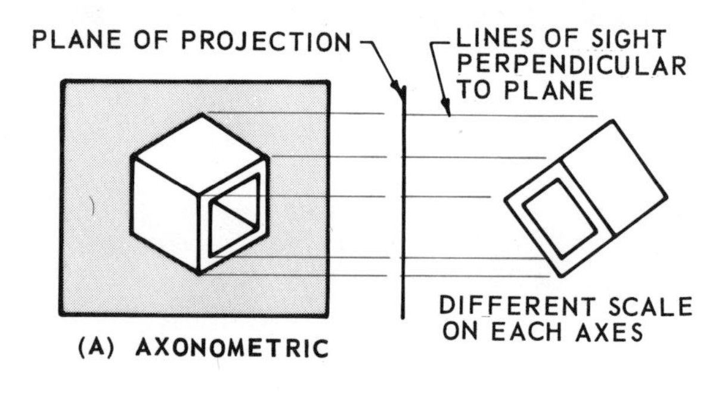

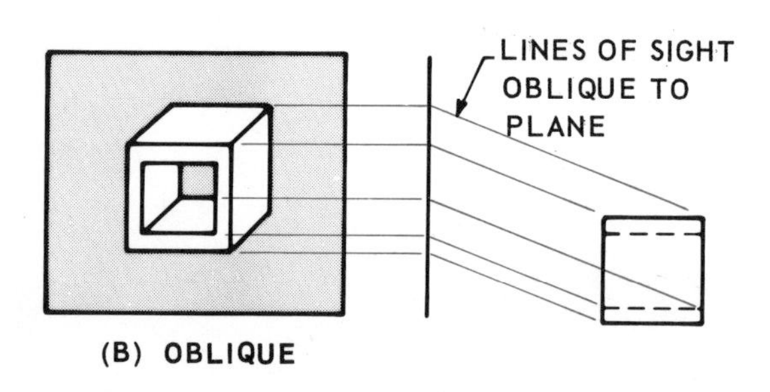

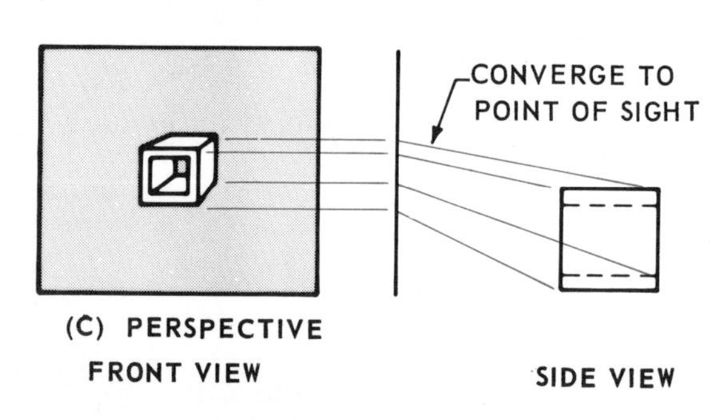

Fig. 9-1-2 Kinds of projection

Axonometric Projection

A projected view in which the lines of sight are perpendicular to the plane of projection, but in which the three faces of a rectangular object are all inclined to the plane of projection, is called an **axonometric projection.** See Figure 9-1-2.

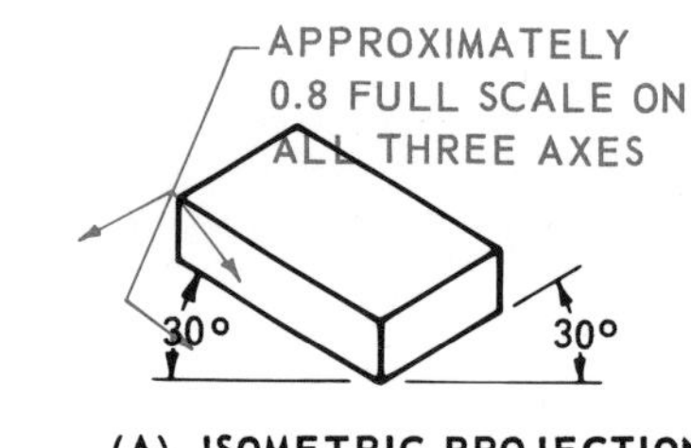

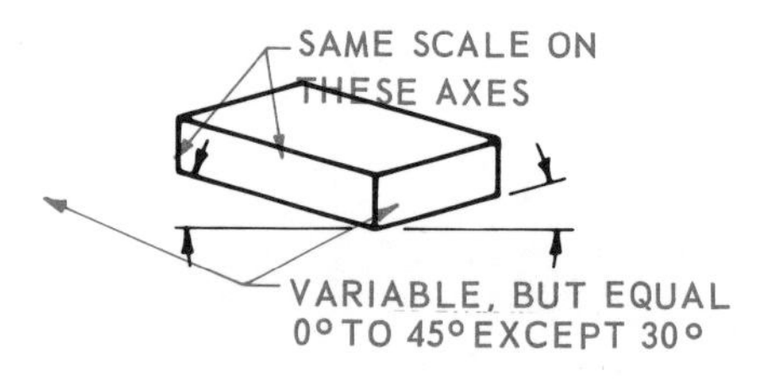

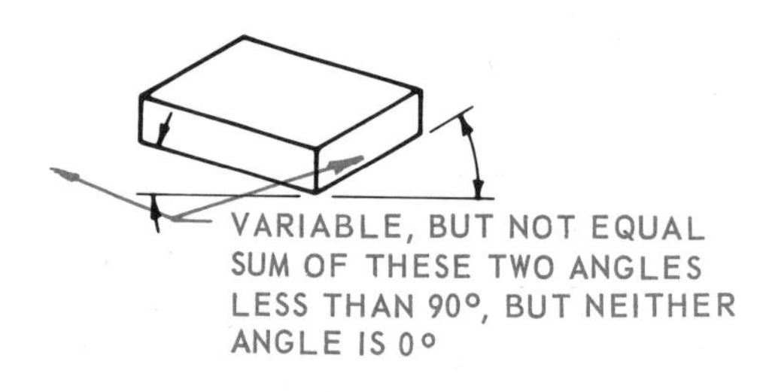

Fig. 9-1-3 Types of axonometric projection

The projections of the three principal axes may make any angle with one another except 90°. Axonometric drawings, as shown in Figure 9-1-3 are classified into three forms: **isometric drawings,** where the three principal faces and axes of the object are equally inclined to the plane of projection; **dimetric drawings,** where two of the three principal faces and axes of the object are equally inclined to the plane of projection; and **trimetric drawings,** where all three faces and axes of the object make different angles with the plane of projection.

The most popular form of axonometric projection is the isometric.

Isometric Drawings

Using this method, the object is revolved at an angle of 45° to the horizontal, so that the front corner is toward the viewer. It is then tipped up or down at an angle of 35° 16′. See Figure 9-1-4.

When this is done to a cube, the three faces visible to the viewer appear equal in shape and size, and the side faces are at an angle of 30° to the horizontal.

If the isometric view were actually projected from a view of the object in the tipped position, the lines in the isometric view would be foreshortened and would, therefore, not be seen in their true length. To simplify the drawing of an isometric view, the actual measurements of the object are used.

Although the object appears slightly larger without the allowance for foreshortening, the proportions are not affected.

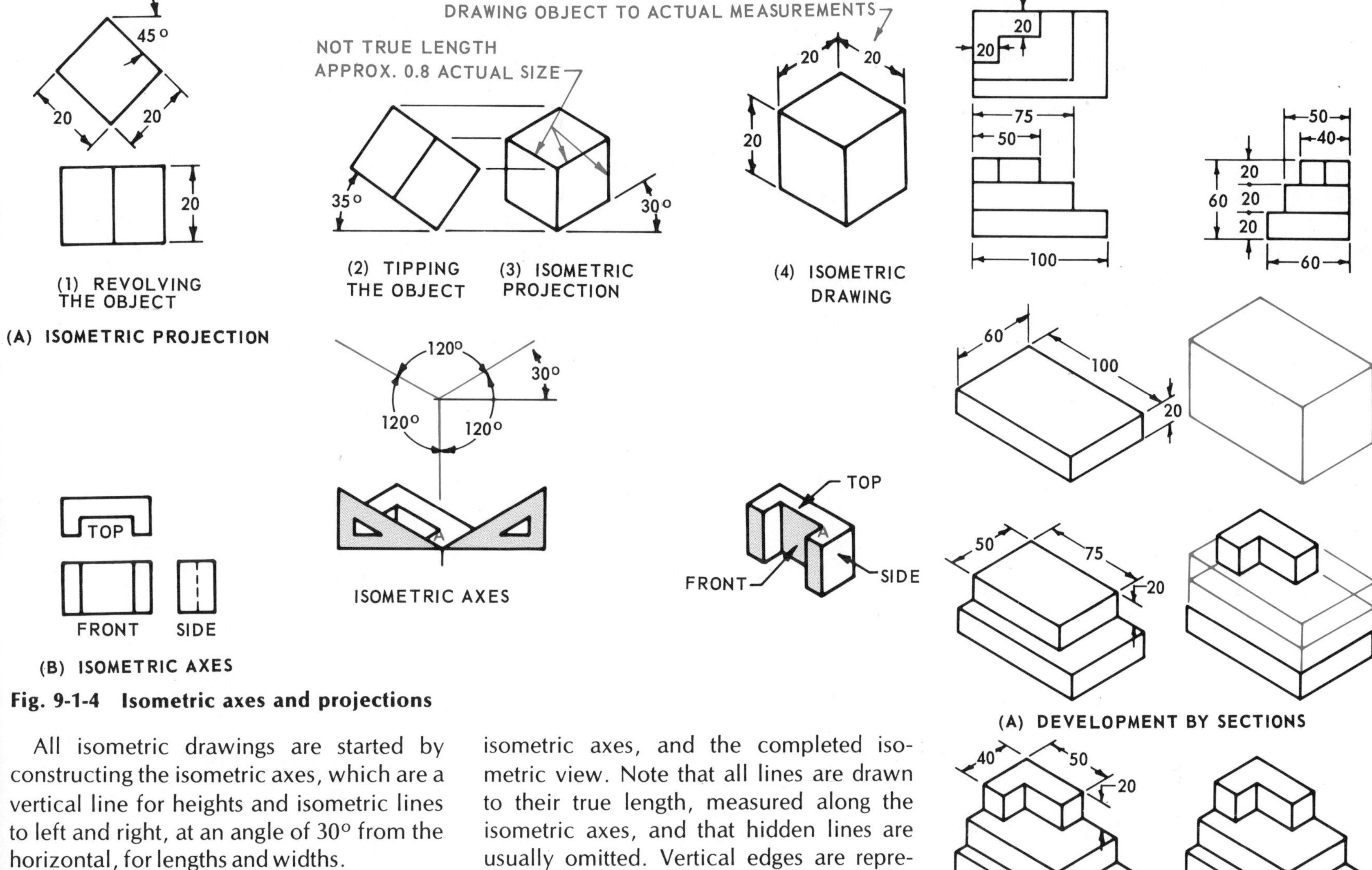

Fig. 9-1-4 Isometric axes and projections

All isometric drawings are started by constructing the isometric axes, which are a vertical line for heights and isometric lines to left and right, at an angle of 30° from the horizontal, for lengths and widths.

The three faces seen in the isometric view are the same faces that would be seen in the normal orthographic views: top, front, and side.

Figure 9-1-4*b* illustrates the selection of the front corner (*A*), the construction of the isometric axes, and the completed isometric view. Note that all lines are drawn to their true length, measured along the isometric axes, and that hidden lines are usually omitted. Vertical edges are represented by vertical lines, and horizontal edges by lines at 30° to the horizontal.

Two techniques can be used for making an isometric drawing of irregularly shaped objects, as illustrated in Figure 9-1-5. In one method the object is divided mentally into

(B) BOX CONSTRUCTION

Fig. 9-1-5 Developing an isometric drawing

a number of sections, and the sections are drawn one at a time in their proper relationship to one another. In the second method, a box is drawn with the maximum height, width, and depth of the object; then the parts of the box that are not part of the object are removed, leaving the pieces that form the total object.

Nonisometric Lines

Many objects have sloping surfaces that are represented by sloping lines in the orthographic views. In isometric drawing, sloping surfaces appear as **nonisometric** lines. To draw them, locate their endpoints, found on the ends of isometric lines, and join them with a straight line. Figures 9-1-6 and 9-1-7 illustrate examples in the construction of nonisometric lines.

Dimensioning Isometric Drawings

At times, an isometric drawing of a simple object may serve as a working drawing. In such cases, the necessary dimensions and specifications are placed on the drawing.

Dimension lines, extension lines, and the line being dimensioned should be in

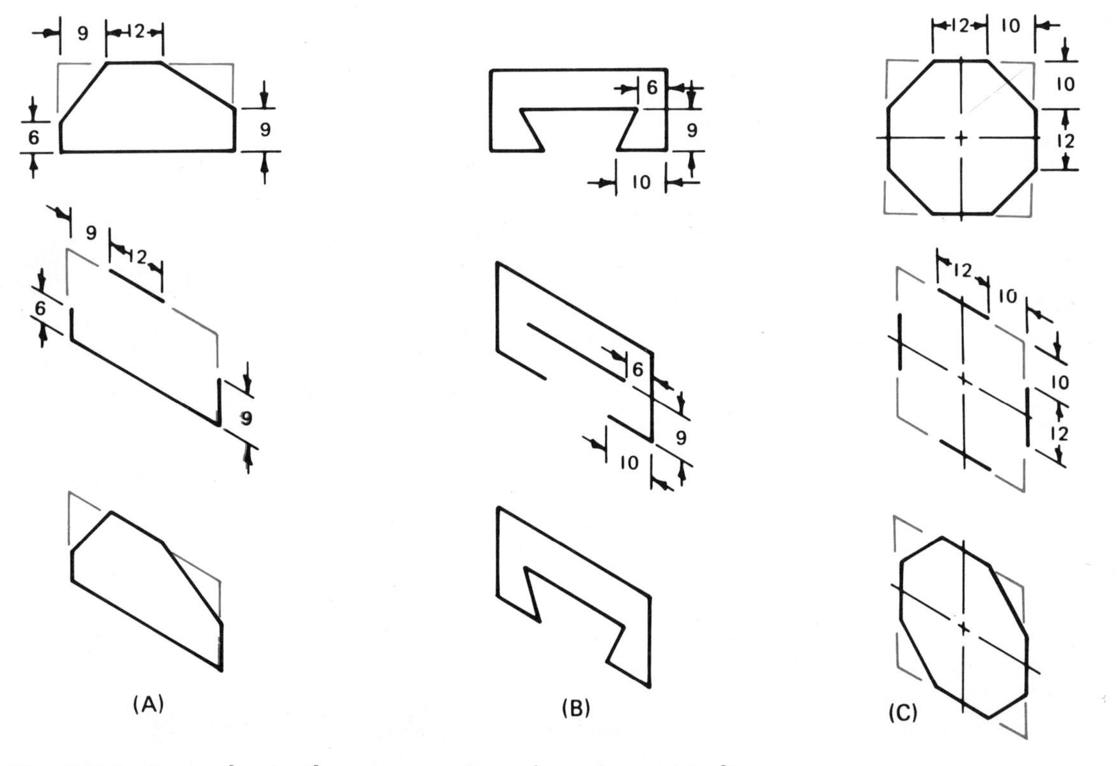

Fig. 9-1-6 Examples in the construction of nonisometric lines

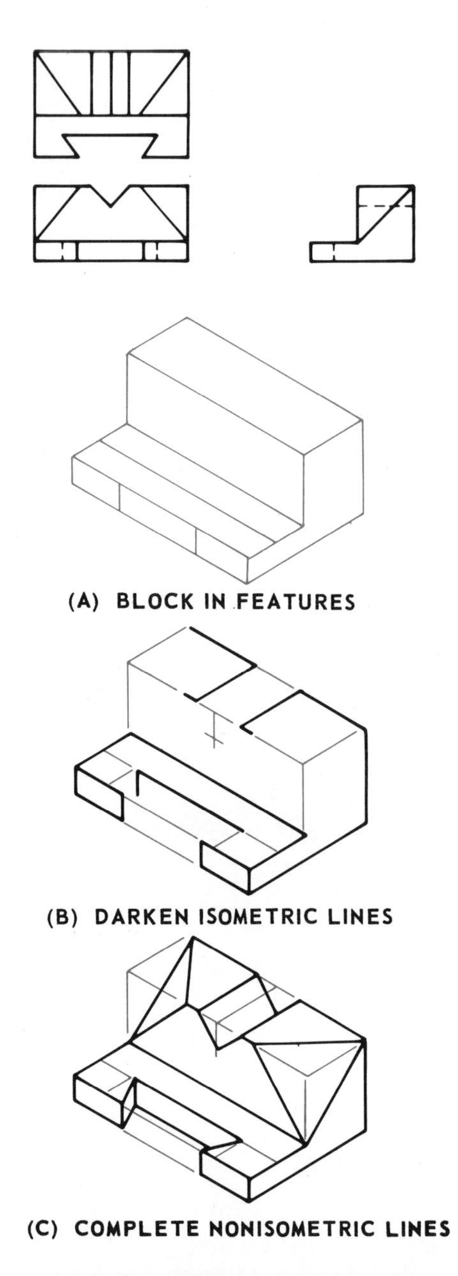

Fig. 9-1-7 Sequence in drawing an object having nonisometric lines

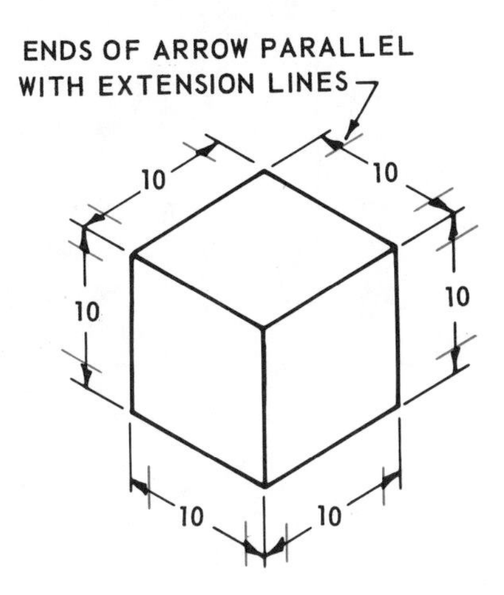

Fig. 9-1-8 Orienting the dimension line, arrowhead, and extension line

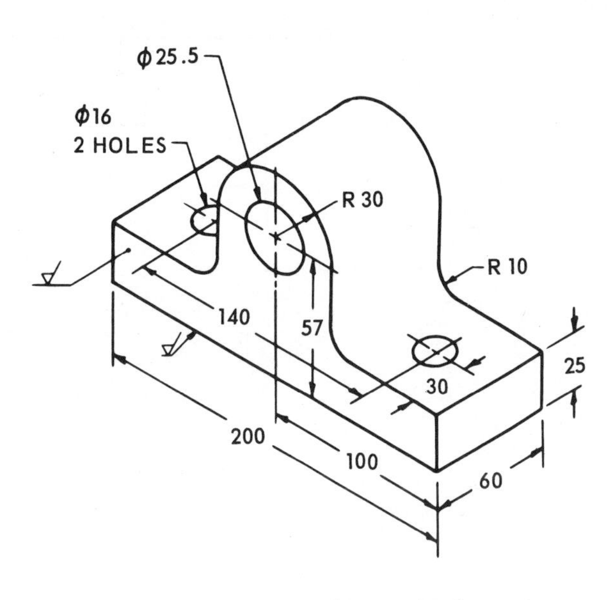

Fig. 9-1-9 Isometric dimensioning

the same plane. Arrowheads, which should be long and narrow, should be in the plane of the dimension and extension lines. See Figure 9-1-8.

Unidirectional dimensioning is used on isometric drawings. The letters and numbers should be vertical and read from the bottom of the sheet. An example of this type of dimensioning is shown in Figure 9-1-9. This is the preferred method.

Since the isometric is a one-view drawing, it is not usually possible to avoid placing dimensions on the view or across dimension lines. However, this practice should be avoided whenever possible.

Isometric Grid Paper

Isometric grid sheets are another time-saving device. Designers and engineers frequently use isometric grid paper on which they sketch their ideas and designs. See Figure 9-1-11. Many companies, such as those which prepare pipe drawings, have large drawing sheets made with non-reproducible isometric grid lines.

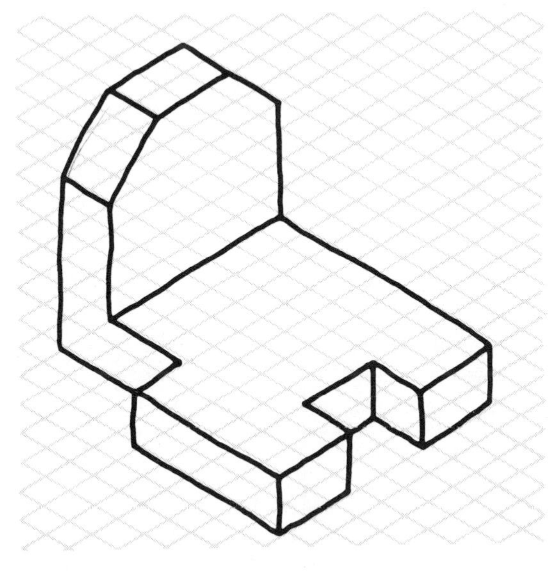

Fig. 9-1-10 Isometric grid paper

Assignments

1. On isometric grid paper sketch the four parts shown in Fig. 9-1-A. Do not show hidden lines. Each square shown on the drawing represents one isometric square on the grid paper.
2. On isometric grid paper sketch the four parts shown in Fig. 9-1-B. Do not show hidden lines. Each square shown on the drawing represents one isometric square on the grid paper.
3. On an A3 sheet, make an isometric drawing, complete with dimensions, of one of the parts shown in Fig. 9-1-C to 9-1-H. Scale is 1:1.

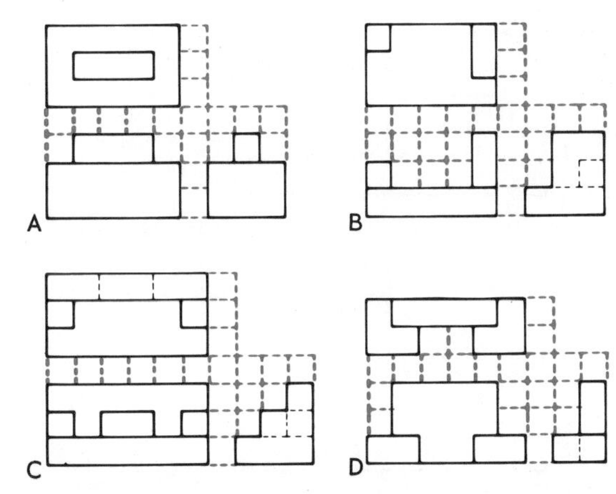

Fig. 9-1-A Sketching problems

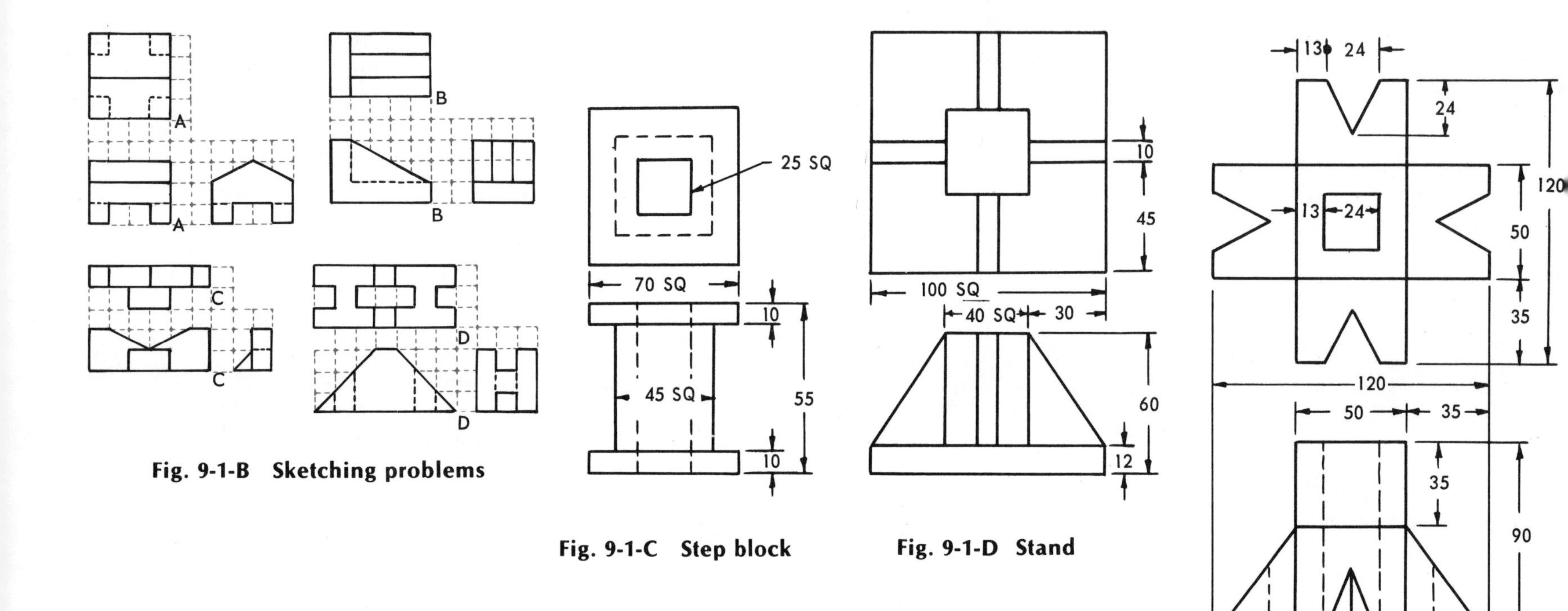

Fig. 9-1-B Sketching problems

Fig. 9-1-C Step block

Fig. 9-1-D Stand

Fig. 9-1-E Base block

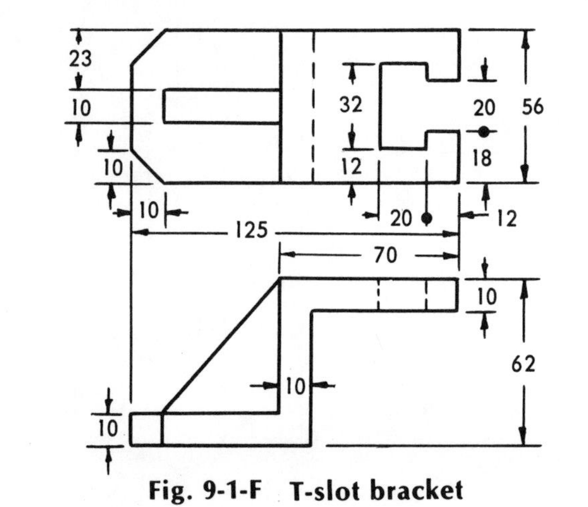

Fig. 9-1-F T-slot bracket

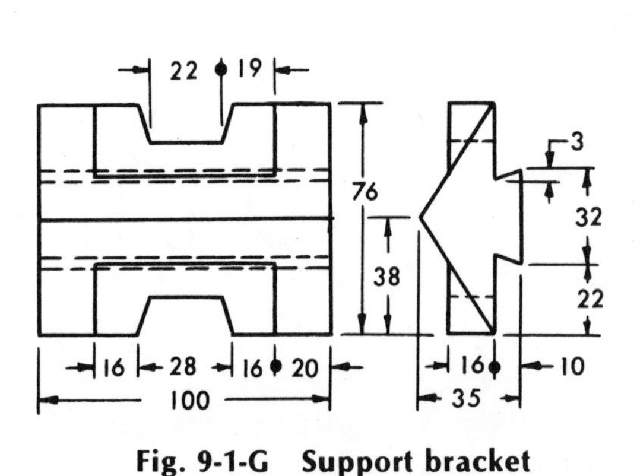

Fig. 9-1-G Support bracket

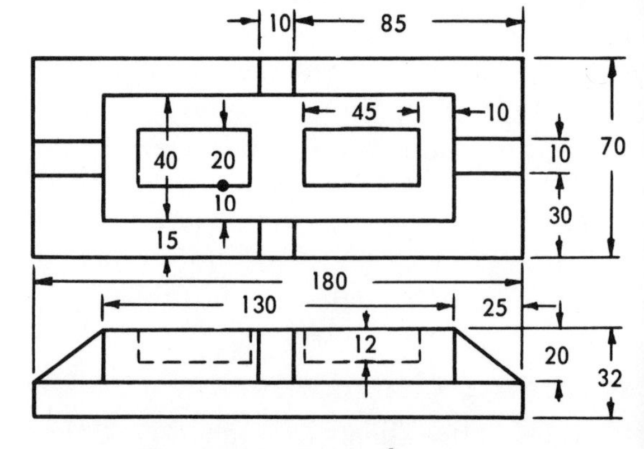

Fig. 9-1-H Base plate

UNIT 9-2
Curved Surfaces in Isometric

Circles and Arcs in Isometric

A circle on any of the three faces of an object drawn in isometric has the shape of an ellipse. See Figure 9-2-1. Figure 9-2-2 illustrates the steps in drawing circular features on isometric drawings.

1. Draw the centre lines and a square, with sides equal to the circle diameter, in isometric.

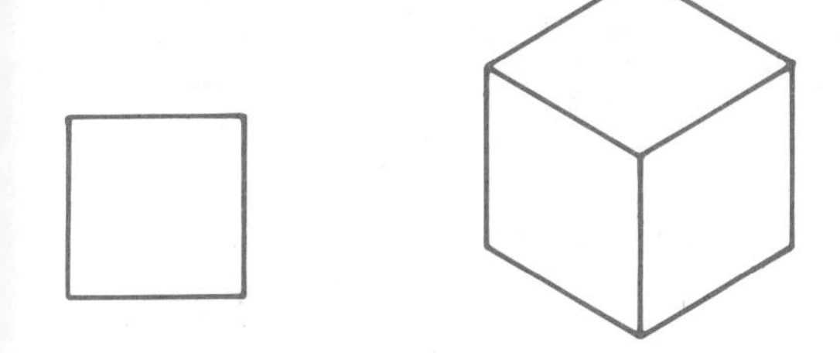

(A) A SQUARE DRAWN IN THE THREE ISOMETRIC POSITIONS

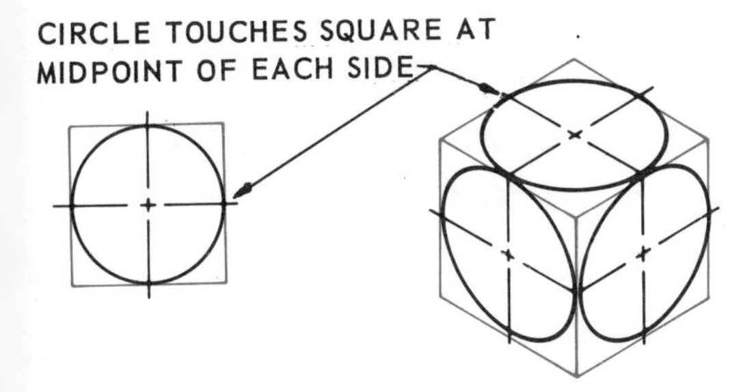

(B) A CIRCLE PLACED INSIDE A SQUARE AND DRAWN IN THE THREE ISOMETRIC POSITIONS

Fig. 9-2-1 Circles in isometric

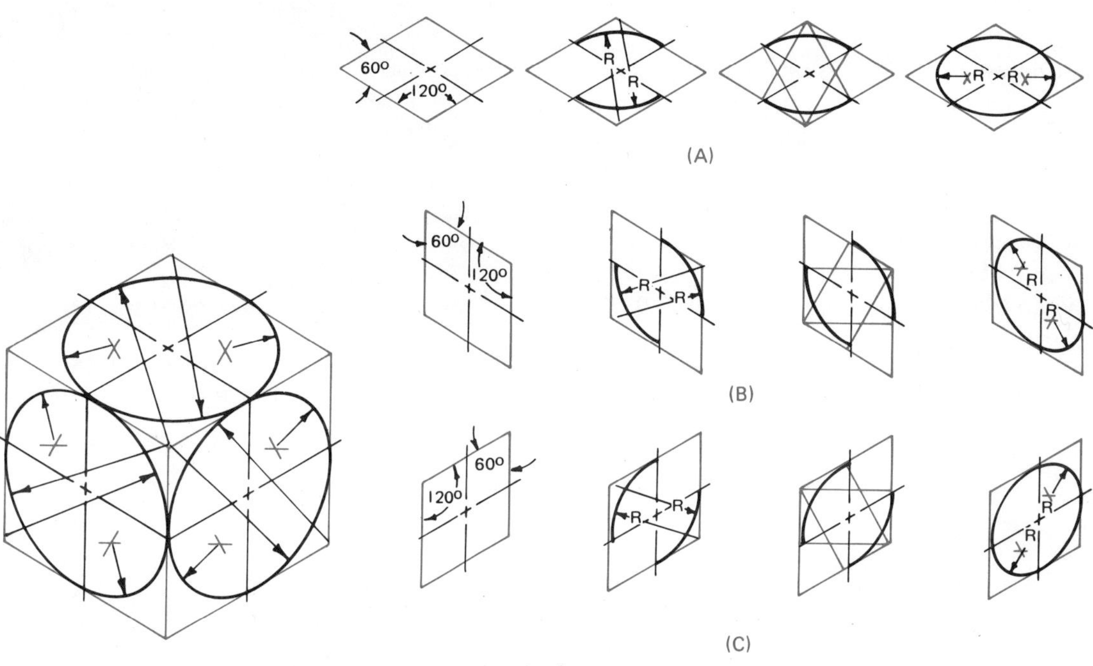

Fig. 9-2-2 Sequence in drawing isometric circles

2. Using the obtuse-angled (120°) corners as centres, draw arcs tangent to the sides forming the obtuse-angled corners, stopping at the points where the centre lines cross the sides of the square.

3. Draw construction lines from these same points to the opposite obtuse-angled corners. The points at which these construction lines intersect are the centres for arcs drawn tangent to the sides forming the acute-angled corners, meeting the first arcs.

In drawing concentric circles, each circle must have its own set of centres for the arcs, as shown in Figure 9-2-3.

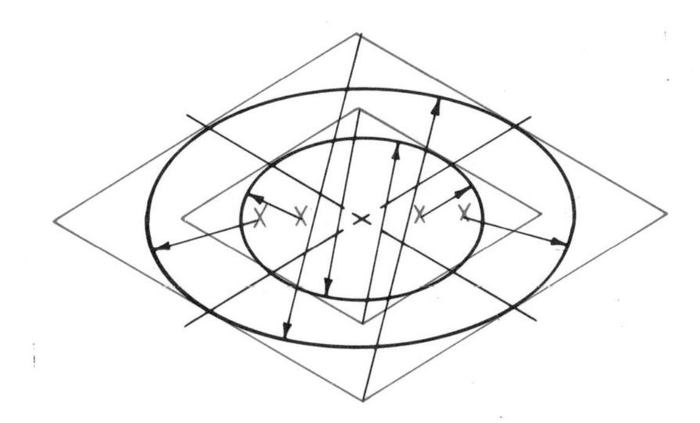

Fig. 9-2-3 Drawing concentric isometric circles

The same technique is used for drawing part-circles (arcs). See Figure 9-2-4. Construct an isometric square with sides equal to twice the radius, and draw that portion of the ellipse necessary to join the two faces. When these faces are parallel, draw half of an ellipse (one long radius and one short radius); when they are at an obtuse angle (120°), draw one long radius; and when they are at an acute angle (60°), draw one short radius.

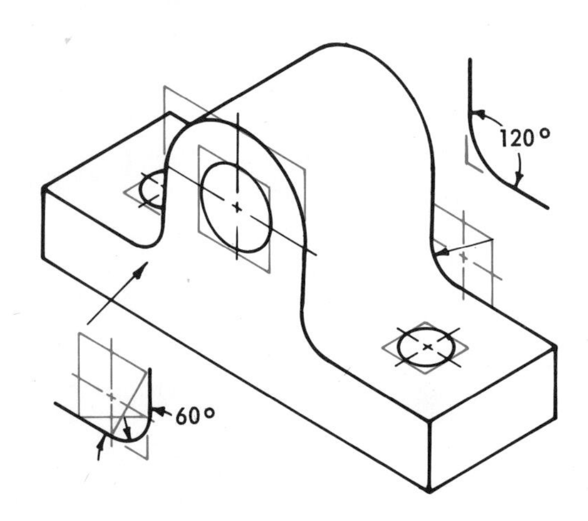

Fig. 9-2-4 Drawing isometric arcs

Isometric Templates

For convenience and time saving, isometric ellipse templates should be used whenever possible. A wide variety of elliptical templates are available. The template shown in Figure 9-2-5 combines ellipse, scales, and angles. Markings on the ellipses coincide with the centre lines of the holes which

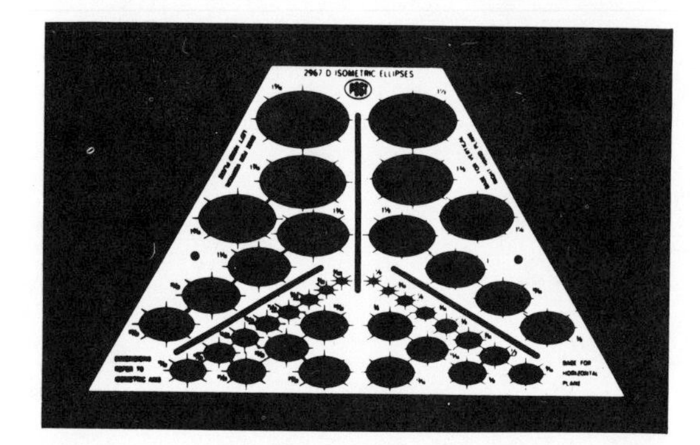

Fig. 9-2-5 Isometric ellipse template

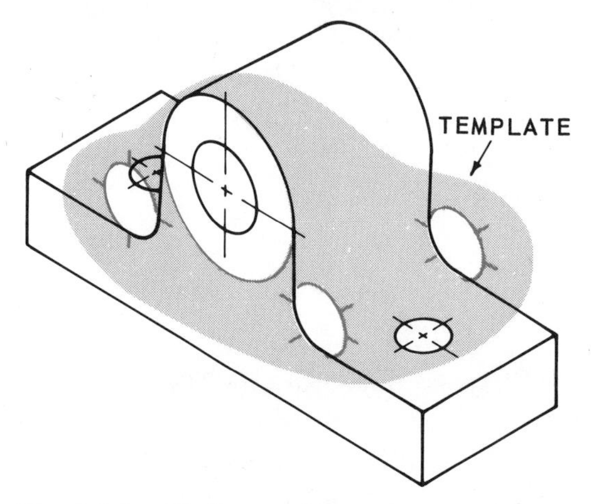

Fig. 9-2-6 Circles and arcs drawn with isometric ellipse template

speeds up drawing circles and arcs. Figure 9-2-6 shows the same part as shown in Figure 9-2-4 but with the arcs and circles being constructed with a template.

Sketching Circles and Arcs

In sketching circles and arcs on isometric grid paper, locate the centre lines first, then lightly sketch in construction boxes (isometric squares) where the circles and arcs should be. See Figure 9-2-7. Sketch the ellipse (isometric circle) just touching the centre of each of the four sides of the square.

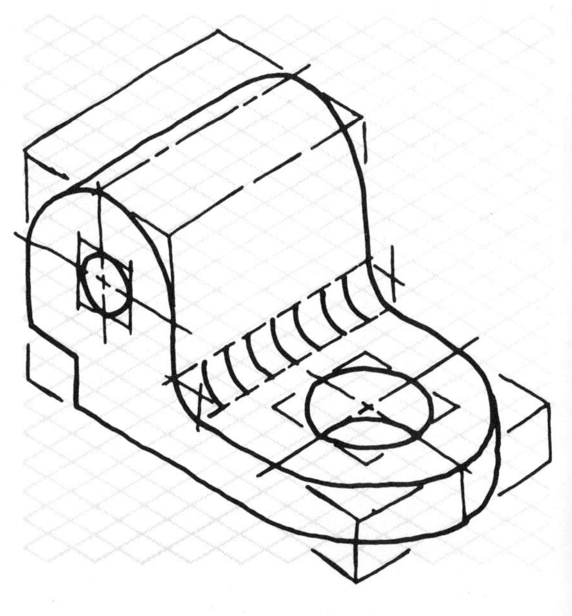

Fig. 9-2-7 Isometric sketching paper

Drawing Irregular Curves in Isometric

To draw curves other than circles or arcs, the plotting method shown in Figure 9-2-8 is used:

1. Draw an orthographic view, and divide the area enclosing the curved line into equal squares.

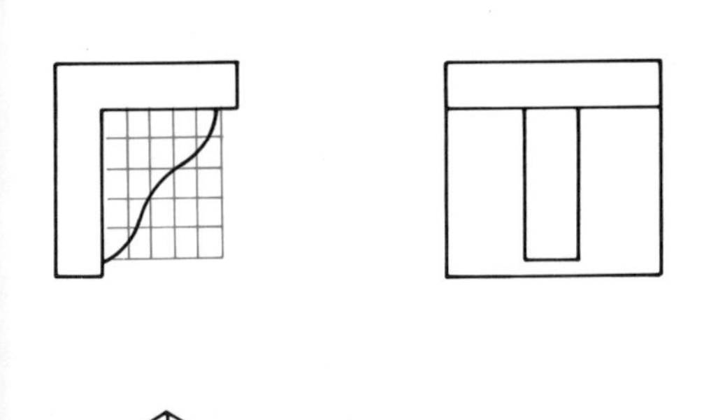

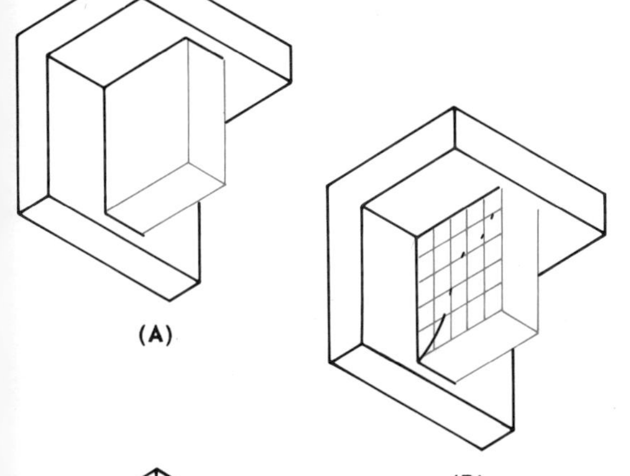

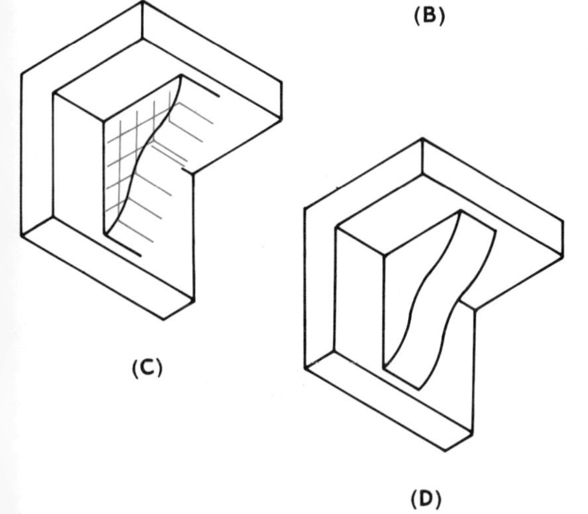

Fig. 9-2-8 Curves drawn in isometric by means of offset measurements

2. Produce an equivalent area on the isometric drawing, showing the offset squares.

3. Take positions relative to the squares from the orthographic view, and plot them on the corresponding squares on the isometric view.

4. Draw a smooth curve through the established points with the aid of an irregular curve.

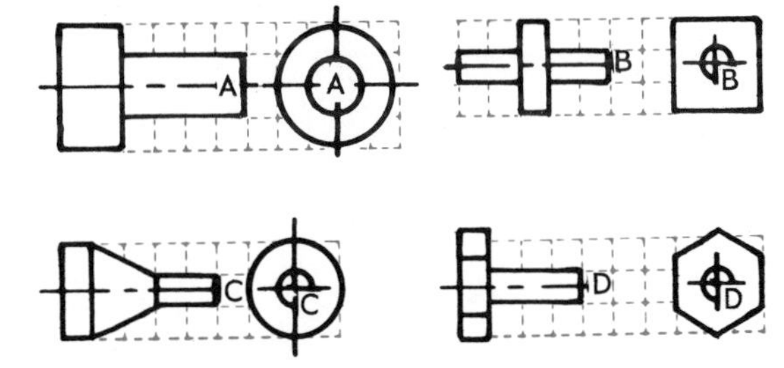

Fig. 9-2-A Sketching problems

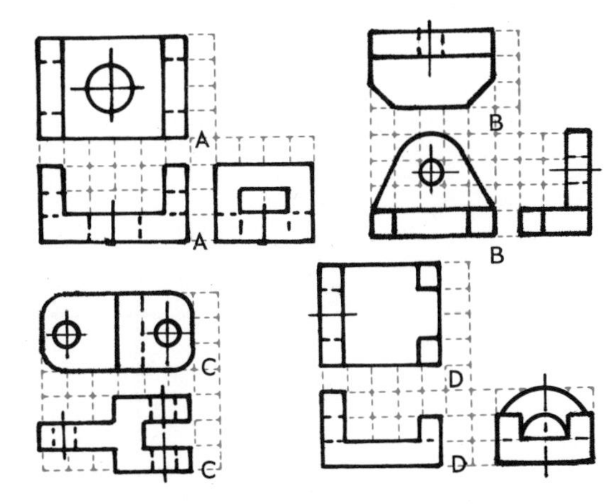

Fig. 9-2-B Sketching problems

Assignments

1. On isometric grid paper sketch the four parts shown in Fig. 9-2-A. Each square shown on the figure represents one square on the isometric grid. Hidden lines may be omitted for clarity.
2. On isometric grid paper sketch the four parts shown in Fig. 9-2-B. Each square shown on the figure represents one square on the isometric grid. Hidden lines may be omitted for clarity.
3. On an A3 sheet, make an isometric drawing complete with dimensions of one of the parts shown in Fig. 9-2-C to 9-2-H. Scale is 1:1

Review for Assignments

Unit 9-1 Dimensioning isometric drawings

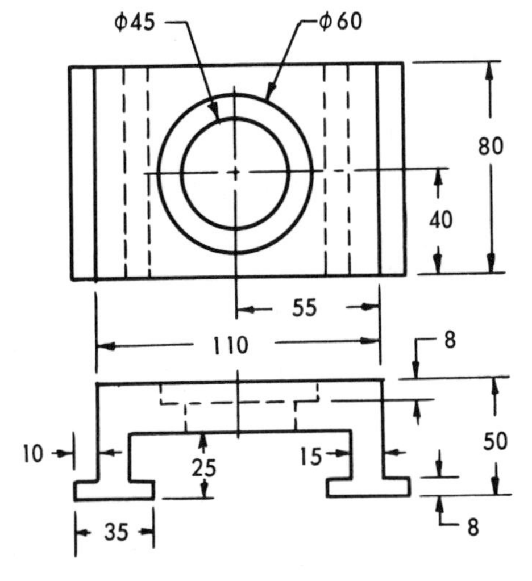

Fig. 9-2-C T-guide

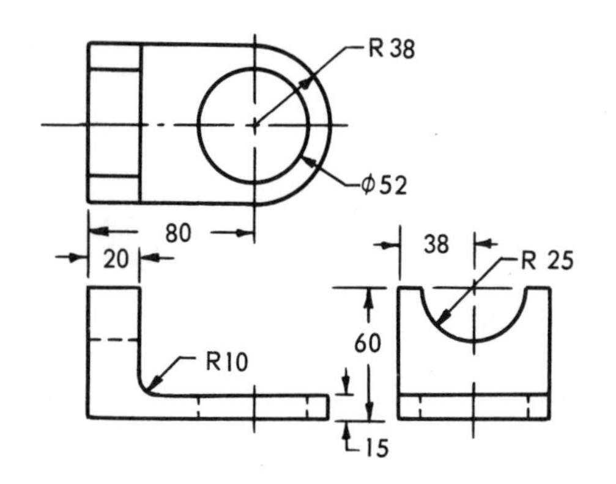

Fig. 9-2-D Cradle bracket

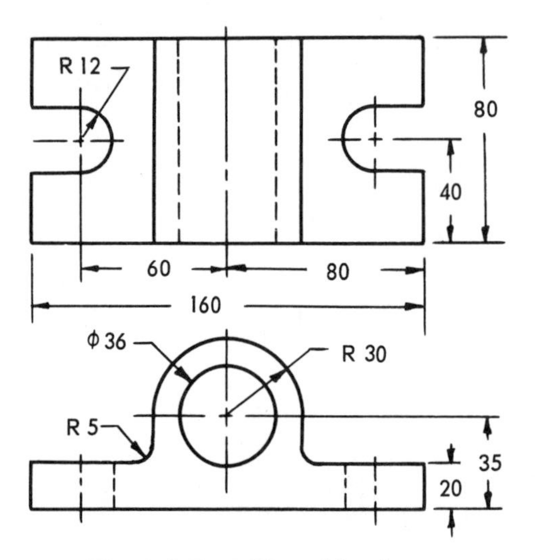

Fig. 9-2-E Pillow block

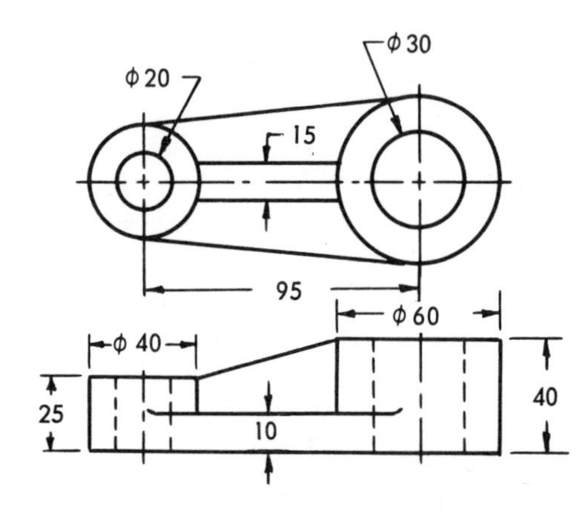

Fig. 9-2-F Link

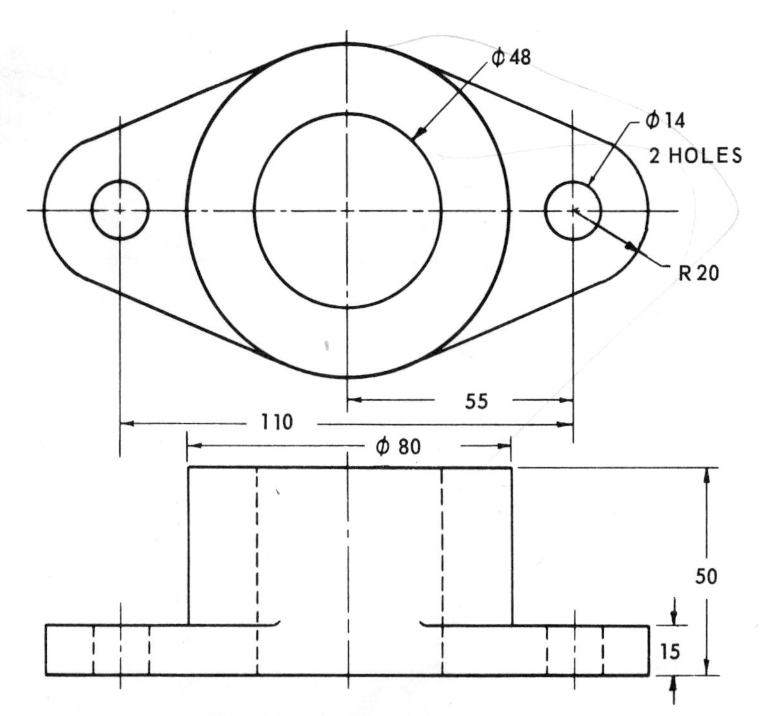

Fig. 9-2-G Bearing support

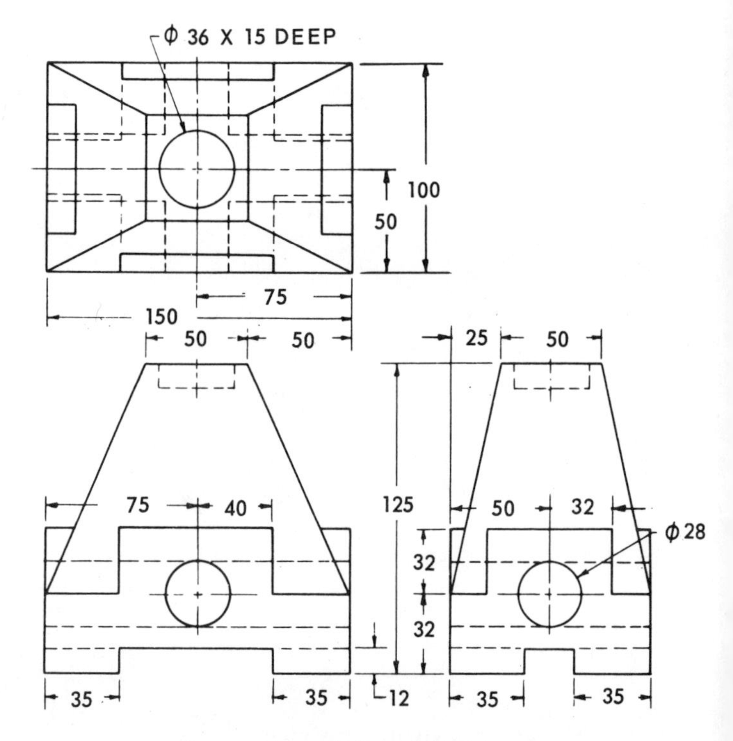

Fig. 9-2-H Sliding block

UNIT 9-3

Common Features in Isometric

Isometric Sectioning

Isometric drawings are generally made as outside views, but sometimes a sectional view is needed. The section is taken on an isometric plane, that is, on a plane parallel to one of the faces of the cube.

Figure 9-3-1 shows isometric full sections taken on a different plane for each of three objects. Note the construction lines indicating the part that has been cut away. Isometric half sections are illustrated in Figure 9-3-2. Notice the outlines of the cut surfaces in *a* and *b*. The cut method is to draw the complete outside view and the isometric cutting plane.

When a section drawing is drawn in isometric, the section lines are drawn at an angle of 60° with the horizontal or in a horizontal position, depending on where the cutting-plane line is located. In half sections, the section lines are sloped in opposite directions, as shown in Figure 9-3-2.

Fillets and Rounds

For most isometric drawings of parts having small fillets and rounds, the adopted practice is to draw the corners as sharp features. However, when it is desirable to represent the part, normally a casting, as having a more realistic appearance, either of the methods shown in Figure 9-3-3 may be used.

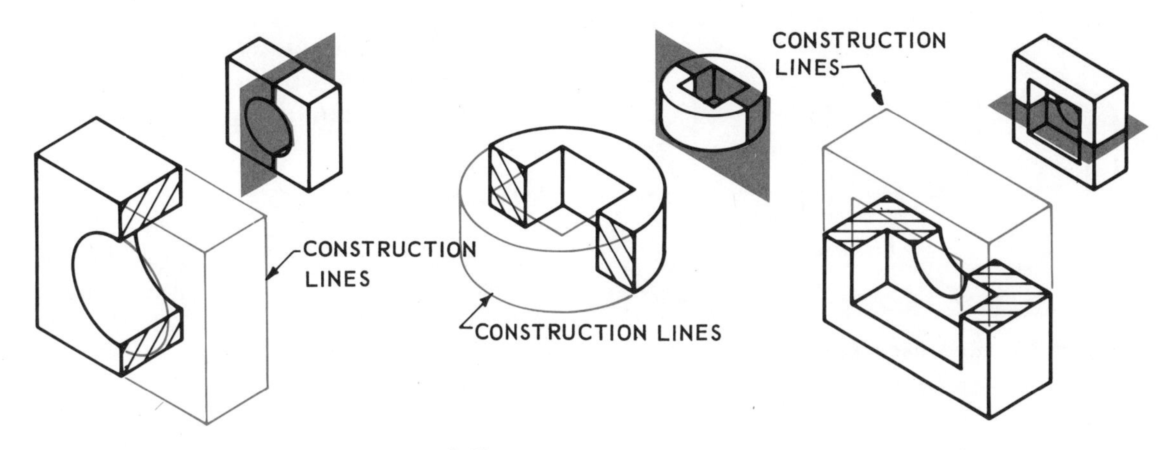

Fig. 9-3-1 Examples of isometric full sections

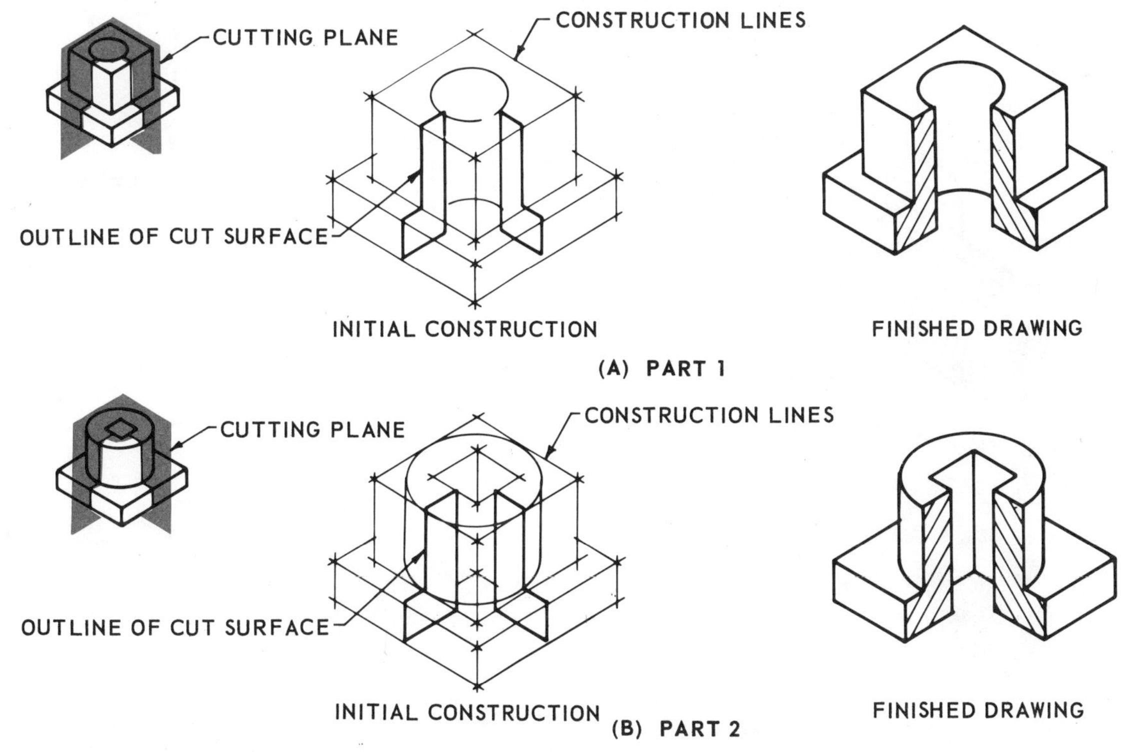

Fig. 9-3-2 Examples of isometric half sections

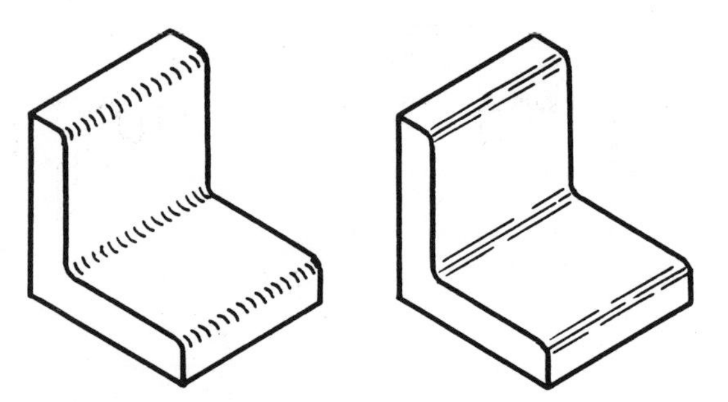

Fig. 9-3-3 Representation of fillets and rounds

Threads

The method for showing threads in isometric is shown in Figure 9-3-4. The threads are represented by a series of ellipses uniformly spaced along the centre line of the thread. The spacing of the ellipses need not be the spacing of the actual pitch.

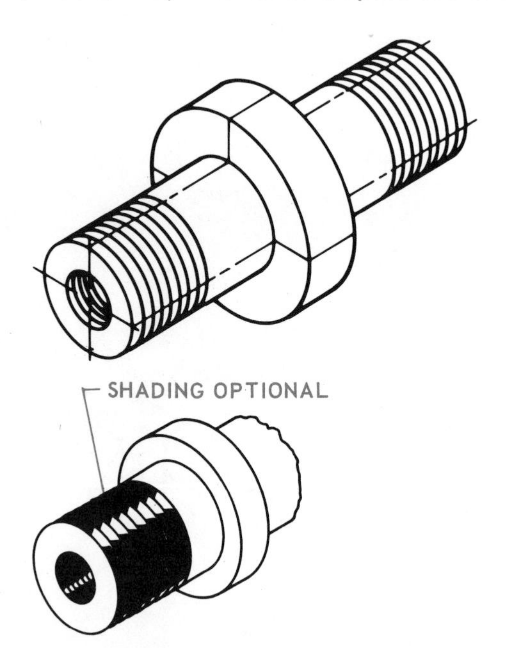

Fig. 9-3-4 Representation of threads in isometric

Break Lines

For long parts, break lines should be used to shorten the length of the drawing. Freehand breaks are preferred, as shown in Figure 9-3-5.

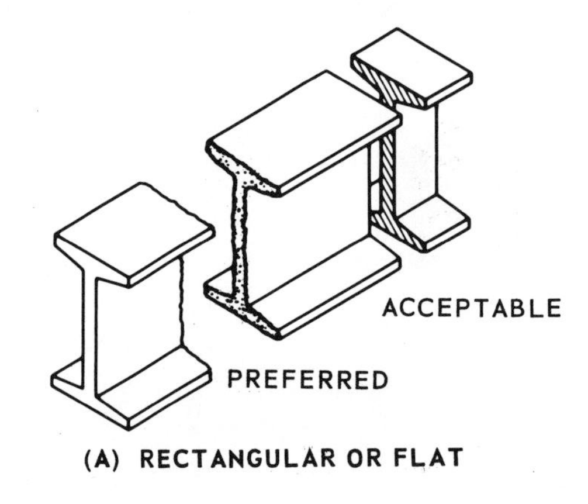

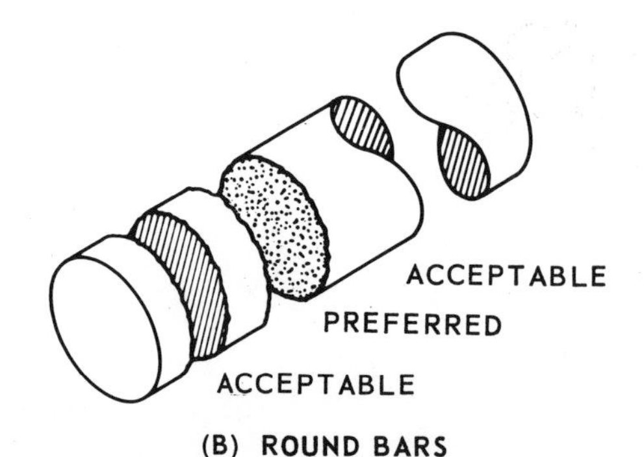

Fig. 9-3-5 Conventional breaks

Isometric Assembly Drawings

Regular or exploded assembly drawings are frequently used in catalogues and sales literature, as illustrated by Figure 9-3-6.

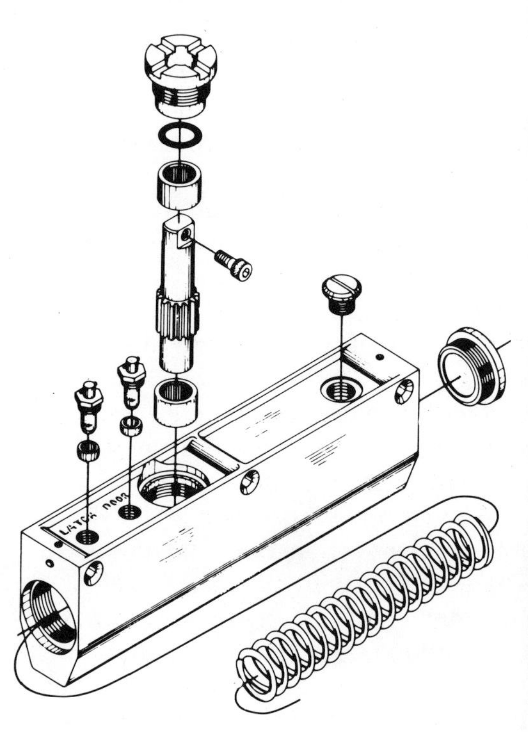

Fig. 9-3-6 Exploded isometric assembly drawing

Assignments

1. On an A3 sheet, draw a section view of one of the parts shown in Fig. 9-3-A or 9-3-B. The section view is to be taken at cutting plane *BB*. Do not dimension. Scale is 1:1.

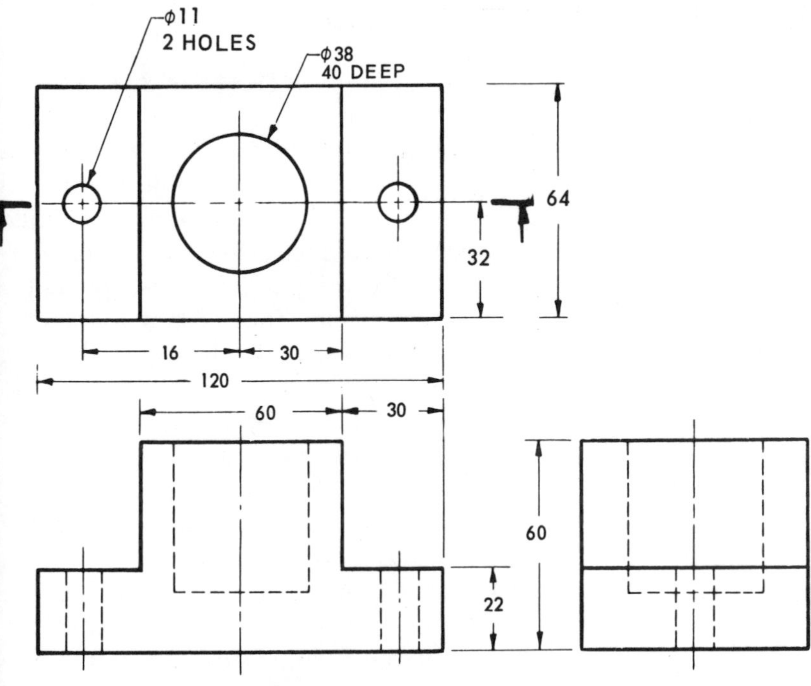

Fig. 9-3-A Pencil holder

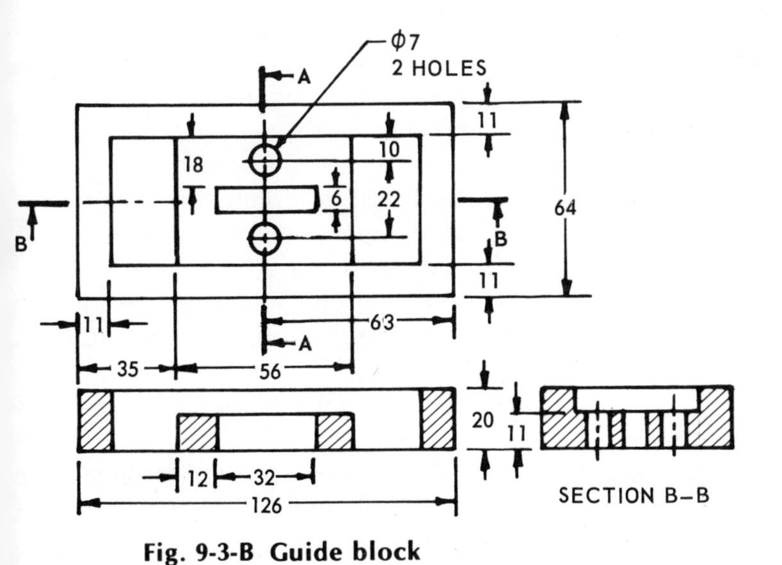

Fig. 9-3-B Guide block

2. On an A3 sheet, draw an isometric assembly drawing of the two-post die set, model 302, shown in Fig. 9-3-C. Allow 50 mm between the top and base. Scale is 1:2. Do not dimension. Include on the drawing a bill of material. Using part numbers, identify the parts on the assembly.

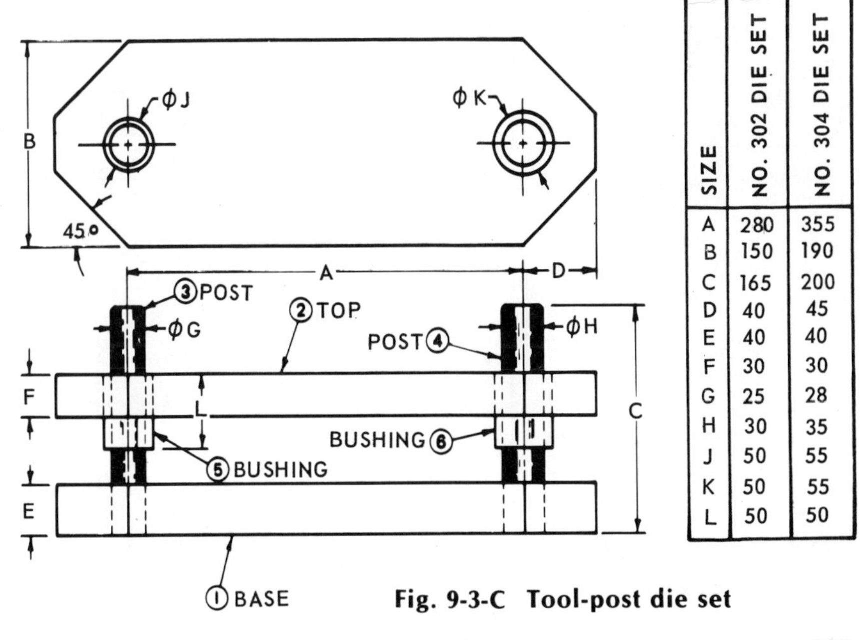

SIZE	NO. 302 DIE SET	NO. 304 DIE SET
A	280	355
B	150	190
C	165	200
D	40	45
E	40	40
F	30	30
G	25	28
H	30	35
J	50	55
K	50	55
L	50	50

Fig. 9-3-C Tool-post die set

3. On an A3 sheet, draw an isometric exploded assembly drawing of the book rack 1 shown in Fig. 9-3-D. Scale is 1:5. Do not dimension. Include on the drawing a bill of material. Using part numbers, identify the parts on the assembly.

4. On an A3 sheet, draw an isometric drawing of the roller shaft shown in Fig. 9-3-E. Use a conventional break to reduce the length of the drawing. Scale 1:1. Do not dimension. Shading the threads is optional.

Review for Assignments

Unit 5-3 Bill of materials and assembly drawings
Unit 7-1 Sections and conventions

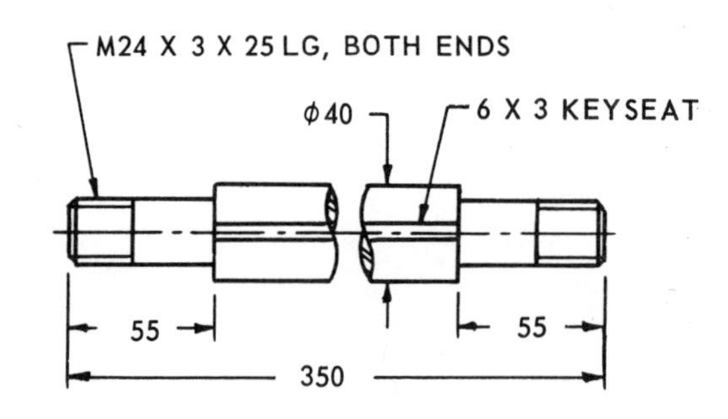

Fig. 9-3-E Shaft

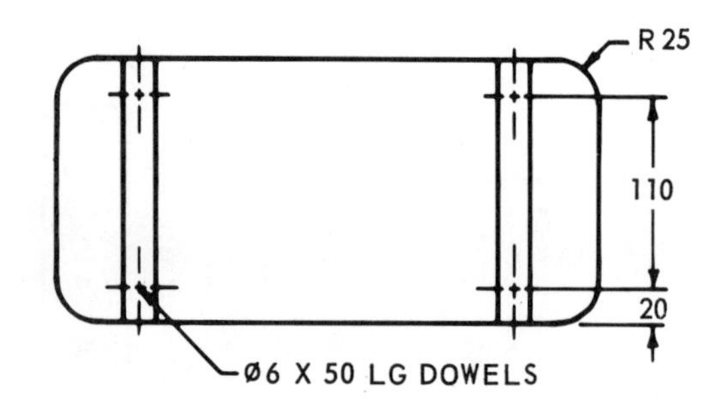

SIZE	NO. 1	NO. 2	NO. 3
A	200	250	300
B	320	370	420

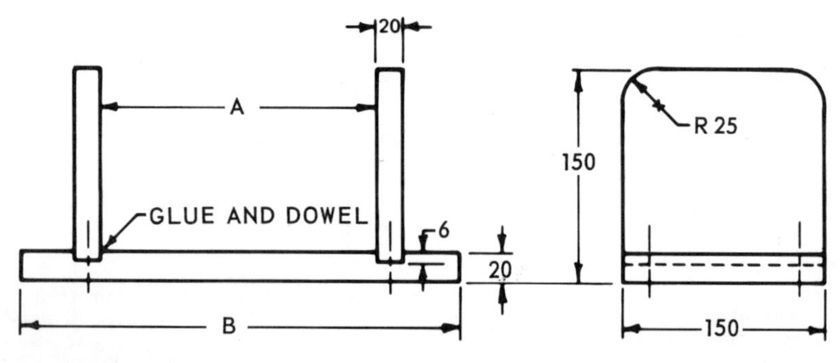

Fig. 9-3-D Book rack

UNIT 9-4
Oblique Projection

This method of pictorial drawing places the object with one face parallel to the frontal plane and the other two faces on oblique (or receding) planes. These planes may recede to left or right, top or bottom, at a convenient angle. The three axes of projection are **vertical, horizontal,** and **receding.**

Figure 9-4-1 illustrates a cube drawn in typical positions with the receding axis at 60, 45, and 30°. This form of projection has the advantage of showing one face of the object without distortion.

The face with the greatest irregularity of outline or contour, or the face with the longest dimension, faces the front. See Figure 9-4-2.

Two types of oblique projection are used extensively. In **cavalier oblique,** all lines are drawn to their true length, measured on the axes of the projection.

In **cabinet oblique,** the lines on the receding axis are shortened by one-half their true length to compensate for distortion and to approximate more closely what the human eye would see.

For this reason, and because of the simplicity of projection, cabinet oblique is a commonly used form of pictorial representation, especially when circles and arcs are to be drawn.

Figure 9-4-3 shows a comparison of cavalier and cabinet oblique. Note that hidden lines are omitted unless required for clarity.

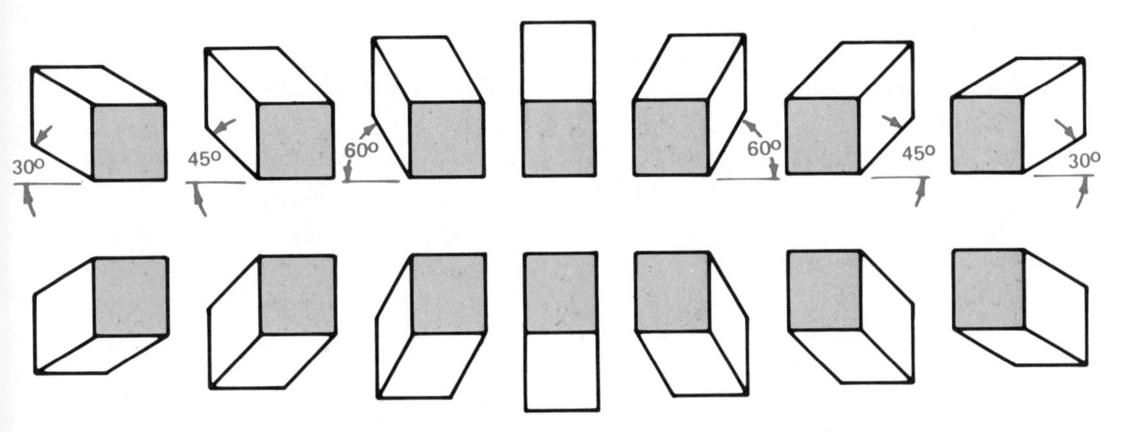

Fig. 9-4-1 Typical positions of receding axes for oblique projections

Many of the drawing techniques for isometric projection apply to oblique projection. Figure 9-4-4 illustrates the construction of an irregularly shaped object by the box method.

Inclined Surfaces

Angles which are parallel to the picture plane are drawn as their true size. Other angles can be laid off by locating the ends of the inclined line.

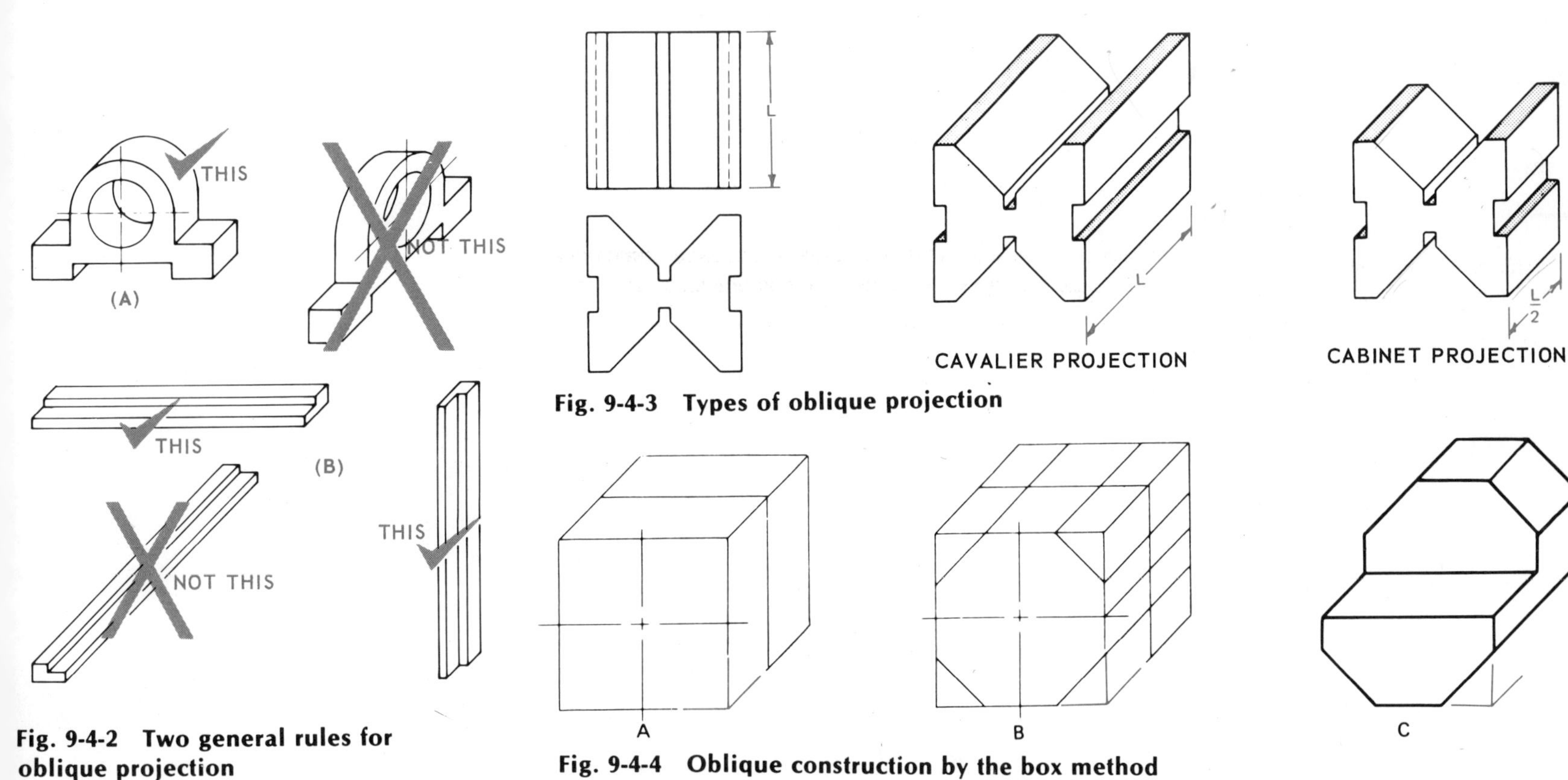

Fig. 9-4-2 Two general rules for oblique projection

Fig. 9-4-3 Types of oblique projection

Fig. 9-4-4 Oblique construction by the box method

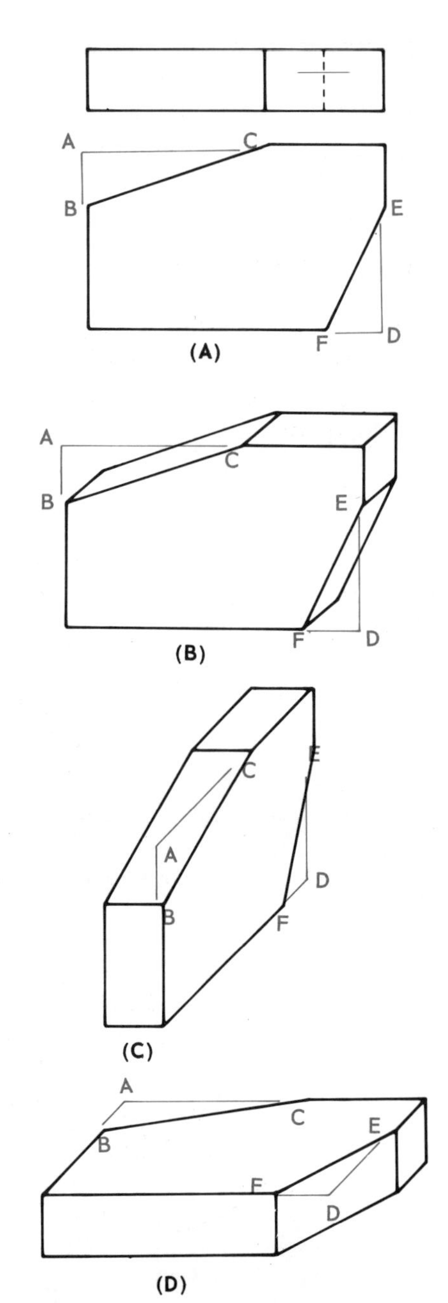

Fig. 9-4-5 Drawing inclined surfaces

A part with notched corners is shown in Figure 9-4-5*a*. An oblique drawing with the angles parallel to the picture plane is shown at *b*. In Figure 9-4-5*c* the angles are parallel to the profile plane.

In each case the angle is laid off by measurement parallel to the oblique axes, as shown by the construction lines. Since the part, in each case, is drawn in cabinet oblique, the receding lines are shortened by one-half their true length.

Oblique Sketching

Specially designed oblique sketching paper with 45° lines is available and, like isometric sketching paper, is used extensively by engineers and drafters. See Figure 9-4-6.

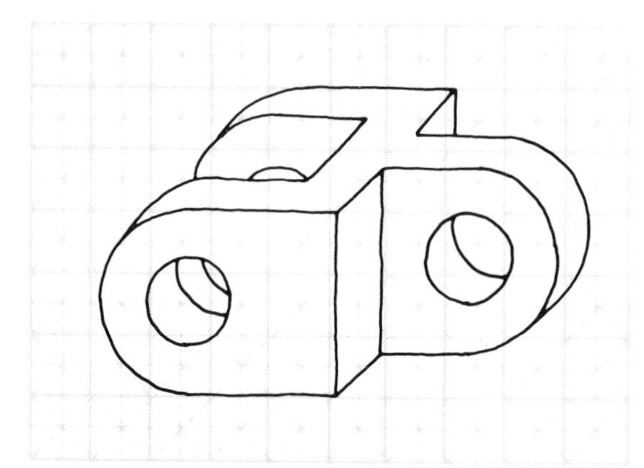

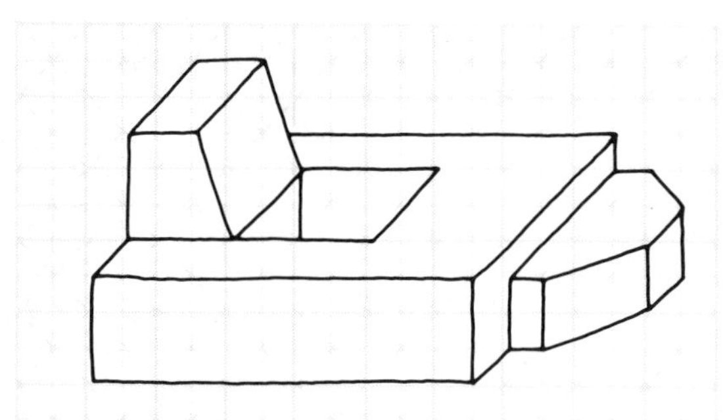

Fig. 9-4-6 Oblique sketching paper

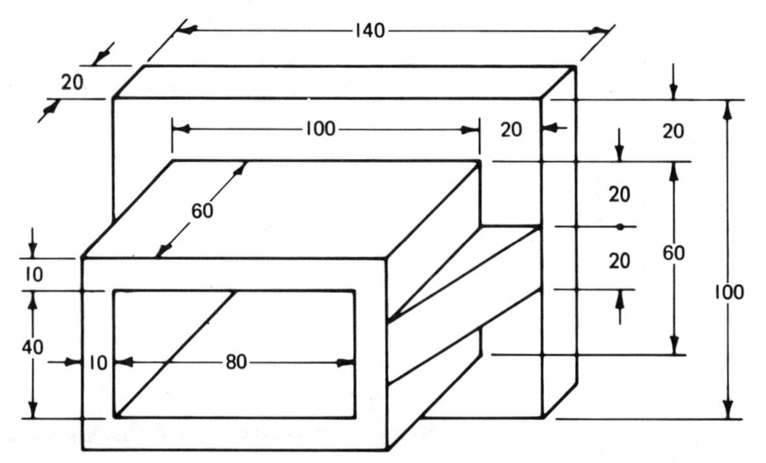

Fig. 9-4-7 Dimensioning an oblique drawing

Dimensioning an Oblique Drawing

Dimension lines are drawn parallel to the axes of projection. Extension lines are projected from the horizontal and vertical object lines whenever possible.

The dimensioning of an oblique drawing is similar to that of an isometric drawing. The recommended method is unidirectional dimensioning, which is shown in Figure 9-4-7. As in isometric dimensioning, usually it is necessary to place some dimensions directly on the view.

Assignments

1. On coordinate grid paper make oblique sketches of the four parts shown in Fig. 9-4-A. Each square shown on the figure represents one square on the grid paper. Hidden lines may be omitted to improve clarity.
2. On coordinate grid paper make oblique sketches of the four parts shown in Fig. 9-4-B. Each square shown on the figure represents one square on the grid paper. Hidden lines may be omitted to improve clarity.
3. On an A3 sheet, make an oblique drawing, complete with dimensions, of one of the parts shown in Fig. 9-4-C or 9-4-D. Scale is 1:1.

Review for Assignments

Unit 9-1 Dimensioning isometric drawings

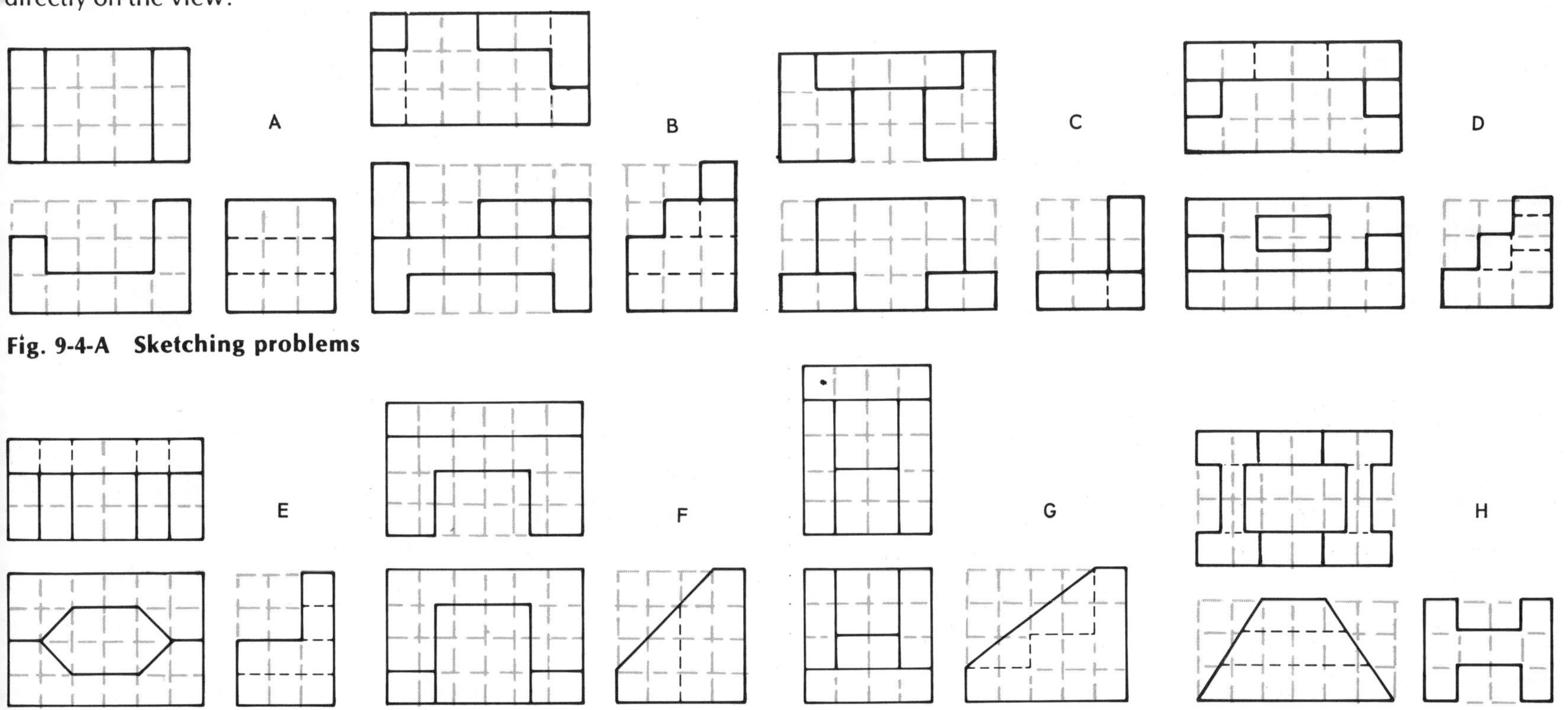

Fig. 9-4-A Sketching problems

Fig. 9-4-B Sketching problems

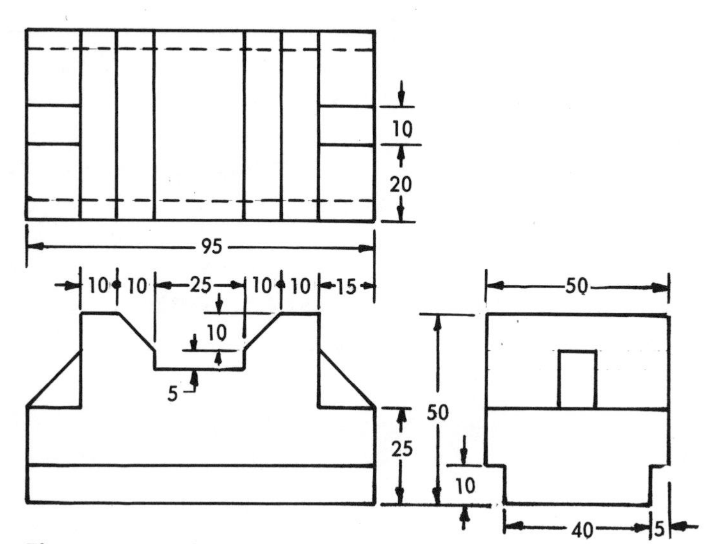

Fig. 9-4-C V-block rest

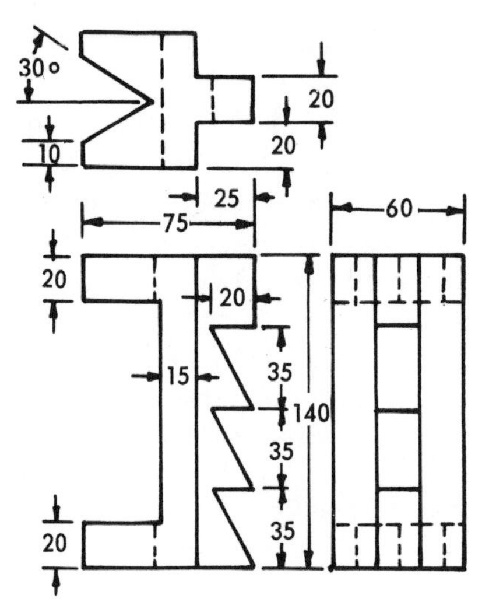

Fig. 9-4-D Ratchet

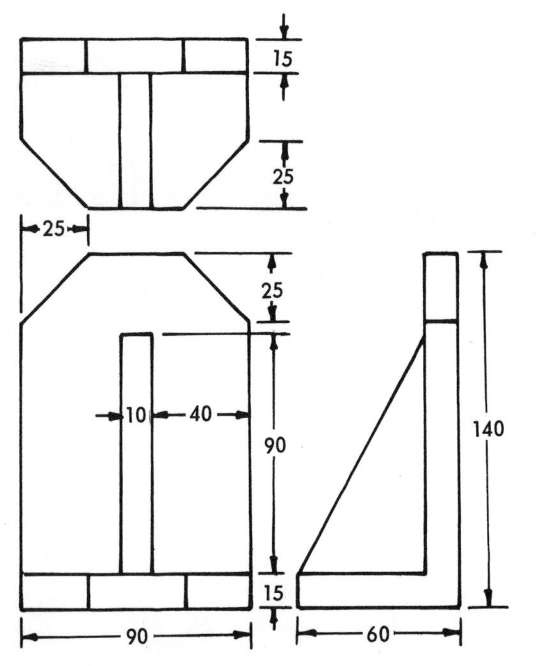

Fig. 9-4-E Brace

UNIT 9-5 Common Features in Oblique

Circles and Arcs

Whenever possible, the face of the object having circles or arcs should be selected as the **front** face, so that such circles or arcs can be easily drawn in their true shape. See Figure 9-5-1.

When circles or arcs must be drawn on one of the oblique faces, the offset measurement method illustrated in Figure 9-5-2 may be used:

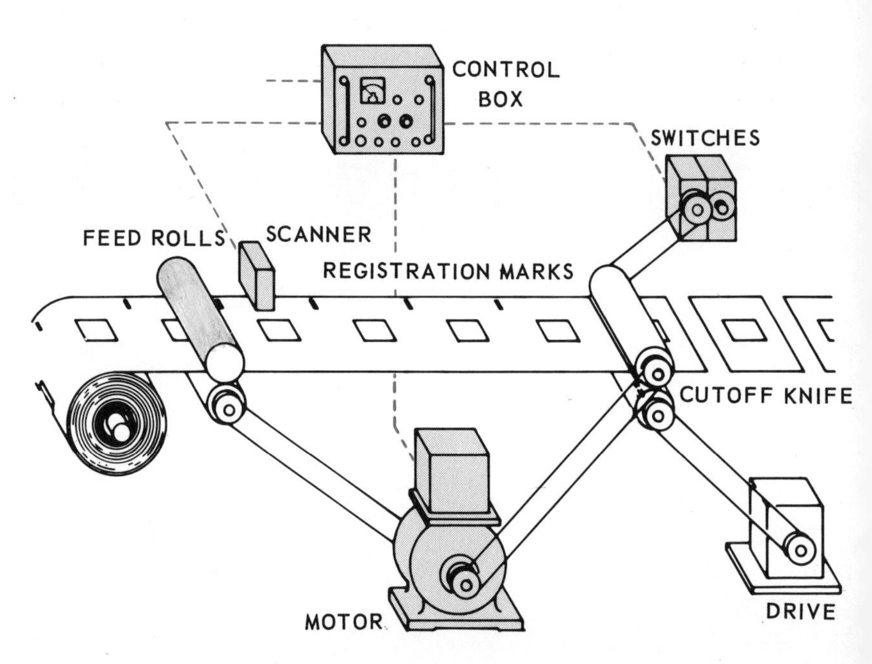

Fig. 9-5-1 Application of oblique drawing

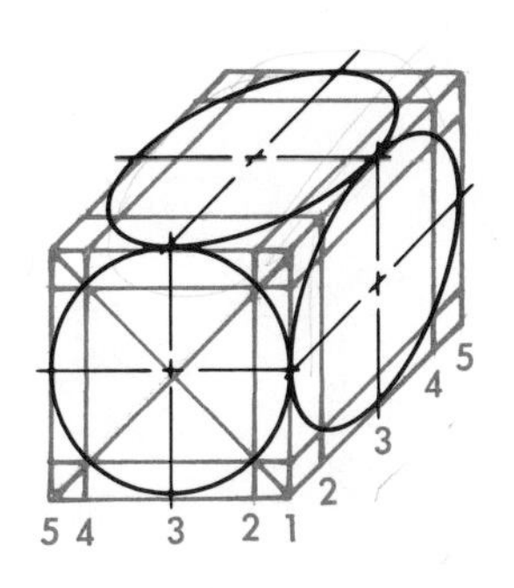

Fig. 9-5-2 Drawing oblique circles by means of offset measurements

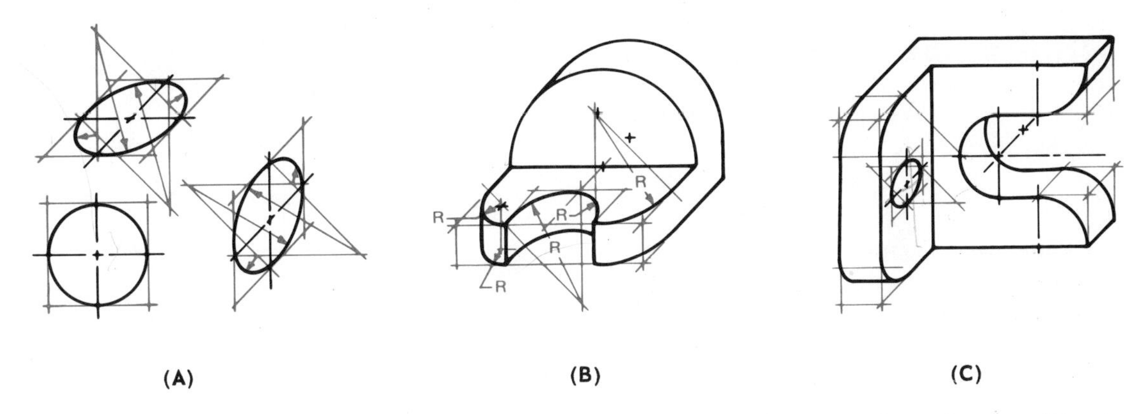

Fig. 9-5-3 Circles parallel to the picture plane are true circles, on other planes, ellipses

1. Draw an oblique square about the centre lines, with sides equal to the diameter.

2. Draw a true circle adjacent to the oblique square, and establish equally spaced points about its circumference.

3. Project these point positions to the edge of the oblique square, and draw lines on the oblique axis from these positions. Similarly spaced lines are drawn on the other axis, forming offset squares and giving intersection points for the oval shape.

Another method used when circles or arcs must be drawn on one of the oblique surfaces is the **four-centre method.** In Figure 9-5-3*a* a circle is shown as it would be drawn on a front plane, a side plane, and a top plane.

In Figure 9-5-3*b*, the oblique drawing has some arcs in a horizontal plane. In Figure 9-5-3*c*, the oblique drawing shown has some arcs in a profile plane.

Circles not parallel to the picture plane when drawn by the approximate method are not pleasing but are satisfactory for some purposes. Ellipse templates, when available, should be used because they reduce drawing time and give much better results. If a template is used, the oblique circle should first be blocked in as an oblique square in order to locate the proper position of the circle. Blocking in the circle first also helps the drafter select the proper size and shape of the ellipse. The construction and dimensioning of an oblique part are shown in Figure 9-5-4.

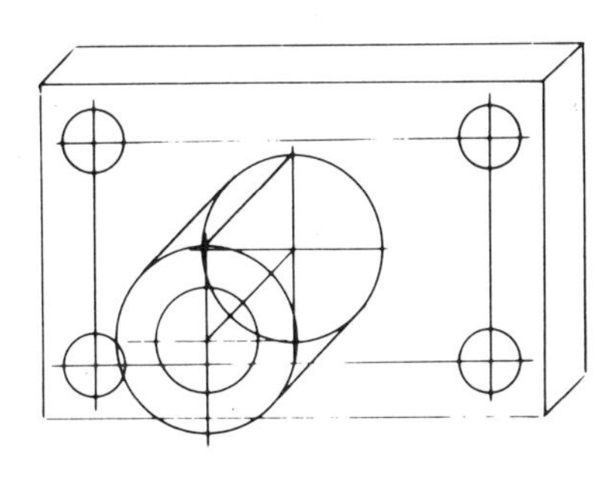

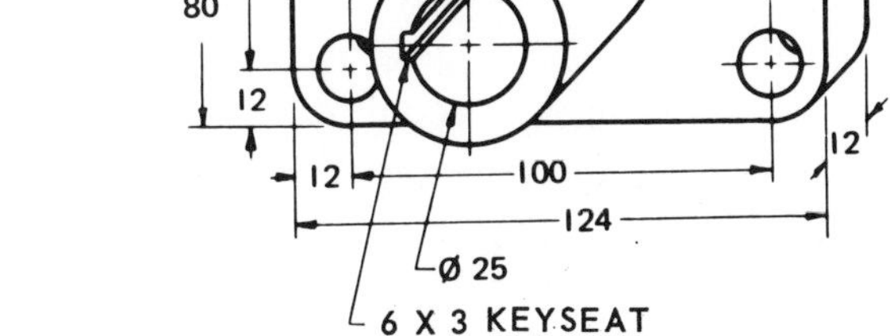

(A) CONSTRUCTING THE OBJECT **(B) DIMENSIONING THE OBJECT**

Fig. 9-5-4 Construction and dimensioning of an oblique object

Oblique Sectioning

Oblique drawings are generally made as outside views, but sometimes a sectional view is necessary. The section is taken on a plane parallel to one of the faces of an oblique cube. Figure 9-9-5 shows an oblique full section and an oblique half section. Note the construction lines which indicate the part that has been cut away.

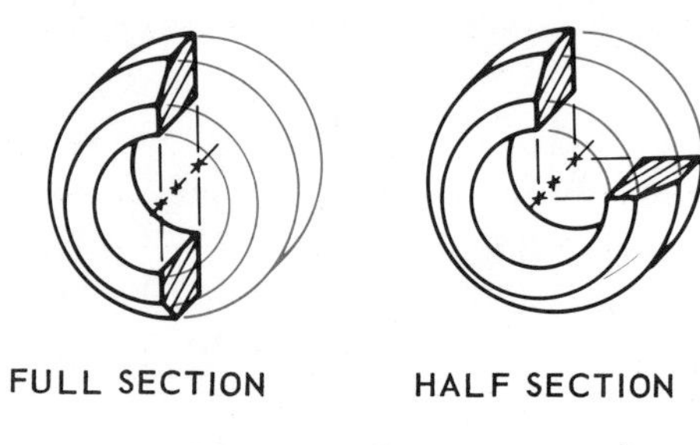

Fig. 9-5-5 Oblique full section and an oblique half section

Treatment of Conventional Features

Fillets and Rounds. Small fillets and rounds normally are drawn as sharp corners. When it is desirable to show the corners rounded, then either of the methods shown in Figure 9-5-6 is recommended.

Threads. The conventional method of showing threads in oblique is shown in Figure 9-5-7. The threads are represented by a series of circles uniformly spaced along the centre of the thread. The spacing of the circles need not be the spacing of the pitch.

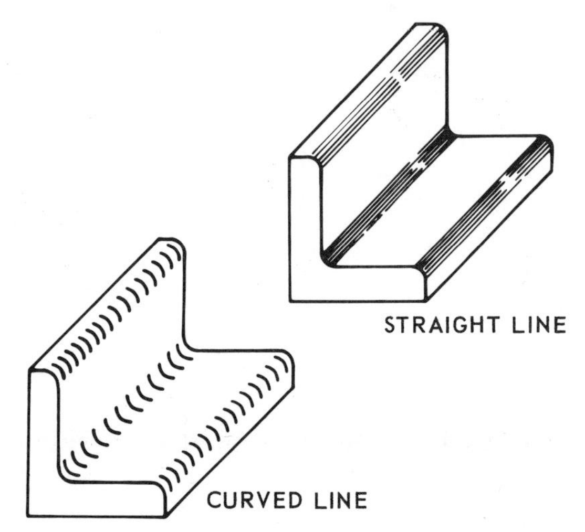

Fig. 9-5-6 Representing rounds and fillets

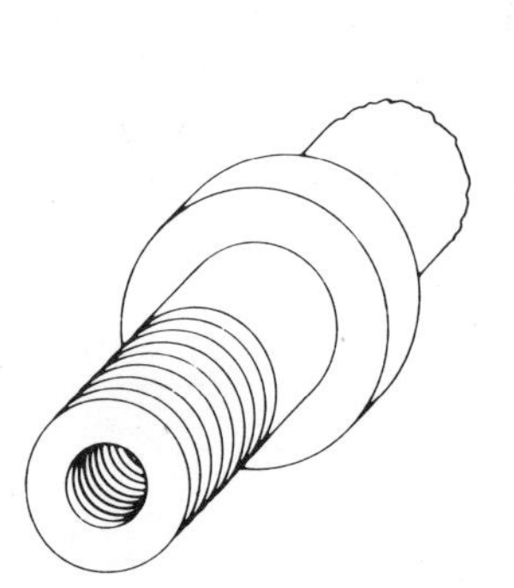

Fig. 9-5-7 Representation of threads in oblique

Breaks. Figure 9-5-8 shows the conventional method for representing breaks.

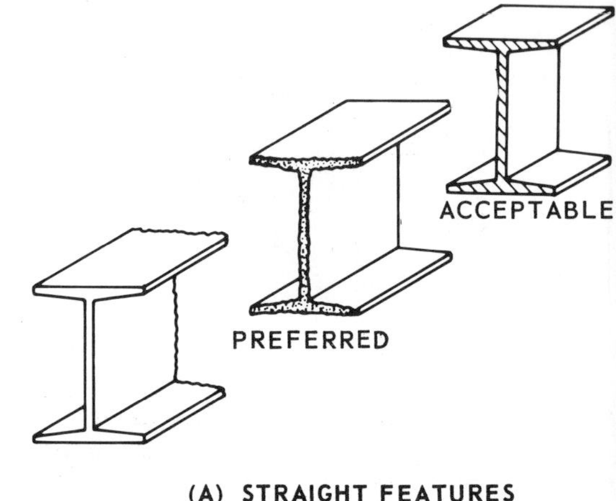

Fig. 9-5-8 Conventional breaks

Assignments

1. On an A3 sheet, make an oblique drawing, complete with dimensions, of one of the parts shown in Fig. 9-5-A to 9-5-D. Scale is 1:2.
2. On an A3 sheet, make an oblique drawing, complete with dimensions, of one of the parts shown in Fig. 9-5-E or 9-5-F. Use the straight-line method of showing the rounds and fillets. Scale is 1:2.

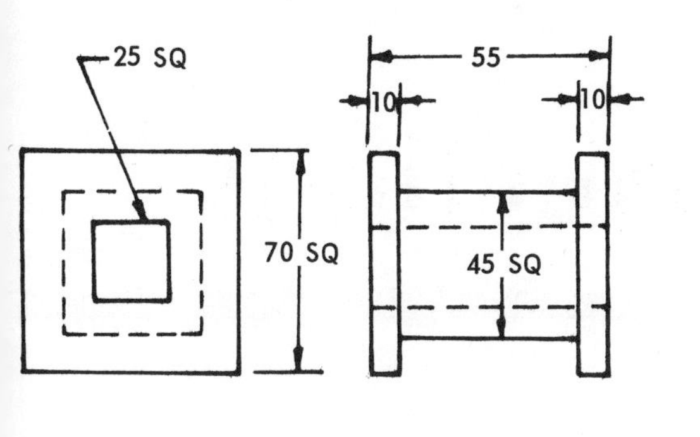

Fig. 9-5-A Guide block

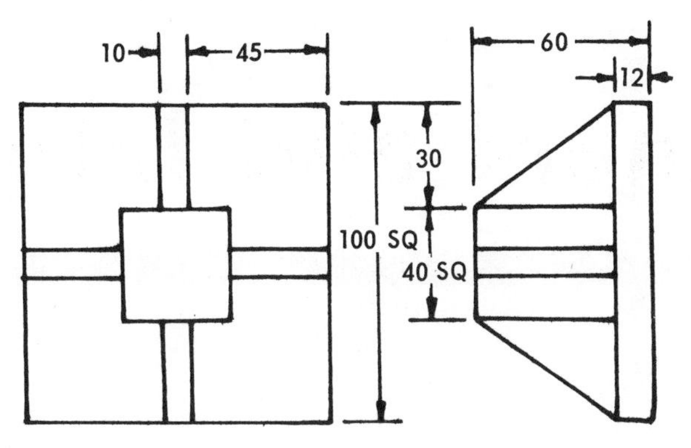

Fig. 9-5-B Lock base

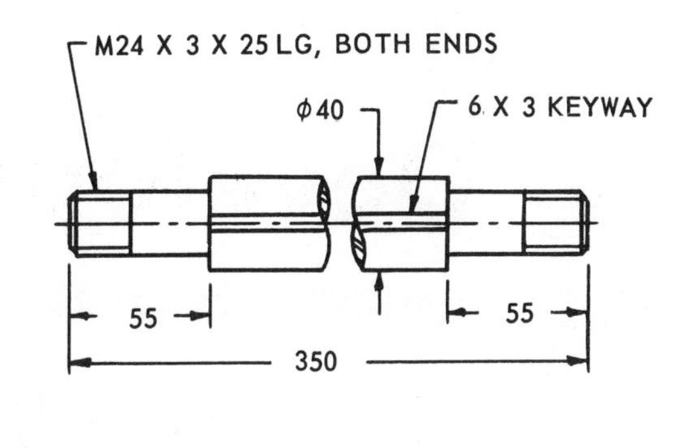

Fig. 9-5-C Shaft

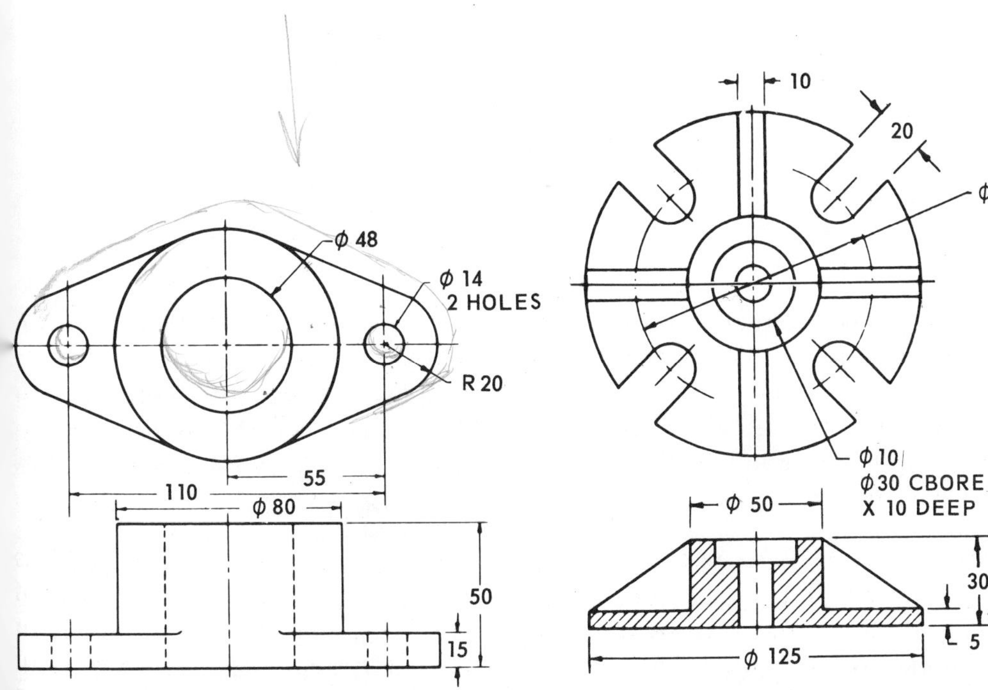

Fig. 9-5-D Bearing support

Fig. 9-5-E Bushing holder

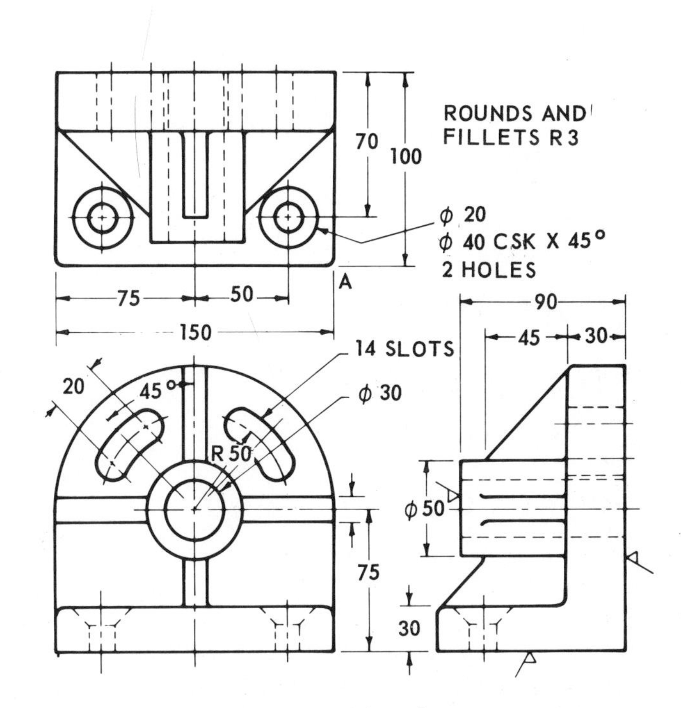

Fig. 9-5-F End bracket

Chapter 10

Development and Intersections

UNIT 10-1

Straight-Line Developments

Surface Developments

Many objects, such as cardboard and metal boxes, tin cans, funnels, cake pans, furnace pipes, elbows, ducts, and eaves troughing, are made from flat sheet material that is cut so that, when folded, formed, or rolled, it will take the shape of an object.

Since a definite shape and size are desired, a regular orthographic drawing of the object, such as shown in Figure 10-1-1, is made first; then a development drawing is made to show the complete surface or surfaces laid out in a flat plane.

Sheet-Metal Development

Surface development drawing is sometimes referred to as **pattern drawing,** because the layout, when made on heavy cardboard, thin metal, or wood, is used as a pattern for tracing out the developed shape on flat material. Such patterns are used extensively in sheet-metal shops.

When making a development drawing of an object which will be constructed of thin metal, such as a tin can or a dust pan, you must be concerned not only with the developed surfaces but also with the joining of the edges of these surfaces and with exposed edges. An allowance must be made for the additional material necessary for such seams and edges. You must also indicate where the material is bent.

Several commonly used methods of rep-

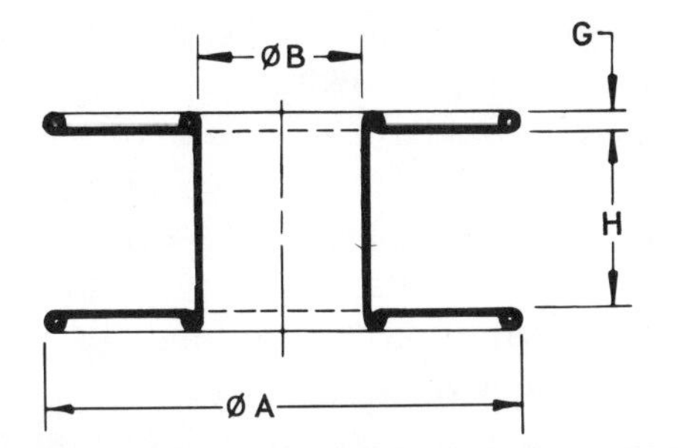

(A) SECTION VIEW THRU ADHESIVE TAPE SPOOL

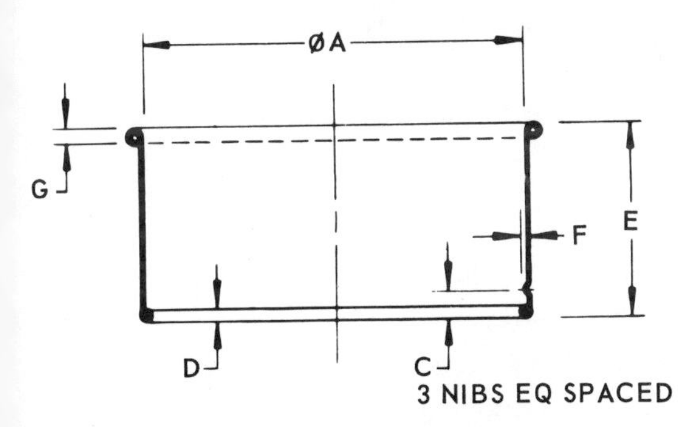

(B) SECTION VIEW THRU ADHESIVE TAPE SHELL

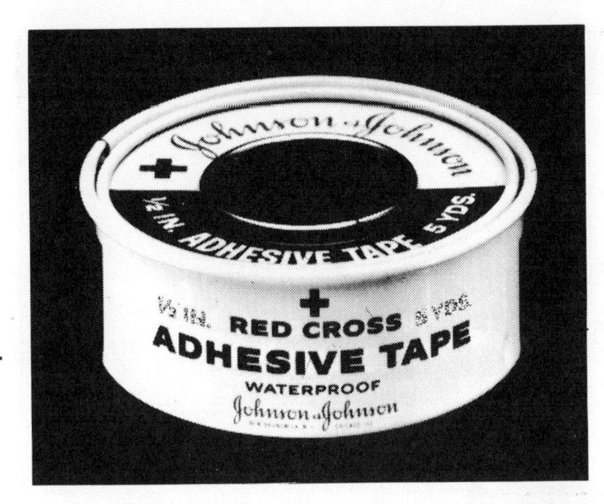

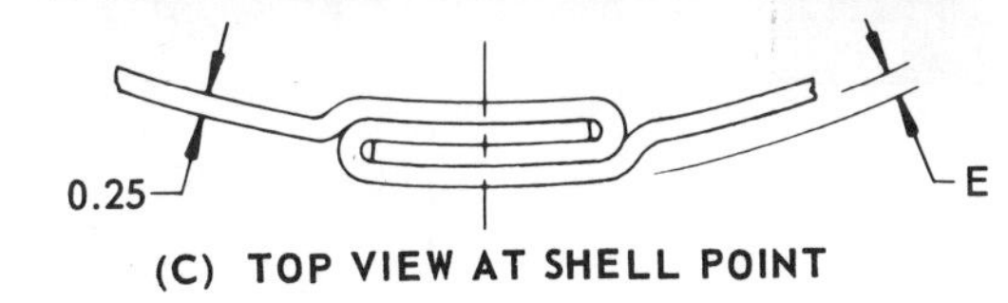

(C) TOP VIEW AT SHELL POINT

Fig. 10-1-1 Sheet metal application

resenting bend lines are seen in Figure 10-1-2. If the finished part is not shown with the development drawing, then bending instructions such as bend up 90°, bend down 180°, bend up 45°, are given beside each bend line.

Figure 10-1-3 shows a number of common methods for seaming and edging. Seams are used to join edges. The seams may be fastened together by lock seams, solder, rivets, adhesive, or welds. Exposed edges are folded or wired to give the edge added strength and to eliminate the sharp point.

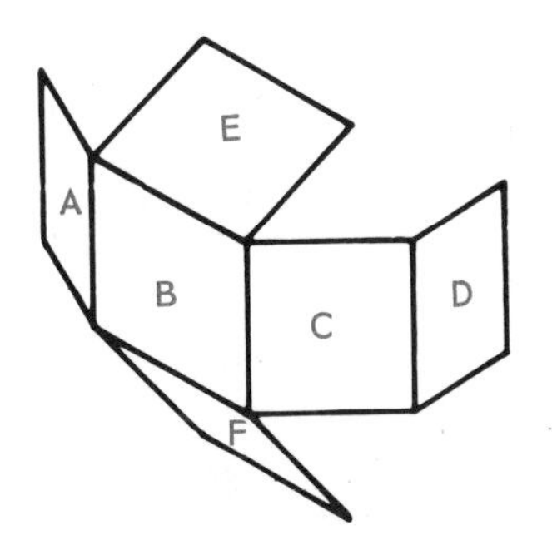

(A) FOLDING THE PART

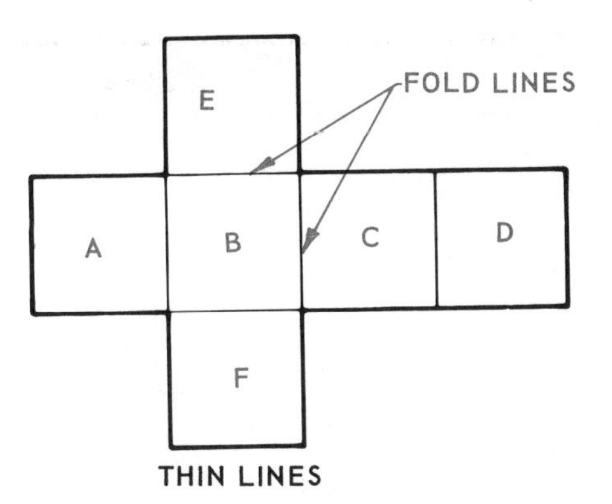

THIN LINES

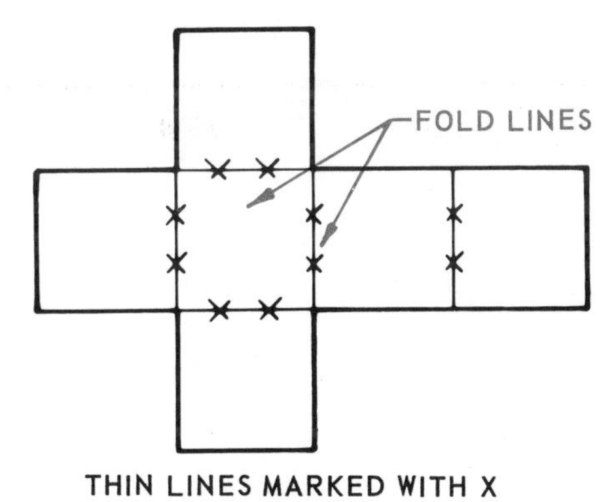

THIN LINES MARKED WITH X

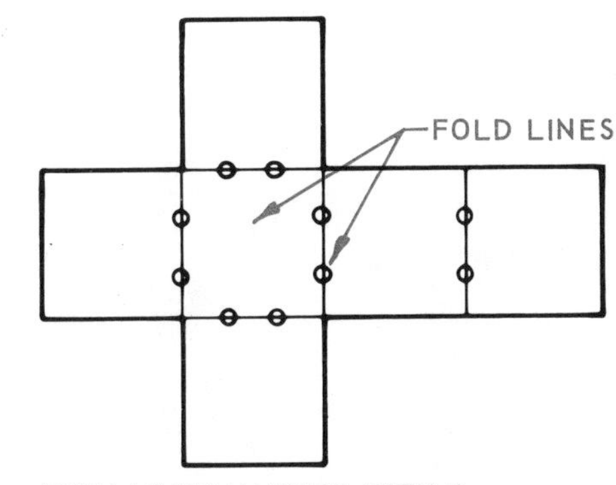

THIN LINES MARKED WITH O

(B) IDENTIFYING THE FOLD LINES ON THE DEVELOPMENT DRAWING

Fig. 10-1-2
Common methods used to identify fold or bend lines on development drawings

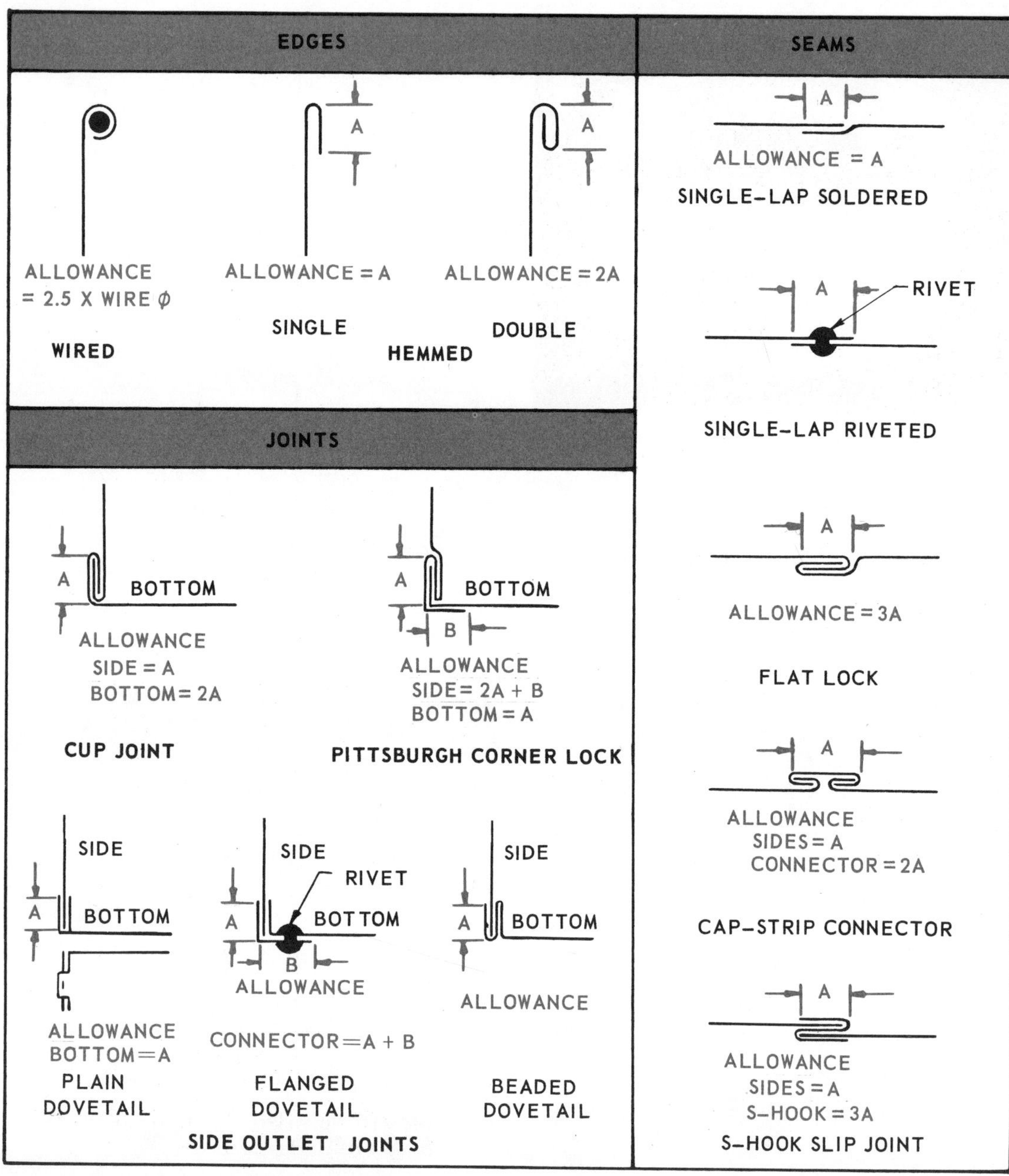

Fig. 10-1-3 Joints, seams and edges

A surface is said to be **developable** if a thin sheet of flexible material, such as paper, can be wrapped smoothly about its surface. Objects that have plane, or flat surfaces or single-curved surfaces are developable; but if a surface is double-curved or warped, approximate methods must be used to develop the surface.

The development of a spherical shape would thus be approximate, and the material would be stretched to compensate for small inaccuracies. For example, the coverings for a football or a basketball are made in segments. Each segment is cut to an approximate developed shape. The segments are then stretched and sewn together to give the desired shape.

Sheet-Metal Sizes

Metal thicknesses up to 6 mm are usually designated by a series of gauge numbers, the more common gauges being shown in Table 8 of the Appendix. Metal 6 mm and over is given in millimetre sizes. In calling for the material size of sheet-metal developments, customary practice is to give the gauge number, type of gauge, and its millimetre equivalent in brackets followed by the developed width and length. See Figure 10-1-4.

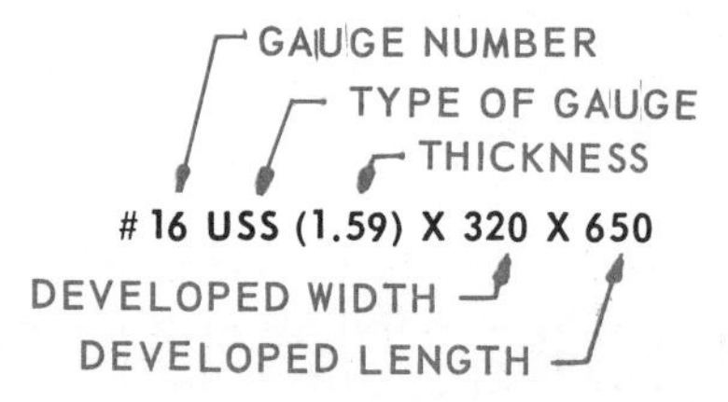

Fig. 10-1-4 Callout of sheet-metal sizes

Straight-Line Development

This is the term given to the development of an object that has surfaces on a flat plane of projection. The true size of each side of the object is known, and these sides can be laid out in successive order.

Figure 10-1-5 shows the development of a simple rectangular box having a bottom and four sides. Note that in the development of the box, an allowance is made for lap seams at the corners and for a folded edge. The fold lines are shown as thin, unbroken lines. Note also that all lines for each surface are straight.

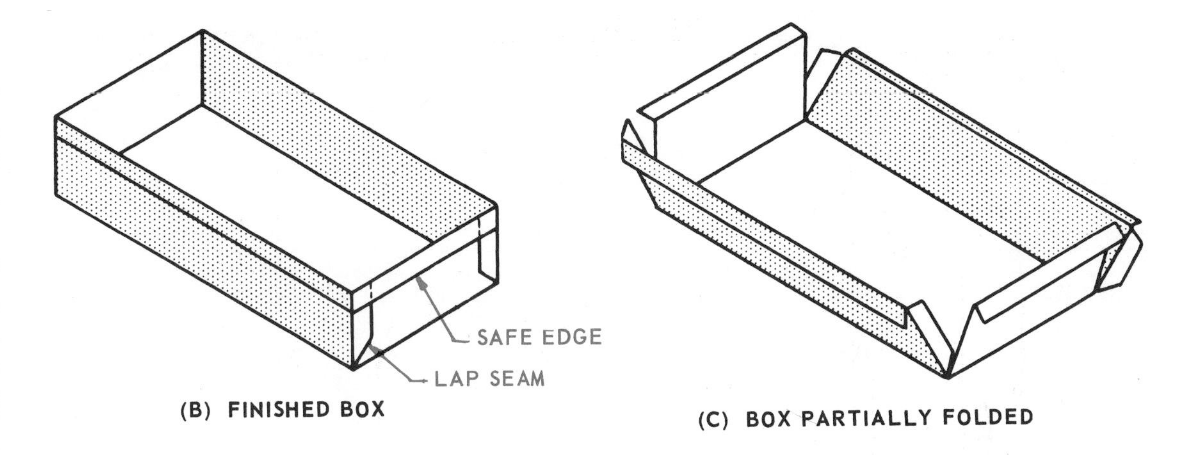

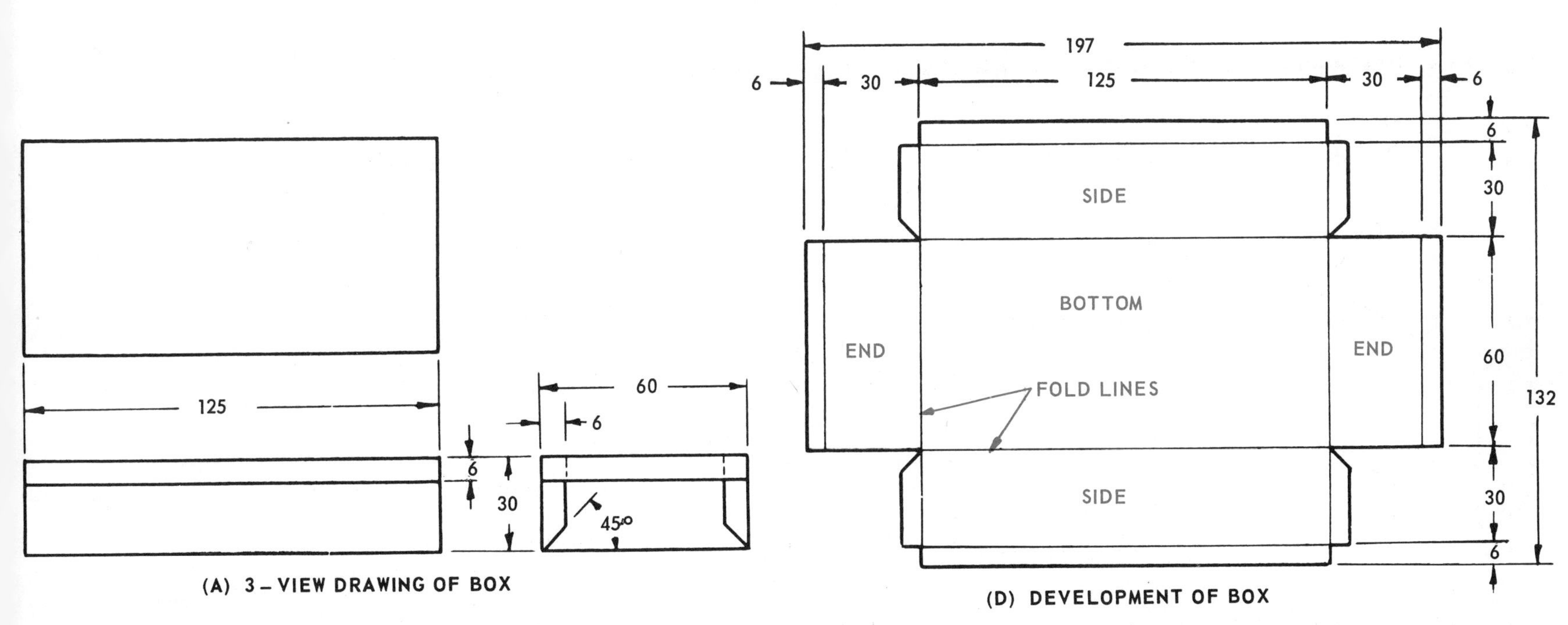

Fig. 10-1-5 Development of a rectangular box

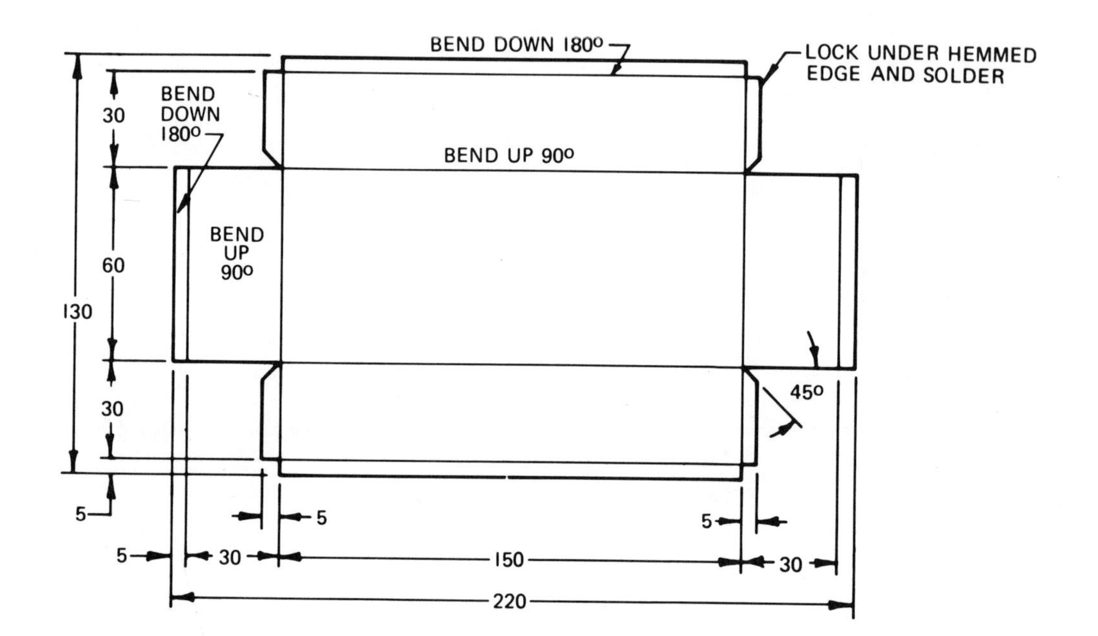

Fig. 10-1-6 Development drawing with a complete set of folding instructions

Figure 10-1-6 shows a development drawing with a complete set of folding instructions. Figure 10-1-7 shows a letter box development drawing where the back is higher than the front surface.

Assignment

On an A3 sheet, make a development drawing complete with bending instructions of one of the parts shown in Figures 10-1-A to 10-1-D. Dimension the development-drawing, showing the distance between bend lines, and show the overall sizes. Scale is 1:1.

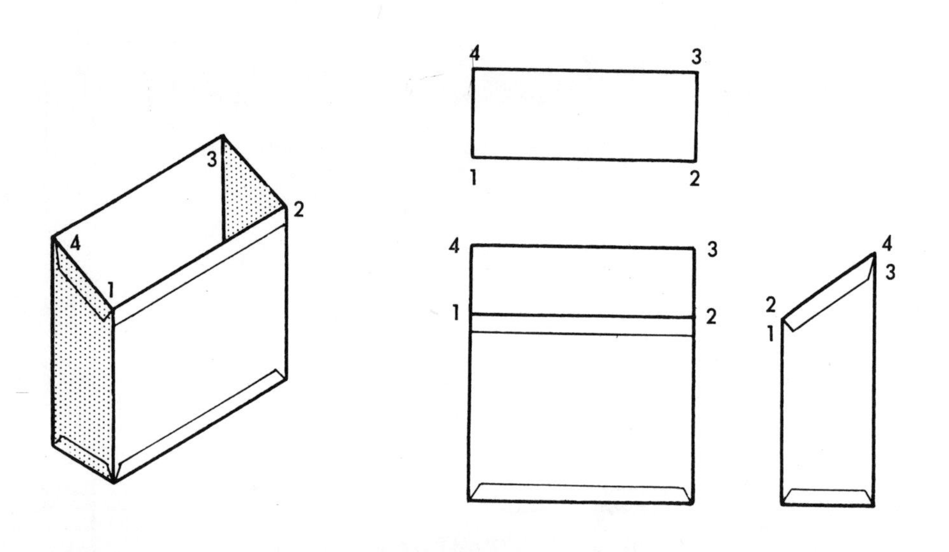

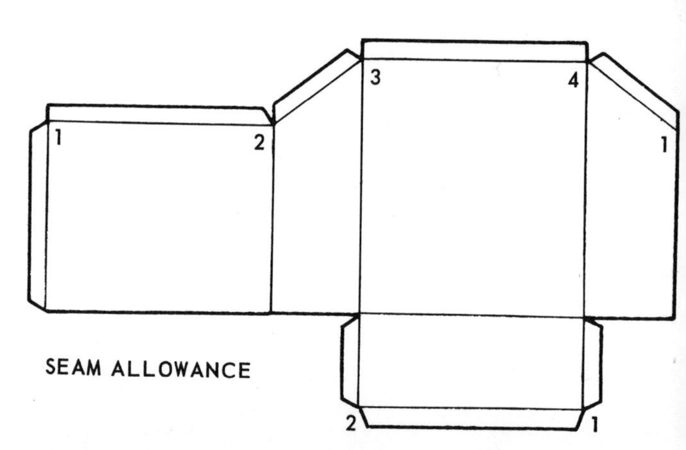

Fig. 10-1-7 Development drawing of a letter box

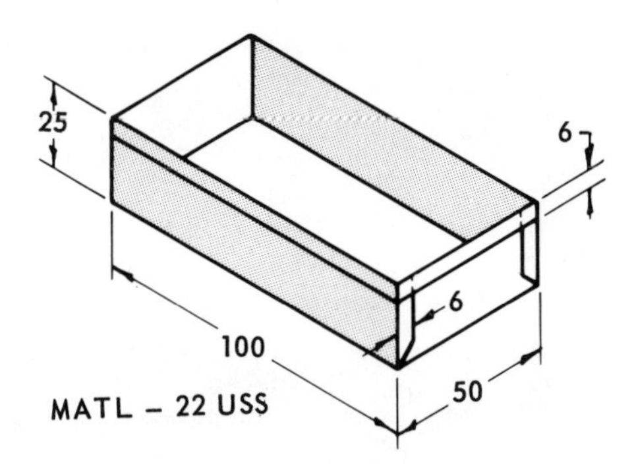

Fig. 10-1-A Nail box

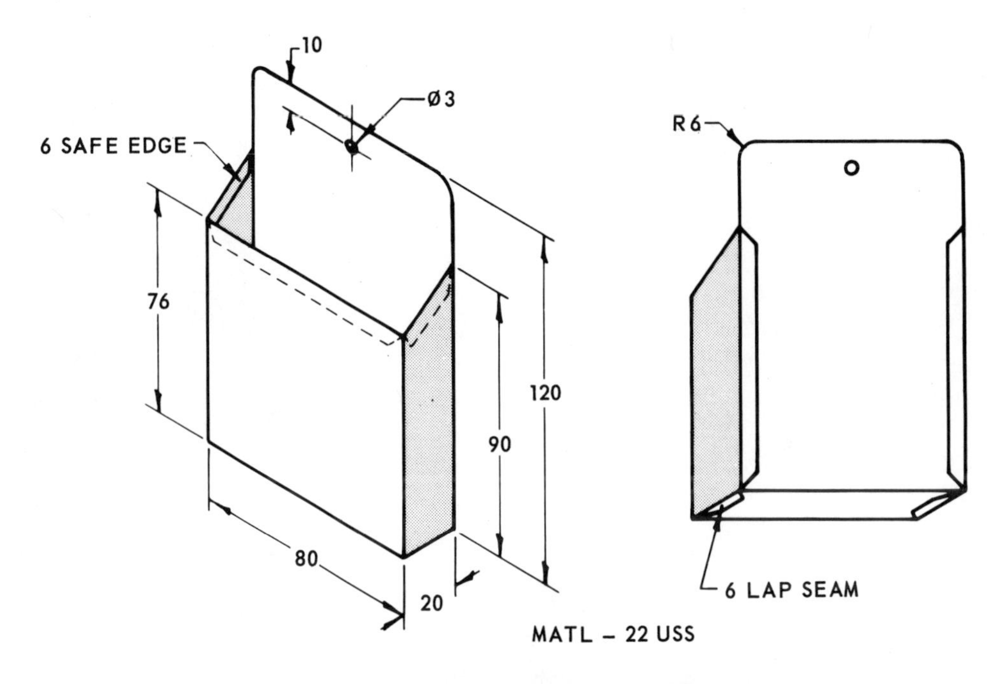

Fig. 10-1-C Memo pad holder

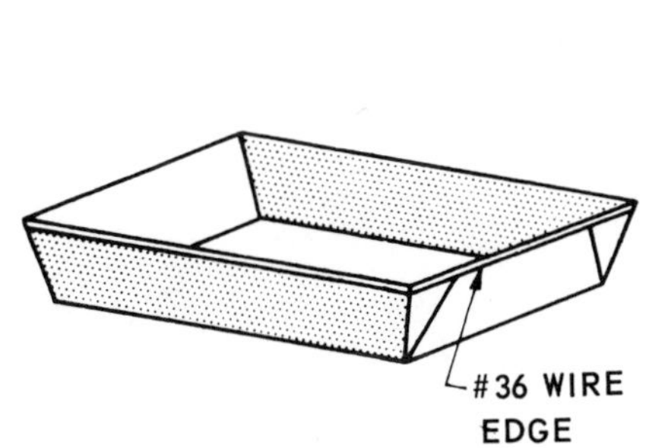

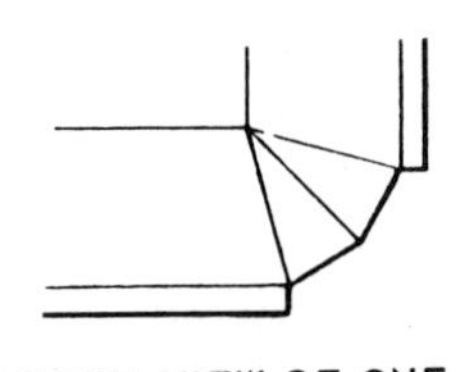

Fig. 10-1-B Cake tray

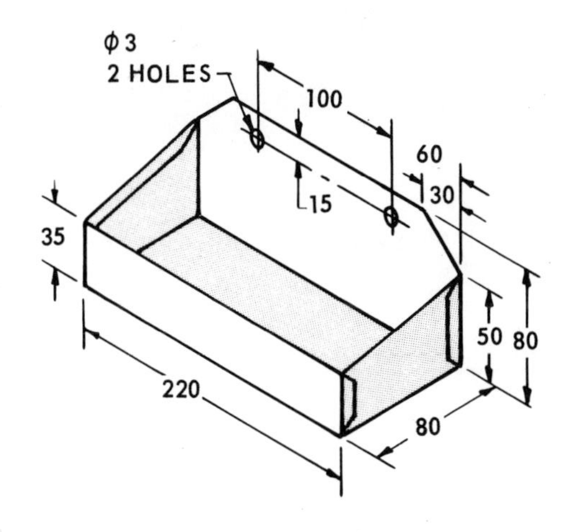

Fig. 10-1-D Wall tray

UNIT 10-2
The Packaging Industry

Packaging, which involves the principles of surface development, is one of the largest and most diversified industries in the world. Most products are packaged in metal, plastic, or cardboard containers. Small products, from candy-coated gum to large television sets, are packaged in cardboard containers. See Figures 10-2-1 and 10-2-2.

Fig. 10-2-1 Typical commercial containers

Such containers, often referred to as **cartons,** in many instances must be attractive as well as functional. They may be designed for sales appeal as well as for protection against contamination, shipping, and handling. They are also designed for temporary or permanent use.

Fig. 10-2-2 A familiar container made by cutting and folding a flat sheet

Many cartons are printed, cut, creased, and sent to the customer in a flat position. See Figure 10-2-3. They take less space to store and ship and are readily assembled. Locking devices such as tabs hold the box together.

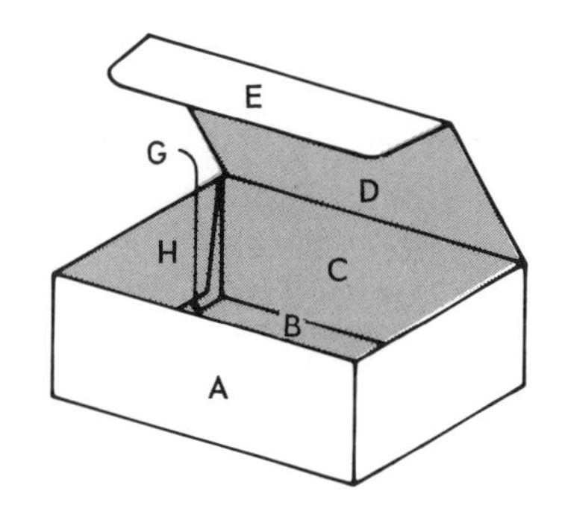

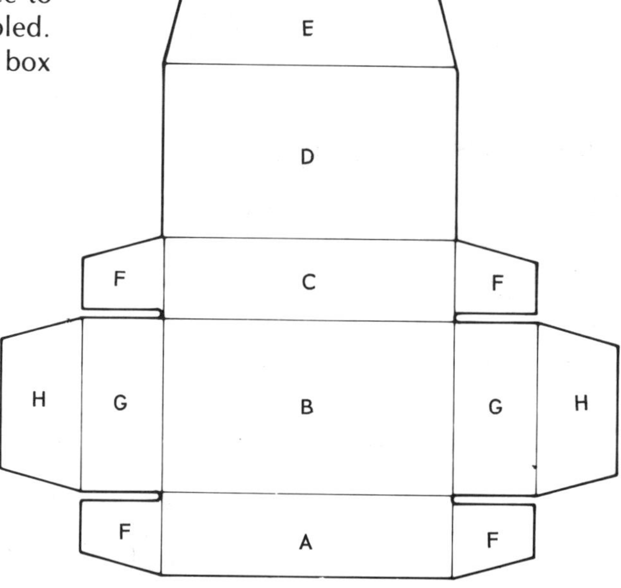

Fig. 10-2-3 Development of a one-piece carton with fold-down corners

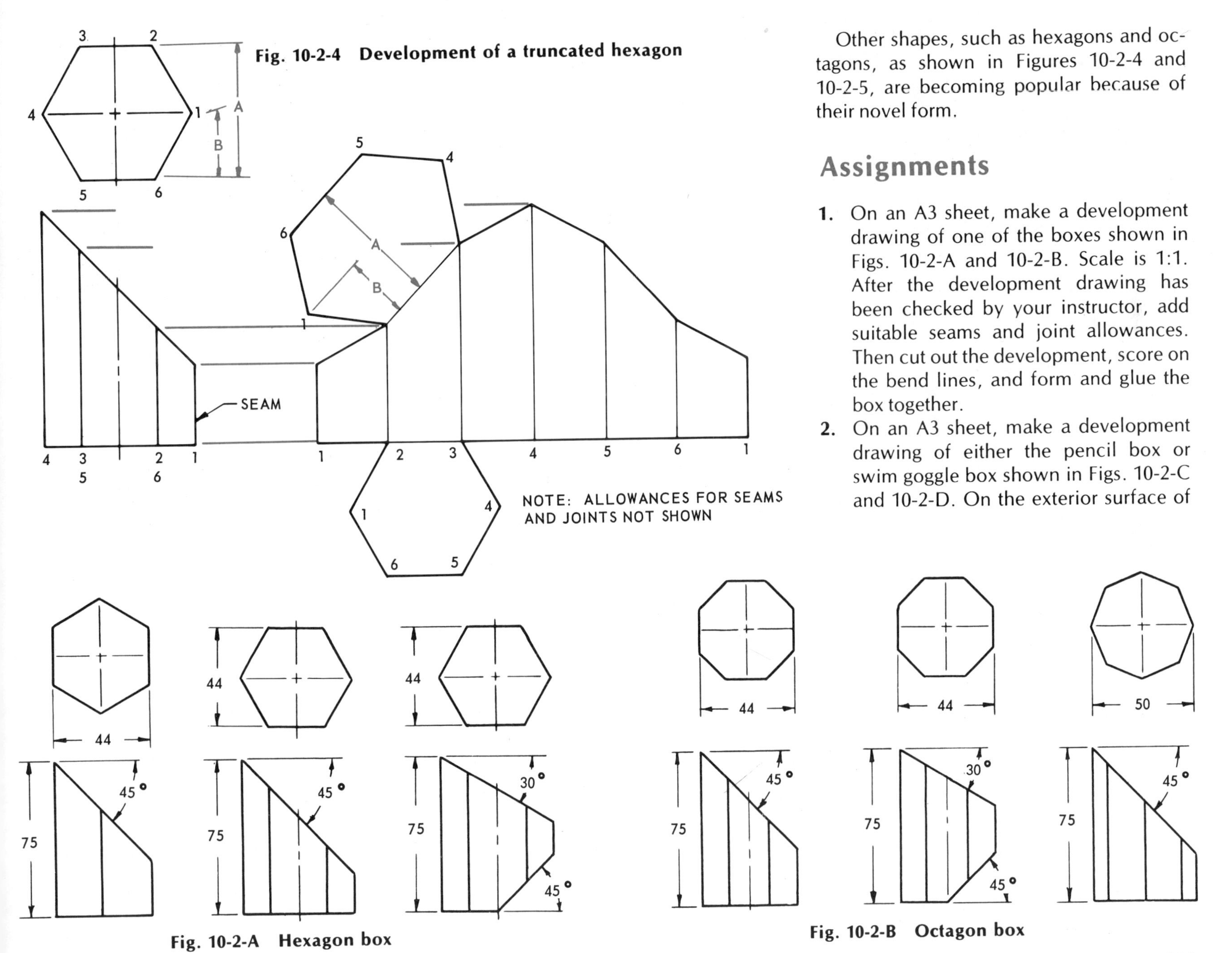

Fig. 10-2-4 Development of a truncated hexagon

Fig. 10-2-A Hexagon box

Fig. 10-2-B Octagon box

Other shapes, such as hexagons and octagons, as shown in Figures 10-2-4 and 10-2-5, are becoming popular because of their novel form.

Assignments

1. On an A3 sheet, make a development drawing of one of the boxes shown in Figs. 10-2-A and 10-2-B. Scale is 1:1. After the development drawing has been checked by your instructor, add suitable seams and joint allowances. Then cut out the development, score on the bend lines, and form and glue the box together.
2. On an A3 sheet, make a development drawing of either the pencil box or swim goggle box shown in Figs. 10-2-C and 10-2-D. On the exterior surface of

the box, lay out a design which has eye appeal (colour can be used) and which contains in the design the name of the item being sold, a company name, a slogan, and any other feature you believe would improve the salability of the article.

Cut out the development, score on the bend lines, and glue together.

Note: With reference to the swim goggle box, the box is completely sealed and must be broken to remove the goggles. Scale is 1:1.

Review for Assignments

Unit 10-1 Surface development
Unit 16-3 Constructing a polygon

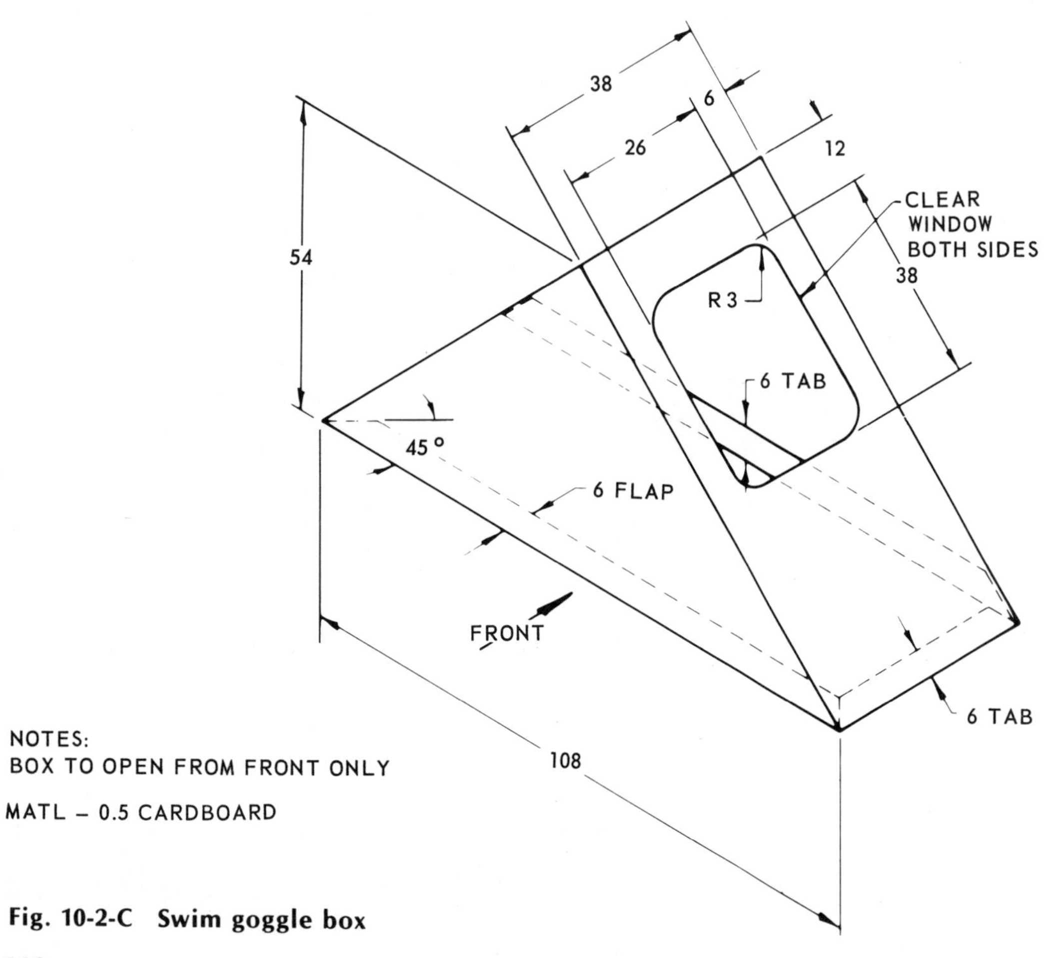

Fig. 10-2-C Swim goggle box

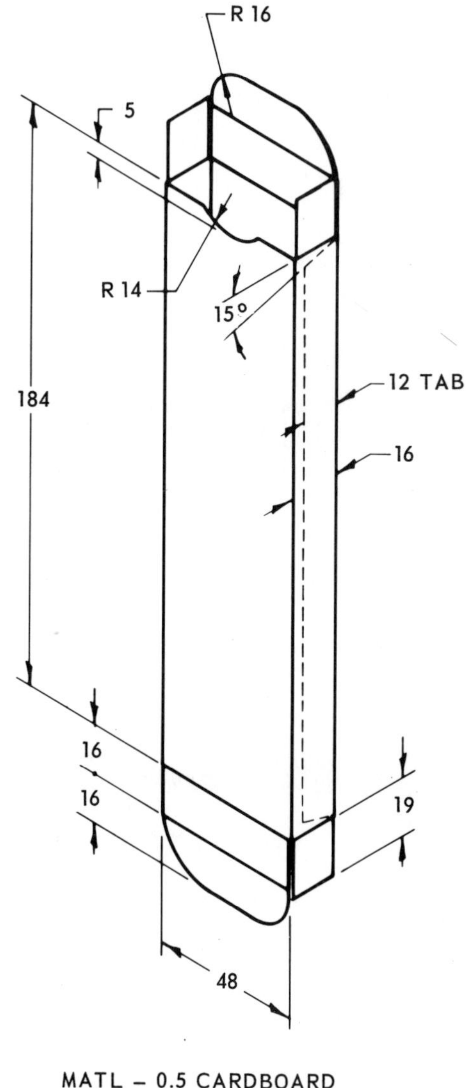

Fig. 10-2-D Pencil box

UNIT 10-3
Radial Line Development of Flat Surfaces

Development of a Right Pyramid with True Length of Edge Lines Shown

See Figure 10-3-1. A right pyramid is a pyramid having all the lateral edges (from vertex to the base) of equal length.

The true length of the lateral edges is shown in the front view (line 0-1 or 0-3) and the top view shows the true lengths of the edges of the base (lines 1-2, 2-3, etc.). The development may be constructed as follows: with 0 as centre (corresponding to the apex) and with a radius equal to the true length of the lateral edges (line 0-1 in the front view), draw an arc as shown. Drop a perpendicular from 0 to intersect the arc at point 3. With a radius equal to the length of the edges of the base (line 1-2 on the top view), start at point 3 and step off distances 3-2, 2-1, 3-4, and 4-1 on the larger arc. Join these points with straight lines. These points are then connected to point 0 by straight lines to complete the development. Lines 0-2, 0-3, and 0-4 are the lines on which the development is folded to shape the pyramid. The base and seam allowances have been omitted for clarity.

In developing a truncated pyramid of this type, the procedure is the same as mentioned above, except that only a portion of lines 0-1, 0-2, 0-3, and 0-4 is required. The positions of points *B* and *D* in the top view are found by projecting lines horizontally from points *B* and *D* in the front view to intersect the true-length line 0-3 at B_1. Project a vertical line from point B_1 to intersect point B_2 in the top view. Rotate B_2 90° from point 0 to intersect line 0-2 at *B* and 0-4 at *D*. It will be noted that only lines A-1 and C-3 appear as their true length in the front view. The true length of lines *B*-2 and *D*-4 may be found by projecting a horizontal line from points *B* and *D* to point *E* on the true-length line 0-1.

To complete the development, step off distances 1-*A* on line 1-0, 2-*B* on line 2-0, 3-*C* on line 3-0, and 4-*D* on line 4-0. Join points *A, B, C, D, A* with straight lines.

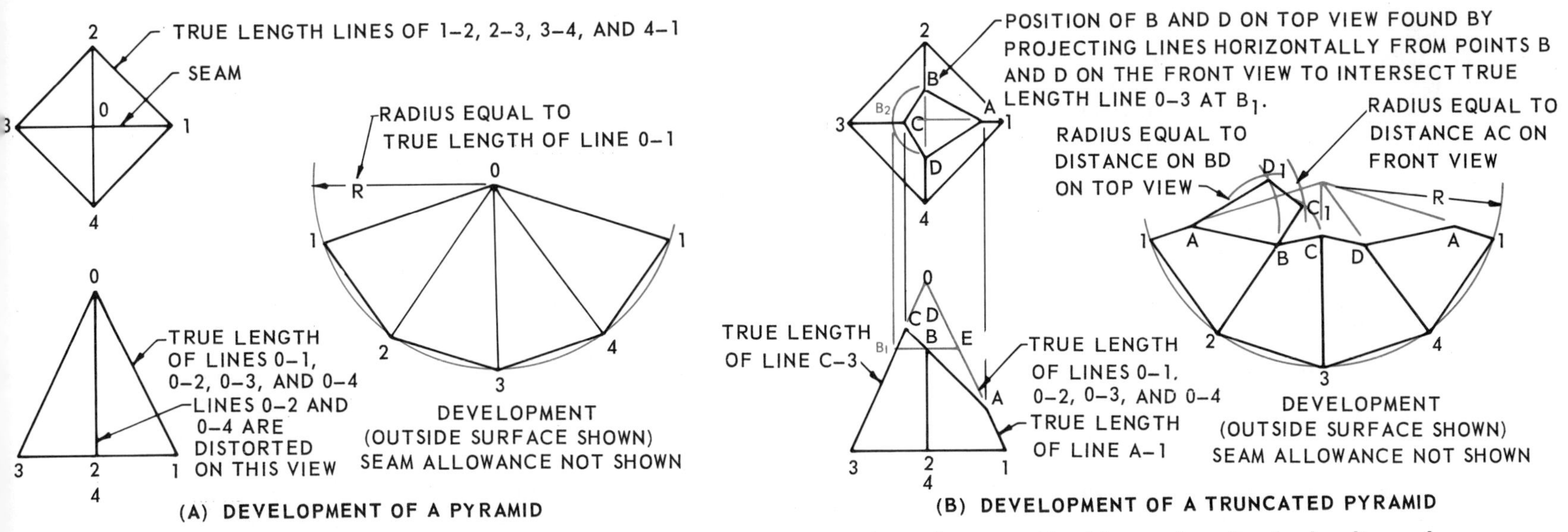

Fig. 10-3-1 Development of a right pyramid with true length of edge lines shown

The top surface of the truncated pyramid may be added to the development as follows: with *A* as centre and with a radius equal to distance *AC* in the front view, swing an arc. With *B* as centre and with a radius equal to line *BC* on the development, swing an arc intersecting the first arc at C_1. Join point *B* to point C_1 with a straight line. With *A* as centre and with a radius equal to line *AB* in the development, swing an arc. With *B* as centre and with a radius equal to line *AB* in the development, swing an arc. With *B* as centre and with a radius equal to distance *BD* in the top view, swing an arc intersecting the first arc at D_1. Join AD_1 and D_1C_1 with straight lines.

Development of a Right Pyramid with True Length of Edge Lines Not Shown

See Figure 10-3-2. In order to construct the development, the true length of the edge lines 0-1, 0-2, etc., must first be found. The true length of the edge lines would be equal to the hypotenuse of a right-angled triangle, having one leg equal in length to the projected edge line in the top view and the other leg equal to the height of the projected edge line in the front view. Since only one true-length line is required, it may be developed directly on the front view rather than by making a separate true-length diagram.

With 0 in the top view as centre and radius equal to distance 0-1 in the top view, swing an arc from point 1 until it intersects the centre line at point 1_1. Project a vertical line down to the front view, intersecting the base line at point 1_1. Line $0\text{-}1_1$ is the true length of the edge lines. The development may now be constructed in a similar manner to that outlined in the previous development.

In developing a truncated pyramid of this type, the procedure is the same except only the truncated edge lines are required. The true length of the truncated edge lines is required and may be found by projecting lines horizontally from points *A*, *B*, *C*, and *D* in the front view to intersect the true-length line $0\text{-}1_1$ at points *F* and *E*, respectively. Line F_1 is the true length of the truncated edge lines *A*1 and *B*1, and line $E1_1$ is

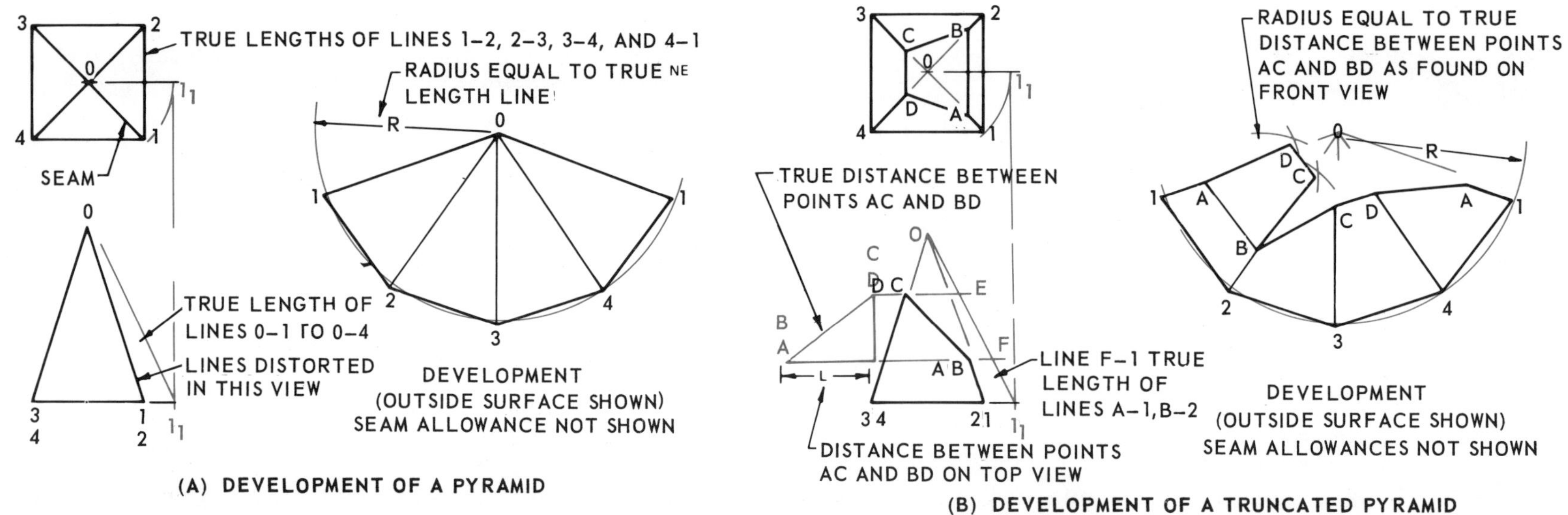

Fig. 10-3-2 Development of a right pyramid with true length of edge lines not shown

the true length of the remaining truncated edge lines *C*3 and *D*4. The sides of the truncated pyramid may now be constructed in the development view.

The top surface of the truncated pyramid may be added to the development as follows: with points *A* and *B* on the development as centres with a radius equal in length to line *BC* on the development, swing light arcs. With a radius equal in length to the true distance between points *A* and *C* or *B* and *D* (this is found on the true-length diagram constructed to the left of the front view) and with centre *B*, swing an arc intersecting the first arc at *C*. Repeat, using point *A* as centre and intersecting the other arc at point *C*. Join points *B*, *C*, *C*, and *A* with straight lines to complete the top surface. The base and seam lines have been omitted for clarity.

Development of a Transition Piece

See Figure 10-3-3. The development of the transition piece is created in a similar manner to that of the development of the right pyramid (Fig.10-3-2).

Fig. 10-3-3 Development of a transition piece

Assignment

On an A3 sheet, make a development drawing of one of the concentric pyramids shown in Figs. 10-3-A to 10-3-C. Add suitable seams. Scale is 1:1.

Review for Assignment

Unit 10-1 Surface development

NOTES:

SEAM IS AT A–1.
TOP IS HINGED AT A–B.
BOTTOM IS HINGED AT 1–2.
MATL – 0.4 CARDBOARD.

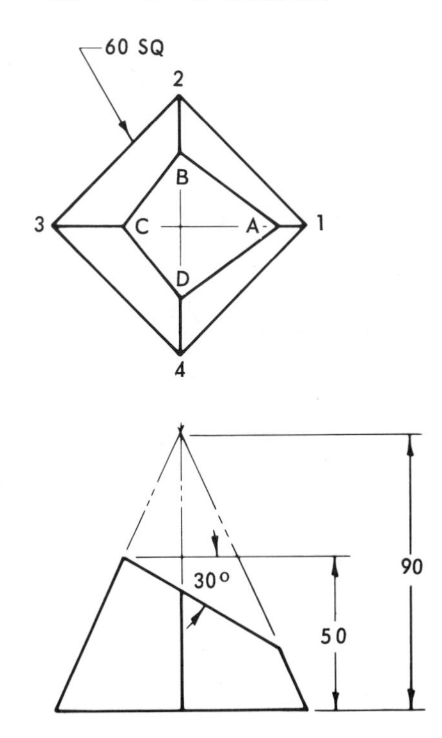

Fig. 10-3-A Truncated concentric pyramid

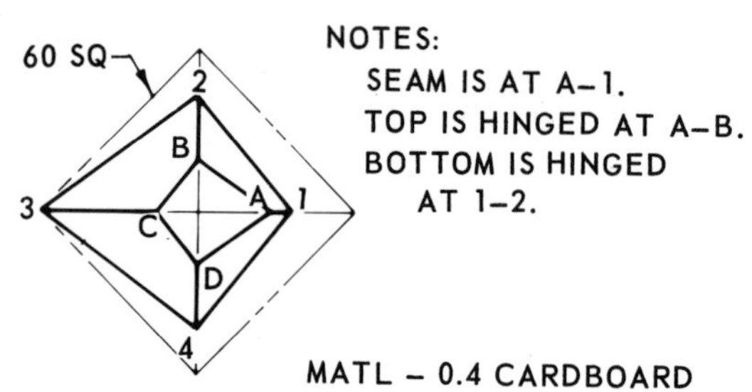

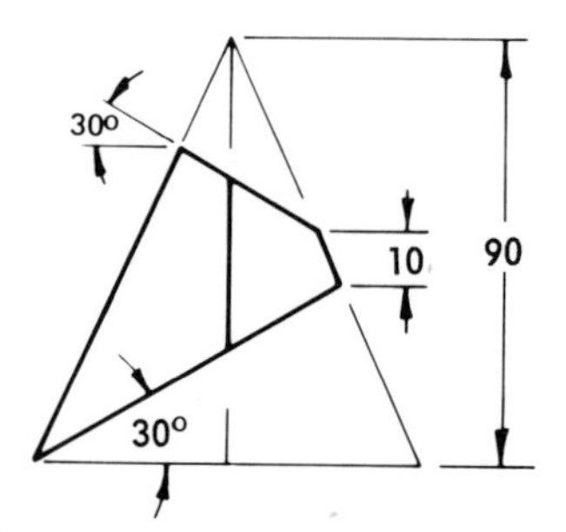

Fig. 10-3-B Truncated concentric pyramid

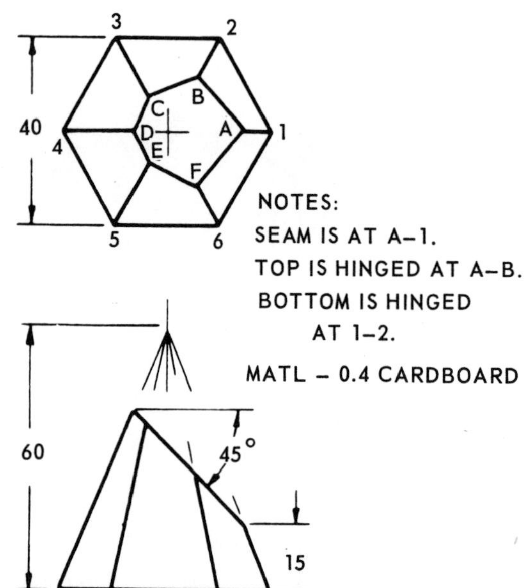

Fig. 10-3-C Truncated concentric pyramid

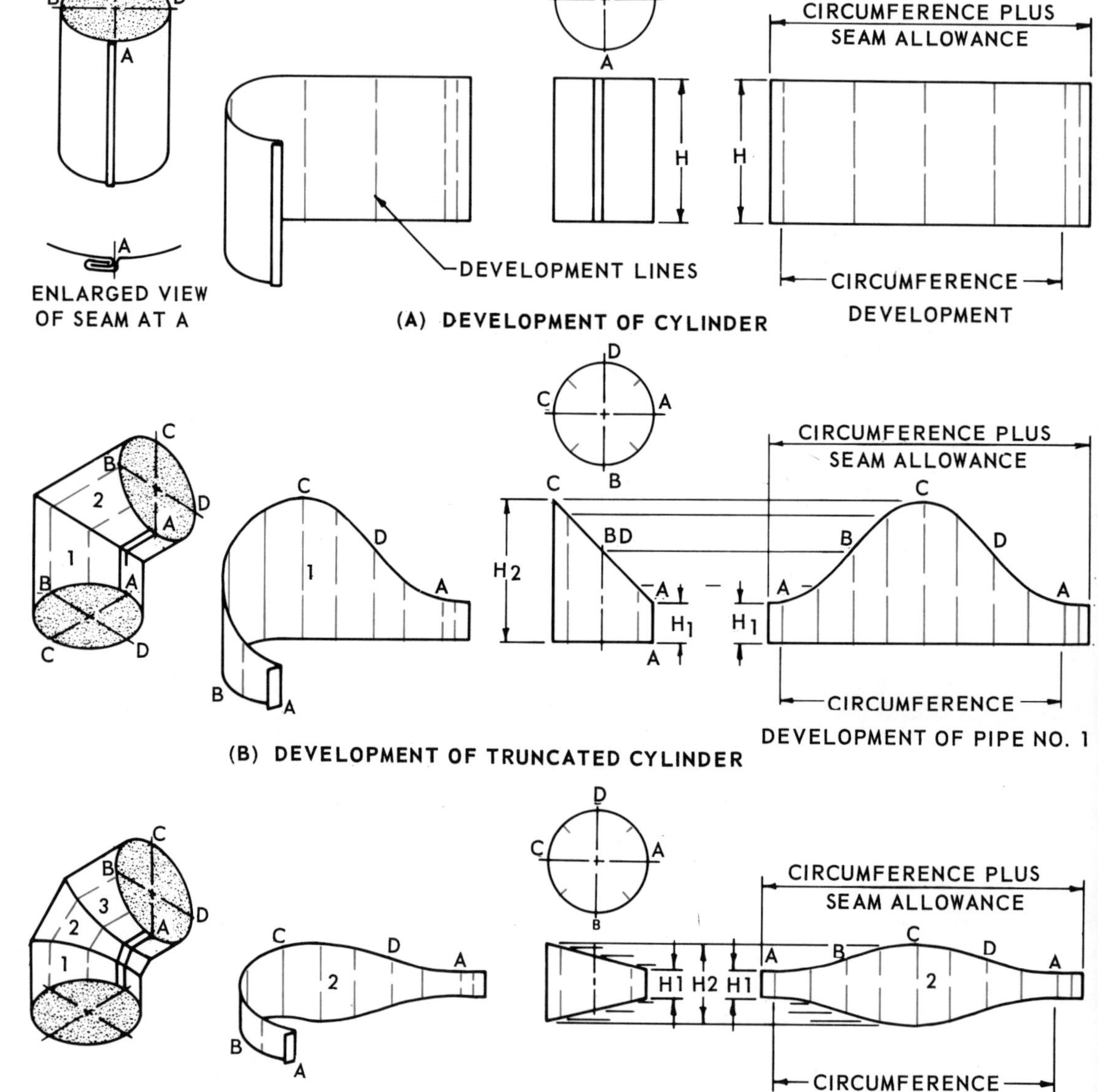

Fig. 10-4-1 Development of cylinders

UNIT 10-4
Parallel Line Development

The lateral, or curved, surface of a cylindrically shaped object, such as a tin can, can be developed because it has a single-curved surface of one constant radius. The development technique used for such objects is called **parallel line development.**

Figure 10-4-1*a* shows the development of the lateral surface of a simple hollow cylinder. The width of the development is equal to the height of the cylinder, and the length of the development is equal to the circumference of the cylinder (*d*) plus the seam allowance.

Figure 10-4-1*b* shows the development of a cylinder with the top truncated at a 45° angle (one-half of the two-piece 90° elbow). Points of intersection are established to give the curved shape on the development. These points are derived from the intersection of a length location, representing a certain distance around the circumference from a starting point, and the height location at that same point on the circumference.

The closer the points of intersection are to one another, the greater the accuracy of the development. An irregular curve is used to connect the points of intersection.

Figure 10-4-1*c* shows the development of the surface of a cylinder with both the top and the bottom truncated at an angle of 22.5° (the centre part of a three-piece elbow).

It is normal practice in sheet-metal work to place the seam on the shortest side. In the development of elbows, however, this practice would result in considerable waste of material, as illustrated by Figure 10-4-2*a*. To avoid this waste and to simplify cutting the pieces, the seams are alternately placed 180° apart, as illustrated by Figure 10-4-2*b* for a two-piece elbow and by Figure 10-4-2*c* for a three-piece elbow. Refer to Figures 10-4-3 and 10-4-4 for complete developments of two- and four-piece elbows.

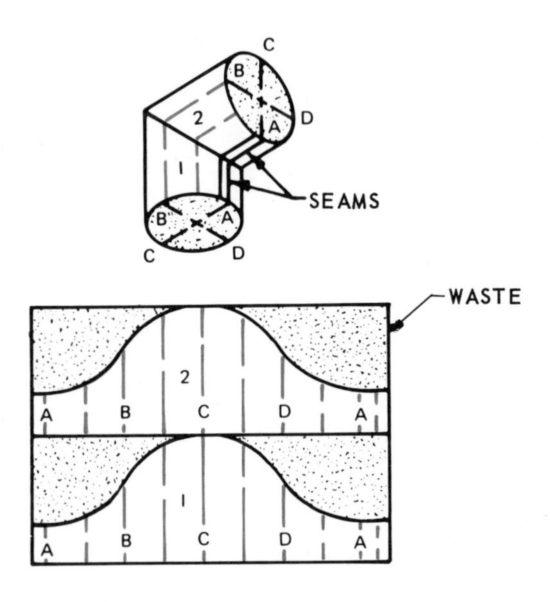

(A) DEVELOPMENT OF A 2–PIECE ELBOW WITH BOTH SEAMS ON LINE A

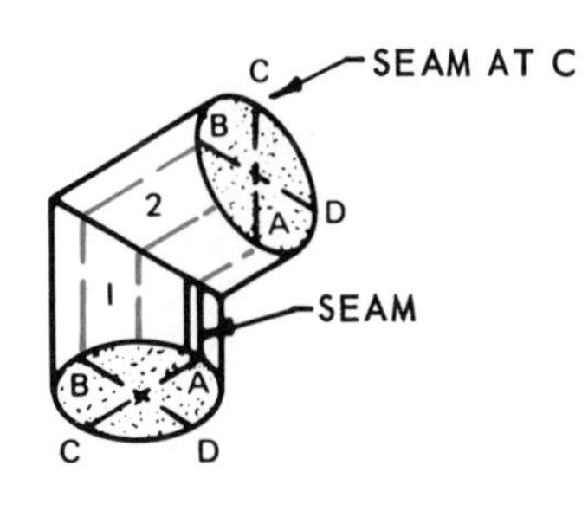

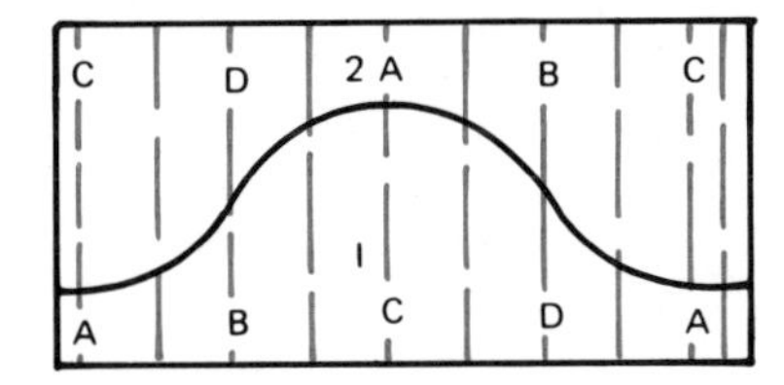

(B) DEVELOPMENT OF A 2–PIECE ELBOW WITH SEAMS ON LINES A AND C

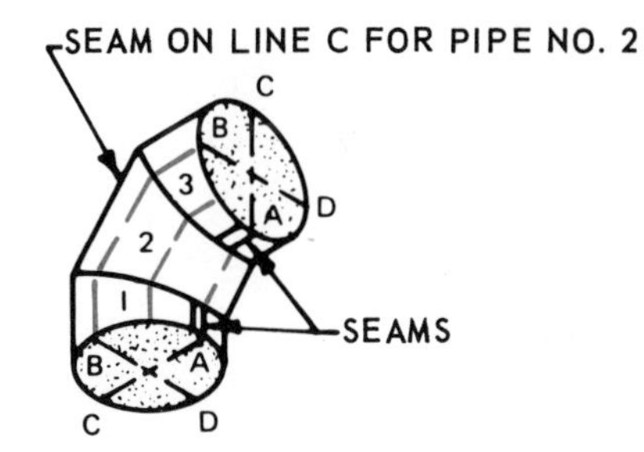

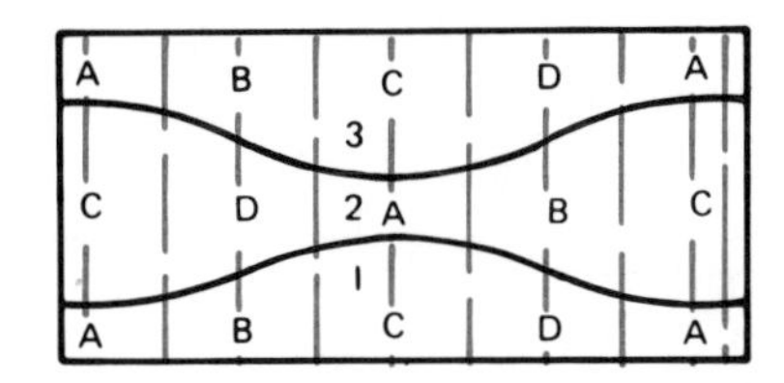

(C) DEVELOPMENT OF A 3–PIECE ELBOW WITH SEAMS ALTERNATED ON LINES A AND C

Fig. 10-4-2 Location of seams on elbows

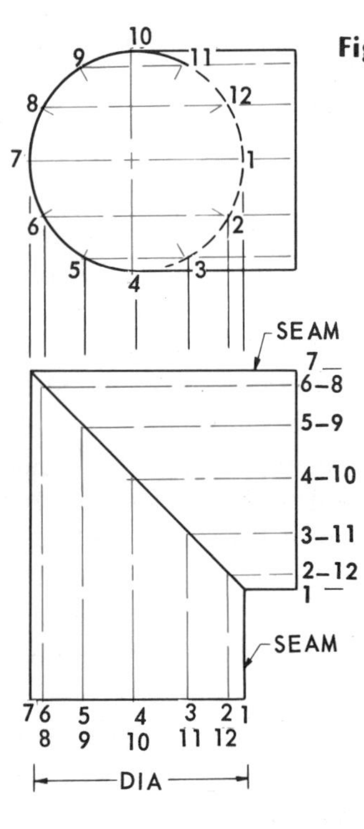

Fig. 10-4-3 Development of a two-piece elbow

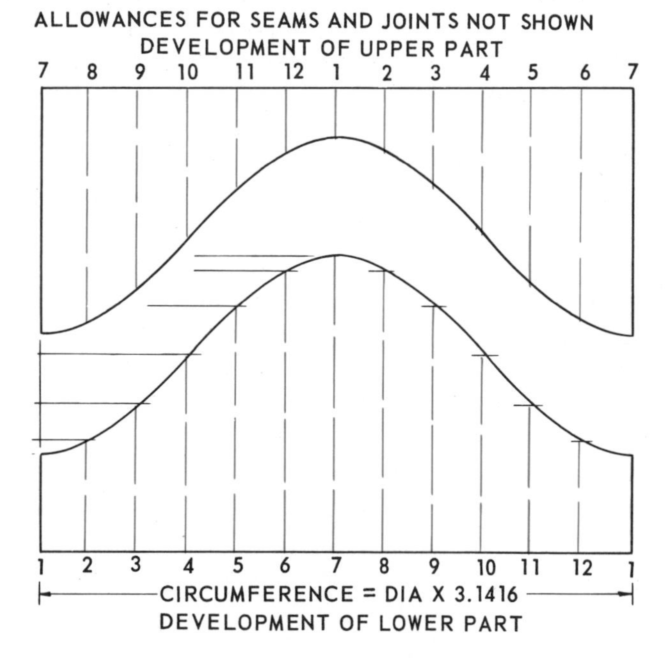

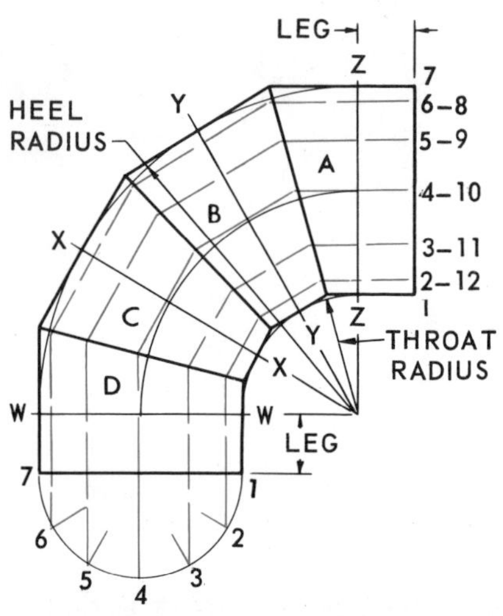

Fig. 10-4-4
Development of a four-piece elbow

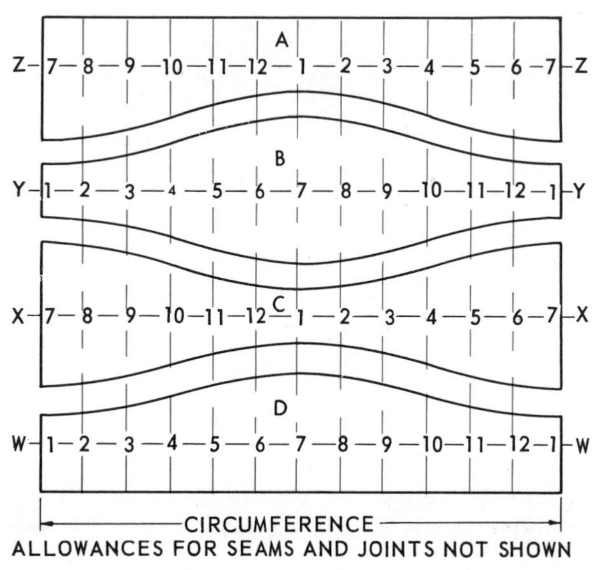

Assignments

1. On an A3 sheet, make a two-view and a development drawing of one of the elbows shown in Figs. 10-4-A to 10-4-C. Scale is 1:1.
2. On an A3 sheet, make a two-view and a development drawing of one of the parts shown in Figs. 10-4-D and 10-4-E. Scale is 1:1.

Review for Assignments

Unit 10-1 Surface development

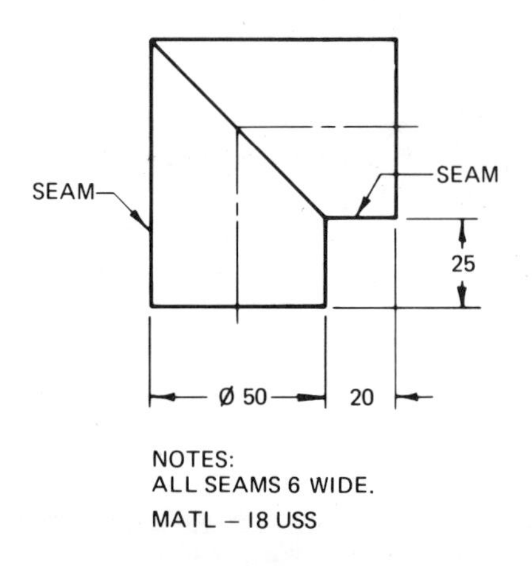

Fig. 10-4-A Two-piece elbow

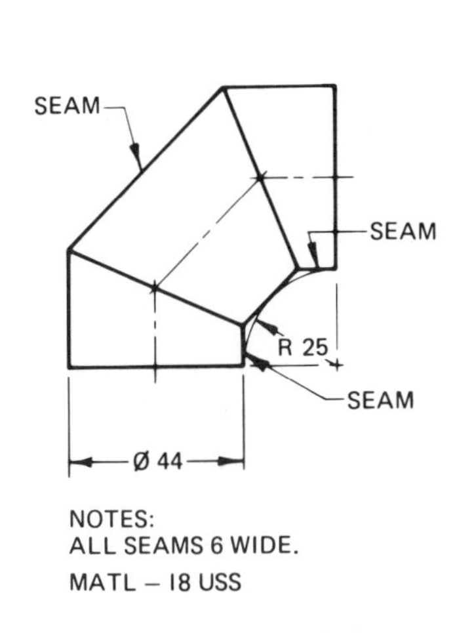

Fig. 10-4-B Three-piece elbow

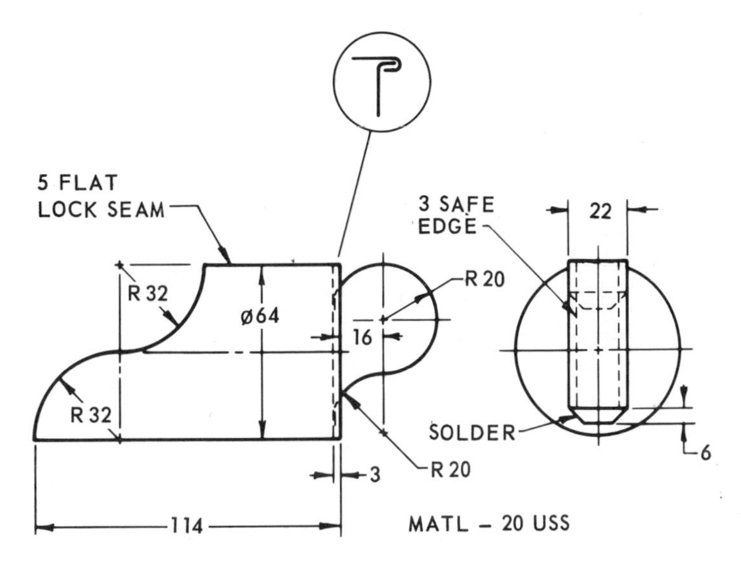

Fig. 10-4-D Sugar scoop

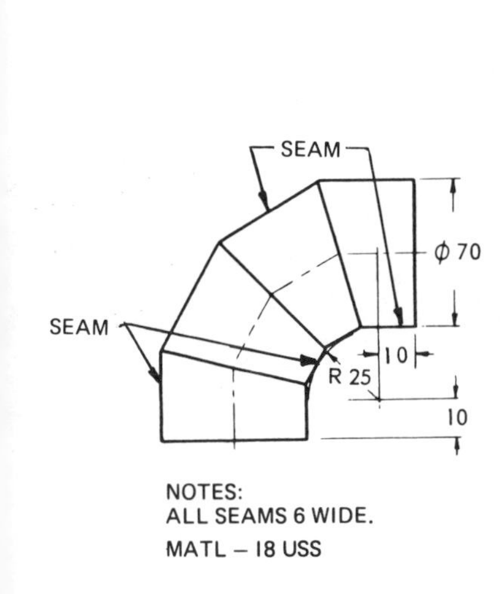

Fig. 10-4-C Four-piece elbow

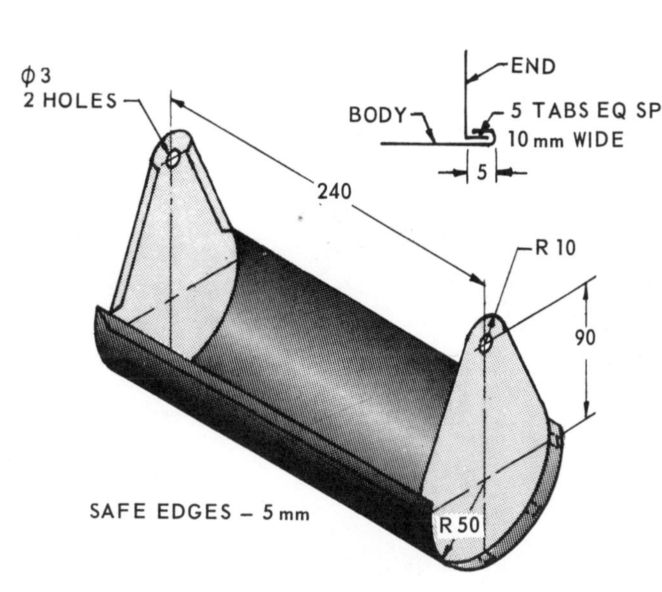

Fig. 10-4-E Planter

UNIT 10-5 Radial Line Development of Conical Surfaces

Development of a Cone

The surface of a cone can be developed because a thin sheet of flexible material can be wrapped smoothly about it. The two dimensions necessary to make the development of the surface are the slant height of the cone and the circumference of its base.

For a right circular cone (symmetrical about the vertical axis), the developed shape is a sector of a circle. The radius for this sector is the slant height of the cone,

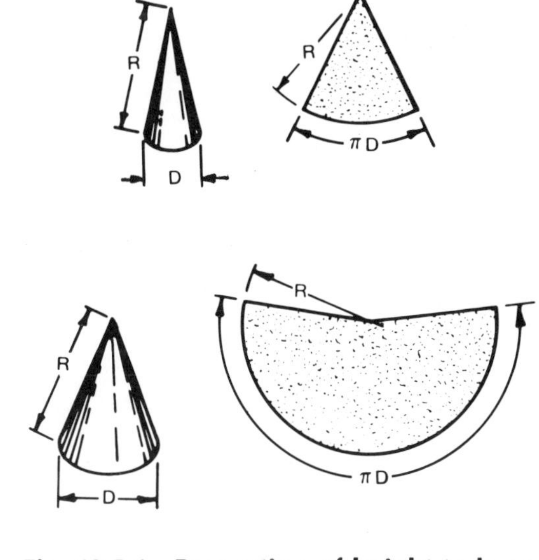

Fig. 10-5-1 Proportion of height to base of a cone

and the length around the perimeter of the sector is equal to the circumference of the base. The proportion of the height to the base diameter determines the size of the sector, as illustrated by Figure 10-5-1.

Figure 10-5-2 shows the steps in the development of a cone. The top view is divided into a convenient number of equal divisions, in this instance 12. The chordal distance between these points is used to step off the length of arc on the development. The radius *R* for the development is seen as the slant height in the front view.

If a cone is truncated at an angle to the base, the inside shape on the development no longer has a constant radius; that is, it is an ellipse, which must be plotted by establishing points of intersection.

The divisions made on the top view are projected down to the base of the cone in the front view. Element lines are drawn from these points to the apex of the cone. These element lines are seen in their true length only when the viewer is looking at right angles to them. Thus the points at which they cross the truncation line must be carried across, parallel to the base, to the outside element line, which is seen in its true length.

The development is first made to represent the complete surface of the cone. Element lines are drawn from the step-off points about the circumference to the centre point. True-length settings for each element line are taken from the front view and marked off on the corresponding element lines in the development. An irregular curve is used to connect these points of intersection, giving the proper inside shape.

Development of a Truncated Cone

The development of a frustum of a cone is the development of a full cone less the development of the part removed, as shown in Figure 10-5-3. Note that, at all times, the radius setting, either R_1 or R_2, is a slant height, a distance taken on the surface of the cone.

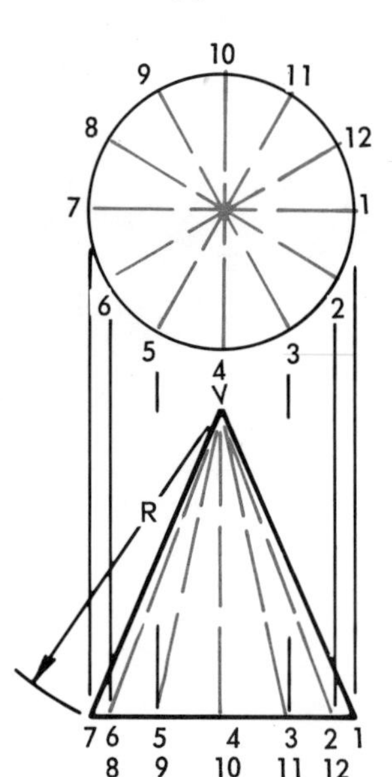

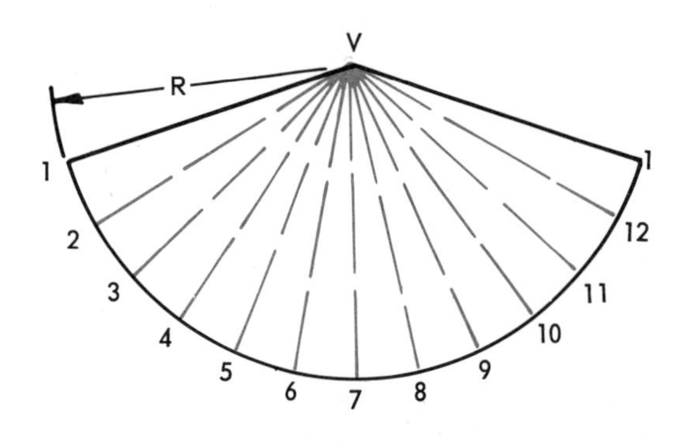

Fig. 10-5-2 Development of a cone

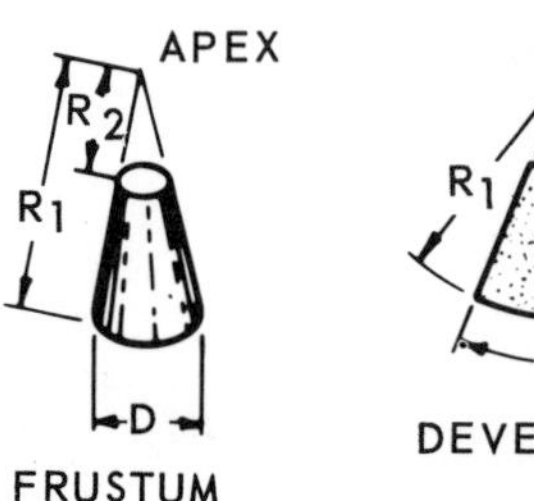

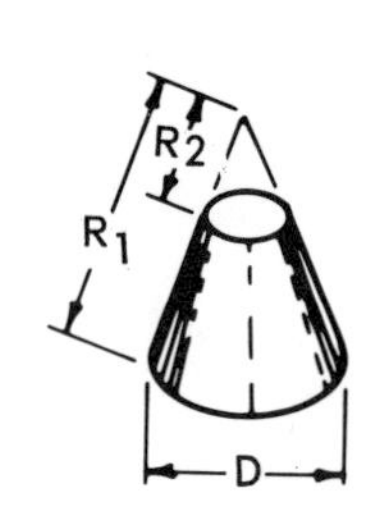

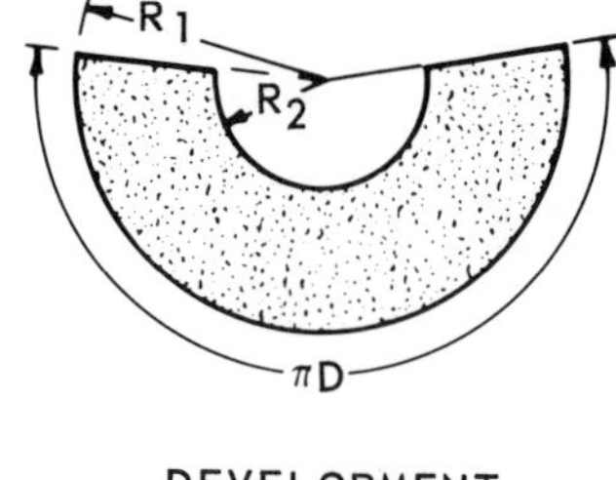

Fig. 10-5-3 Frustum of a cone

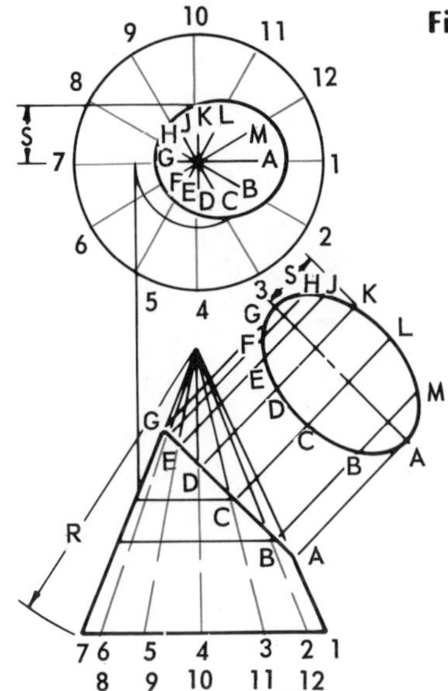

Fig. 10-5-4 **Development of a truncated cone**

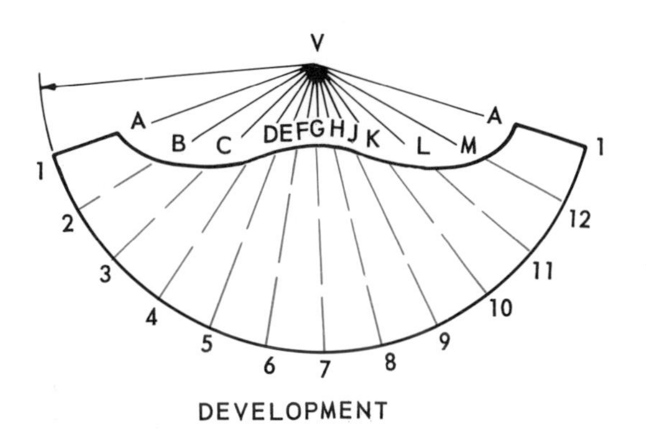

the triangulation method. The base of the cone is divided into a convenient number of equal parts and elements; 0-1, 0-2, etc., are drawn in the top view and projected down and drawn in the front view. The true lengths of the elements are not shown in either the top or front view but would be equal in length to the hypotenuse of a right-angle triangle, having one leg equal in length to the projected element in the top view and the other leg equal to the height of the projected element in the front view.

When it is necessary to find the true length of a number of edges, or elements, then a true-length diagram is drawn adjacent to the front view. This prevents the front view from being cluttered with lines.

When the top of a cone is truncated at an angle to the base, the top surface will not be seen as a true circle. This shape must also be plotted by establishing points of intersection. True radius settings for each element line are taken from the front view and marked off on the corresponding element line in the top view. These points are connected with an irregular curve to give the correct oval shape for the top surface. If the development of the sloping top surface is required, an auxiliary view of this surface shows its true shape.

Development of an Oblique Cone

See Figure 10-5-5. The development of an oblique cone is generally accomplished by

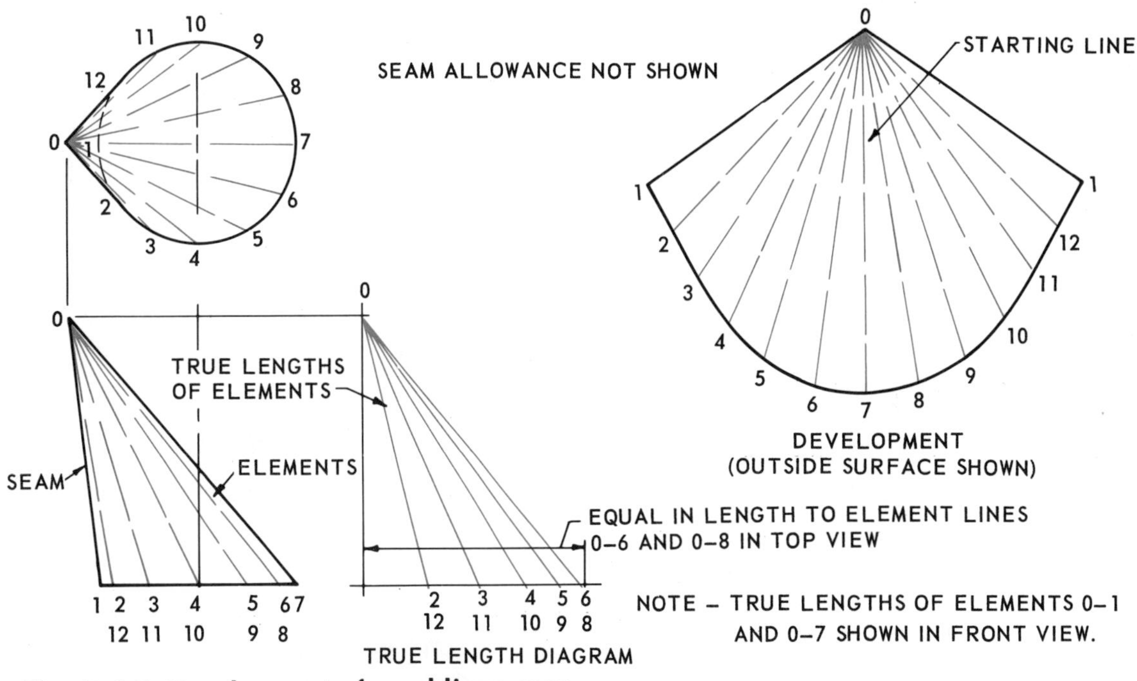

Fig. 10-5-5 **Development of an oblique cone**

Since the development of the oblique cone will be symmetrical, the starting line will be element 0-7. The development is constructed as follows: with 0 as centre and radius equal to the true length of element 0-6, draw an arc. With 7 as centre and radius equal to distance 6-7 in the top view, draw a second arc intersecting the first at point 6. Draw element 0-6 on the development. With 0 as centre and the radius equal to the true length of element 0-5, draw an arc. With 6 as centre and the radius equal to the true length of element 0-5, drawn an arc. With 6 as centre and the radius equal to distance 5-6 in the top view, draw a second arc intersecting the first at point 5. Draw element 0-5 on the development. This is repeated until all the element lines are located on the development view. No seam allowance is shown on the development.

Assignment

On an A3 sheet, make a development drawing of one of the assembled parts shown in Figs. 10-5-A and 10-5-B. Use your judgment for the other views required and add suitable seams. Scale is 1:1.

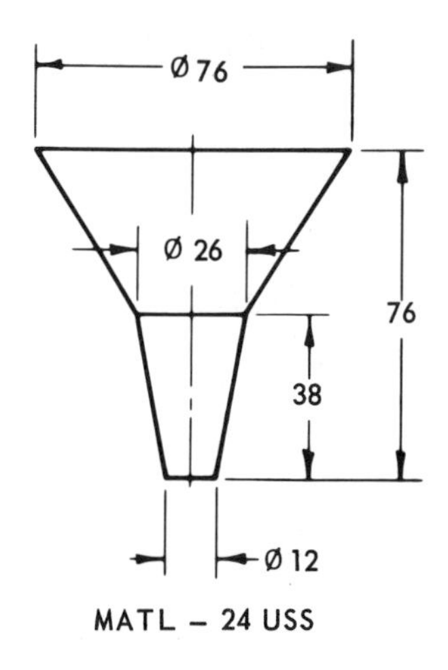

Fig. 10-5-A Funnel

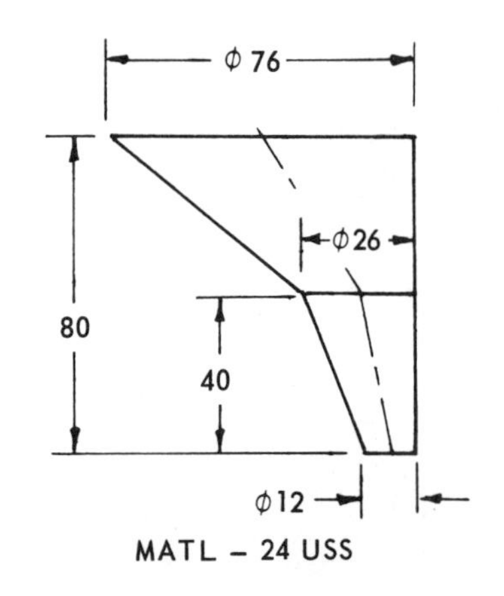

Fig. 10-5-B Offset funnel

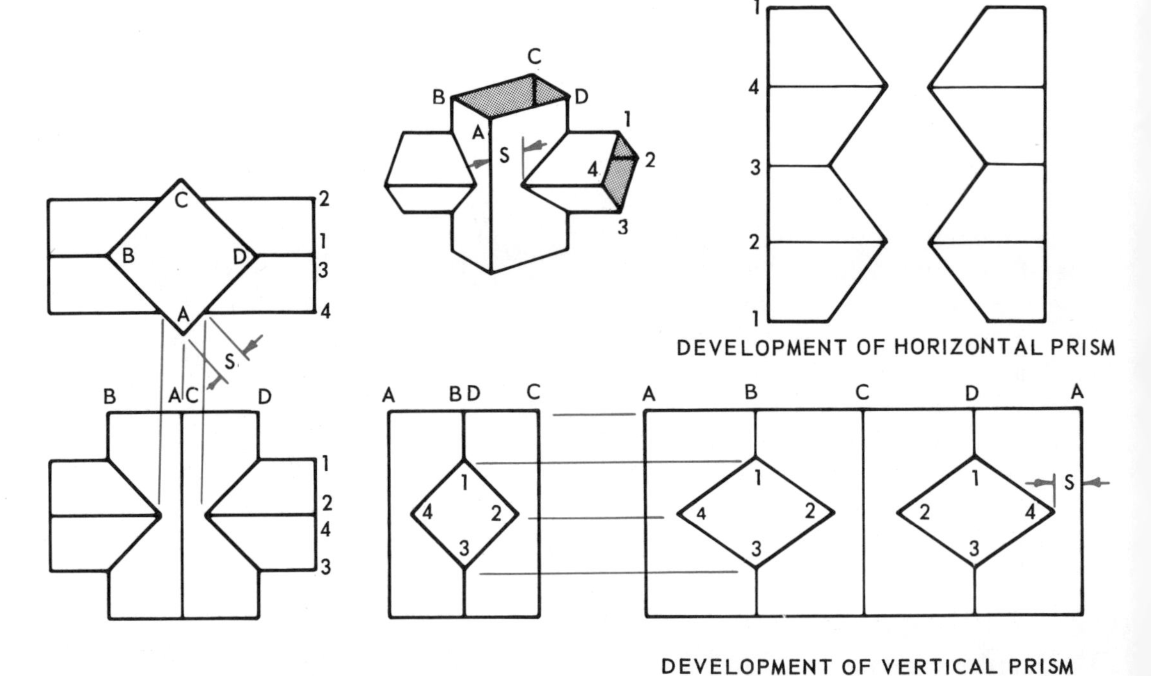

Fig. 10-6-1 Plotting lines of intersection and making development drawings of intersecting prisms

UNIT 10-6
Intersection of Flat Surfaces

Whenever two surfaces meet, there is a line common to both called the **line of intersection.** In making the orthographic drawing of objects that comprise two or more intersecting parts, the lines of intersection of these parts must be plotted on the orthographic views. Figure 10-6-1 and 10-6-2 illustrate this plotting technique for the intersection of flat-sided prisms. Figure 10-6-1 shows the development of the parts.

A numbering technique is very valuable in plotting lines of intersection.

In the illustrations shown, the lines of intersection appear in the front view. The end points for these lines are established by projecting the height position from the right side view to intersect the corresponding length position projected from the top view. When the prisms are flat-sided, the lines of intersection are straight, and the lines in the development are straight.

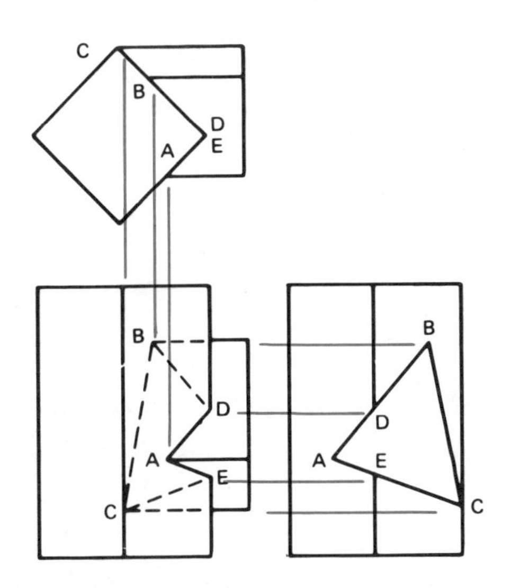

Fig. 10-6-2 Intersecting prisms at right angles

Assignment

On an A3 sheet, select one of the assembled parts shown in Figs. 10-6-A to 10-6-C, complete the views, and make a development drawing of the vertical parts. In each case assume that the horizontal part passes through the vertical support. Add suitable seams. Scale is 1:1.

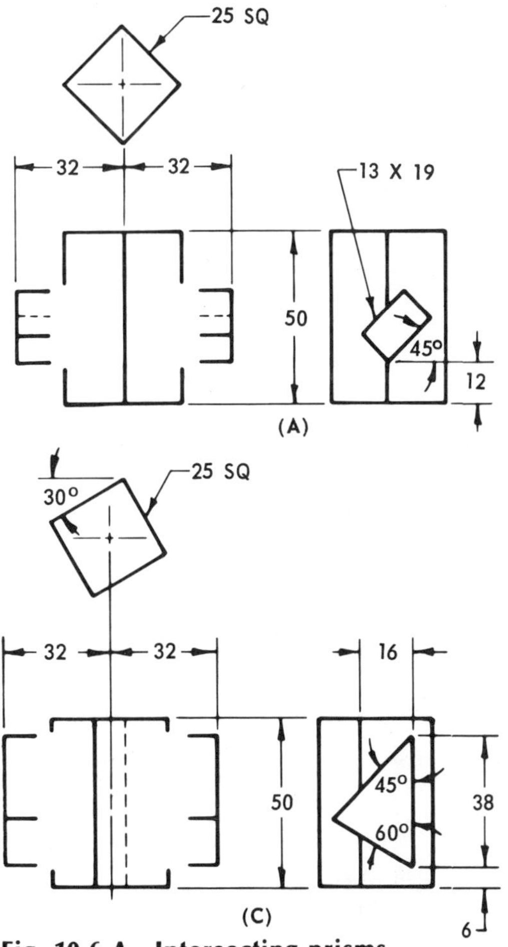

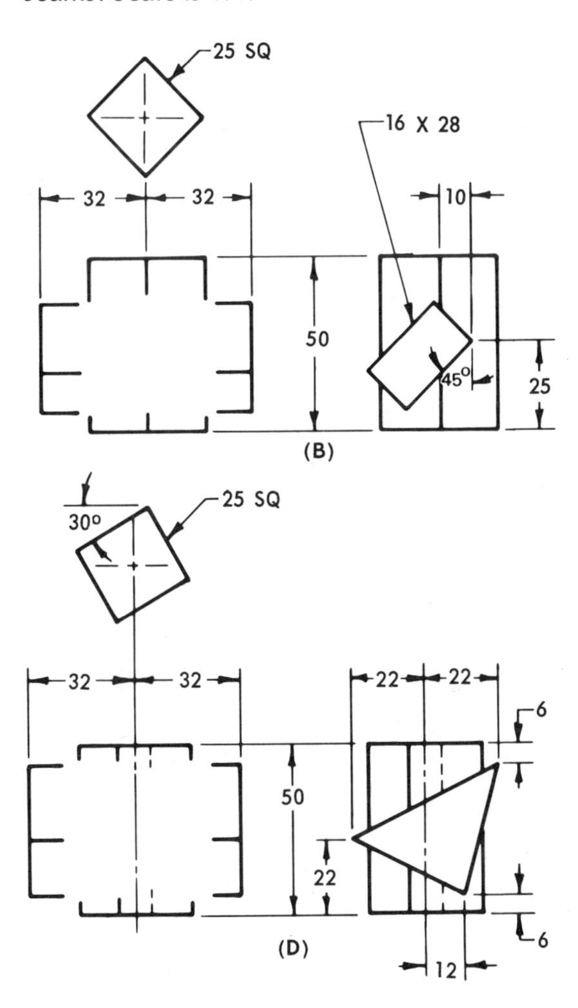

Fig. 10-6-A Intersecting prisms

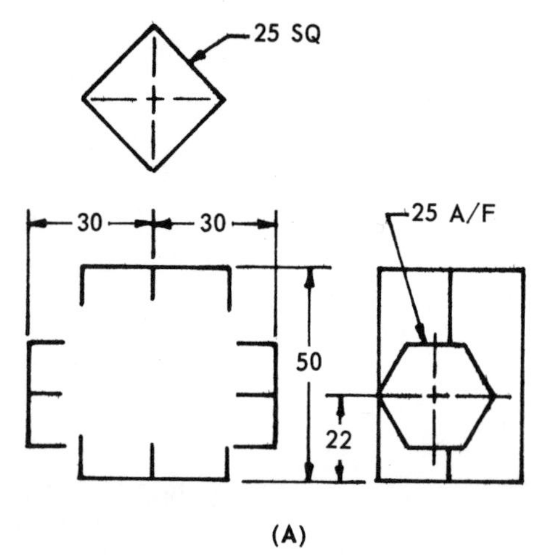

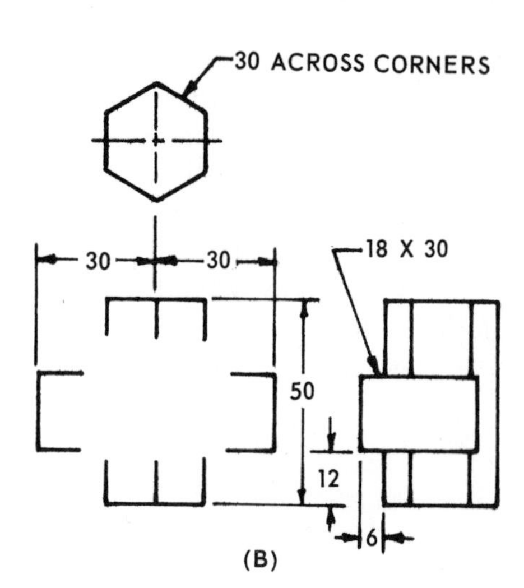

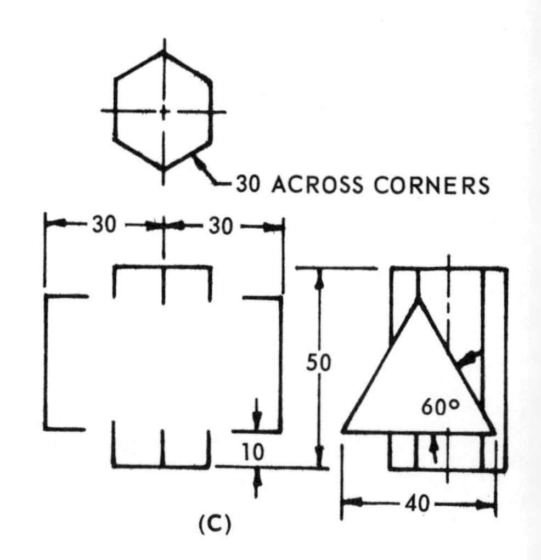

Fig. 10-6-B Intersecting prisms

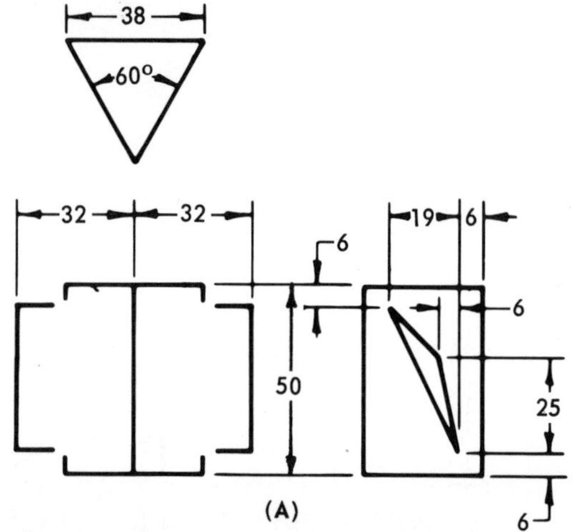

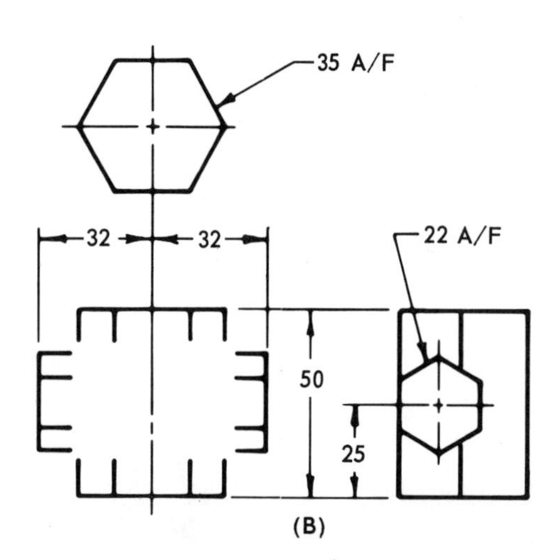

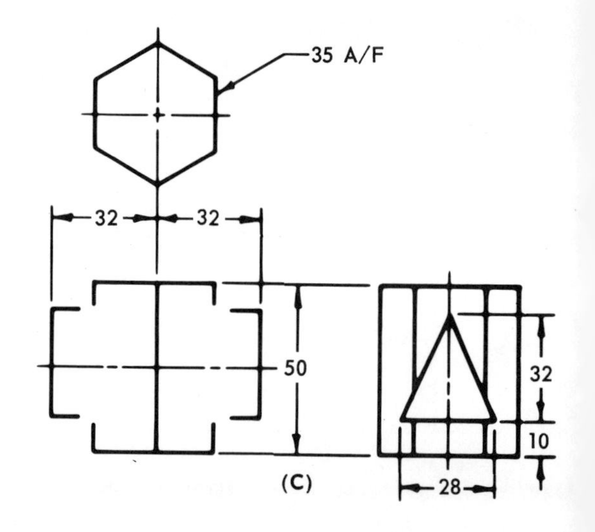

Fig. 10-6-C Intersecting prisms

UNIT 10-7
Intersection of Cylindrical Surfaces

90° Reducing Tee

See Figure 10-7-1. This figure illustrates the plotting technique for the intersection of cylinders. Because there are no edges on the cylinders, element lines of reference are established about the cylinders in their orthographic views.

In the top view, the element lines for the small cylinder are drawn to touch the surface of the large cylinder; for example, line 2 touches at *E*. This point location is then projected down to the front view to intersect the corresponding element line, establishing the height at that point.

The points of intersection thus established are connected by an irregular curve to produce the line of intersection. The same points of reference used to establish the line of intersection are used to draw the development.

Assignment

On an A3 sheet, select one of Figs. 10-7-A or 10-7-B and draw the assembly. Complete the views, and make a development drawing of the vertical parts. In each case assume the horizontal part passes through the horizontal support. Add the suitable seams. Scale is 1:1.

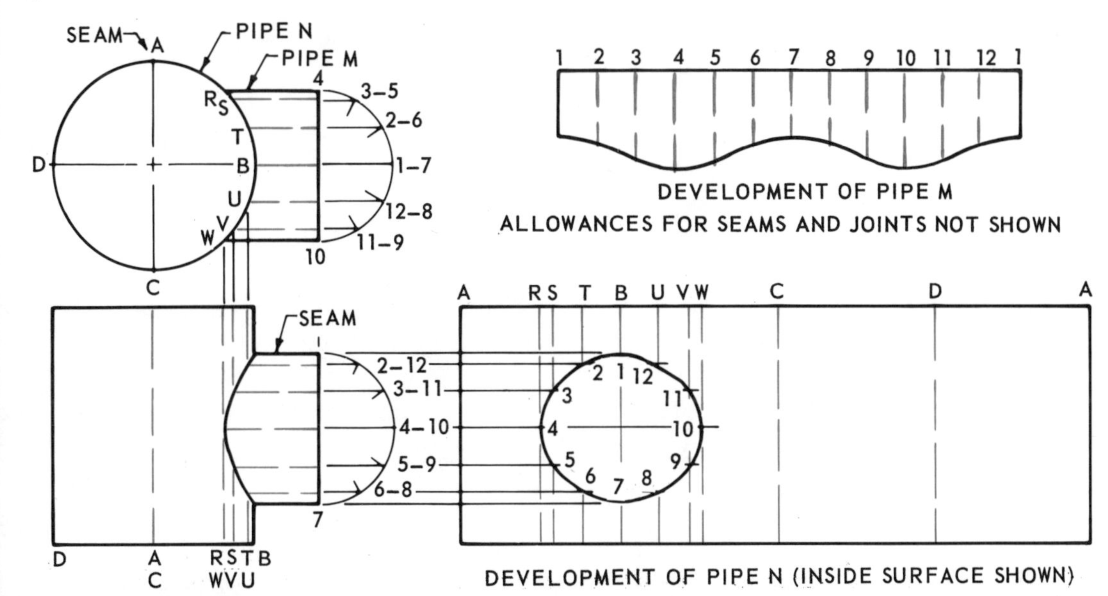

Fig. 10-7-1 Plotting lines of intersection and making development drawings for a 90° reducing tee

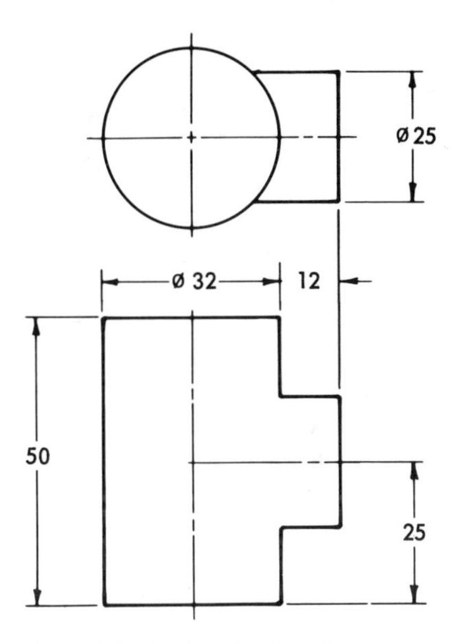

Fig. 10-7-A Reducing tee

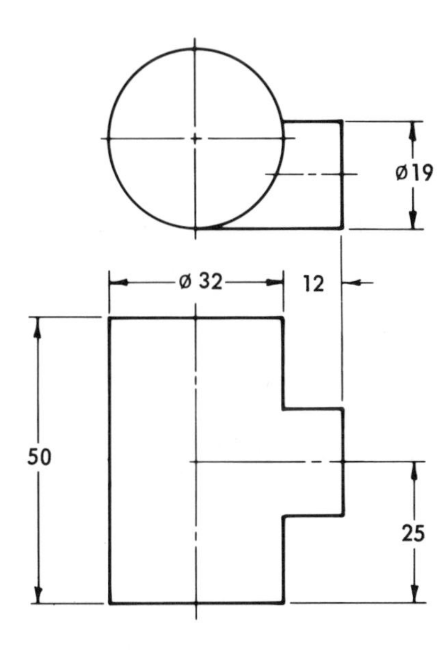

Fig. 10-7-B Offset reducing tee

CHAPTER 11
Materials and Manufacturing Processes

UNIT 11-1
Basic Metallurgy

In addition to drafting skills, a drafter must have basic knowledge of the materials used in the manufacture of component parts, the common manufacturing processes for the production of parts, and the various machines and shop operations involved in manufacturing.

Ferrous Metals

Metals that contain iron and are magnetic are called ferrous metals. The three general classes of ferrous metals are cast iron, wrought iron, and steel.

The first step in the manufacture of any iron or steel is the conversion of the iron ore into pig iron by mixing the iron ore, coke fuel, and limestone (to carry off impurities), in a blast furnace.

The iron melts from the ore and drops into a pool at the bottom of the furnace. Then it is drawn off into an iron ladle and poured into molds called **pigs.** The limestone combines with the impurities to form a scum called **slag.** The slag, which is lighter than the molten iron, floats to the top and is drawn off into a slag ladle.

About 93% of pig iron is pure iron and about 3% to 5% is carbon. The rest is silicon, phosphorus, sulphur, and other

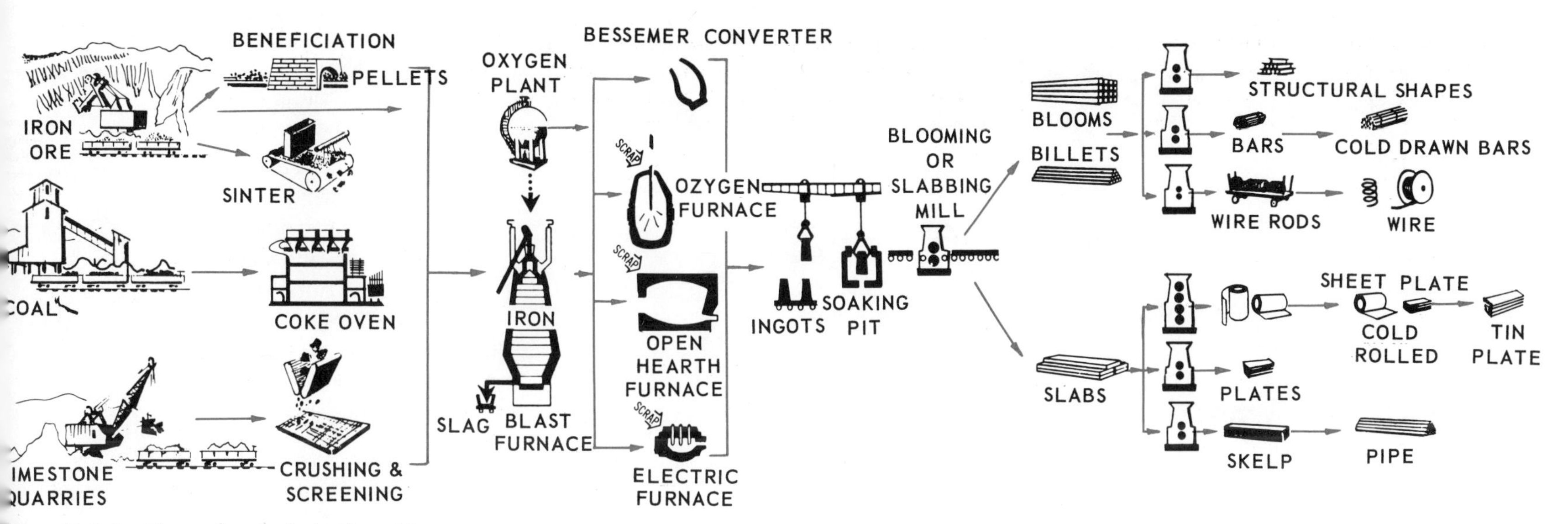

Fig. 11-1-1 Flow chart of steel making

elements. Pig iron is used in foundries for the manufacture of iron castings.

Carbon content is a very important factor in the classification and uses of ferrous metals.

Cast Iron

Cast iron, used in foundries for the making of various castings, has a carbon content of up to 5%. It is used extensively for machine frames and bases, and for a wide variety of machine parts such as gears, levers, and pulleys.

Wrought Iron

Wrought Iron has most of the carbon removed, and is therefore, very ductile. It is made up into common bar stock and sheet sizes, and is used mainly for ornamental and structural shapes and for making pipe and wire.

Steel

Steel is a name given to many iron based metals that differ greatly from each other in their chemical and physical qualities. The basic ingredient for all steel is pig iron. Before molten pig iron can be converted into steel, some of the impurities must be burned out, and other ingredients added, to give the metal the desired chemical composition. About 90% of all steel is now produced in open hearth furnaces. Electric furnaces are used primarily to make fine alloy and tool steel. Carbon content has a great influence on the classification and uses of the various kinds of steel. See Figure 11-1-1 for the basic sequence in producing steel.

Low carbon steel, commonly called machinery or mild steel, contains from 0.1 to 0.3% carbon. This steel, which can be easily forged, welded, and machined, is used for making such things as rivets, shafting, and chains.

Medium carbon steel contains from 0.3 to 0.6% carbon and is commonly used for axles, rails, and heavy forgings.

High carbon steel, commonly called tool steel, contains from 0.6 to 1.5% carbon and can be hardened and tempered. It is primarily used for making tools such as drills, reamers, hammers, and crowbars.

Alloy steels are made by adding metals such as chromium, nickel, tungsten, and vanadium to give the steel certain new and special characteristics such as resistance to rust, corrosion, heat, shock, and fatigue.

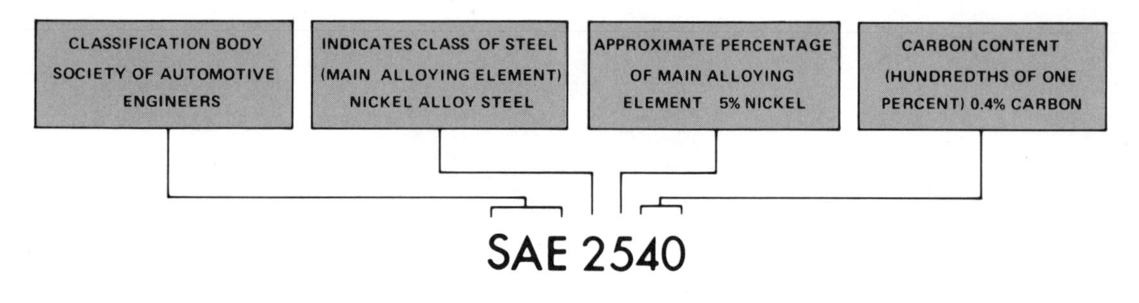

Fig. 11-1-2 Steel designation system

Steel Designation System

Several ways are used to identify a specific steel: by chemistry, by mechanical properties, by its ability to meet a standard specification or industry-accepted practice, or by its ability to be fabricated into an identified part.

SAE and AISI — Systems of Steel Identification. The specifications for steel are based on a code that indicates the composition of each type of steel covered. They include both plain carbon and alloy steels. The code is a 4-number system. See Figures 11-1-2 and 11-1-3. Each figure in the number has the following specific function: the first or left-side figure indicates the major class of steel, the second figure represents a subdivision of the major class; for example, the series having **one** as the left-hand figure covers the carbon steels. The second figure breaks this class up into normal low sulphur steels, the high sulphur free machining grades, and another grade having higher than normal manganese.

Class 1	
Carbon steels	1xxx
Basic open hearth and acid Bessemer carbon steels nonsulphurized and nonphosphorized	10xx
Basic open hearth and acid Bessemer carbon steels, sulphurized but not phosphorized	11xx
Basic open hearth carbon steels phosphorized	12xx

Originally this second figure indicated the percentage of the major alloying element present, and this is true of many of

TYPE OF CARBON STEEL	NUMBER SYMBOL	PRINCIPAL PROPERTIES	COMMON USES
—Plain Carbon	10XX		
—Low-Carbon Steel (0.6% to 0.20% Carbon)	1006 to 1020	Toughness and Less Strength	Chains, Rivets, Shafts, Pressed Steel Products
—Medium-Carbon Steel (0.20% to 0.50% Carbon)	1020 to 1050	Toughness and Strength	Gears, Axles, Machine Parts, Forgings, Bolts and Nuts
—High-Carbon Steel (Over 0.50% Carbon)	1050 and over	Less Toughness and Greater Hardness	Saws, Drills, Knives, Razors, Finishing Tools, Music Wire
—Sulphurized (Free Cutting)	11XX	Improves Machinability	Threads, Splines, Machined Parts
—Phosphorized	12XX	Increases Strength and Hardness but Reduces Ductility	
—Manganese Steels	13XX	Improves Surface Finish	

Fig. 11-1-3 Steel designations, properties and use of carbon steel

TYPE OF STEEL	ALLOY SERIES	APPROXIMATE ALLOY CONTENT	PRINCIPAL PROPERTIES	COMMON USES
Manganese Steel	13XX	Mn 1.6-1.9	Improve Surface Finish	
Molybdenum Steel	40XX	Mo 0.15-0.3	High Strength	Axles, Forgings, Gears, Cams, Mechanical Parts
	41XX	Cr 0.4-1.1; Mo 0.08-0.35		
	43XX	Ni 1.65-2; Cr 0.4-0.9; Mo 0.2-0.3		
	44XX	Mo 0.45-0.6		
	46XX	Ni 0.7-2; Mo 0.15-0.3		
	47XX	Ni 0.9-1.2; Cr 0.35-0.55; Mo 0.15-0.4		
	48XX	Ni 3.25-3.75; Mo 0.2-0.3		
Chromium Steels	50XX	Cr 0.3-0.5	Hardness, Great Strength and Toughness	Gears, Shafts, Bearings, Springs, Connecting Rods
	51XX	Cr 0.7-1.15		
	E51100	C 1.0; Cr 0.9-1.15		
	E52100	C 1.0; Cr 0.9-1.15		
Chromium Vanadium Steel	61XX	Cr 0.5-1.1; V 0.1-0.15	Hardness and Strength	Punches and Dies, Piston Rods, Gears, Axles
Nickel-Chromium-Molybdenum Steels	86XX	Ni 0.4-0.7; Cr 0.4-0.6; Mo 0.15-0.25	Rust Resistance, Hardness and Strength	Food containers, Surgical Equipment
	87XX	Ni 0.4-0.7; Cr 0.4-0.6; Mo 0.2-0.3		
	88XX	Ni 0.4-0.7; Cr 0.4-0.6; Mo 0.3-0.4		
Silicon-Manganese Steels	92XX	Si 1.8-2.2	Springiness and Elasticity	Springs

Fig. 11-1-4 Steel designations, properties and uses of alloy steel

the alloy steels. However, this had to be varied in order to care for all the steels that are available.

The third and fourth figures indicate carbon content in hundredths of 1%, thus the figure xx15 indicates 0.15 of 1% carbon.

Example. SAE 2335 is a nickel steel containing 3.5% nickel and 0.35 of 1% carbon.

Non-Ferrous Metals

Metals that contain little or no iron and are non-magnetic and resistant to corrosion are called non-ferrous metals.

Aluminum

Aluminum is made from **bauxite** ore, and is used extensively in aircraft manufacture because it has only one third the mass of steel. Because it is very soft in its pure state, aluminum is usually alloyed with other metals to increase its strength and stiffness.

Bronze

Bronze is an alloy of copper, tin, and zinc. Other ingredients such as lead, phosphorus, manganese, and aluminum are often added to the alloy to give it special qualities of strength, toughness, ductility, and resistance to corrosion. Bronze is used extensively for machine bearings, propellers, gears, marine hardware, and weather-stripping.

Brass

Brass is an alloy of about 66% copper and 33% zinc. A small quantity of lead (about 3%), is sometimes added to the alloy to make it easier to machine. Brass is commonly used for plumbing fittings, small bushings, radiator parts, hardware, cartridge shells, and many cast parts.

Copper

Copper is a soft, ductile, tough metal, with the special properties of resistance to corrosion and excellent conductivity of electricity and heat. Some of its many uses include wiring, tubin, screening, roofing, radio parts, electric contacts, and piping.

Babbitt

Babbitt is an alloy of tin, copper and antimony. It is easily machined and is primarily used for the linings of bearings in automobiles and machines.

Assignments

Answer the following questions.

1. Name the three general classes of ferrous metal.
2. How does wrought iron differ from cast iron?
3. What is the basic ingredient of steel?

UNIT 11-2
Plastics

The widespread and growing use of plastics is evident in almost every product manufactured for our use. Non-natural fibres and materials, often referred to as "synthetics," are being used in combination with, or to replace, many of nature's materials.

Because the base materials of plastics are manufactured, the specific ingredients can be carefully controlled to produce materials and products which have a unique range of physical properties. They can be soft or hard, transparent or opaque and coloured, flexible or rigid, smooth or textured. Plastic products are resistant to corrosion, rusting, and extreme temperatures, and can be manufactured accurately and at low cost.

Plastics are usally classified as either **thermoplastic** or **thermosetting.** Some other materials may be formed in a similar manner but are not normally referred to as plastics. These include rubber, glass, some forms of ceramics, and compressed powders.

Fig. 11-2-1 A variety of plastic parts

NAME OF PLASTIC	PROPERTIES	USES
ABS (Acrylonitrile Butadiene-Styrene)	Strong, tough, good electrical properties.	Pipe, wheels, football helmets, battery cases, radio cases, children's skates, tote boxes.
ACETAL RESIN	Rigid without being brittle, tough, resistant to extreme temperatures, good electrical properties.	Automobile instrument clusters, gears, bearings, bushings, door handles, plumbing fixtures, threaded fasteners, cams.
ACRYLICS	Exceptional clarity and good light transmission. Strong, rigid, and resistant to sharp blows. Excellent insulator, colourless or full range of transparent, translucent, or opaque colours.	Airplane canopies and windows, television and camera viewing lenses, combs, costume jewelry, salad bowls, trays, lamp bases, scale models, automobile tail lights, outdoor signs.
CELLULOSICS		
(A) Cellulose Acetate		Spectacle frames, toys, lamp shades, combs, shoe heels.
(B) Cellulose Acetate Butyrate	Among the toughest of plastics. Retains a lustrous finish under normal wear. Transparent, translucent, or opaque in wide variety of colours and in clear transparent. Good insulators.	Steering wheels, radio cases, pipe and tubing, tool handles, playing cards.
(C) Cellulose Propionate		Appliance housing, telephone hand sets, pens and pencils.
(D) Ethyl Cellulose		Edge moldings, flashlights, electrical parts.
(E) Cellulose Nitrate		Shoe heel covers, fabric coating.
FLUOROCARBONS	Low coefficient of friction, resistant to extreme heat and cold. Strong, hard, and good insulators.	Valve seats, gaskets, coatings, linings, tubings.
NYLON (Polyamides)	Resistant to extreme temperatures. Strong and long-wearing range of soft colours.	Tumblers, faucet washers, gears. As a filament, it is used as brush bristles, fishing line.
POLYCARBONATE	High impact strength, resistant to weather, transparent.	Parts for aircraft, automobiles, business machines, gauges, safety-glass lenses.
POLYETHYLENE	Excellent insulating properties, moisture proof. Clear, transparent, translucent.	Ice cube trays, tumblers, dishes, bottles, bags, balloons, toys, moisture barriers.
POLYSTYRENE	Clear, transparent, translucent, or opaque. All colours. Water and weather resistant. Resistance to heat or cold.	Kitchen items, food containers, wall tile, toys, instrument panels.
POLYPROPYLENES	Good heat resistance. High resistance to cracking. Light range of colour.	Thermal dishware, washing machine agitators, pipe and pipe fittings, wire and cable insulation, battery boxes, packaging film and sheets.
URETHANES	Tough and shock resistant for solid materials. Flexible for foamed material, can be foamed in place.	Mattresses, cushioning, padding, toys, rug underlays, crashpads, sponges, mats, adhesion, thermal insulation, industrial tires.
VINYLS	Strong and abrasion-resisting. Resistant to heat and cold. Wide colour range.	Raincoats, garment bags, inflatable toys, hose, records, floor and wall tile, shower curtains, draperies, pipe, paneling.

Fig. 11-2-2 Thermoplastics

(The Society of Plastics Industry, Inc.)

Thermoplastics

These materials soften, or liquefy, and flow when heat is applied. Removal of the heat causes these materials to set or solidify. They may be reheated and reformed or reused. In this group fall the acrylics, the cellulosics, nylon (polyamide), polyethylene, polystyrene or styrene, polyfluorocarbons, the vinyls, polyvinylidene, ABS, acetal resin, polypropylene, and polycarbonates. See Figure 11-2-2.

Thermosetting

These materials undergo an irreversible chemical change when heat is applied or when a catalyst or reactant is added. They become hard, insoluble, and infus-

NAME OF PLASTIC	PROPERTIES	USES
Alkyds	Excellent dielectric strength, heat resistance, and resistance to moisture.	Light switches, electric motor insulator and mounting cases, television tuning devices and tube supports. Enamels and lacquers for automobiles, refrigerators, and stoves are typical uses for the liquid form.
Allylics	Excellent dielectric strength and insulation resistance. No moisture absorption; stain resistant. Full range of opaque and transparent colours.	Electrical connectors, appliance handles, knobs, etc. Laminated overlays or coatings for plywood, hardboard, and other laminated materials needing protection from moisture.
Amino (melamine and urea)	Full range of translucent and opaque colours. Very hard, strong, but not unbreakable. Good electrical qualities.	Melamine—tableware, buttons, distributor cases, tabletops, plywood adhesive, and as a paper and textile treatment. Urea—scale housing, radio cabinets, electrical devices, appliance housings, stove knobs in resin form as baking enamel coatings, plywood adhesive and as a paper and textile treatment.
Casein	Excellent surface polish. Wide range of near transparent and opaque colours. Strong, rigid, affected by humidity and temperature changes.	Buttons, buckles, beads, game counters, knitting needles, toys, and adhesives.
Cold-Molded 3 Types: Bitumin Phenolic Cement-Asbestos	Resistance to high heat, solvents, water, and oil.	Switch bases and plugs, insulators, small gears, handles and knobs, tiles, jigs and dies, toy building blocks.
Epoxy	Good electrical properties; water and weather resistance.	Protective coating for appliances, cans, drums, gymnasium floors, and other hard-to-protect surfaces. They firmly bond metals, glass, ceramics, hard rubber and plastics, printed circuits, laminated tools and jigs, and liquid storage tanks.
Phenolics	Strong and hard. Heat and cold resistant; excellent insulators.	Radio and TV cabinets, washing machine agitators, juke box housings, jewelry, pulleys, electrical insulation.
Polyesters (Fibreglass)	Strong and tough, bright and pastel colours, high dielectric qualities.	Used to impregnate cloth or mats of glass fibres, paper, cotton, and other fibres in the making of reinforced plastic for use in boats, automobile bodies, luggage.
Silicones	Heat resistant, good dielectric properties.	Coil forms, switch parts, insulation for motors, and generator coils.

Fig. 11-2-3 Thermosetting plastics

PRODUCTION REPORT					
PART	MATERIAL 1ST CHOICE	MATERIAL 2ND CHOICE	REASON FOR SELECTION	MACHINING REQUIRED	COLOUR
TELEPHONE CASE	ABX	CELLULOSICS	GOOD IMPACT STRENGTH	NONE	GREEN
			GOOD RANGE OF COLOURS		BLUE
			LIGHT MASS		WHITE
			EXCELLENT SURFACE FINISH		TAN
			GOOD ELECTRICAL PROPERTIES		RED
			VARIETY OF FORMING METHODS		

ible, and they do not soften upon reapplication of heat. Thermosetting plastics include phenolics, amino plastics (melamine and urea), cold-molded polyesters, epoxies, silicones, alkyds, allylics, and casein. See Figure 11-2-3.

Forming Methods

The most common methods of forming plastic parts are by the application of heat and pressure, casting, and machining. Heat and pressure may be applied by different methods, the most common of which are compression molding, injection molding, transfer molding, extrusion, blow and vacuum forming, and embossing. Other methods include laminating or layup and cold forming and embossing. The method of forming is governed by the material, part, part design, and cost.

Machining

Practically all thermoplastics and thermosets can be satisfactorily machined on standard equipment with adequate tooling. The nature of the plastic will determine whether heat should be applied, as in some laminates, or avoided, as in buffing some thermoplastics. Standard machining operations can be used, such as turning, drilling, tapping, milling, blanking, and punching.

Assignments

1. Why do plastics have a unique range of physical properties?
2. Name ten physical properties that plastics may have.
3. Explain the main difference between thermoplastics and thermosetting.

Fig. 11-2-4 Selection of material

UNIT 11-3 Manufacturing Processes

Most machine parts are produced by casting, forging, machining from standard stock, welding, or forming from sheet stock.

Casting

Castings are made by pouring molten metal into cavities of the desired shape.

Sand-Casting

When the cavity for a casting is made in a mixture of sand and damp clay, the casting is called a sand casting. Patterns are used to produce the cavity in the sand mold. These are usually made of wood or metal.

The pattern, normally made in two parts, is slightly larger in every dimension than the part to be cast to allow for shrinkage when the metal cools.

Draft or **slight angles** are placed on the pattern to allow for easy withdrawal from the sand mold. Additional metal, known as machining or finish allowance, is also provided on the casting, where a surface is to

(A) CASTING REQUIRED

(B) PATTERN

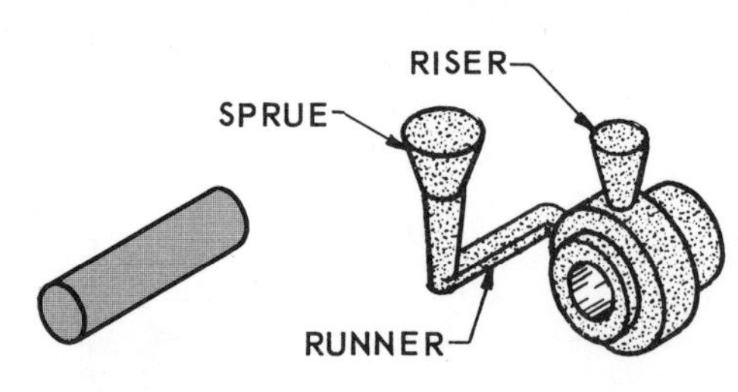

(C) CORE

(D) CASTING AS REMOVED FROM MOLD

Fig. 11-3-2 Sand casting parts

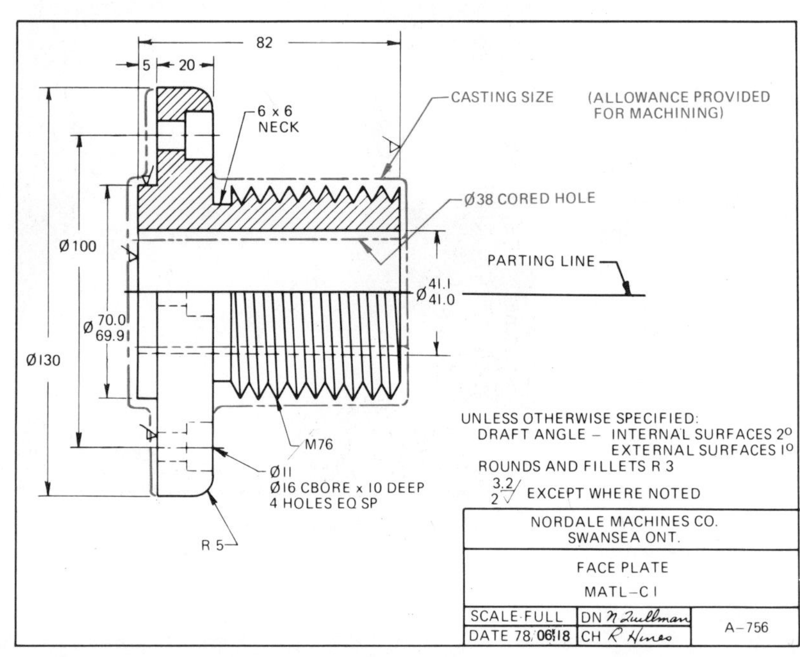

(A) WORKING DRAWING OF A CAST PART

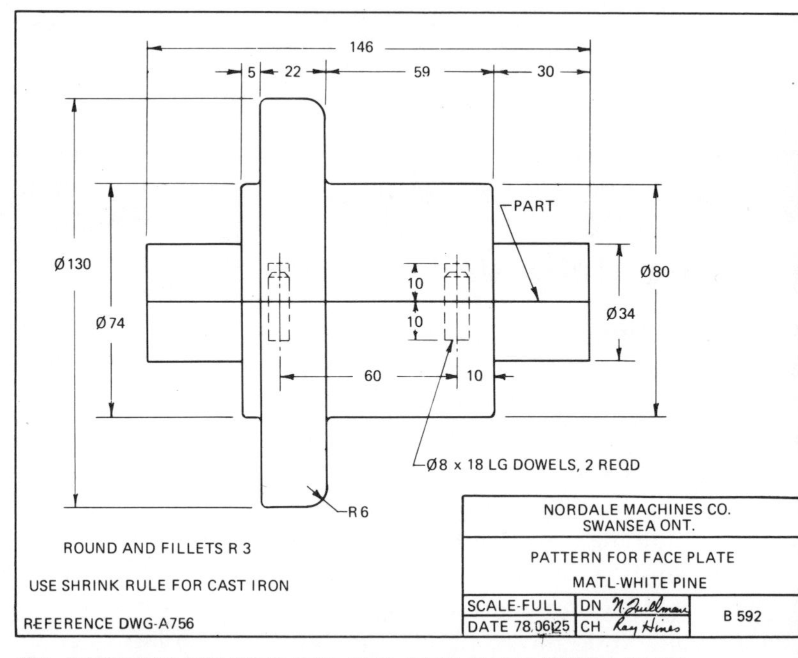

(B) PATTERN DRAWING FOR THE CAST PART SHOWN IN (A)

Fig. 11-3-1 Cast part drawings

be finished. Between 2 and 4 mm, depending on the material being cast, is usually allowed on small castings for each surface that requires finishing.

A typical sand mold is illustrated in Figure 11-3-3. Sand molds consist of two or more sections: bottom (drag), top (cope), and intermediate sections (cheeks) when required. The sand is contained in flasks equipped with pins and lugs to ensure alignment of cope and drag.

Molten metal is poured into the sprue (passage for molten metal). Connecting runners provide flow channels for the metal to enter the mold cavity through gates. Riser cavities are located over the heavier sections of the casting to allow for the addition of liquid metal to the casting during solidification.

The whole gating system not only provides for the molten metal to enter the mold, but also functions as a venting system for the removal of gases.

When the metal has hardened, the sand is broken and the casting removed. The excess metal, gates and risers, are then removed and remelted. In sand casting, a new mold must be made for each part.

Die Casting

This fairly new process uses a permanent metal mold or die. The die is usually made of steel and is in two parts, with a vertical parting line. The stationary half of the die is called the cover die; the movable half is called the ejector die.

Molten metal is forced into the die cavity under pressure. After the molten metal has

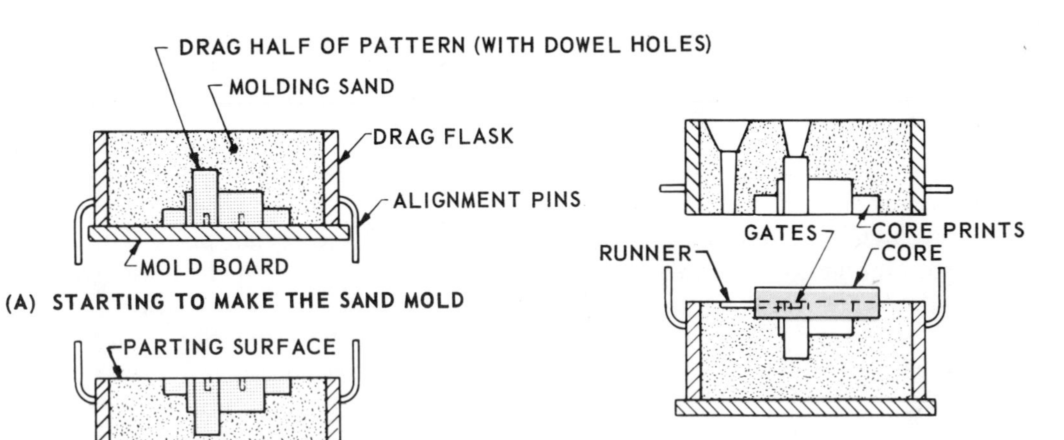

(A) STARTING TO MAKE THE SAND MOLD

(B) AFTER ROLLING OVER THE DRAG

(E) PARTING FLASKS TO REMOVE PATTERN AND TO ADD CORE AND RUNNER

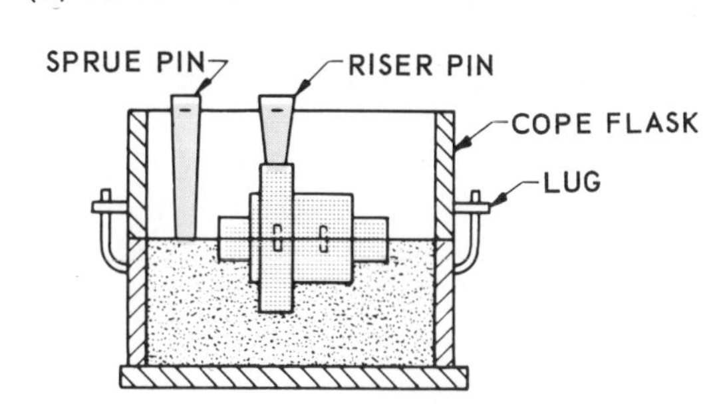

(C) PREPARING TO RAM MOLDING SAND IN COPE

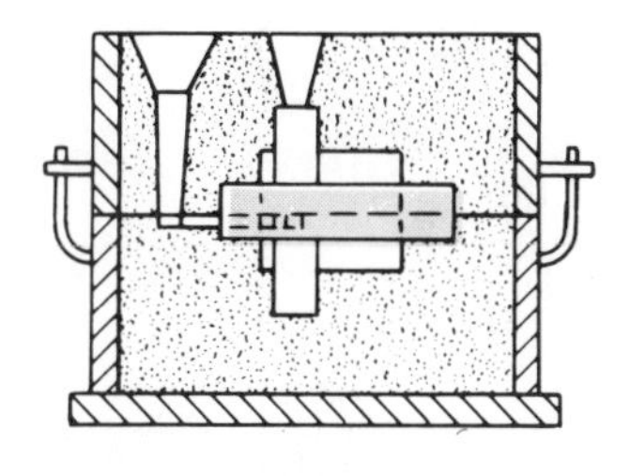

(F) SAND MOLD READY FOR POURING

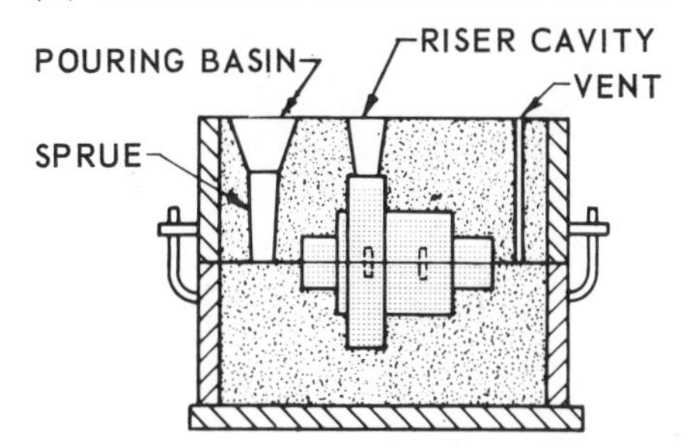

(D) REMOVING RISER AND GATE SPRUE PINS AND ADDING POURING BASIN

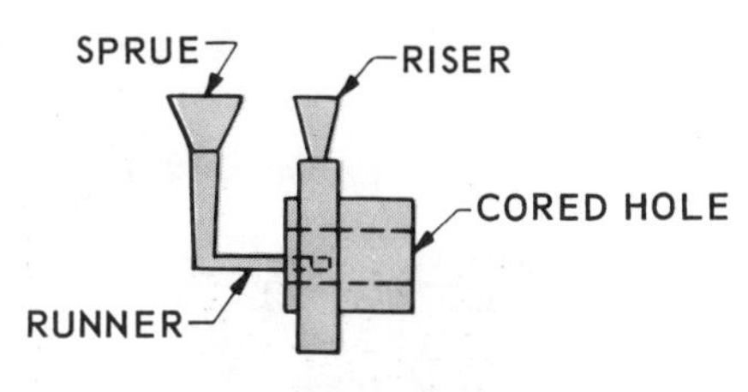

SPRUE, RISER, AND RUNNER TO BE REMOVED FROM CASTING

(G) CASTING AS REMOVED FROM THE MOLD

Fig. 11-3-3 Sequence in preparing a sand casting

solidified, the die is opened and the casting is ejected from the ejector die.

Parts produced by this method are very accurate in size, and are often completely finished when taken from the die. This factor eliminates or reduces machining costs.

Large production quantities are necessary to justify the large capital expenditure for machines and dies. At the present time, only non-ferrous alloys can economically be die cast, because of the lack of a suitable die material to withstand the higher temperatures required for casting steel and iron.

The die casting process is also used for the assembly of small parts. The parts are placed in an assembly die either by hand or automatically from vibratory feeders or magazines. The die is then closed, locating all parts in exact position. Molten metal, usually a lead, tin, or zinc alloy, is injected into a suitably arranged cavity to lock the parts together. The gate or sprue is sheared as the die opens, and the assembly is removed from the die.

Forming from Sheet Stock

Many parts can be produced from a standard thickness of sheet or strip stock. The stock is first cut to size **in the flat** and then is bent, punched, or formed to the desired shape.

For limited quantities of simple parts, a template is made to control the shape in the flat, and the piece is formed using standard sheet metal equipment. See Chapter 10 for information on development techniques and examples of formed parts.

When very large quantities of a part are required, special dies are made and the parts are produced in punch presses. Such parts are often referred to as **stampings.**

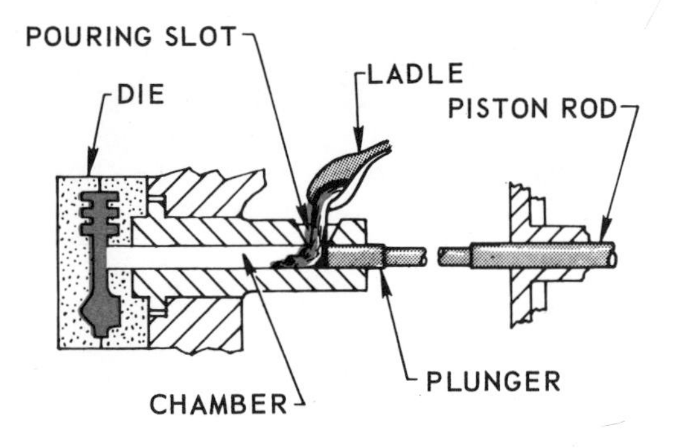

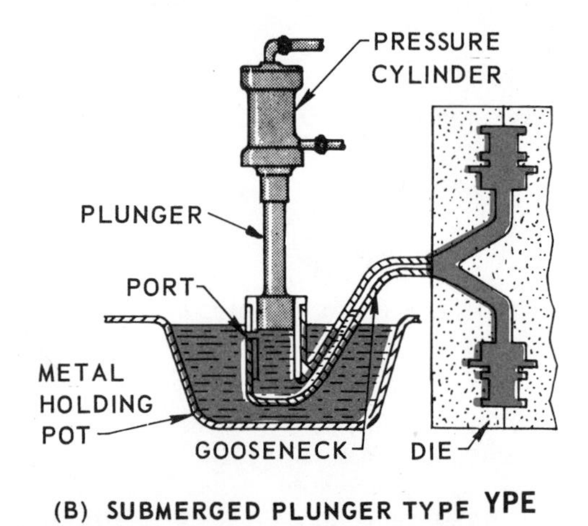

Fig. 11-3-4 Die casting machines

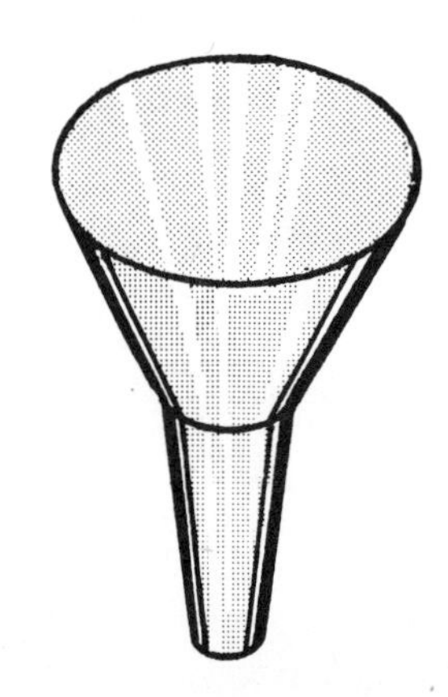

Fig. 11-3-5 Forming from sheet stock

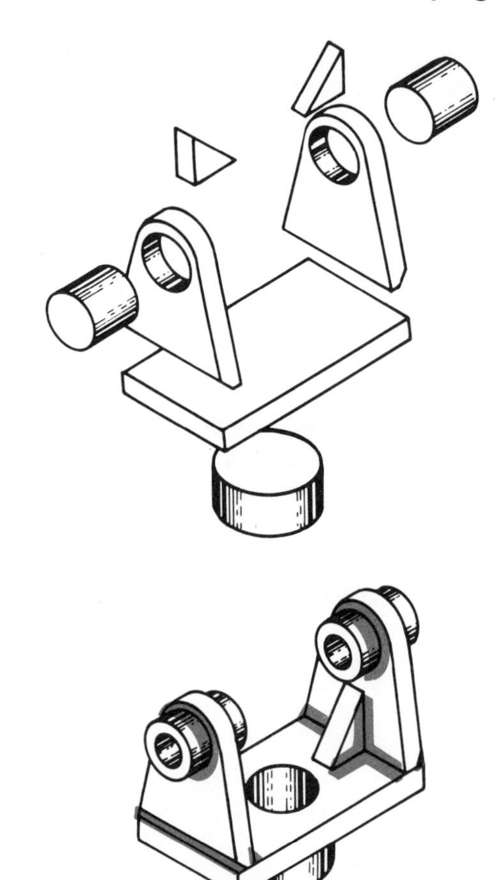

Fig. 11-3-6 Welded part

(A) BILLET

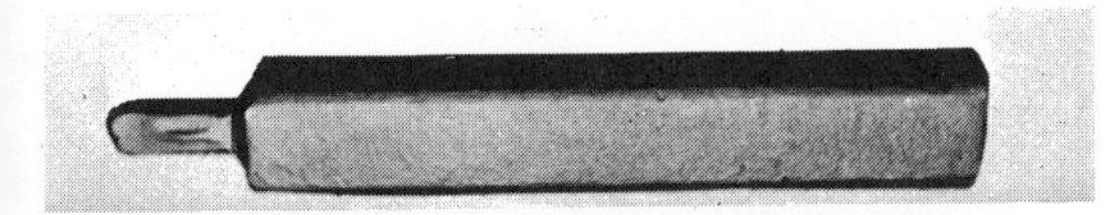

(B) TONGHOLD IS FORGED

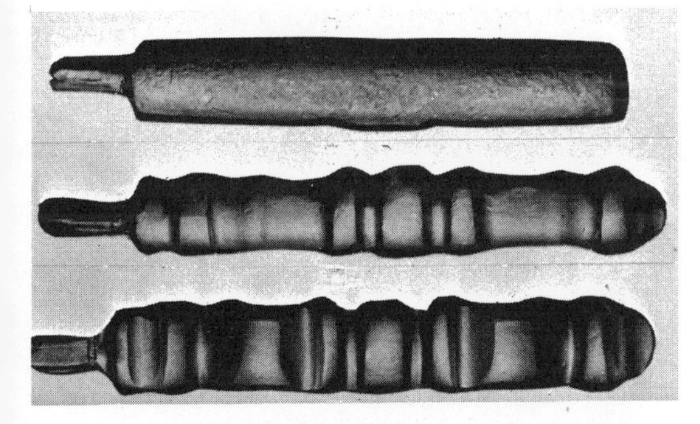

(C) THREE–STEP FORMING IMPRESSION

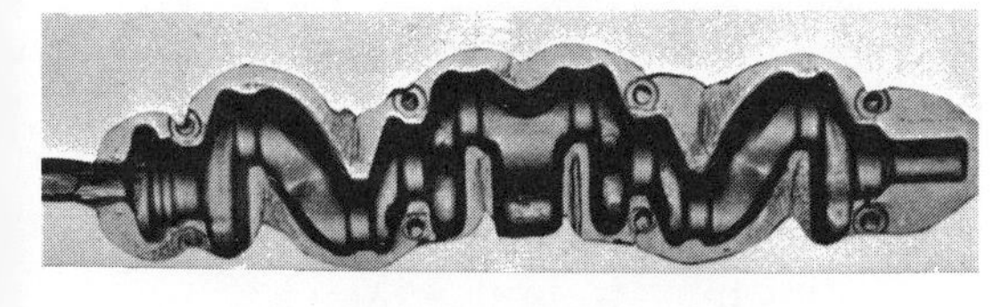

(D) BLOCKING AND FINISHING

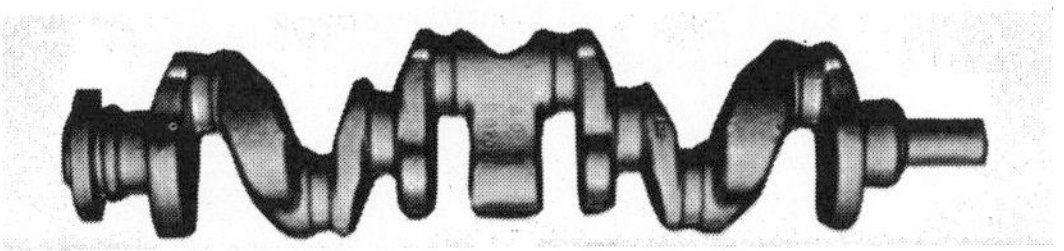

(E) AFTER TRIMMING CRANKS ARE TWISTED INTO POSITION

Fig. 11-3-7 Forging a crankshaft

Welding

Parts can be produced by welding simple pieces cut from standard rod, bar, or plate stock. Such parts are called welded fabrications. For many purposes this method provides a lighter and stronger part than can be obtained by casting or forging, often at less cost, particularly when the quantity of parts to be produced is not large enough to warrant the expense of making patterns or forging dies.

As with castings and forgings, some machining is usually necessary on the completed welded fabrications.

Forging

Forging is a process of squeezing or hammering a bar or billet into a desired shape. Materials to be forged must be ductile, and are usually heated to make them more plastic.

When a large quantity of parts is to be forged, a special steel die is used. The lower die is held on the bed of the drop hammer, and the upper die is fastened to the hammer mechanism. The hot billet is placed on the lower die, and the upper die is dropped several times, causing the metal to flow into the cavities of the dies.

The small amount of excess material forms a thin web, or **flash,** surrounding the part between the two die faces. The flash is then removed in a trimming die. A generous draft angle must be provided for easy release of the forging from the die.

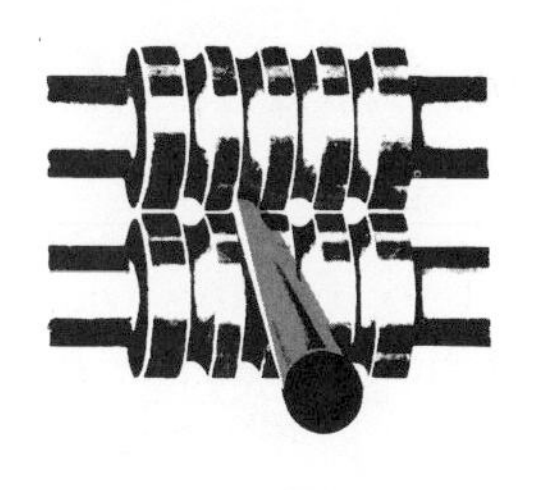

ROUNDS

SQUARES

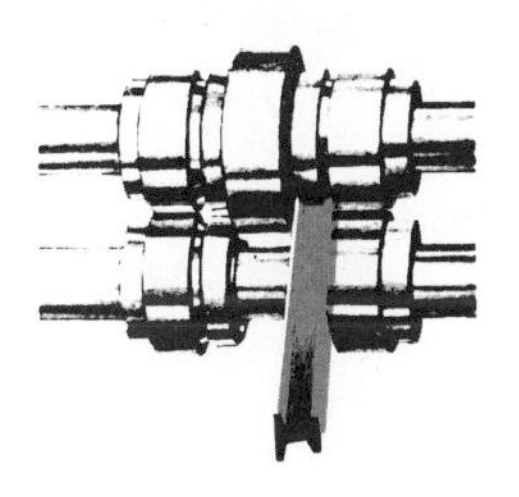

RAILS

CHANNELS

H COLUMNS

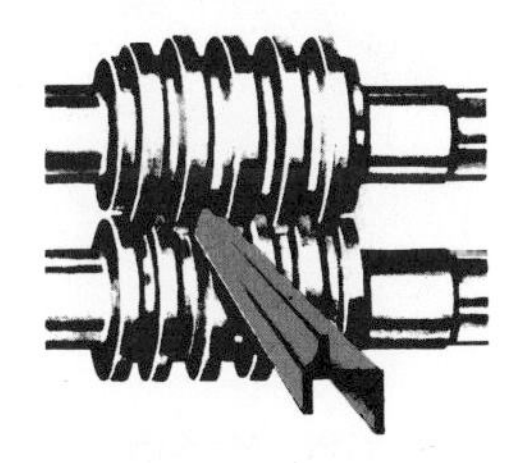

SPECIAL SECTIONS

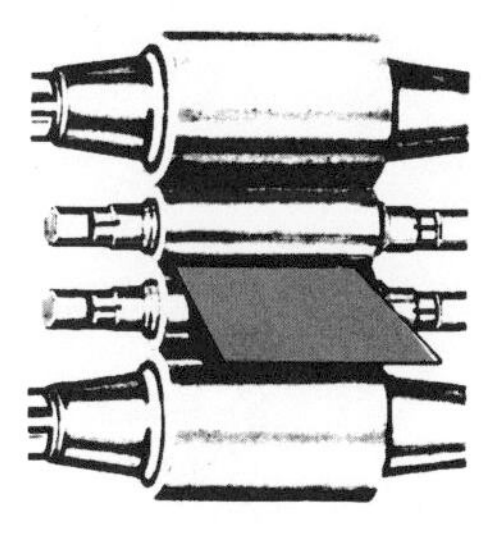

SHEETS

ZEE BARS

Fig. 11-3-8 The forming of standard shapes

Machining from Standard Stock

Many parts are designed so that they can be made from standard stock material by machining. Materials available in standard sizes include solid round, square, and rectangular bars; round and square pipe and tubing; a variety of structural steel shapes such as angle and channel; and plate and sheet.

All machining operations remove metal for one or more of the following reasons:

1. To produce a smooth surface or an accurate size.
2. To give a desired shape.
3. To produce holes, slots, threads, keyseats, or other special features.

Machining operations to remove soft metal are performed by hardened steel or carbide cutting tools. Diamond cutting tools or abrasive wheels are necessary to remove metal that has been hardened.

The machines commonly used for the production of parts from standard stock, as well as machining operations on castings, forgings, welded and formed parts, are the lathe, horizontal and vertical milling machines, shaper, drill press, cylindrical and surface grinders, boring machine, planer, and band saw.

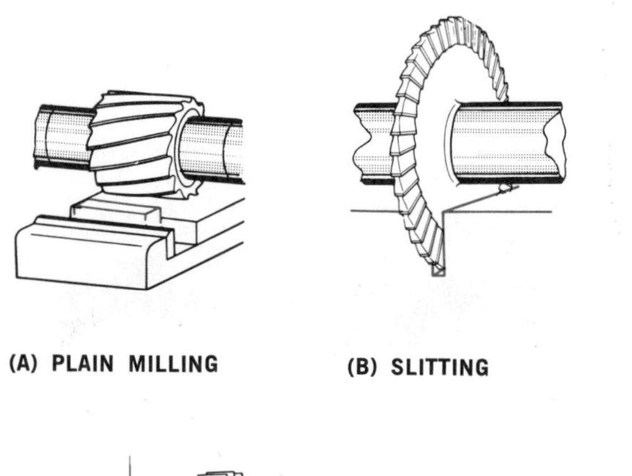

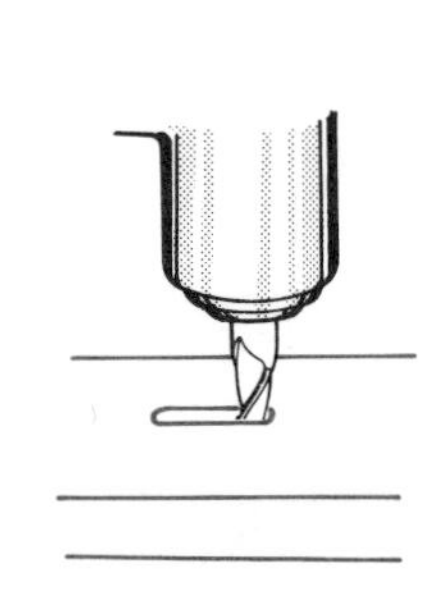

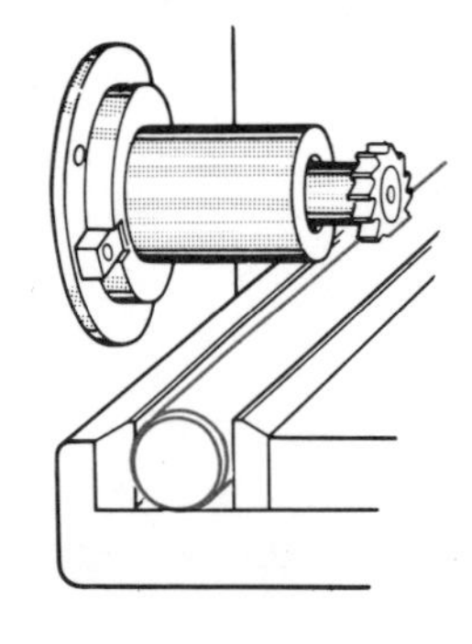

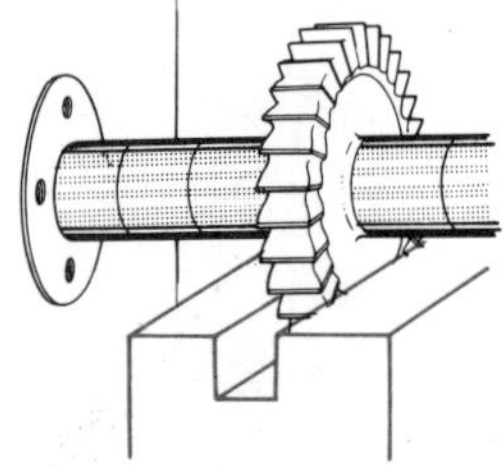

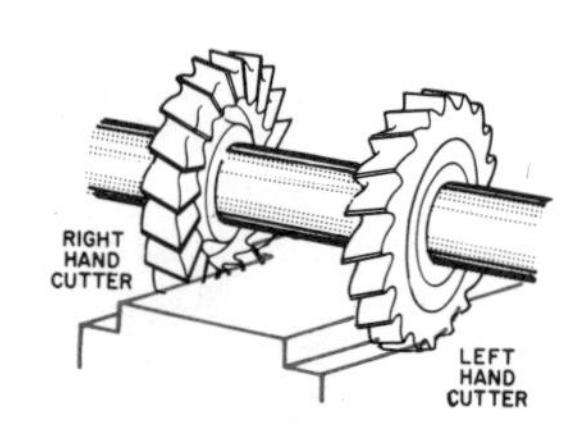

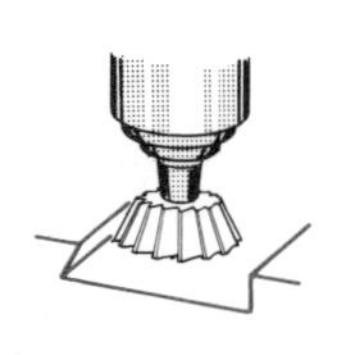

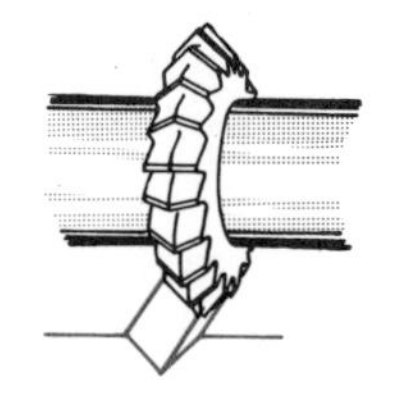

Fig. 11-3-9 Milling machine operations

Assignments

Review Questions

1. What is the purpose of a pattern in sand casting?
2. Give two reasons for a pattern being larger than the finished part.
3. How do **die cast** parts differ from **sand cast** parts?
4. Why must materials be ductile to be forged?
5. Why is a generous draft angle provided on forging dies?
6. Give three reasons for machining operations to remove metal.
7. What cutting tools are used to remove metal that has been hardened?
8. What are some of the advantages of welded fabrications?
9. What is the purpose of a template in making parts from sheet stock?
10. How are stampings produced?

Drawing Assignments

1. Cast pivot arm, Fig. 11-3-A, sheet size-A4, scale 1:1. Make a two-view working drawing.
2. Forged wrench, Fig. 11-3-B, sheet size-A4, scale 1:1. Make a two-view working drawing.

3. Formed bracket, Fig. 11-3-C, sheet size-A4, scale 1:1. Make a three-view working drawing.
4. Make detail or assembly drawings of the parts shown in Figs. 11-3-D or 11-3-E.

Review for Assignments

Unit 5-3 Detail and assembly drawings
Unit 6-3 Common threaded fasteners
Unit 7-1 Sections and conventions
Unit 7-5 Assemblies in section

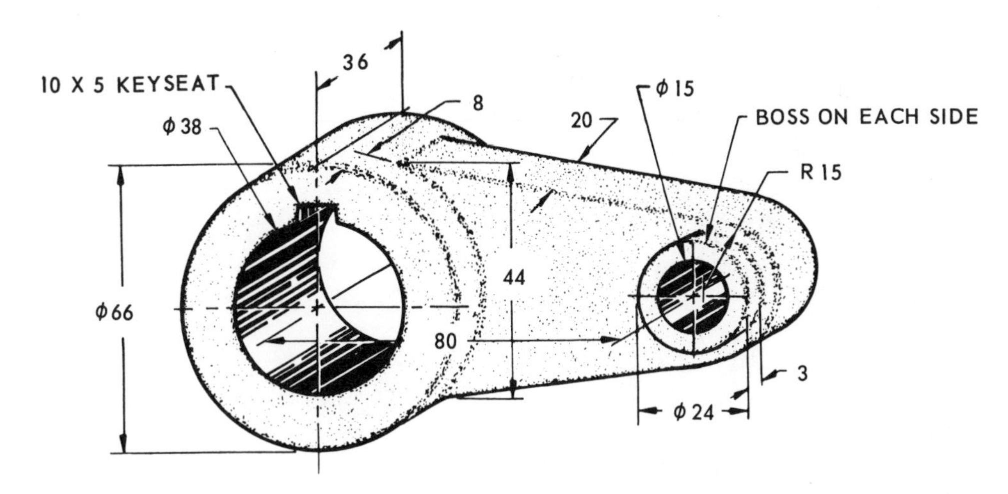

Fig. 11-3-A Pivot arm - casting

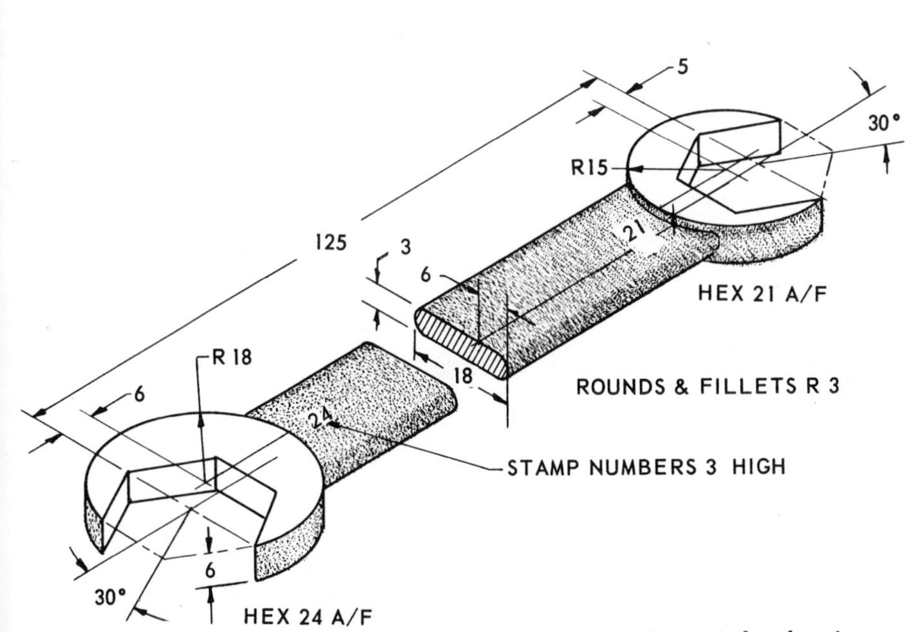

Fig. 11-3-B Open-end wrench - forging

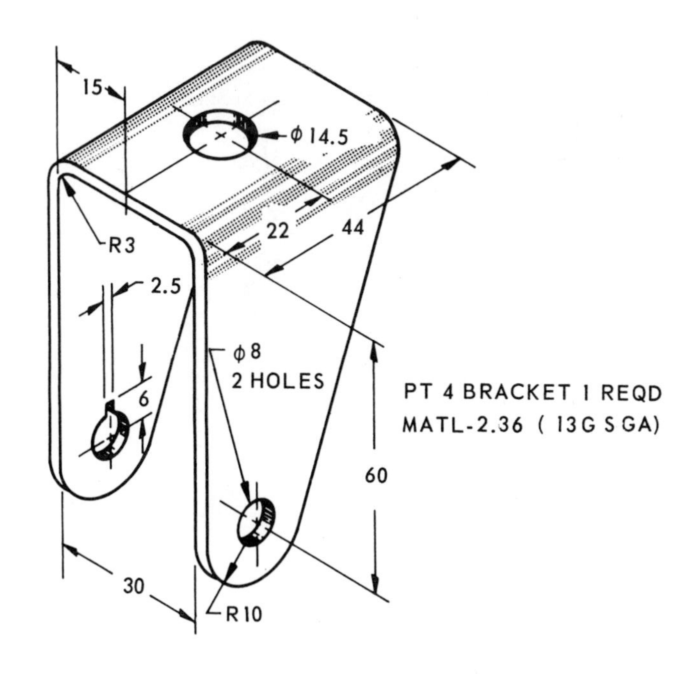

Fig. 11-3-C Bracket - formed part

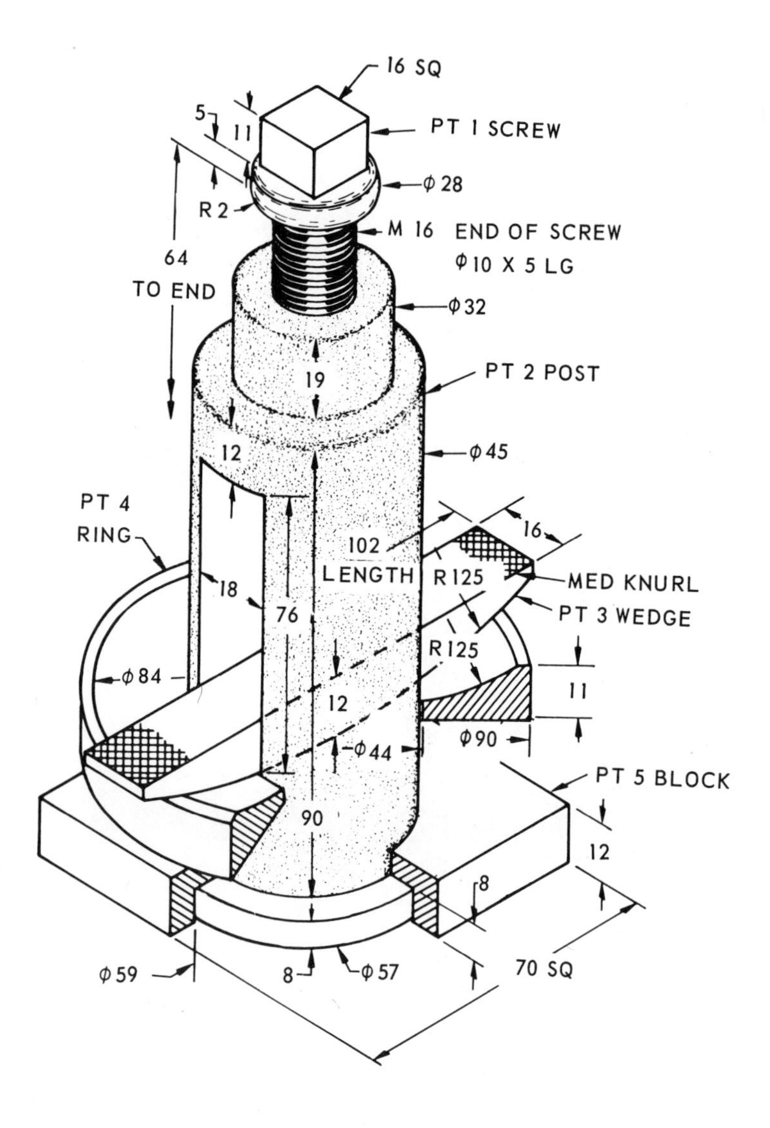

Fig. 11-3-D Tool Post Holder

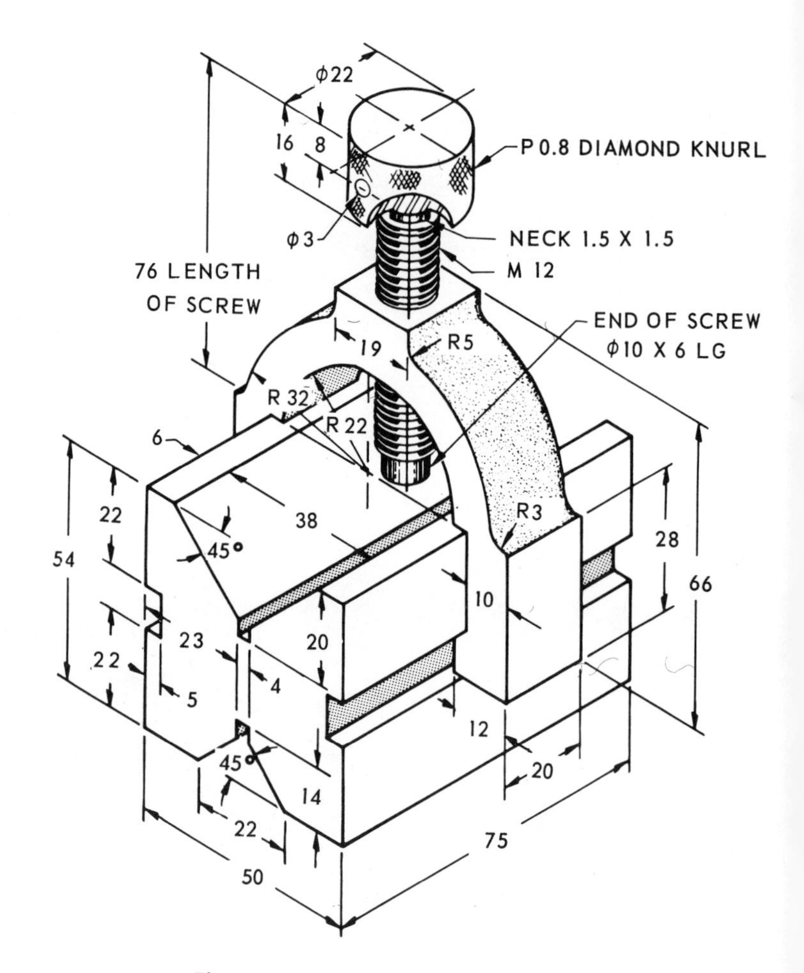

Fig. 11-3-E V-Block clamp

Chapter 12
Modern Engineering Tolerancing

UNIT 12-1
Limits and Tolerances

Although drawings have been the means of communication of engineering information for about 6000 years, it was not until early in the twentieth century that the concept of tolerances and allowances was introduced to the dimensioning of mating parts.

Prior to this time, skilled craftsmen produced parts that fit together properly, but each part was not necessarily made to a specific size. To a large extent, most mechanisms were handcrafted, and the parts from one could not likely be interchanged with a duplicate mechanism.

As manufacturing equipment and methods improved, and better measuring instruments were developed, it became possible to make parts to controlled sizes, or tolerances.

The development of a system of tolerances and allowances has made possible the great advances in the economical manufacture of interchangeable parts, and mass production techniques.

Tolerances are the permissible variations in the specified form, size, or location of individual features of a part from that shown on the drawing. For example, a dimension given as 40 ± 0.1 mm means that the manufactured part can be anywhere between 39.9 mm and 40.1 mm and that the tolerance permitted on this dimension is 0.2 mm. The largest and smallest permissible sizes (40.1 and 39.9 mm, respectively) are known as the **limits.**

Greater accuracy costs more money; and since economy in manufacturing would not permit all dimensions to be held to the same

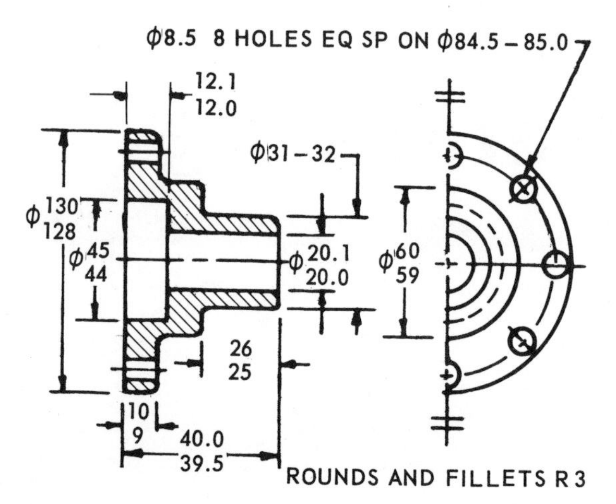

Fig. 12-1-1 A working drawing

accuracy, a system for dimensioning must be used (Fig. 12-1-1). Generally, most parts require only a few dimensions to be accurate.

In order that assembled parts may function properly and to allow for interchangeable manufacturing, it is necessary to permit only a certain amount of tolerance on each of the mating parts and a certain amount of allowance between them.

In order to calculate limit dimensions, the following definitions should be clearly understood (refer to Fig. 12-1-2).

Basic Size. The basic size of a dimension is the theoretical size from which the limits for that dimension are derived, by the application of the allowance and tolerance.

Limits of Size. These limits are the maximum and minimum sizes permissible for a specific dimension, to give a desired fit (Fig. 12-1-3).

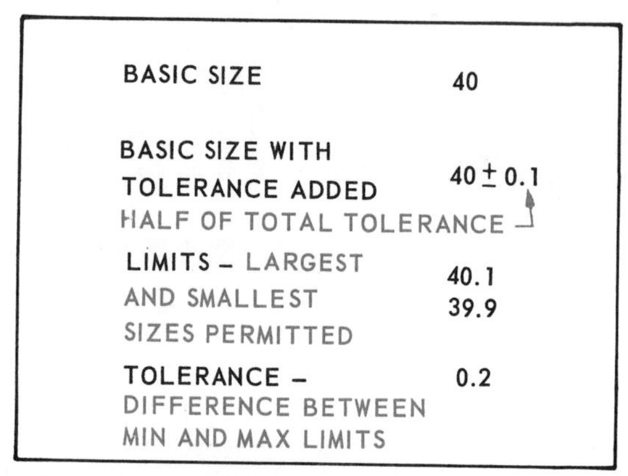

BASIC SIZE	40
BASIC SIZE WITH TOLERANCE ADDED HALF OF TOTAL TOLERANCE	40 ± 0.1
LIMITS – LARGEST AND SMALLEST SIZES PERMITTED	40.1 39.9
TOLERANCE – DIFFERENCE BETWEEN MIN AND MAX LIMITS	0.2

Fig. 12-1-2 Limit and tolerance terminology

Tolerance. The tolerance on a dimension is the total permissible variation in its size. The tolerance is the difference between the limits of size.

Allowance. An allowance is an intentional difference in correlated dimensions of mating parts. It is the minimum clearance (positive allowance) or maximum interference (negative allowance) between such parts.

Maximum Material Size. The maximum material size is that limit of size of a feature which results in the part containing the maximum amount of material. Thus it is the maximum limit of size for a shaft or an external feature, or the minimum limit of size for a hole or internal feature.

Tolerancing

All dimensions requried in the manufacture of a product have a tolerance. Tolerances

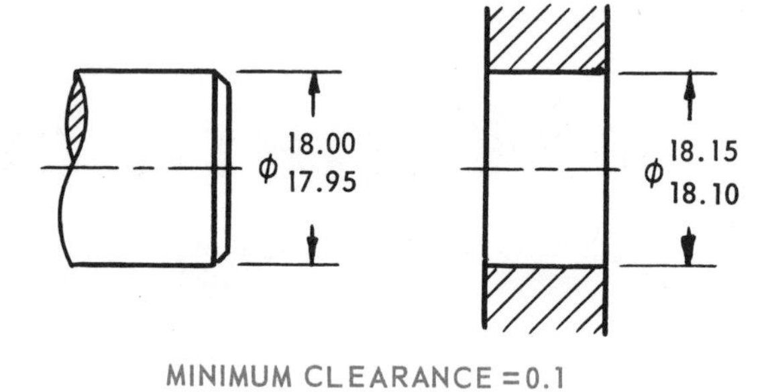

MINIMUM CLEARANCE = 0.1
MAXIMUM CLEARANCE = 0.2

(A) CLEARANCE FIT

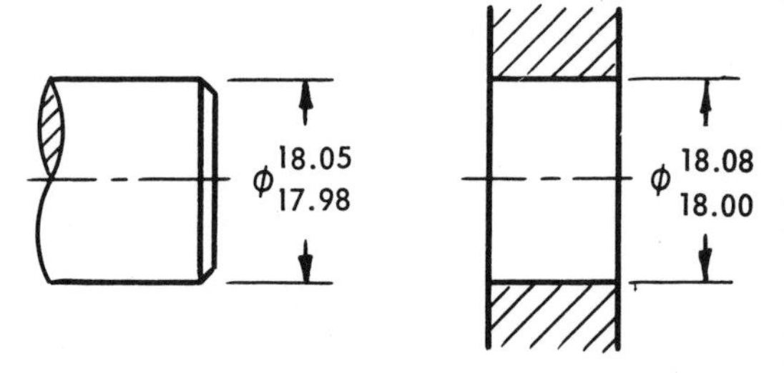

MAXIMUM CLEARANCE = 0.1
MAXIMUM INTERFERENCE = 0.05

(B) TRANSITION FIT

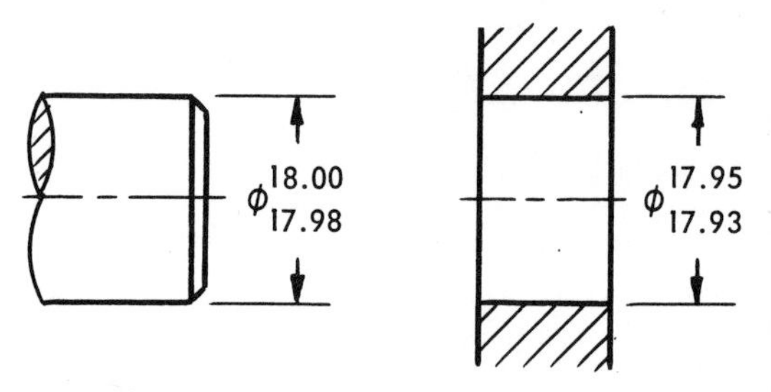

MINIMUM INTERFERENCE = 0.03
MAXIMUM INTERFERENCE = 0.07

(C) INTERFERENCE FIT

Fig. 12-1-3 Basic types of fits

may be expressed in one of the following ways:

- As specified limits or tolerances shown directly on the drawing for a specified dimension.
- In a general tolerance note, referring to all dimensions on the drawing for which tolerances are not otherwise specified.
- In the form of a special note referring to specific dimensions.

Tolerances on dimensions that locate features may be applied directly to the locating dimensions or by the positional tolerancing method described in Unit 12-2.

Tolerancing Methods

A tolerance applied directly to a dimension may be expressed in two ways.

Limit Dimensioning. For this method, the high limit (maximum value) is placed above the low limit (minimum value). When it is expressed in a single line, the low limit should precede the high limit and they should be separated by a dash, as shown in Figures 12-1-4(A) and 12-1-5.

Where limit dimensions are used and where either the maximum or minimum dimension has digits to the right of the decimal point, the other value should have the zeros added so that both the limits of size are expressed to the same number of decimal places.

Plus and Minus Tolerancing. (Refer to Fig. 12-1-4). For this method the dimension of the specified size is given first and is followed by a plus or minus expression

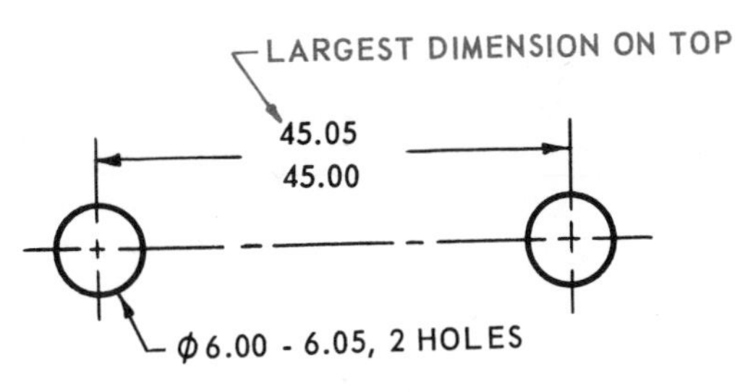

TOLERANCES ON DIMENSIONS 0.05

(A) LIMIT DIMENSIONING

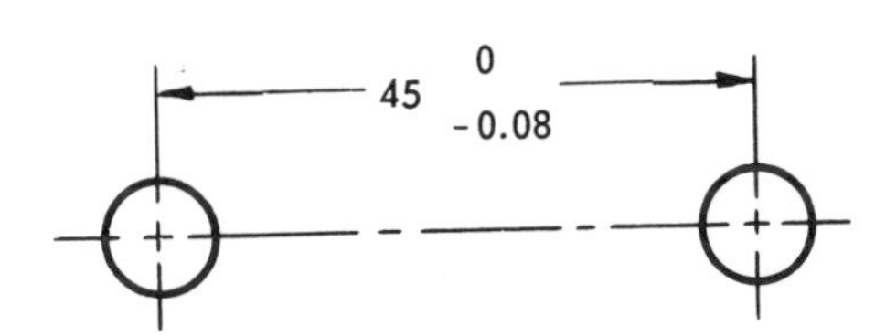

TOLERANCE ON DIMENSION 0.08

NOTE: + OR − SIGN NOT REQUIRED FOR A ZERO TOLERANCE

(B) UNILATERAL TOLERANCING

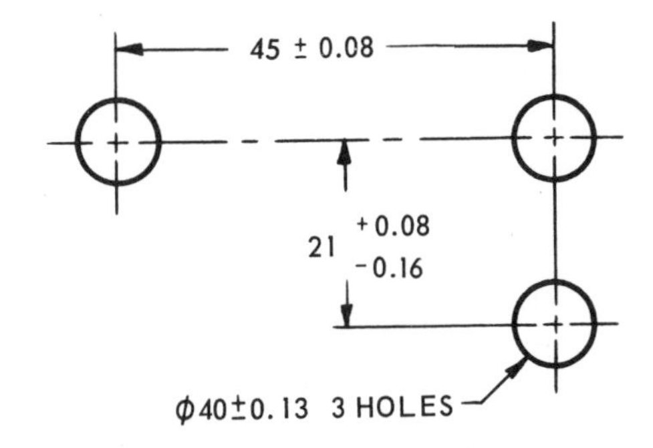

TOLERANCE ON DIMENSION A = 0.16

TOLERANCE ON DIMENSION B = 0.24

TOLERANCE ON DIMENSION C = 0.26

(C) BILATERAL TOLERANCING

Fig. 12-1-4 Tolerancing methods

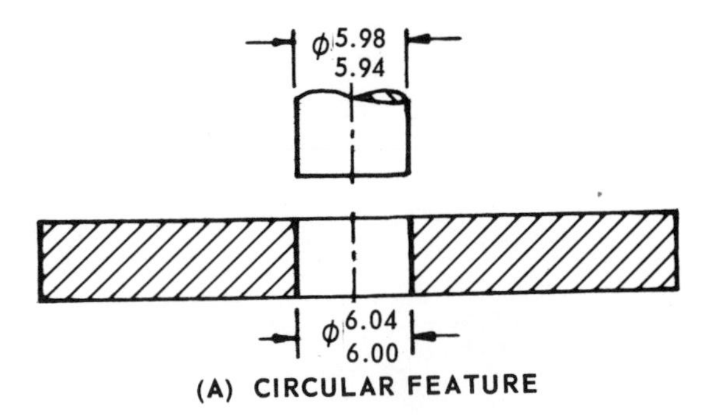

(A) CIRCULAR FEATURE

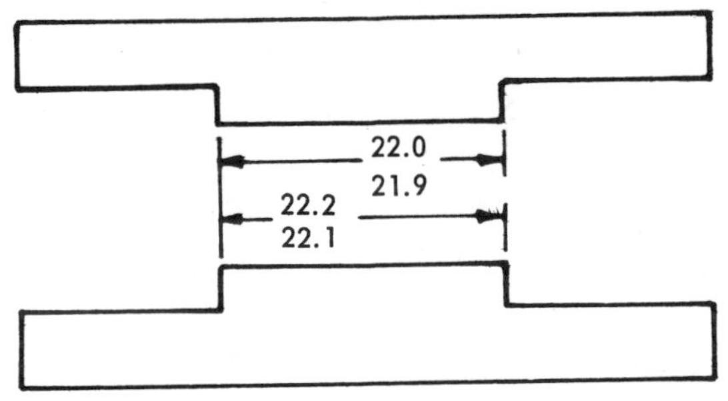

(B) FLAT FEATURE

Fig. 12-1-5 Limit dimensioning application

of tolerancing. The plus value should be placed above the minus value. This type of tolerancing can be broken down into bilateral and unilateral tolerancing. In a bilateral tolerance, the plus and minus tolerances should generally be equal, but special design considerations may sometimes dictate unequal values (Fig. 12-1-6). The specified size is the design size, and the tolerance represents the desired control of quality and appearance. The dimension need not be shown to the same number of decimal places as its tolerance. For example:

1.5±0.04 *not* 1.50±0.04
10±0.1 not 10.0±0.1

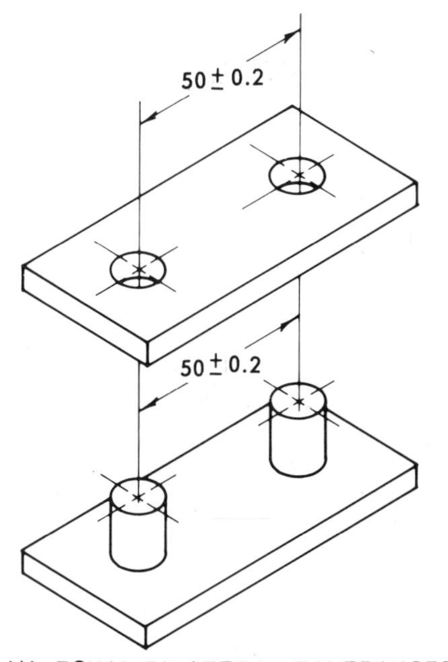

(A) EQUAL BILATERAL TOLERANCES

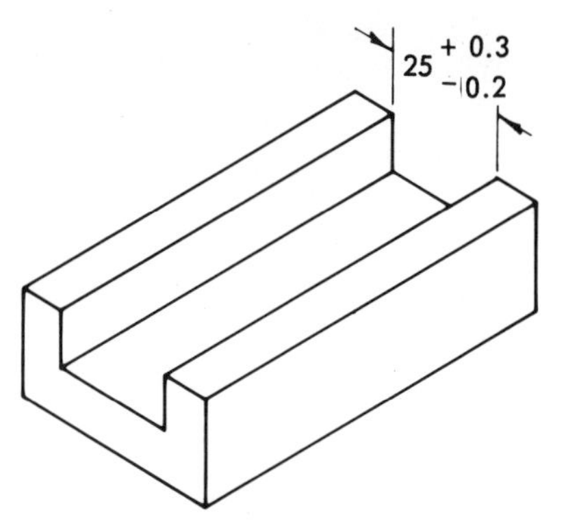

(B) UNEQUAL BILATERAL TOLERANCES

Fig. 12-1-6 Application of bilateral tolerances

The unilateral tolerance is generally used to establish the position of a feature, as shown in Figure 12-1-7, whenever the ideal position of a feature is midway in the allowable tolerance range.

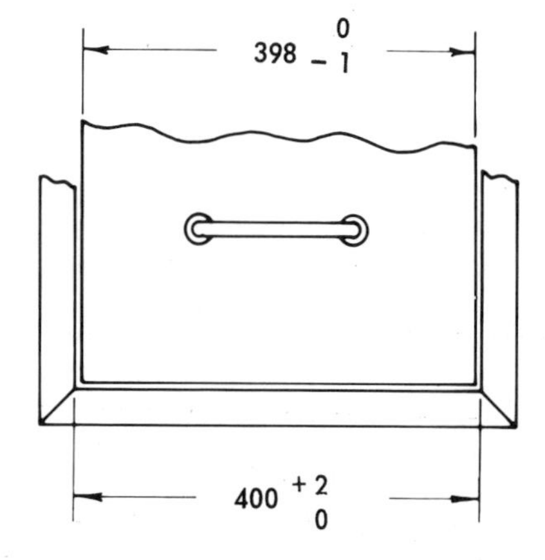

Fig. 12-1-7 Application of unilateral tolerances

General Tolerance Notes. The use of general tolerance notes greatly simplifies the drawing and saves much layout in its preparation. The following examples illustrate the wide field of application of this system. The values given in the examples are typical.

Example 1.
EXCEPT WHERE STATED OTHERWISE, TOLERANCES ON FINISHED DECIMAL DIMENSIONS ±0.1.

Example 2.
EXCEPT WHERE STATED OTHERWISE, TOLERANCES ON FINISHED DIMENSIONS TO BE AS FOLLOWS:

Dimension (mm)	Tolerance
Up to 100	±0.1
From 101 to 300	±0.2
From 301 to 600	±0.5
Over 600	±1

A comparison between the tolerancing methods described is shown in Figure 12-1-8.

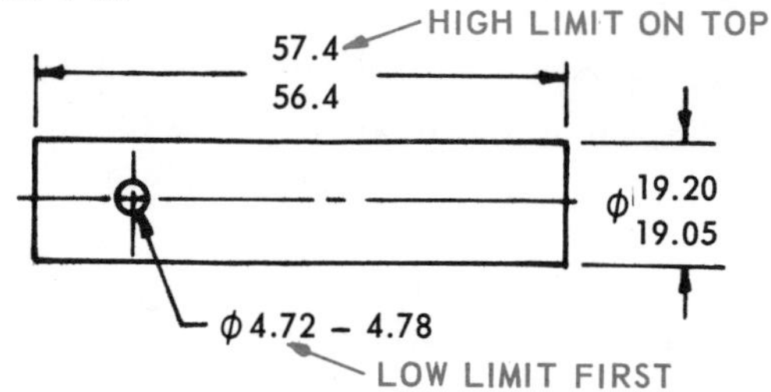

(A) LIMIT DIMENSIONING

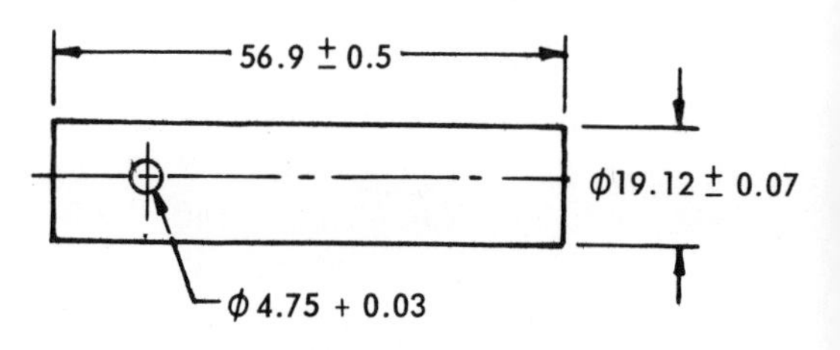

(B) BILATERAL TOLERANCING

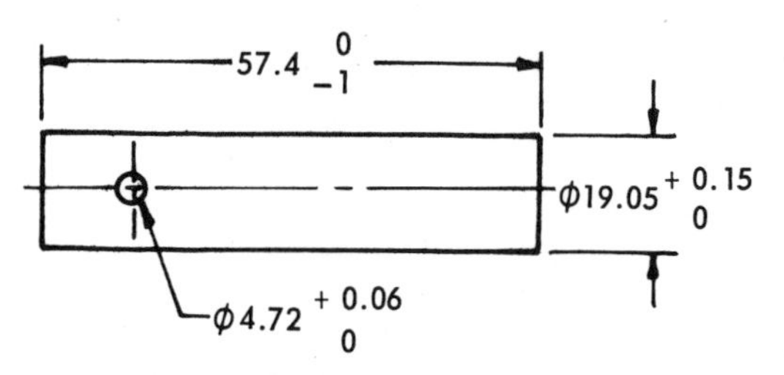

(C) UNILATERAL TOLERANCING

Fig. 12-1-8 A comparison of the tolerancing methods

Assignments

1. Calculate the maximum and minimum clearances and interferences of the features in Fig. 12-1-A.
2. Calculate the sizes and tolerances for the parts shown in Figs. 12-1-B and C.

FEATURE		LIMITS OF SIZE		CLEARANCE OR INTERFERENCE	
		MAX	MIN	MAX	MIN
1	HOLE	10.015	10.000		
	SHAFT	10.024	10.015		
2	HOLE	6.018	6.000		
	SHAFT	5.990	5.978		
3	HOLE	12.018	12.000		
	SHAFT	12.012	12.001		
4	HOLE	16.027	16.000		
	SHAFT	15.984	15.966		
5	HOLE	20.021	20.000		
	SHAFT	20.035	20.022		
6	HOLE	8.015	8.000		
	SHAFT	8.000	7.991		
7	HOLE	5.012	5.000		
	SHAFT	5.000	4.992		
8	HOLE	8.035	8.013		
	SHAFT	8.000	7.985		
9	HOLE	18.020	18.000		
	SHAFT	18.049	18.037		

Fig. 12-1-A Clearance and interference problems

PROBLEM	HOLE	SHAFT	HOLE LIMITS	SHAFT LIMITS	MIN CLEARANCE OR MAX INTERFERENCE	MAX CLEARANCE OR MIN INTERFERENCE
1	$10^{+0.09}_{0}$	$9.20^{0}_{-0.09}$				
2	$16^{+0.018}_{0}$	$16.001^{+0.011}_{0}$				
3	8 ± 0.02	$7.96^{0}_{-0.02}$				
4	$50^{+0.016}_{0}$	$49.87^{0}_{-0.16}$				
5	$50^{+0.025}_{0}$	$50^{0}_{-0.016}$				
6	$25^{+0.021}_{0}$	$24.980^{+0.013}_{0}$				
7	$40^{+0.05}_{-0.03}$	$39.9^{+0.02}_{0}$				
8	$30^{+0.021}_{0}$	$30.048^{0}_{-0.013}$				
9	36 ± 0.05	$36^{+0.02}_{-0.01}$				
10	56 ± 0.1	55.5 ± 0.1				

Fig. 12-1-B Limits and tolerance problems

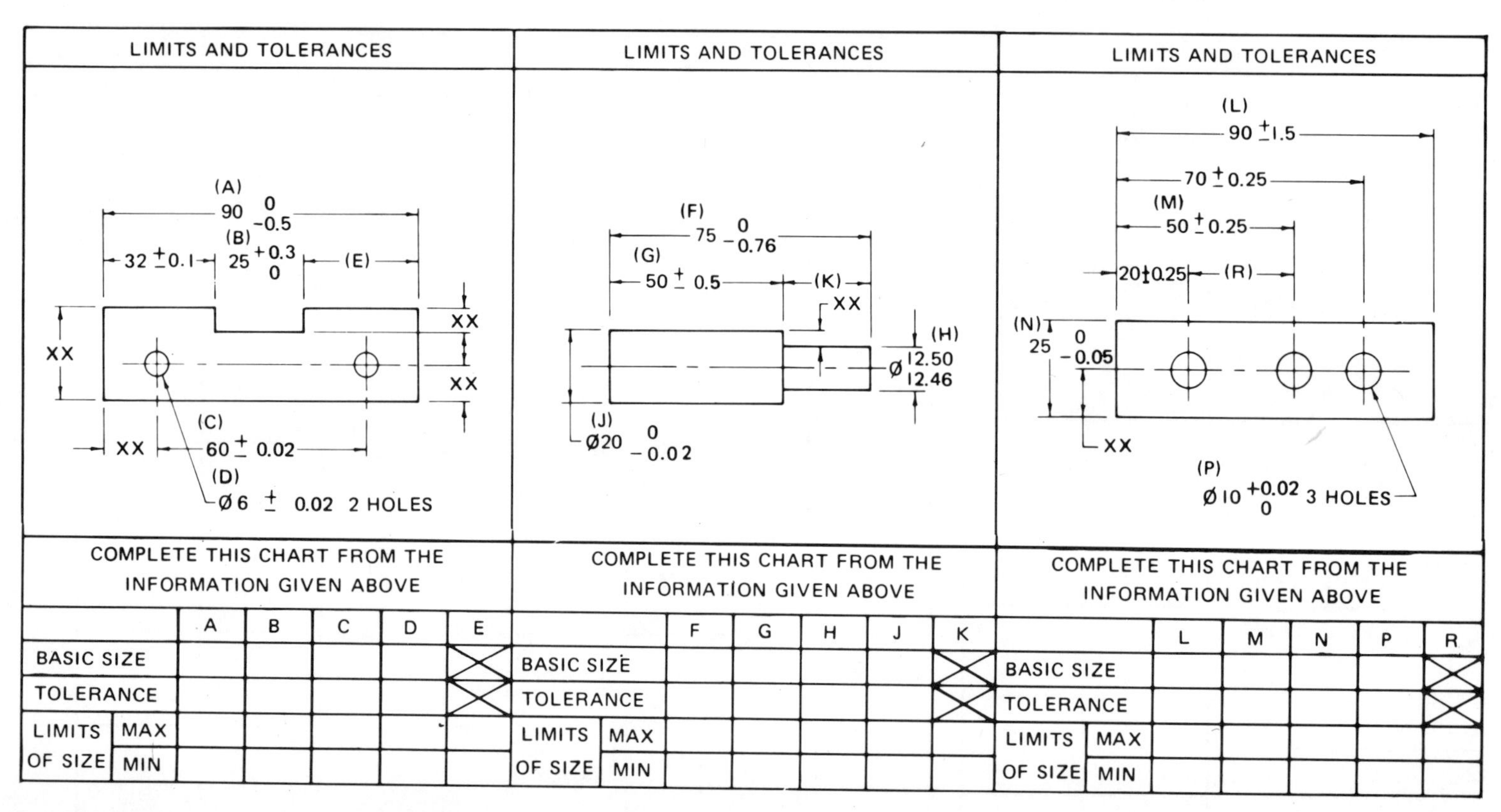

COMPLETE THIS CHART FROM THE INFORMATION GIVEN ABOVE

		A	B	C	D	E
BASIC SIZE						
TOLERANCE						
LIMITS OF SIZE	MAX					
	MIN					

COMPLETE THIS CHART FROM THE INFORMATION GIVEN ABOVE

		F	G	H	J	K
BASIC SIZE						
TOLERANCE						
LIMITS OF SIZE	MAX					
	MIN					

COMPLETE THIS CHART FROM THE INFORMATION GIVEN ABOVE

		L	M	N	P	R
BASIC SIZE						
TOLERANCE						
LIMITS OF SIZE	MAX					
	MIN					

Fig. 12-1-C Limits and tolerance assignment

UNIT 12-2
Positional Tolerancing

The location of features is one of the most frequently used applications of dimensions on technical drawings. Tolerancing may be accomplished either by **coordinate tolerances** applied to the dimensions or by **geometric (positional) tolerancing.**

Positional tolerancing is especially useful as this method meets functional requirements in most cases and permits assessment with simple gauging procedures.

It is necessary, however, first to look at the widely used method of coordinate tolerancing in order to explain and understand the advantages and disadvantages of the positional tolerancing methods.

Coordinate dimensions and tolerances may be applied to the location of a single hole, as shown in Figure 12-2-1. They indicate the location of the hole axis and result in a rectangular or wedge-shaped tolerance zone within which the axis of the hole must lie.

If the two coordinate tolerances are equal, the tolerance zone formed will be a square. Unequal tolerances result in a rectangular tolerance zone. Where one of the locating dimensions is a radius, polar dimensioning gives a circular ring section tolerance zone. For simplicity, square tolerance zones are used in the analyses of most of the example in this section.

It should be noted that the tolerance zone extends for the full depth of the hole, that is, the whole length of the axis. This is illustrated in Figure 12-2-2. In most of the illustrations, tolerances will be analyzed as they apply at the surface of the part, where the axis is represented by a point.

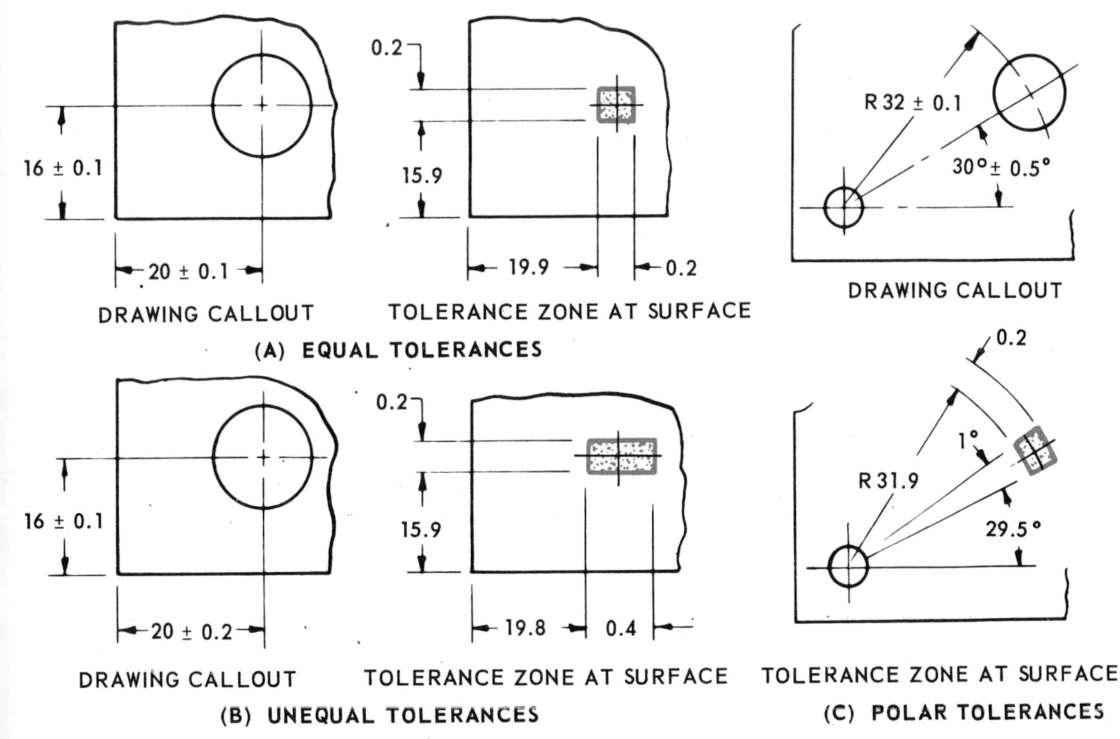

Fig. 12-2-1 Tolerance zones for coordinate tolerances

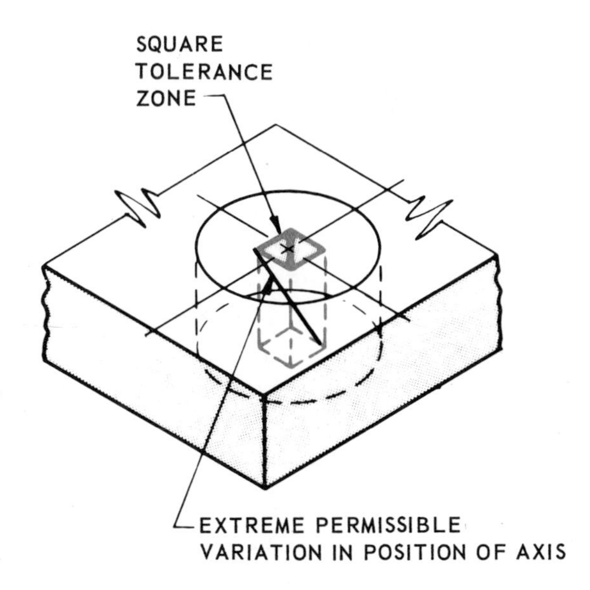

Fig. 12-2-2 Tolerance zone extending through part

Maximum Permissible Error in Coordinate Tolerancing

The actual position of the feature axis may be anywhere within the rectangular tolerance zone. For square tolerance zones, the maximum allowable variation from the desired position occurs in a direction of 45° from the direction of the coordinate dimensions. See Figure 12-2-3.

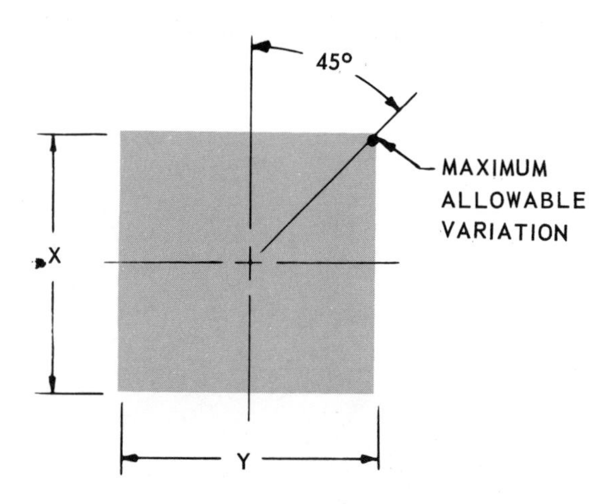

Fig. 12-2-3 Square tolerance zone

For rectangular tolerance zones this maximum tolerance is the square root of the sum of the squares of the individual tolerances, or expressed mathematically

$$\sqrt{X^2 + Y^2}$$

For the examples shown in Figure 12-2-1 the tolerance zones are shown in Figure 12-2-4 and the maximum tolerance values are as shown in the following examples.

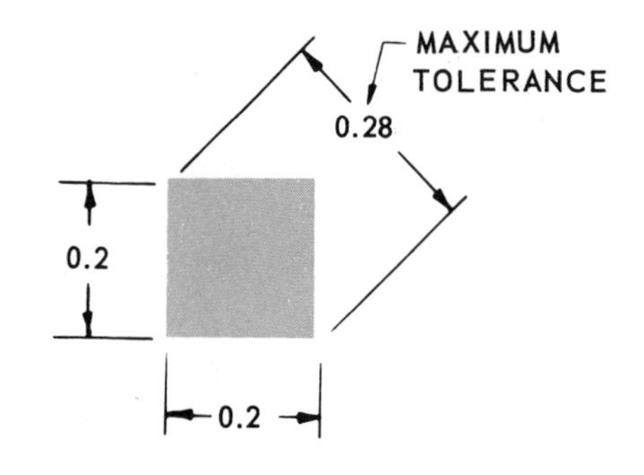

EXAMPLE 1

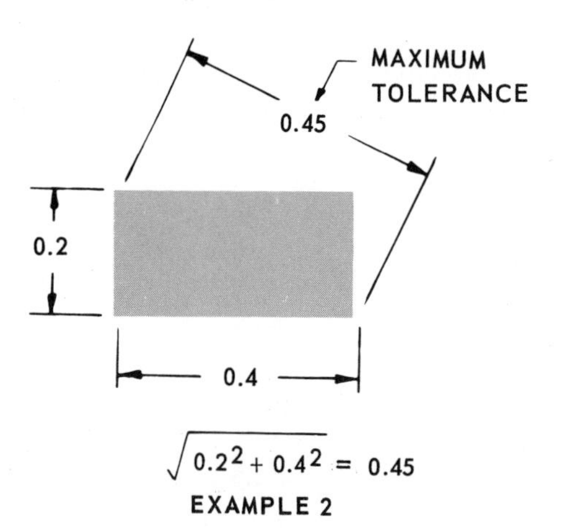

$$\sqrt{0.2^2 + 0.4^2} = 0.45$$

EXAMPLE 2

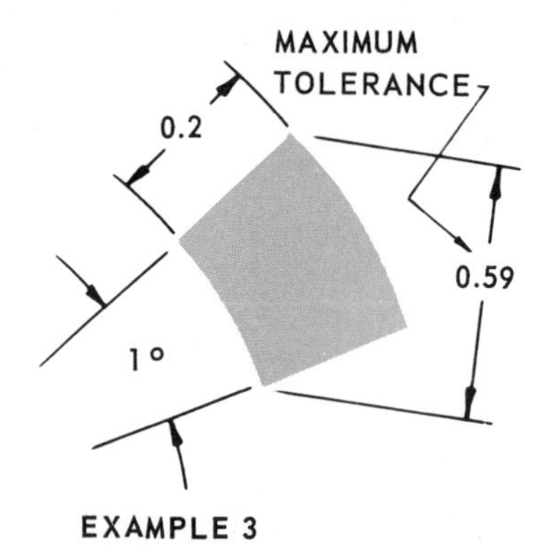

EXAMPLE 3

Fig. 12-2-4 Tolerance zones for parts shown in Fig. 12-2-1

Example 1.

$$\sqrt{0.2^2 + 0.2^2} = 0.28$$

Example 2.

$$\sqrt{0.2^2 + 0.4^2} = 0.45$$

Example 3. For polar coordinates the extreme variation is

$$\sqrt{A^2 + T^2}$$

where **A** = **R** Tan **a**
T = tolerance on radius
R = mean radius
a = angular tolerance

Thus, the extreme variation in the third example is

$$\sqrt{(32 \times 0.017\,45)^2 \times 0.2^2} = 0.59$$

Use of Chart

A quick and easy method of finding the maximum positional error permitted with coordinate tolerancing, without having to calculate squares and square roots, is by use of a chart like that shown in Figure 12-2-5.

In the first example shown in Figure 12-2-1, the tolerance in both directions is 0.2 mm. The extensions of the horizontal and vertical lines of 0.2 in the chart intersect at point *A*, which lies between the radii of 0.275 and 0.3 mm. When interpolated and rounded to two decimal places, a maximum permissible variation of position is 0.28 mm.

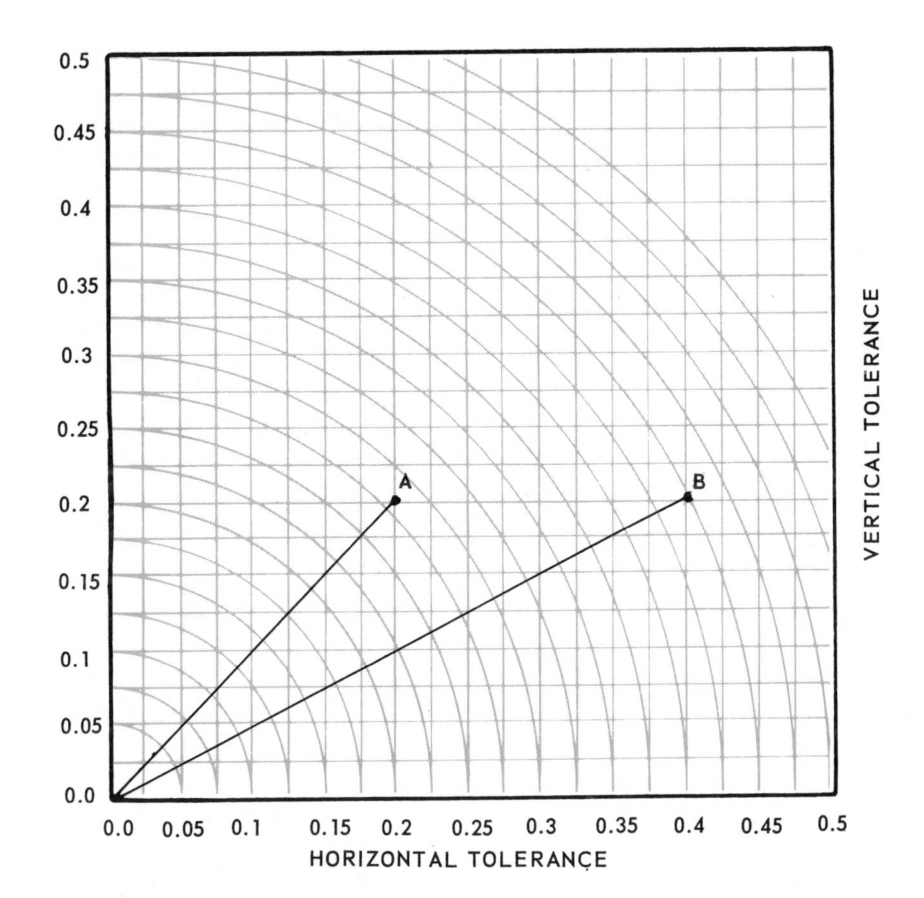

Fig. 12-2-5 Chart for calculating maximum tolerance using coordinate tolerancing

In the second example shown in Figure 12-2-1, the tolerances are 0.2 mm in one direction and 0.4 mm in the other. The extensions of the vertical and horizontal lines at 0.2 and 0.4 mm respectively in the chart intersect at point *B*, which lies between the radii of 0.425 and 0.45 mm. When interpolated and rounded to two decimal places, a maximum variation of position is 0.45 mm.

Advantages of Coordinate Tolerancing

The advantages claimed for direct coordinate tolerancing are as follows.

1. It is simple and easily understood, and therefore it is commonly used.

2. It permits direct measurements to be made with standard instruments and does not require the use of special-purpose functional gauges or other calculations.

Disadvantages of Coordinate Tolerancing

There are a number of disadvantages to the direct tolerancing method:

1. It results in a square or rectangular tolerance zone within which the axis must lie. For a square zone this permits a variation in a 45° direction of approximately 1.4 times the specified tolerance. This amount of variation may necessitate the specification of tolerances which are only 70% of those that are functionally acceptable.

2. It may result in an undesirable accumulation of tolerances when several features are involved, especially when chain dimensioning is used.

3. It is more difficult to assess clearances between mating features and components than when positional tolerancing is used.

Positional Tolerancing

Positional tolerancing is part of the system of geometric tolerancing. In this system the location of features is shown either by coordinate dimensions or by polar and angular dimensions, except that the dimensions are shown without direct tolerances. These dimensions represent the basic sizes and are commonly known as **true-position** dimensions. Each such dimension is enclosed in a rectangular frame to indicate that it represents an exact value, to which tolerances shown in the general tolerance note do not apply. See Figure 12-2-6. The frame size need not be any larger than that necessary to

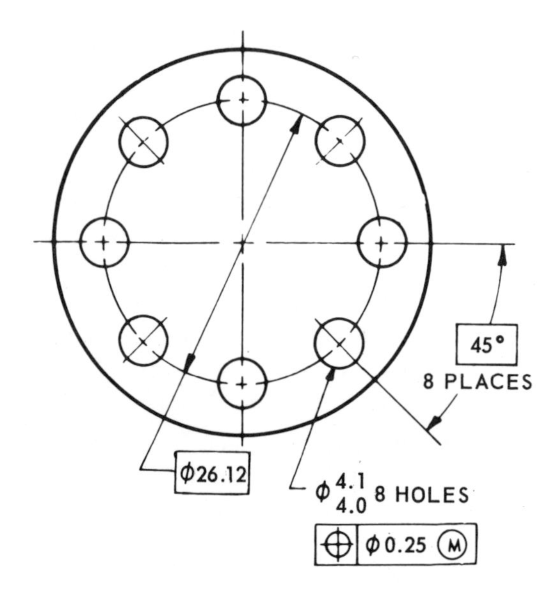

Fig. 12-2-6 Identifying true position dimensions

enclose the dimension. Permissible deviations from true position are then indicated by a positional tolerance as described in this unit.

Symbol for Position

The geometric characteristic symbol for position is a circle and two solid centre lines, as shown in Figure 12-2-7.

As positional tolerancing controls the position of the axis of the hole, the feature control symbol is attached to the size of the feature, as shown in Figure 12-2-8.

The positional tolerance represents the diameter of a cylindrical tolerance zone, located at true position as determined by the true-position dimensions on the drawing, within which the axis or centre line of the hole must lie.

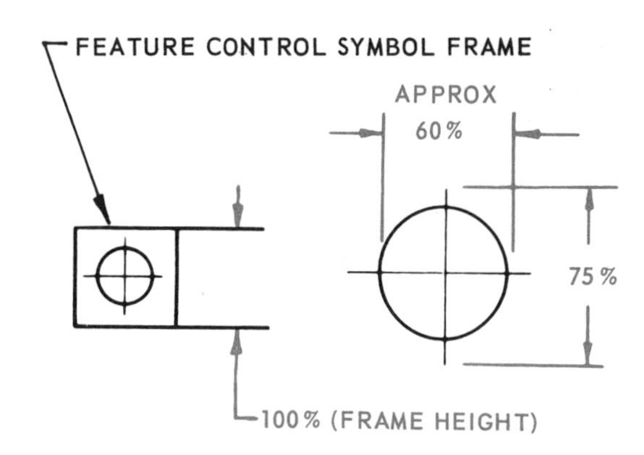

Fig. 12-2-7 Position symbol

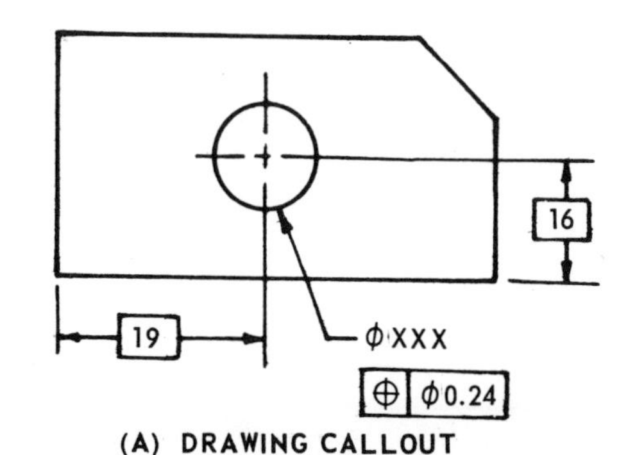

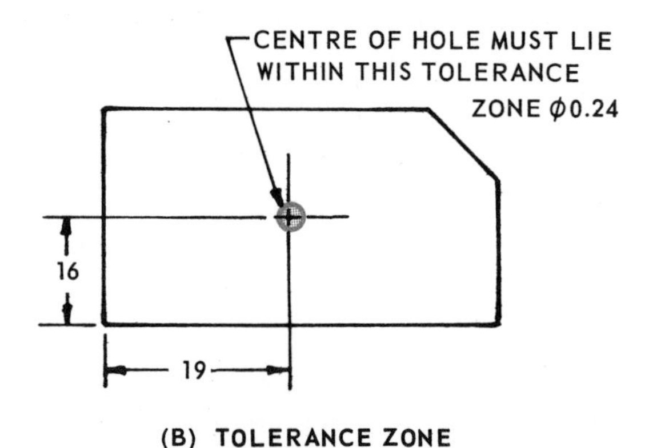

Fig. 12-2-8 Positional tolerancing - regardless of feature size

Except for the fact that the tolerance zone is circular instead of square, a positional tolerance on this basis has exactly the same meaning as direct coordinate tolerancing with equal tolerances in both directions.

It has already been shown that with rectangular coordinate tolerancing the maximum permissible error in location is not the value indicated by the horizontal and vertical tolerances, but rather is equivalent to the length of the diagonal between the two tolerances. For square tolerance zones this amount is 1.4 times the specified tolerance values. If the same tolerance as shown for a square coordinate tolerance is specified for a positional tolerance (circular), the total area of the tolerance zone is increased and

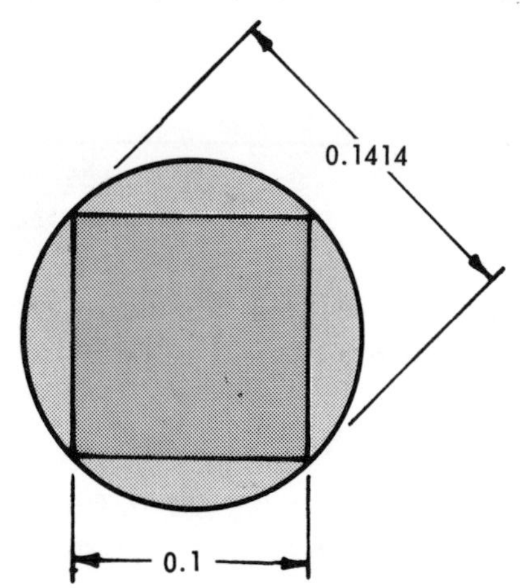

Fig. 12-2-9 Relationship of tolerance zones

the diameter of the tolerance zone is equal to the diagonal of the coordinate tolerance.

This does not affect the clearance between the hole and its mating part, yet it offers 57% more tolerance area, as shown in Figure 12-2-9. Such a change would most likely result in a reduction in the number of parts rejected for positional errors.

Larger variations in hole positions can be obtained if the tolerance is based on the Maximum Material Condition, the theory of which is beyond the scope of this text.

Assignment

On an A3 sheet, prepare a detail drawing of the Guide Block shown in Fig. 12-2-A. Replace the coordinate tolerances for the holes with positional tolerancing. Scale 1:1.

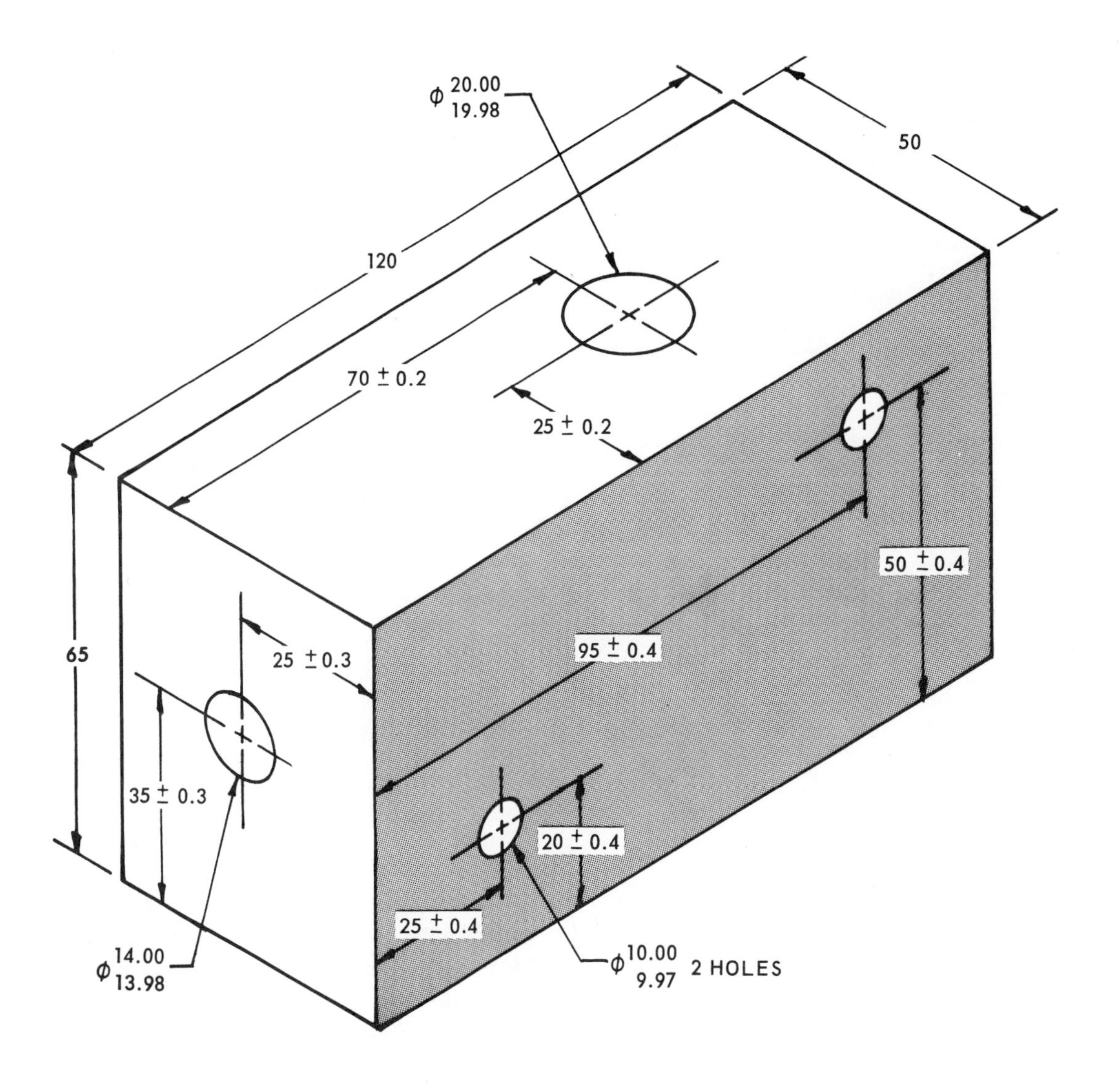

Fig. 12-2-A Guide block

Chapter 13
Drawing for Numerical Control

UNIT 13-1
Two-Axis Control Systems

In the 1980 decade, and those to follow, Canadian industry will face not only new opportunities, but also increased competition for current and future markets, both at home and abroad.

In this context, the rapidly emerging use of Computer Aided Design and Computer Aided Manufacturing (CAD/CAM) technology is of special importance.

The technical and economic factors associated with CAD/CAM systems will cause the technology to be adopted in industry at a rapid rate. There is, therefore, a degree of urgency, for competitive reasons, to include this important technology in the drafting courses of secondary schools and college programmes.

Numerical control is a means of directing some of or all the functions of a machine automatically from instructions. The instructions are generally stored on tape and are fed to the controller through a tape reader. The controller interprets the coded instructions and directs the machine through the required operations.

It has been established that because of the consistent, high accuracy of numerically controlled machines, and because human errors have been almost entirely eliminated, scrap has been considerably reduced.

Because both setup and tape preparation times are short, numerically controlled machines produce a part faster than manually controlled machines.

When changes become necessary on a part, they can easily be implemented by changing the original tapes. The process takes very little time and expense in comparison to the alterations of a jig or fixture.

Another area where numerically controlled machines are better is in the quality or accuracy of the work. In many cases a numerically controlled machine can produce parts more accurately at no additional cost, resulting in reduced assembly time and better interchangeability of parts. This latter fact is especially important when spare parts are required.

Acceptance of numerical control processes is being accelerated by the development of computer-aided design and computer-aided manufacturing techniques.

Dimensioning for Numerical Control

Common guidelines have been established that enable dimensioning and tolerancing practices to be used effectively in delineating parts for both numerical control and conventional fabrication.

Coordinate System

The numerical control concept is based on the system of rectangular or cartesian coordinates in which any position can be described in terms of distance from an origin point along either two or three mutually perpendicular axes. Two dimensional coordinates (*X*, *Y*) define points in a plane. See Figure 13-1-1.

The *X* axis is horizontal and is considered the first and basic reference axis. Distances to the right of the zero *X* axis are considered positive *X* values and to the left of the zero *X* axis hasnegative *X* values.

The *Y* axis is vertical and perpendicular to the *X* axis in the plane of a drawing showing *XY* relationships. Distances above the zero *Y* axis are considered positive *Y* values and below the zero *Y* axis as negative *Y* values. The position where the *X* and *Y* axes cross is called the **origin,** or **zero point.**

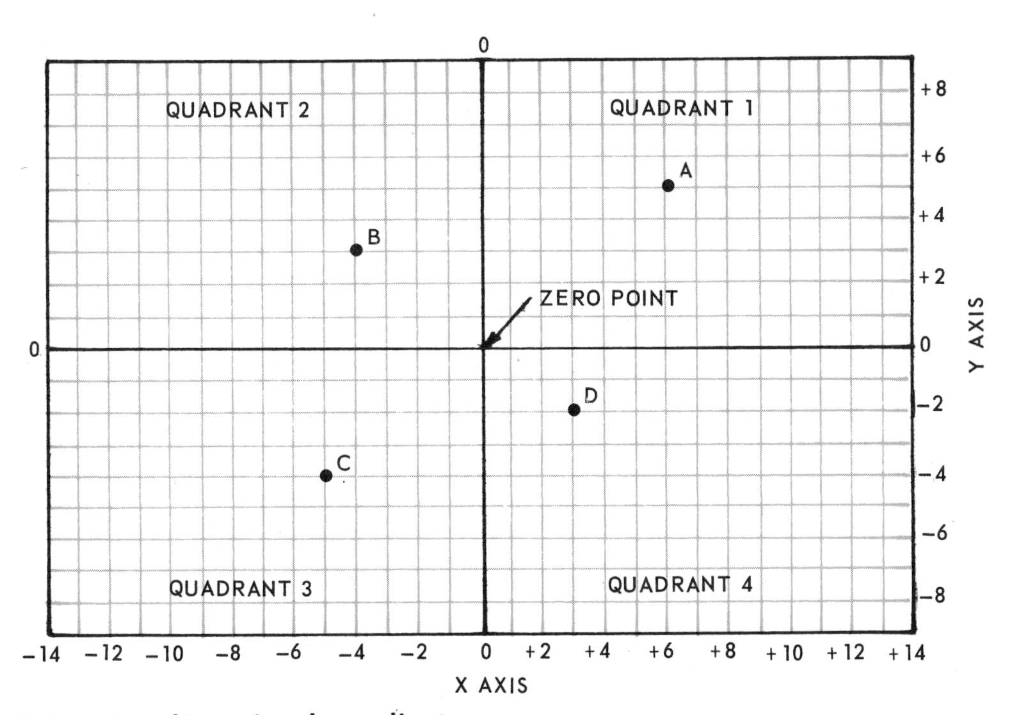

Fig. 13-1-1 Two dimensional coordinates

For example, four points lie in a plane, as shown in Figure 13-1-1. The plane is divided into four quadrants. Point *A* lies in quadrant 1 and is located at position (6, 5), with the *X* coordinate first, followed by the *Y* coordinate. Point *B* lies in quadrant 2 and is located at position (−4, 3). Point *C* lies in quadrant 3 and is located at position (−5, −4). Point *D* lies in quadrant 4 and is located at position (3, −2).

Designing for numerical control would be greatly simplified if all work were done in the first quadrant because all the values would be positive and the plus and minus signs would not be required. However, any of the four quadrants may be used, and, as such, programming in any of the quadrants should be understood.

Some numerically controlled machines are designed for locating points in only the *X* and *Y* directions. These are called **two-axis machines.**

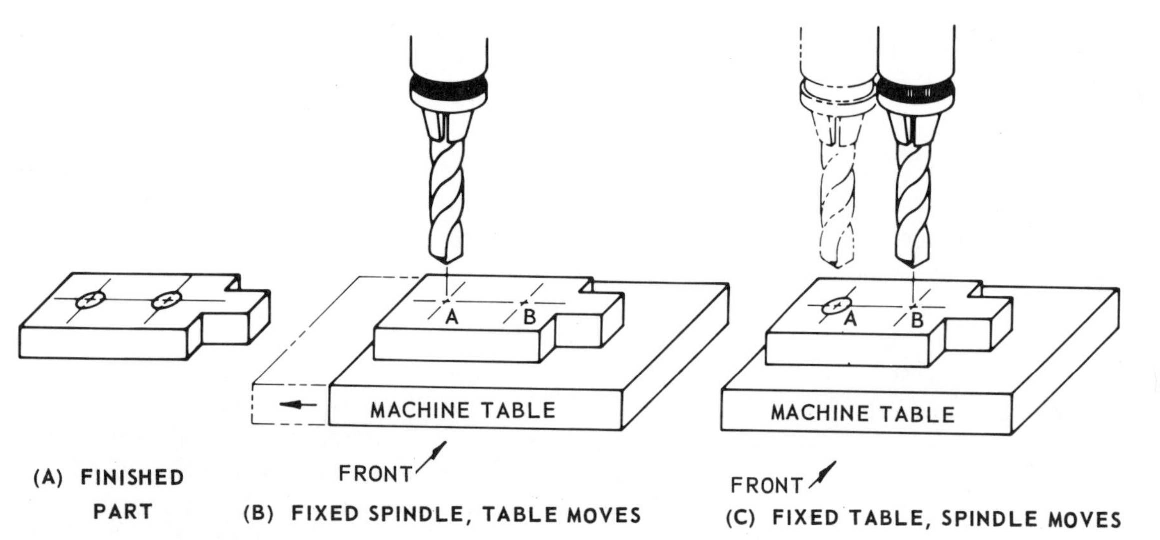

Fig. 13-1-2 Positioning the work

The function of these machines is to move the machine table or tool to a specified position in order to perform work, as shown in Figure 13-1-2. With the fixed spindle and movable table as shown in Figure 13-1-2*b*, hole *A* is drilled. Then the table moves to the left, positioning point *B* below the drill. This is the most frequently used method. With the fixed table and movable spindle as shown in Figure 13-1-2*c*, hole *A* is drilled. Then the spindle moves to the right, positioning the drill above point *B*. This changes the direction of the motion, but the movement of the cutter as related to the work remains the same.

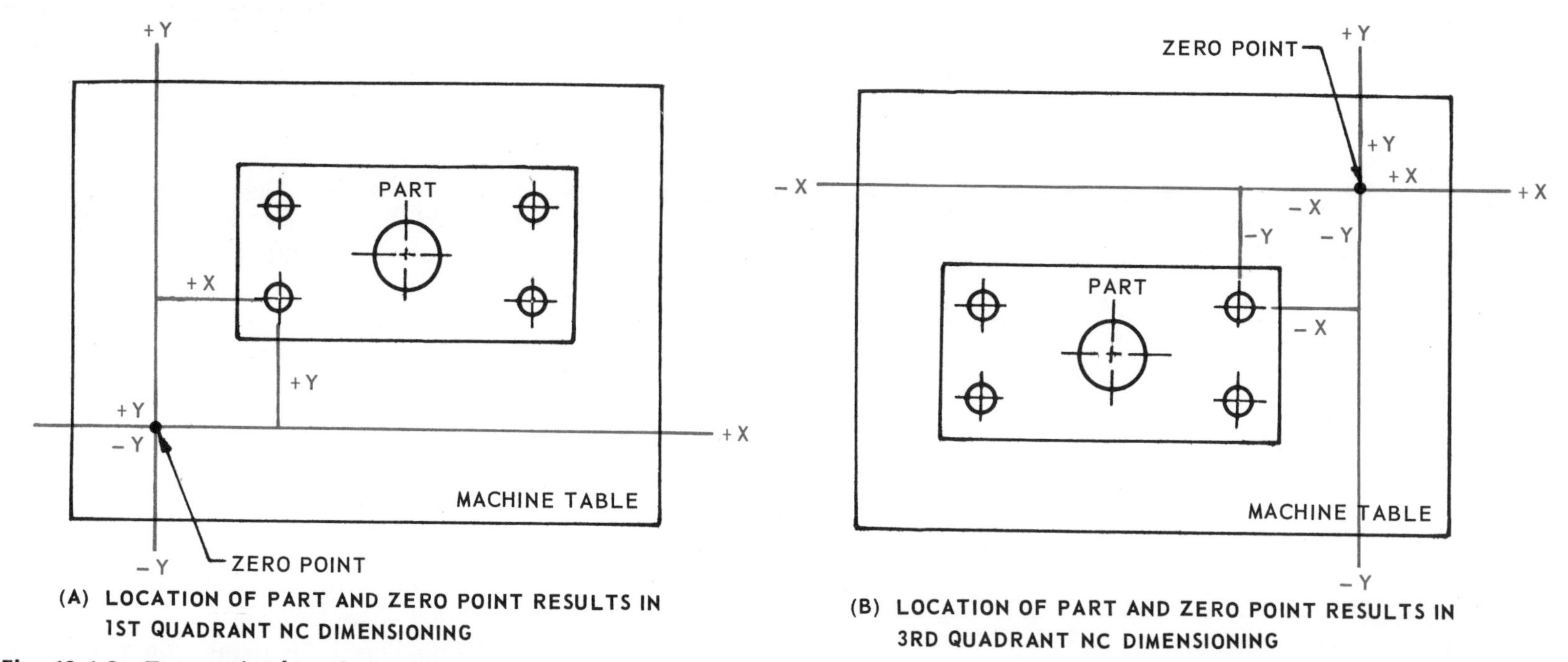

Fig. 13-1-3 Zero point location

Zero Point

As previously mentioned, the zero point is the point where the *X* and *Y* axes intersect. It is the point where all coordinate dimensions are measured. Many machines have a fixed zero point built in.

Two examples of fixed zero points on machine tables are shown in Figure 13-1-3. In Figure 13-1-3*a* all points are located in the first quadrant, resulting in positive *X* and *Y* values. In Figure 13-1-3*b* all points are located in the third quadrant, resulting in negative *X* and *Y* values.

Setup Point

The setup point is located on the part or the fixture holding the part. It may be the intersection of two finished surfaces, the centre of a previously machined hole in the part, or a feature of the fixture. It must be accurately located in relation to the zero point, as shown in Figure 13-1-4.

Point-to-Point Programming

Point-to-point programming is the most common type of positioning system. With this system each new positioning is given from the last position. To compute the next position wanted, it is necessary to establish the sequence in which the work is to be done.

An example of this type of dimensioning is shown in Figure 13-1-4*a*. The distance between the left edge of the part and hole 1 is given as 20 mm. From hole 1 to hole 4 the dimension shown is 120 mm (*X* axis), and from hole 1 to hole 2 the dimension shown is 40 mm (*Y* axis). These dimensions give the distance from the last hole drilled to the next drilled hole. Assume the holes are to be drilled in the sequence shown in the figure. Hole 1 is located (70, 70) from the zero point. After hole 1 has been drilled, the drill spindle is positioned above the centre of hole 2. Hole 2 has the same *X*-coordinate dimension as hole 1, making the *X* increment zero. Since the vertical distance between holes 1 and 2 is 40 mm, the *Y* increment becomes +40.

After hole 2 is drilled, the drill spindle is positioned above the centre of hole 3. Since the horizontal distance between holes 2 and 3 is 120 mm, the *X* increment is +120. Hole 3 has the same *Y*-coordinate dimension as hole 2, making the *Y* increment zero.

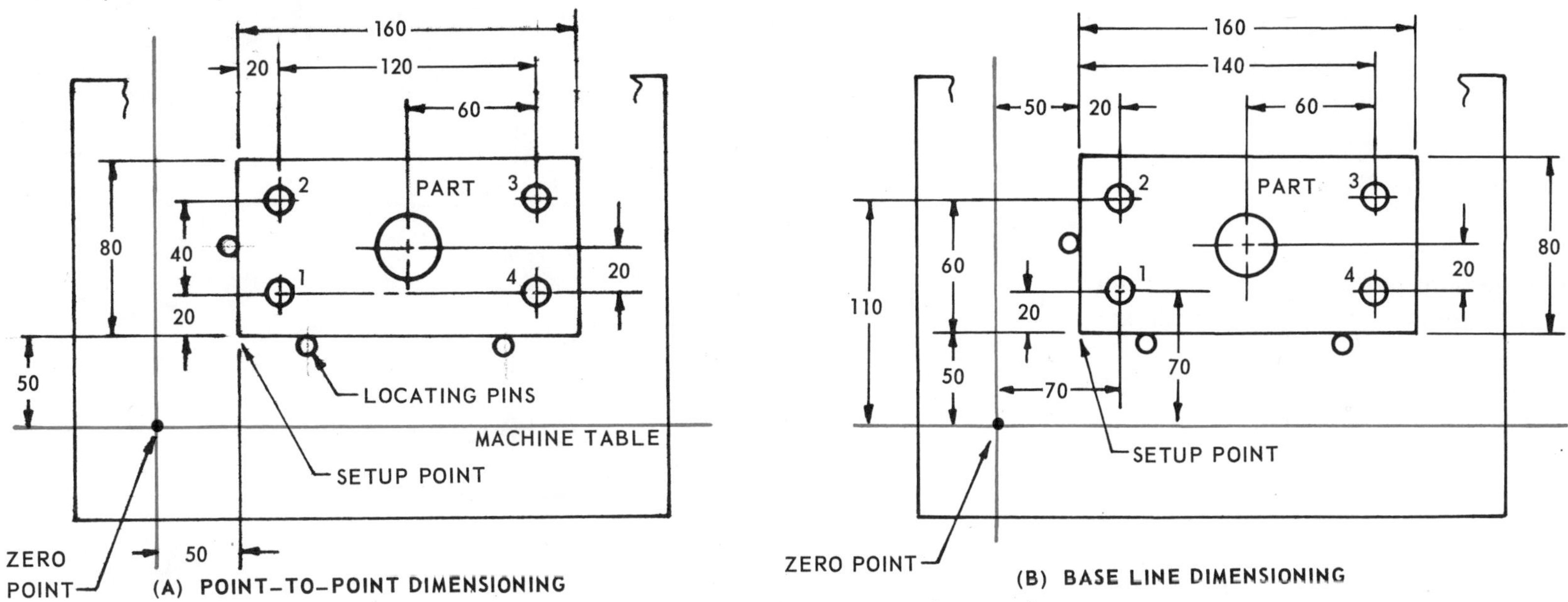

Fig. 13-1-4 Dimensioning for numerical control

From hole 3 the drill spindle is positioned above the centre of hole 4. Hole 4 has the same *X*-coordinate dimension as hole 3, making the *X* increment zero. Since the vertical distance between hole 3 and hole 4 is 40 mm, the *Y* increment is −40.

Figure 13-1-5 lists the distance between holes and indicates the direction of motion by plus and minus signs. It can be seen that each pair of coordinates is the distance between the two locations.

HOLE	X	Y
1	+ 70	+ 70
2	0	+ 40
3	+ 120	0
4	0	− 40

Fig. 13-1-5 Point-to-point dimensioning of holes shown in Fig. 13-1-4

Coordinate Programming

Many machines use coordinate programming instead of the point-to-point method of dimensioning. With this type of dimensioning all dimensions are taken from the zero point; as such, base-line dimensioning, as shown in Figure 13-1-4*b*, is used. For example, after hole 1 is drilled, then the drill spindle has to be positioned above the centre of hole 2. The coordinates for hole 2 are (70, 110). Figure 13-1-6 shows the coordinate dimensions of the holes shown in Figure 13-1-4.

HOLE	X	Y
1	+ 70	+ 70
2	+ 70	+ 110
3	+ 190	+ 110
4	+ 190	+70

Fig. 13-1-6 Coordinate dimensioning of holes shown in Fig. 13-1-4

Assignments

1. On an A3 size sheet, prepare two drawings of the cover plate shown in Fig. 13-1-A. One drawing is to use point-to-point dimensioning for the 10 holes; the other drawing is to use coordinate dimensioning. Only the dimensions locating the holes need be shown. The radial and angular dimensions are to be replaced with coordinate dimensions and taken to two decimal places. Below

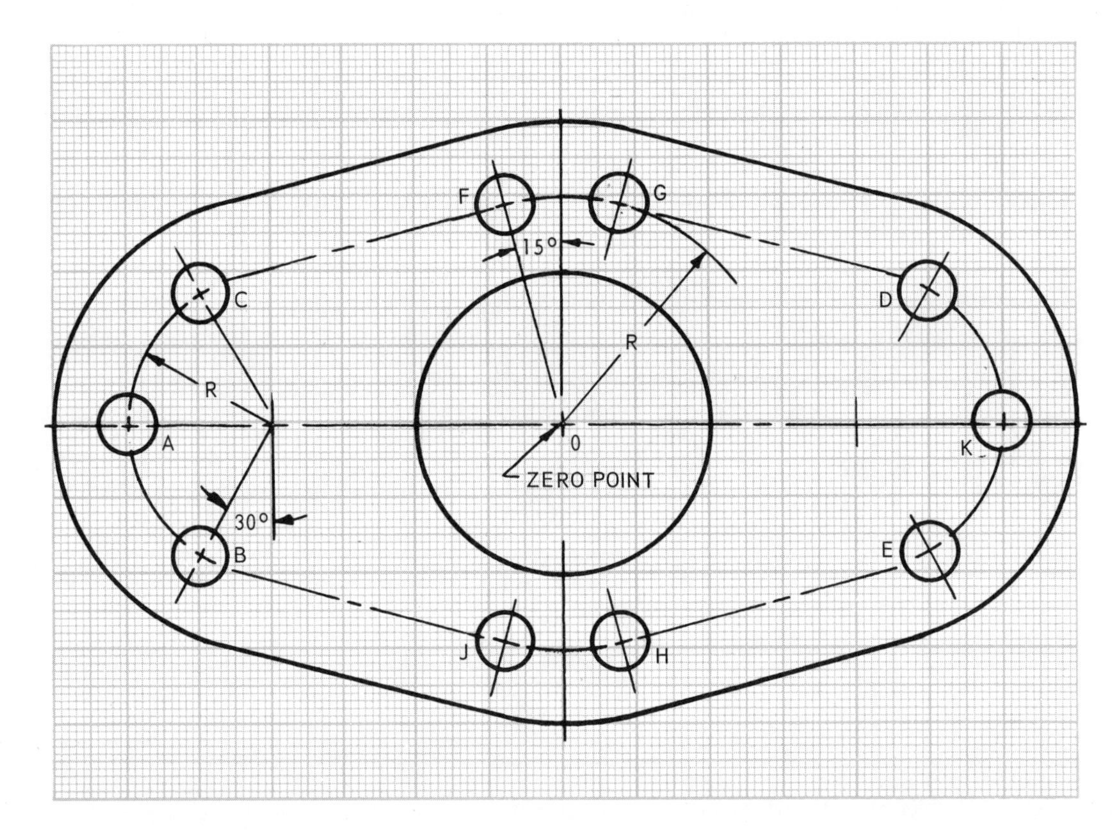

Fig. 13-1-A Cover plate

each drawing prepare a chart listing each hole and their X and Y coordinates. The letters shown on the holes show the sequence in which they are to be drilled. The centre of the part is the zero point. Scale is 1:1.

2. On An A3 size sheet, prepare a chart listing the X and Y (coordinate) locations and the quadrant for the points A to V shown on Fig. 13-1-B. The grid is 10 × 10 to the centimetre.

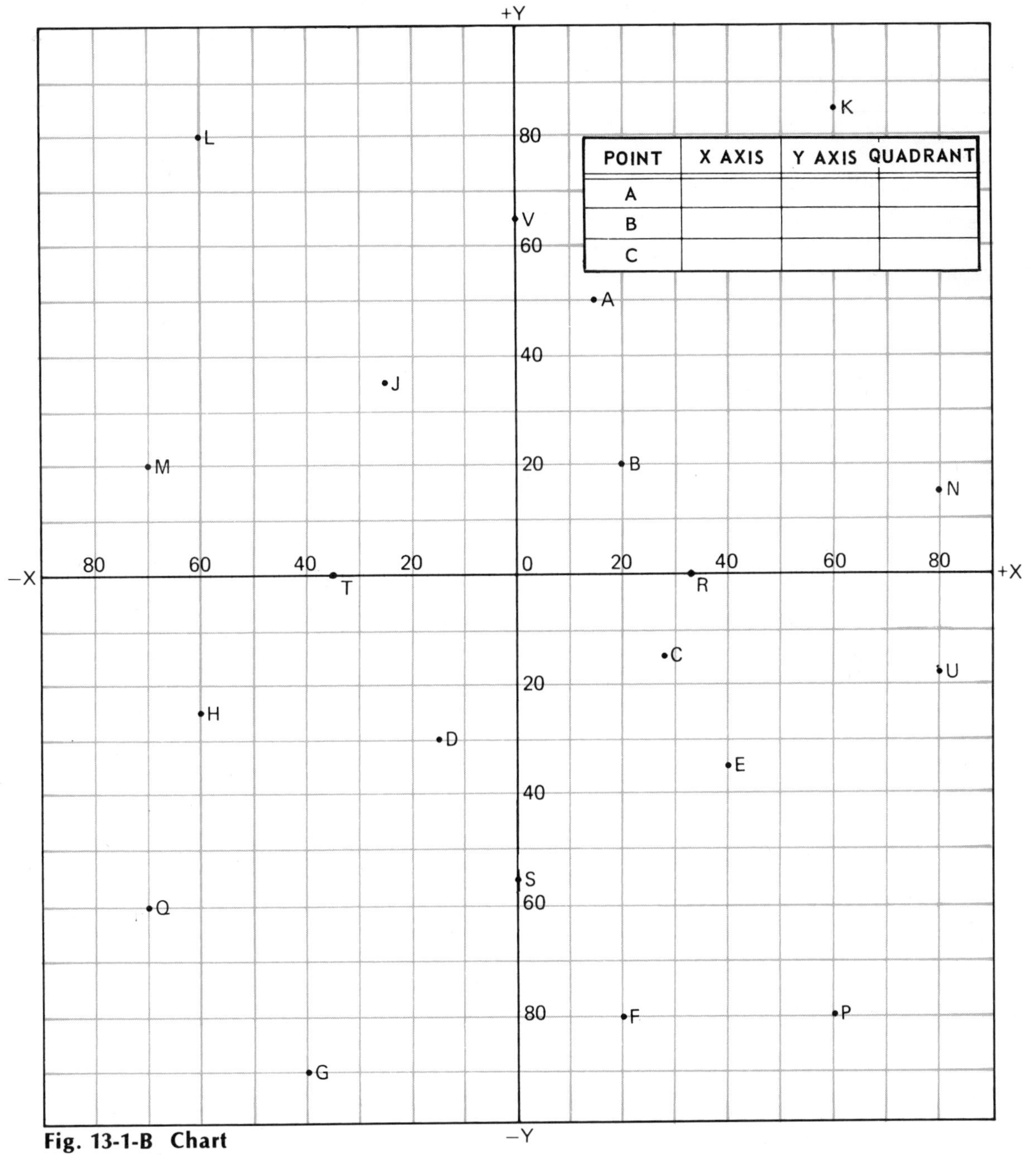

POINT	X AXIS	Y AXIS	QUADRANT
A			
B			
C			

Fig. 13-1-B Chart

UNIT 13-2
Three-Axis Control Systems

Many numerically controlled machines operate in three directions, the table and carriage moving in the *X* and *Y* directions, as explained in Unit 13-1, and the tool spindle, such as a turret drill, traveling in an up-and-down direction. A vertical line taken through the centre of the machine spindle is referred to as the *Z* axis and is perpendicular to the plane formed by the *X* and *Y* axes. See Figure 13-2-1.

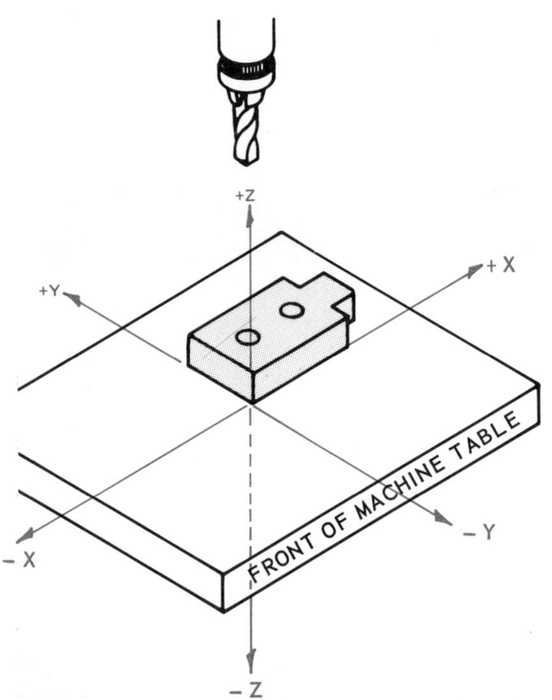

Fig. 13-2-1 X, Y, and Z axis

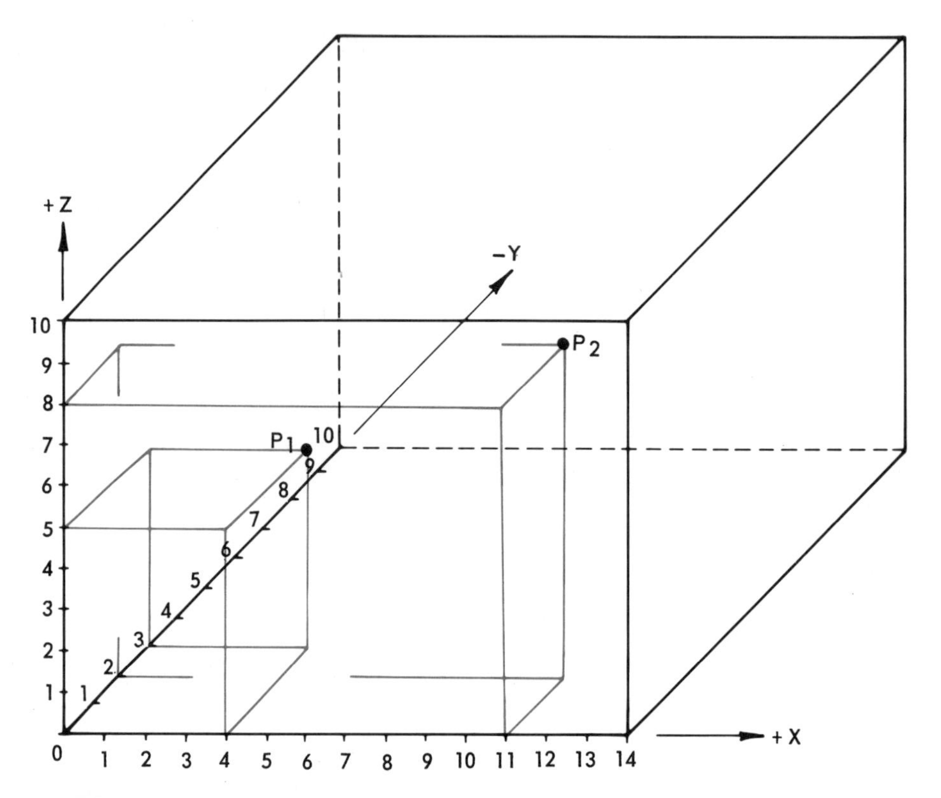

Fig. 13-2-2 Points in space

Thus, a point in space can be described by its *X*, *Y*, and *Z* coordinates. For example, P_1 in Fig. 13-2-2 can be described by its (*X*, *Y*, *Z*) coordinates as (4, 3, 5) and P_2 as (11, 2, 8).

A popular system used on many numerically controlled machines, such as the turret drill, is to establish the *Z* zero reference plane above the workpiece. Each tool is then adjusted and calibrated to the *Z* zero reference plane.

For example, Figure 13-2-3 shows a part requiring three drilled holes. As the centre hole is drilled through, the part is raised by gauge blocks so that the drill does not touch the machine table. The height of the gauge blocks is determined by the distance the drill passes through the workpiece plus clearance, or $2 + 0.3D + 4$. See Figure 13-2-4. If a 20 mm drill were used, the gauge block height would be $2 + 6 + 4 = 12$ mm.

If the distance from the top of the workpiece to the *Z* zero reference plane is set at 20 mm, the *Z* coordinates for the three holes shown are $-(20 + A)$, $-(20 + B)$, and $-(20 + C)$.

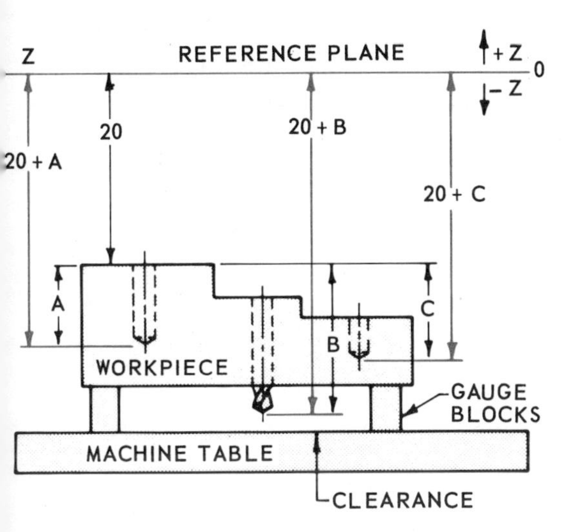

Fig. 13-2-3 Calculating Z distance

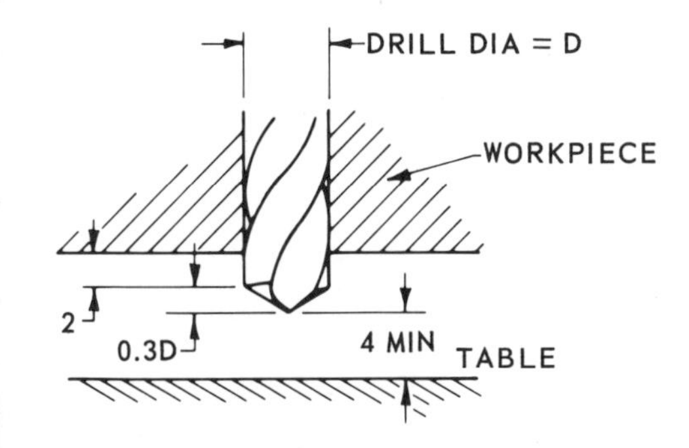

Fig. 13-2-4 Determining gauge block distances

Dimensioning and Tolerancing

Recommended guidelines for dimensioning and tolerancing practices for use in defining parts for numerical control fabrication are as follows:

1. When the basic coordinate system is established, the setup point should be placed at an appropriate location on the part itself.

2. Any number of subcoordinate systems may be used to define features of a part as long as these systems can be related to the basic coordinate system of the given part.

3. Define part surfaces in relation to three mutually perpendicular reference planes. Establish these planes along part surfaces which parallel the machine axes if these axes can be predetermined.

4. Dimension the part precisely so that the physical shape can be readily determined. Dimension to points on the part surfaces.

5. Regular geometric contours such as ellipses, parabolas, hyperbolas, etc., may be defined on the drawing by mathematical formulas. The numerically controlled machinery can easily be programmed to approximate these curves by linear interpolation, that is, as a series of short, straight lines whose endpoints are close enough together to ensure meeting the required tolerances for contour. Consideration should be given to the number of points needed to define the curve; however, one should keep in mind the fact that the tighter the tolerance or the smaller the radius of curvature, the closer together the points should be.

6. Holes in a circular pattern should be located with coordinate dimensions preferably.

7. Express angular dimensions, where possible, relative to the *X* axis in degrees and decimal parts of a degree.

8. Use plus and minus tolerances, not limit dimensions. Preferably, the tolerance should be equally divided bilaterally.

9. Tolerances are specified only on the basis of actual design requirements. The accuracy capability of numerically controlled equipment is not a basis for specifying more restrictive tolerances than functionally required.

Assignments

1. On an A3 size sheet make a two-view drawing of the end plate shown in Fig. 13-2-A. Point-to-point dimensioning is to be used, and only the dimensions locating the holes need be shown. Below the drawing prepare a chart listing each hole and its *X*, *Y*, and *Z* coordinates. The letters on the holes show the sequence in which they are to be drilled. Calculating the *Z* coordinate is to be done in the same manner as for the part shown in Fig. 13-2-3. Scale is 1:1.

   Note: Programming will be for the tap-drill holes and the six through holes shown. Zero point for the *X* and *Y* coordinates is the centre of the end plate.
2. On an A4 size sheet accurately lay out the chart shown in Fig. 13-2-B to the scale 1:1. Locate points *P*1 to *P*10 on the chart.

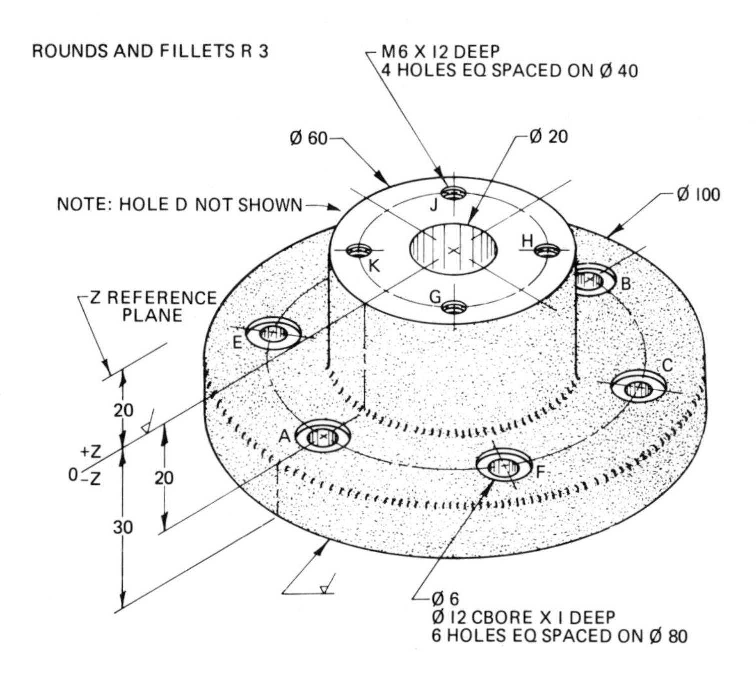

Fig. 13-2-A End plate

3. Lay out the chart shown in Fig. 13-2-C. Points *P*1 to *P*10 are located, but only one of their coordinates is known. Accurately lay out the missing coordinates on the chart and record their values in the table. Use the scale shown on the chart.

NOTE: ALL COORDINATES ARE +

POINT	X AXIS	Y AXIS	Z AXIS
P1	10	20	60
P2	20	70	70
P3	0	20	30
P4	100	60	75
P5	40	30	20
P6	40	60	10
P7	70	20	0
P8	50	50	50
P9	85	65	30
P10	60	65	15

Fig. 13-2-B Chart

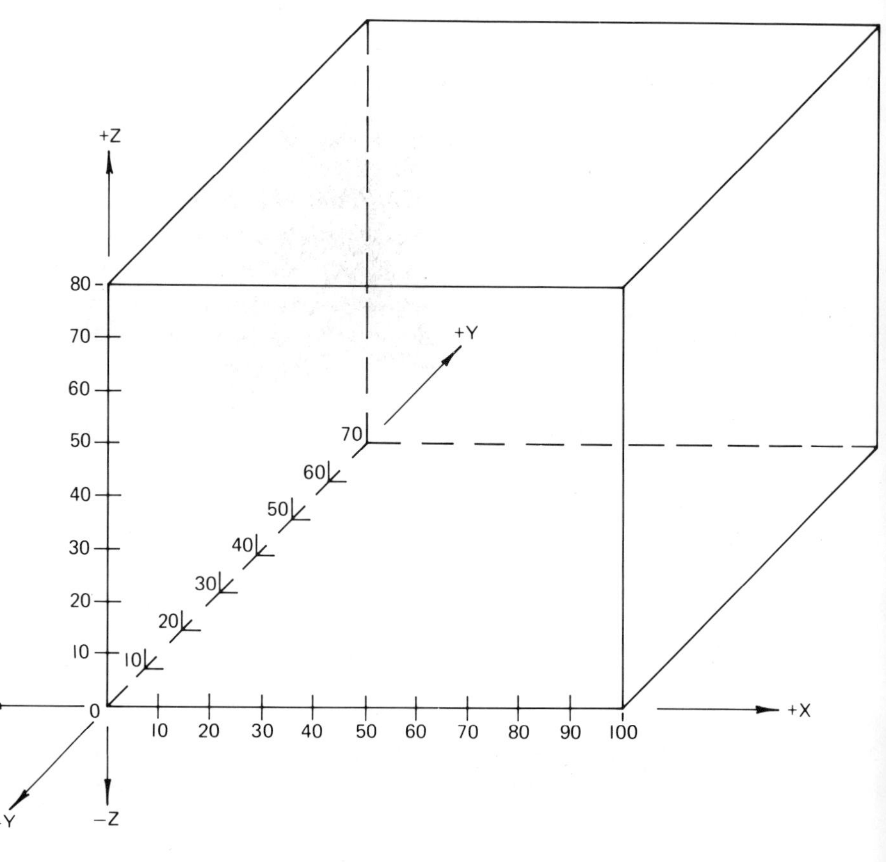

NOTE: ALL COORDINATES ARE +

POINT	X AXIS	Y AXIS	Z AXIS
PI	20		
P2			55
P3		60	
P4	30		
P5		30	
P6			0
P7			40
P8	65		
P9		30	
PI0			55

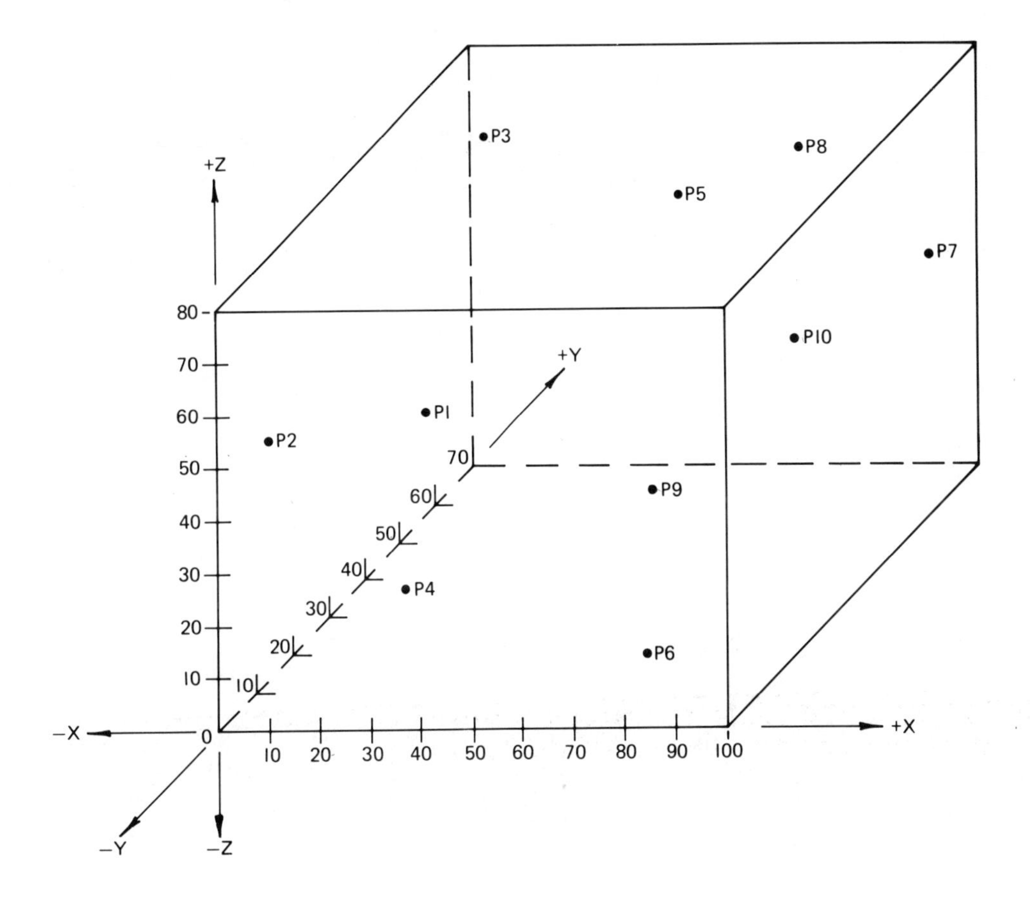

Fig. 13-2-C Chart

Chapter 14
Functional Drafting

UNIT 14-1
Drawing Aids and Drawing Practices

The basic function of the drafting department is to provide sufficient information to produce or assemble parts. Functional drafting, therefore, must embrace every possible means to communicate this information in the least expensive manner.

Functional drafting also applies to any method which would lower the cost of producing the part. New technological developments have provided many new ways of producing drawings at lower costs and/or in less time. This means that the drafting office must be prepared to discard some of the old, traditional methods in lieu of these newer means of communication.

There are many ways in which to reduce the drafting time in preparing a drawing. These drawing shortcuts, when collectively used, are of prime importance in an effective drafting system.

Drawing Aids

Numerous time-saving devices are available: templates for every application, "pens" for easier linework, and more application of tape to artwork, transfer-type lettering, etc. Since drafting applications vary so widely, only the drafting supervisor can determine which devices will increase the drafting production.

Drafting aids are designed to facilitate the making of drawings by removing or reducing some of the more tedious aspects of drafting.

Templates. Templates play an important part in functional drafting, for they save a great deal of time in drawing common shapes of details such as rounds, squares, hexagons, and ellipses. In addition to common shapes, templates have been made for standard parts such as nuts, bolt heads, electrical symbols, outline of tools and equipment, and many other outlines which are often repeated.

Mechanical Lettering. The results of a recent survey disclosed that mechanical lettering is generally replacing hand lettering. When mechanical lettering is required, it should be performed, whenever possible, by a subordinate. Mechanical lettering is particularly effective when it is used for general notes, particularly when preprinted standard notes on adhesive material are used.

Drawing Practices

The cost of a project is, to some extent, directly related to the number of drawings which must be prepared. Therefore, careful planning to reduce the number of drawings required can result in significant savings. Some ways to reduce the number of drawings are:

Using Standard and Existing Drawings. Every year numerous drawings of parts are prepared which are repetitions of existing drawings. If the drafter were to incorporate in the new design parts that were already drawn, many drawing hours would be saved. Good drawing application records and an efficient multiple-use drawing system can eliminate a great deal of duplication. Standard tabulated drawings may be used to eliminate hundreds of drawings. See Figure 14-1-1.

Detail Assembly Drawings. Detail assembly drawings, in which parts are detailed in place on the assembly (see Fig. 14-1-2), and multidetail assembly drawings, in which there are separate detail views for the assembly and each of its parts, will reduce the number of drawings required. However, they must be used with extreme care. They can easily become too complicated and confusing to be an effective means of communication.

QTY	PART	MATL	DESCRIPTION	PT NO
2	CABLE SUPPORT	MAPLE	A–5374 PT 1	1
2	CABLE SUPPORT	MAPLE	A–5374 PT 2	2
3	CABLE SUPPORT	MAPLE	A–5374 PT 4	3

(A) DRAWING CALLOUT

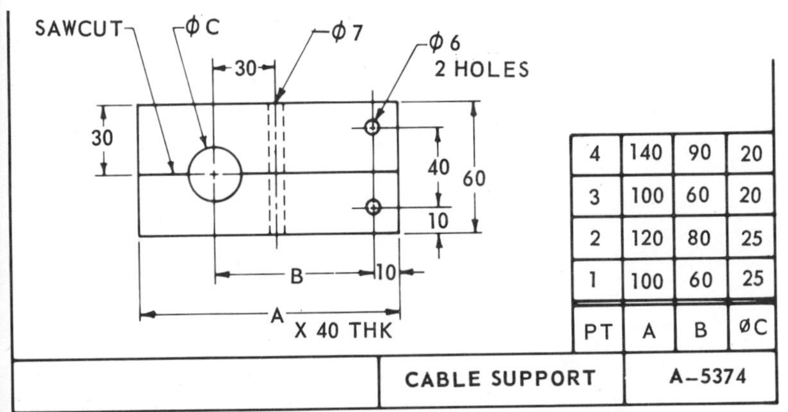

PT	A	B	ØC
4	140	90	20
3	100	60	20
2	120	80	25
1	100	60	25

(B) STANDARD PART

Fig. 14-1-1 Standard tabulated drawings

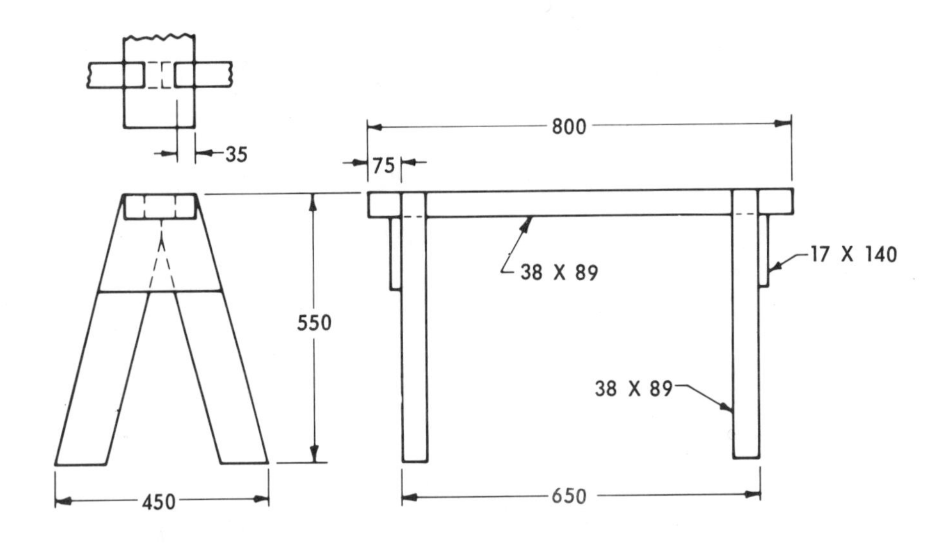

Fig. 14-1-2 Detail and assembly drawing of a sawhorse

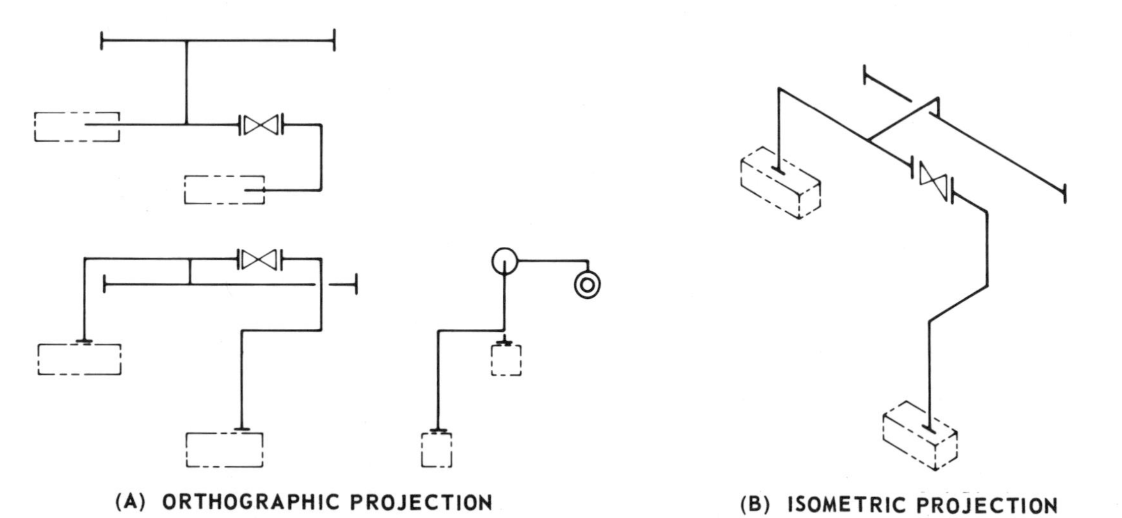

Fig. 14-1-3 Selecting the most suitable type of projection

Selecting the Most Suitable Type of Projection to Describe the Part. The selection of the type of projection (orthographic, isometric, or oblique) can greatly increase the ease with which some drawings can be read and, in many cases, reduce drafting time. For example, a single-line piping drawing drawn in isometric projection simplifies an otherwise difficult drawing problem in orthographic projection. See Figure 14-1-3.

Assignments

1. After the number of drawings made over the last 6 months was reviewed, it was discovered that a great number of cable straps, shown in Fig. 14-1-A, were being

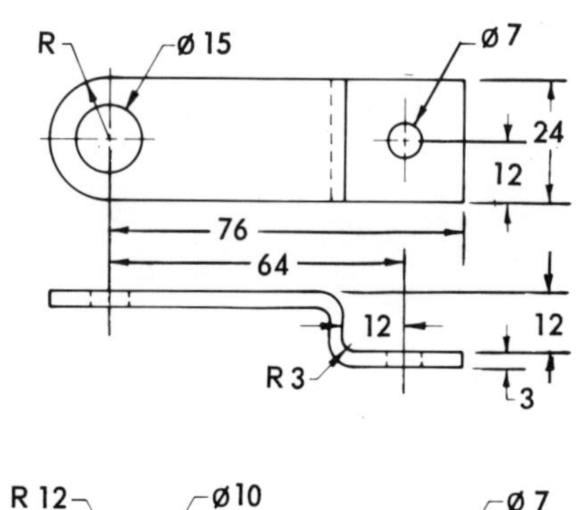

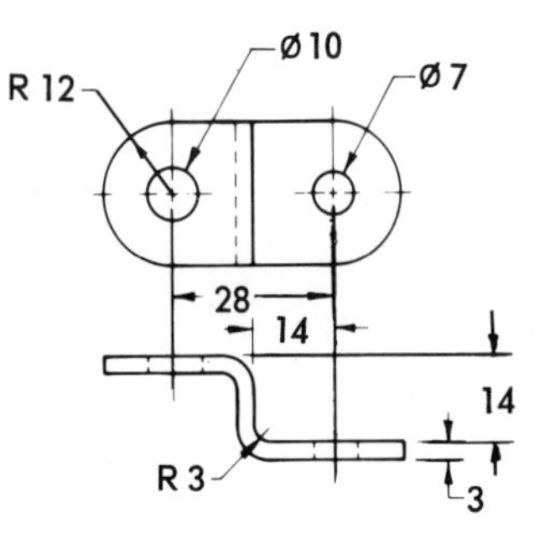

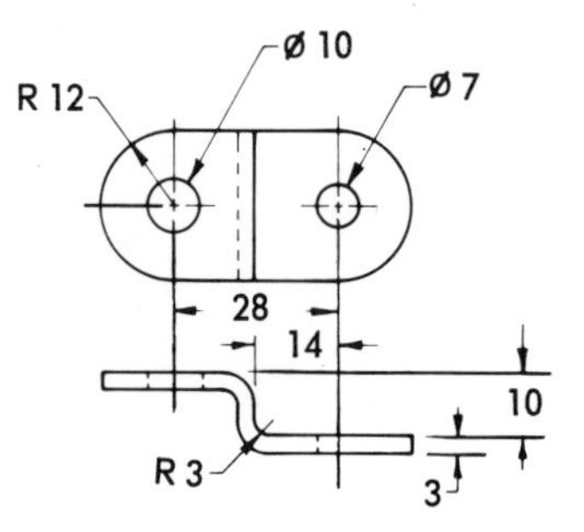

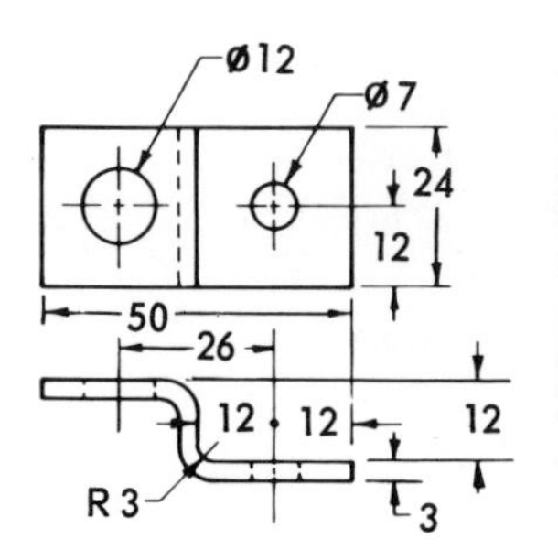

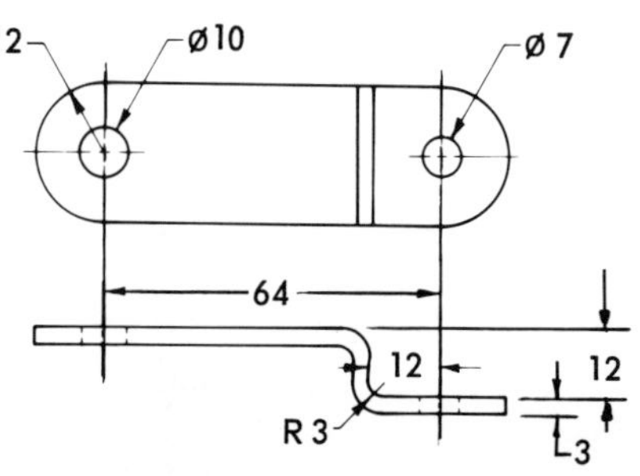

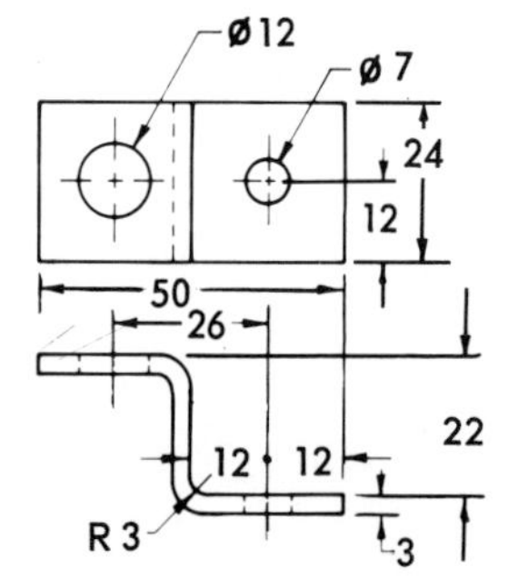

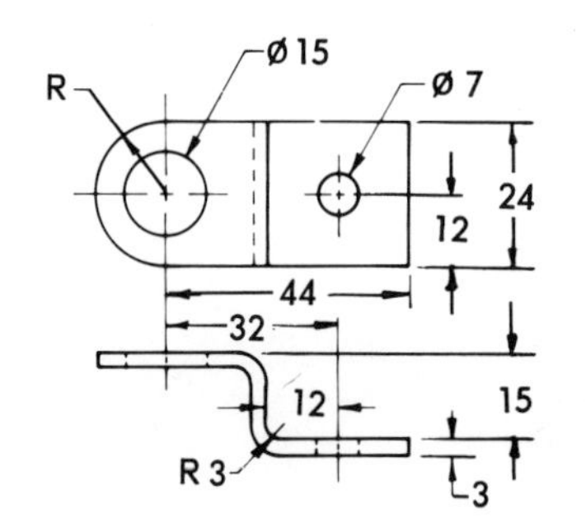

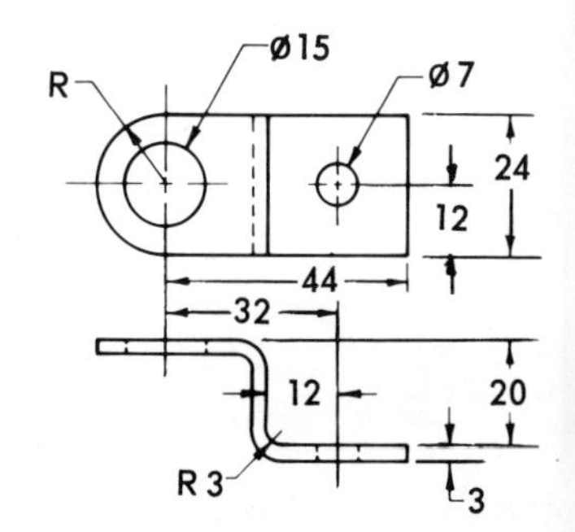

Fig. 14-1-A Cable straps

made which were similar in design. On an A3 size sheet, prepare a standard tabulated drawing similar to Fig. 14-1-1, reducing the number of standard parts to 4. Scale is 1:1.

2. On an A3 sheet, the rod guide shown in Fig. 14-1-B is to be drawn twice and the drawing time for each recorded. First, on plain paper make an isometric drawing of the part, using a compass to draw the circles and arcs. Next, repeat the drawing, only this time use isometric grid paper and a template for drawing the circle and arcs. From the drawing times recorded, state in percentage the time saved by the use of grid paper and templates. Scale is 1:1. Do not dimension.

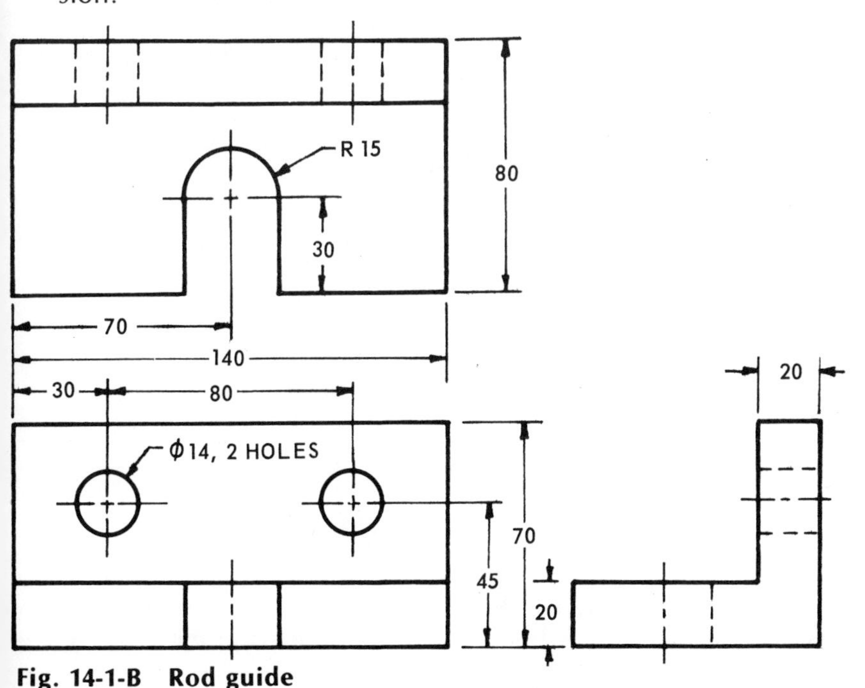

Fig. 14-1-B Rod guide

UNIT 14-2 Freehand Sketching

Most shops care little whether the drawing is freehand, whether one view is shown, or whether the drawing is to scale, as long as the proportions are approximate. They are interested in having the necessary information clearly shown. Freehand sketches and drawings made with instruments can be shown on one sheet. However, it must be clearly understood that the use of freehand sketching does not give the drafter a licence to turn out sloppy work.

Savings as high as 30% in the preparation of working drawings have been attributed to the use of freehand sketches as opposed to instrument-produced drawings. However, freehand sketching has its limitations. It is highly effective on simple detailed parts, for small radii, such as rounds and fillets, and for small holes. In many cases the term **freehand** is not entirely correct. For instance, templates may be used to draw circles, resistors, or other common features, or a straightedge may be used to produce long lines since it is faster and more accurate than freehand sketching. But short lines are drawn faster freehand.

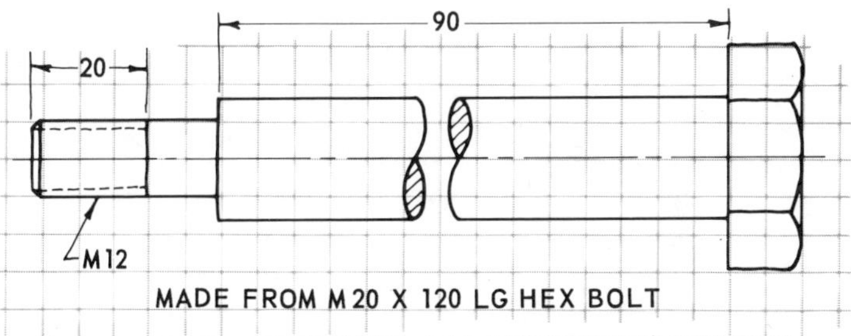

(A) COMBINED FREEHAND AND INSTRUMENT DRAWING

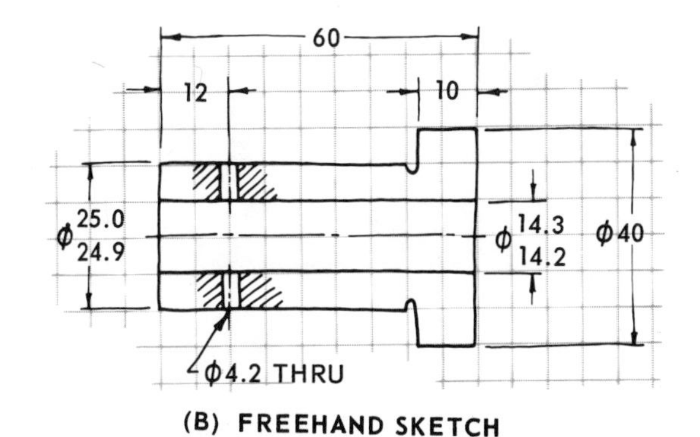

(B) FREEHAND SKETCH

Fig. 14-2-1 Sketching parts on coordinate paper

Drawing paper with nonreproducible grid lines is ideal for freehand sketching. See Figure 14-2-1. For this reason many companies have their drawing paper made with nonreproducible grid lines over the entire drawing area. Other advantages of having the grid lines on the paper are that (1) they may serve as guidelines in lettering notes and dimensions and (2) they may be used for measuring distances, thereby reducing the number of times the scale is used for measuring.

Assignment

On a sheet of coordinate graph paper sketch two views of the shaft bracket shown in Fig. 14-2-A. Scale is 1:1. Add dimensions.

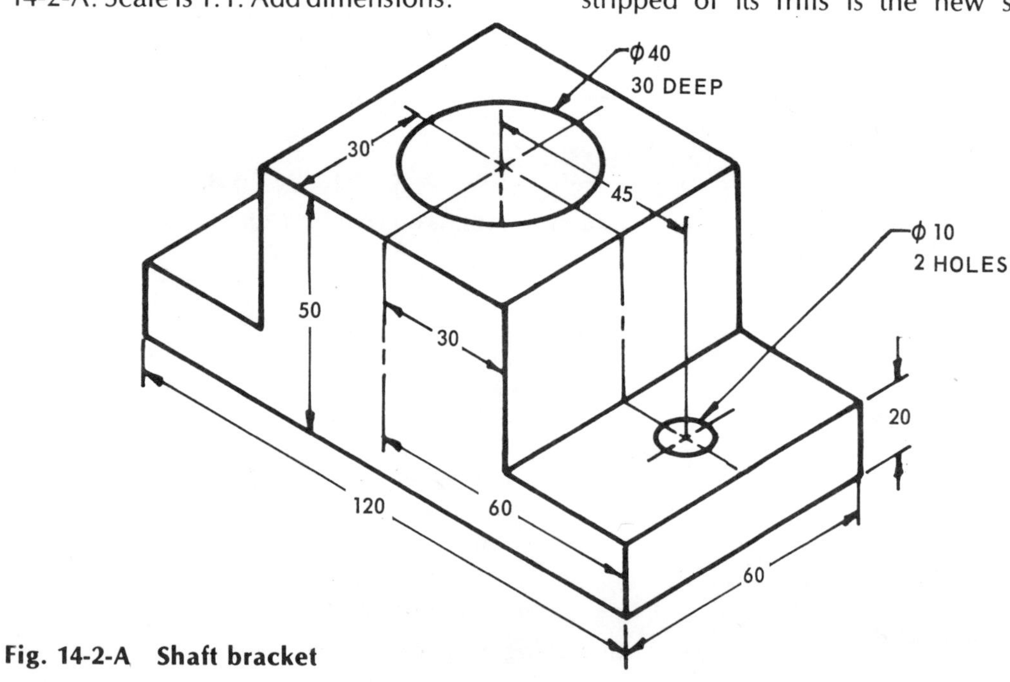

Fig. 14-2-A Shaft bracket

UNIT 14-3
Simplified Drafting

The challenge of modern industry is to produce more and better goods at competitive prices. Drafting, like all other branches of industry, must share in the responsibility for making this increased productivity possible. The old concept of drafting — that of producing an elaborate and beautiful drawing, complete with all the lines, projected views, and sections — must give way to a simplified method. The new, simplified method of drafting must embrace many modern economical drafting practices but surrender nothing in either clarity of presentation or accuracy of dimensioning. Drafting stripped of its frills is the new standard for production drawings (drawings used only one or two times). See Figure 14-3-1.

Although simplified drafting was pursued largely to reduce drafting costs and to im-

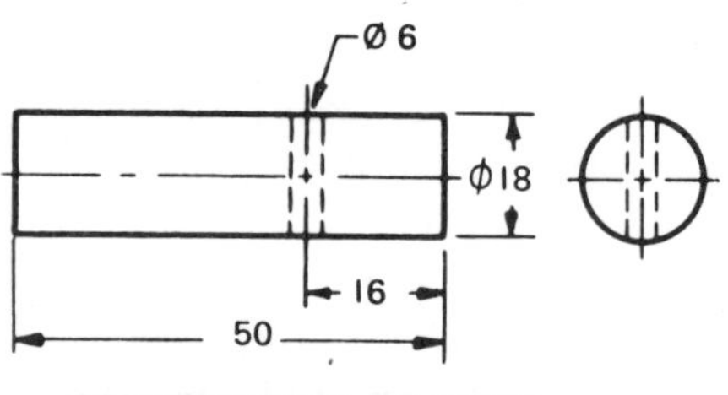

CONVENTIONAL DRAWING

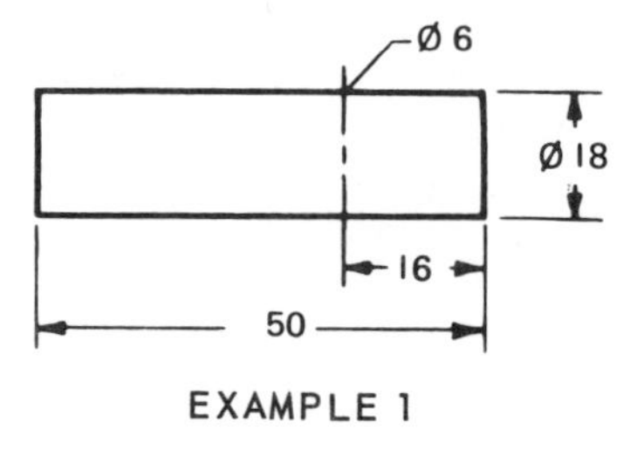

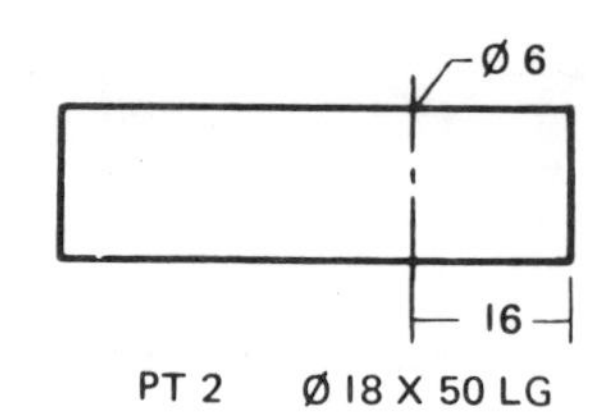

EXAMPLE 2

NOTE PT 2 Ø 18 X 50 LG
Ø 6 HOLE – 16 FROM END

PART DESCRIBED BY A NOTE

EXAMPLE 3

Fig. 14-3-1 Simplified drafting practices for detailed parts

prove the drafting product, it had a beneficial side effect when the drawing had to be microfilmed. Even more important, the worker on the shop floor has an "easier to understand" print. Much of the fine, difficult-to-produce detail, such as unnecessary frills and curlicues, has been eliminated and replaced by simplified or symbolic presentation. Drawings cluttered by repetitive detail are now shown simplified and are more suitable for making quality microfilm reproduction and reduced-size drawings.

Three most effective practices used in simplified drafting are

- Simplification of dimensioning.
- Simplification of detail drawing.
- Extensive use of freehand sketching.

Simplification of Dimensioning

Simplification of dimensioning not only reduces drafting time but also avoids cluttering a drawing with unnecessary lines, thereby making it easier to read.

No other single factor has more influence on the use of drawings than dimensioning.

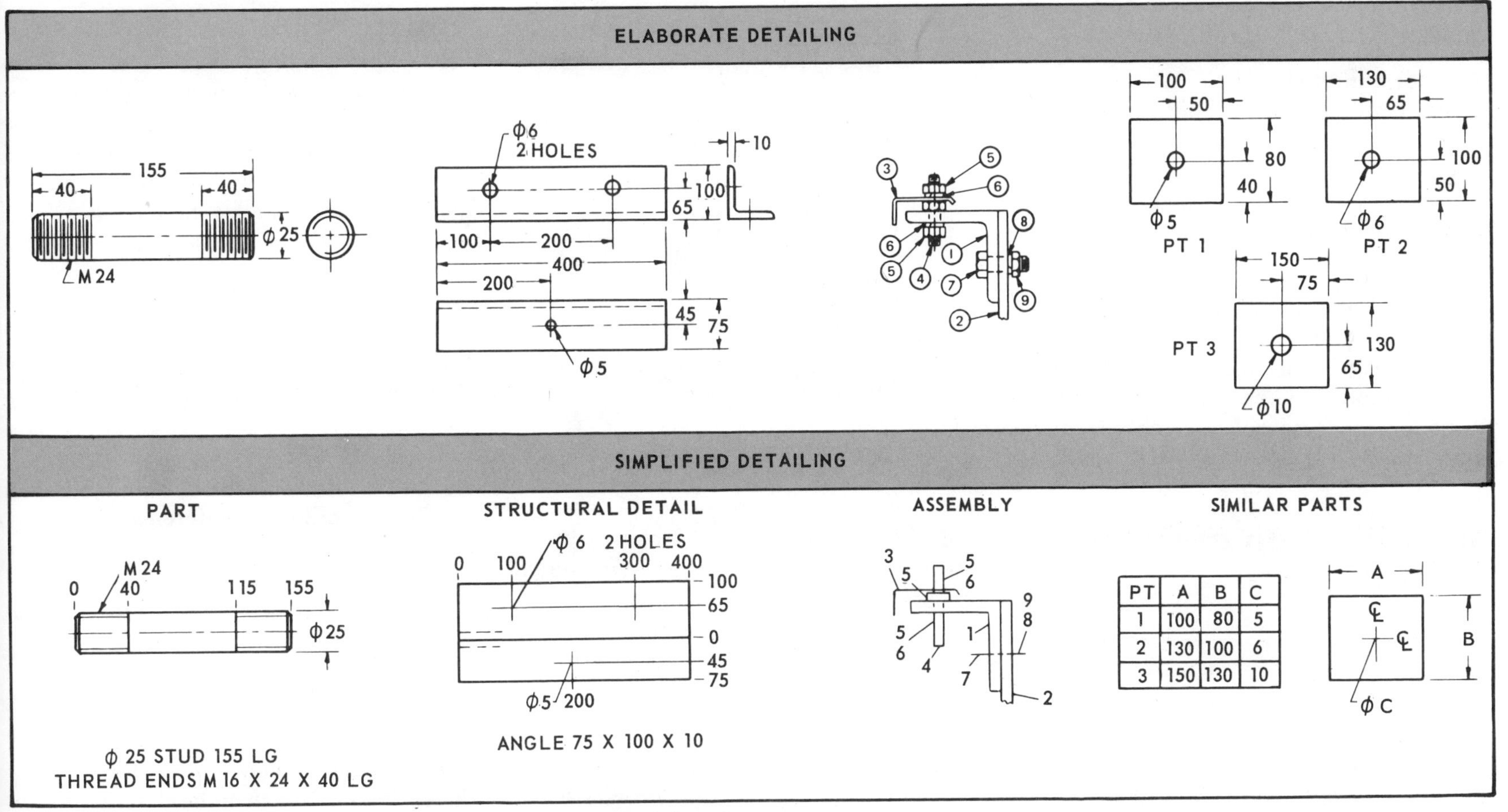

PT	A	B	C
1	100	80	5
2	130	100	6
3	150	130	10

Fig. 14-3-2 Comparison between elaborate and simplified drawings

It is not sufficient to have dimensions numerically correct; it is equally important to dimension a drawing properly so that computation of sizes is unnecessary.

- Use arrowless dimensioning.
- Use tabular dimensioning.
- Use abbreviations and symbols.

Arrowless Dimensioning. To avoid having a large number of dimensions extending away from the part, arrowless or ordinate dimensioning may be used. See Unit 4-4.

Tabular Dimensioning. When there are a very large number of holes or repetitive features, such as in a chassis or a printed circuit board, and where the multitude of centre lines would make a drawing difficult to read, tabular dimensioning is recommended. See Unit 4-4.

Abbreviations and Symbols. Abbreviations and symbols are shortened forms of words or expressions used to conserve drafting time and drawing space. Refer to the Appendix for commonly used abbreviations and symbols.

Simplification of Detail Drawing

1. Complicated parts are best described by means of a drawing. However, explanatory notes can complement the drawing, thereby eliminating views that are time-consuming to draw. See Figure 14-3-2.

2. Use simplified drawing practices as described throughout this text, especially on threads and common features.

3. Avoid unnecessary views. In many cases one or two views are sufficient to explain the part fully.

4. Show only partial views of symmetrical objects. See Figure 14-3-3.

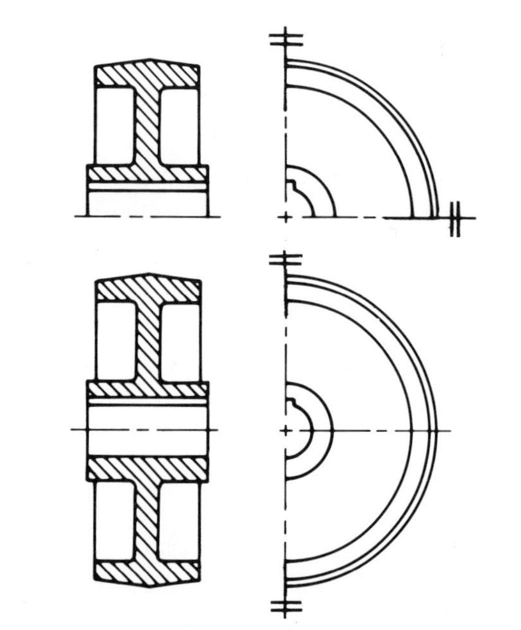

Fig. 14-3-3 Partial views

5. The use of the symbol ╪ indicates that all dimensions are symmetrical about the line indicated.

6. A simplified assembly drawing should be for assembly purposes only. Some means of simplification are:

- Standard parts such as nuts, bolts, and washers need not be drawn.
- Reference part circles and arrowheads on leaders can be omitted.
- Small fillets and rounds on cast parts need not be shown.
- Phantom outline of complicated details can often be used.

7. Use templates where possible.

8. Within limits a small drawing is made more quickly than a large drawing.

9. Eliminate hidden lines which do not add clarification.

10. When a large number of holes of similar size are to be made in a part, there is a chance that the person producing the part may misinterpret a conventional drawing. To simplify the drawing and reduce the chance of error, hole symbols such as shown in Figures 14-3-4 and 14-3-5 are recommended.

Fig. 14-3-4 Recommended hole symbols in order of preference

11. Avoid the use of elaborate pictorial or repetitive detail.

12. Eliminate repetitive data by use of general notes or phantom lines.

13. Eliminate views where the shape or dimension can be given by description, for example, hex, sq., ϕ, thk, etc.

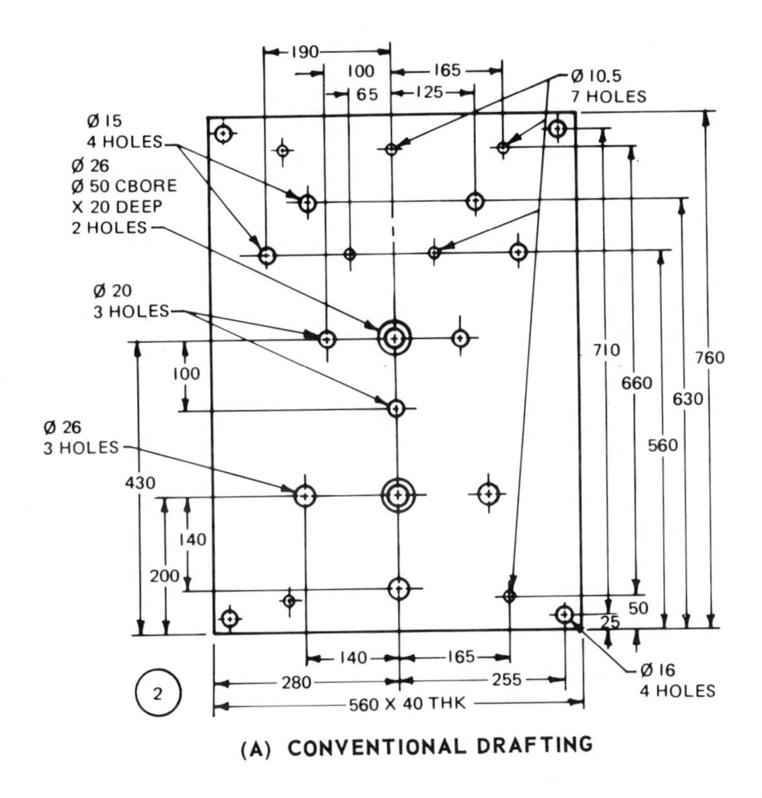

(A) CONVENTIONAL DRAFTING

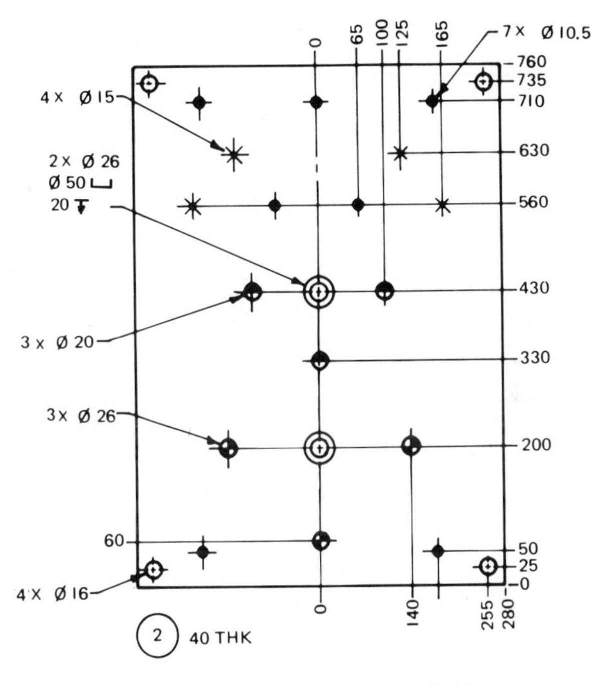

(B) SIMPLIFIED DRAFTING

Fig. 14-3-5 Application of hole symbols and arrowless dimensioning

Assignments

1. On an A3 sheet, redraw the two parts shown in Fig. 14-3-A using arrowless dimensioning and simplified drawing practices. Scale is 1:2. For the cover plate use the bottom and left-hand edge for the datum surfaces. For the back plate use the bottom and the centre of the part for the datum surfaces.

2. On an A2 sheet, make simplified drawings of the parts shown in Fig. 14-3-B. Scale to suit.

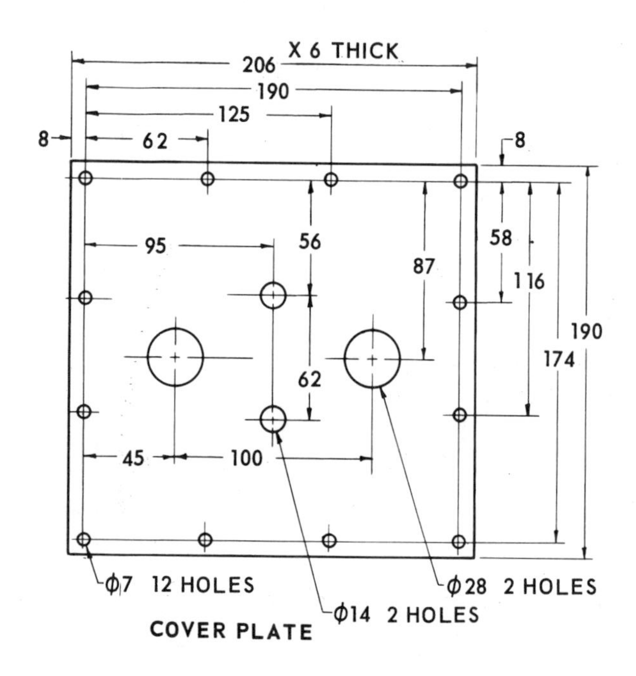

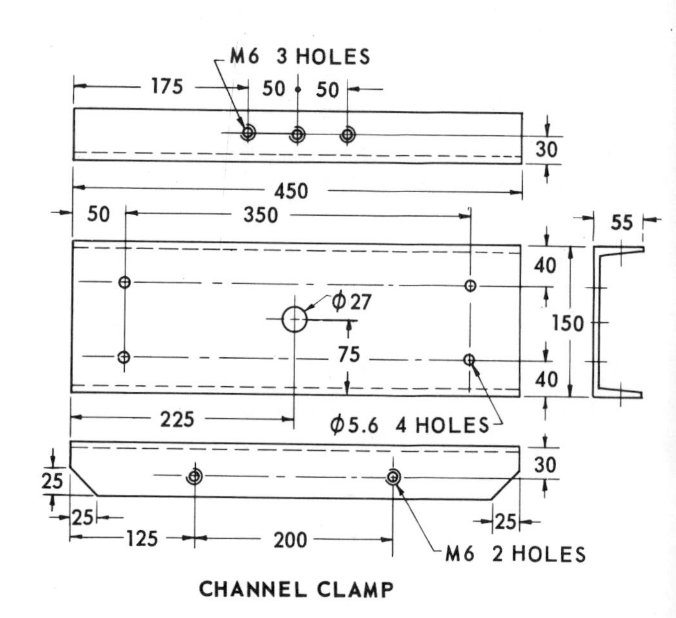

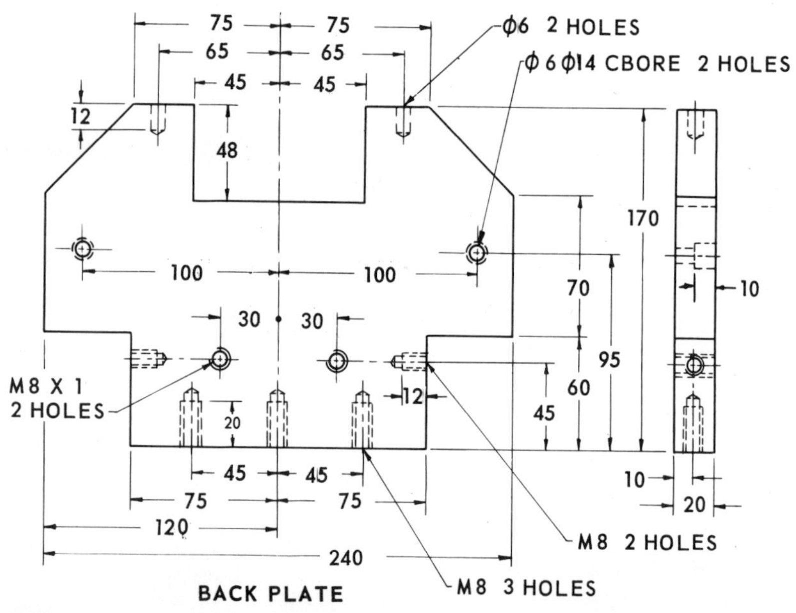

Fig. 14-3-A Arrowless dimensioning assignment

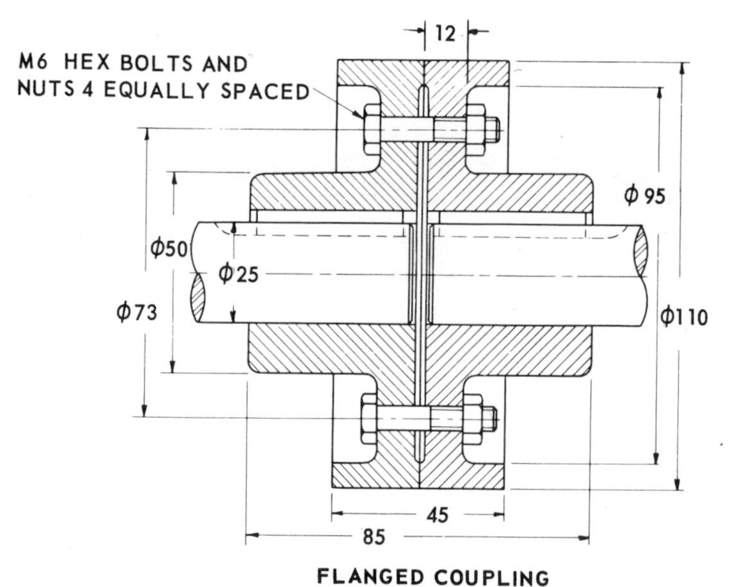

Fig. 14-3-B Simplification of detail assignment

UNIT 14-4 Cut-And-Paste Drafting

Using Parts of Existing Drawings

No matter how original a design may be, a great number of part features are repetitive. With the aid of modern reproduction methods, drawings can be created by using unchanged portions of existing drawings. Transferring them from one drawing to the next is accomplished by cut-and-paste drafting. It provides a way of using all or parts of existing drawings, notes, charts, drawing forms, to revise existing drawings and to create new drawings.

Through the utilization of existing drawings much valuable drafting time is freed for creative design drafting rather than hand copying.

Finished prints can be made on paper, acetate, or vellum. They can be the same size or reduced to five different sizes, depending on the reproduction equipment being used.

Another important advantage of cut-and-paste drafting is that materials copied from existing drawings do not have to be rechecked minutely, as must be done with new drawings. Rechecking time is reduced. See Figure 14-4-1.

Appliqués

One of the most successful methods of reducing drawing time is the use of appliqués. When parts, shapes, symbols, or notes are used repeatedly, then appliqués should be considered. These pressure-sensitive overlays may be printed on plain, transparent or translucent sheets with an adhesive backing.

Appliqués are available in a great variety of standard symbols or patterns (Fig. 14-4-2) and in blank (unprinted) sheets. A matte surface on the blank sheet will accept typewriter copy as well as pencil or ink lines.

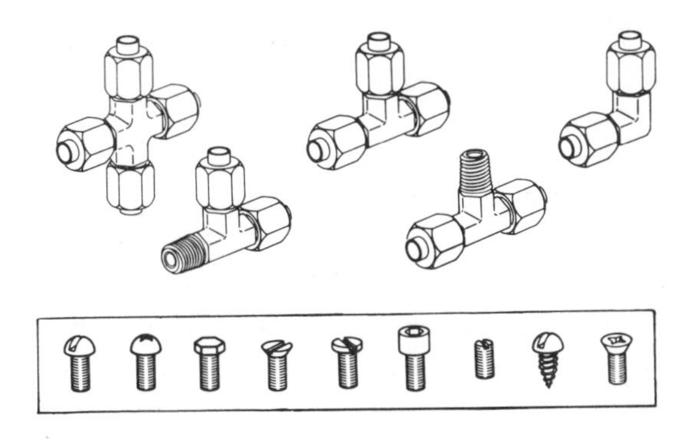

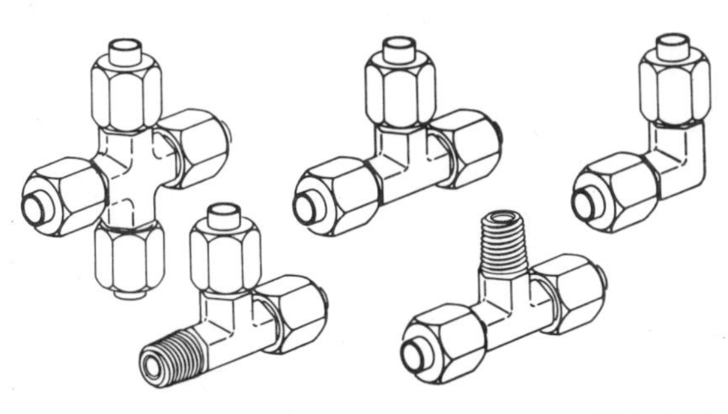

Fig. 14-4-2 A variety of shapes and sizes of appliqués

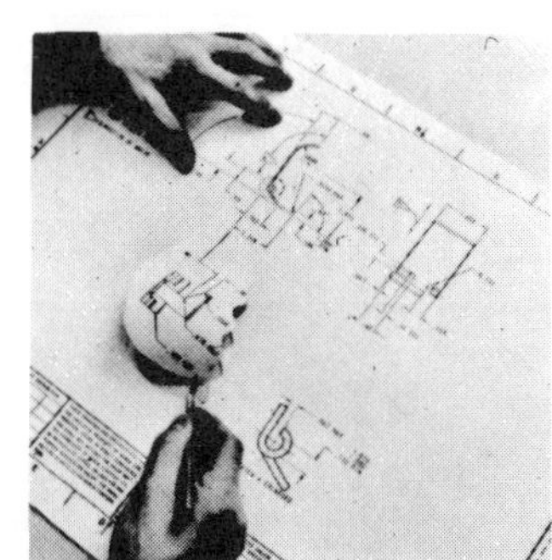

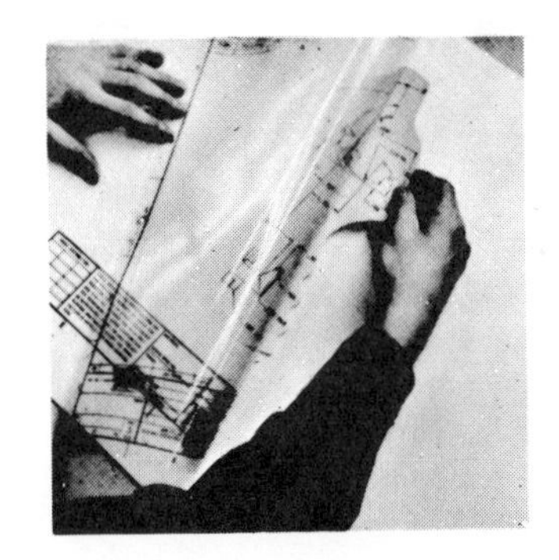

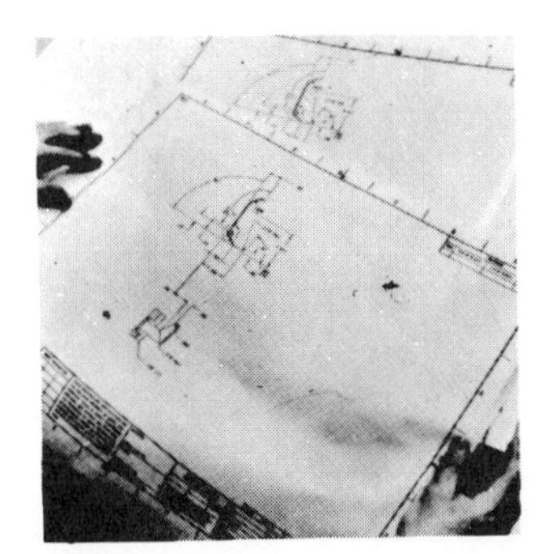

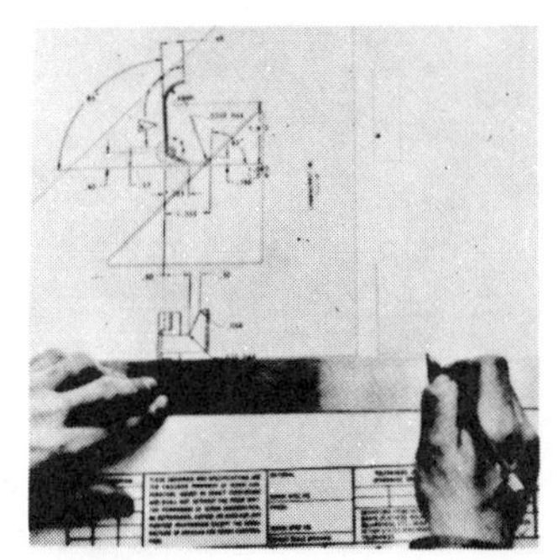

Fig. 14-4-1 Cut and paste drafting

This material is often used for making corrections on drawings and for adding materials lists or detailed notes which can be typed faster than they can be lettered. Figure 14-4-3 shows a drawing which used many of the appliqués shown in Figure 14-4-2.

Appliqués are available in two basic types: **cutout** and **transfer.** Cutout appliqués are applied by positioning the desired image in the correct position on the drawing, burnishing (rubbing) the image area, and cutting around it to remove the portion not wanted. The transfer-type pressure-sensitive appliqué works on a somewhat different principle. The carrier is removed from the translucent image sheet, and the area to be transferred is placed in position on the drawing. The image to be transferred is then rubbed over the top surface of the transfer sheet with a burnishing stick or other blunt instrument. When the transfer sheet is lifted, the image remains on the drawing sheet. After the image has been transferred, the carrier sheet is placed over the image and reburnished.

The combined use of cut-and-paste drafting and appliqués for new drawings is found extensively in industry, especially in the electronics and piping fields.

Assignment

On an A3 sheet, make the electronic layout drawing shown in Fig. 14-4-A. Appliqués of the electronic components are not available, make your own by photostating Fig. 14-4-3 and cutting them out and gluing them to your drawing. There is no scale.

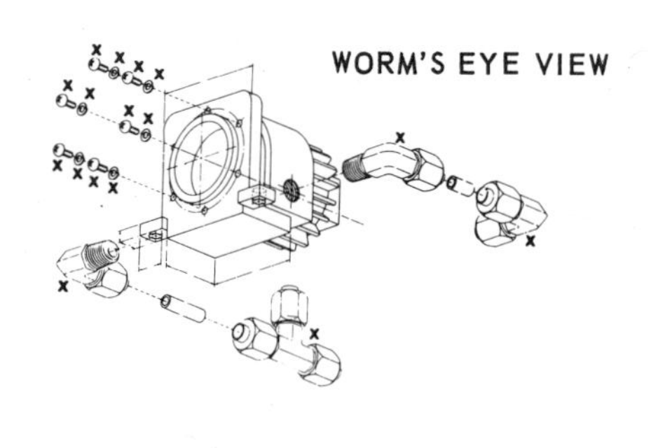

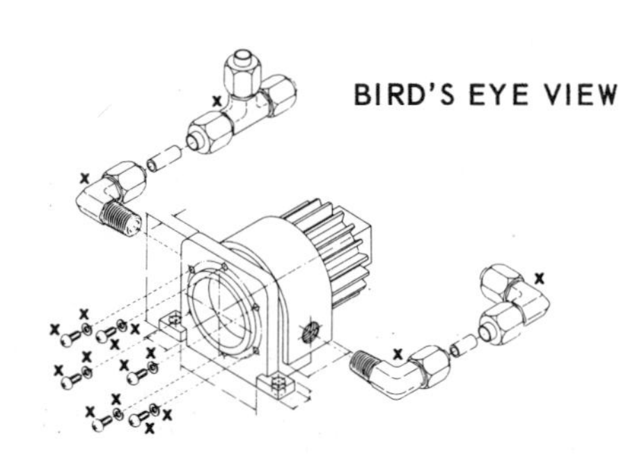

(A) APPLICATION OF SYMBOLS SHOWN IN FIG. 14–4–2

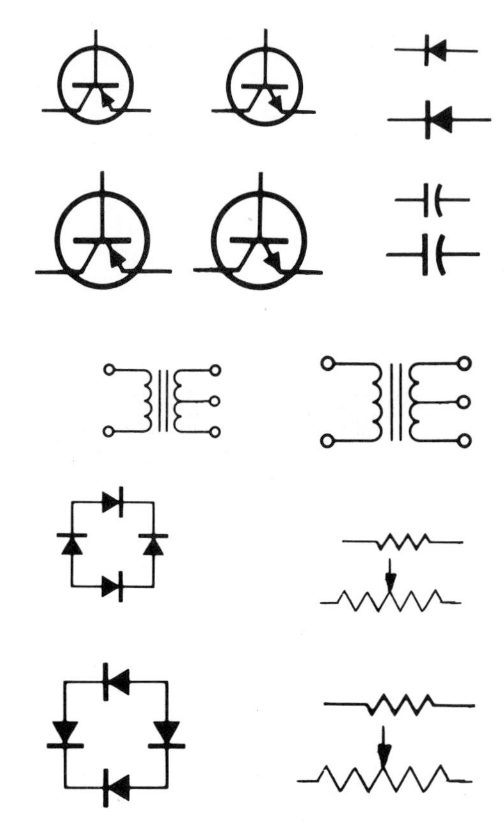

(B) ELECTRONIC APPLIQUES

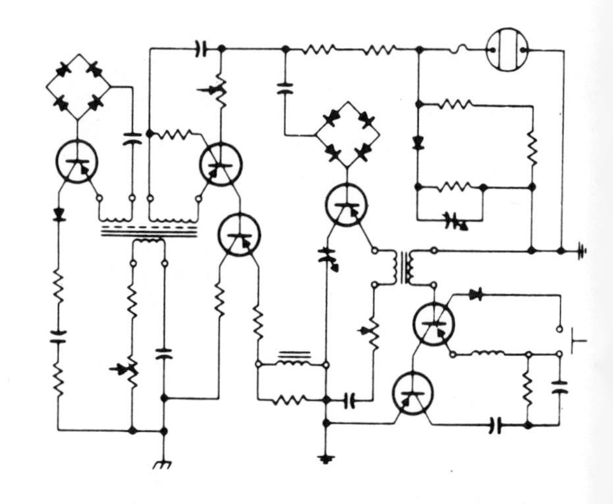

(C) APPLICATION OF ELECTRONIC APPLIQUES

Fig. 14-4-3 Application of appliqués

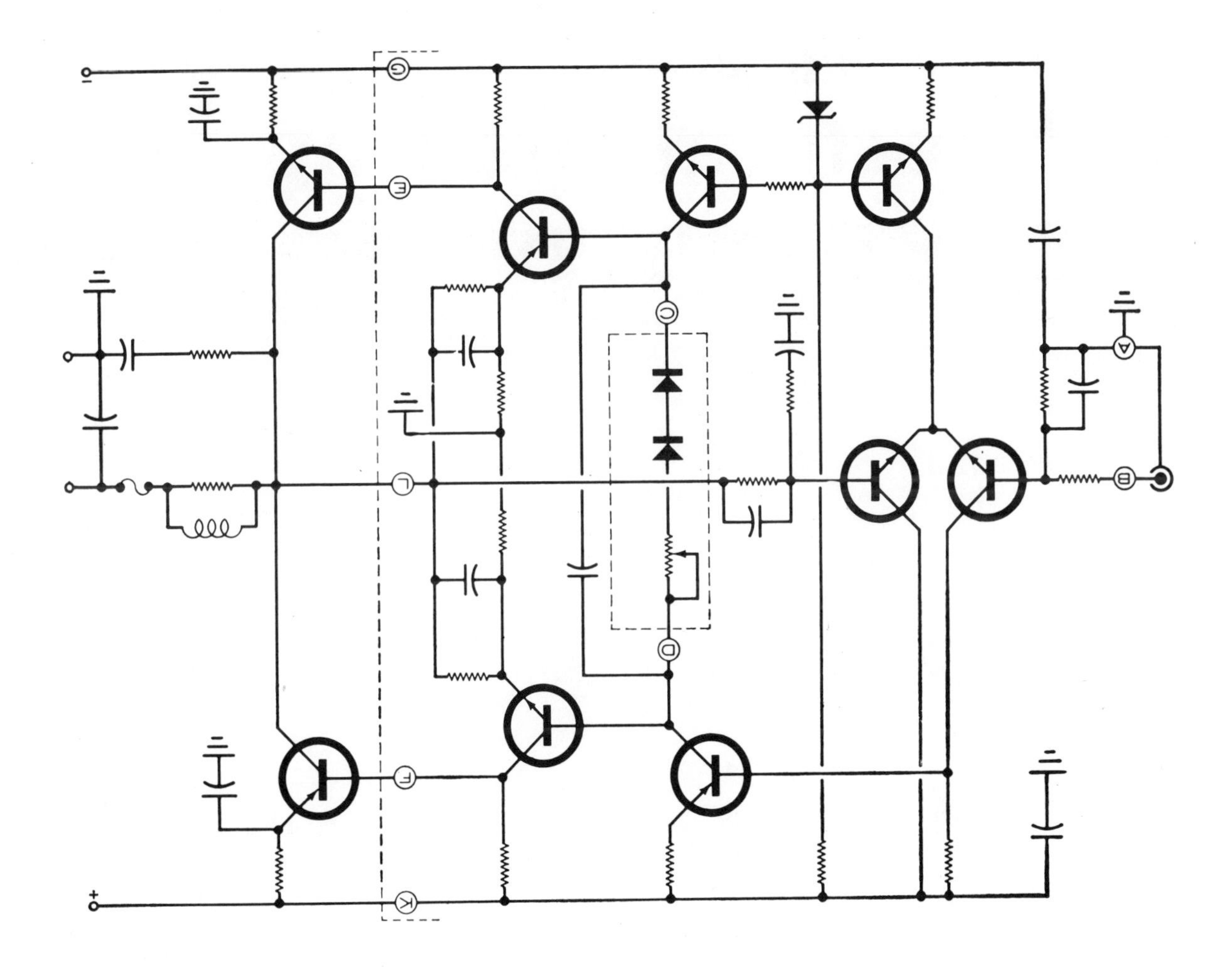

Fig. 14-4-A   Electronic drawing with appliqués

Chapter 15
Architectural Drafting

UNIT 15-1
Presentation Drawings

Architectural drafting is concerned with the design, representation, and specifications for a variety of buildings and structures. The principles of architectural drawing are basically those used in other technical drawings. There are, however, many conventional symbols and practices peculiar to architectural drawing because of the nature of the work and the smaller scale used. This chapter will deal with the basic principles of architectural drawing as related to home design.

There are five basic house types:

1. The one-storey or bungalow
2. The 1½-storey
3. The two-storey
4. The split-level
5. The bi-level

Information on these house types is found in Figures 15-1-1 and 15-1-2.

Architectural drawings are of two types: **presentation drawings** and **construction** (working) **drawings.**

Presentation Drawings

Presentation drawings are prepared by architectural drafters for use in periodicals, magazines and other publications read by the general public. These drawings are pictorial in form, showing only the basic room arrangements and exterior features. They are drawn omitting the complicated building detail so that the general public may readily understand them. A floor plan is the central feature of such drawings, and a pictorial view and elevations (views) are usually shown also (Fig. 15-1-3).

Landscaping and rendering is used on

ONE–STOREY PITCHED ROOF AND FLAT CEILING	ONE–STOREY PITCHED ROOF AND SLOPING CEILING – FLAT ROOF	ONE AND ONE–HALF STOREY TWO LIVING LEVELS – VARYING SECOND FLOOR AREA AND CEILING	TWO–STOREY TWO LIVING LEVELS – VARYING ROOF AND CEILING TYPES
LIVING	LIVING LIVING	LIVING LIVING	LIVING LIVING
Home building statistics reveal that one-story houses, with and without basements, are built in larger numbers each year than any other basic house type. There are more size, shape, and design variations in one-story houses than in any other type. Despite these variables, the construction simplicity of most one-story designs provides an excellent basis for studies of construction methods which may result in important cost savings. From a livability standpoint, one-level houses are advantageous. Families of all ages, including the elder retirement group, favor the convenience of one-story houses. Many multi-level houses are designed with a variety of one-story additions as a basic feature of the composite architectural design.	The roof design and load-bearing elements of houses with flat roofs and those with continuous sloping roofs and ceilings are similar. Both types are significantly different, in structural design, from houses with trusses or with rafters and ceiling joists. Good architectural detail, in the functional design of these basic house types, will result in exciting examples of contemporary living. These same roof construction details are often used in multi-level houses.	The traditional homes of New England have been popular for two centuries. Probably the most familiar of these is the 1½-story "Cape Cod." The basic simplicity of the Cape Cod design should be retained, in proportion and detail, if an approach to historical authenticity is desired. The 1½-story basic shape permits a wide design variety other than traditional. The steep, sloping roof can be the basis for many outstanding contemporary designs. The second-floor living area, which varies in size with the house dimensions and use of dormers, provides for flexible planning.	Like 1½-story houses, two-story house types are rich in traditional heritage. They, too, should be designed with great respect for proportion and detail, regardless of their architectural style. The box-like, two-story form provides maximum living area at the least cost. A wide range of roof types can be used to vary the design characteristics of the two-story house. Lower- and upper-level walls can be in the same plane or, in some cases, the upper-level walls may be projected to gain more floor area and break the high wall appearance. Various types of additions, to the two-story box, often enhance overall design composition.

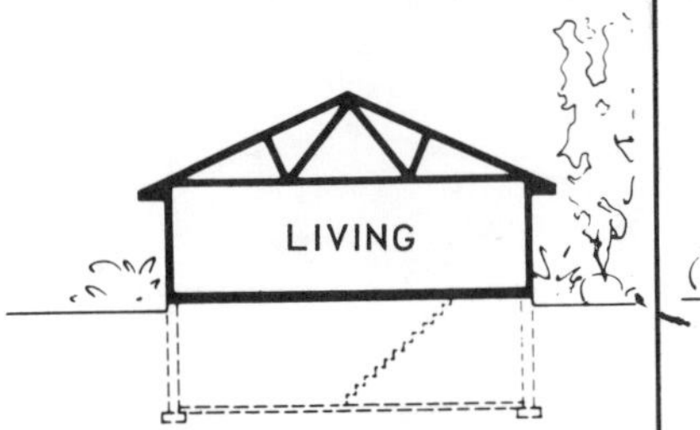

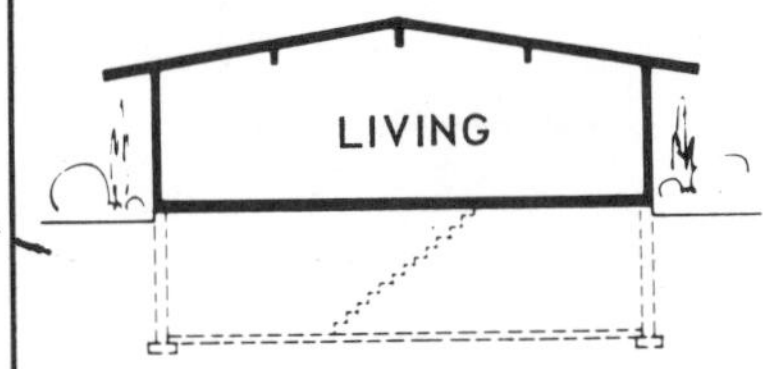

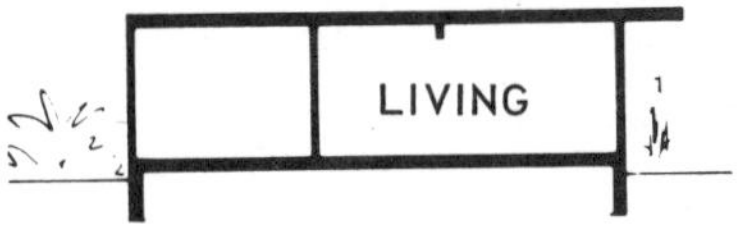

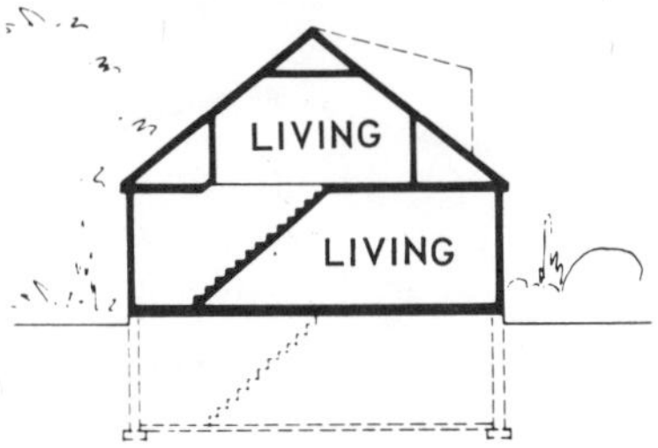

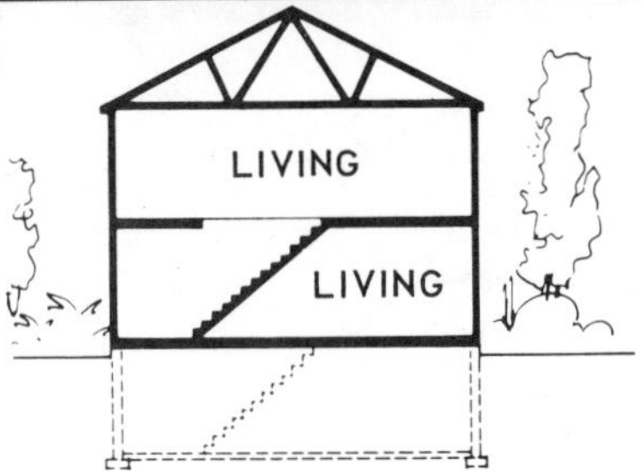

Canada Mortgage and Housing Corp.

Fig. 15-1-1 House design for level lots

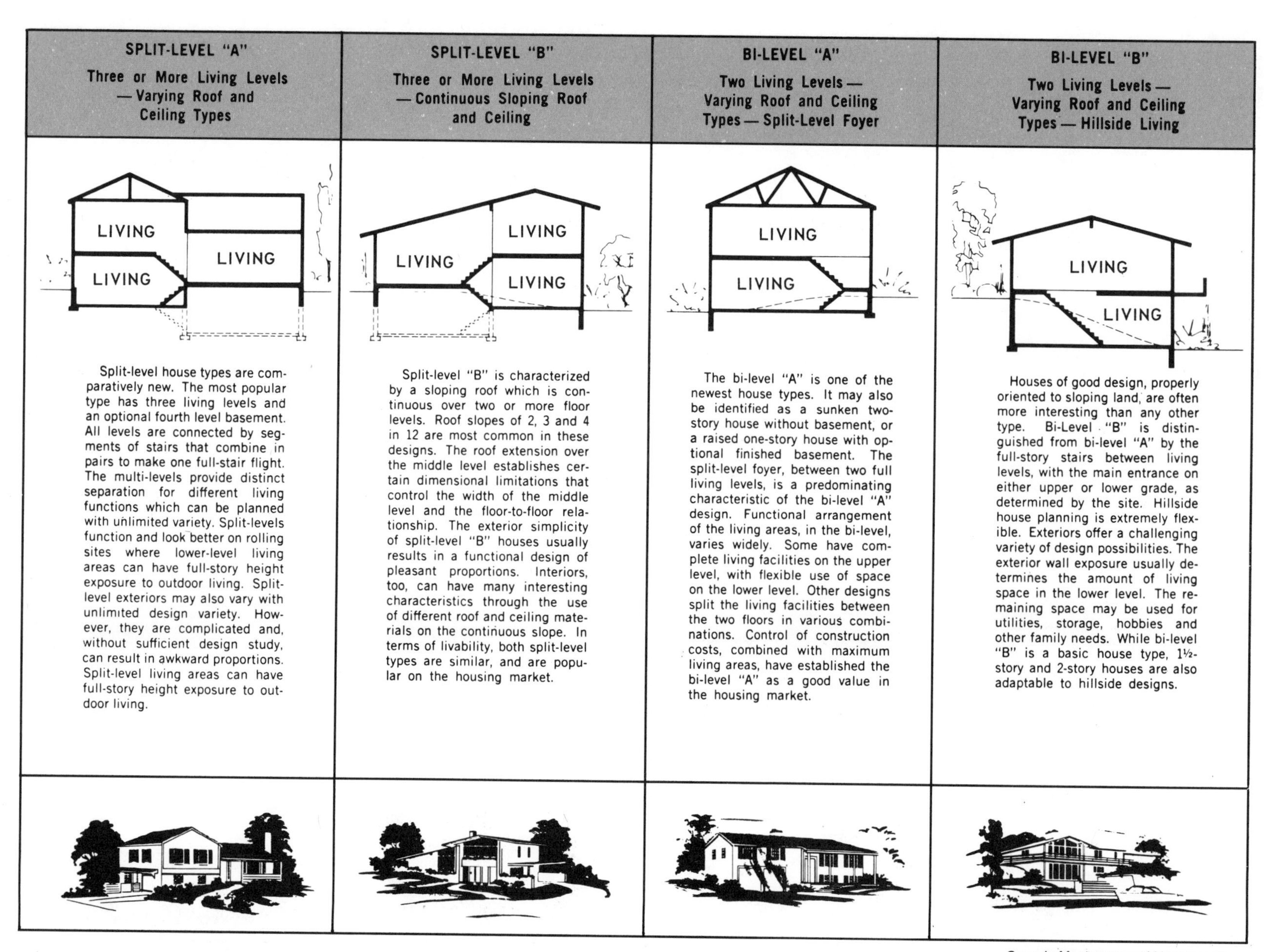

Canada Mortgage and Housing Corp.

Fig. 15-1-2 House designs for sloped lots

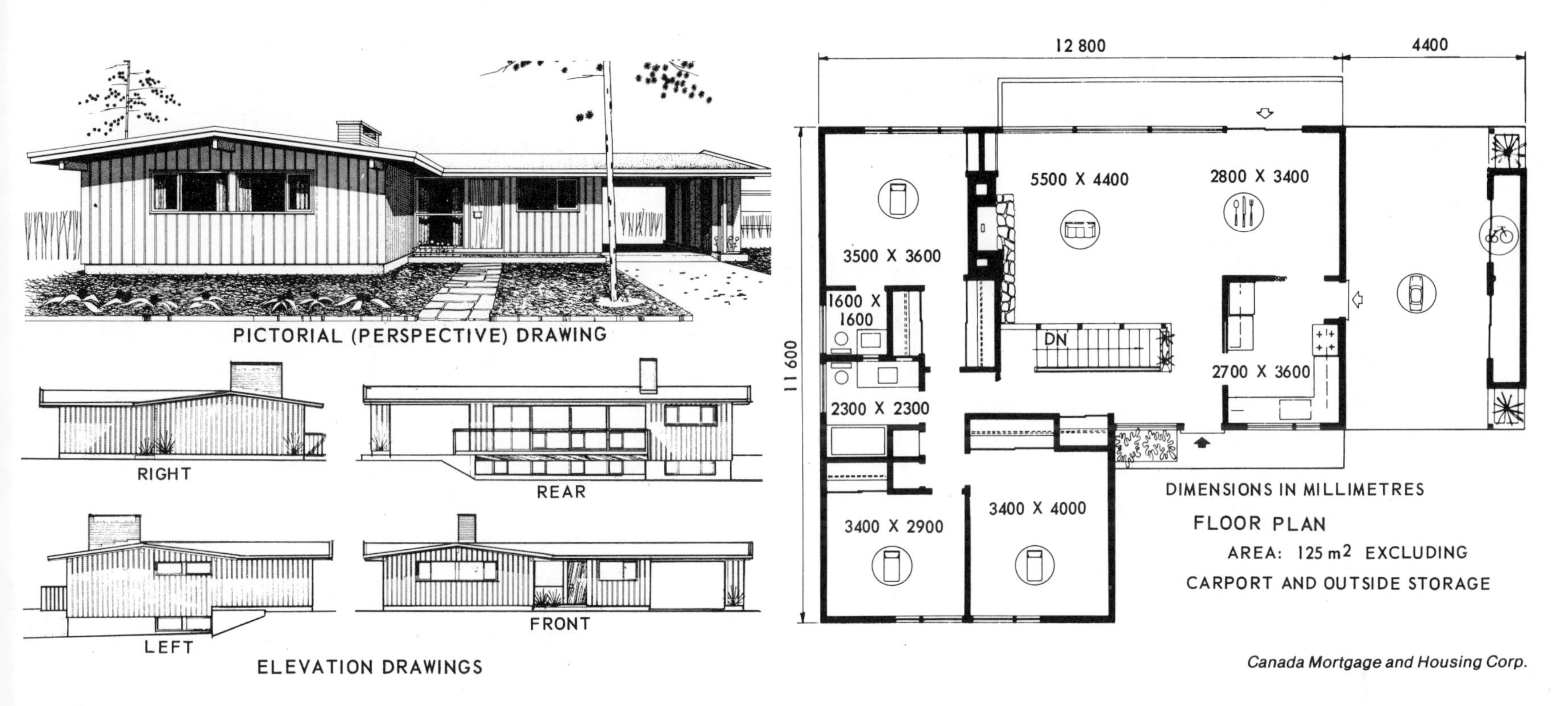

Fig. 15-1-3 Presentation drawing of a house

presentation drawings to make the house more attractive.

Only the overall house size is shown by the means of dimension and extension lines. Finished room sizes are given along with the room names, the first dimension being the horizontal distance across the drawing, the second dimension being the vertical distance on the drawing. The overall floor area in square metres is shown beside the floor plan.

The floor plan represents a horizontal slice through the house. Figure 15-1-4 shows the slice and the resulting floor-plan drawing. The floor plan gives the information that a person would need to know when selecting a house design. It gives the location of doors, windows, fireplaces, closets, and other features of the building. Also the floor plan shows the relationships of the rooms to one another. Figure 15-1-5 illustrates the symbols used for presentation floor-plan drawings. A knowledge of these symbols is necessary for a full understanding of the architect's design.

Review Questions

1. Name the five basic house types and give the distinguishing features of each.
2. What is the purpose of presentation drawings?
3. What are side, front, and back views called in architectural drawings?
4. What information does the presentation floor plan drawing give?

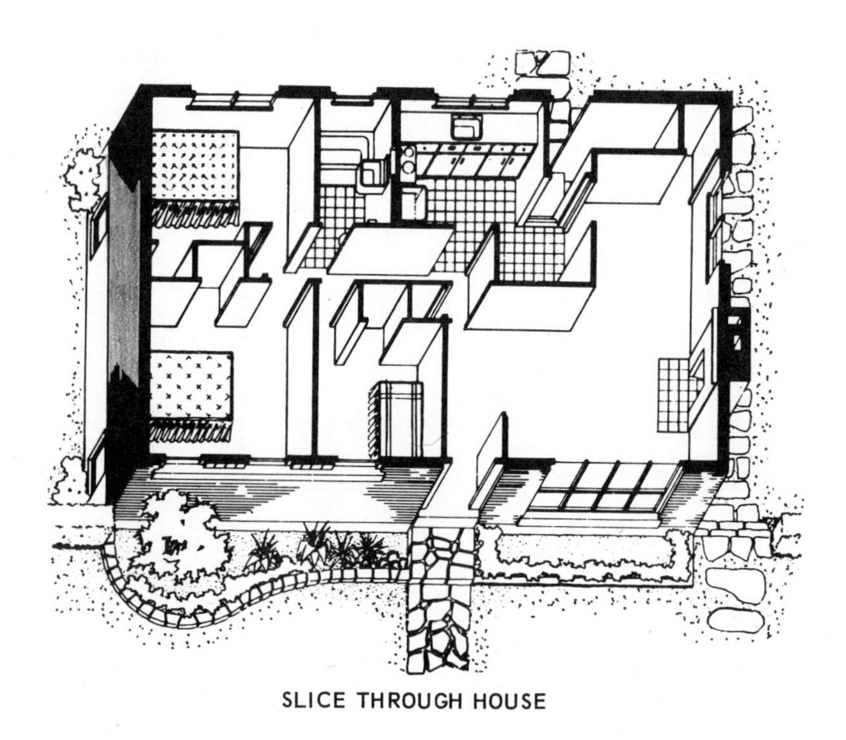

SLICE THROUGH HOUSE

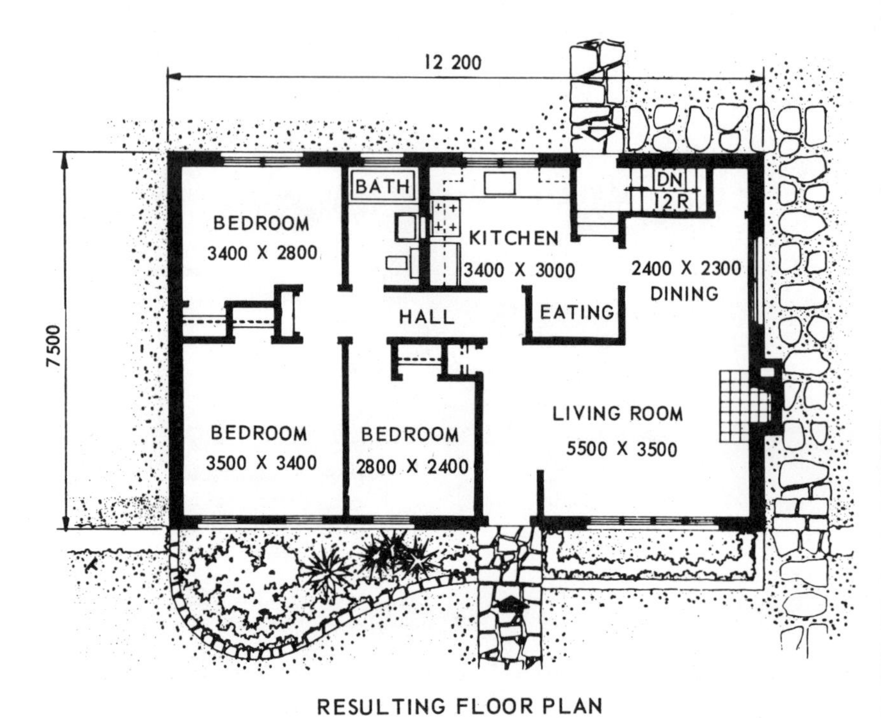

RESULTING FLOOR PLAN

Canada Mortgage and Housing Corp.

Fig. 15-1-4 How to read presentation floor plan drawings

Assignments

1. Make a presentation plan drawing of the summer cottage shown in Figure 15-2-A.
2. Make a presentation plan drawing of bungalow #1 shown in Figure 15-2-B.
3. A prospective customer is interested in having a 4300 × 6700 mm garage built. The garage is to have two double hung windows, 600 × 900 mm; one door, 900 × 2030 mm; and one garage door 2800 mm wide and 2200 mm high. Make a presentation plan drawing of the garage.
4. Mr. Hines is interested in building a 16-unit motel. Construction is to be of 200 mm concrete block except for the front wall, which will be 90 mm angel stone backed by 140 mm block. The maximum width of the motel is to be 5500 mm. The units will each have a 3-piece bath, a telephone, and a television set, and will be suitable to accommodate two double beds. Make a plan drawing, complete with dimensions, of a unit which you would present to Mr. Hines. (Scale 1:50).
5. Make a presentation drawing of the floor plan of a summer cottage having the following details: floor area between 75 and 100 m^2, 3 bedrooms, kitchen with eating area, bathroom with basin and water closet, and a large living room. Other details may be selected by the student. (Scale 1:50).

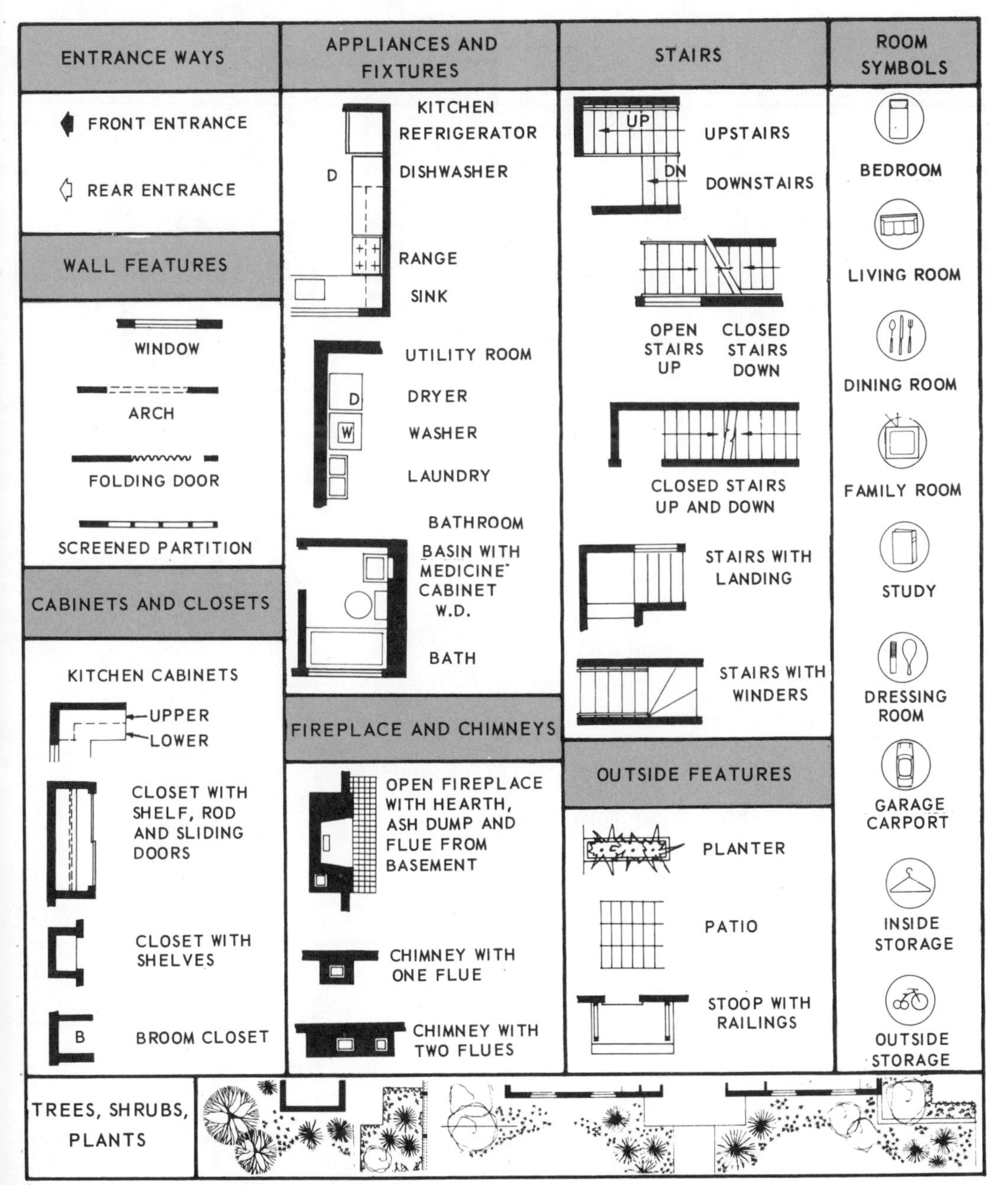

Fig. 15-1-5 Floor plan symbols used on presentation drawings

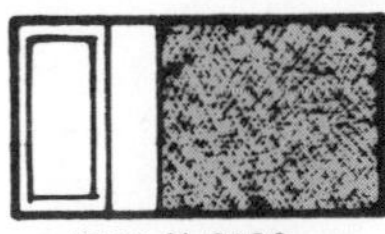

1000 X 2000
SINGLE BED

2000 X 850
CHESTERFIELD

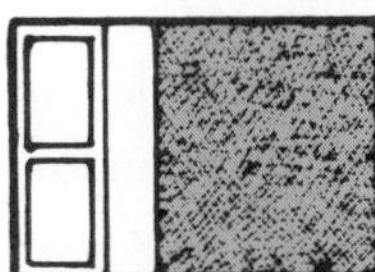

1350 X 2000
DOUBLE BED

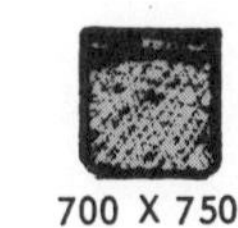

700 X 750
OCCASIONAL CHAIR

800 X 850
ARMCHAIR

1200 X 450
DOUBLE

750 X 450
SINGLE

DRESSERS

500 X 400
BEDSIDE TABLE

900 X 450
DESK

1200 X 900
TABLE

400 X 650
END TABLE

1200 X 450
COFFEE TABLE

1200 X 450
BUFFET

Fig. 15-1-6 Basic furniture size and symbols

UNIT 15-2
Construction Drawings

Construction (Working) drawings contain the information required by the builder to erect the building. In order to understand and read construction drawings, it is necessary to become familiar with architectural symbols and terms. Figures 15-2-1 and 15-2-2 illustrate the framing of a bungalow and give the architectural terms for the principal parts of the house.

1. PREFABRICATED LADDER FRAME FOR VERGE
2. PREFABRICATED GABLE END UNIT
3. PLY OR ALUMINUM SOFFIT
4. 19 mm FASCIA
   OVERHANG TIED TO FIRST TRUSS AND BEARING ON GABLE END UNIT
5. MOULDING
6. GYPSUM BOARD INTERIOR FINISH
7. WALL SHEATHING (INSULATED IF REQUIRED)
8. VAPOUR BARRIER
9. 38 X 140 mm STUDS AT 600 mm O.C. MAX.
10. INSULATION
11. AIR SPACE BEHIND BRICK
12. VERTICAL WOOD OR ALUMINUM SIDING FINISH
13. BRICK VENEER
14. FLASHING
15. STEEL ANGLE LINTEL
16. BUILDING PAPER
17. WINDOW FRAME
18. 38 X 140 mm SOLE PLATE
19. SUBFLOOR
20. DOUBLE 38 X 140 mm PLATES
21. 38 X 140 mm STUDS AT 600 mm O.C. MAX.
22. FLASHING
23. BRICK TIES
24. WEEP HOLES
25. FLOOR FINISH ON UNDERLAY

FRONT ELEVATION

26. VERTICAL WOOD OR ALUMINUM SIDING EXTERIOR FINISH
27. LINTEL
28. FINISHED GRADE
29. CONCRETE CORBEL TO SUPPORT PILASTER
30. GYPSUM BOARD INTERIOR FINISH
31. VAPOUR BARRIER
32. RIGID INSULATION IN TOP SECTION ONLY
33. DAMPPROOFING TO GRADE
34. 200 mm POURED CONCRETE FOUNDATION WALL
35. CRUSHED STONE
36. WEEPING TILE
37. PREFABRICATED ROOF TRUSS 600 mm O.C. MAX.
38. ROOF INSULATION
39. LINTEL
40. 38 X 38 mm SOFFIT BEARER
41. TOP CHORD OF TRUSS EXTENDED TO SUPPORT FASCIA
42. PLY OR ALUMINUM SOFFIT WITH CONTINUOUS VENTING
43. 19 mm FASCIA ON 38 X 89 mm BACKING
44. ROOF SHEATHING
45. ROOF SHEATHING TRIMMED FOR CONTINUOUS RIDGE VENTING
46. RIDGE
47. 38 X 38 mm STRAPPING
48. CROSS BRIDGING
49. GYPSUM BOARD CEILING FINISH
50. FLOOR JOISTS
51. STRINGER
52. TREAD
53. RISER
54. HEADER

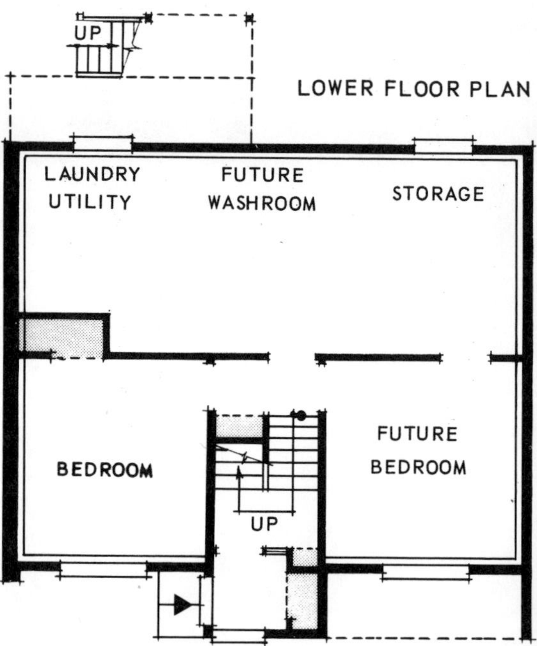

Fig. 15-2-1 House framing

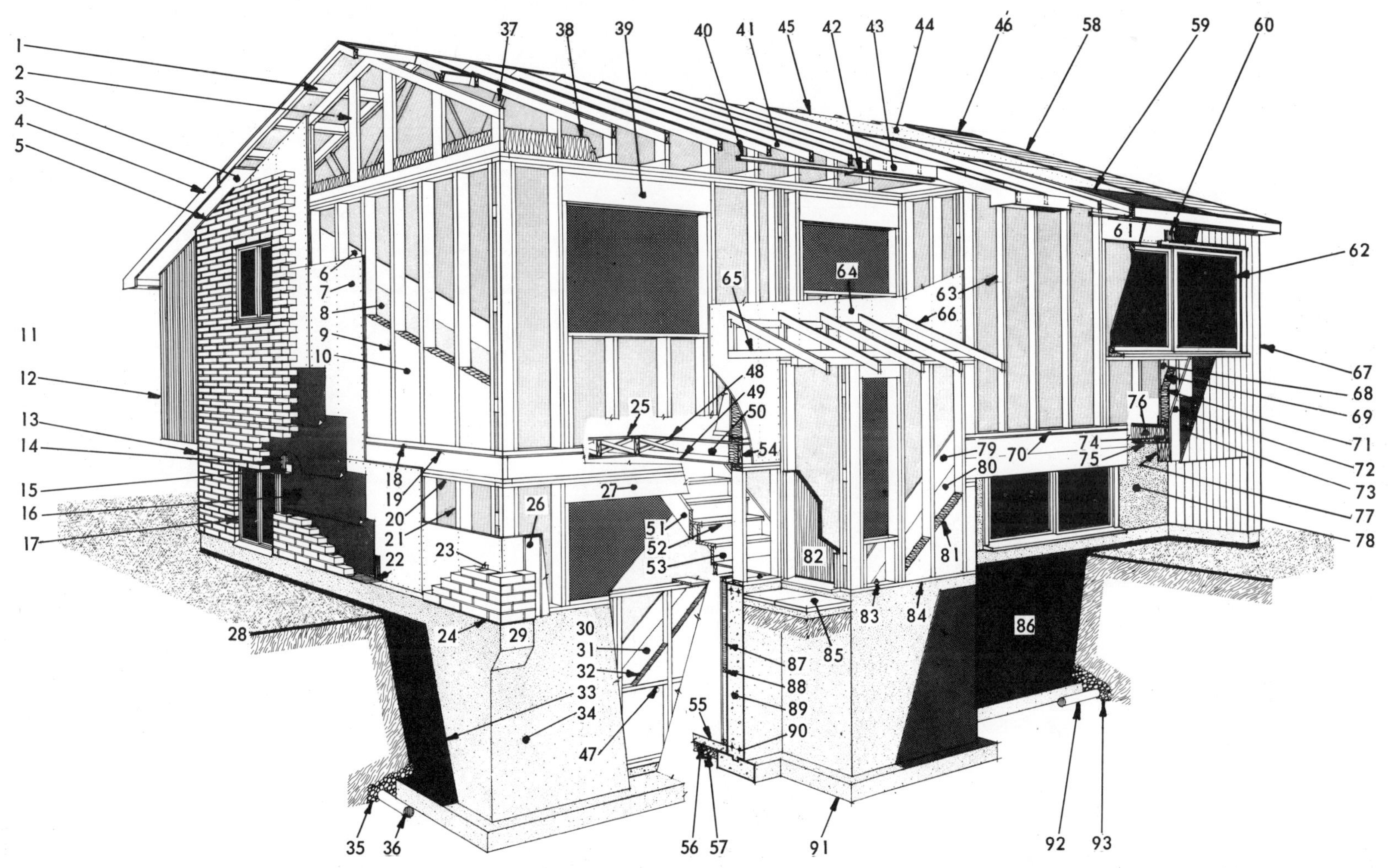

57. 25 mm GRANULAR FILL
58. ASPHALT SHINGLES
59. EAVE PROTECTION
60. FLASHING
61. LINTEL
62. WINDOW FRAME
63. 38 X 140 mm STUDS AT 600 mm O.C. MAX.
64. WALL SHEATHING (INSULATED IF REQUIRED)
65. 38 X 89 mm RAIL
66. 38 X 89 mm FRAMED RAFTERS AND CEILING JOISTS OVER ENTRANCE VESTIBULE
67. VERTICAL WOOD OR ALUMINUM SIDING EXTERIOR FINISH
68. GYPSUM BOARD INTERIOR FINISH
69. VAPOUR BARRIER
70. 38 X 140 mm SOLE PLATE
71. INSULATION
72. WALL SHEATHING (INSULATED IF REQUIRED)
73. BUILDING PAPER
74. HEADER
75. PLY SOFFIT – INVERTED
76. INSULATION TO FULL DEPTH OF JOIST IN OVERHANG
77. 3 – 38 X 235 mm BEAM
78. STUCCO FINISH
79. GYPSUM BOARD INTERIOR FINISH
80. VAPOUR BARRIER
81. INSULATION
82. ENTRANCE DOOR
83. ANCHOR BOLTS AT 2400 mm O.C. MAX.
84. 38 X 140 mm SILL PLATE
85. 600 X 760 mm PRECAST PATIO SLABS ON 50 mm SAND
86. DAMPPROOFING TO GRADE
87. RIGID INSULATION IN TOP SECTION ONLY
88. 38 X 38 mm STRAPPING
89. 200 mm POURED CONCRETE FOUNDATION WALL
90. 10 m REINFORCING BARS TOP AND BOTTOM
91. 200 X 400 mm MIN. CONCRETE FOOTING
92. WEEPING TILE
93. CRUSHED STONE

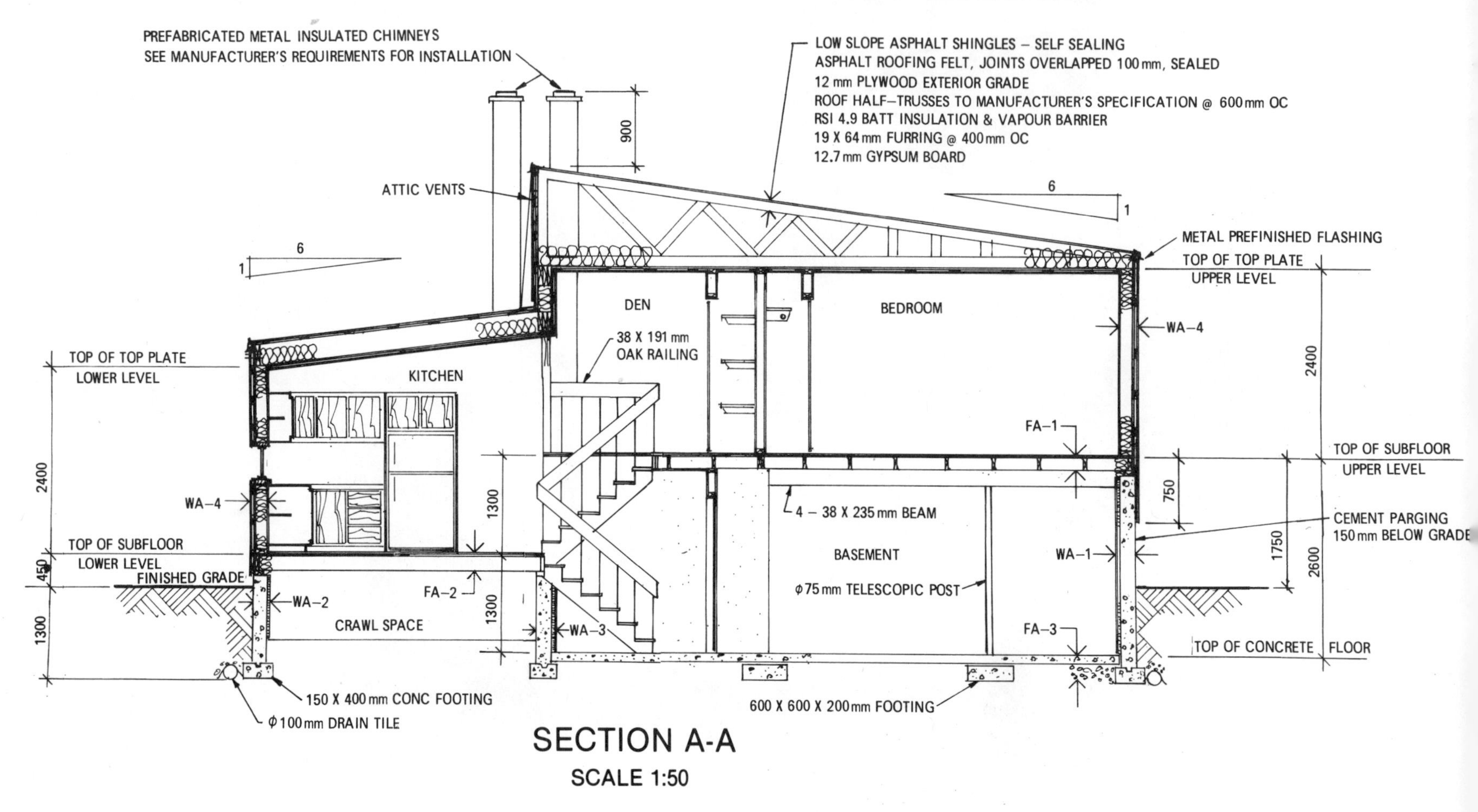

SECTION A-A

SCALE 1:50

FLOOR ASSEMBLY SCHEDULE

FA–1 CARPET
6 mm PLYWOOD UNDERLAY
15 mm PLYWOOD SUBFLOOR
38 X 184 mm FLOOR JOISTS @ 400 mm OC
38 X 184 mm SOLID BRIDGING
12.7 mm GYPSUM BOARD

FA–2 FINISHED FLOORING
6 mm PLYWOOD UNDERLAY
15 mm PLYWOOD SUBFLOOR
38 X 184 mm FLOOR JOISTS @ 300 mm OC
38 X 184 mm SOLID BRIDGING
VAPOUR BARRIER
RSI 4.9 FRICTION BATT INSULATION
WIRE LATH

FA–3 75 mm CONCRETE SLAB
0.15 POLYETHYLENE
150 mm COMPACTED GRAVEL

WALL ASSEMBLY SCHEDULE

WA–1 CEMENT PARGING 150 mm BELOW FINISH GRADE
200 mm CONCRETE WALL
50 mm RIGID INSULATION AND VAPOUR BARRIER
12.7 mm GYPSUM BOARD
DAMPPROOF ALL CONCRETE BELOW GRADE

WA–2 CEMENT PARGING 150 mm BELOW FINISH GRADE
200 mm CONCRETE WALL
50 mm RIGID INSULATION AND VAPOUR BARRIER
6.35 mm ASBESTOS BOARD

WA–3 200 mm CONCRETE WALL
50 mm RIGID INSULATION AND VAPOUR BARRIER
12.7 mm GYPSUM BOARD

WA–4 15 mm RUFF–SAWN PLYWOOD OR METAL SIDING
19 X 64 mm HORIZONTAL FURRING @ 400 mm OC
#15 BUILDING PAPER
9.5 mm FIBREBOARD SHEATHING
RSI 3.5 BATT INSULATION & VAPOUR BARRIER
38 X 140 mm STUDS @ 600 mm OC
12.7 mm GYPSUM BOARD

WA–5 12.7 mm GYPSUM BOARD
38 X 89 mm STUDS @ 400 mm OC
12.7 mm GYPSUM BOARD

Fig. 15-2-2 Section through a house

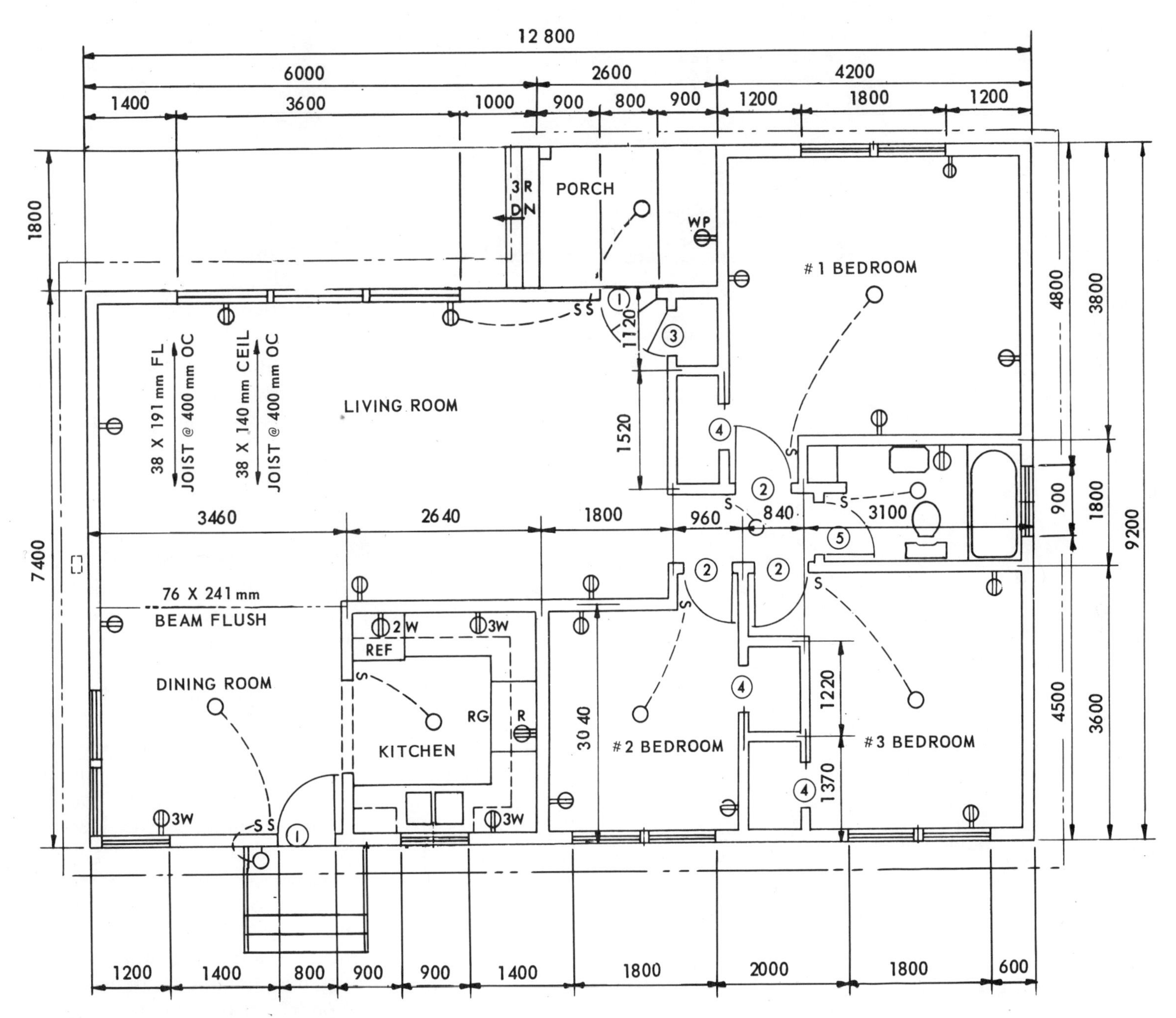

Fig. 15-2-3 Floor plan of a summer cottage

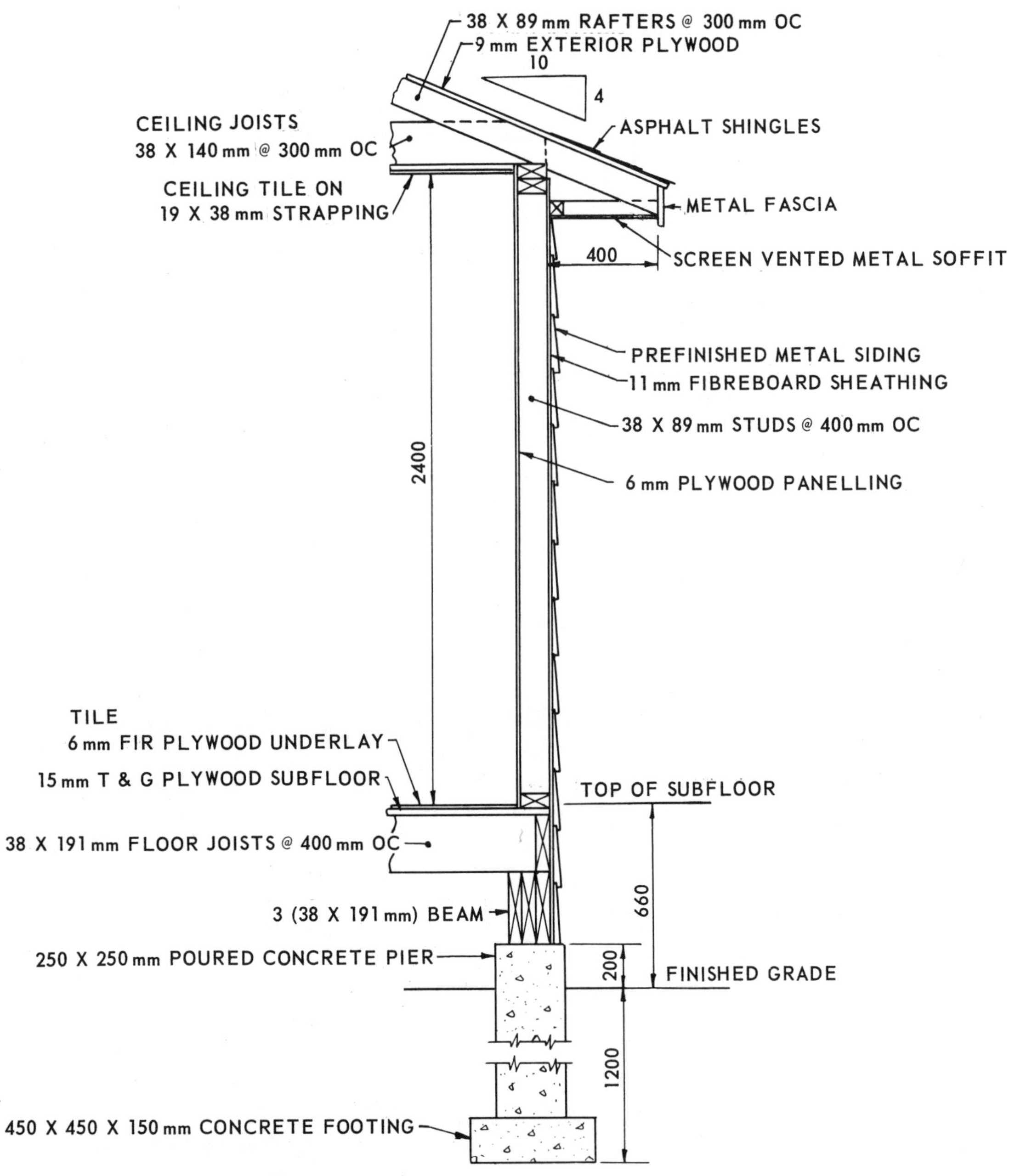

Fig. 15-2-4 Exterior wall section of summer cottage

A typical set of construction drawings for a house includes: floor plans for each level, a basement plan, a typical section through a wall, and elevations of the front, the back, and the two sides of the house. The floor plan and typical section view for a summer cottage are shown in Figures 15-2-3 and 15-2-4.

Drawing Scales

Floor plans and elevations for houses are usually drawn to the scale 1:50. Section views which show close detail are drawn to a larger scale such as 1:10.

Lettering

Lettering should be used on drawings to convey information that is not readily or clearly indicated by the drawings alone. The combination of lettering and drawing should fully and concisely define the object being constructed.

The most important requirements for lettering are legibility, reproducibility and ease of execution. These are particularly important due to the increased use of microfilming which requires optimum clarity and adequate size of all details and lettering.

The requirements for lettering are best met by the style of lettering known as standard upper case roman (simple block without serif or scroll), which should be used exclusively on drawings, except for the metric symbols requiring lower case lettering, e.g., millimetres (mm) and metres (m). Condensed or extended styles are not recommended.

Except for the main headings and drawing number, the height of numbers and letters on architectural drawings should be approximately 3 mm.

Drawing Symbols

Figures 5-2-5 to 5-2-12 show the more common symbols found on construction drawings.

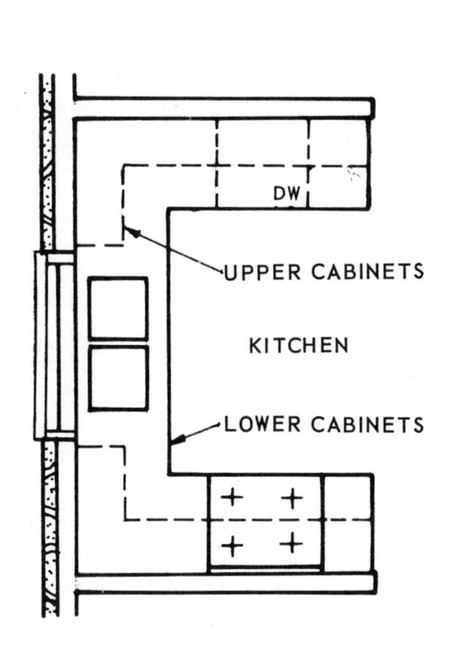

Fig. 15-2-5 Kitchen cabinets

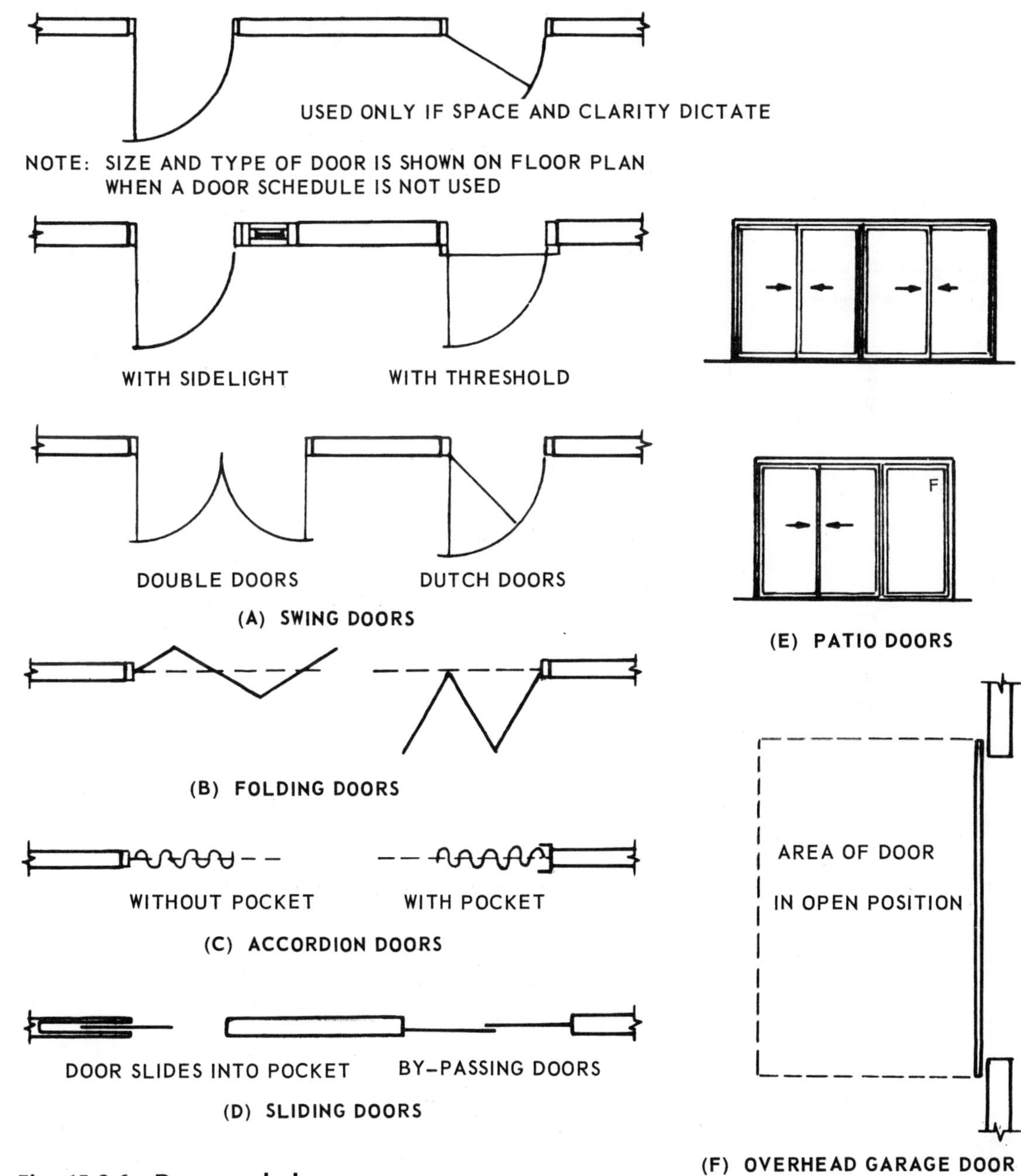

Fig. 15-2-6 Door symbols

PLAN AND SECTION	ELEVATION	PLAN AND SECTION	ELEVATION
POURED CONCRETE		RIGID INSULATION	
REINFORCED CONCRETE		BATT INSULATION	
CONCRETE OR CINDER BLOCK		EARTH	
BRICK MASONRY		GRAVEL	
CUT STONE MASONRY		PLASTER, GYPSUM BOARD CEMENT	
NATURAL STONE		GLASS	
ARTIFICIAL STONE		GLASS BLOCK	
WOOD FRAMING FOR NEW WORK FOR ALTERATION WORK		STRUCTURAL STEEL	

Fig. 15-2-7 Material symbols

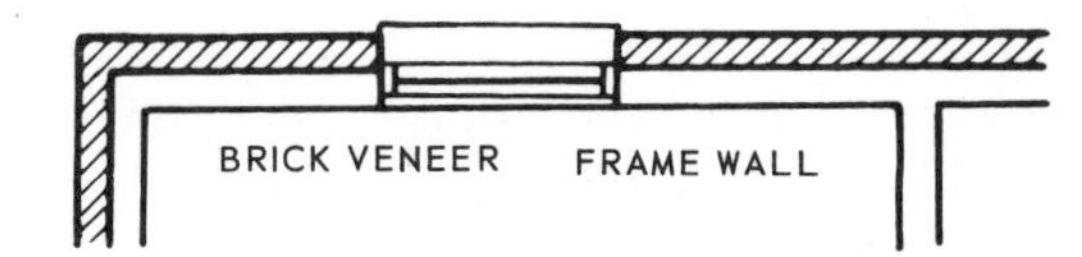

(A) BRICK VENEER ON FRAME WALL

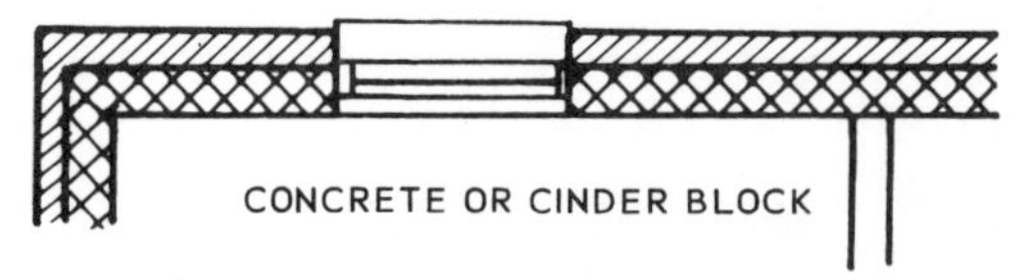

(B) BRICK VENEER ON CONCRETE OR CINDER BLOCK

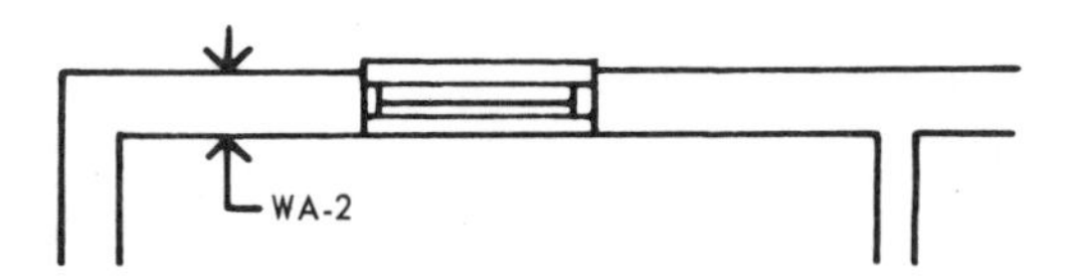

(C) WALL SYMBOLS NEED NOT BE SHOWN WHEN WALL CONSTRUCTION DETAILS ARE SHOWN OR GIVEN IN A WALL SCHEDULE

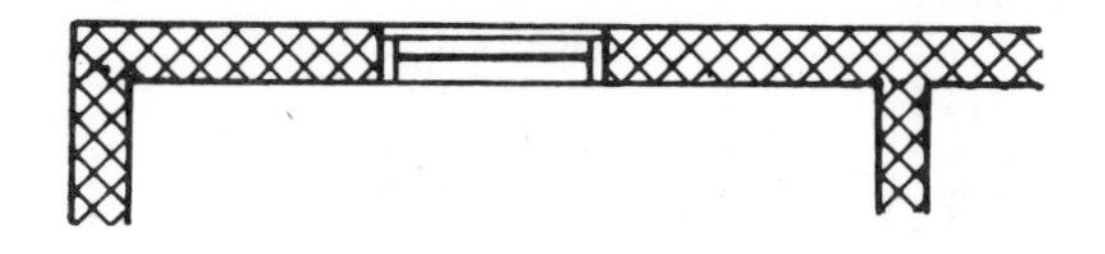

(D) BLOCK FOUNDATION

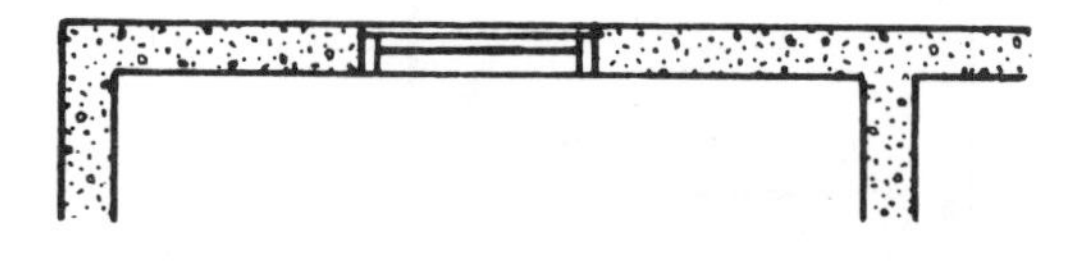

(E) POURED CONCRETE FOUNDATION

Fig. 15-2-8 Wall symbols

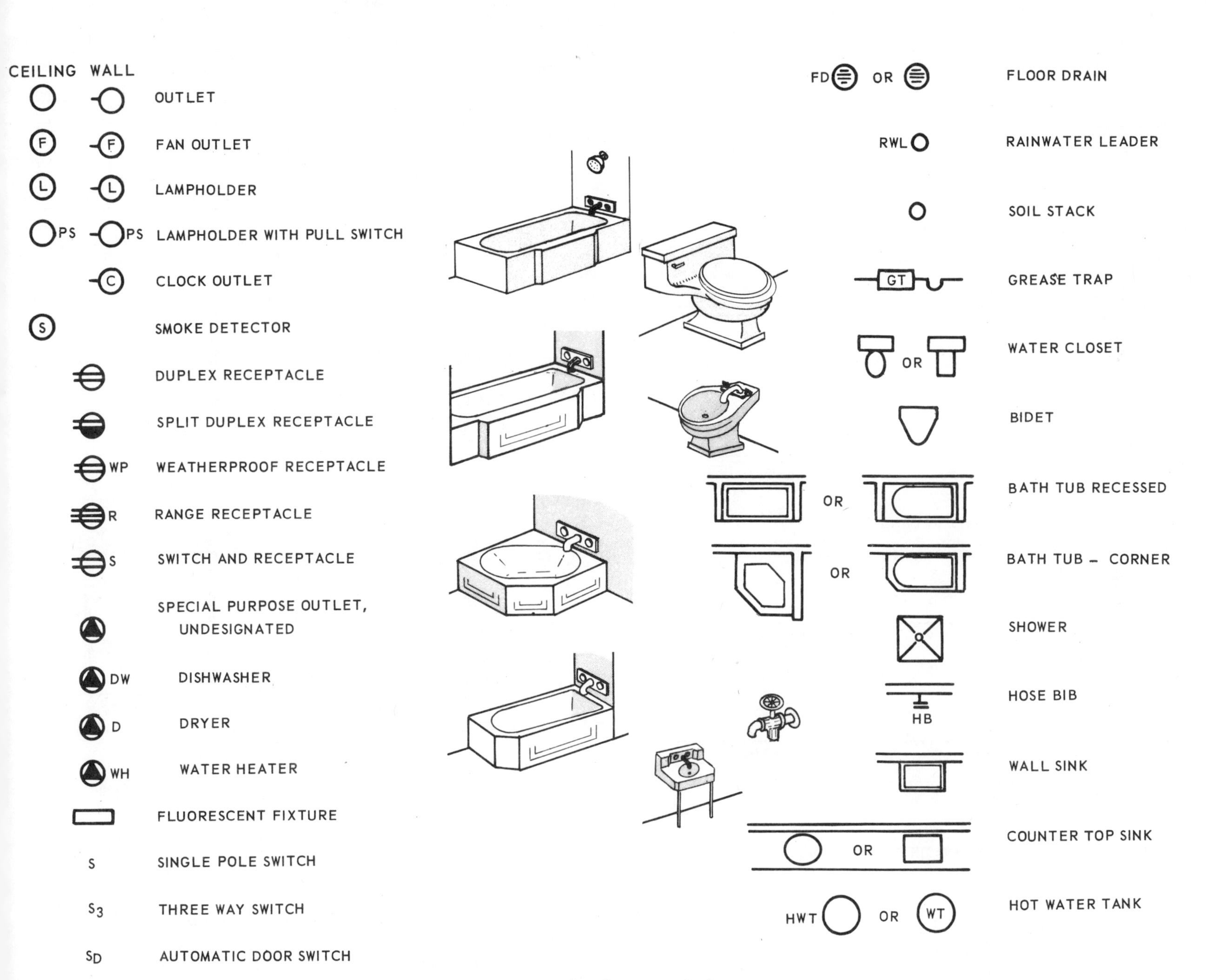

Fig. 15-3-9 Electrical symbols

Fig. 15-2-10 Plumbing symbols

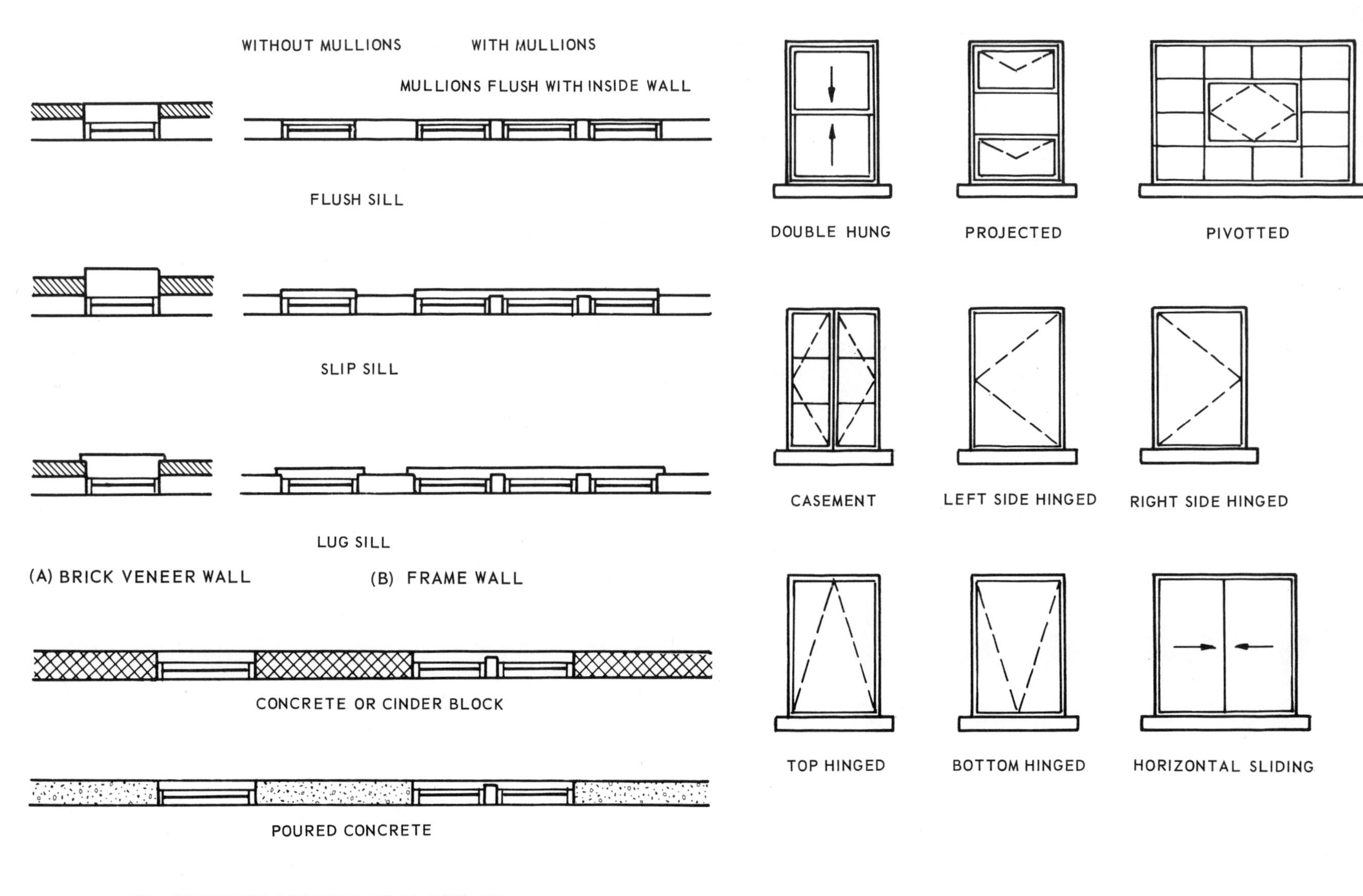

Fig. 15-2-11 Window symbols

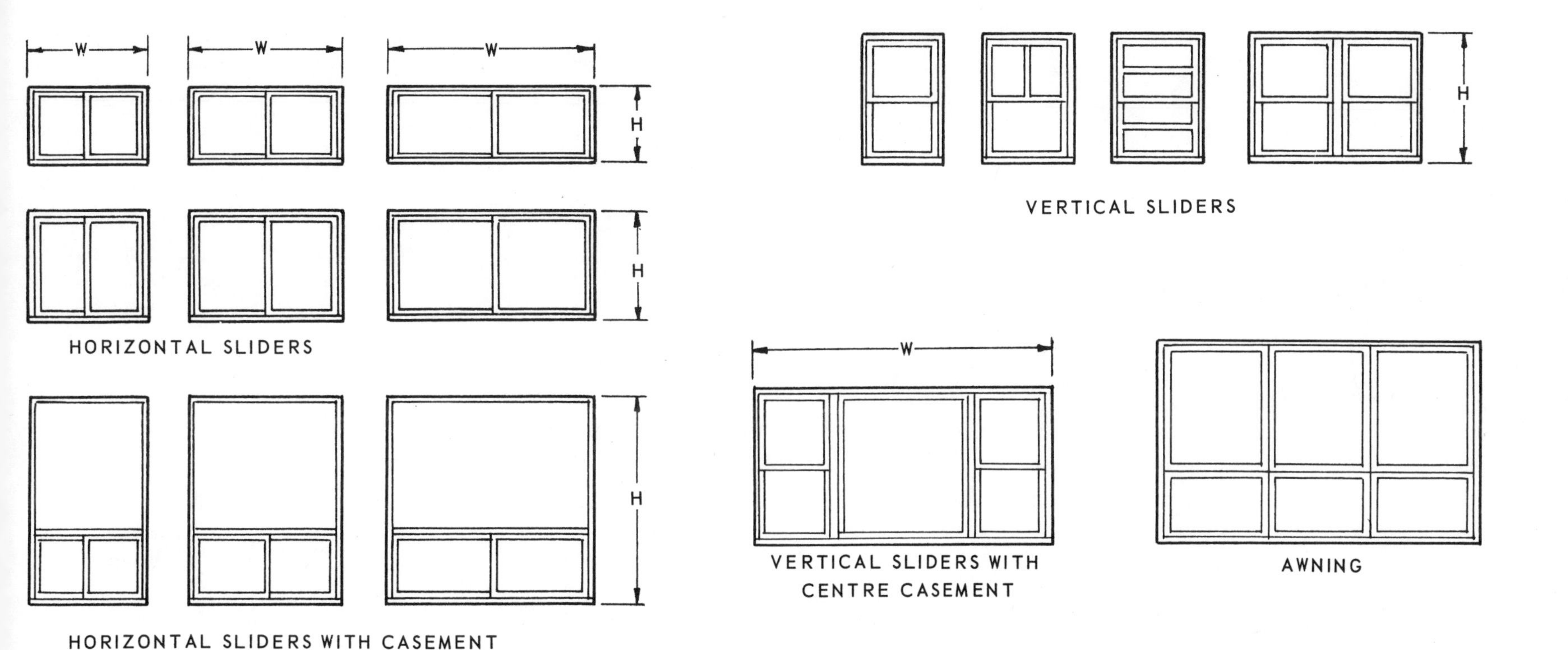

CASEMENT

HEIGHT	WIDTH							NOTES
	600	900	1200	1500	1800	2100	2400	
2400	1	2	1	2	1	2	1	1. FIRST PREFERENCE
2100	1	2	1	2	1	2	1	2. SECOND PREFERENCE
1900	1	2	1	2	1	2	1	
1800	1	2	1	2	1	2	1	
1600	1	2	1	2	1	2	1	
1500	1	2	1	2	1	2	1	
1200	1	2	1	2	1	2	1	SIZES SHOWN
900	1	2	1	2	1	2	1	INCLUDE FRAME
600	1	2	1	2	1	2	1	AND CLEARANCE

Fig. 15-2-12 Window details

Dimensioning

Dimensioning for architectural construction drawings differs from dimensioning for mechanical drafting in several ways.

1. The linear units of measurement used in architectural drawings are the millimetre and the metre.

2. The smallest unit of linear measurement is one millimetre. Thus the decimal point and parts of a millimetre are not used with linear millimetre dimensioning. However, due to fire regulations, when specifying the thickness of building products, such as wall sheathing and plywood, the thickness, where required, may be shown to one or two decimal places. Examples are

12.7 mm GYPSUM BOARD
0.15 mm POLYETHYLENE FILM (VAPOUR BARRIER)
15.5 mm PLYWOOD SUBFLOOR

3. Metres and decimal parts of the metre are used to a limited extent on some architectural drawings. Thus, a dimension having a decimal point is a dimension in metres. When the decimal point is used on construction drawings the dimension must be shown to three places beyond the decimal marker. For presentation drawings it is not necessary to take the dimension to three places beyond the decimal marker.

4. Metre and millimetre dimensioning on one drawing, where possible, should be avoided.

5. Dimension lines are not broken and the dimension is placed above the dimension line.

6. Aligned dimensioning is preferred.

7. An oblique 45° slash crossing the intersection of the dimension and extension line is more commonly used than an arrowhead. See Figure 15-2-13.

8. All stud walls should be dimensioned to the face of the stud, rather than to the centre line of the wall.

Dimensioning Systems

There are two basic methods of dimensioning the location of features, such as doors and windows on working drawings. See Figure 15-2-14. One method is to dimension the opening in which the component fits. The second method is to dimension to the centre of the feature. Once the manufacturers of building components—doors, windows, blocks, etc., establish standard modular sizes (dimensional co-ordination) for their products, then the dimensioning to the centre of the feature will become obsolete.

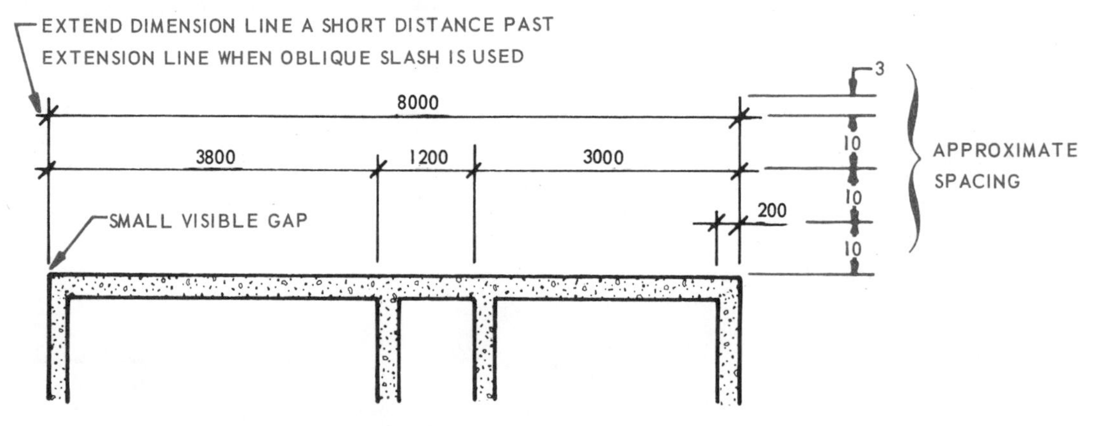

(A) OBLIQUE LINE TO INDICATE EXTENT OF DIMENSION

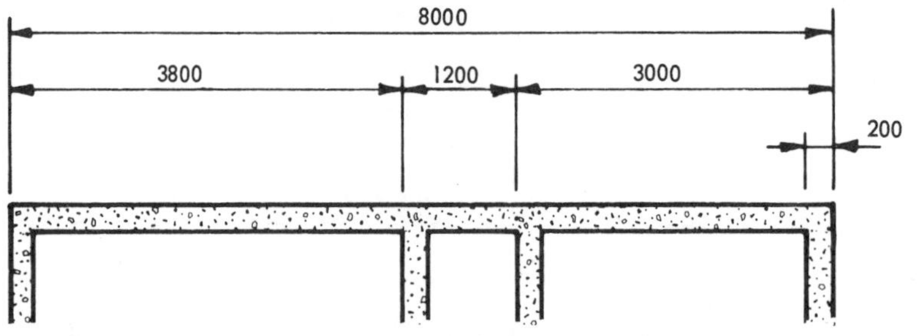

(B) ARROWHEAD TO INDICATE EXTENT OF DIMENSION

Fig. 15-2-13 Linear dimensioning

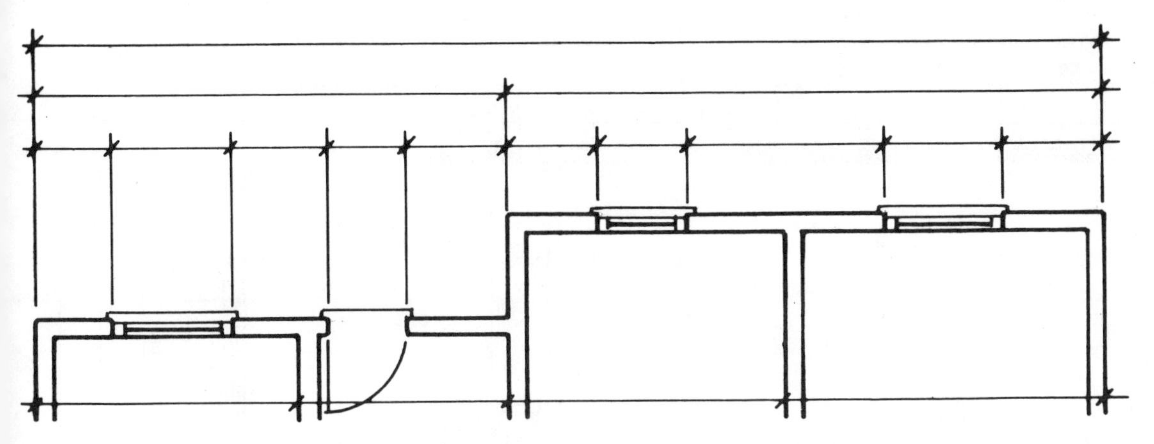

DIMENSIONING INTERIOR PARTITIONS AND TO THE OPENING IN WHICH DOORS AND WINDOWS FITS

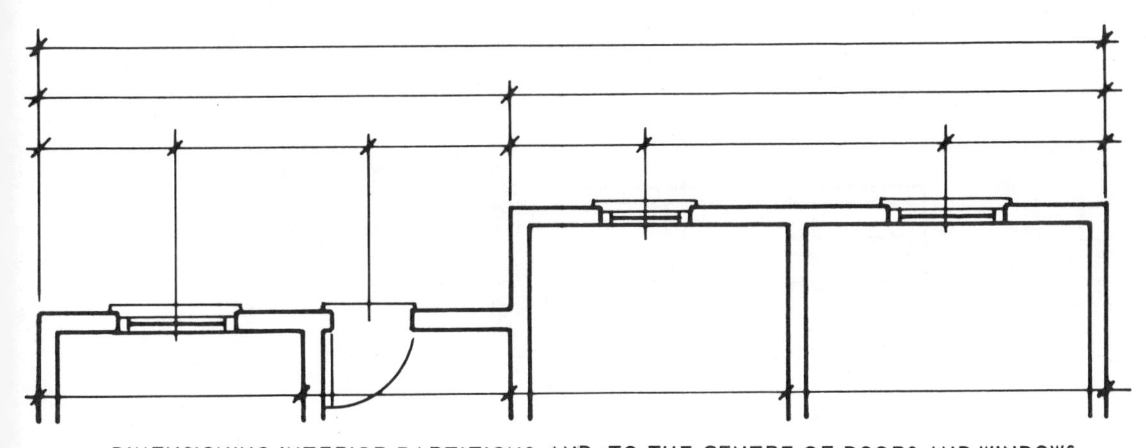

DIMENSIONING INTERIOR PARTITIONS AND TO THE CENTRE OF DOORS AND WINDOWS

Fig. 15-2-14 Dimensioning the location of components

Basic dimensioning rules illustrated by the numbered arrows in Figure 15-2-15 are:

1. The space between parallel dimension lines and the object line should not be less than 10 mm.
2. Interior stud walls are dimensioned to the face of the stud, and except for the one outside wall, should all be dimensioned to the same side. This practice assists the carpenter when erecting and locating the walls. Lines indicating the position of the walls are marked off on the sub floor and the walls are positioned and erected along these lines.
3. Overall building dimensions are located outside the other dimensions.
4. Overall dimensions of the wood framing are measured to the outside face of the stud walls.
5. When an offset occurs on the outer walls, the lengths of the offset (masonry and stud wall) are aligned in a straight line.
6. All of the dimensions locating the openings in an outside wall are shown aligned in a straight line and must add up to the overall dimensions.
7. In dimensioning stairs, the number of risers and the word UP or the abbreviation DN are shown with an arrow indicating the direction.
8. The dimensions of the brick and stone veneer are shown on the floor plan and are normally in line with the overall wall dimension.
9. Curved leaders are sometimes used to eliminate confusion with other lines or when there is a limited space.
10. Linear dimensions that cannot be seen on the floor plan or those too small to be placed on the object are placed on leaders for easier reading.
11. Direction, size and spacing of joists are given as a note and symbol.
12. Direction and size of beams are given on the floor plan and section drawing.

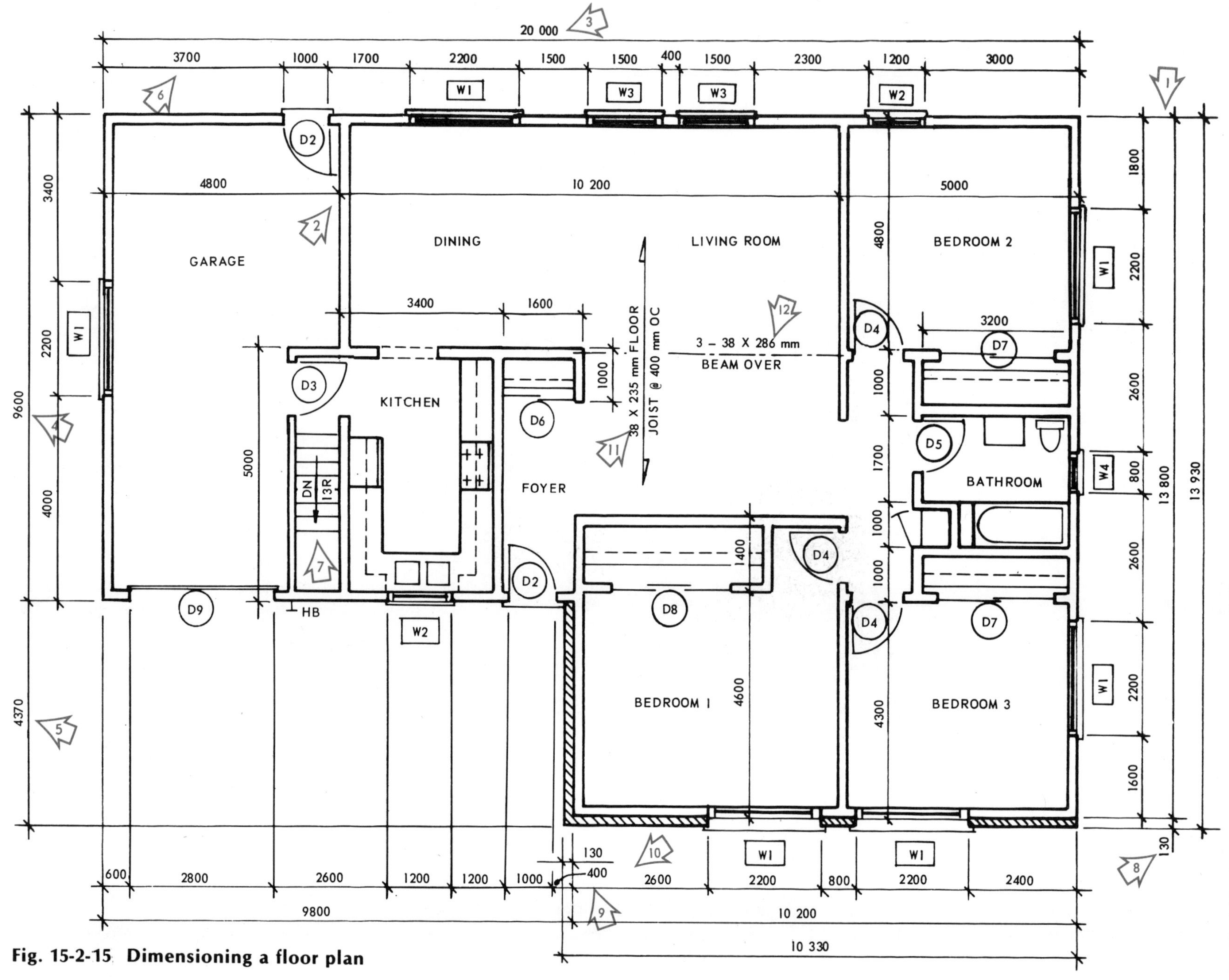

Fig. 15-2-15 **Dimensioning a floor plan**

Review Questions

1. How are location dimension of doors and windows given on a working drawing?
2. How are location dimensions for interior partitions given?
3. What does a typical set of *construction* drawings for a house include?

Assignments

1. Make the construction floor plan drawing of the cottage shown in Fig. 15-2-A.

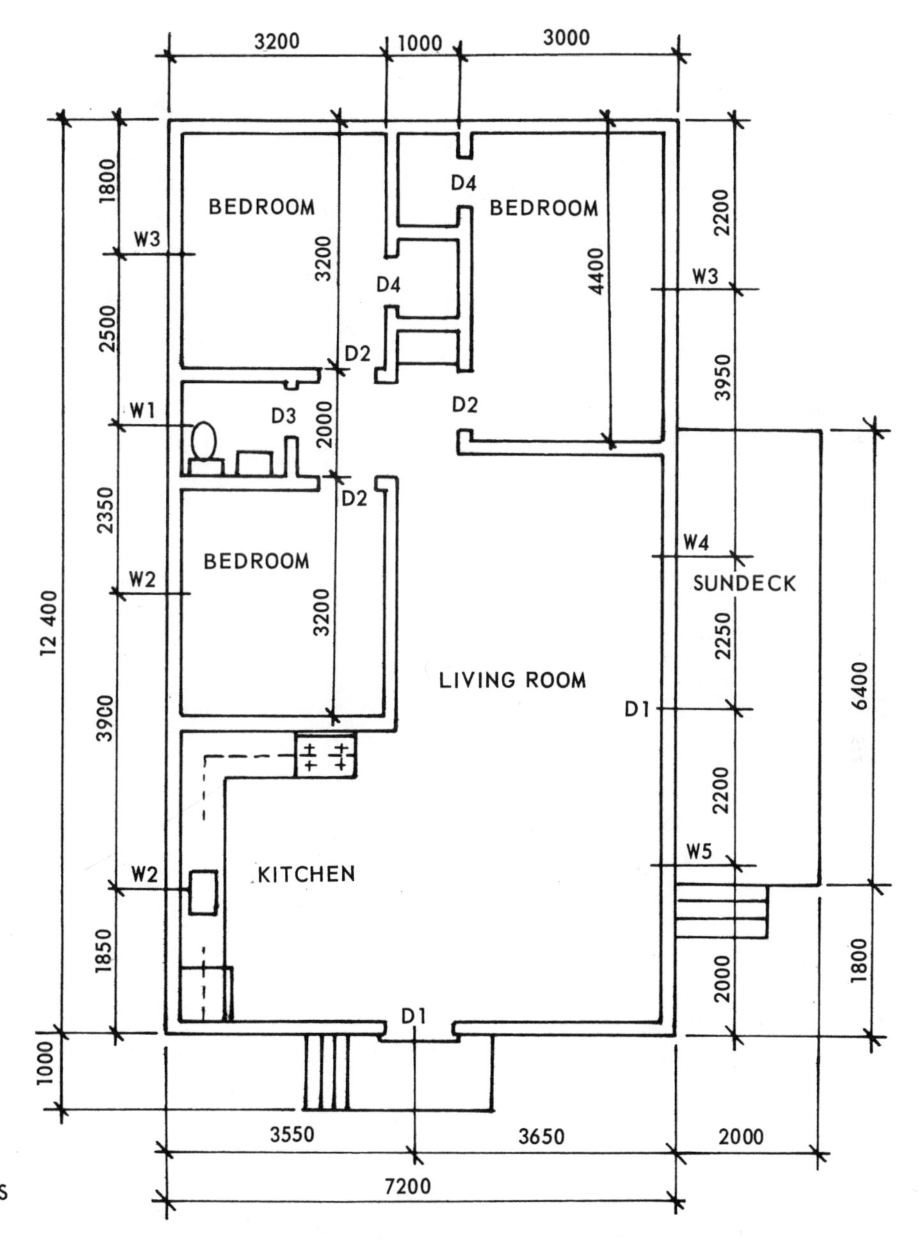

DOOR SCHEDULE

SWING

D1 900 WIDE

D2 800 WIDE

D3 700 WIDE

FOLDING

D4 700 WIDE

WINDOW SCHEDULE

HORIZONTAL SLIDERS

W1 600 WIDE

W2 900 WIDE

W3 1200 WIDE

HORIZONTAL SLIDER WITH CASEMENT

W4 1500 WIDE

W5 1800 WIDE

DIMENSION TO THE OUTSIDE OF THE DOORS AND WINDOWS

Fig. 15-2-A Summer cottage

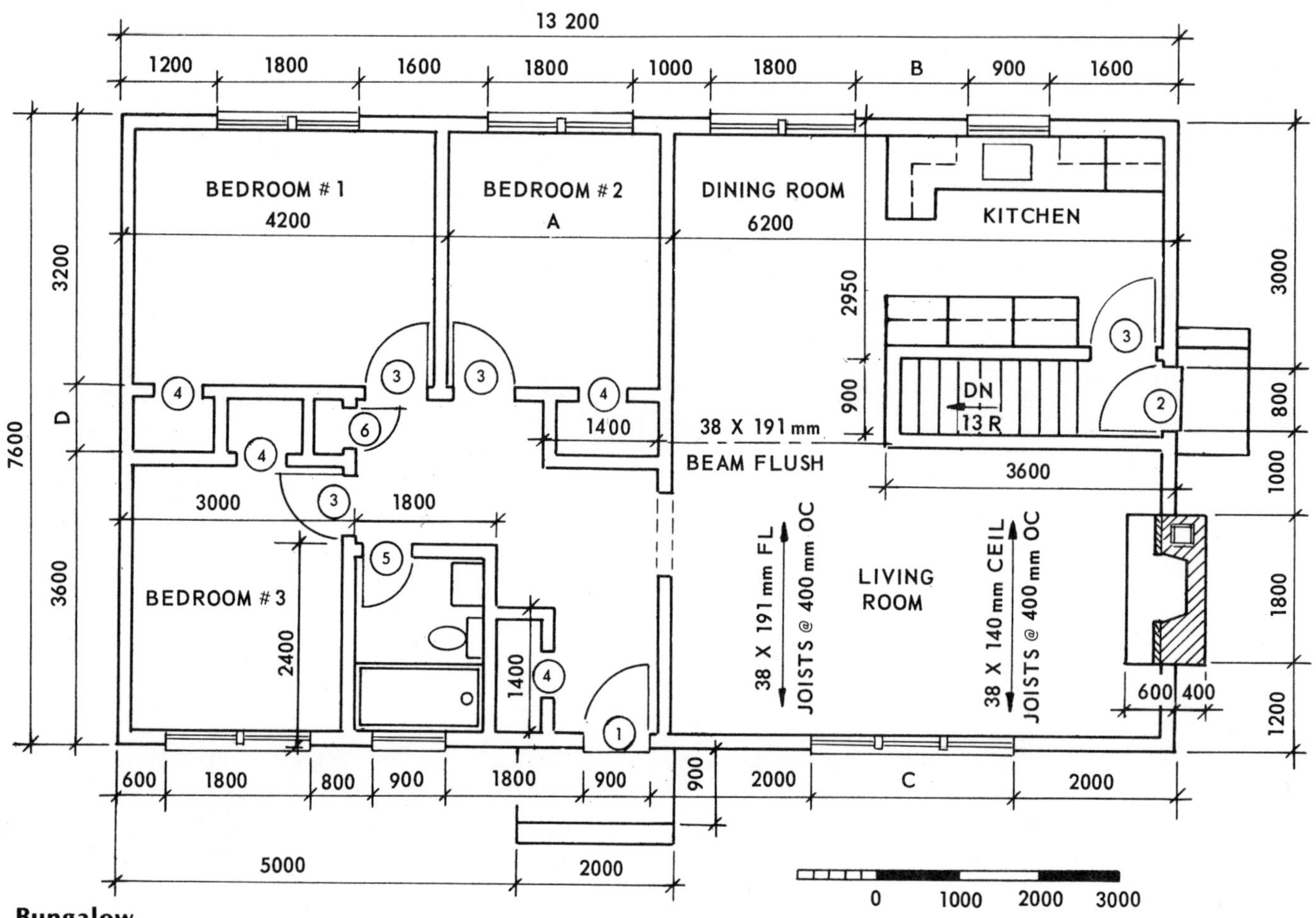

Fig. 15-2-B Bungalow

2. With reference to Fig. 15-2-B

A. Calculate dimensions A to D.

B. What type of exterior wall is shown?

C. What are the overall dimensions of the house?

D. Calculate the size of (1) the #3 bedroom, (2) the front hall closet, (3) the front porch, (4) the chimney, (5) the combined chimney and hearth.

E. What is the size and spacing of (1) the floor joists, (2) the ceiling joists?

F. How many closets are there?

G. What type of door is used on the bedroom closets?

H How many swinging doors are there?.

J. What is the length of the living room?

K. Where is the archway located?

L. Using the scale on the drawing, what is the total length of the kitchen counters? (Use counter front measurements. Do not include the refrigerator or range frontage).

3. Make the construction floor plan drawing of the house shown in Fig. 15-2-B.

UNIT 15-3
Developing a House Plan

How much space is required for the furniture in a small house? This is an important consideration in trying to devise an economical house plan with reasonable minimum dimensions. In buying furniture, it is important to select pieces that do not occupy too much floor space. Bulky old-fashioned furniture wastes space. Furniture is well designed if it is comfortable, convenient, compact, light, easily cleaned, and assists in giving the effect of free floor space.

Bedrooms

The three-bedroom house is most popular because it provides for a typical family of parents with children of both sexes. Each individual in a family spends more time in the bedroom than in any other room in the house. The bedroom should be tailored to suit the individual, to offer opportunity for reasonable self-expression and provide a haven from the rest of the family. A well designed bedroom gives the maximum in pleasure, comfort, leisure, and quiet.

Youngsters are generally active and noisy, so that, while their rooms should be close by those of the parents, there should be some form of sound barrier, possibly a bank of closets, separating them. Children's bedrooms should also be near the bathroom. Bedrooms vary in size, usually between 7.5

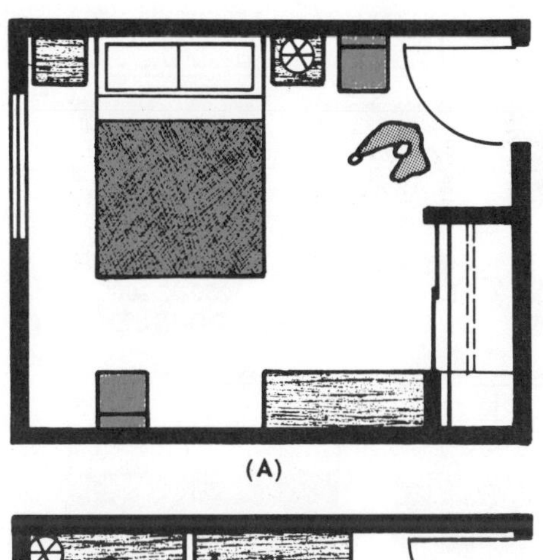

(A)

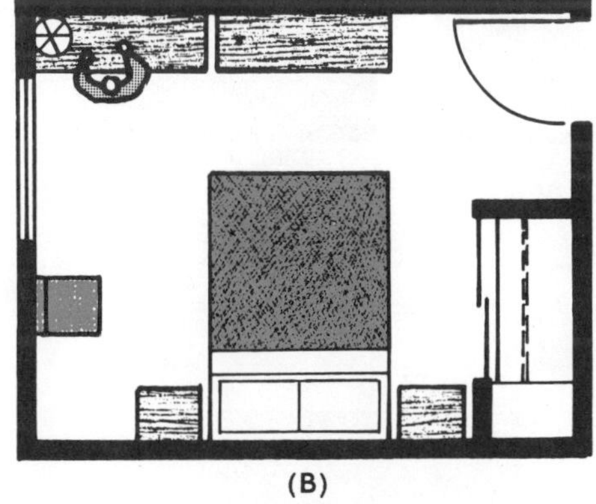

(B)

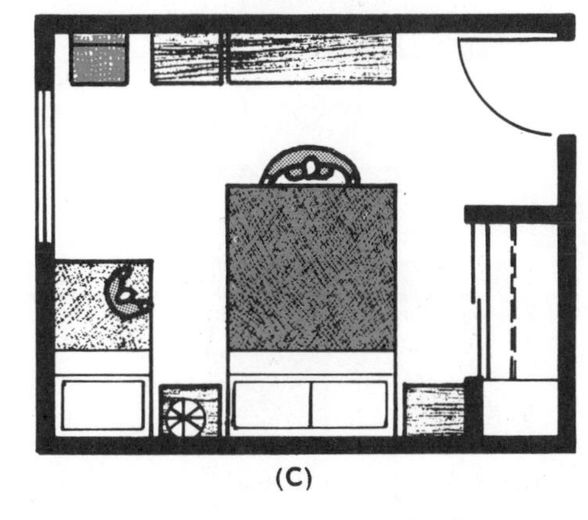

(C)

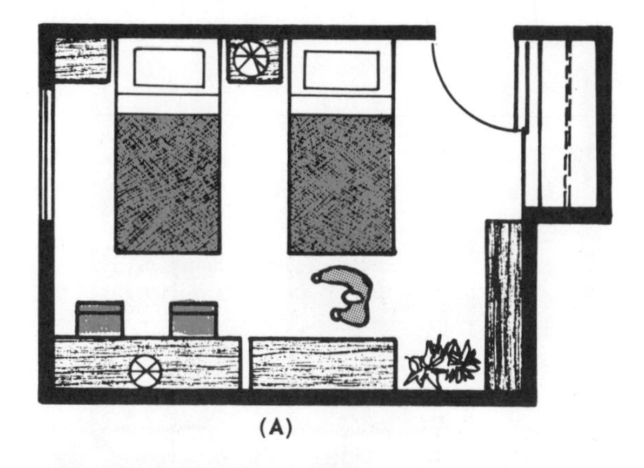

(A)

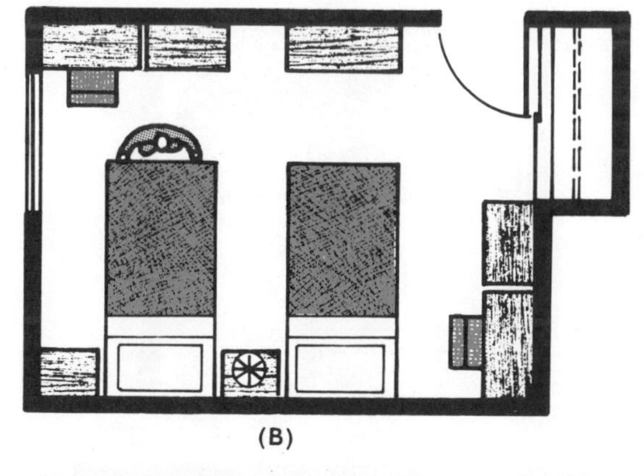

(B)

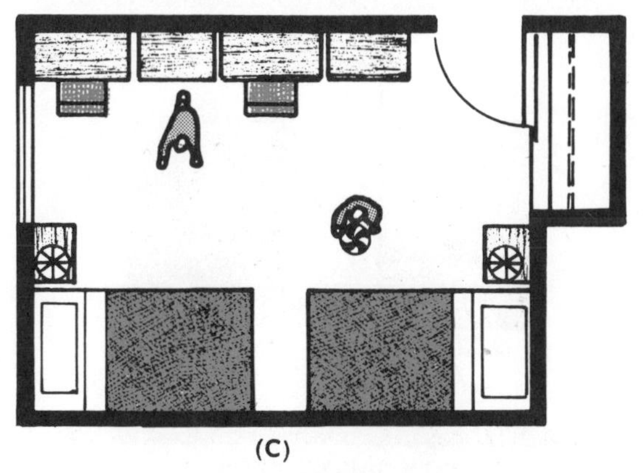

(C)

Fig. 15-3-1 Two-person bedrooms

Canada Mortgage and Housing Corp.

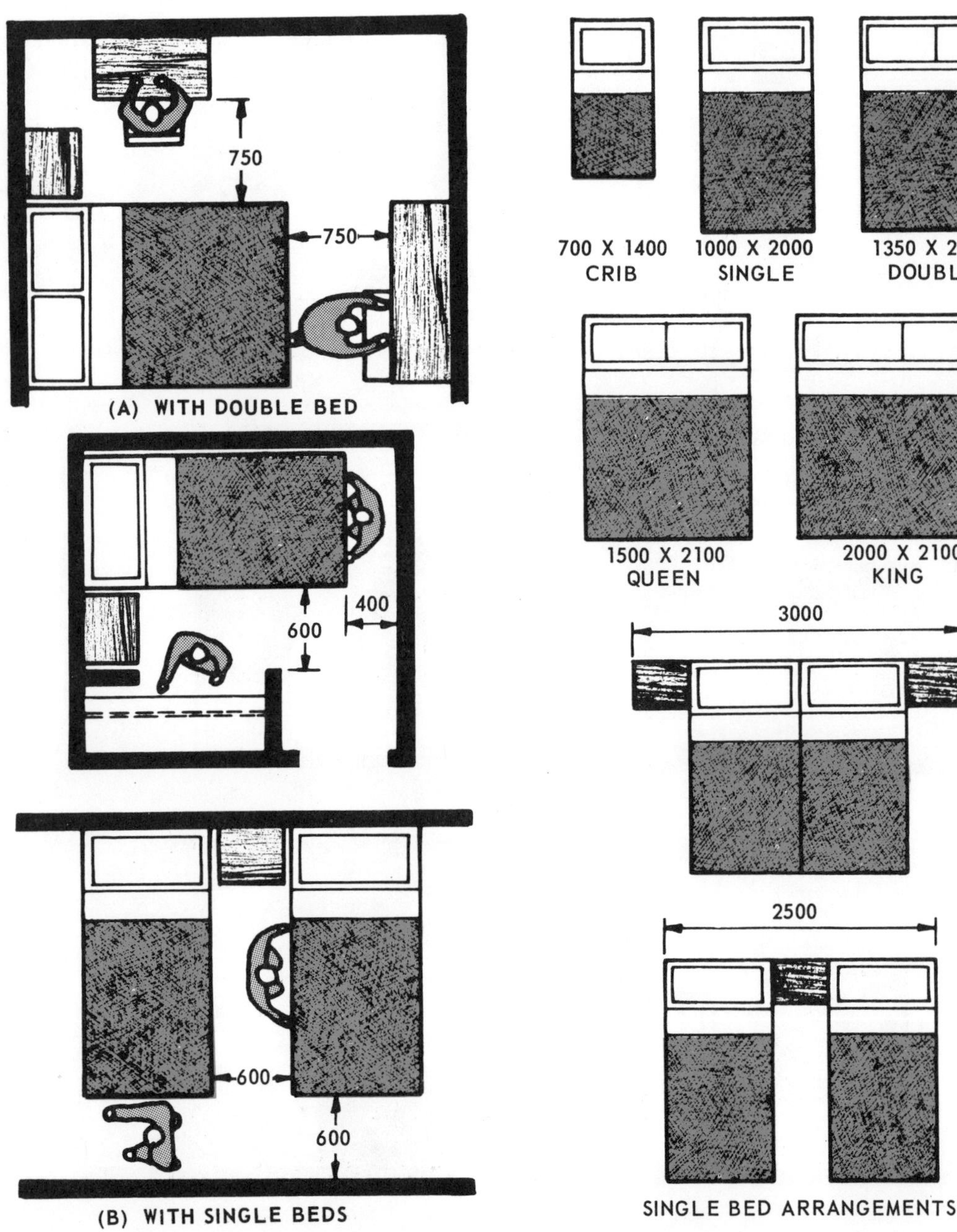

Fig. 15-3-2 Bedroom area clearances

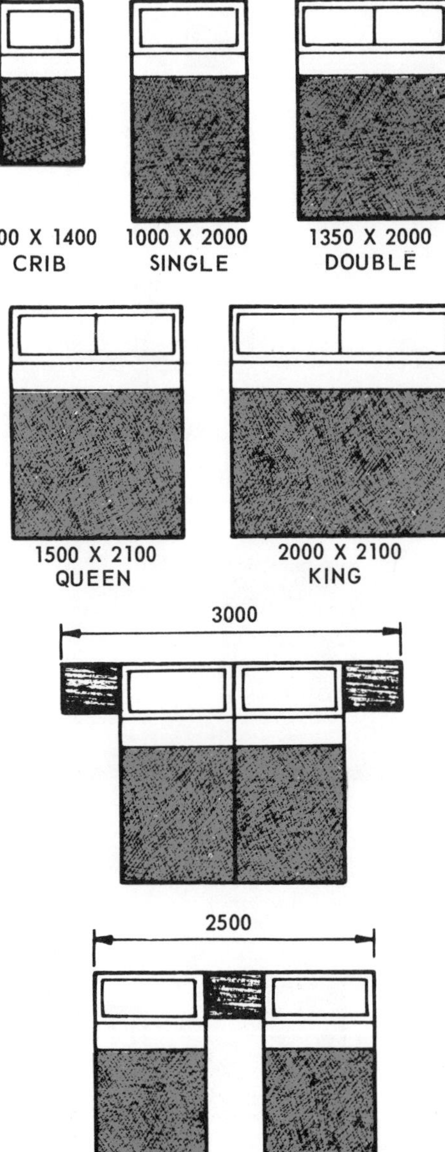

Fig. 15-3-3 Bed sizes and arrangements

and 14 m². Furniture is likely to occupy almost half of the floor area. The remaining space is necessary for dressing, to give access to drawers and cupboards, and to make beds and clean with mop and vacuum cleaner.

BEDROOM	FURNITURE REQUIREMENTS
PARENTS	1 DOUBLE BED 2 BEDSIDE TABLES 1 DOUBLE DRESSER OR 2 SINGLE DRESSERS
TWO-PERSON	2 SINGLE BEDS 2 BEDSIDE TABLES OR 1 BEDSIDE TABLE BETWEEN PARALLEL BEDS 1 DOUBLE DRESSER OR 2 SINGLE DRESSERS 2 WORK SURFACES
ONE-PERSON	1 SINGLE BED 1 BEDSIDE TABLE 1 SINGLE DRESSER 1 WORK SURFACE

Fig. 15-3-4 Bedroom furniture requirements

Bathrooms

In houses with bedrooms on two floors, for example, a two-storey home, it is a good idea to install bathroom facilities on both floors. Also, where the family is a large one, some duplication of bathroom facilities is desirable, although it is not necessary to duplicate all the fixtures. The most useful combination is to have a standard, four-fixture bathroom supplemented elsewhere by a compartment containing an additional wash basin and toilet. See Figure 15-3-2.

The main family bathroom should open off the hall serving the bedrooms. This hall should be accessible to all parts of the

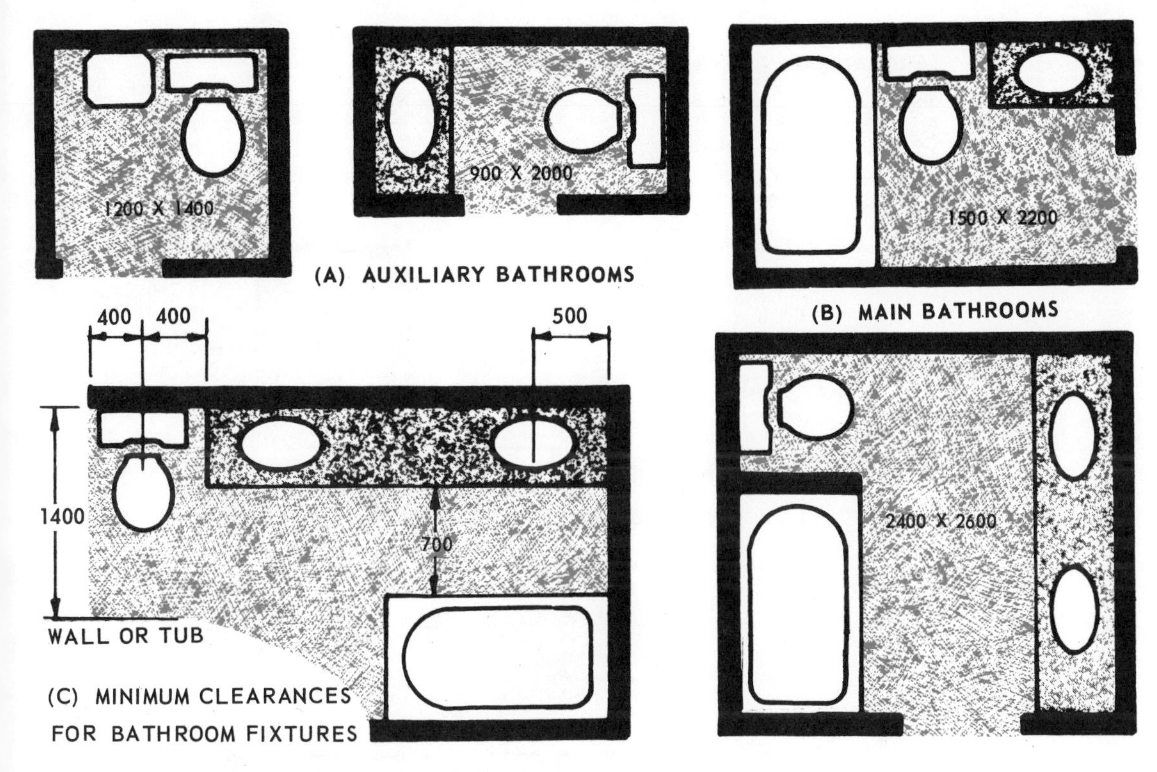

Fig. 15-3-5 Bathroom layouts and clearances

house without an individual's having to pass through any room, such as the living room, to reach it.

The auxiliary bathroom may be located off the master bedroom, in the basement if there are recreational facilities offered there, or off the family room. The latter location is ideally situated to provide scrub-up and toilet facilities for outdoor activities.

Kitchens

The kitchen serves both as a food preparation centre and as a food store. Often, food is eaten in the kitchen. Apart from regular meal preparation, other activities such as canning, baking, and pickle-making are done here. The kitchen is also used as a "home base" for cleaning the house, for doing laundry work, and for ironing. Children's home-work and hobbies are sometimes done in the kitchen. The cleaning of items such as shoes and silver and even minor repair jobs also take place in the kitchen. In fact, the kitchen is the hardest used and most useful room in the house.

A regular sequence of actions takes place in preparing a meal. This sequence is:

1. The assembling of the foods,
2. Preparation of food,
3. Cooking,
4. Serving up, and
5. Washing up and storage.

To make the handling of meals easy, this sequence of food preparation should be provided for in the layout of the kitchen and its equipment. Three work centres are necessary. They are:

1. Mix Centre
2. Sink Centre
3. Cook Centre

Supplies and equipment for use at each centre should be stored there, close at hand.

There is a direct path between these centres forming a work triangle by which the efficiency of the kitchen may be gauged. The sum of the distances between these centres should not be less than 3600 mm and not more than 6000 mm. The arrangement of these centres, together with that of the doors and windows, determines the type of kitchen. Some of the common types are shown in Figure 15-3-6. Note that in all but one case the work centres are placed side by side with a continuous counter connecting them.

Most kitchens require two doors; one to the dining area and one to the service entrance. Unnecessary doors break up the assembly of the work centres and thus decrease the efficiency of the kitchen.

A serving hatch from the cook centre counter through to the dining room is often a great asset and makes the serving of meals easier.

NUMBER OF PERSONS	SINK WIDTH	RANGE WIDTH *	REFRIGERATOR WIDTH *	A	B
1 – 2	510	660	660	760	380
3 – 4	510	810	810	1070	380
5 – 6	510	810	810	1220	460
7 – 8	610	810	810	1370	460

* INCLUDES 50 mm TOTAL CLEARANCE BETWEEN APPLIANCE AND BASE CABINET

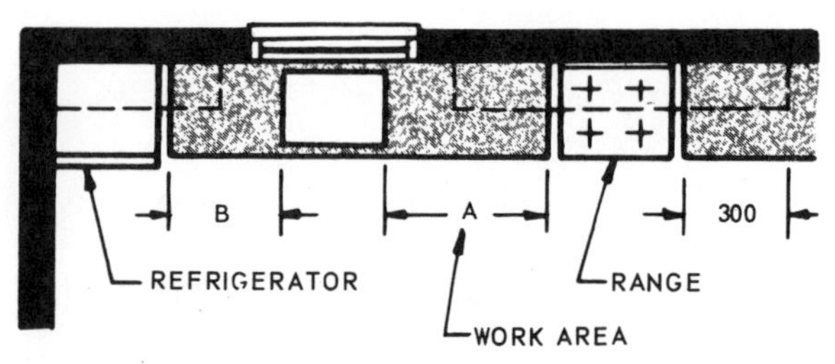

(A) MINIMUM COUNTER SPACE ADJACENT TO APPLIANCES

The following figures are suggested for kitchen storage requirements:

Wall cabinets, 2500-3000 mm in length
Base cabinets, 2500-3000 mm in length

Additional storage space is needed for cleaning equipment. There should be counter-top work surfaces:

1. Beside the refrigerator (note which way the door opens). (Mix Centre)
2. At both sides of the sink. (Sink Centre)
3. At least on one side of the stove. (Cook Centre)
4. On one side of a washing machine and built-in ironer, if this equipment is included.

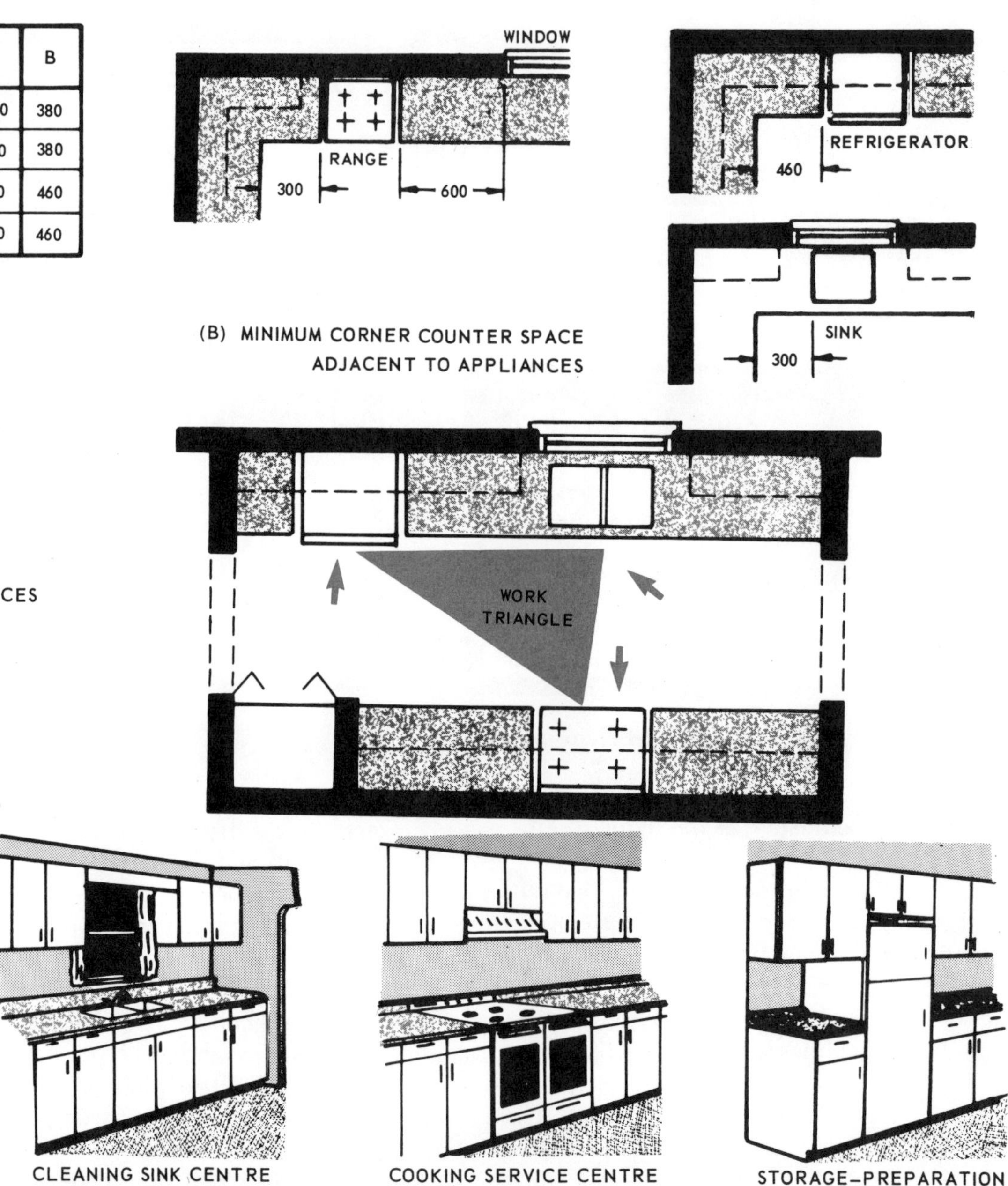

(B) MINIMUM CORNER COUNTER SPACE ADJACENT TO APPLIANCES

Fig. 15-3-6 Kitchen cabinet design and work centres

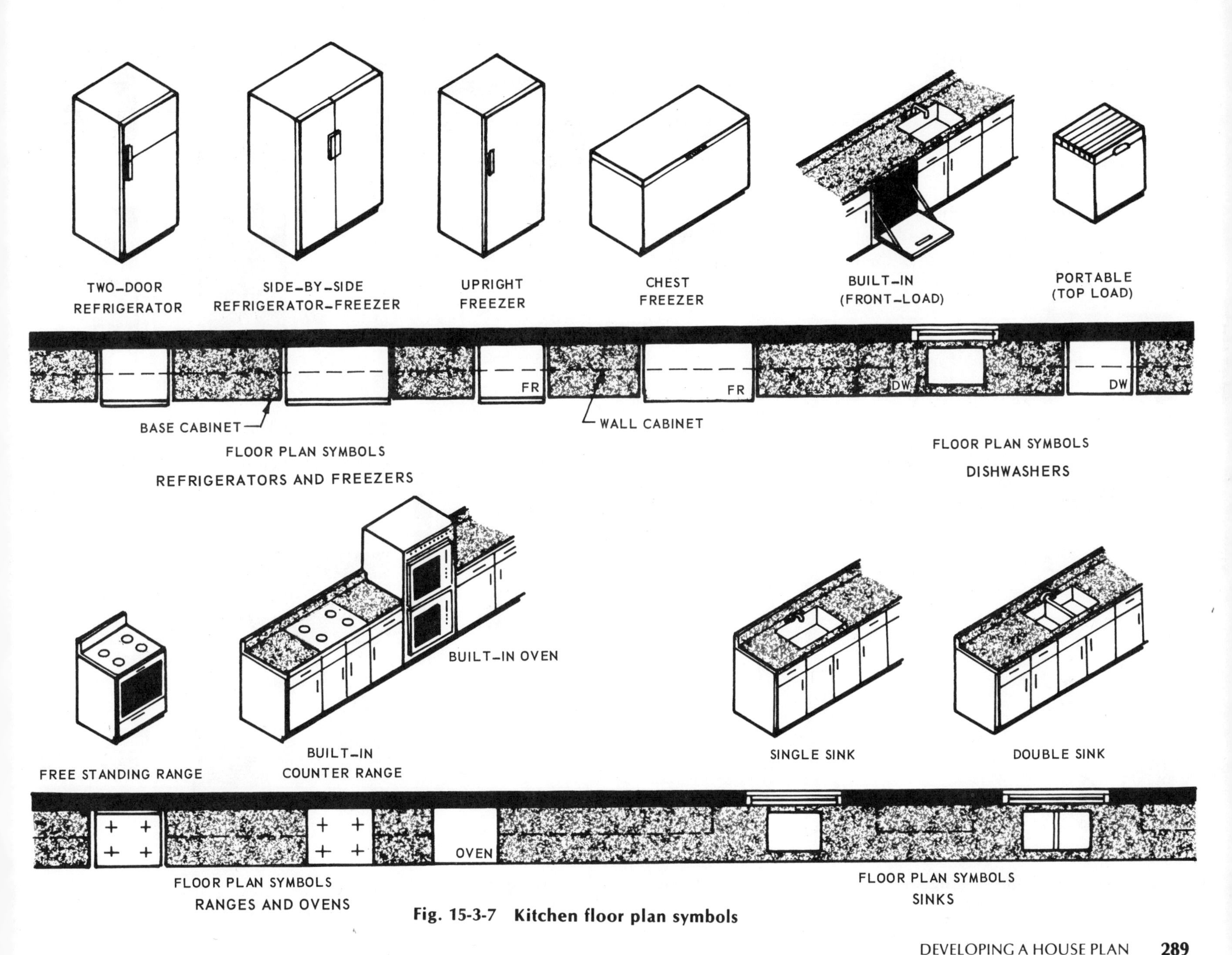

Fig. 15-3-7 Kitchen floor plan symbols

Living and Dining Rooms

This is the room where the character of the family is expressed and where friends are received. It should combine the requirements for quiet relaxation, for active entertainment, for study, and for children's play.

Unlike other rooms, which are enclosed by four walls and a door, the living room is often the open central core of the house.

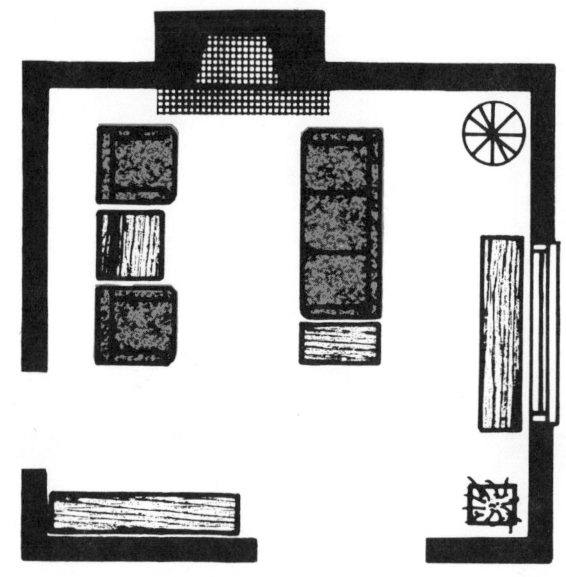

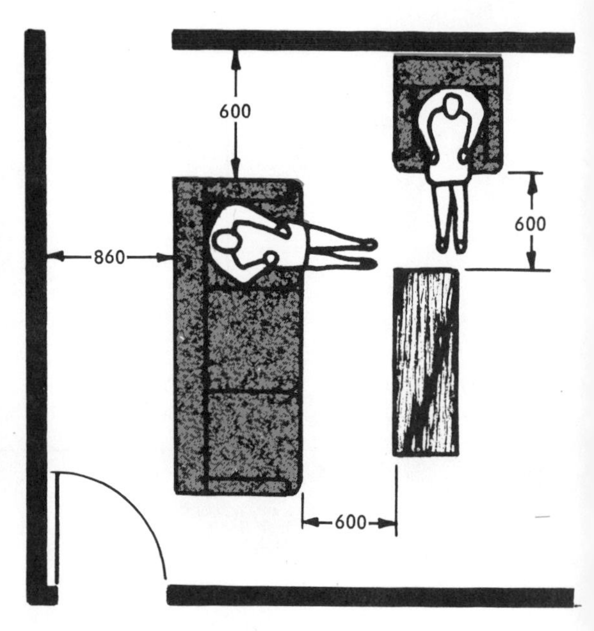

Fig. 15-3-10 **Living area clearances**

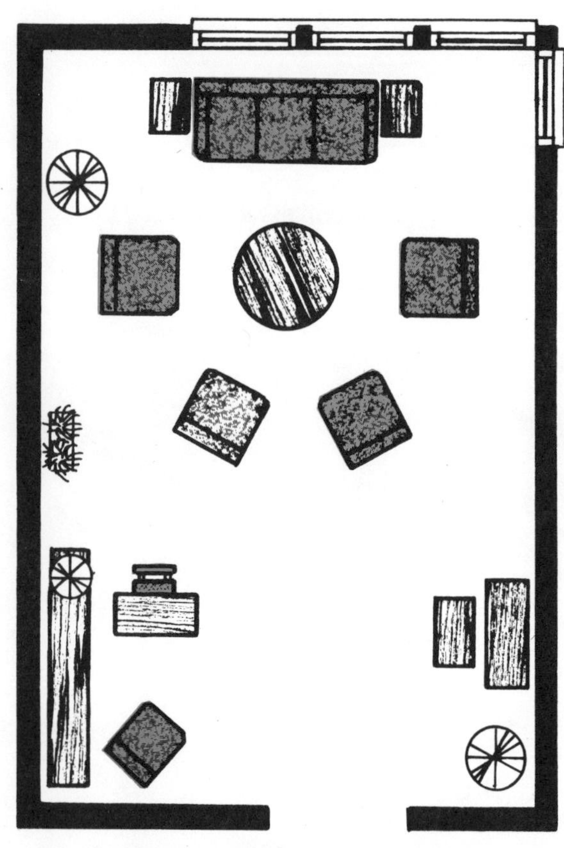

Fig. 15-3-8 **Furniture arrangement for entertainment**

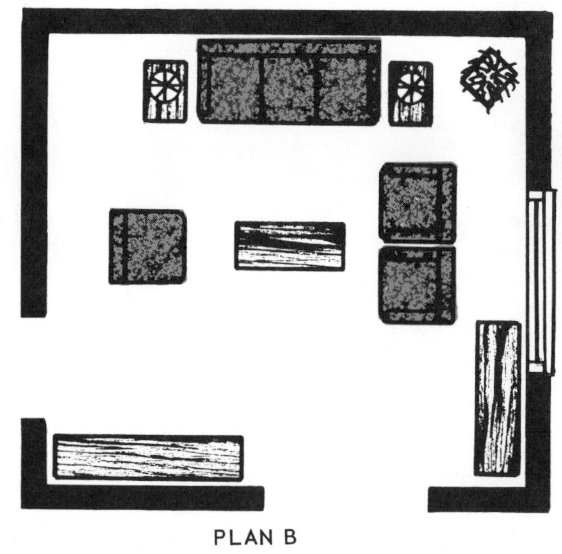

Fig. 15-3-9 **Furniture arrangement for conversation**

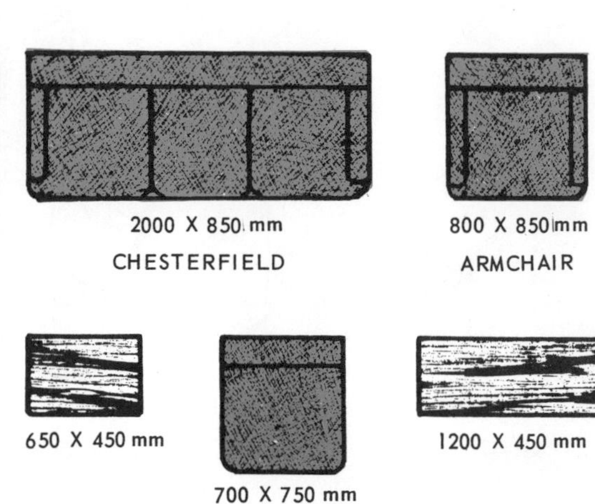

Fig. 15-3-11 **Living room furniture sizes**

FURNITURE	NUMBER OF PIECES OF FURNITURE REQUIRED					
NUMBER OF PERSONS	1	2	3	4	5	6
CHESTERFIELD	1	1	1	1	1	1
ARMCHAIR	1	1	1	2	2	2
OCCASIONAL CHAIR	–	–	–	–	–	1
END TABLE	2	2	2	2	2	2
COFFEE TABLE	1	1	1	1	1	1
TELEVISION SET	1	1	1	1	1	1

Fig. 15-3-12 Minimum living room furniture requirements

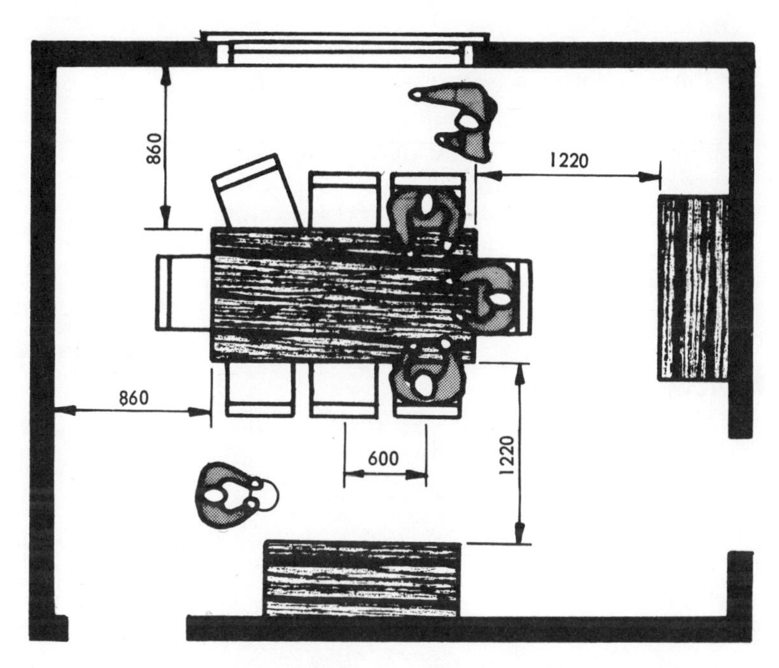

Fig. 15-3-14 Dining area clearances

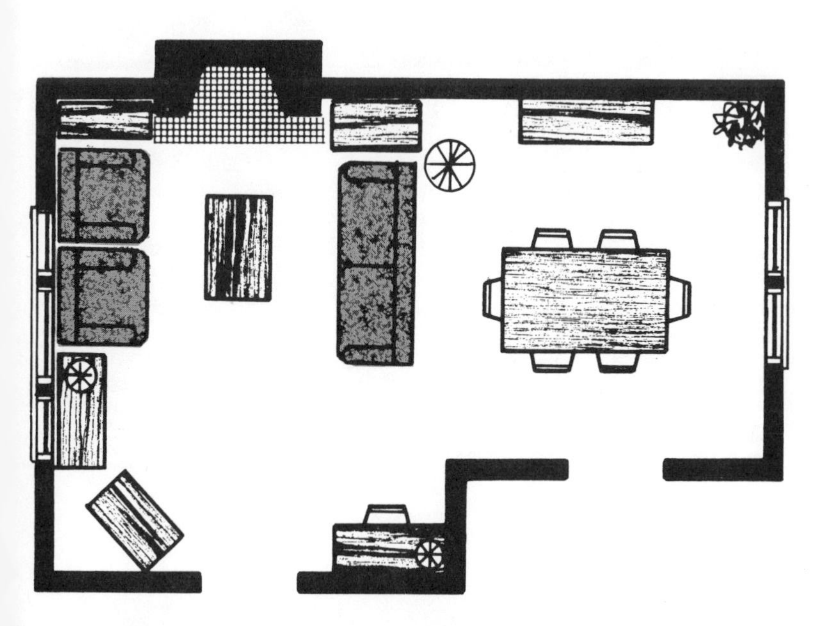

Fig. 15-3-13 Combination living-dining room

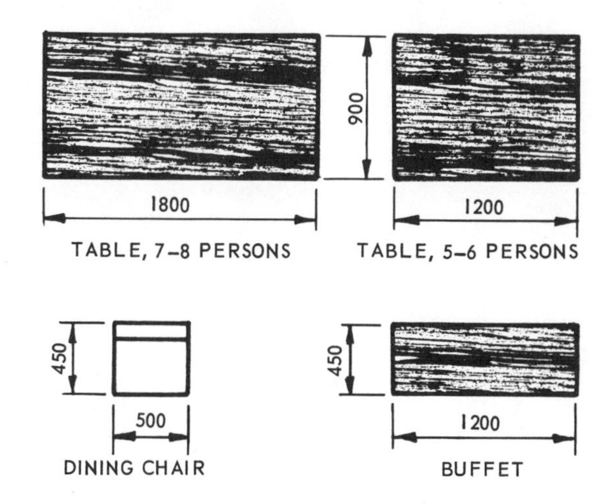

Fig. 15-3-15 Dining room furniture sizes

In contemporary houses it is often of irregular shape with openings revealing the other parts of the house and adding to its sense of freedom, accessibility and informality. For this reason the selection and placing of furniture is an essential part of the planning and design of the house. This determines the real shape and uses of the space within the living room.

Review Questions

1. What room in the house receives the hardest use, and why?
2. What is meant by a work *triangle* in the kitchen?
3. Where should counter-top work surfaces be located?

Assignments

Draw one of the floor plans shown in Fig. 15-3-A to the scale 1:50 and draw in the furniture that you recommend each room should have. Wall thicknesses:

Exterior wall - 240 mm
Interior walls - 110 mm.

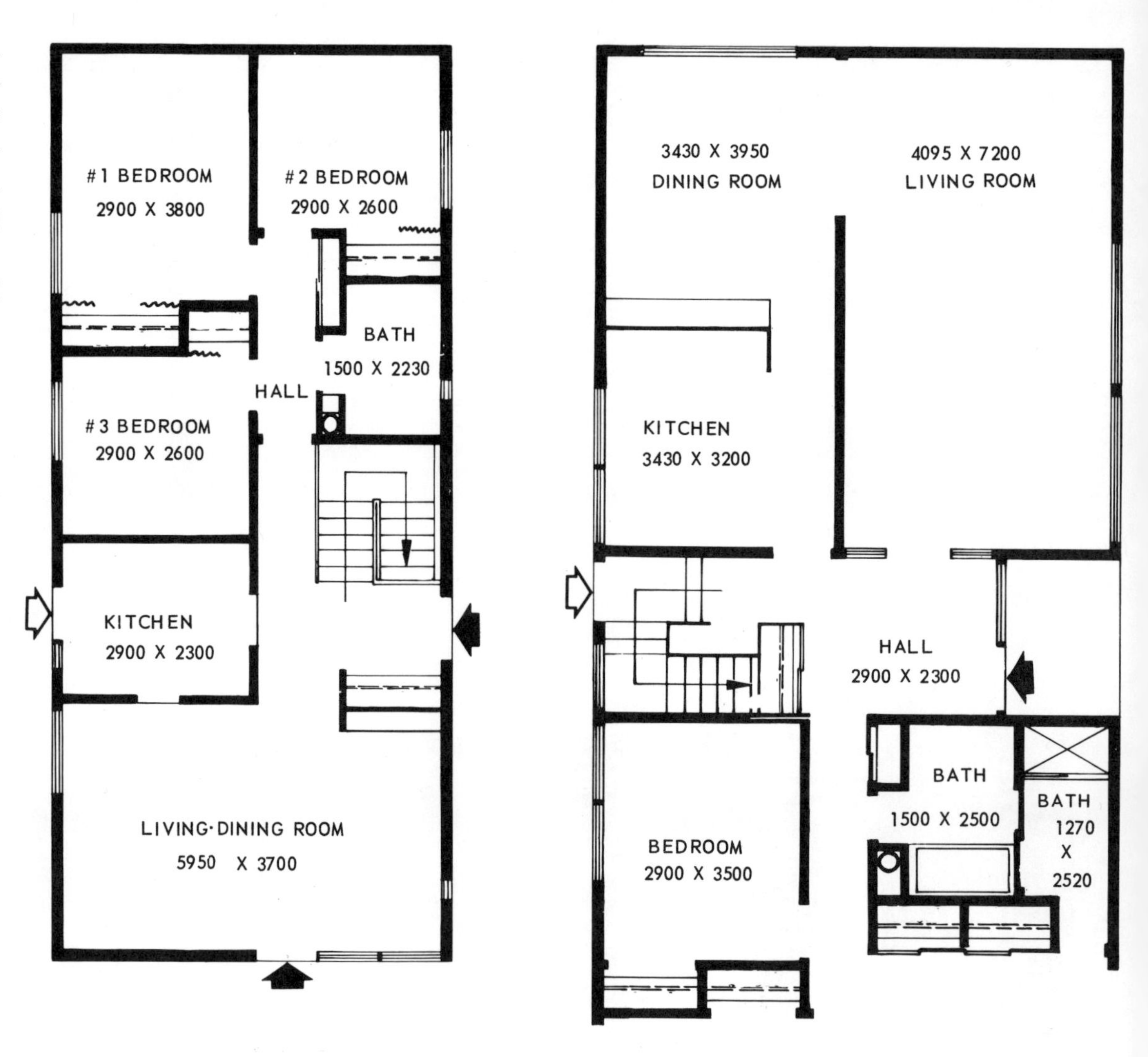

Fig. 15-3-A Furniture layout assignments

Chapter 16

Applied Geometry

UNIT 16-1
Straight Lines

Most of the lines forming the views on mechanical drawings can be drawn using the instruments and equipment described in Chapter 2. However, geometric constructions have important uses, both in making drawings and in solving problems by graphs and diagrams.

Sometimes it is necessary to use geometric constructions, particularly if you do not have the advantages afforded by a drafting machine, an adjustable set square, or templates for drawing hexagonal and elliptical shapes.

To Draw a Line or Lines Parallel To and At a Given Distance From an Oblique Line

1. Given line *AB* (Fig.16-1-1), erect a perpendicular *CD* to *AB*.

2. Space the given distance from the line *AB* by scale measurement or by an arc

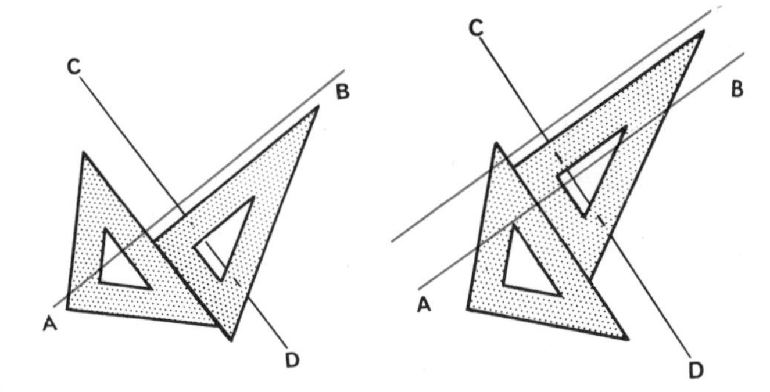

Fig. 16-1-1 Drawing parallel lines with the side of set squares

3. Position a set square, using a second set square or a T square as base, so that one side of the set square is parallel with the given line.

4. Slide this set square along the base to the point at the desired distance from the given line, and draw the required line.

To Draw a Straight Line Tangent to Two Circles

Place a T square or straightedge so that the top edge just touches the edges of the circles, and draw the tangent line (Fig. 16-1-2). Perpendiculars to this line from the centres of the circles give the tangent points T_1 and T_2.

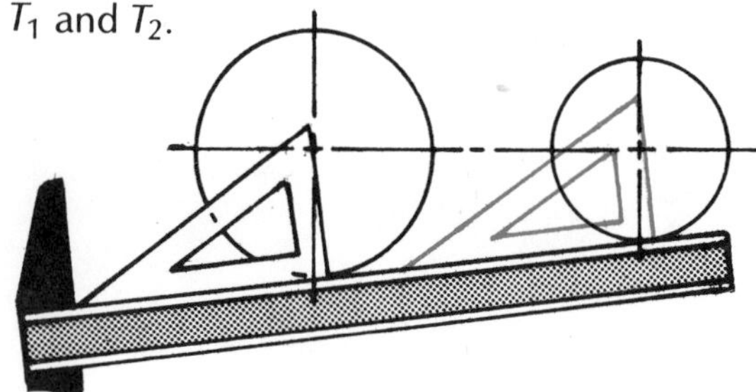

Fig. 16-1-2 Drawing a straight line tangent to two circles

To Bisect a Straight Line

1. Given line *AB* (Fig. 16-1-3), set the compass to a radius greater than 0.5 *AB*.

2. Using centres at *A* and *B*, draw intersecting arcs above and below line *AB*. A line *CD* drawn through the intersections will divide *AB* into two equal parts and will be perpendicular to line *AB*.

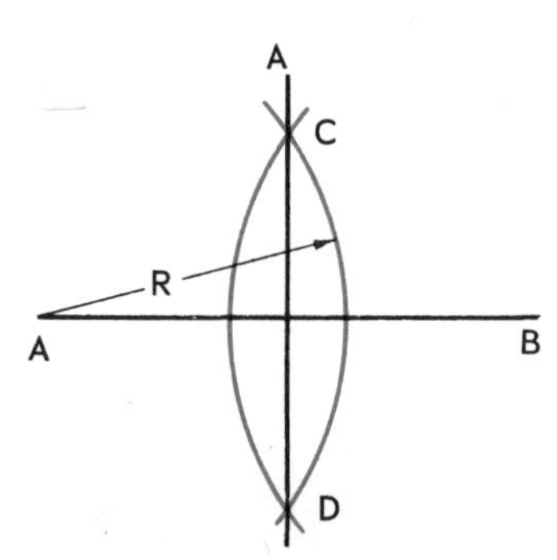

Fig. 16-1-3 Bisecting a line

To Bisect an Arc

1. Given arc *AB* (Fig. 16-1-4), set the compass to a radius greater than 0.5 *AB*.

2. Using points *A* and *B* as centres, draw intersecting arcs above and below arc *AB*. A line drawn through the intersections *C* and *D* will bisect the arc *AB* into two equal

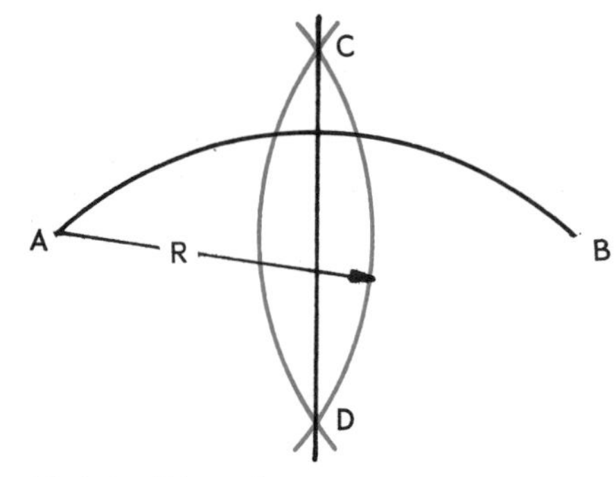

Fig. 16-1-4 Bisecting an arc

To Bisect an Angle

1. Given angle *ABC* with centre *B* and a suitable radius (Fig. 16-1-5), draw an arc to cut *BC* at *D* and *BA* at *E*.

2. With centres *D* and *E* and equal radii, draw arcs to intersect at *F*.

3. Join *BF* and produce to *G*. Line *BG* is the required bisector.

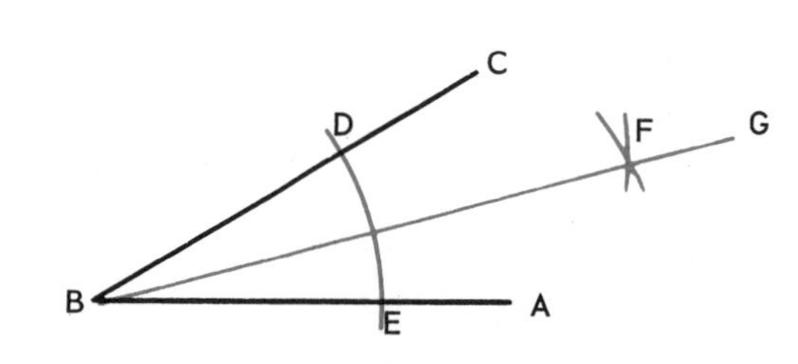

Fig. 16-1-5 Bisecting an angle

To Divide a Line Into a Given Number of Equal Parts

1. Given line *AB* and the number of equal divisions desired (12, for example), draw a perpendicular from *A*.

2. Place the scale so that the desired number of equal divisions is conveniently included between *B* and the perpendicular. Then mark these divisions, using short vertical marks from the scale divisions as shown in Figure 16-1-6.

3. Draw perpendiculars to line *AB* through the points marked, dividing the line *AB* as required.

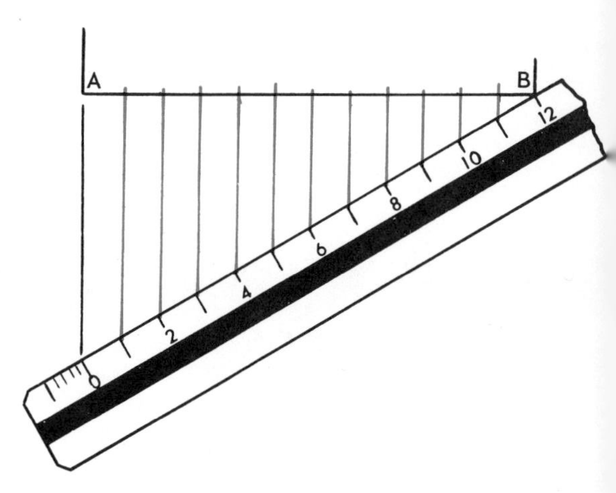

Fig. 16-1-6 Dividing a straight line into equal parts

Assignment

Divide an A3-size sheet as shown in Fig. 16-1-A. In the designated areas draw the geometric constructions.

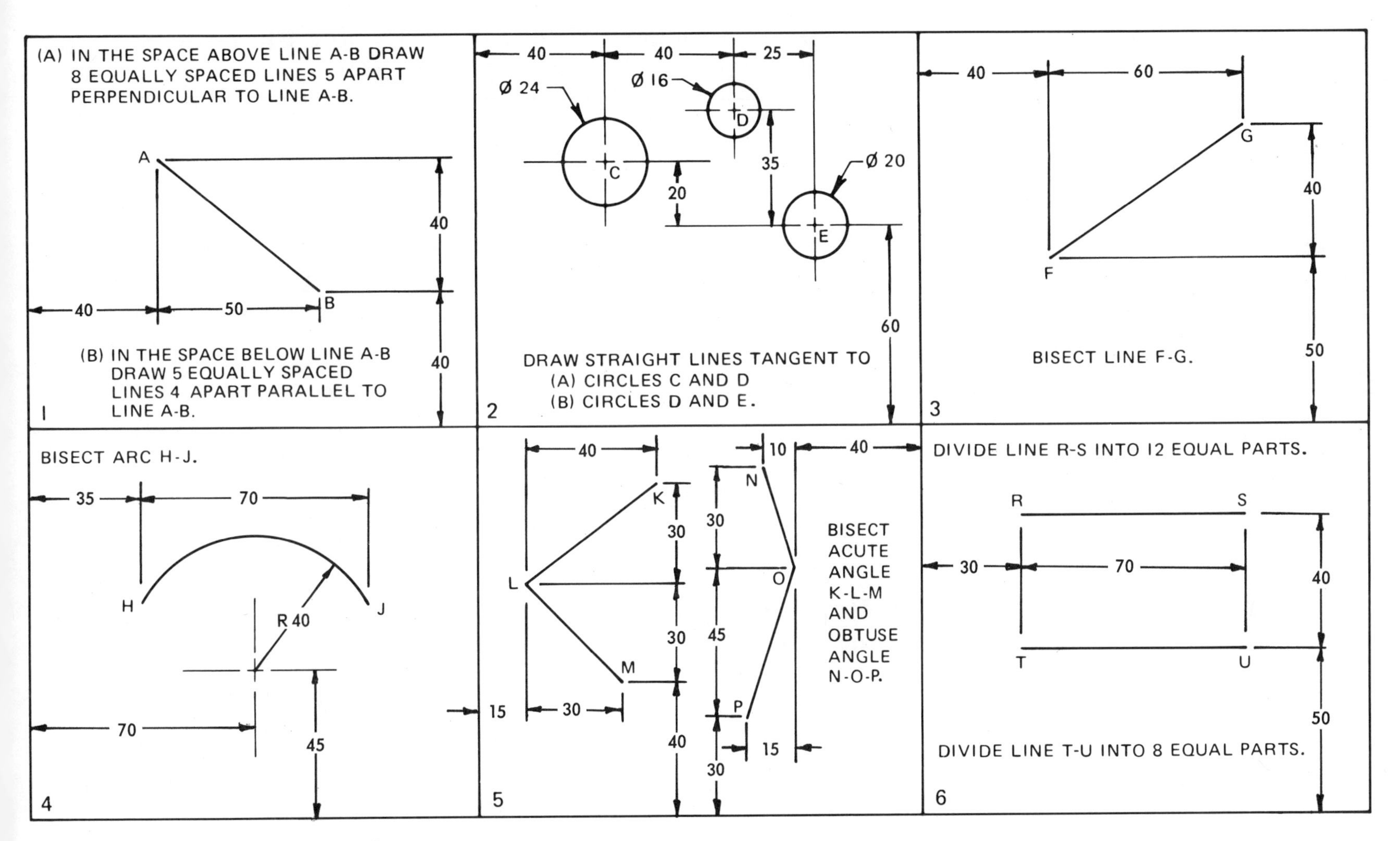

Fig. 16-1-A Straight line constructions

UNIT 16-2
Arcs and Circles

To Draw an Arc Tangent to Two Lines at Right Angles to Each Other

Given radius *R* of the arc (Fig. 16-2-1).

1. Draw an arc having radius *R* with centre at *B*, cutting the lines *AB* and *BC* at *D* and *E* respectively.

2. With *D* and *E* as centres and with the same radius *R*, draw arcs intersecting at 0.

3. With centre 0 draw the required arc. The tangent points are *D* and *E*.

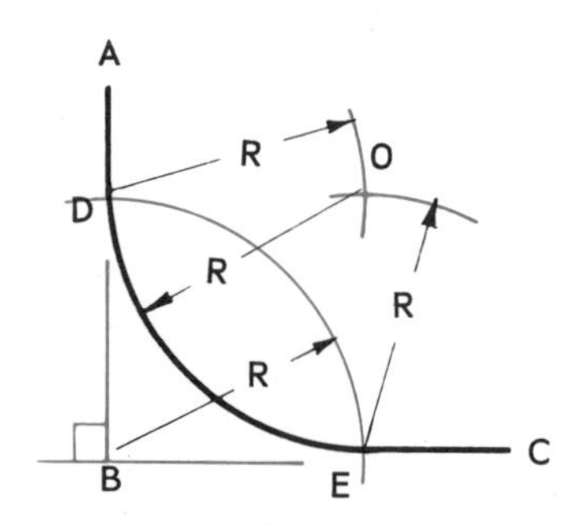

Fig. 16-2-1 Arc tangent to two lines at right angles to each other

To Draw an Arc Tangent to the Sides of an Acute Angle

Given radius *R* of the arc (Fig. 16-2-2).

1. Draw lines inside the angle, parallel to the given lines, at distance *R* away from the given lines. The centre of the arc will be at *C*.

2. Set the compass to radius *R*, and with centre *C* draw the arc tangent to the given sides. The tangent points *A* and *B* are found by drawing perpendiculars through point *C* to the given lines.

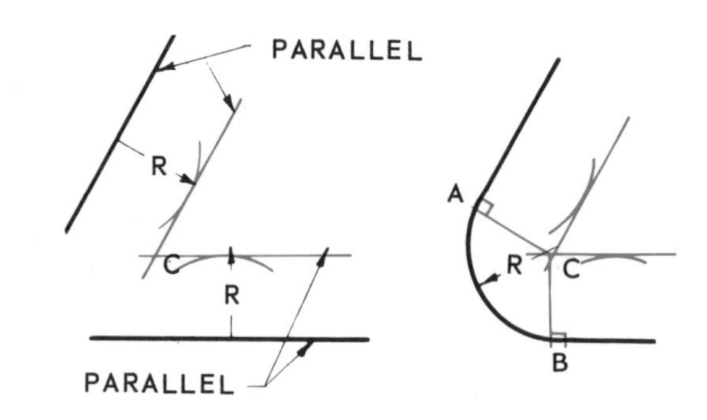

Fig. 16-2-2 Drawing an arc tangent to the sides of an acute angle

To Draw an Arc Tangent to Two Sides of an Obtuse Angle

Follow the same procedure as for an acute angle (Fig. 16-2-3).

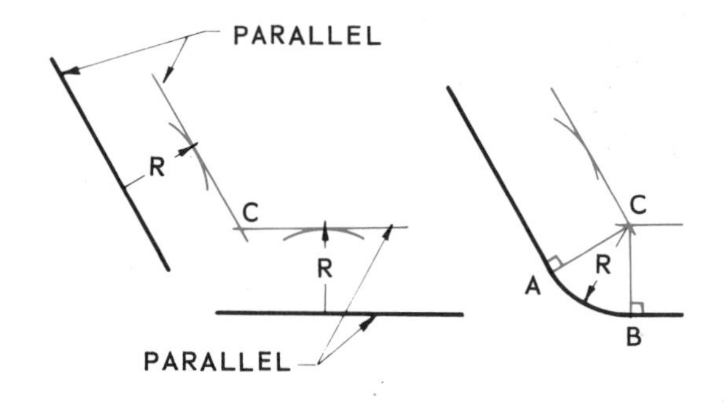

Fig. 16-2-3 Drawing an arc tangent to the sides of an acute angle

To Draw a Circle on a Regular Polygon

1. Given the size of the polygon (Fig. 16-2-4), bisect any two sides; for example, *BC* and *DE*. The centre of the polygon is where bisectors *F*0 and *G*0 intersect at point 0.

2. The inner circle radius is 0*H*, and the outer circle radius is 0*A*.

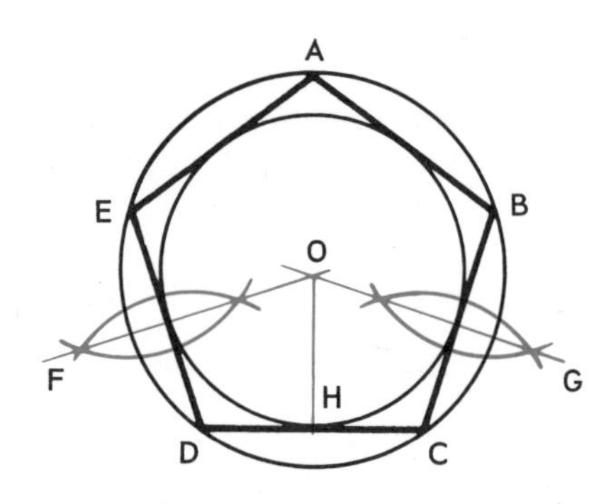

Fig. 16-2-4 Drawing a circle on a regular polygon

To Draw a Reverse, or Ogee, Curve Connecting Two Parallel Lines

1. Given two parallel lines *AB* and *CD* and distances *X* and *Y* (Fig. 16-2-5), join points *B* and *C* with a line.

2. Erect a perpendicular to *AB* and *CD* from points *B* and *C* respectively.

3. Select point *E* on line *BC* where the curves are to meet.

4. Bisect *BE* and *EC*.

5. Points *F* and *G* where the perpendiculars and bisectors meet are the centres for the arcs forming the ogee curve.

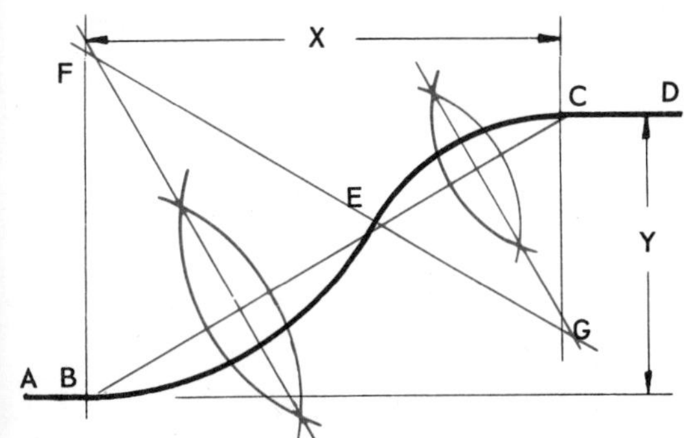

Fig. 16-2-5 Drawing a reverse (ogee) curve connecting two parallel lines

To Draw an Arc Tangent to a Given Circle and Straight Line

1. Given R, the radius of the arc (Fig. 16-2-6), draw a line parallel to the given straight line between the circle and the line at distance R away from the given line.

2. With the centre of the circle as centre and radius R_1 (radius of the circle plus R), draw an arc to cut the parallel straight line at C.

3. With centre C and radius R, draw the required arc tangent to the circle and the straight line.

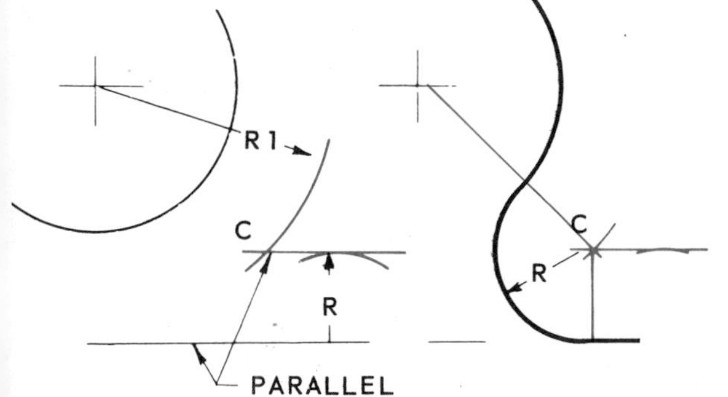

Fig. 16-2-6 Drawing an arc tangent to a circle and a straight line

To Draw an Arc Tangent to Two Circles (Fig. 16-2-7)

Figure 16-2-7*a*

1. Given the radius of arc R, with the centre of circle A as centre and radius R_2 (radius of circle A plus R), draw an arc in the area between the circles.

2. With the centre of circle B as centre and radius R_2 (radius of circle B plus R), draw an arc to cut the other arc at C.

3. With centre C and radius R, draw the required arc tangent to the given circles.

Figure 16-2-7*b*

1. Given radius of arc R, with the centre of circle A as centre and radius $R - R_2$, draw an arc in the area between the circles.

2. With the centre of circle B as centre and radius $R - R_3$, *draw an arc to cut the other arc at C.*

3. With centre C and radius R, draw the required arc tangent to the given circles.

To Draw an Arc or Circle Through Three Points Not in a Straight Line

1. Given points A, B, and C (Fig. 16-2-8), join points A, B, and C as shown.

2. Bisect lines AB and BC and extend bisecting lines to intersect at 0. Point 0 is the centre of the required circle or arc.

3. With centre 0 and radius $0A$ draw an arc.

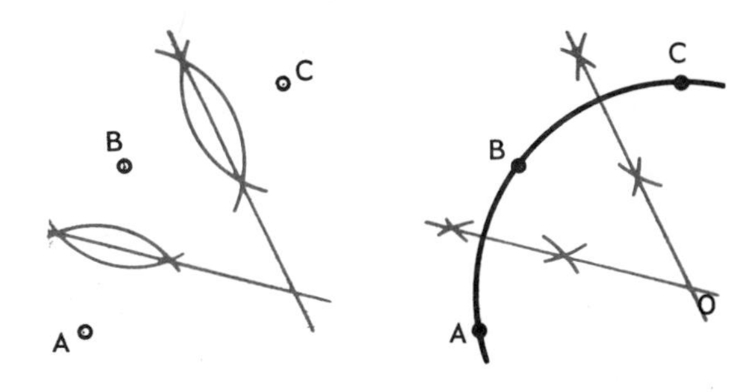

Fig. 16-2-8 Drawing an arc or circle through three points not in a straight line

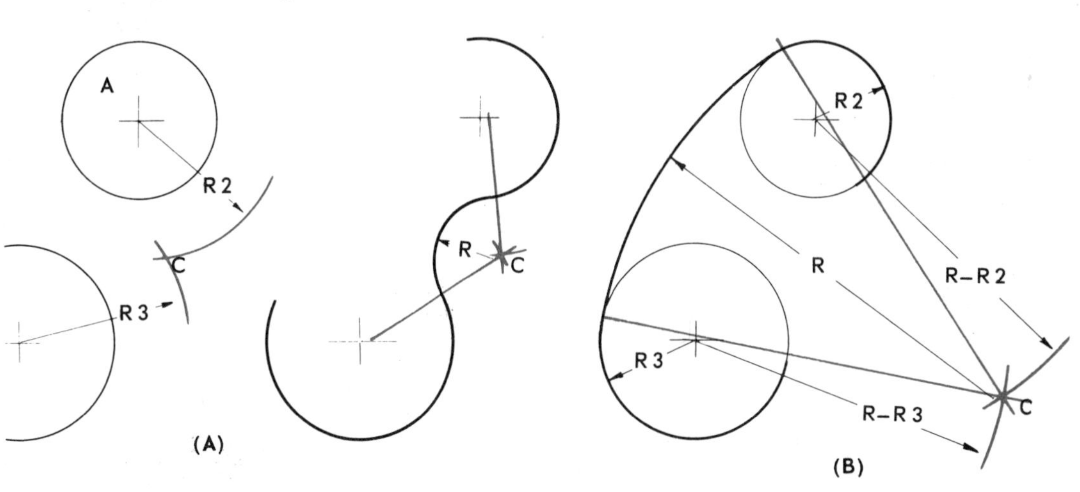

Fig. 16-2-7 Drawing an arc tangent to two circles

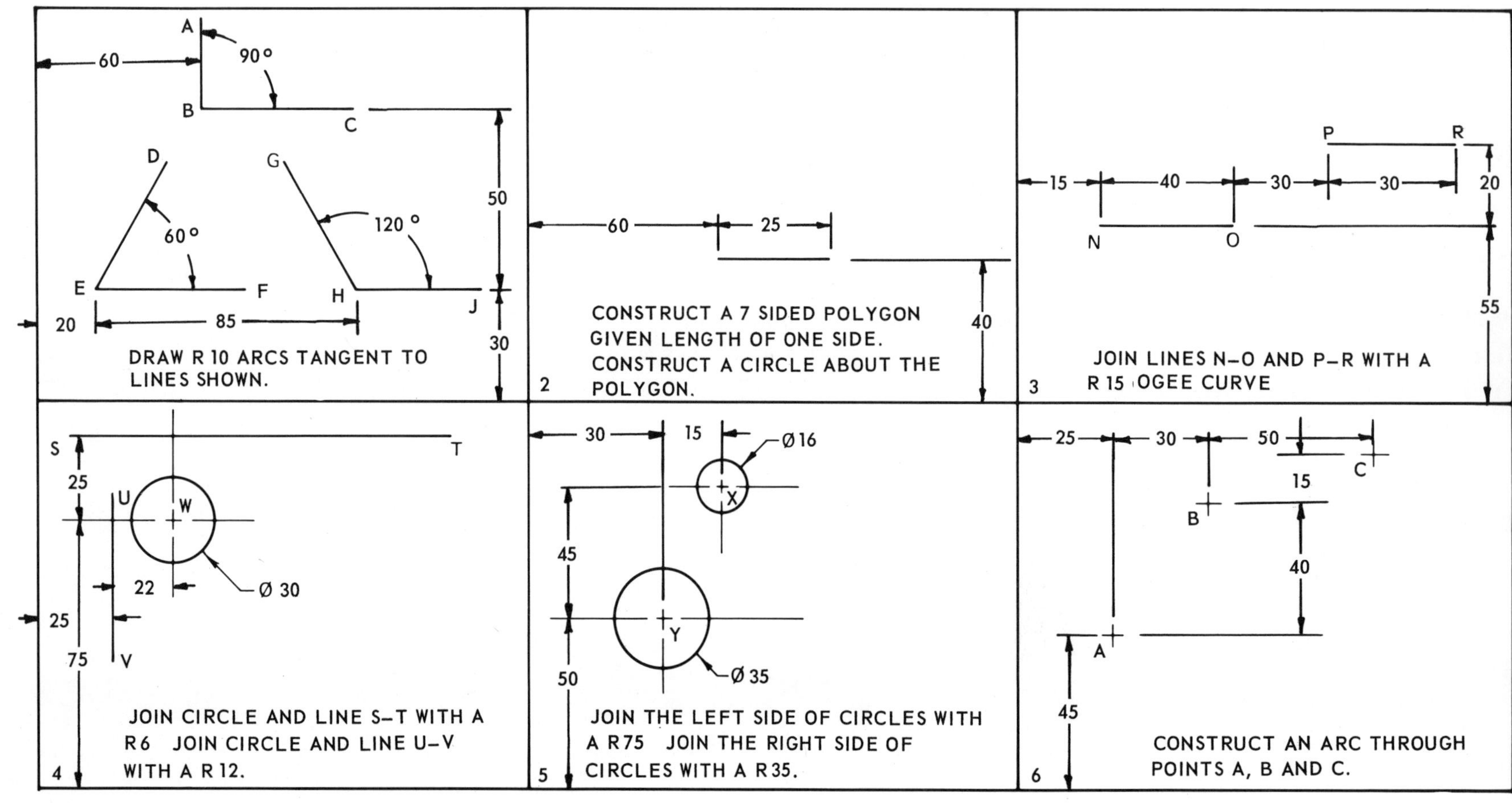

Fig. 16-2-A Circle and arc constructions

Assignment

Divide an A3 sheet as shown in Fig. 16-2-A. In the designated areas draw the geometric constructions.

UNIT 16-3
Polygons

A regular polygon is a plane figure bounded by straight lines of equal length and contains angles of equal size.

To Draw a Hexagon (Fig. 16-3-1), Given the Distance Across the Flats

1. Establish horizontal and vertical centre lines for the hexagon.

2. Using the intersection of these lines as centre, with radius one-half the distance across the flats, draw a light construction circle.

3. Using the 60° set square, draw six straight lines, equally spaced, passing through the centre of the circle.

4. Draw tangents to these lines at their intersection with the circle.

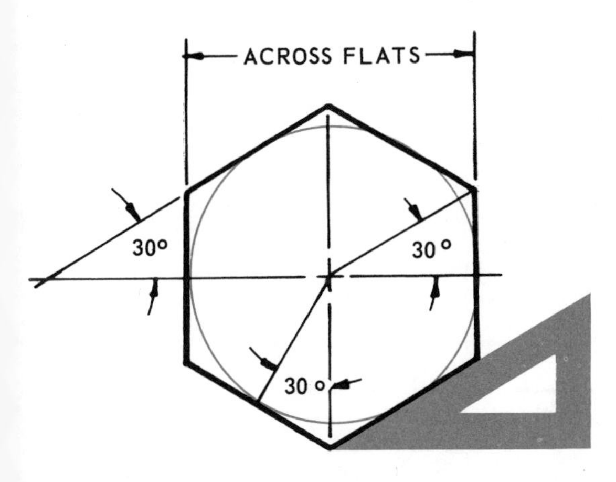

Fig. 16-3-1 Constructing a hexagon, given distance across flats

To Draw a Hexagon, Given the Distance Across the Corners (Fig. 16-3-2)

1. Establish horizontal and vertical centre lines, and draw a light construction circle with radius one-half the distance across the corners.

2. With a 60° set square, establish points on the circumference 60° apart.

3. Draw straight lines connecting these points.

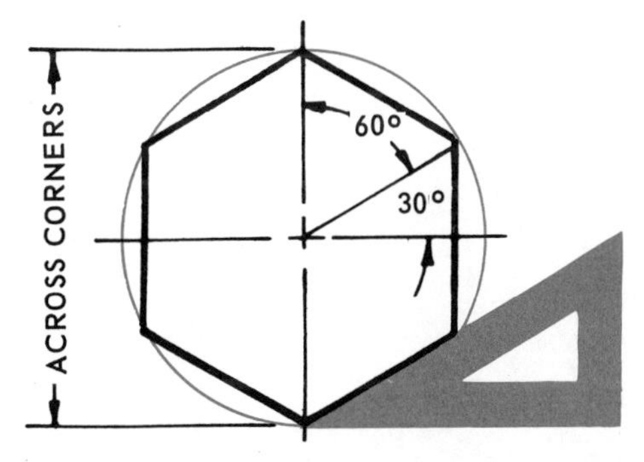

Fig. 16-3-2 Constructing a hexagon, given distance across corners

To Draw an Octagon, Given the Distance Across the Flats (Fig. 16-3-3)

1. Establish horizontal and vertical centre lines and draw a light construction circle with radius one-half the distance across the flats.

2. Draw horizontal and vertical lines tangent to the circle.

3. Using the 45° set square, draw lines tangent to the circle at a 45° angle from the horizontal.

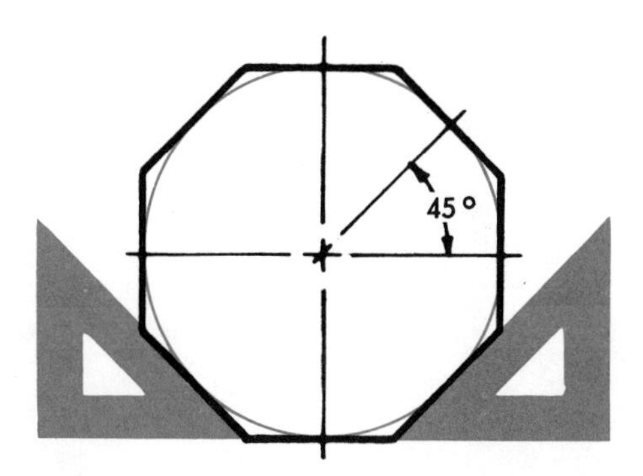

Fig. 16-3-3 Constructing a hexagon, given distance across flats

To Draw an Octagon, Given the Distance Across the Corners (Fig. 16-3-4)

1. Establish horizontal and vertical centre lines and draw a light construction circle with radius one-half the distance across the corners.

2. With the 45° set square, establish points on the circumference between the horizontal and vertical centre lines.

3. Draw straight lines connecting these points to the points where the centre lines cross the circumference.

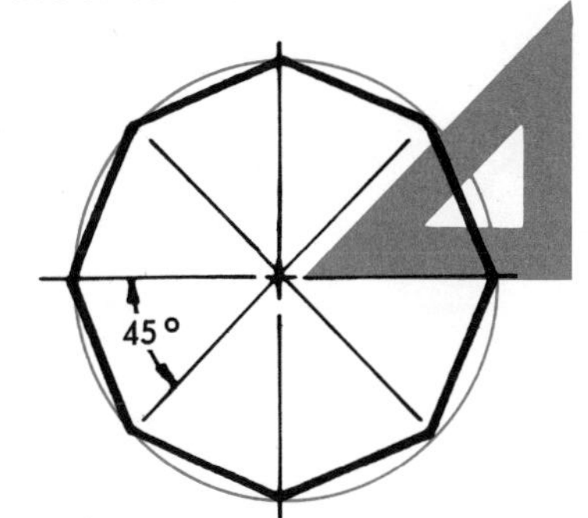

Fig. 16-3-4 Constructing a hexagon, given distance across corners

To Draw a Regular Polygon, Given the Length of the Sides

Let the polygon have seven sides.

1. Given the length of side *AB* (Fig. 16-3-5), with radius *AB* and *A* as centre, draw a semicircle and divide it into seven equal parts using a protractor.

2. Through the second division from the left, draw radial line *A2*.

3. Through points 3, 4,5, and 6 extend radial lines as shown.

4. With *AB* as radius and *B* as centre, cut line *A6* at *C*. With the same radius and *C* as centre, cut line *A5* at *D*. Repeat at *E* and *F*.

5. Connect these points with straight lines.

These steps can be followed in drawing a regular polygon with any number of sides.

To Inscribe a Regular Pentagon in a Given Circle

1. Given circle with centre 0 (Fig. 16-3-6), draw the circle with diameter *AB*.

2. Bisect line 0*B* at *D*.

3. With centre *D* and radius *DC*, draw arc *CE* to cut the diameter at *E*.

4. With *C* as centre and radius *CE*, draw arc *CF* to cut the circumference at *F*. Distance *CF* is one side of the pentagon.

5. With radius *CF* as a chord, mark off the remaining points on the circle. Connect the points with straight lines.

Assignment

Divide an A3-size sheet as shown in Fig. 16-3-A. In the designated areas, draw the geometric constructions.

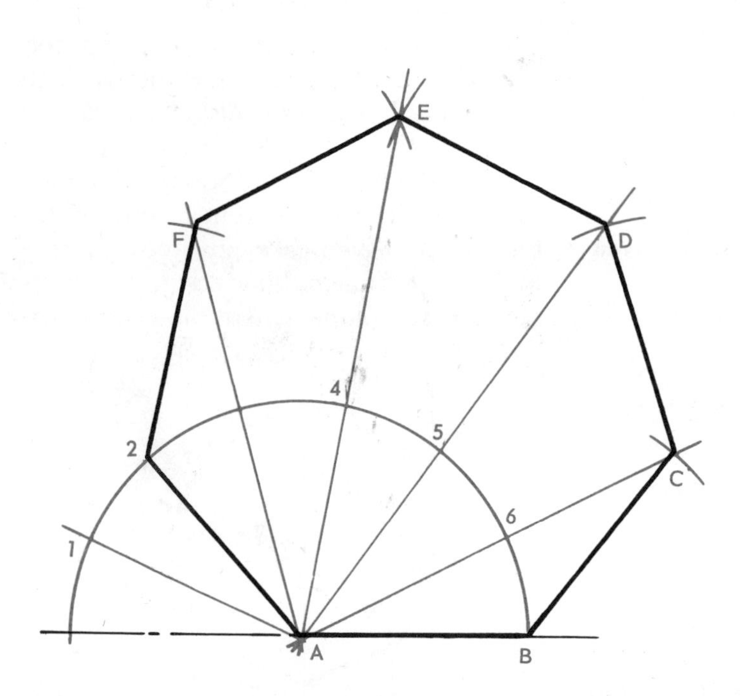

Fig. 16-3-5 Constructing a regular polygon, given length of one side

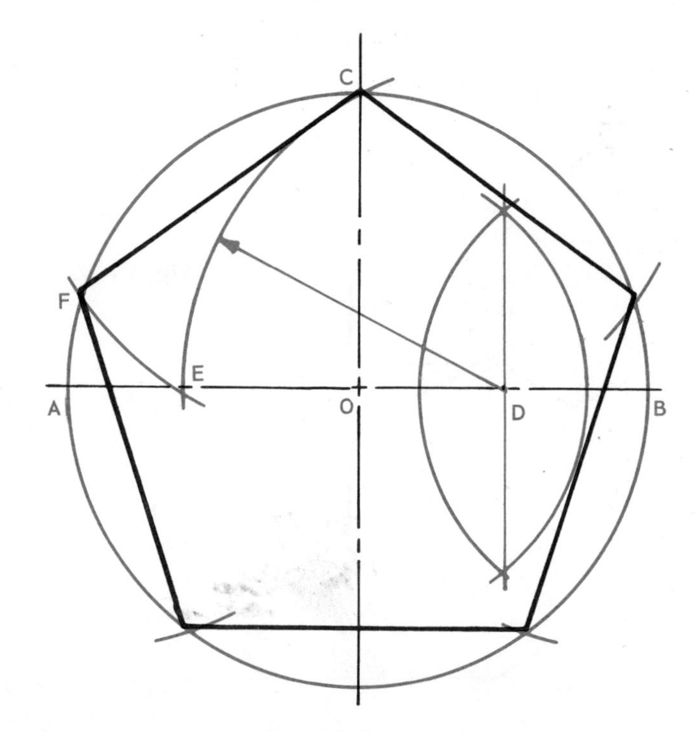

Fig. 16-3-6 Inscribing a regular pentagon in a given circle

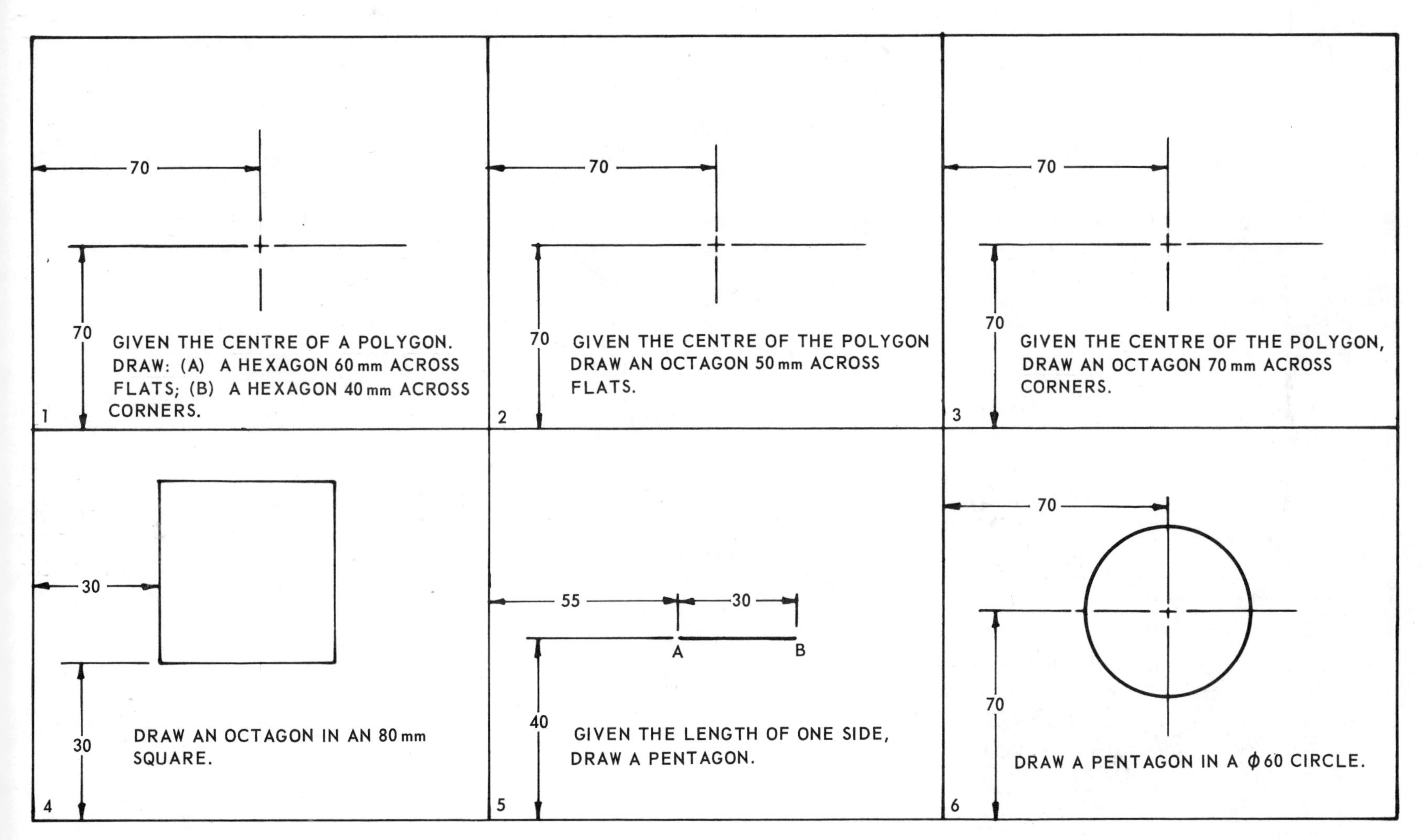

Fig. 16-3-A Polygon constructions

UNIT 16-4
The Ellipse

The **ellipse** is the plane curve made by a point moving so that the sum of the distances from any point on the curve to two fixed points, called foci, is a constant.

Often you are called upon to draw oblique and inclined holes and surfaces which take the form of an ellipse. Several methods, true and approximate, are used for its construction. The terms **major diameter** and **minor diameter** will be used in place of **major axis** and **minor axis** so you won't become confused with the mathematical *X* and *Y* axes.

To Draw an Ellipse — Two Circle Method

1. Given the major and minor diameters (Fig. 16-4-1), construct two concentric circles with diameters equal to *AB* and *CD*.

2. Divide the circles into a convenient number of equal parts. Figure 16-4-1 shows 12.

3. Where the radial lines intersect the outer circle, as at 1, draw lines parallel to line *CD* inside the outer circle.

4. Where the same radial line intersects the inner circle, as at 2, draw a line parallel to axis *AB* away from the inner circle. The intersection of these lines, as at 3, gives points on the ellipse.

5. Draw a smooth curve through these points.

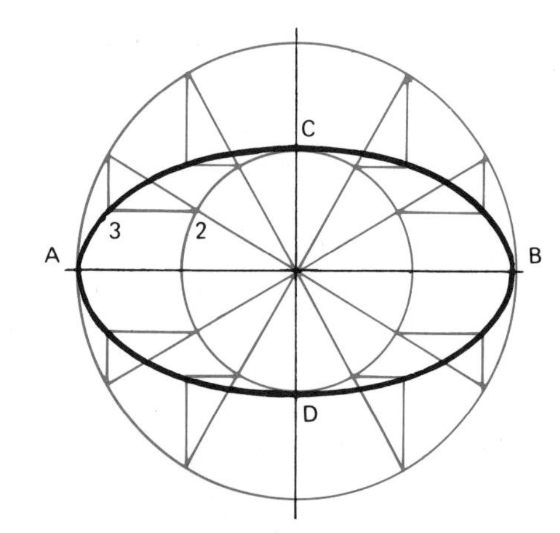

Fig. 16-4-1 Drawing an ellipse - two circle method

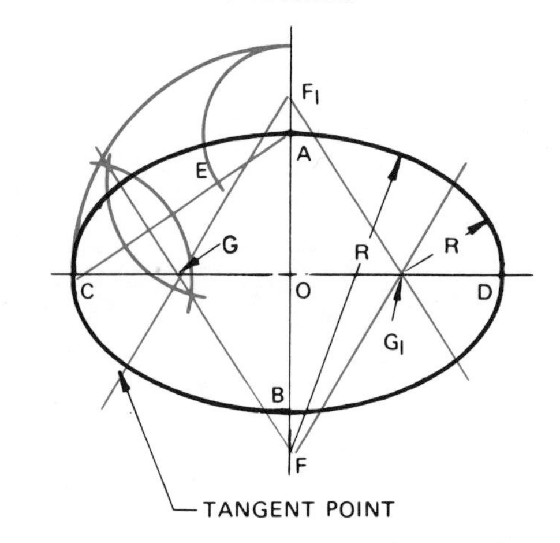

Fig. 16-4-2 Drawing an ellipse - four centre method

To Draw an Ellipse — Four-Centre Method

1. Given the major diameter *CD* and the minor diameter *AB* (Fig. 16-4-2), join points *A* and *C* with a line.

2. Lay off *AE* equal to *CO-AO*.

3. Draw the right bisector of *CE* locating point *G* on line *CO* and point *F* on line *AB*. (Line *AB* may have to be extended.)

4. Make *OF* equal to OF_1 and *OG* equal to OG_1.

5. Points *F*, F_1, *G*, and G_1 are the centres for the two large and two small arcs forming the ellipse.

To Draw an Ellipse — Parallelogram Method

1. Given the major diameter *CD* and minor diameter *AB* (Fig. 16-4-3), construct a parallelogram.

2. Divide *CO* into a number of equal parts. Divide *CE* into the same number of equal parts. Number the points from *C*.

3. Draw a line from *B* to point 1 on line *CE*. Draw a line from *A* through point 1 on *CO*, intersecting the previous line. The point of intersection will be one point on the ellipse.

4. Proceed in the same manner to find other points on the ellipse.

5. Draw a smooth curve through these points.

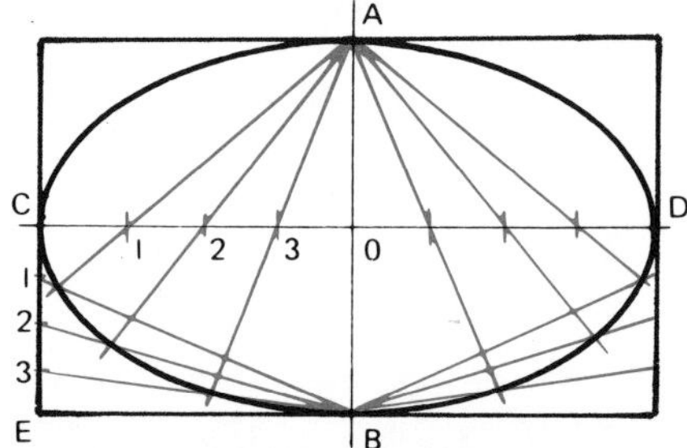

Fig. 16-4-3 Drawing an ellipse - parallelogram method

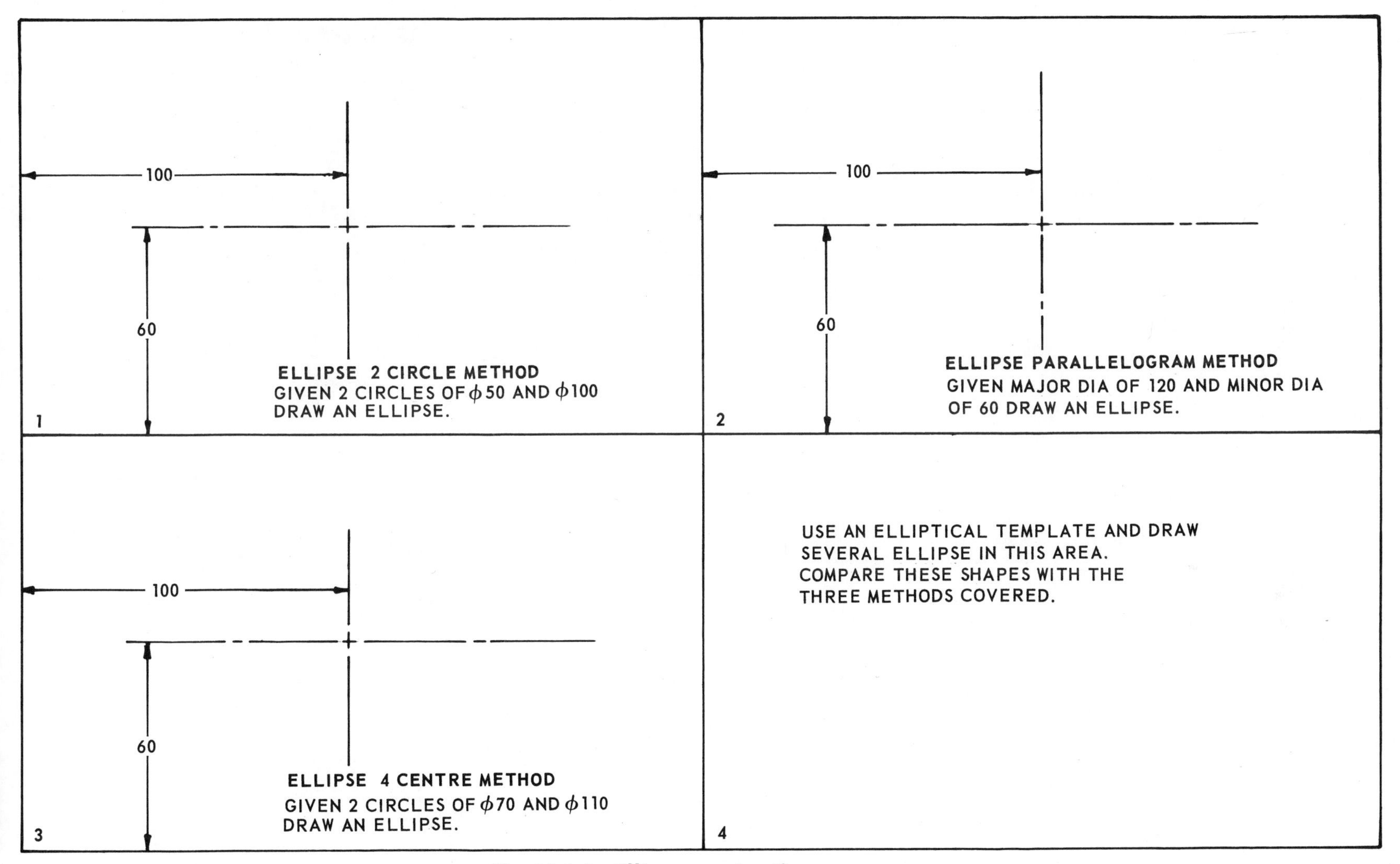

Fig. 16-4-A Ellipse constructions

Assignment

Divide an A3-size sheet as shown in Fig. 16-4-A. In the designated areas draw the geometric constructions.

UNIT 16-5
The Helix

The **helix** is the curve made by a point that revolves uniformly around and up or down the surface of a cylinder. The **lead** is the vertical distance that the point rises or drops in one complete revolution.

To Draw a Helix

1. Given the diameter of the cylinder and the lead (Fig. 16-5-1), draw the top and front views.

2. Divide the circumference (top view) into a convenient number of parts (use 12) and label them.

3. Project lines down to the front view.

4. Divide the lead into the same number of equal parts and label them as shown in Fig. 16-5-1.

5. The points of intersection of lines with corresponding numbers lie on the helix. **Note:** Since points 8 to 12 lie on the back portion of the cylinder, the helix curve starting at point 7 and passing through points 8, 9,10, 11, 12 to point 1 will appear as a hidden line.

6. If the development of the cylinder is drawn, the helix will appear as a straight line on the development.

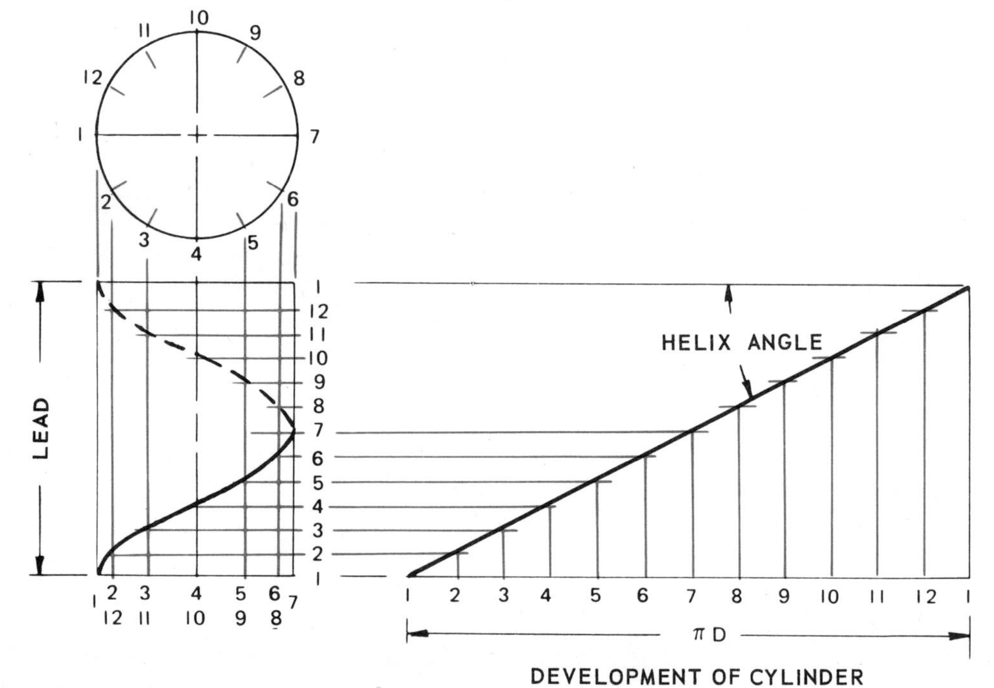

Fig. 16-5-1 Drawing a cylindrical helix

Assignment

On an A4 sheet draw the helix shown in Figure 16-5-A

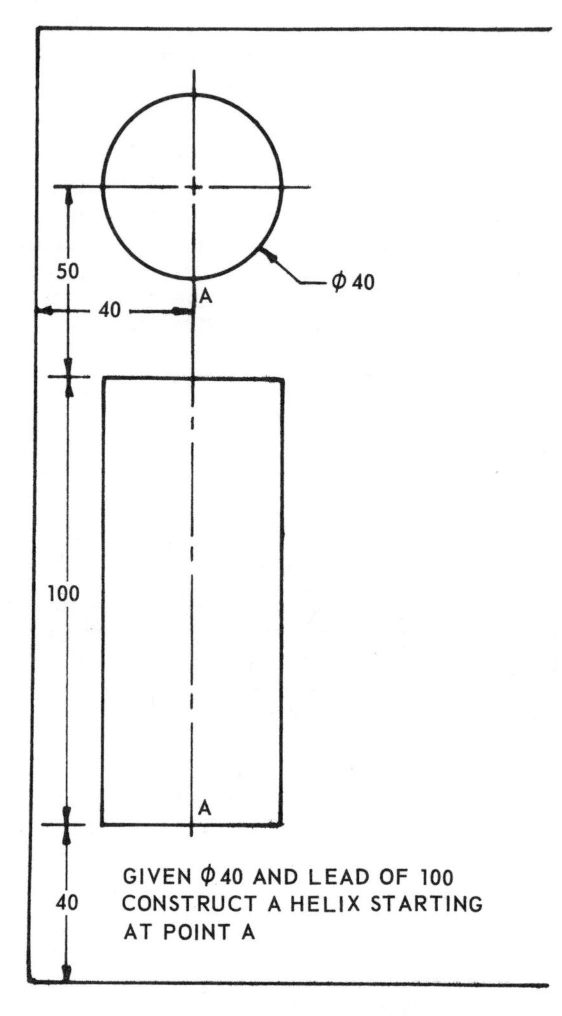

Fig. 16-5-A Helix construction

Appendix

Table 1. CSA and Ansi Publications, **306**
Table 2. Abbreviations and symbols, p, **307**
Table 3. Conversion of decimals of an inch to millimetres, **308**
Table 4. Conversion of fractions of an inch to millimetres, **308**
Table 5. Function of numbers, **309**
Table 6. Trigonometric functions, **310**
Table 7. Wire and sheet-metal gauges and thicknesses, **311**
Table 8. Metric twist drill sizes, **312**
Table 9. Number and letter drill sizes, **313**
Table 10. Metric screw threads, **314**
Table 11. Inch screw threads, **315**
Table 12. Common cap screws, **316**
Table 13. Hexagon-head bolts and cap screws, **316**
Table 14. Twelve-spline flange screws, **316**
Table 15. Setscrews, **317**
Table 16. Common washer sizes, **317**
Table 17. Hexagon-head nuts, **318**
Table 18. Hex flange nuts, **318**
Table 19. Prevailing - torque insert - type nuts, **318**
Table 20. Square and flat stock keys, **319**
Table 21. Woodruff keys, **319**
Table 22. Taper pins, **320**
Table 23. Spring pins, **320**
Table 24. Groove pins, **320**
Table 25. Grooved studs, **321**
Table 26. Cotter Pins, **321**
Table 27. Clevis pins, **321**

CSA AND ANSI PUBLICATIONS

CSA—Canadian Standards

CSA	B1.1	Unified and American Screw Threads
CSA	B19.1	Plain Washers
CSA	B33.1	Square and Hexagon Bolts and Nuts, Studs and Wrench Openings
CSA	B35.1	Machine Screws, Stove Bolts and Associated Nuts
CSA	B78.1	Drawing Standard—General Principles
CSA	B78.2	Drawing Standard—Dimensioning and Tolerancing
CAN3	B78.3-M77	Building Drawings
CSA	B95	Surface Texture
CSA	B97.1	Limits and Fits for Engineering and Manufacturing
CSA	Z85	Abbreviations for Scientific and Engineering Terms

ANSI—American National Standards Institute

ANSI	Y14.1	Size and Format
*ANSI	Y14.2	Line Conventions, Sectioning and Lettering
*ANSI	Y14.3	Projections
*ANSI	Y14.4	Pictorial Drawing
*ANSI	Y14.5	Dimensioning and Tolerancing for Engineering Drawings
*ANSI	Y14.6	Screw Threads
ANSI	Y14.7	Gears, Splines and Serrations
ANSI	Y14.9	Forgings
ANSI	Y14.10	Metal Stampings
ANSI	Y14.11	Plastics
ANSI	Y14.14	Mechanical Assemblies
ANSI	Y14.15	Electrical and Electronics Diagrams
ANSI	B1.1	Unified Screw Threads
ANSI	B18.2.1	Square and Hex Bolts and Screws
ANSI	B18.2.2	Square and Hex Nuts
ANSI	B18.3	Socket Cap, Shoulder and Setscrews
ANSI	B18.6.3	Machine Screws and Machine Screw Nuts
ANSI	B17.2	Woodruff Key and Keyslot Dimensions
ANSI	B17.1	Keys and Keyseats
ANSI	B18.21.1	Lock Washers
ANSI	B27.2	Plain Washers
ANSI	B46.1	Surface Texture

The above standards may be purchased from the

CANADIAN STANDARDS ASSOCIATION
178 REXDALE BOULEVARD
REXDALE, ONTARIO, CANADA, M9W 1R3

Table 1 CSA and ANSI Publications.

And	&
Across Flats	A/F
American National Standards Institute	ANSI
Angular	ANG
Approximate	APPROX
Arc Dimension	**R105**
Assembly	ASSY
Basic	BASIC
Basic Dimension	**[110]**
Bill of Material	B/M
Bolt Circle	BC
Brass	BR
Brown and Sharpe Gauge	B & S GA
Bushing	BUSH
Canada Standards Institute	CSI
Casting	CSTG
Cast Iron	CI
Centimetre	cm
Center Line	℄
Center to Center	C to C
Chamfered	CHAM
Circularity	CIR
Cold-Rolled Steel	CRS
Concentric	CONC
Counterbore	⌴ CBORE
Countersink	⌵ ... CSK
Cubic Centimetre	cm^3
Cubic Metre	m^3
Datum	ISO [A] .. ANSI [-A-] .. DATUM
Deep/Depth Symbol	↧ **DEEP**
Degree (Angle)	° or DEG
Diameter	ϕ or DIA
Diametral Pitch	DP
Dimension	DIM
Drawing	DWG
Eccentric	ECC
Figure	FIG
Finish All Over	FAO
Gage	GA
Heat Treat	HT TR
Head	HD
Heavy	HY
Hexagon	HEX
Hydraulic	HYD
Inside Diameter	ID
International Organization for Standardization	ISO
Iron Pipe Size	IPS
Kilogram	kg
Kilometre	km
Large End	LE
Left Hand	LH
Litre	L
Machined	√ or ∀
Machine Steel	MS or MACH ST
Material	MATL
Maximum	MAX
Maximum Material Condition	M or MMC
Metre	m
Metric Thread	M
Micrometre	μm
Millimetre	mm
Minimum	MIN
Minute (Angle)	MIN
Newton	N
Nominal	NOM
Not to Scale	75 (underlined)
Number	NO
Number of Times	**8X**
Outside Diameter	OD
Parallel	PAR
Pascal	Pa
Perpendicular	PERP
Pitch	P
Pitch Circle Diameter	PCD
Pitch Diameter	PD
Plate	PL
Position	⌖
Radian	rad
Radius	R
Radius, Common	**CR20**
Radius, Spherical	**SR20**
Reference or Reference Dimension	REF
Revolutions per Minute	rev/min
Right Hand	RH
Second (Arc)	(")
Second (Time)	S
Section	SECT
Slotted	SLOT
Socket	SOCK
Spherical	SPHER
Spotface	⌴ .. SFACE
Square	□ or SQ
Square Centimetre	cm^2
Square Metre	m^2
Steel	ST
Straight	STR
Surface Texture	
Symmetrical	SYM
Taper - Conical	⊲
Taper - Flat	◺
Thread	THD
Through	THRU
Tolerance	TOL
True Profile	TP
Undercut	UCUT
U.S. Sheet-Metal Gage	USS GA
Watt	W
Wrought iron	WI

Table 2 Abbreviations and symbols.

ONE TENTH OF AN INCH INCREMENTS TO TWENTY INCHES										
Inch	0	.10	.20	.30	.40	.50	.60	.70	.80	.90
0	0.0	2.5	5.1	7.6	10.2	12.7	15.2	17.8	20.3	22.9
1	25.4	27.9	30.5	33.0	35.6	38.1	40.6	43.2	45.7	48.3
2	50.8	53.3	55.9	58.4	61.0	63.5	66.0	68.6	71.1	73.7
3	76.2	78.7	81.3	83.8	86.4	88.9	91.4	94.0	96.5	99.1
4	101.6	104.1	106.7	109.2	111.8	114.3	116.8	119.4	121.9	124.5
5	127.0	129.5	132.1	134.6	137.2	139.7	142.2	144.8	147.3	149.9
6	152.4	154.9	157.5	160.0	162.6	165.1	167.6	170.2	172.7	175.3
7	177.8	180.3	182.9	185.4	188.0	190.5	193.0	195.6	198.1	200.7
8	203.2	205.7	208.3	210.8	213.4	215.9	218.4	221.0	223.5	226.1
9	228.6	231.1	233.7	236.2	238.8	241.3	243.8	246.4	248.9	251.5
10	254.0	256.5	259.1	261.6	264.2	266.7	269.2	271.8	274.3	276.9
11	279.4	281.9	284.5	287.0	289.6	292.1	294.6	297.2	299.7	302.3
12	304.8	307.3	309.9	312.4	315.0	317.5	320.0	322.6	325.1	327.7
13	330.2	332.7	335.3	337.8	340.4	342.9	345.4	348.0	350.5	353.1
14	355.6	358.1	360.7	363.2	365.8	368.3	370.8	373.4	375.9	378.5
15	381.0	383.5	386.1	388.6	391.2	393.7	396.2	398.8	401.3	403.9
16	406.4	408.9	411.5	414.0	416.6	419.1	421.6	424.2	426.7	429.3
17	431.8	434.3	436.9	439.4	442.0	444.5	447.0	449.6	452.1	454.7
18	457.2	459.7	462.3	464.8	467.4	469.9	472.4	475.0	477.5	480.1
19	482.6	485.1	487.7	490.2	492.8	495.3	497.8	500.4	502.9	505.5
20	508.0	510.5	513.1	515.6	518.2	520.7	523.2	525.8	528.3	530.9

ONE HUNDREDTH OF AN INCH INCREMENTS TO ONE INCH										
Inches	.00	.01	.02	.03	.04	.05	.06	.07	.08	.09
.00	0.00	0.25	0.51	0.76	1.02	1.27	1.52	1.78	2.03	2.29
.10	2.54	2.79	3.05	3.30	3.56	3.81	4.06	4.32	4.57	4.83
.20	5.08	5.33	5.59	5.84	6.10	6.35	6.60	6.86	7.11	7.37
.30	7.62	7.87	8.13	8.38	8.64	8.89	9.14	9.40	9.65	9.91
.40	10.16	10.41	10.67	10.92	11.18	11.43	11.68	11.94	12.19	12.45
.50	12.70	12.95	13.21	13.46	13.72	13.97	14.22	14.48	14.73	14.99
.60	15.24	15.49	15.75	16.00	16.26	16.51	16.76	17.02	17.27	17.53
.70	17.78	18.03	18.29	18.54	18.80	19.05	19.30	19.56	19.81	20.07
.80	20.32	20.57	20.83	21.08	21.34	21.59	21.84	22.10	22.35	22.61
.90	22.86	23.11	23.37	23.62	23.88	24.13	24.38	24.64	24.89	25.15

Table 3 Conversion of decimals of an inch to millimetres.

IN.	0	1/16	1/8	3/16	1/4	5/16	3/8	7/16	1/2	9/16	5/8	11/16	3/4	13/16	7/8	15/16
0	.0	1.6	3.2	4.8	6.4	7.9	9.5	11.1	12.7	14.3	15.9	17.5	19.1	20.6	22.2	23.8
1	25.4	27.0	28.6	30.2	31.8	33.3	34.9	36.5	38.1	39.7	41.3	42.9	44.5	46.0	47.6	49.2
2	50.8	52.4	54.0	55.6	57.2	58.7	60.3	61.9	63.5	65.1	66.7	68.3	69.9	71.4	73.0	74.6
3	76.2	77.8	79.4	81.0	82.6	84.1	85.7	87.3	88.9	90.5	92.1	93.7	95.3	96.8	98.4	100.0
4	101.6	103.2	104.8	106.4	108.0	109.5	111.1	112.7	114.3	115.9	117.5	119.1	120.7	122.2	123.8	125.4
5	127.0	128.6	130.2	131.8	133.4	134.9	136.5	138.1	139.7	141.3	142.9	144.5	146.1	147.6	149.2	150.8
6	152.4	154.0	155.6	157.2	158.8	160.3	161.9	163.5	165.1	166.7	168.3	169.9	171.5	173.0	174.6	176.2
7	177.8	179.4	181.0	182.6	184.2	185.7	187.3	188.9	190.5	192.1	193.7	195.3	196.9	198.4	200.0	201.6
8	203.2	204.8	206.4	208.0	209.6	211.1	212.7	214.3	215.9	217.5	219.1	220.7	222.3	223.8	225.4	227.0
9	228.6	230.2	231.8	233.4	235.0	236.5	238.1	239.7	241.3	242.9	244.5	246.1	247.7	249.2	250.8	252.4
10	254.0	255.6	257.2	258.8	260.4	261.9	263.5	265.1	266.7	268.3	269.9	271.5	273.1	274.6	276.2	277.8
11	279.4	281.0	282.6	284.2	285.8	287.3	288.9	290.5	292.1	293.7	295.3	296.9	298.5	300.0	301.6	303.2
12	304.8	306.4	308.0	309.6	311.2	312.7	314.3	315.9	317.5	319.1	320.7	322.3	323.9	325.4	327.0	328.6
13	330.2	331.8	333.4	335.0	336.6	338.1	339.7	341.3	342.9	344.5	346.1	347.7	349.3	350.8	352.4	354.0
14	355.6	357.2	358.8	360.4	362.0	363.5	365.1	366.7	368.3	369.9	371.5	373.1	374.7	376.2	377.8	379.4

Table 4 Conversion of fractions of an inch to millimetres.

NUMBER	SQUARE	SQUARE ROOT	CIRCLE CIRCUMFERENCE	AREA OF CIRCLE
1	1	1	3.14	0.78
2	4	1.41	6.28	3.14
3	9	1.73	9.43	7.07
4	16	2.00	12.57	12.57
5	25	2.34	15.71	19.64
6	36	2.4495	18.85	28.27
7	49	2.6458	21.99	38.48
8	64	2.8284	25.13	50.27
9	81	3.0000	28.27	63.62
10	100	3.1623	31.46	78.54
11	121	3.3166	34.56	95.03
12	144	3.4641	37.70	113.09
13	169	3.6056	40.84	132.73
14	196	3.7417	43.98	153.94
15	225	3.8730	47.12	176.72
16	256	4.0000	50.27	201.06
17	289	4.1231	53.41	226.98
18	324	4.2426	56.55	254.47
19	361	4.3589	59.69	283.53
20	400	4.4721	62.83	314.16
21	441	4.5826	65.97	346.36
22	484	4.6904	69.12	380.13
23	529	4.7958	72.26	415.48
24	576	4.8990	75.39	452.39
25	625	5.0000	78.54	490.87
26	676	5.0990	81.68	530.93
27	729	5.1962	84.82	572.56
28	784	5.2915	87.97	615.75
29	841	5.3852	91.11	660.52
30	900	5.4772	94.25	706.86
31	961	5.5678	97.39	754.77
32	1024	5.6569	100.53	804.25
33	1089	5.7446	103.67	855.30
34	1156	5.8310	106.81	907.92
35	1225	5.9161	109.96	962.113
36	1296	6.0000	113.10	1017.88
37	1369	6.0828	116.24	1075.21
38	1444	6.1644	119.38	1134.11
39	1521	6.2450	122.52	1194.59
40	1600	6.3246	125.66	1256.64
41	1681	6.4031	128.81	1320.25
42	1764	6.4807	131.95	1385.44
43	1849	6.5574	135.09	1452.20
44	1936	6.6332	138.23	1520.53
45	2025	6.7082	141.37	1590.43
46	2116	6.7823	144.51	1661.90
47	2209	6.8557	147.65	1734.94
48	2304	6.9282	150.80	1809.56
49	2401	7.0000	153.94	1885.74
50	2500	7.0711	157.08	1963.50
51	2601	7.1414	160.22	2042.82
52	2704	7.2111	163.36	2123.72
53	2809	7.2801	166.50	2206.18
54	2916	7.3485	169.65	2290.22
55	3025	7.4162	172.79	2375.83
56	3136	7.4833	175.93	2463.01
57	3249	7.5498	179.07	2551.76
58	3364	7.6158	182.21	2642.08
59	3481	7.6811	185.35	2733.97
60	3600	7.7460	188.50	2827.43
61	3721	7.8102	191.64	3922.47
62	3844	7.8740	194.78	3019.07
63	3969	7.9373	197.92	3117.25
64	4096	8.0000	201.06	3216.99
65	4225	8.0623	204.20	3318.31
66	4356	8.1240	207.35	3421.19
67	4489	8.1854	210.49	3525.65
68	4624	8.2462	213.63	3631.68
69	4761	8.3066	216.77	3739.28
70	4900	8.3666	219.91	3848.50
71	5041	8.4261	223.05	3959.19
72	5184	8.4853	226.19	4071.50
73	5329	8.5440	229.34	4185.39
74	5476	8.6023	232.48	4300.84
75	5625	8.6603	235.62	4417.88
76	5776	8.7178	238.76	4536.47
77	5929	8.7750	241.90	4656.64
78	6084	8.8318	245.04	4778.37
79	6241	8.8882	248.19	4901.68
80	6400	8.9443	251.33	5026.56
81	6561	9.0000	254.47	5183.01
82	6724	9.0554	257.61	5281.03
83	6889	9.1104	260.75	5410.62
84	7056	9.1652	263.89	5541.78
85	7225	9.2200	267.04	5674.52
86	7396	9.2736	270.18	5808.82
87	7569	9.3274	273.32	5944.69
88	7744	9.3808	276.46	6082.14
89	7921	9.4340	279.60	6221.15
90	8100	9.4868	282.74	6361.74
91	8281	9.5393	285.89	6503.90
92	8464	9.5917	289.03	6647.63
93	8649	9.6437	292.17	6792.92
94	8836	9.6954	295.31	6939.79
95	9025	9.7468	298.45	7088.24
96	9216	9.7979	301.59	7238.25
97	9409	9.8489	304.74	7389.83
98	9604	9.8995	307.88	7542.98
99	9801	9.9509	311.02	7697.71
100	10 000	10.000	314.16	7854.00
101	10 201	10.0499	317.30	8011.87
102	10 404	10.0995	320.44	8171.30
103	10 609	10.1489	323.58	8332.31
104	10 816	10.1980	326.73	8494.89
105	11 025	10.2470	329.87	8659.04

Table 5 Function of numbers.

ANGLE	SINE	COSINE	TAN	COTAN	ANGLE
0°	.0000	1.0000	.0000	θ	90°
1°	0.0175	0.9998	0.0175	57.290	89°
2°	0.0349	0.9994	0.0349	28.636	88°
3°	0.0523	0.9986	0.0524	19.081	87°
4°	0.0698	0.9976	0.0699	14.301	86°
5°	0.0872	0.9962	0.0875	11.430	85°
6°	0.1045	0.9945	0.1051	9.5144	84°
7°	0.1219	0.9925	0.1228	8.1443	83°
8°	0.1392	0.9903	0.1405	7.1154	82°
9°	0.1564	0.9877	0.1584	6.3138	81°
10°	0.1736	0.9848	0.1763	5.6713	80°
11°	0.1908	0.9816	0.1944	5.1446	79°
12°	0.2079	0.9781	0.2126	4.7046	78°
13°	0.2250	0.9744	0.2309	4.3315	77°
14°	0.2419	0.9703	0.2493	4.0108	76°
15°	0.2588	0.9659	0.2679	3.7321	75°
16°	0.2756	0.9613	0.2867	3.4874	74°
17°	0.2924	0.9563	0.3057	3.2709	73°
18°	0.3090	0.9511	0.3249	3.0777	72°
19°	0.3256	0.9455	0.3443	2.9042	71°
20°	0.3420	0.9397	0.3640	2.7475	70°
21°	0.3584	0.9336	0.3839	2.6051	69°
22°	0.3746	0.9272	0.4040	2.4751	68°
23°	0.3907	0.9205	0.4245	2.3559	67°
24°	0.4067	0.9135	0.4452	2.2460	66°
25°	0.4226	0.9063	0.4663	2.1445	65°
ANGLE	COSINE	SINE	COTAN	TAN	ANGLE

ANGLE	SINE	COSINE	TAN	COTAN	ANGLE
0°	.0000	1.0000	.0000	θ	90°
26°	0.4384	0.8988	0.4877	2.0503	64°
27°	0.4540	0.8910	0.5095	1.9626	63°
28°	0.4695	0.8829	0.5317	1.8807	62°
29°	0.4848	0.8746	0.5543	1.8040	61°
30°	0.5000	0.8660	0.5774	1.7321	60°
31°	0.5150	0.8572	0.6009	1.6643	59°
32°	0.5299	0.8480	0.6249	1.6003	58°
33°	0.5446	0.8387	0.6494	1.5399	57°
34°	0.5592	0.8290	0.6745	1.4826	56°
35°	0.5736	0.8192	0.7002	1.4281	55°
36°	0.5878	0.8090	0.7265	1.3764	54°
37°	0.6018	0.7986	0.7536	1.3270	53°
38°	0.6157	0.7880	0.7813	1.2799	52°
39°	0.6293	0.7771	0.8098	1.2349	51°
40°	0.6428	0.7660	0.8391	1.1918	50°
41°	0.6561	0.7547	0.8693	1.1504	49°
42°	0.6691	0.7431	0.9004	1.1106	48°
43°	0.6820	0.7314	0.9325	1.0724	47°
44°	0.6947	0.7193	0.9657	1.0355	46°
45°	0.7071	0.7071	0.0000	1.0000	45°
ANGLE	COSINE	SINE	COTAN	TAN	ANGLE

Table 6 Trigonometric functions.

NORTH AMERICAN GAUGES												EUROPEAN GAUGES								
Ferrous metals, such as galvanized steel, tin plate						Nonferrous metals, such as copper, brass, aluminum			Steel and iron wire and bare copper piano wire			Galvanized steel, tin plate, copper, strip steel and steel, copper and aluminum tubes						Nonferrous		
U.S. Standard (USS)			U.S. Standard (Revised) Formerly Manufactures Standard			American Standard or Brown and Sharpe (B & S)			United States Steel Wire Gauge			Birmingham (BWG)			New Birmingham (BG)			Imperial Wire Gauge Imperial Standard (SWG)		
Gauge	mm	in.	Gauge	mm	in.	Gauge	mm	in.	Gauge	mm	in.	Gauge	mm	in.	Gauge	mm	in.	Gauge	mm	in.
			3	6.01	.240	3	5.83	.229												
4	5.95	.234	4	5.70	.224	4	5.19	.204	4	5.72	.225	4	6.05	.238	4	6.35	.250	4	5.89	.232
5	5.56	.219	5	5.31	.209	5	4.62	.182	5	5.26	.207	5	5.59	.220	5	5.65	.223	5	5.39	.212
6	5.16	.203	6	4.94	.194	6	4.12	.162	6	4.88	.192	6	5.16	.203	6	5.03	.198	6	4.88	.192
7	4.76	.188	7	4.55	.179	7	3.67	.144	7	4.50	.177	7	4.57	.180	7	4.48	.176	7	4.47	.176
8	4.37	.172	8	4.18	.164	8	3.26	.129	8	4.11	.162	8	4.19	.165	8	3.99	.157	8	4.06	.160
9	3.97	.156	9	3.80	.149	9	2.91	.114	9	3.77	.148	9	3.76	.148	9	3.55	.140	9	3.66	.144
10	3.57	.141	10	3.42	.135	10	2.59	.102	10	3.43	.135	10	3.40	.134	10	3.18	.125	10	3.25	.128
11	3.18	.125	11	3.04	.120	11	2.30	.091	11	3.06	.121	11	3.05	.120	11	2.83	.111	11	2.95	.116
12	2.78	.109	12	2.66	.105	12	2.05	.081	12	2.68	.106	12	2.77	.109	12	2.52	.099	12	2.64	.104
13	2.38	.094	13	2.78	.090	13	1.83	.072	13	2.32	.092	13	2.41	.095	13	2.24	.088	13	2.34	.092
14	1.98	.078	14	1.90	.075	14	1.63	.064	14	2.03	.080	14	2.11	.083	14	1.99	.079	14	2.03	.080
15	1.79	.070	15	1.71	.067	15	1.45	.057	15	1.83	.072	15	1.83	.072	15	1.78	.070	15	1.83	.072
16	1.59	.063	16	1.52	.060	16	1.29	.051	16	1.63	.063	16	1.65	.065	16	1.59	.063	16	1.63	.064
17	1.43	.056	17	1.37	.054	17	1.15	.045	17	1.37	.054	17	1.47	.058	17	1.41	.056	17	1.42	.056
18	1.27	.050	18	1.21	.048	18	1.02	.040	18	1.21	.048	18	1.25	.049	18	2.58	.050	18	1.22	.048
19	1.11	.044	19	1.06	.042	19	0.91	.036	19	1.04	.041	19	1.07	.042	19	1.19	.044	19	1.02	.040
20	0.95	.038	20	0.91	.036	20	0.81	.032	20	0.88	.035	20	0.89	.035	20	1.00	.039	20	0.91	.036
21	0.87	.034	21	0.84	.033	21	0.72	.029	21	0.81	.032	21	0.81	.032	21	0.89	.035	21	0.81	.032
22	0.79	.031	22	0.76	.030	22	0.65	.025	22	0.73	.029	22	0.71	.028	22	0.79	.031	22	0.71	.028
23	0.71	.028	23	0.68	.027	23	0.57	.023	23	0.66	.026	23	0.64	.025	23	0.71	.028	23	0.61	.024
24	0.64	.025	24	0.61	.024	24	0.51	.020	24	0.58	.023	24	0.56	.022	24	0.63	.025	24	0.56	.022
25	0.56	.022	25	0.53	.021	25	0.46	.018	25	0.52	.020	25	0.51	.020	25	0.56	.022	25	0.51	.020
26	0.48	.019	26	0.46	.018	26	0.40	.016	26	0.46	.018	26	0.46	.018	26	0.50	.020	26	0.46	.018
27	0.44	.017	27	0.42	.016	27	0.36	.014	27	0.44	.017	27	0.41	.016	27	0.44	.017	27	0.42	.016
28	0.40	.016	28	0.38	.015	28	0.32	.013	28	0.41	.016	28	0.36	.014	28	0.40	.016	28	0.38	.015
29	0.36	.014	29	0.34	.014	29	0.29	.011	29	0.38	.015	29	0.33	.013	29	0.35	.014	29	0.35	.014
30	0.32	.013	30	0.31	.012	30	0.25	.010	30	0.36	.014	30	0.31	.012	30	0.31	.012	30	0.32	.012
31	0.28	.011	31	0.27	.011	31	0.23	.009	31	0.34	.013	31	0.25	.010	31	0.28	.011			
32	0.26	.010	32	0.25	.010	32	0.20	.008	32	0.33	.013	32	0.23	.009				32	0.27	.011
33	0.24	.009	33	0.23	.009	33	0.18	.007	33	0.30	.012	33	0.20	.008	33	0.22	.009	33	0.25	.010
34	0.22	.009	34	0.21	.008	34	0.16	.006	34	0.26	.010	34	0.18	.007	34	0.20	.008	34	0.23	.009
									35	0.24	.010	35	0.13	.005	35	0.18	.007	35	0.21	.008
36	0.18	.007	36	0.17	.007	36	0.13	.005	36	0.23	.009	36	0.10	.004	36	0.16	.006			
									37	0.22	.008							37	0.17	.007
38	0.16	.006	38	0.15	.006	38	0.10	.004	38	0.20	.008				38	0.12	.005	38	0.15	.006
									39	0.19	.008									
									40	0.18	.007				40	0.10	.004	40	0.12	.005
									41	0.17	.007							42	0.10	.004

Note: Metric standards governing gage sizes were not available at the time of publication. The sizes given in the above chart are "soft conversion" from current inch standards and are not meant to be representative of the precise metric gage sizes which may be available in the future. Conversions are given only to allow the student to compare gage sizes readily with the metric drill sizes.

Table 7 Wire and sheet-metal gauges and thicknesses.

METRIC DRILL SIZES		Reference Decimal Equivalent (Inches)
Preferred	Available	
—	0.40	.0157
—	0.42	.0165
—	0.45	.0177
—	0.48	.0189
0.50	—	.0197
—	0.52	.0205
0.55	—	.0217
—	0.58	.0228
0.60	—	.0236
—	0.62	.0244
0.65	—	.0256
—	0.68	.0268
0.70	—	.0276
—	0.72	.0283
0.75	—	.0295
—	0.78	.0307
0.80	—	.0315
—	0.82	.0323
0.85	—	.0335
—	0.88	.0346
0.90	—	.0354
—	0.92	.0362
0.95	—	.0374
—	0.98	.0386
1.00	—	.0394
—	1.03	.0406
1.05	—	.0413
—	1.08	.0425
1.10	—	.0433
—	1.15	.0453
1.20	—	.0472
1.25	—	.0492
1.3	—	.0512
—	1.35	.0531
1.4	—	.0551
—	1.45	.0571
1.5	—	.0591
—	1.55	.0610
1.6	—	.0630
—	1.65	.0650
1.7	—	.0669
—	1.75	.0689
1.8	—	.0709
—	1.85	.0728
1.9	—	.0748
—	1.95	.0768
2.0	—	.0787
—	2.05	.0807
2.1	—	.0827
—	2.15	.0846
2.2	—	.0866
—	2.3	.0906
2.4	—	.0945
2.5	—	.0984
2.6	—	.1024
—	2.7	.1063
2.8	—	.1102
—	2.9	.1142
3.0	—	.1181
—	3.1	.1220
3.2	—	.1260
—	3.3	.1299
3.4	—	.1339
—	3.5	.1378
3.6	—	.1417
—	3.7	.1457
3.8	—	.1496
—	3.9	.1535
4.0	—	.1575
—	4.1	.1614
4.2	—	.1654
—	4.4	.1732
4.5	—	.1772
—	4.6	.1811
4.8	—	.1890
5.0	—	.1969
—	5.2	.2047
5.3	—	.2087
—	5.4	.2126
5.6	—	.2205
—	5.8	.2283
6.0	—	.2362
—	6.2	.2441
6.3	—	.2480
—	6.5	.2559
6.7	—	.2638
—	6.8	.2677
—	6.9	.2717
7.1	—	.2795
—	7.3	.2874
7.5	—	.2953
—	7.8	.3071
8.0	—	.3150
—	8.2	.3228
8.5	—	.3346
—	8.8	.3465
9.0	—	.3543
—	9.2	.3622
9.5	—	.3740
—	9.8	.3858
10	—	.3937
—	10.3	.4055
10.5	—	.4134
—	10.8	.4252
11	—	.4331
—	11.5	.4528
12	—	.4724
12.5	—	.4921
13	—	.5118
—	13.5	.5315
14	—	.5512
—	14.5	.5709
15	—	.5906
—	15.5	.6102
16	—	.6299
—	16.5	.6496
17	—	.6693
—	17.5	.6890
18	—	.7087
—	18.5	.7283
19	—	.7480
—	19.5	.7677
20	—	.7874
—	20.5	.8071
21	—	.8268
—	21.5	.8465
22	—	.8661
—	23	.9055
24	—	.9449
25	—	.9843
26	—	1.0236
—	27	1.0630
28	—	1.1024
—	29	1.1417
30	—	1.1811
—	31	1.2205
32	—	1.2598
—	33	1.2992
34	—	1.3386
—	35	1.3780
36	—	1.4173
—	37	1.4567
38	—	1.4361
—	39	1.5354
40	—	1.5748
—	41	1.6142
42	—	1.6535
—	43.5	1.7126
45	—	1.7717
—	46.5	1.8307
48	—	1.8898
50	—	1.9685
—	51.5	2.0276

Table 8 Metric twist drill sizes.

NUMBER OR LETTER SIZE DRILL	SIZE		NUMBER OR LETTER SIZE DRILL	SIZE		NUMBER OR LETTER SIZE DRILL	SIZE		NUMBER OR LETTER SIZE DRILL	SIZE	
	mm	INCHES		mm	INCHES		mm	INCHES		mm	INCHES
80	0.343	.014	50	1.778	.070	20	4.089	.161	K	7.137	.821
79	0.368	.015	49	1.854	.073	19	4.216	.166	L	7.366	.290
78	0.406	.016	48	1.930	.076	18	4.305	.170	M	7.493	.295
77	0.457	.018	47	1.994	.079	17	4.394	.173	N	7.671	.302
76	0.508	.020	46	2.057	.081	16	4.496	.177	O	8.026	.316
75	0.533	.021	45	2.083	.082	15	4.572	.180	P	8.204	.323
74	0.572	.023	44	2.184	.086	14	4.623	.182	Q	8.433	.332
73	0.610	.024	43	2.261	.089	13	4.700	.185	R	8.611	.339
72	0.635	.025	42	2.375	.094	12	4.800	.189	S	8.839	.348
71	0.660	.026	41	2.438	.096	11	4.851	.191	T	9.093	.358
70	0.711	.028	40	2.489	.098	10	4.915	.194	U	9.347	.368
69	0.742	.029	39	2.527	.100	9	4.978	.196	V	9.576	.377
68	0.787	.031	38	2.578	.102	8	5.080	.199	W	9.804	.386
67	0.813	.032	37	2.642	.104	7	5.105	.201	X	10.084	.397
66	0.838	.033	36	2.705	.107	6	5.182	.204	Y	10.262	.404
65	0.889	.035	35	2.794	.110	5	5.220	.206	Z	10.490	.413
64	0.914	.036	34	2.819	.111	4	5.309	.209			
63	0.940	.037	33	2.870	.113	3	5.410	.213			
62	0.965	.038	32	2.946	.116	2	5.613	.221			
61	0.991	.039	31	3.048	.120	1	5.791	.228			
60	1.016	.040	30	3.264	.129	A	5.944	.234			
59	1.041	.041	29	3.354	.136	B	6.045	.238			
58	1.069	.042	28	3.569	.141	C	6.147	.242			
57	1.092	.043	27	3.658	.144	D	6.248	.246			
56	1.181	.047	26	3.734	.147	E	6.350	.250			
55	1.321	.052	25	3.797	.150	F	6.528	.257			
54	1.397	.055	24	3.861	.152	G	6.629	.261			
53	1.511	.060	23	3.912	.154	H	6.756	.266			
52	1.613	.064	22	3.988	.157	I	6.909	.272			
51	1.702	.067	21	4.039	.159	J	7.036	.277			

Table 9 **Number and letter size drills.**

NOMINAL SIZE DIA (mm)		SERIES WITH GRADED PITCHES				SERIES WITH CONSTANT PITCHES																	
		COARSE		FINE		4		3		2		1.5		1.25		1		0.75		0.5		0.35	
Preferred		Thread Pitch	Tap Drill Size	Thread Pitch	Tap Drill Size	Thread Pitch	Tap Drill Size	Thread Pitch	Tap Drill Size	Thread Pitch	Tap Drill Size	Thread Pitch	Tap Drill Size	Thread Pitch	Tap Drill Size	Thread Pitch	Tap Drill Size	Thread Pitch	Tap Drill Size	Thread Pitch	Tap Drill Size	Thread Pitch	Tap Drill Size
1.6		0.35	1.25																				
	1.8	0.35	1.45																				
2		0.4	1.6																				
	2.2	0.45	1.75																				
2.5		0.45	2.05																			0.35	2.15
3		0.5	2.5																			0.35	2.65
	3.5	0.6	2.9																			0.35	3.15
4		0.7	3.3																	0.5	3.5		
	4.5	0.75	3.7																	0.5	4.0		
5		0.8	4.2																	0.5	4.5		
6		1	5.0															0.75	5.2				
8		1.25	6.7	1	7.0											1	7.0	0.75	7.2				
10		1.5	8.5	1.25	8.7									1.25	8.7	1	9.0	0.75	9.2				
12		1.75	10.2	1.25	10.8							1.5	10.5	1.25	10.7	1	11						
	14	2	12	1.5	12.5							1.5	12.5	1.25	12.7	1	13						
16		2	14	1.5	14.5							1.5	14.5			1	15						
	18	2.5	15.5	1.5	16.5					2	16	1.5	16.5			1	17						
20		2.5	17.5	1.5	18.5					2	18	1.5	18.5			1	19						
	22	2.5	19.5	1.5	20.5					2	20	1.5	20.5			1	21						
24		3	21	2	22					2	22	1.5	22.5			1	23						
	27	3	24	2	25					2	25	1.5	25.5			1	26						
30		3.5	26.5	2	28					2	28	1.5	28.5			1	29						
	33	3.5	29.5	2	31					2	31	1.5	31.5										
36		4	32	3	33					2	34	1.5	34.5										
	39	4	35	3	36					2	37	1.5	37.5										
42		4.5	37.5	3	39	4	38	3	39	2	40	1.5	40.5										
	45	4.5	39	3	42	4	41	3	42	2	43	1.5	43.5										
48		5	43	3	45	4	44	3	45	2	46	1.5	46.5										

Table 10 Metric screw threads.

THREADS PER INCH AND TAP DRILL SIZES													
SIZE INCHES		Graded Pitch Series						Constant Pitch Series					
		Coarse UNC		Fine UNF		Extra Fine UNEF		8 UN		12 UN		16 UN	
Number or Fraction	Deci-mal	Threads per Inch	Tap Drill Dia.	Threads per Inch	Tap Drill Dia.	Threads per Inch	Tap Drill Dia.	Threads per Inch	Tap Drill Dia.	Threads per Inch	Tap Drill Dia.	Threads per Inch	Tap Drill Dia.
0	.060	—	—	80	3/64	—	—	—	—	—	—	—	—
2	.086	56	No. 50	64	No. 49	—	—	—	—	—	—	—	—
4	.112	40	No. 43	48	No. 42	—	—	—	—	—	—	—	—
5	.125	40	No. 38	44	No. 37	—	—	—	—	—	—	—	—
6	.138	32	No. 36	40	No. 33	—	—	—	—	—	—	—	—
8	.164	32	No. 29	36	No. 29	—	—	—	—	—	—	—	—
10	.190	24	No. 25	32	No. 21	—	—	—	—	—	—	—	—
1/4	.250	20	7	28	3	32	.219	—	—	—	—	—	—
5/16	.312	18	F	24	1	32	.281	—	—	—	—	—	—
3/8	.375	16	.312	24	Q	32	.344	—	—	—	—	UNC	—
7/16	.438	14	U	20	.391	28	Y	—	—	—	—	16	V
1/2	.500	13	.422	20	.453	28	.469	—	—	—	—	16	.438
9/16	.562	12	.484	18	.516	24	.516	—	—	UNC	—	16	.500
5/8	.625	11	.531	18	.578	24	.578	—	—	12	.547	16	.562
3/4	.750	10	.656	16	.688	20	.703	—	—	12	.672	UNF	—
7/8	.875	9	.766	14	.812	20	.828	—	—	12	.797	16	.812
1	1.000	8	.875	12	.922	20	.953	UNC	—	UNF	—	16	.938
1 1/8	1.125	7	.984	12	1.047	18	1.078	8	1.000	UNF	—	16	1.062
1 1/4	1.250	7	1.109	12	1.172	18	1.188	8	1.125	UNF	—	16	1.188
1 3/8	1.375	6	1.219	12	1.297	18	1.312	8	1.250	UNF	—	16	1.312
1 1/2	1.500	6	1.344	12	1.422	18	1.438	8	1.375	UNF	—	16	1.438
1 5/8	1.625	—	—	—	—	18	—	8	1.500	12	1.547	16	1.562
1 3/4	1.750	5	1.562	—	—	—	—	8	1.625	12	1.672	16	1.688
1 7/8	1.875	—	—	—	—	—	—	8	1.750	12	1.797	16	1.812
2	2.000	4.5	1.781	—	—	—	—	8	1.875	12	1.922	16	1.938
2 1/4	2.250	4.5	2.031	—	—	—	—	8	2.125	12	2.172	16	2.188
2 1/2	2.500	4	2.250	—	—	—	—	8	2.375	12	2.422	16	2.438
2 3/4	2.750	4	2.500	—	—	—	—	8	2.625	12	2.672	16	2.688
3	3.000	4	2.750	—	—	—	—	8	2.875	12	2.922	16	2.938
3 1/4	3.250	4	3.000	—	—	—	—	8	3.125	12	3.172	16	3.188
3 1/2	3.500	4	3.250	—	—	—	—	8	3.375	12	3.422	16	3.438
3 3/4	3.750	4	3.500	—	—	—	—	8	3.625	12	3.668	16	3.688
4	4.000	4	3.750	—	—	—	—	8	3.875	12	3.922	16	3.938

Table 11 Inch screw threads.

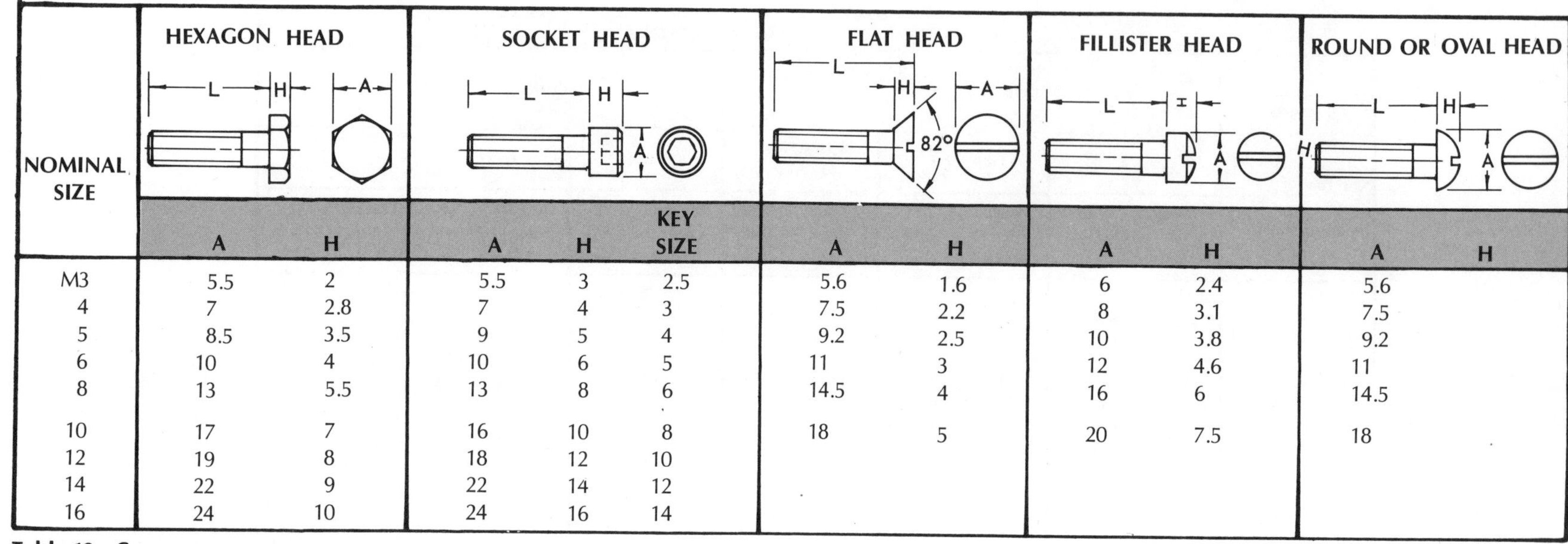

NOMINAL SIZE	HEXAGON HEAD		SOCKET HEAD			FLAT HEAD		FILLISTER HEAD		ROUND OR OVAL HEAD	
	A	H	A	H	KEY SIZE	A	H	A	H	A	H
M3	5.5	2	5.5	3	2.5	5.6	1.6	6	2.4	5.6	
4	7	2.8	7	4	3	7.5	2.2	8	3.1	7.5	
5	8.5	3.5	9	5	4	9.2	2.5	10	3.8	9.2	
6	10	4	10	6	5	11	3	12	4.6	11	
8	13	5.5	13	8	6	14.5	4	16	6	14.5	
10	17	7	16	10	8	18	5	20	7.5	18	
12	19	8	18	12	10						
14	22	9	22	14	12						
16	24	10	24	16	14						

Table 12 Common cap screws.

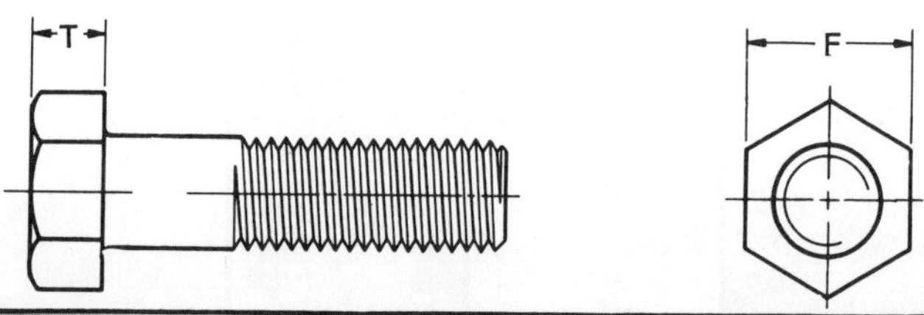

NOMINAL BOLT SIZE AND THREAD PITCH	WIDTH ACROSS FLATS F	THICKNESS T
M6 x 1	10	4.7
M8 x 1.25	13	5.7
M10 x 1.5	15	6.8
M12 x 1.75	18	8
M14 x 2	21	9.3
M16 x 2	24	10.5
M20 x 2.5	30	13.1
M24 x 3	36	15.6
M30 x 3.5	46	19.5
M36 x 4	55	23.4

Table 13 Hexagon-head bolts and cap screws.

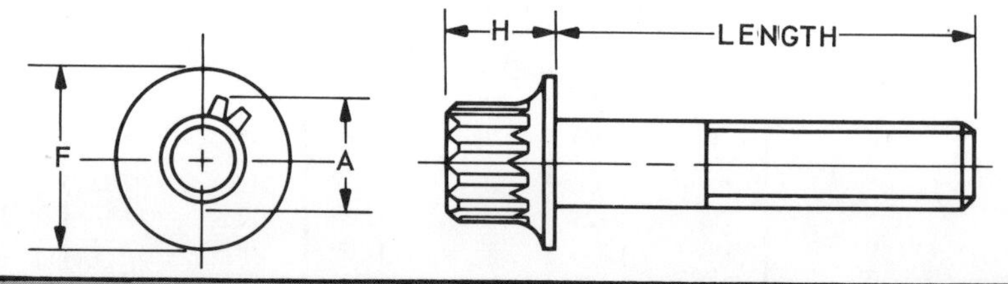

NOMINAL BOLT SIZE AND THREAD PITCH	HEAD SIZES		
	F	A	H
M5 x 0.8	9.4	5.9	5
M6 x 1	11.8	7.4	6.3
M8 x 1.25	15	9.4	8
M10 x 1.5	18.6	11.7	10
M12 x 1.75	22.8	14	12
M14 x 2	26.4	16.3	14
M16 x 2	30.3	18.7	16
M20 x 2.5	37.4	23.4	20

Table 14 Twelve-spline flange screws.

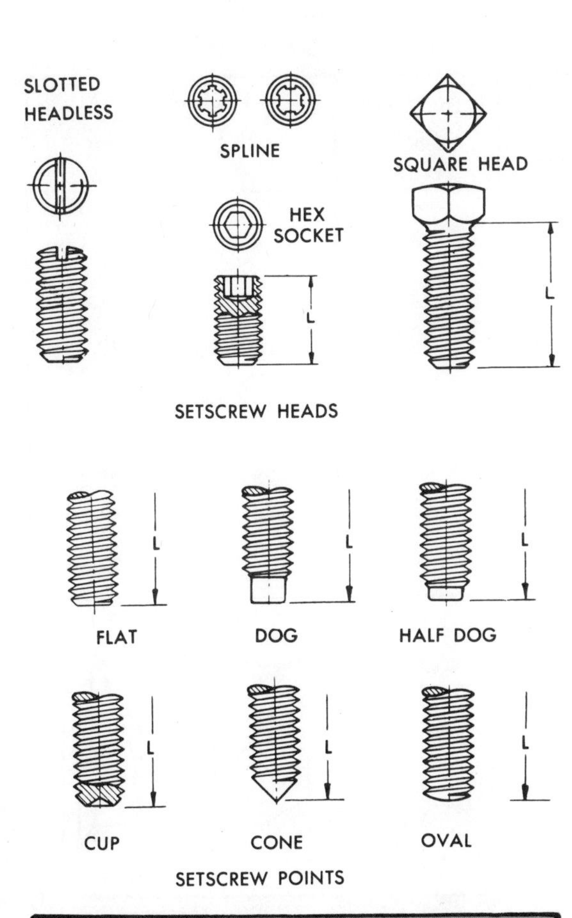

NOMINAL SIZE	KEY SIZE
M1.4	0.7
2	0.9
3	1.5
4	2
5	2
6	3
8	4
10	5
12	6
16	8

Table 15 Setscrews.

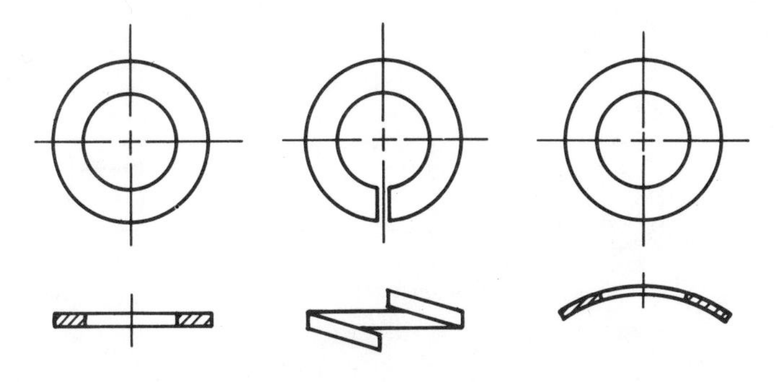

BOLT SIZE	FLAT WASHERS			LOCKWASHERS			SPRING LOCKWASHERS		
	ID	OD	THICK	ID	OD	THICK	ID	OD	THICK
2	2.2	5.5	0.5	2.1	3.3	0.5			
3	3.2	7	0.5	3.1	5.7	0.8			
4	4.3	9	0.8	4.1	7.1	0.9	4.2	8	0.3 0.4
5	5.3	11	1	5.1	8.7	1.2	5.2	10	0.4 0.5
6	6.4	12	1.5	6.1	11.1	1.6	6.2	12.5	0.5 0.7
7	7.4	14	1.5	7.1	12.1	1.6	7.2	14	0.5 0.8
8	8.4	17	2	8.2	14.2	2	8.2	16	0.6 0.9
10	10.5	21	2.5	10.2	17.2	2.2	10.2	20	0.8 1.1
12	13	24	2.5	12.3	20.2	2.5	12.2	25	0.9 1.5
14	15	28	2.5	14.2	23.2	3	14.2	28	1.0 1.5
16	17	30	3	16.2	26.2	3.5	16.3	31.5	1.2 1.7
18	19	34	3	18.2	28.2	3.5	18.3	35.5	1.2 2.0
20	21	36	3	20.2	32.2	4	20.4	40	1.5 2.25
22	23	39	4	22.5	34.5	4	22.4	45	1.75 2.5
24	25	44	4	24.5	38.5	5			
27	28	50	4	27.5	41.5	5			
30	31	56	4	30.5	46.5	6			

Table 16 Common washer sizes.

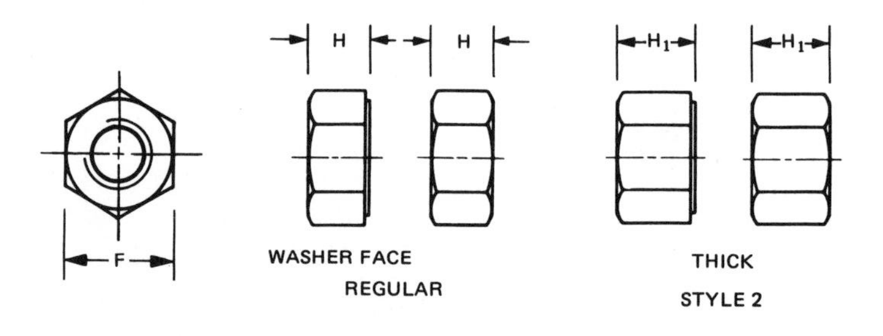

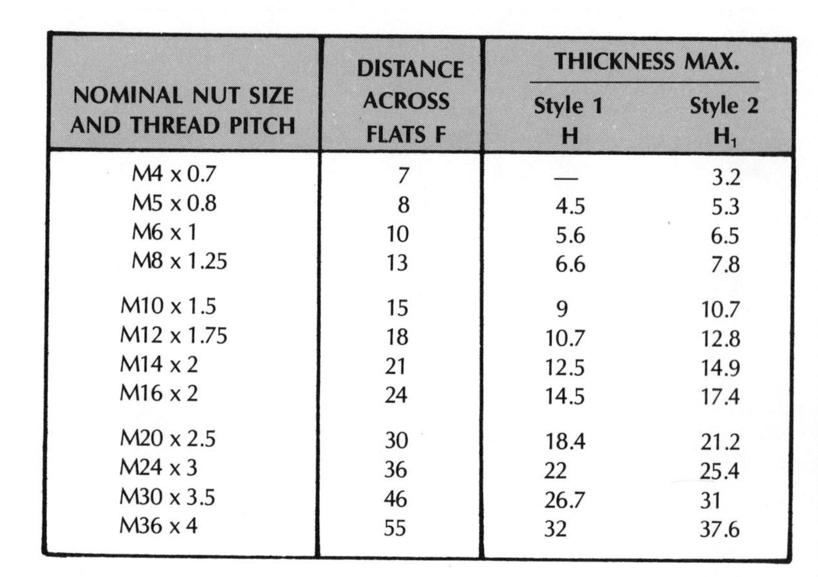

NOMINAL NUT SIZE AND THREAD PITCH	DISTANCE ACROSS FLATS F	THICKNESS MAX. Style 1 H	Style 2 H_1
M4 x 0.7	7	—	3.2
M5 x 0.8	8	4.5	5.3
M6 x 1	10	5.6	6.5
M8 x 1.25	13	6.6	7.8
M10 x 1.5	15	9	10.7
M12 x 1.75	18	10.7	12.8
M14 x 2	21	12.5	14.9
M16 x 2	24	14.5	17.4
M20 x 2.5	30	18.4	21.2
M24 x 3	36	22	25.4
M30 x 3.5	46	26.7	31
M36 x 4	55	32	37.6

Table 17 Hexagon-head nuts.

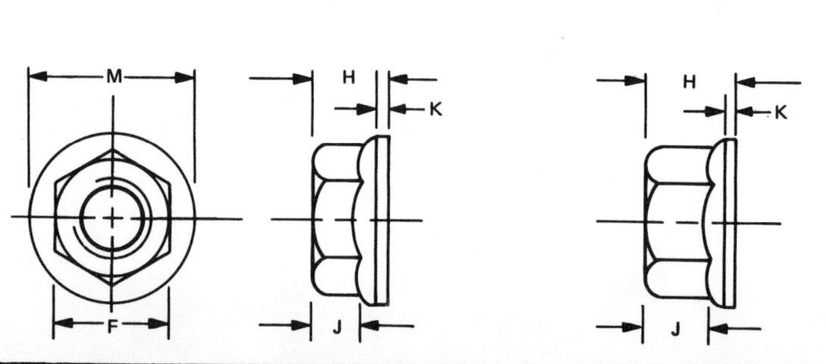

NOMINAL NUT SIZE AND THREAD PITCH	WIDTH ACROSS FLATS F	STYLE 1 H	J	K	M	STYLE 2 H	J
M6 x 1	10	5.8	3	1	14.2	6.7	3.7
M8 x 1.25	13	6.8	3.7	1.3	17.6	8	4.5
M10 x 1.5	15	9.6	5.5	1.5	21.5	11.2	6.7
M12 x 1.75	18	11.6	6.7	2	25.6	13.5	8.2
M14 x 2	21	13.4	7.8	2.3	29.6	15.7	9.6
M16 x 2	24	15.9	9.5	2.5	34.2	18.4	11.7
M20 x 2.5	30	19.2	11.1	2.8	42.3	22	12.6

Table 18 Hex flange nuts.

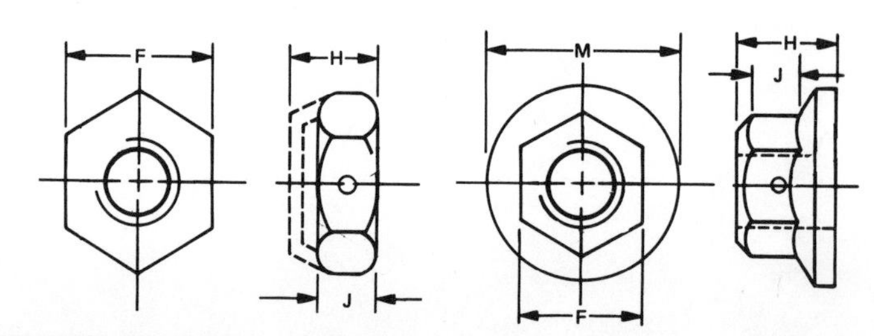

NOMINAL NUT SIZE AND THREAD PITCH	WIDTH ACROSS FLATS F	HEX NUTS Style 1 H	J	Style 2 H	J	HEX FLANGE NUTS Style 1 H	J	M	Style 2 H	J
M5 × 0.8	8.0	6.1	2.3	7.6	2.9					
M6 × 1	10	7.6	3	8.8	3.7	7.6	3	14.2	8.8	3.7
M8 × 1.25	13	9.1	3.7	10.3	4.5	9.1	3.7	17.6	10.3	4.5
M10 × 1.5	15	12	5.5	14	6.7	12	5.5	21.5	14	6.7
M12 × 1.75	18	14.2	6.7	16.8	8.2	14.4	6.7	25.6	16.8	8.2
M14 × 2	21	16.5	7.8	18.9	9.6	16.6	7.8	29.6	18.9	9.6
M16 × 2	24	18.5	9.5	21.4	11.7	18.9	9.5	34.2	21.4	11.7
M20 × 2.5	30	23.4	11.1	26.5	12.6	23.4	11.1	42.3	26.5	12.6

Table 19 Prevailing-torque insert-type nuts.

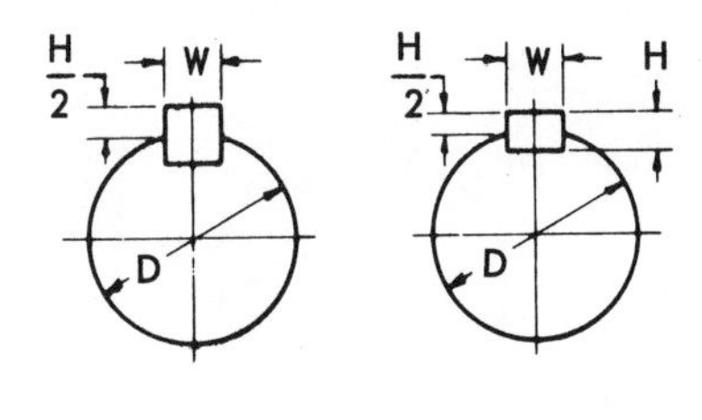

Diameter of Shaft		Square Key		Flat Key	
		Nominal Size		Nominal Size	
Over	Up To	W	H	W	H
6	8	2	2		
8	10	3	3		
10	12	4	4		
12	17	5	5		
17	22	6	6		
22	30	7	7	8	7
30	38	8	8	10	8
38	44	9	9	12	8
44	50	10	10	14	9
50	58	12	12	16	10

Table 20 Square and flat stock keys.

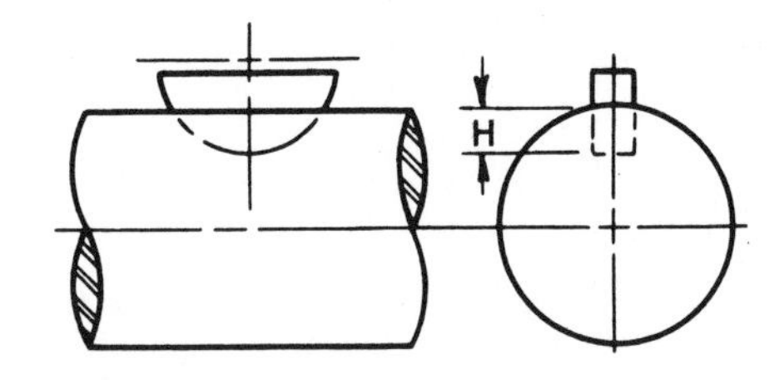

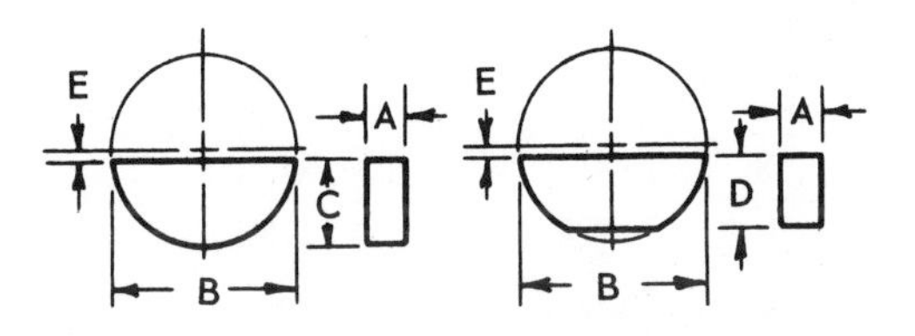

Key No.	Nominal Size	Key			Keyseat
	A × B	E	C	D	H
204	1.6 × 12.7	1.5	5.1	4.8	4.3
304	2.4 × 12.7	1.3	5.1	4.8	3.8
305	2.4 × 15.9	1.5	6.4	6.1	5.1
404	3.2 × 12.7	1.3	5.1	4.8	3.6
405	3.2 × 15.9	1.5	6.4	6.1	4.6
406	3.2 × 19.1	1.5	7.9	7.6	6.4
505	4.0 × 15.9	1.5	6.4	6.1	4.3
506	4.0 × 19.1	1.5	7.9	7.6	5.8
507	4.0 × 22.2	1.5	9.7	9.1	7.4
606	4.8 × 19.1	1.5	7.9	7.6	5.3
607	4.8 × 22.2	1.5	9.7	9.1	7.1
608	4.8 × 25.4	1.5	11.2	10.9	8.6
609	4.8 × 28.6	2.0	12.2	11.9	9.9
807	6.4 × 22.2	1.5	9.7	9.1	6.4
808	6.4 × 25.4	1.5	11.2	10.9	7.9

Table 21 Woodruff keys.

NUMBER	6/0	5/0	4/0	3/0	2/0	0	1	2	3	4	5	6	7	8
SIZE (LARGE END)	2	2.4	2.8	3.2	3.6	4	4.4	4.9	5.6	6.4	7.4	8	10.4	12.5
LENGTH 12	x	x	x	x	x	x								
16	x	x	x	x	x	x								
20	x	x	x	x	x	x	x	x	x					
22				x	x	x	x	x	x					
25		x	x	x	x	x	x	x	x	x	x			
30					x	x	x	x	x	x	x	x		
40						x	x	x	x	x	x	x		
45							x	x	x	x	x	x		
50							x	x	x	x	x	x	x	x
55								x	x	x	x	x	x	x
65								x	x	x	x	x	x	x
70									x	x	x	x	x	x

Table 22 Taper pins.

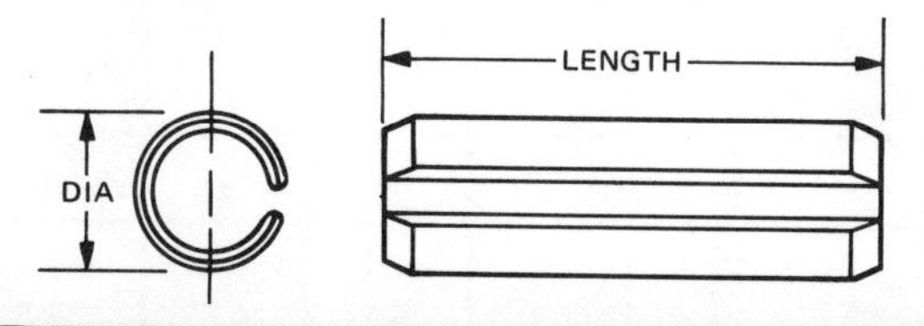

Length	PIN DIAMETER									
	1.5	2	2.5	3	4	5	6	8	10	12
10	x	x	x	x						
15	x	x	x	x	x	x				
20	x	x	x	x	x	x	x			
25	x	x	x	x	x	x	x	x		
30		x	x	x	x	x	x	x	x	x
35		x	x	x	x	x	x	x	x	x
40		x	x	x	x	x	x	x	x	x
45				x	x	x	x	x	x	x
50				x	x	x	x	x	x	x
55					x	x	x	x	x	x
60					x	x	x	x	x	x
70							x	x	x	x
75							x	x	x	x

Table 23 Spring pins.

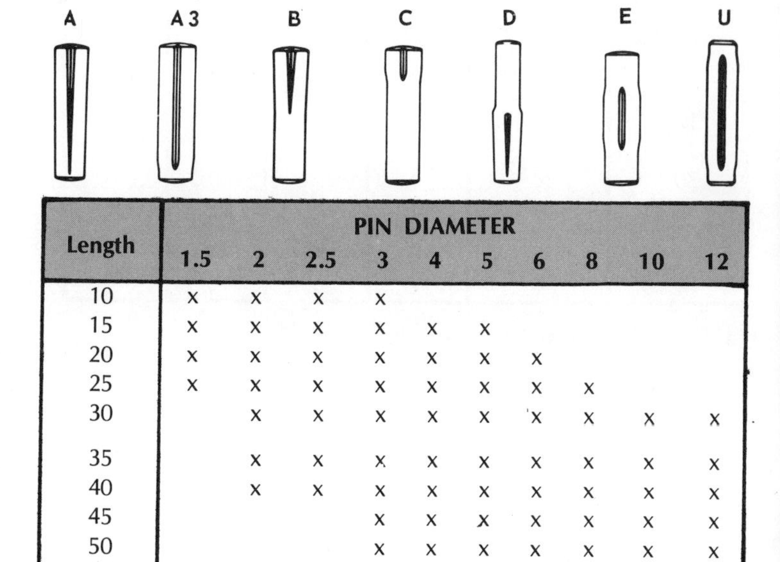

Length	PIN DIAMETER									
	1.5	2	2.5	3	4	5	6	8	10	12
10	x	x	x	x						
15	x	x	x	x	x	x				
20	x	x	x	x	x	x	x			
25	x	x	x	x	x	x	x	x		
30		x	x	x	x	x	x	x	x	x
35		x	x	x	x	x	x	x	x	x
40		x	x	x	x	x	x	x	x	x
45				x	x	x	x	x	x	x
50				x	x	x	x	x	x	x
55					x	x	x	x	x	x
60					x	x	x	x	x	x
70							x	x	x	x
75							x	x	x	x

Table 24 Groove pins.

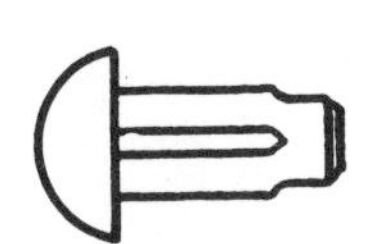

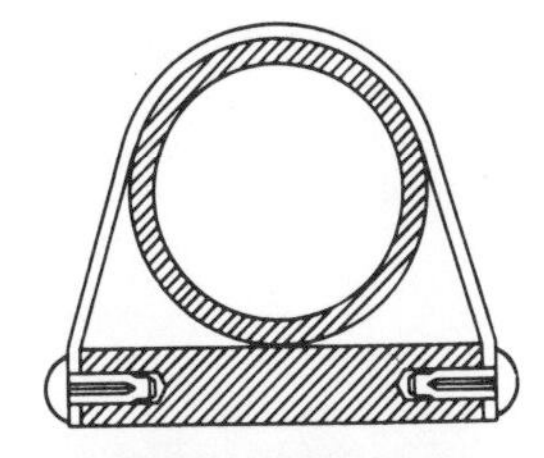

WIDELY USED FOR FASTENING BRACKETS

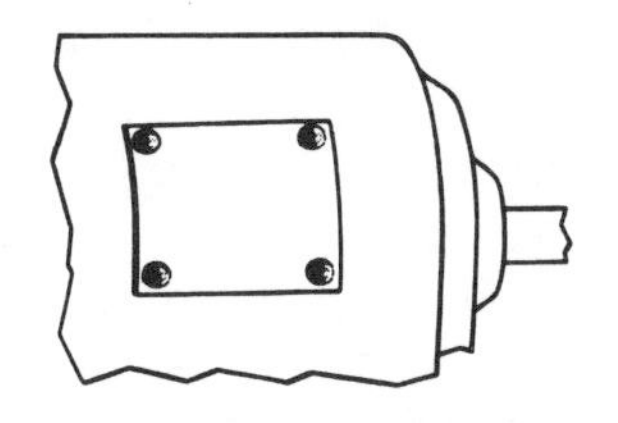

ATTACHING NAMEPLATES, INSTRUCTION PANELS

STUD NUMBER	SHANK DIA.	RECOMMENDED DRILL SIZE	MAXIMUM HEAD DIA.	MAXIMUM HEAD HEIGHT	STANDARD LENGTHS					
					4	6	8	10	12	14
0	1.7	1.7	3.3	1.3	x	x	x			
2	2.2	2.2	4.1	1.8	x	x	x			
4	2.6	2.6	5.4	2.2		x	x	x		
6	3.0	3.0	6.6	2.6			x	x	x	
7	3.4	3.4	7.8	3.0				x	x	x
8	3.8	3.8	7.8	3.0					x	x
10	4.1	4.1	9.1	3.5					x	x
12	5.0	5.0	10.4	3.9						x
14	5.6	5.6	11.6	4.3						x
16	6.3	6.3	12	4.4						x

Table 25 Grooved studs.

Nominal Bolt or Thread-Size Range	Nominal Cotter-Pin Size	Cotter-Pin Hole	Min. End Clearance*
3.5–4.5	1.0	1.2	2.0
4.5–5.5	1.2	1.4	2.5
5.5–7.0	1.6	1.8	2.5
7.0–9.0	2.0	2.2	3.0
9.0–11	2.5	2.8	3.5
11–14	3.2	3.6	5
14–20	4	4.5	6
20–27	5	5.6	7
27–39	6.3	6.7	10
39–56	8.0	8.5	15
56–80	10	10.5	20

*End of bolt to centre of hole

Table 26 Cotter pins.

Pin Dia. A	B	C	Min. D	E	Drill Size F
4	6	1	16	2.2	1
6	10	2	20	3.2	1.6
8	14	3	24	3.5	2
10	18	4	28	4.5	3.2
12	20	4	36	5.5	3.2
16	25	4.5	44	6	4
20	30	5	52	8	5
24	36	6	66	9	6.3

Table 27 Clevis pins.

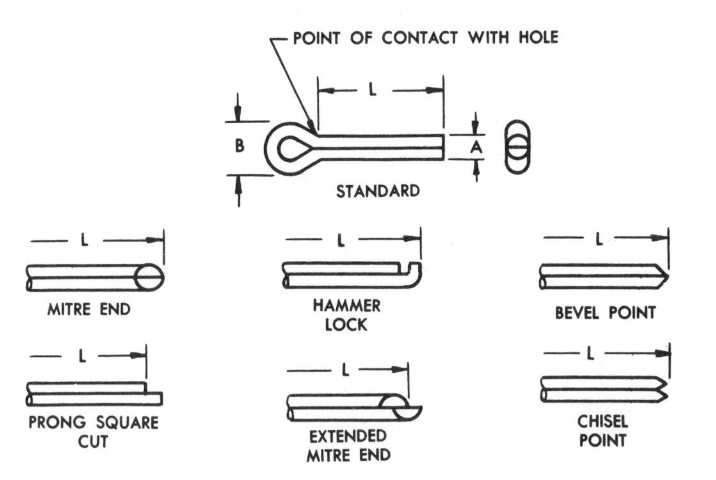

Index

A

Abbreviations 63
Alternative thread representation 114
Aluminum 219
Angular units of measurements 62
Applied geometry 293-304
  arcs and circles 296
  ellipse 302
    four-centre method 302
    parallelogram method 302
    two-circle method 302
  helix 304
  polygons 299-301
    draw a hexagon 299
    draw an octagon 299
  straight lines 293
Appliques 261
Architectural drafting 264-292
Arcs and circles 296
Arms in section 162
Arrangement of views 33-34
Arrowheads 57
Arrowless dimensioning 80
Artistic drawing 3
Assemblies in section 150
Assembly drawings 99
Axonometric Projection 174

B

Babbitt 220
Basic metallurgy 216-220
Basic rules for dimensioning 63
Bathrooms 286
Bedrooms 285
Bills of materials 102
Bolts 119
Brass 220
Break lines (geometric) in isometrics 184
Broken - out sections 163
Bronze 219
Brushes 10

C

Cabinet oblique 186
Calculators 11
Cap screws 119
Captive or self-retaining nuts 129
Cast iron 217
Casting 224-225
Cavalier oblique 186
Centre lines 21, 44, 58, 60
Chain dimensionning 80
Chamfers 74
Chemical locking 127
Chordal dimensioning 80
Circle and arc features in oblique 190
Circles and arcs in isometric drawings 179
Circular features 44
Circular features in auxiliary projection 169-170
Circular tapers 73
Clevis pins 134
Common-point dimensioning 81
Compasses 8
Conical washer 123
Construction Drawings 270-282
Construction lines
Conventional breaks
Conventional features in oblique
Conventional representation of common
  features 88-93
Conventional thread representation 113-119
Coordinate programming 246
Coordinate system 243
Coordinate tolerancing 238
Copper 220
Cotter pins 134
Counterbore 70
Countersink 70
Cut-and-paste drafting 261-262
Cutting-plane lines 140
Cylindrical holes 68
Cylindrical intersections 92

D

Datum or common-point dimensioning 81
Decimal inch system 61
Detail assembly drawings 101
Detail drawings 99
Developing a house plan 285-292
  bathrooms 286
  bedrooms 285
  kitchens 287
  living and dining rooms 290-292
Development of a cone 209
Diameters 66
Diazo (whiteprint) 104
Die-casting 225
Dimension lines 57
Dimensioning (architectural drafting) 280
Dimensioning (definition) 56
Dimensioning an oblique drawing 189
Dimensioning auxiliary views 167
Dimensioning for numerical control 243
Dimensioning isometric drawings 176
Dimensioning methods 79-81
  arrowless 80
  chain 80
  chordal 80
  datum or common-point 81
  polar coordinate 79
  rectangular coordinate 79
  tabular 80
  true position 133
Dimensioning of keyseats 133
DImensioning systems (architctural drafting) 280
Dining rooms 290-292
Dividers 8
Dowel pins 134
Drafting careers 4
Drafting instruments and equipment 5-11
Drafting leads 9
  drafting pencils 9
  lead pointers 9
Drafting machines 8
Drafting office 4
Drawing aids and drawing practises 252-254
Drawing circles and arcs 21-22
  centre lines 21
Drawing curves in isometric 180
Drawing paper 15
Drawing practises 253
Drawing reproduction 104-106
  diazo (whiteprint) 104
  microfilm 104
  photo-reproduction process 106
Drawing scales
Drawing straight lines 13
Drawings 1, 3

E

Ellipse 302
Enlarged views 95
Extension lines 57, 59

F

Fasteners, specifying *see Specifying fasteners*
Fastening devices 107
Ferrous metals 216-219
  cast iron 217
  steel 217
    steel designation system 218
  wrought iron 217
Fillets and rounds 76
Fillets and rounds in isometrics 183
Flat tapers 74
Flat washers 123
Foot and inch system 62
Foreshortened projection 93
Forging 227
Formed parts 76
Forming from sheet stock 226
Forming methods, plastics 223
Four-centre method 302
Free-spinning 127
Free-spinning locknuts 129
Freehand sketching 255
Full sections 140
Functional drafting 252-262

G

General Notes 60
Geometry *see Applied geometry*

Gothic 14
Grooved straight pins 135
Guide lines 15

H

Half sections 146
Hatching lines 141
Head styles 120
Helical spring washer 123
Helix 108, 304
Hexagon, draw a 299
Hidden lines 18, 38
Hidden sections 165
Holes in sections 156
Holes revolved to show true distance from centre 93
Hollow spring pins 136
House plan, developing a see *Developing a house plan*

I

Inch threads 117
Inch units 61-62
  decimal inch system 61
  foot and inch system 62
Inclined surfaces 41, 187
Inserts 130
Instant lettering 10
Intersection of cylindrical surfaces 215
Intersection of flat surfaces 213
Intersection of unfinished surfaces 88
Irregular curves 10
Isometric drawings 174-184
  breaklines (geometric) 184
  circles and arcs 179
  dimensioning isometric drawings 176
  drawing curves in isometric 180
  fillets and rounds 183
  isometric assembly drawings 184
  isometric grid paper 177
  isometric sectioning 183
  isometric templates 180
  non-isometric lines 176
  sketching circles and arcs 180
  threads 184

K

Keys 132-133
  dimensioning of keyseats 133
Kitchens 287
Knurls 75

L

Lead 108
Lead pointers 9
Leaders 60
Left-handed threads 112
Lettering 274
Lettering 10
  instant lettering 10
  lettering sets 10
Lettering (styles) 14
  gothic 14
  microfont 14
Lettering heights
Limit Dimensioning 233
Limits and tolerances 231-234
Line widths 12
Linear units or measurements 60-62
  inch units 61
    decimal inch system 61
    foot and inch system 62
  SI Metric units 61
Lines 12
Living rooms 290-292
Local notes 60
Locknuts 129
  free-spinning locknuts 129
  prevailing-torque locknuts 129
Lugs in sections 156

M

Machine pins 134
Machine screws 119
Machining, removal of material by see *Removal of material by machining*
Machining from standard stock 228
Manufacturing processes 224-228
  casting 224
    die-casting 225
    sandcasting 224
  forging 227
  forming from sheet stock 226
  machining from standard stock 228
  welding 227
Material removal allowance 84
Material removal prohibited 84
Metallurgy, basic 216-220
Metric scales 7
Metric threads 116
Microfilm 105
Microfont 14
Micrometre 84
Mitre line 34
Multi-auxiliary view drawings 171
Multi-view drawings 30
Multiple threads 111

N

Non-Ferrous metals 219-220
  aluminum 219
  babbitt 220
  brass 220
  bronze 219
  copper 220
Non-isometric lines 176
Not-to-scale dimensions 63
Notes 60
  general 60
  local 60
Numerical control 242-249
  three-axis control systems 248
  two-axis control systems 242
Nuts 122

O

Oblique projection 186-192

cabinet oblique 186
cavalier oblique 186
circle and arc features in oblique 190
conventional features in oblique 192
dimensioning an oblique drawing 189
inclined surfaces 192
oblique sectioning 192
oblique sketching 188
Oblique surface 46
Octagon, draw an 299
Offset sections 153
Oneview drawings 29
Operational names 64
Opposite-hand views 96
Orthographic projection 29-30

P

Packaging industry 200
Parallel line development 207
Parallelogram method 302
Partial or broken-out sections 163
Partial views 96
Phantom or hidden section 165
Photo-reproduction 106
Pictorial drawings 173-192
Pictorial thread presentation 110
Pictorial views 29
Pin fasteners 134-136
semipermanent pins 134
machine pins 134
·clevis pins 134
dowel pins 134
taper pins 134
radial locking pins 134
grooved straight pins 135
hollow spring pins 136
Pipe threads 117
Pitch 108
Placement of sectional views 160
Placement of views 95
Plastics 220-223
forming methods 223
thermoplastics 222
thermosetting 222
Plus and minus tolerancing 233
Point designs 122
Point-to-point programming 245
Polar coordinate dimensioning 79
Polygons 299-301
Positional tolerancing 237-241
Presentation drawings 264-267
Prevailing-torque 127
Prevailing-torque nuts 129
Primary auxiliary views 166-168
dimensioning auxiliary views 167

R

Radial line development of conical surfaces 209-212
development of a cone 209
Radial line development of flat surfaces 203-205
Radial locking pins 134
Radii 66
Reading direction 62
Rear views 96
Rectangular coordinate dimensioning 79
Reference dimension 63
Removal of material by machining 83
Repetitive features and dimensions 76
Revolved and removed sections 159-160
Ribs in sections 156
Right- and left-handed threads 112
Roughness 85
Roughness-height value 85
Roughness spacing 85
Rounds 76
Rounds, (in isometrics) 183

S

Sand-casting 224
Scales 7
metric scales 7
Screw threads 108
Sealing fasteners 131
Sectional views 139-142
Sectional views, placement of see *Placement of sectional views*
Selection of views 30
Self-retaining nuts 129
Semipermanent pins 134
Set squares 6
Setscrews 124
Shape description 28-32
Sheet-metal sizes 196
SI metric units 61
Simplification of detail drawing 258
Simplification of dimensioning 257
Simplified drafting 256-259
Single and multiple threads 111
Single-thread engaging nuts 130
Sketching 25-26
Sketching circles and arcs in isometrics 180
Slotted holes 69
Spacing the views 33
Special fasteners 127-131
Special purpose washer 124
Specifying fasteners 126
Spherical features 68
Spokes and arms in section 162
Spotface 70
Spring washer 124
Standard drawing sheet sizes 15
Steel 217
Steel designation system 218
Straight-line developments 194-198
Straight lines 293
Straightedges 5
Studs 119
Surface developments 194
Surface texture 82-85
Surface texture characteristics 84-85
micrometre 84
roughness 85
roughness-height value 85
roughness spacing 85
waviness 85
Surface texture control 84
Surface texture symbol 82
Symmetrical outlines 76

T

T squares 6
Tabular dimensioning 80
Taper pins 134
Tapers 73-74
  circular tapers 73
  flat tapers 74
Technical drawing 3
Templates 10
Thermoplastics 222
Thermosetting 222
Thread designation 116-117
  inch threads 117
  metric threads 116
Thread forms 107-112
Thread grades and classes 116
Thread standards 115
Threaded assemblies 114, 148
Threads, pictorial presentation see *Pictorial threads presentation*
Threads in isometrics 184
Threads in section 148-149
Three-axis control systems 248-249
Title block 15
Title strips 15
Tolerances 231
Tolerancing 232-234
  methods 233
    limit dimensioning 233
    plus and minus tolerancing 233
Tooth lock washer 123
True position dimensioning 79
Two-axis control systems 242-246
  coordinate programming 246
  coordinate system 243
  dimensioning for numerical control 243
  point-to-point programming 245
Two-circle method 302
Two-view drawings 30

U

Undercuts 75

W

Washers 123-124
  conical 123
  flat 123
  helical spring 123
  special purpose 124
  spring 124
  tooth lock 123
Waviness 85
Welding 227
Working drawings 56-64
Wrought iron 217